Essential
Developmental
Biology

3rd *Edition*

Essential Developmental Biology

Jonathan M.W. Slack **MA, PhD, FMedSci**

Stem Cell Institute
University of Minnesota
Minneapolis
MN, USA

Professor Emeritus
University of Bath, UK

✦WILEY-BLACKWELL

A John Wiley & Sons, Ltd., Publication

Registered office: John Wiley & Sons, Ltd, The Atrium, Southern Gate, Chichester, West Sussex, PO19 8SQ, UK

Editorial offices: 9600 Garsington Road, Oxford, OX4 2DQ, UK
The Atrium, Southern Gate, Chichester, West Sussex, PO19 8SQ, UK
111 River Street, Hoboken, NJ 07030-5774, USA

For details of our global editorial offices, for customer services and for information about how to apply for permission to reuse the copyright material in this book please see our website at www.wiley.com/wiley-blackwell.

Library of Congress Cataloging-in-Publication Data

Slack, J. M. W. (Jonathan Michael Wyndham), 1949–
    Essential developmental biology / Jonathan M.W. Slack. – 3rd ed.
       p. cm.
    Includes bibliographical references and index.
    ISBN 978-1-118-02286-3 (hardback : alk. paper)
    ISBN 978-0-470-92351-1 (paperback : alk. paper)
1. Developmental biology–Laboratory manuals.   I. Title.
    QH491.S6 2012
    571.8'6–dc23
                        2012008542

A catalogue record for this book is available from the British Library.

Wiley also publishes its books in a variety of electronic formats. Some content that appears in print may not be available in electronic books.

Cover design by Design Deluxe

Set in 9.5/12pt Minion by Toppan Best-set Premedia Limited, Hong Kong

Printed and bound in Singapore by Markono Print Media Pte Ltd

1   2013

Cover image: The cover shows a double posterior limb of an axolotl (a type of salamander), together with a normal limb. The double posterior limb arose after grafting a strip of limb rudiment tissue into the flank of a host embryo. This means that a polarizing region develops on both sides, leading to the double posterior symmetry (see Chapter 15). The specimen was made by the author using the technique first described in Slack, JMW. *Determination of polarity in the amphibian limb, Nature* 1976: 261:44–46. It was stained and photographed by Drs Ying Chen and Gufa Lin. Red indicates bone and blue indicates cartilage.

Contents

Preface

This book presents the basic concepts and facts relating to the developmental biology of animals. It is designed as a core text for undergraduate courses from the second to the fourth year, and also for first year graduate students. The first and second editions were "road tested" by myself and by many other instructors, and were found suitable for both biologically based and medically oriented courses. A basic knowledge of cell and molecular biology is assumed, but no prior knowledge of development, animal structure, or histology should be necessary.

The book is arranged in four sections and the order of topics is intended to represent a logical progression. The first section introduces the basic concepts and techniques. The second covers the six main "model organisms," *Xenopus*, zebrafish, chick, mouse, *Drosophila*, and *Caenorhabditis elegans*, describing their early development to the stage of the general body plan. The third deals with organ development, mostly of vertebrates but including also *Drosophila* imaginal discs. The fourth deals with some topics of high contemporary interest: growth and stem cells, regeneration, regenerative medicine, and evolution. To assist readers unfamiliar with the families of genes and molecules that are important in development, they are listed in the Appendix in the context of a short revision guide to basic molecular and cell biology.

Like the previous editions, the new version of *Essential Developmental Biology* differs from its main competitors in four important respects, all of which I feel are essential for effective education.

- It keeps the model organisms separate when early development is discussed. This avoids the muddle that arises all too often when students think that knockouts can be made in *Xenopus*, or that bindin is essential for mammalian fertilization.
- It avoids considerations of history and experimental priority because students do not care who did something first if it all happened 20 years ago.
- It does, however, explain *why* we believe what we do. Understanding does not come from simply memorizing long lists of gene names, so I continue to explain how to investigate developmental phenomena and what sorts of evidence are needed to prove a particular type of result.

- The work is highly focused. In order to keep the text short and concise I have not wandered off into areas such as the development of plants or lower eukaryotes, that may be very interesting but are really separate branches of biology.

The first two editions were very well received by both users and reviewers and I hope that the third edition will make this book an even more popular choice for undergraduate and graduate level teaching around the world.

The curse of detail

The principal challenge today is that of exponentially increasing detail. The molecular life sciences seem to double in workforce and output every 10 years or so, and developmental biology has grown at least this fast since its re-founding as a molecular subject in the 1980s. But students' brains do not double in size every 10 years, nor do their courses lengthen. So very serious thought has to be given to what to include and what to leave out.

I have taken the view that the value of model organisms is that they enable experiments to be performed whose results illuminate some general principle of development. However, a lot of recent research has diverged from this strategy and has focused on an ever more detailed molecular analysis of every possible process and organ system in every model organisms. Although the principles of development are common to all animals, it is inevitable that the harder you look, the more differences between organisms you will find. My view is that for a textbook at this level it is not appropriate to cover all the detailed differences between, say, myogenesis in mouse, chicken, *Xenopus*, and zebrafish, so I have focused on just the main themes. In the general field of organogenesis it is also not feasible to cover all of the organs in the vertebrate body. Accordingly, I have selected for inclusion those that are classic developmental biology topics, such as limb development, and those of the greatest medical importance, such as the nervous system, the heart, or the pancreas. Likewise, when considering regeneration, it is not possible to cover every structure in the living world that regenerates to some extent, so I have focused on three classic systems: the planarian, insect limbs, and vertebrate limbs.

New features

Two new chapters in the third edition are devoted to Techniques for Studying Organogenesis and Postnatal Development and Applications of Pluripotent Stem Cells. The Techniques chapter mostly deals with the very important genetic methods in the mouse used for conditional knockouts, induction of transgene activity, and labeling of specific cell lineages. The Applications chapter deals with human embryonic stem cells and induced pluripotent stem cells and their current and future applications. The two chapters on Tissue Organization and Stem Cells and on Growth, Aging, and Cancer, areas of enormous current interest, are both very substantially revised, and place tissue-specific stem cells firmly into the context of normal postembryonic development and cell turnover.

Otherwise, the text has all been rewritten and updated, the grouping of topics has been reorganized to some extent, the further reading has been updated, and errors have been removed. I have also taken the opportunity of the third edition to introduce many photographs. This helps to communicate the remarkable beauty of specimens, which is what motivates many developmental biologists to do their work.

Obstacles to learning

Students sometimes consider developmental biology to be a difficult subject, but this need not be the case so long as certain obstacles to understanding are identified at an early stage. The names and relationships of embryonic body parts are generally new to students so in this book the number of different parts mentioned is kept to the minimum required for understanding the experiments, and a consistent nomenclature is adopted (e.g. "anterior" is used throughout rather than "rostral" or "cranial").

The competitor texts mix up species and, for example, would typically consider sea urchin gastrulation, *Xenopus* mesoderm induction, and chick somitogenesis in quick succession. This leaves the student unsure about which processes occur in which organisms. In order to avoid confusion, I have kept separate the model organism species in Section 2, and for Sections 3 and 4 it is made clear to which organisms particular findings apply.

Although most students do understand genetics in its simple Mendelian form, they do not necessarily appreciate certain key features prominent in developmental genetics. Among these are the fact that one gene can have several mutant alleles with different properties (e.g. loss of function, constitutive, or dominant negative), or that the name of a gene often corresponds to its loss-of-function phenotype rather than its normal function (e.g. the normal function of the *dorsal* gene in *Drosophila* is to promote ventral development!). Furthermore, pathways with inhibitory steps, such as the Wnt pathway, cause considerable trouble because of the difficulty of representing the lack of something in a diagram. Here, these issues are fully explained in the early chapters, with appropriate reinforcement later on. I provide charts showing the state of each component of inhibitory pathways such that the consequences of altering a particular component can be seen at a glance. I also always distinguish clearly between loss-of-function, gain-of-function, and dominant negative mutations.

Gene nomenclature is an awkward problem for a textbook because there are different conventions in use for different model organisms and between those genes discovered through mutation as opposed to those discovered via biochemistry of the protein product. Here, the species-specific conventions are followed where the text relates to a particular species, but if the text relates a gene in more than one species, I use a generic convention with the name italicized and an initial capital letter.

Students usually fail to distinguish between genes and gene products, and should hopefully be encouraged to do so by the use of italics for gene symbols and regular type for proteins. It is also necessary to understand the difference between increasing the expression of a gene product and activating the biochemical function of a product that is already present. Here, I refer respectively to "upregulation" and "activation" for these two situations, and to "repression" and "inhibition" for the situation where expression or activity is reduced.

I am careful not to adopt the colloquial joining of name and function, as for example in "Notch receptor." This all too easily suggests a receptor *for* Notch rather than the Notch molecule itself, and is a style best avoided.

Finally, I have tried to keep the overall level of detail, in terms of the number of genes, signaling systems, and other molecular components, to the minimum required to explain the workings of a particular process. This often means that various parallel or redundant components are not mentioned, and the latest detail published in *Cell* is omitted.

Learning outcomes

When students have completed a course corresponding to the content of this book they should be able to understand the main principles and methods of the subject. If they wish to enter graduate school, they should be very well prepared for a graduate program in developmental biology. If they go to work in the pharmaceutical industry, they should be able to evaluate assays based on developmental systems where these are used for the purposes of drug screening or drug development. If they become high school teachers, they should be able to interpret the increasing flow of stories in the media dealing with developmental topics, which are sometimes inaccurate and often sensationalized. Whether the story deals with miracle stem cell cures, human cloning, four-legged chickens, or headless frogs, the teacher should be able to understand and explain the true nature of the results and the real motivation behind the work. It is in all our interests to ensure that the results of scientific research are disseminated widely, but also that they are a source of enlightenment and not of sensation.

Acknowledgements

I am grateful to the Wiley-Blackwell editorial staff, Karen Chambers and Mindy Okura-Marszicky, for their assistance; to Debbie Maizels of Zoobotanica for her excellent artwork; to Kathy Syplywczak for her able production management; to Mike O'Connor and Anne Rougvie of the University of Minnesota who read chapters for me; to Abeer Yakout who helped eliminate some errors; to James Dutton, Lucas Greder, Kevin Lin, and many past lab members for providing photographs; to all my current lab members, who have tolerated the many hours I spend reading and writing; and to my family who have provided unstinting support.

Jonathan M.W. Slack
Minnesota

About the companion website

There is a companion website available for this book at
www.essentialdevelopmentalbiology.com

On the site you will find:
- Animations
- Figures from the book
- Self-test questions and answers
- Useful website links
and more

The website is sign-posted throughout the book. Look out for these icons:

Section 1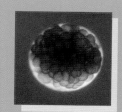
Groundwork

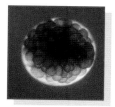
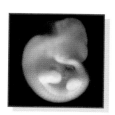
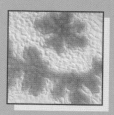

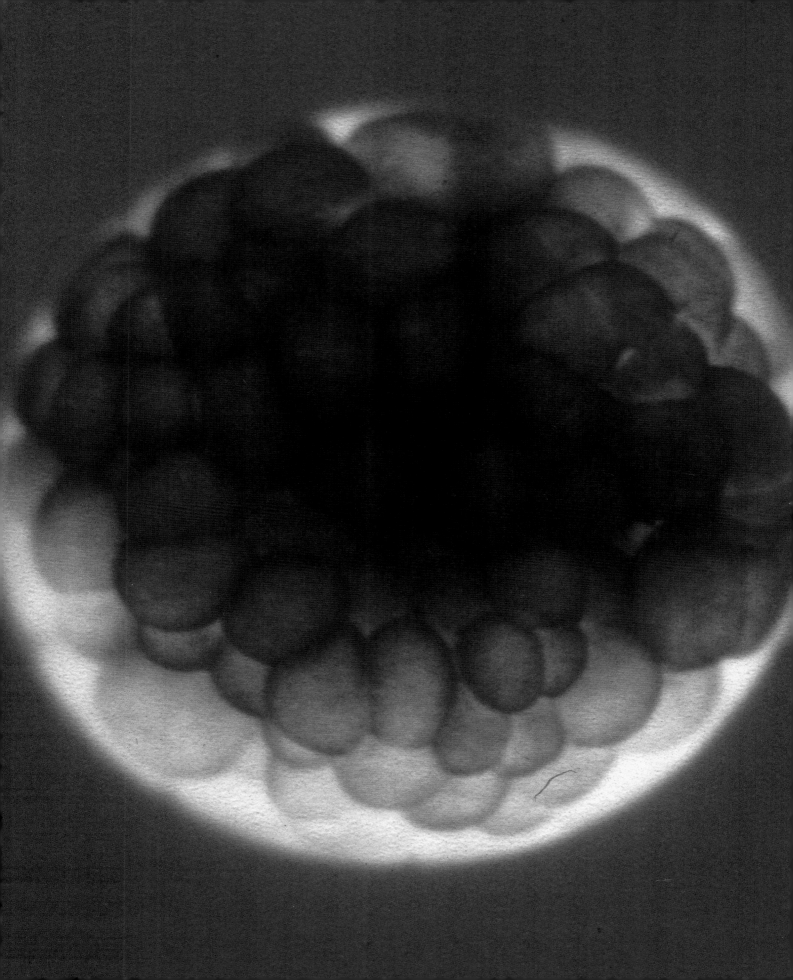

Chapter 1

The excitement of developmental biology

Developmental biology is the science of how biological form changes in time. Development occurs most obviously in the embryo, where the fertilized egg develops into a complex animal containing many cell types, tissues, and body parts. But development also occurs in other contexts, for example during regeneration of missing body parts, during metamorphosis of larval animals to the adult form, and even within our own bodies as the continuous differentiation of new functional cells from stem cells.

Developmental biology occupies a unique central position in modern biology. This is because it unites the disciplines of molecular/cellular biology, genetics, and morphology. Molecular and cell biology tell us about how individual genes and cells work. In development this means inducing factors, their receptors, signal transduction pathways, and transcription factors. Genetics tells us directly about the function of an individual gene and how it relates to the activities of other genes. Morphology, or anatomical structure, is both a consequence and a cause of the molecular events. The first processes of development create a certain simple morphology, which then serves as the basis on which further rounds of signaling and responses can occur, creating a progressively more complex morphology.

So developmental biology is a synthetic discipline, involving contributions from these three areas of science. When thinking about developmental problems it is necessary to be able to use concepts from these three areas simultaneously because they are all required to achieve a complete picture.

Where the subject came from

One of the most amazing conclusions of modern biological research is that the mechanisms of development are very similar for all animals, including humans. This fact has only been known since it has become possible to examine the molecular basis of developmental processes. Before 1980, we knew virtually nothing of these mechanisms but 30 years later we know a lot and it is possible to write undergraduate textbooks on the subject. Over this period, developmental biology has been one of the most exciting areas of biological research. The dramatic advances came from three main traditions that became fused together into a single world view: experimental embryology, developmental genetics, and molecular biology.

Experimental embryology had been going since the beginning of the twentieth century, when it consisted mainly of microsurgical experiments on embryos of frogs and sea urchins. These demonstrated the existence of **embryonic induction**: chemical signals that controlled the pathways of development of cells within the embryo. The experiments showed where and when these signals operated, but they could not identify the signals nor the molecular nature of the responses to them.

Developmental genetics has also existed for a long time, but it really flowered in the late 1970s when mass genetic screens were carried out on the fruit fly *Drosophila*, in which thousands of mutations affecting development were examined. These **mutagenesis screens** resulted in the identification of a high proportion of the genes that control development, not just in *Drosophila*, but in all animals.

Molecular biology had started with the discovery of the three dimensional structure of DNA in 1953, and became a practical science of gene manipulation in the 1970s. The key technical innovations were methods for **molecular cloning** to enable single genes to be amplified to a chemically useful quantity, methods for **nucleic acid hybridization** to enable the identification of DNA or RNA samples, and methods for **DNA sequencing** to determine the primary structures of genes and their

Essential Developmental Biology, Third Edition. Jonathan M.W. Slack.
© 2013 John Wiley & Sons, Ltd. Published 2013 by John Wiley & Sons, Ltd.

protein products. Once this toolkit had been assembled it could be applied to a whole range of biological problems, including those of development. It was used initially to clone the developmental genes of *Drosophila*. This turned out to be of enormous importance because most of the key *Drosophila* genes were found to exist also in other animals, and frequently to be controlling similar developmental processes. Molecular biological methods were also applied directly to vertebrate embryos and used to identify the previously mysterious inducing factors and the genes regulated by them.

The application of molecular biology techniques meant that the mechanisms of development could for the first time be understood in molecular detail. It also meant that the path of development could be experimentally altered by the introduction of new genes, or the selective removal of genes, or by an alteration of the regulatory relationships between genes. It also showed that all animals use very similar mechanisms to control their development. This is particularly exciting because it means that we really can learn about human development by understanding how it happens in the fruit fly, zebrafish, frog, or mouse.

Impact of developmental biology

Some areas of developmental biology have had a significant impact on society in recent decades. *In vitro* fertilization (IVF) is now a routine procedure and has enabled millions of previously infertile couples to have a baby. It is estimated that as many as 2–3% of births in developed countries now arise from IVF. Its variants include artificial insemination by donor (AID), egg donation, and storage of fertilized eggs by freezing. In 2011, Robert Edwards received the Nobel Prize in Physiology/Medicine for introducing this technique. It is less widely appreciated that AID, IVF, embryo freezing, and embryo transfer between mothers are also very important for farm animals. These techniques have been used for many years in cattle to increase the reproductive potential of the best animals.

Developmental biology also led to the understanding that human embryos are particularly sensitive to damage during the period of **organogenesis** (i.e. after the general body plan is formed, and while individual organs are being laid down). The science of **teratology** studies the effects of environmental agents such as chemicals, viral infection, or radiation on embryos. This has led to an awareness of the need to protect pregnant women from the effects of these agents. For example the statin drugs, used to lower cholesterol levels, can compromise the cholesterol modification of the signaling molecule Sonic hedgehog. This can potentially lead to a variety of defects in systems dependent on hedgehog signaling during development: the central nervous system (CNS), limbs, and vertebrae. Although normal doses of statins are probably not teratogenic in humans, this provides a good reason to avoid them during early pregnancy.

Developmental biology is responsible for an understanding of the genetic or chromosomal basis of many human **birth defects**. In particular Down's syndrome is due to the presence of an extra chromosome, and there are a number of relatively common abnormalities of the sex chromosomes. These can be detected in cells taken from the amniotic fluid and form the basis of the **amniocentesis** tests taken by millions of expectant mothers every year. They can also be detected in samples of **chorionic villi**, which may be taken in the early stages of pregnancy. Many more birth defects are due to mutations in genes that control development. It is now possible to screen for some of these, either in the DNA of the parents or that of the embryo or chorionic villi, using molecular biology techniques.

Developmental biology research has also led to the identification of several new growth regulatory substances, some of which have entered clinical practice. For example the hematopoietic growth factors erythropoietin and granulocyte–macrophage colony-stimulating factor (GM-CSF) have both been used for some years to treat patients whose blood cells are depleted by cancer chemotherapy, or for other reasons. Some other growth factors, such as the fibroblast growth factors (FGFs) have been used to assist the healing of wounds.

Developmental biology has also impacted in a major way on other areas of science. This is especially true of the methods for making genetically modified mice, which are now commonly used as **animal models** of human diseases, enabling more detailed study of pathological mechanisms and the testing of new experimental therapies. These are by no means limited to models for human genetic disease as often a targeted mutation in the mouse can mimic a human disease that arises from non-mutational causes.

Developmental biology has also been the "midwife" of **stem cell** biology. Embryonic stem cells were discovered by developmental biologists and the methods for directed differentiation of these cells depends on the understanding of the normal sequence of embryonic inductions which has been built up by developmental biologists. Stem cell biology has now become a huge science in its own right, with many potential medical applications.

Future impact

Although the past impact of developmental biology is significant, the future impact will certainly be much greater. Some of the benefits are indirect and not immediately apparent. Some, particularly those involving human genetic manipulation, may cause some serious ethical and legal problems. These problems will have to be resolved by society as a whole and not just the scientists who are the current practitioners of the subject. For this reason it is important that an understanding of developmental biology becomes as widespread as possible, because only with an appreciation of the science will people be able to make informed choices.

The first main area of practical significance is that an understanding of developmental mechanisms will assist the pharmaceutical industry in designing new drugs effective against cancer or against degenerative diseases such as diabetes, arthritis, and neurodegeneration, conditions that continue to cause enormous suffering and premature death. The processes that fail in degenerative diseases are those established in the course of embryonic development, particularly its later stages. Understanding which genes and signaling molecules are involved has provided a large number of potential new **therapeutic targets** for possible intervention. Once the targets have been identified by developmental biology, the new powerful techniques of **combinatorial chemistry** are applied by pharmaceutical chemists to create drugs that can specifically augment or inhibit their action.

Secondly, and as a quite separate contribution to the work of the pharmaceutical industry, various developmental model systems are important as assays. The *in vivo* function of many **signal transduction pathways** can be visualized in *Xenopus* or zebrafish or *Drosophila* or *Caenorhabditis elegans*, and can be used to assay substances that interfere with them using simple dissecting microscope tests. Genetically modified mice have become very important as models for specific human diseases. Because they are looking at the whole organism these assays are more powerful than biochemical assays on cells in tissue culture.

Thirdly, there is the extension of the existing **prenatal screening** to encompass the whole variety of single-gene disorders. Although this is welcome as a further step in the elimination of human congenital defects, it also presents a problem. The more tests are performed on an individual's genetic makeup, the more likely they are to be denied insurance or particular career opportunities because they have some susceptibility to some disease or other. This is a problem that society as a whole will have to resolve.

Fourthly, there will be a widening application of our understanding of growth and regeneration processes for therapy. For example factors may be developed that could make pancreatic β cells grow, which would be very useful for the treatment of diabetes, or something that could promote neuronal regeneration, which would be useful in treating a variety of neurodegenerative disorders.

Fifthly, there is the application of developmental biology to the production of human cells, tissues, or organs for **transplantation**. This is usually called **cell therapy** if cells or tissues are transplanted, rather than whole organs. At present all types of transplantation are seriously limited by the availability of donors and the ability to make cells, tissues, or even organs on demand has been the principal public justification for funding of stem cell research. The route to replacement envisages their growth from human **pluripotent stem cells**. In a dramatic recent discovery it has been found possible to reprogram normal fibroblasts or other cell types to become pluripotent stem cells (**iPS cells**) through the upregulation of a small number of genes already known to be important for the properties of **embryonic stem cells**. Pluripotent stem cells have the ability to grow without limit in tissue culture, and, when placed in the appropriate environment, to differentiate into all or most of the cell types in the body. Especial interest is shown in methods for making pancreatic β-**cells** for treatment of diabetes, **dopaminergic neurons** for treatment of Parkinson's disease, and **cardiomyocytes** for treatment of heart disease. Not only do pluripotent stem cells hold out the promise of creating these cell types on demand, but it is in principle possible to grow "personalized" cell lines, which would be a perfect immunological match for the patient to be treated.

The new stem cell technology is likely to become fused with the methods for **tissue engineering** which can potentially generate more complex tissues and organs starting with the constituent cell types. This involves the production of novel types of three-dimensional extracellular matrix, or **scaffold**, on which the cells grow and with which they interact. Tissue engineering will need more input from developmental biology in order to be able to create tissues containing several interacting cell types, or tissues with appropriate vascular and nerve supplies.

Finally, we should not overlook the likely applications of developmental biology to agriculture. With farm animals the possibilities are likely to be limited by a public wish to retain a "traditional" appearance for their cows, pigs, sheep, and poultry, but already technologies have been developed to produce pharmaceuticals in the milk of sheep, or vaccines in eggs, and other opportunities will doubtless present themselves in the future.

Further reading

Useful web sites

Society for Developmental Biology: Education section
http://www.sdbonline.org/
index.php?option=com_content&task=section&id=6&Itemid=62
The virtual embryo:
http://www.ucalgary.ca/UofC/eduweb/virtualembryo/

Textbooks, mainly descriptive

Gilbert, S.F. & Raunio, A.M., eds. (1997) *Embryology: Constructing the Organism.* Sunderland, MA: Sinauer Associates.
Hildebrand, M. & Goslow, G.E. (2001) *Analysis of Vertebrate Structure,* 5th edn. New York: John Wiley & Sons.
Carlson, B.M. (2004) *Human Embryology and Developmental Biology,* 4th edn. Philadelphia: Mosby Elsevier.
Schoenwolf, G., Bleyl, S., Brauer, P. & Francis-West, P. (2008) *Larsen's Human Embryology,* 4th edn. New York: Churchill Livingstone.

Textbooks, mainly analytical

Gilbert, S.F. (2010) *Developmental Biology,* 9th edn. Sunderland, MA: Sinauer Associates.
Wolpert, L. & Tickle, C.A. (2010) *Principles of Development,* 4th edn. Oxford: Oxford University Press.

Reproductive technology, teratology, ethics

Braude, P. (2001) Preimplantation genetic diagnosis and embryo research – human developmental biology in clinical practice. *International Journal of Developmental Biology* 45, 607–611.

Maienschein, J. (2003) *Whose View of Life? Embryos, Cloning and Stem Cells.* Cambridge, MA: Harvard University Press.

Gilbert, S.F., Tyler, A. & Zackin, E. (2005) *Bioethics and the New Embryology: Springboards for Debate.* Sunderland, MA: Sinauer Associates.

Ferretti, P., Copp, A., Tickle, C. & Moore, G. (2006) *Embryos, Genes and Birth Defects.* Chichester, England: Wylie.

Pearson, H. (2008) Making babies: the next 30 years. *Nature* **454**, 260–262.

Gearhart, J. & Coutifaris, C. (2011) In vitro fertilization, the Nobel Prize, and human embryonic stem cells. *Cell Stem Cell* **8**, 12–15.

Stem cells and associated technologies

Zandonella, C. (2003) The beat goes on. *Nature* **421**, 884–886.

Verma, I.M. & Weitzman, M.D. (2005) Gene therapy: twenty-first century medicine. *Annual Review of Biochemistry* **74**, 711–738.

Atala, A. (2006) Recent developments in tissue engineering and regenerative medicine. *Current Opinion in Pediatrics* **18**, 167–171.

Gardner, R.L. (2007) Stem cells and regenerative medicine: principles, prospects and problems. *Comptes Rendus Biologies* **330**, 465–473.

Lutolf, M.P., Gilbert, P.M. & Blau, H.M. (2009) Designing materials to direct stem-cell fate. *Nature* **462**, 433–441.

Kay, M.A. (2011) State-of-the-art gene-based therapies: the road ahead. *Nature Review Genetics* **12**, 316–328.

Rossant, J. (2011) The impact of developmental biology on pluripotent stem cell research: successes and challenges. *Developmental Cell* **21**, 20–23.

Slack, J.M.W. (2012) *Stem Cells. A Very Short Introduction.* Oxford: Oxford University Press.

Chapter 2

How development works

Some of the basic processes and mechanisms of embryonic development are now quite well understood, others are not. This chapter will give a summary of how development works to the extent that we currently understand it. Evidence for why we believe that the mechanisms are like this, and many examples of developmental processes in specific organisms, will be presented in later chapters. Further information about the genes and molecules that are mentioned by name will be found in the Appendix.

Embryonic development involves the conversion of a single cell, the fertilized egg, into a complex organism consisting of many anatomical parts. We can break down the complexity of what happens by considering it as five types of process:

1 Regional specification deals with how pattern appears in a previously similar population of cells. For example, most early embryos pass through a stage called the **blastula** or **blastoderm** at which they consist of a featureless ball or sheet of cells. The cells in different regions need to become programmed to form different body parts such as the head, trunk, and tail. The initial steps usually involve regulatory molecules deposited in particular positions within the fertilized egg (**determinants**). The later steps usually involve intercellular signaling events, known as **embryonic inductions**, which lead to the upregulation of different combinations of developmental control genes in each zone of cells.

2 Cell **differentiation** refers to the mechanism whereby different sorts of cells arise. There are more than 200 different specialized cell types in a vertebrate body, ranging from epidermis to thyroid epithelium, lymphocyte, or neuron. Each cell type owes its special character to particular proteins coded by particular genes. The study of cell differentiation deals with the way in which these genes are upregulated and how their activity is subsequently maintained. Cell differentiation continues throughout life in regions of persistent cell turnover fed by stem cells.

3 Morphogenesis refers to the cell and tissue movements that give the developing organ or organism its shape in three dimensions. This depends on the dynamics of the **cytoskeleton** and on the mechanics and viscoelastic properties of cells. Some morphogenetic processes persist into adult life in regions of tissue renewal.

4 Growth refers both to the overall increase of size of the organism, and to the control of proportion between body parts. Although more familiar to the lay person than other aspects of development, it is less well understood in terms of molecular mechanisms.

5 Somehow the component processes of development are coordinated in time. But **developmental time** remains the most mysterious aspect of the process. We know that different species develop at different rates but we do not know why. In this area there are serious gaps in our knowledge.

Ultrashort summary

The following provides a quick summary of how development works. The remainder of this chapter will explore some basic developmental processes in more detail. Later chapters will explain how these processes work in specific model organisms, or situations of organ development, and provide experimental evidence for the basic model.

Male and female **gametes** develop and undergo **meiosis**, thus halving their **chromosome** number to one copy of each chromosome. The male and female gametes fuse in the process of **fertilization** to form a fertilized egg, or **zygote**. This undergoes a period of **cleavage** divisions to form a ball or sheet of similar

Essential Developmental Biology, Third Edition. Jonathan M.W. Slack.
© 2013 John Wiley & Sons, Ltd. Published 2013 by John Wiley & Sons, Ltd.

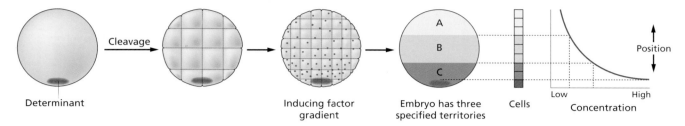

Fig. 2.1 Generation of complexity from a simple beginning. This embryo has a cytoplasmic determinant at the vegetal end which acts as the source of a morphogen. Genes controlling the formation of two territories, B and C, are upregulated at appropriate threshold concentrations. Territory A is the default that arises in the absence of any inducing factor.

cells called a **blastula** or **blastoderm**. Cleavage divisions are typically rapid and involve no growth so the daughter cells are half the size of the mother cell and the whole embryo stays about the same size. A series of morphogenetic movements, called **gastrulation**, converts the original cell mass into a three-layered structure consisting of multicellular sheets called **ectoderm**, **mesoderm**, and **endoderm**, which are known as **germ layers**. During cleavage and gastrulation the first regional specification events occur. In addition to the formation of the three germ layers themselves, these often generate **extraembryonic** structures, needed for support and nutrition, and establish differences of commitment between future **anteroposterior** body regions (head, trunk, and tail).

Regional specification is initiated by the presence of **cytoplasmic determinants** in one part of the zygote, which become inherited by the cells that form from this region. This region becomes a signaling center and its cells emit an inducing factor (Fig. 2.1). Because it is produced in one place, diffuses away, and decays, the inducing factor forms a concentration gradient, high near the source cells and low further away. The remaining cells of the embryo, which do not contain the determinant, are competent to respond to different concentrations of the inducing factor by upregulating particular developmental control genes. So a series of zones becomes set up, arranged at progressively greater distance from the signaling center established by the determinant. In each zone a different combination of developmental control genes becomes upregulated. These encode **transcription factors**, which upregulate new combinations of gene activity in each region. Some of these regions will eventually become new signaling centers, emitting inducing factors different from that emitted by the first center.

It may seem that all embryos would have to be radially symmetrical, around the axis defined by the direction of diffusion of the first inducing factor. Some embryo types do have radial symmetry, but bilateral symmetry is more usual in animal development. This arises from the fact that there is normally another determinant, off center from the first, and so the initial subdivision of the cells arises from two nonparallel signals. This naturally generates an initially bilateral pattern of territories

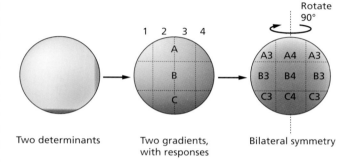

Fig. 2.2 Generation of bilateral symmetry with two determinants. Two gradients partition the embryo into territories along two axes. The resulting embryo has territories arranged symmetrically around a medial plane.

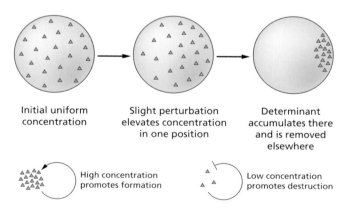

Fig. 2.3 Localization of a determinant by a symmetry breaking process.

(Fig. 2.2). Determinants occur in all animal zygotes. In some cases they consist of specific RNA or protein deposited in some part of the egg during oogenesis. In other cases they become localized as a result of a **symmetry breaking** process that segregates some substances to one region of the zygote and other substances to other regions (Figs 2.3, 2.4). It is this process of symmetry breaking that explains why a spherically symmetrical

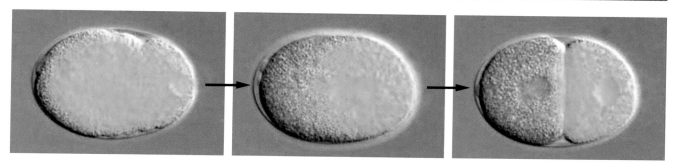

Fig. 2.4 Localization of the protein PIE-1 to the posterior during the first cell cycle of a *C. elegans* embryo. The PIE-1 is a fusion protein with green fluorescent protein (GFP), making it fluoresce green. Reproduced from Gönczy (2008), with permission from Nature Publishing Group.

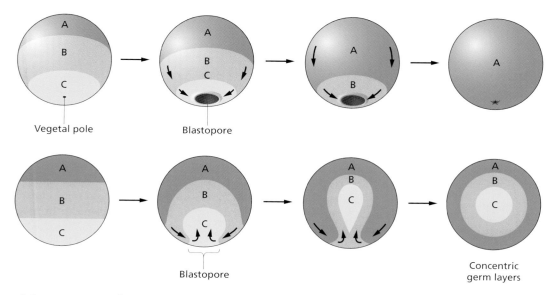

Fig. 2.5 Gastrulation movements: surface view above and sections below. In this very simple example, the C territory invaginates through the vegetal pole, followed by the B territory, so that the three territories end up in a concentric arrangement.

or radially symmetrical egg can nonetheless initiate the establishment of an internal pattern of structures.

Among the earliest developmental commitments are those responsible for the formation of the three germ layers, **ectoderm**, **mesoderm**, and **endoderm**. These are each associated with the expression of specific transcription factors. Among other functions, these transcription factors upregulate expression of genes conferring specific adhesive and motility properties on the cells in which they are active. Because of these different morphogenetic properties, the cells of each germ layer move to form sheets such that the ectoderm ends up on the outside, mesoderm in the middle, and endoderm on the inside (Fig. 2.5). The morphogenetic movements that result in the positioning of the three germ layers are collectively called **gastrulation**. Morphogenetic movements not only change the

shape and structure of the embryo, but by bringing cell sheets into new spatial relationships they also make possible new cycles of signaling and response between these cell populations.

In the development of a typical embryo there will be many more cycles of the same types of event, involving an inductive signal and responses to it. Each new territory of cells, with a specific combination of genes active, has a specific competence to respond to inductive signals because of the presence of particular cell surface receptors, particular signal transduction pathways, and particular developmental control genes poised for upregulation once the signal has been received. The number of different types of inducing factor is relatively small but because competence may change in time the cell populations can respond in different ways at different times. For this reason the final complexity that can be built up by successive rounds of inductive signals and responses, together with morphogenetic movements, is very large.

Growth is mostly autonomous. For each territory of cells the growth rate is controlled by the combination of genes that are active. Free-living embryos do not grow in mass as they have no external food supply. But embryos fed by a placenta or extraembryonic yolk supply can grow very fast, and changes to relative growth rate between parts in these organisms help to produce the final overall anatomy.

The whole process needs to be coordinated in time. For example it will not work if the secretion and diffusion of inducing factors is not on a time scale consistent with the perception of the signals and the activation of target genes. How this is controlled is not understood. There may be a master clock able to communicate with all parts of the embryo that controls the course of events, or timing may depend simply on local causal sequences of events.

Gametogenesis

Sexual reproduction really starts with the formation of the gametes. By definition the male gamete is small and motile and called a spermatozoon (sperm), and the female gamete is large and immotile and called an **egg** or ovum. Each gamete contributes a **haploid** (1n) chromosome set so the zygote is **diploid** (2n), containing a maternal- and a paternal-derived copy of each chromosome.

The gametes are formed from **germ cells** in the embryo. The germ cells are referred to collectively as the **germ line**, consisting of cells that will or can become the future gametes, and all other cells are referred to as the **somatic** tissues or soma. The importance of the germ line is that its genetic information can be passed to the next generation, while that of the soma cannot. This means that a germ-line mutation is one occurring in the DNA of germ cells, which may be carried to the next generation. By contrast, a somatic mutation may occur in a cell at any stage of development and may be important in the life of the individual animal, but it cannot affect the next generation.

It is often the case that the future germ cells become committed to their fate at an early stage of animal development. In some cases there is a cytoplasmic determinant present in the egg that programs cells that inherit it to become germ cells. This is associated with a visible specialization of the cytoplasm called **germ plasm**. It occurs in *Caenorhabditis elegans* where the cells inheriting the **polar granules** become the P lineage and thereafter the germ cells. It occurs in *Drosophila*, where cells inheriting the **pole plasm** become pole cells and later germ cells. It also occurs in *Xenopus* where there is a vegetally localized germ plasm rich in mitochondria. In other species there may be no visible germ plasm in the egg but germ cells still appear at a relatively early stage of development.

During embryonic development, germ cells undergo a period of multiplication and will also often undergo a migration from the site of their formation to the **gonad**, which may be some distance away. The gonad arises from mesoderm and is initially composed entirely of somatic tissues. After the germ cells arrive they become fully integrated into its structure, and in postembryonic life undergo gamete formation or **gametogenesis**. At some stage in mid-development the key decision of **sex determination** is made and the gonad is determined to become either an ovary or a testis. The molecular mechanism of this is, somewhat surprisingly, different for each of the principal experimental model species, so it will not be described in this chapter. But the upshot is that in the male the germ cells will need to become sperm and in the female they will need to become eggs. Unlike the other model organisms, *C. elegans* is normally a **hermaphrodite** and the germ cells will produce both sperm and eggs in the same individual. However there are also male individuals of *C. elegans*, and the sex determination mechanism controls the male–hermaphrodite decision rather than a male–female decision.

Meiosis

The critical cellular event in gamete production is **meiosis**. This is a modified type of cell cycle in which the number of chromosomes is reduced by half (Fig. 2.6). As in mitosis, meiosis is also preceded by an S-phase in which each chromosome becomes replicated to form two identical sister **chromatids**, so the process starts with the nucleus possessing a total DNA content of four times the haploid complement. In mitosis the sister chromatids segregate into two identical diploid daughter cells. But meiosis involves two successive cell divisions. In the first the homologous chromosomes, which are the equivalent chromosome derived from mother and father, pair with each other. At this stage the chromosomes are referred to as **bivalents**, and each consists of four chromatids, two maternal-derived and two paternal-derived. **Crossing over** can occur between these chromatids, bringing about recombination of the alleles present at different loci. Hence, alleles present at two different loci on the same chromosome of one parent may become separated into different gametes and be found in different offspring. The frequency with which alleles on the same chromosome are separated by recombination is roughly related to the physical separation of the loci, and this is why the measurement of recombination frequencies is the basis of genetic mapping. Recombination can also occur between sister chromatids but here the loci should all be identical because they have just been formed by DNA replication, so there are no genetic consequences.

In the first meiotic division the four-stranded bivalent chromosomes separate into homologous pairs, which are segregated to the two daughter cells. There is no further DNA replication and in the second meiotic division the two chromatids of each chromosome become separated into individual gametes.

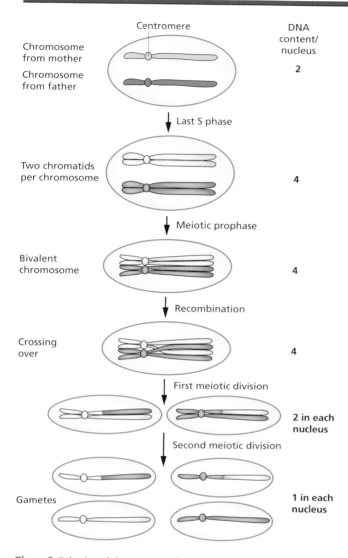

Centromere

DNA content/ nucleus

Chromosome from mother

Chromosome from father

2

↓ Last S phase

Two chromatids per chromosome

4

↓ Meiotic prophase

Bivalent chromosome

4

↓ Recombination

Crossing over

4

↓ First meiotic division

2 in each nucleus

Second meiotic division

↓

Gametes

1 in each nucleus

Fig. 2.6 Behavior of chromosomes during meiosis.

It should be noted that the terms **haploid** (1n) and **diploid** (2n) are normally used to refer to the number of homologous chromosome sets in the nucleus rather than the actual amount of DNA. After DNA replication a nucleus contains twice as much DNA as before, but retains the same ploidy designation.

Oogenesis

The process of formation of eggs is called **oogenesis** (Fig. 2.7). Following sex determination to female, the germ cells become **oogonia**, which continue mitotic division for a period. After the final mitotic division, the germ cell becomes known as an **oocyte**. It is called a **primary oocyte** until completion of the first meiotic division, and a **secondary oocyte** until completion

of the second meiotic division. After this it is known as an unfertilized **egg** or ovum. In all the vertebrate organisms considered in this book, fertilization occurs before completion of the second division, so it is technically an **oocyte** rather than an **egg** that is being fertilized. However, the term "egg" is often used rather loosely to refer to oocytes, fertilized ova, and even early embryos.

Eggs are larger than sperm and the process of oogenesis involves the accumulation of materials in the oocyte. Usually, the primary oocyte is a rather long-lived cell that undergoes a considerable increase in size. Its growth may be assisted by the absorption of materials from the blood, such as the **yolk** proteins of fish or amphibians that are made in the liver. It may also be assisted by direct transfer of materials from other cells. This is seen in *Drosophila* where the last four mitoses of each oogonium produces an egg chamber containing one oocyte and 15 **nurse cells**. The nurse cells then produce materials that are exported to the oocyte. Animals that produce a lot of eggs usually maintain a pool of oogonia throughout life capable of generating more oocytes. Mammals differ from this pattern as they produce all their primary oocytes before birth. In humans no more oocytes are produced after the seventh month of gestation, and the primary oocytes then remain dormant until puberty.

Ovulation refers to the resumption of the meiotic divisions and the release of the oocyte from the ovary. It is provoked by hormonal stimulation and involves a breakdown of the oocyte nucleus (the **germinal vesicle**) and the migration of the cell-division spindle to the periphery of the cell. The meiotic divisions do not divide the oocyte into two halves, but instead result in the budding off of small **polar bodies**. The first meiotic division divides the primary oocyte into the secondary oocyte and the first polar body, which is a small projection containing a replicated haploid chromosome set (i.e. 1n information content and 2x DNA content). The second meiotic division divides the secondary oocyte into an egg and a second polar body, which consists of another small projection enclosing a haploid chromosome set (in this case 1n information content and 1x DNA content). The polar bodies soon degenerate and play no further role in development.

Spermatogenesis

If the process of sex determination yields a male then the germ cells undergo spermatogenesis (Fig. 2.7). Mitotic germ cells in the testis are known as **spermatogonia**. Some of these are **stem cells** that can both produce more of themselves and also produce **progenitor cells**, which divide a number of times before differentiation into sperm. After the last mitotic division the male germ cell is known as a primary spermatocyte. Meiosis is equal, the first division yielding two secondary spermatocytes and the second division yielding four spermatids, which mature to become motile spermatozoa.

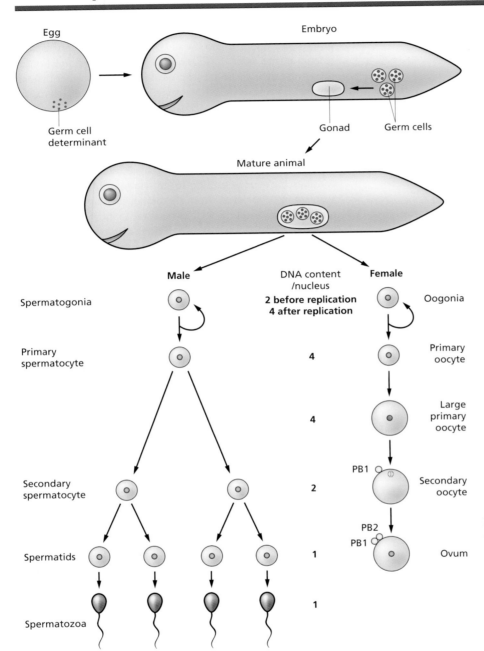

Fig. 2.7 Typical sequence of gametogenesis. The germ cells are initially formed from a cytoplasmic determinant and during development they migrate to enter the gonad. Spermatogenesis generally results in the production of four haploid sperm per meiosis. Oogenesis generally results in the formation of one egg and two polar bodies (PB1 has the same chromosome number but twice the DNA content as PB2).

Early development

Fertilization

The process of fertilization differs considerably between animal groups but there are a few common features. When the sperm fuses with the egg there is a fairly rapid change in egg structure that excludes the fusion of any further sperm. This is called a block to **polyspermy**. Fusion activates the inositol trisphosphate signal transduction pathway resulting in a rapid increase in intracellular calcium. This causes exocytosis of **cortical granules** whose contents form, or contribute to, a fertilization membrane; and also triggers the metabolic activation of the egg, increasing the rate of protein synthesis and, in vertebrates, starting the second meiotic division. The calcium may, in addition, trigger cytoplasmic rearrangements that position determinants that are important for the future regional specification of the embryo. For example dorsal localization of components of the

Wnt pathway in *Xenopus*, or polar granule segregation in *C. elegans* occur in this manner. The sperm and egg pronuclei fuse to form a single diploid nucleus and at this stage the fertilized egg is known as a **zygote**.

Cleavage

A generalized sequence of early development is shown in Fig. 2.8. A typical zygote of an animal embryo is small, spherical, and polarized along the vertical axis. The upper hemisphere,

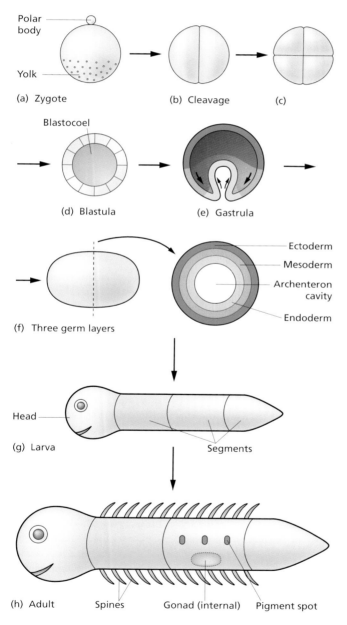

(a) Zygote   (b) Cleavage   (c)

(d) Blastula   (e) Gastrula

(f) Three germ layers

Ectoderm
Mesoderm
Archenteron cavity
Endoderm

(g) Larva   Segments

(h) Adult   Spines   Gonad (internal)   Pigment spot

Fig. 2.8 A generalized sequence of early development.

usually carrying the polar bodies, is called the **animal hemisphere**, and the lower hemisphere, rich in yolk, the **vegetal hemisphere**. The early cell divisions are called **cleavages**. They differ from normal cell division in that there is no growth phase between successive divisions. So each division partitions the mother cell into two half-size daughters (Fig. 2.9). The products of cleavage are called **blastomeres**. Cell division without growth can proceed for a considerable time in free-living embryos without an extracellular yolk mass. Embryos that do have some form of food supply, either mammals that are nourished by the mother, or egg types with a large yolk mass such as birds and reptiles, only undergo a limited period of cleavage at the beginning of development. In many species, the embryo's own genome remains inactive during part or all of the cleavage phase, and protein synthesis is directed by messenger RNA transcribed during oogenesis (maternal mRNA). This is the stage of genetic **maternal effects** because the properties of the cleavage-stage embryo depends entirely on the genotype of the mother and not on that of the embryo itself (see Chapter 3).

Different animal groups display different types of cleavage (Fig. 2.10) and this is controlled to a large extent by the amount of yolk in the egg. Where there is a lot of yolk, as in an avian egg, the cytoplasm is concentrated near the animal pole and only this region cleaves into blastomeres, with the main yolk mass remaining acellular. This type of cleavage is called **meroblastic**. Where cleavage is complete, dividing the whole egg into blastomeres, it is called **holoblastic**. Holoblastic cleavages are often somewhat unequal, with the blastomeres in the yolk-rich vegetal hemisphere being larger (**macromeres**), while those in the animal hemisphere are smaller (**micromeres**). Each animal class or phylum tends to have a characteristic mode of early cleavage and these can be classified by the arrangement of the blastomeres into such categories as **radial** (echinoderms), bilateral (ascidians), and rotational (mammals). An important type is the **spiral** cleavage shown by most annelid worms, molluscs, and flatworms. Here, the macromeres cut off successive tiers of micromeres, first in a right-handed sense when viewed from

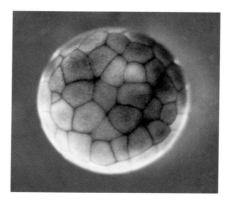

Fig. 2.9 A cleaving embryo of an axolotl.

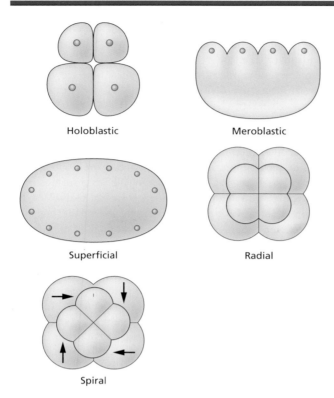

Holoblastic

Meroblastic

Superficial

Radial

Spiral

Fig. 2.10 Cleavage types.

netic movements of gastrulation can vary quite a lot even between related animal groups, but the outcome is similar (Fig. 2.11). The three tissue layers formed during gastrulation are called **germ layers**, but these should not be confused with **germ cells**. Conventionally, the outer layer is known as the **ectoderm**, and later forms the skin and nervous system; the middle layer is the **mesoderm** and later forms the muscles, connective tissue, excretory organs, and gonads; and the inner layer is the **endoderm**, later forming the epithelial tissues of the gut. The **germ cells** have usually appeared by the stage of gastrulation and are not regarded as belonging to any of the three germ layers.

After the completion of the major body morphogenetic movements, most types of animal embryo have reached the general **body plan** stage at which each major body part is present as a region of committed cells, but is yet to differentiate internally. This stage is often called the **phylotypic stage**, because it is the stage at which different members of an animal group, not necessarily a whole phylum, show maximum similarity to each other (see Chapter 22). For example all vertebrates show a phylotypic stage at the **tailbud** stage when they have a notochord, neural tube, paired somites, branchial arches, and tailbud. All insects show a phylotypic stage at the **extended germ band** when they show six head segments, three appendage-bearing thoracic segments, and a variable number of abdominal segments.

above, then another tier in a left-handed sense, and so on. Most insects and some crustaceans show a special type of cleavage called **superficial cleavage**. Here only the nuclei divide and there is no cytoplasmic cleavage at the early stages. Thus, the early embryo becomes a **syncytium** consisting of many nuclei suspended within the same body of cytoplasm. At a certain stage the nuclei migrate to the periphery and shortly afterwards cell membranes grow in from the outer surface of the embryo and surround the nuclei to form an epithelium.

During the cleavage phase a cavity usually forms in the center of the ball of cells, or sheet of cells in the case of meroblastic cleavage. This expands due to uptake of water and becomes known as the **blastocoel**. At this stage of development the embryo is called a **blastula** or **blastoderm**. The cells often adhere tightly to one another, being bound by cadherins, and will usually have a system of **tight junctions** forming a seal between the external environment and the internal environment of the blastocoel.

Gastrulation

Following the formation of the blastula, all animal embryos show a phase of cell and tissue movements called **gastrulation**, that converts the simple ball or sheet of cells into a three-layered structure known as the **gastrula**. The details of the morphoge-

Axes and symmetry

In order that specimens can be oriented in a consistent way, it is necessary to have terms for describing embryos (Fig. 2.12). If the egg is approximately spherical with an animal and vegetal pole then the line joining the two poles is the animal–vegetal axis. Unfertilized eggs are usually radially symmetrical around this axis, but after fertilization there is often a cytoplasmic rearrangement that breaks the initial radial symmetry and generates a bilateral symmetry. In some organisms, such as *Drosophila*, this may occur earlier, in the oocyte; in others, such as mammals, it may occur later, at a multicellular stage. But even animals such as sea urchins, which are radially symmetrical as adults, or gastropods, which are asymmetrical as adults, still have bilaterally symmetrical early embryos. The change of symmetry means that the animal now has a distinct **dorsal** (upper) and **ventral** (lower) side.

If the animal and vegetal poles are at the top and bottom, then the equatorial plane is the horizontal plane dividing the egg into **animal** and **vegetal** hemispheres, just like the equator of the Earth. Any vertical plane, corresponding to circles of longitude, is called a meridional plane. Once the embryo has acquired its bilateral symmetry then there is a particularly important meridional plane, the **medial** (= **sagittal**) plane, separating the right and left sides of the body. This is often, but not always, the plane of the first cleavage. The **frontal** plane is

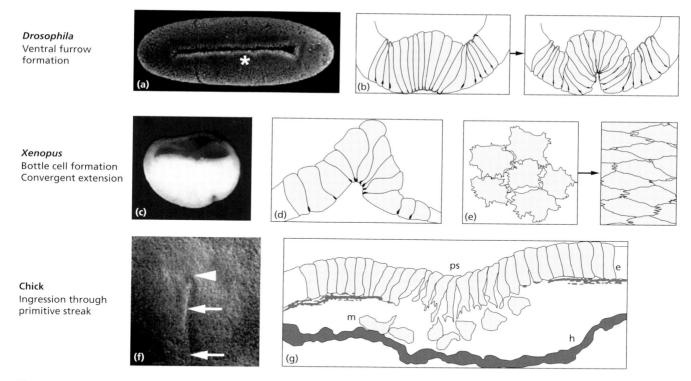

Fig. 2.11 Different processes during gastrulation. (a,b) Ventral furrow formation in *Drosophila*, * indicates the ventral furrow; (c–e) *Xenopus* gastrulation; (c) a bisected embryo, * indicates the dorsal lip; (d) bottle cells appearing at the blastopore; (e) intercalation of cells leading to axial elongation (convergent extension). (f,g) Ingression of cells through the primitive streak of a chick embryo. Reproduced from Hammerschmidt and Wedlich (2008), with permission from Company of Biologists Ltd.

the meridional plane at right angles to the medial plane, and is often, but not necessarily, the plane of the second cleavage.

Following gastrulation most animals become elongated. The head end is the **anterior**, the tail end is the **posterior**, so the head-to-tail axis is called **anteroposterior** (= craniocaudal or rostrocaudal). The top–bottom axis is called **dorsoventral** and the left–right axes are called **mediolateral**. In human anatomy, because we stand upright on two legs, the term anteroposterior is normally synonymous with dorsoventral, but the term "craniocaudal" remains acceptable for the head-to-tail axis. The terms **proximal** and **distal** are usually used in relation to appendages, proximal meaning "near the body" and distal meaning "further away from the body."

Generally, the principal body parts will become visible some time after completion of gastrulation. Some phyla, including annelids, arthropods, and chordates, show prominent **segmentation** of the anteroposterior axis. To qualify as segmented an organism should show repeated structures that are similar or identical to each other, are principal rather than minor body parts, and involve contributions from all the germ layers.

Although most animals have an overriding bilateral symmetry, this is not exact and there are systematic deviations which make right and left sides slightly different. For example in mammals the cardiac apex, stomach, and spleen are on the left and the liver, vena cava, and greater lung lobation are on the right. This asymmetrical arrangement is known as ***situs solitus***. If the arrangement is inverted, as occurs in some mutants or experimental situations, it is called ***situs inversus*** (Fig. 2.12). If the parts on the two sides are partly or wholly equivalent, it is called an **isomerism**.

Developmental control genes

The state of commitment of different body parts is controlled by the expression of specific sets of developmental control genes. These are sometimes called **homeotic** genes, or **selector** genes. The expression of these genes in embryos is controlled by cytoplasmic determinants or by inducing factors.

Developmental control genes virtually all encode transcription factors, whose function is regulating the activity of other genes. It is important to note that for such genes as much information is encoded by the "off" state as by the "on" state because the absence of a repressor can be equivalent to the presence of an activator. The existence of two discrete states of gene activity is a natural way of ensuring a sharp and discontinuous

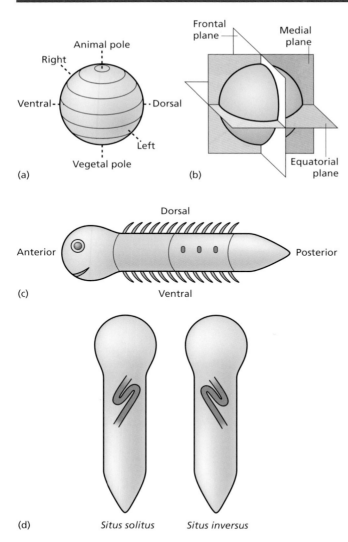

(a)

(b)

(c)

Anterior   Posterior

Dorsal

Ventral

(d)   *Situs solitus*   *Situs inversus*

Fig. 2.12 Axes and symmetry. (a) Axes of a fertilized egg after it has acquired a dorsoventral asymmetry. (b) Anatomical planes of an early embryo. (c) Principal axes of an animal viewed from the left side. (d) Ventral view of an animal showing deviation from bilateral symmetry.

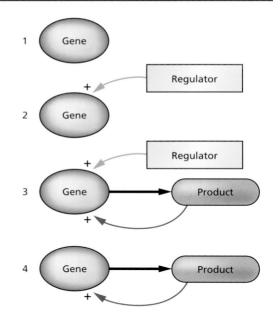

Fig. 2.13 Operation of a bistable switch. The figure depicts a temporal sequence: in step 2 the gene is upregulated by a regulator; in step 3 it is also upregulated by its product; in step 4 it remains on because of the product even though the regulator is now gone.

memory of exposure to the regulator. This is because the gene remains on permanently despite its transient exposure to the regulator. Thirdly, bistable switches are kinetic phenomena. This means that they depend on the continuous production and removal of substances, in this case the gene product. The simple equilibrium thermodynamic properties of binding and dissociation cannot create sharp thresholds or memory.

Although sharp thresholds of gene activity are very common, they are in most cases maintained by much more complex mechanisms than this simple positive-feedback loop, and involve many gene products.

Morphogen gradients

The properties of inductive signals explain how it is possible for an embryo to increase in complexity. There needs to be at least two regions to begin with: one emitting the signal and the other responding to it. Even if there is just one threshold response to the signal, the responding tissue will be partitioned into two territories so there is an increase of territories from two to three. Often, however, there is more than one threshold response to an inducing factor, at different concentrations of the signal. If the inducing factor is distributed in a concentration **gradient** then multiple threshold responses can bring about the formation of a complex pattern in one step. Such a gradient controls the types of territories, their sequence in space, and the overall orientation or polarity of the series of new structures. When

threshold response to a determinant or inductive signal. One way of ensuring that there are just two discrete states of activity for developmental control genes is to have a positive-feedback regulation, as shown in Fig. 2.13. This type of system is called a **bistable switch**, because it has two stable states. The gene is off if both the regulator and the gene product are absent. It is initially turned on by the regulator, which might be a cytoplasmic determinant or a signal transduction pathway activated by an inducing factor. Once the gene product has accumulated, the gene remains on even if the regulator is removed. This model shows three critically important features of gene regulation in development. Firstly, it can yield a sharp and discontinuous **threshold** in response to the regulator. Secondly, the system has

inducing factors are present as gradients with multiple responses, they are called **morphogens**.

A stable concentration gradient cannot be produced simply by releasing a pulse of the morphogen. Such a pulse would spread out by diffusion and eventually the concentration would become uniform all over the embryo. A concentration gradient is only set up in a situation where the morphogen is continuously produced in one region (a **source**) and destroyed in another (a **sink**). This will produce a gradient of concentration, with a flux of material from the source to the sink. A model that seems to fit the behavior of many natural systems maintains a constant concentration of morphogen in the signaling region, and has destruction at a rate proportional to local concentration throughout the responding tissue. This produces a gradient which is approximately of exponential form when it reaches the steady state.

A concentration gradient has two important properties. It can subdivide the competent zone of cells into several states of commitment by means of threshold responses, and it automatically imparts a **polarity** as well as a pattern to the responding tissue. In Fig. 2.14 are shown three examples to illustrate basic properties of this type of system. The competent field of cells is subdivided into four territories by the activation of three developmental control genes in a nested pattern. Because there are just the two states of gene activity, they are represented by binary digits where "1" means "on" and "0" means "off." The action of the gradient will subdivide the field into four territories (here head and three body segments) with the codings 000, 001, 011, 111. In the second example, suppose that a graft has been carried out to place a second source at the other end of the field. With a source at both ends and destruction throughout the central region, the concentration gradient will become U-shaped. The same threshold responses will now produce a different pattern 111, 011, 001, 011, 111. This is a type of structure called a mirror-symmetrical duplication, because it consists of two similar halves joined by a plane of **mirror symmetry**. The polarity is normal in the right half but inverted in the left half of the field. Mirror-symmetrical duplications arise quite often in embryological experiments following the grafting of signaling centers, for example the double-dorsal *Xenopus* embryo arising from the creation of a second organizer (see Chapter 7), or the double posterior limbs produced by ZPA grafting (Chapter 15). In the lower panel is shown another type of experiment involving the insertion of an impermeable barrier that interrupts the passage of the morphogen. Because no morphogen is being produced on the left-hand side, the concentration soon falls to zero, while on the right-hand side it actually piles up to a higher concentration than normal. This is because the size of the sink has been reduced relative to the source. As the rate of destruction is proportional to concentration, the overall concentration has to increase to re-establish the steady state. Operation of the same threshold responses shows that not only has the left half of the pattern been lost, but so has part of the right half because the elevation of concentration has

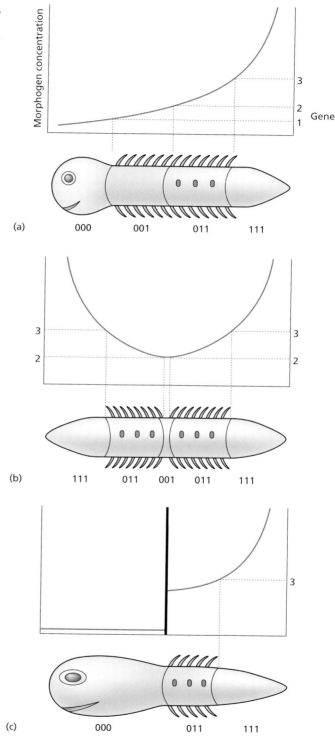

Fig. 2.14 Properties of morphogen gradients. (a) Normal development of an animal with head and three segments. (b) Graft of the posterior source to the anterior causes formation of a U-shaped gradient and produces a double-posterior animal. (c) Insertion of an impermeable barrier causes formation of a large gap in the pattern.

expanded the size of the most posterior territories. This example shows how the properties of developmental systems may not be obvious at first sight. They may sometimes be counterintuitive, and interpretation of experimental results may require an understanding of the properties of the underlying dynamic mechanisms as well as a familiarity with the genes and signaling molecules at work.

Homeotic mutations

A homeotic mutation is one that can convert one body part into another. How this can happen is explained by our simple model. Suppose that the organism of Fig. 2.15 is a mutant and that the second gene in the series is permanently inactive and cannot be turned on. In this case the codings will be 000, 001, 001, and 101. In other words the second body segment has now been turned into another copy of the first body segment (Fig. 2.15a,b). What happens to the third, tail, segment cannot be predicted without knowing more about the logical circuitry, because this

now has a novel coding not found in the normal organism. This example illustrates the behavior of a loss-of-function homeotic mutation. Such mutations are genetically **recessive** because function would be restored by a good copy of the gene on either chromosome.

On the other hand, suppose that the mutation in gene 2 caused constitutive activity, in other words the gene is on all the time everywhere. Now the sequence of codings would become 010, 011, 011, 111. Here the first body segment has become a copy of the second body segment and the head has an abnormal coding (Fig. 2.15c). This is a gain-of-function mutation. It is genetically **dominant** because inappropriate activation will occur with only one copy of the mutated gene.

This example represents, in a highly oversimplified manner, the operation of the Hox gene system for anteroposterior patterning of animals. In general, loss-of-function mutations produce anteriorizations, as in Fig. 2.15b, and gain-of-function mutations produce posteriorizations, as in Fig. 2.15c. It also illustrates the **combinatorial** nature of regional specification in development. Although striking anatomical transformations may sometimes be produced by the manipulation of individual genes, the identity of body parts is actually controlled in normal development by combinations of several genes.

Homeotic, homeobox, and Hox

Considerable confusion arises about the relationship between the definition of homeotic behavior and the molecular identity of homeotic genes. In 1984, a common DNA sequence was discovered in the genes of the Bithorax and Antennapedia complexes of *Drosophila* (see Classic experiments box "Discovery of the homeobox" in Chapter 11). Because these genes were homeotic, this motif was called the "**homeobox**." The proteins encoded by these genes are all transcription factors and the homeobox encodes a sequence of 60 basic amino acids which form their DNA binding domain, called the homeodomain. Subsequently, many homeobox-containing genes were found in all types of eukaryotic organisms. All homeobox-containing genes encode homeodomain-containing transcription factors. Many, but not all, of these are concerned with development, but rather few of them are homeotic when mutated or misexpressed.

The **Hox genes** are a family of genes, found in animals but not in other eukaryotes, that are responsible for specifying **anteroposterior** identity to body levels. They are a subset of the general class of homeobox genes and are specifically the homologues of the *Drosophila* Bithorax and Antennapedia gene clusters. In many animals the Hox genes form a single gene cluster, with the different genes adjacent to each other on the chromosome.

The Hox genes are activated at an early stage in body-plan formation and are generally maximally expressed around the **phylotypic** stage of the group in question. Each gene in the

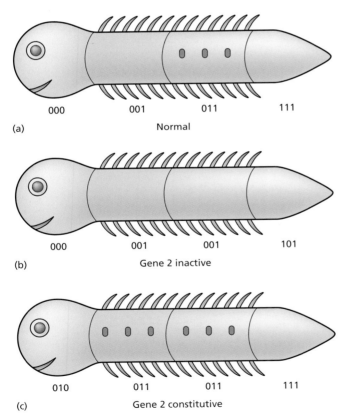

(a)   000   001   011   111   Normal

(b)   000   001   001   101   Gene 2 inactive

(c)   010   011   011   111   Gene 2 constitutive

Fig. 2.15 Homeotic mutants. (a) Normal genotype and phenotype. (b) Loss-of-function mutation of gene 2 causes second body segment to resemble the first. (c) Gain-of-function mutation of gene 2 causes first body segment to resemble the second. This example assumes that the abnormal codings (010 and 101) do not produce homeotic effects.

cluster is expressed at a particular anteroposterior level, running from a sharp anterior boundary to fade out gradually in the posterior. They are expressed in both central nervous system and mesoderm. Remarkably, the spatial order of expression of Hox genes in the body, from anterior to posterior, is often the same as the order of the arrangement of the genes on the chromosome. Invertebrates have just one cluster of Hox genes but vertebrates have four or more clusters, each situated on a different chromosome.

Morphogenetic processes

Cell shape changes and movements are fundamental to early development as the embryo needs to convert itself from a simple ball or sheet of cells into a multilayered and elongated structure. The processes by which this is achieved are called **gastrulation**. Although the details of gastrulation can differ substantially between even quite similar species, the basic cellular processes are common. At later stages of development the same repertoire of processes is re-used repeatedly in the morphogenesis of individual tissues and organs.

From a morphological point of view, most embryonic cells can be regarded as epithelial or mesenchymal (Fig. 2.16). These terms relate to cell shape and behavior rather than to embryonic origin, as epithelia can arise from all three germ layers and mesenchyme can arise from both ectoderm and mesoderm. An **epithelium** is a sheet of cells, arranged on a **basement membrane**, each cell joined to its neighbors by specialized junctions, and showing a distinct apical–basal polarity. **Mesenchyme** is a descriptive term for scattered cells of stellate appearance embedded in loose extracellular matrix. It fills up much of the embryo and later forms fibroblasts, adipose tissue, smooth muscle, and skeletal tissues. Epithelia and mesenchyme are further described in Chapter 18.

Cell movement

Many morphogenetic processes depend on the movement of individual cells. This is most apparent in the case of migrations, for example of the neural crest cells or the germ cells, which may move long distances as individuals. But shorter-range movements are also important for processes such as differential adhesion or shape changes in cell sheets. The mechanism of cell movement is most apparent in fibroblasts moving on a substratum (Fig. 2.17). They extend a flat process called a **lamellipodium** (plural **lamellipodia**) which is rich in microfilaments. This attaches to the substratum by focal attachments containing integrins, which are connected to the microfilament bundles by a complex of actin-associated proteins. The body of the cell is then drawn forward by a process involving active contraction in which myosin molecules migrate towards the plus end of microfilaments. Cells in embryos are thought to move in essentially the same way. Instead of a large flat lamellipodium they may extend multiple thin **filopodia** to form the contacts. Individual cell movement often occurs up a concentration gradient of a substance emitted by a signaling center. Very frequently this involves the chemokine SDF1 and its receptor CXCR4.

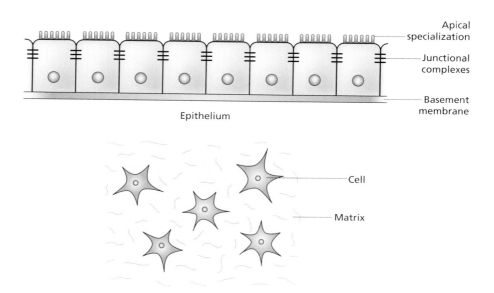

Fig. 2.16 Epithelium and mesenchyme.

Epithelium

Mesenchyme

Apical specialization

Junctional complexes

Basement membrane

Cell

Matrix

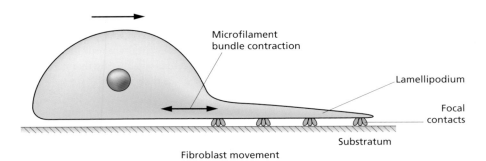

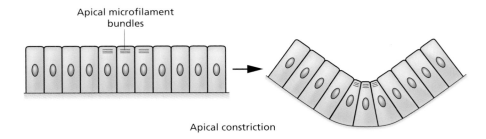

Fig. 2.17 Types of cell movement.

Nerve axons also elongate by a similar mechanism. At the tip is a flattened structure called a **growth cone**, which emits the filopodia. The occurrence and specificity of nerve connections are also controlled by secreted substances, of which there are several groups including ephrins, netrins, semaphorins, and slits.

Microfilament bundles and their associated motor proteins also control cell shape. An apical constriction in a group of epithelial cells will reduce the apical surface area and increase the length of the cells (Fig. 2.17). This is often a preliminary to an **invagination** movement in which the cells leave their epithelium and enter the space below.

Cell adhesion

Vertebrate epithelial cells are bound together by tight junctions, adherens junctions and desmosomes, the latter two types involving cadherins as major adhesion components. Mesenchymal cells may also adhere by means of cadherins but usually more loosely. The adhesion of early embryo cells is usually dominated by the cadherins, as shown by the fact that most types of early embryo can be fully disaggregated into single cells by removal of calcium from the medium.

There is some qualitative specificity to cell adhesion. Cadherin-based adhesion is **homophilic** and so cells carrying E-cadherin will stick more strongly to each other than to cells bearing N-cadherin. The calcium-independent immunoglobulin superfamily-based adhesion systems such as N-CAM (neural cell adhesion molecule), particularly important on developing

neurons and glia, are different again, and also promote adhesion of similar cells. This qualitative specificity of adhesion systems provides a mechanism for the assembly of different types of cell aggregate in close proximity, and also prevents individual cells wandering off into neighboring domains. If cells with different adhesion systems are mixed they will sort out into separate zones, eventually forming a dumbbell-like configuration or even separating altogether (Fig. 2.18a).

In addition to the qualitative aspect of specificity, it is also known that cell-sorting behavior can result simply from different strengths of adhesion of the same system. If cell type A is more adhesive than cell type B then B will eventually surround A (Fig. 2.18b). The process is based on the existence of small, random movements of the cells in the aggregates and the same final configuration will be reached from any starting configuration, for example from an intimate mixture of the two types, or from blocks of the two types pressed together. An experimental example is shown in Fig. 2.19.

Classification of morphogenetic processes

Similar repertoires of morphogenetic processes are re-used repeatedly in different developmental contexts (Fig. 2.20). They are driven partly by shape changes of cells due to the contraction of microfilament complexes, partly by changes in gene expression of the cell adhesion molecules: cadherins, CAMs, and integrins, and partly by local changes in cell division rates. These events are in turn under the control of regional specifica-

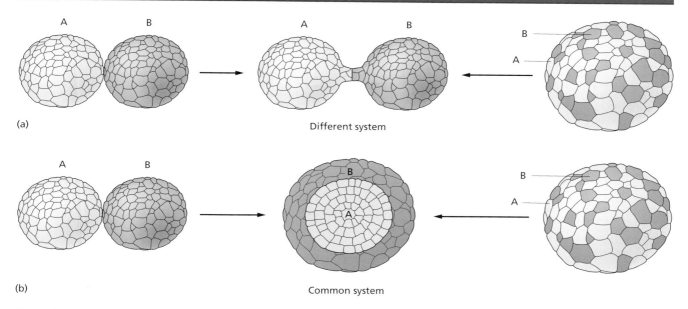

(a)

Different system

(b)

Common system

Fig. 2.18 Cell-sorting processes. (a) Cell types A and B each have their own adhesion mechanism, so adhesion A–B<< adhesion A–A or B–B. (b) Cell types A and B have the same adhesion mechanism, but adhesion A–A > B–B and adhesion A–B has the average strength: (A–A + B–B)/2. The center image in (b) represents a section while the others represent surface views.

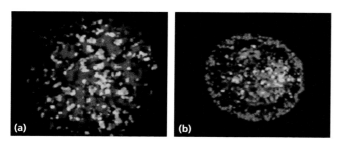

Fig. 2.19 An experiment to demonstrate cell sorting by differential adhesion. The red and green cells are colored with vital dyes. They are identical except for their level of production of N-cadherin, from a transfected gene, which is 2.4 times greater in the green cells. The cell types are mixed together in an aggregate, viewed as a confocal optical section. After 24 hours they have sorted out, with the more adhesive green cells in the middle. Reproduced from Foty and Steinberg. *Dev Biol* 2005; 278: 255–263 with permission from Elsevier.

tion mechanisms: the inductive signals and regulatory genes (see Animation 1: Morphogenetic processes).

Condensation of cells to form an aggregate (Fig. 2.20a) is often a prelude to the formation of structures, for example the somites, or the skeletal elements of the limbs. Condensation arises partly by increased local cell division, partly by reduction of local matrix secretion, and partly by increased cell–cell adhesion.

Invagination and **involution** (Fig. 2.20b) are processes that generate multilayered structures from a simple epithelium. They are found in gastrulation, in neurulation, and in the for-

mation of many other structures such as glands, sense organs, and insect appendages. They are usually initiated from a localized apical constriction that arises from a contraction of the terminal microfilament complexes. In an **invagination** the constriction causes the cell sheet to buckle so that the constricted region of cells forms a protrusion into the interior. Internalization of this protrusion can then occur by several different mechanisms. They will normally involve differential adhesion between the invaginating cells and the surroundings, for example the expression of N-CAM by the forming neural plate and L-CAM (leucocyte cell adhesion molecule) by the epidermis helps to ensure closure and complete invagination of the neural tube. In a distinct mechanism also depending on cell adhesion, the forming archenteron of the sea-urchin embryo is drawn into the embryo by filopodia thrown from cells of the initial invagination across to the far side of the gastrula. If the initial entry of cells arises not from mechanical buckling of the sheet but from a migration of cells around the edge of the constricted surface then it is called an **involution**. This will involve the formation of a free edge in the involuting tissue. Again the process depends on a differential cell adhesion between the involuting cells and their surroundings. The cell movement may arise from individual migration or by convergent extension (see below) or both.

Invagination is one way of generating a hollow ball or tube of cells. Alternatively, a fluid-filled space can be hollowed out from a solid mass of cells by **cavitation** (Fig. 2.20c). This may occur either by cell rearrangement, as in secondary neurulation, or by apoptosis of the cells in the interior, as found for example in formation of the mouse egg cylinder. However a cavity is

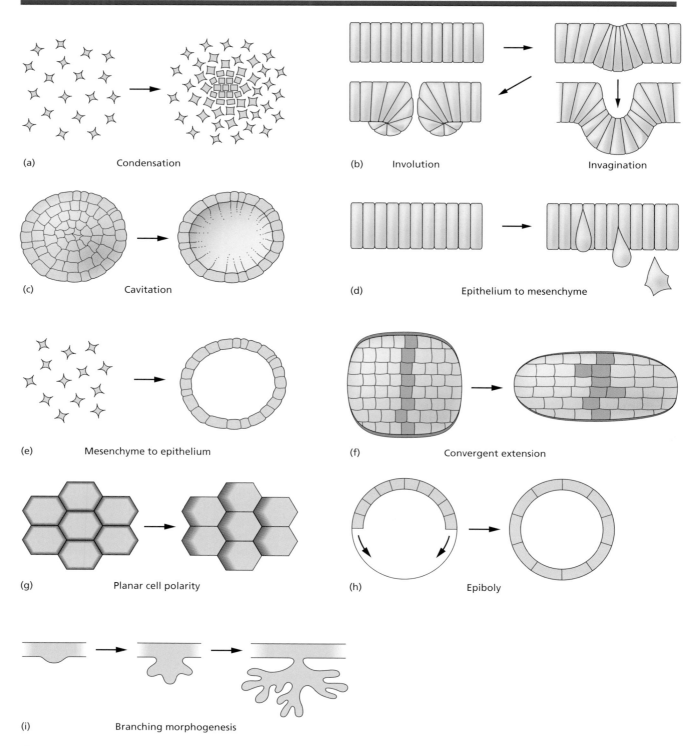

(a)   Condensation

(b)   Involution   Invagination

(c)   Cavitation

(d)   Epithelium to mesenchyme

(e)   Mesenchyme to epithelium

(f)   Convergent extension

(g)   Planar cell polarity

(h)   Epiboly

(i)   Branching morphogenesis

Fig. 2.20 Classification of morphogenetic processes. (a) Mesenchymal condensation. (b) Invagination and involution. (c) Cavitation. (d) Epithelium to mesenchyme transition. (e) Mesenchyme to epithelium transition. (f) Convergent extension. Here a typical file of cells is labeled and the intercalation movements of the cells rearrange the file so that some cells are excluded. (g) Planar cell polarity. (h) Epiboly. (i) Branching morphogenesis.

produced, it may later be referred to as a **lumen**, and the cells abutting it as being on the lumenal (also spelt luminal) surface.

Epithelial-to-mesenchymal transitions (Fig. 2.20d) occur whenever cells leave an epithelium and move off as individuals or as a mesenchymal mass, also referred to as **delamination**. This happens in the ingression of cells from the chick epiblast to form the hypoblast, or in the formation of the neural crest from the dorsal neural tube. It requires a local reduction in cell–cell adhesion in the cells that separate out from the epithelium. The reverse mesenchyme-to-epithelium transition (Fig. 2.20e) is also found, for example in the formation of the coelomic lining epithelium or of kidney tubules.

Cell sheets can change their shape quite dramatically as a result of active cell rearrangement. This usually takes the form of **convergent extension** (Fig. 2.20f), in which individual cells intercalate in between each other, causing a constriction of the sheet in the direction of intercalation and an elongation of the sheet at right angles to the intercalation. The intercalation movements depend on contractile activity at the termini of the cells, which exert traction on the extracellular matrix and force the cells in between one another. The overall polarity of the process is controlled by boundary regions at which the contractile activity is inhibited. These boundaries will eventually become the long edges of the elongated cell sheet.

During convergent extension the cells all need to be polarized in the same direction. This is an instance of another general process called **planar cell polarity** (Fig. 2.20g), referring to the acquisition of a polarity by cells in an epithelium in the direction of the plane of the epithelium (in addition to the apical–basal polarity found in all epithelial cells). This requires a long-range signal to orient the direction of polarization, and also on local communication that enables the cellular asymmetry to be propagated from one cell to the next. Planar cell polarity is found in most epidermal structures of *Drosophila*, and in several vertebrate tissues including the epidermis and the cochlea.

Also found in gastrulation is **epiboly** (Fig. 2.20h), in which a sheet of cells expands to surround and enclose another population. The mechanisms of this may be somewhat diverse. *Xenopus* and zebrafish gastrulation both show epiboly of the animal hemisphere, which expands to cover the whole embryo. However, in *Xenopus* the process is driven by radial intercalation of cells producing a thinner sheet of greater area, thus showing some similarity to convergent extension. By contrast, in the zebrafish it is driven by the expansion of the acellular yolk syncytial layer by a microtubule-dependent process.

A type of morphogenetic behavior that is characteristic of organogenesis, rather than early development, is **branching morphogenesis**. Typically, an epithelial bud grows into a mesenchymal cell mass and as it does so the number of growing points progressively increases to generate a branched structure (Fig. 2.20i). This process may depend mainly on cell movement, as in the formation of the *Drosophila* tracheal system, or by differential growth at the tips, as is more usual in formation of vertebrate organs such as the lung or kidney.

Growth and death

Growth usually involves cell division and the molecular biology of the animal cell cycle is quite well understood. A typical cell cycle is shown in Fig. 2.21 and some typical patterns of cell division in Fig. 2.22. The cell cycle is conventionally described as consisting of four phases: M indicates the phase of mitosis, S indicates the phase of DNA replication, and G1 and G2 are the intervening phases. For growing cells, the increase in mass is continuous around the cycle and so is the synthesis of most of the cell's proteins. Normally, the cell cycle is coordinated with the growth rate. If it were not, cells would increase or decrease in size with each division, as happens for example during embryonic cleavage divisions. There are various internal controls built into the cycle to ensure that mitosis does not start before DNA replication is completed. These controls operate at **checkpoints** around the cycle at which the process stops unless the appropriate conditions are fulfilled.

Control of the cell cycle depends on a metabolic oscillator comprising a number of proteins called cyclins and a number of cyclin-dependent protein kinases (Cdks). In order to pass the M checkpoint and enter mitosis, a complex of cyclin and Cdk (called M-phase promoting factor, or **maturation promoting factor, MPF**) has to be activated. This phosphorylates and

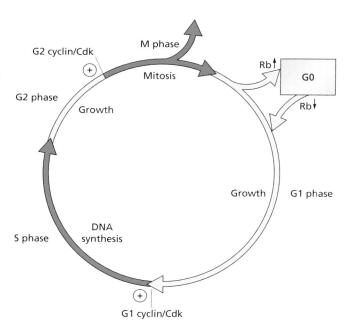

Fig. 2.21 The cell cycle of a typical eukaryotic cell. ⊕ Indicates checkpoints at the entry to M phase and S phase.

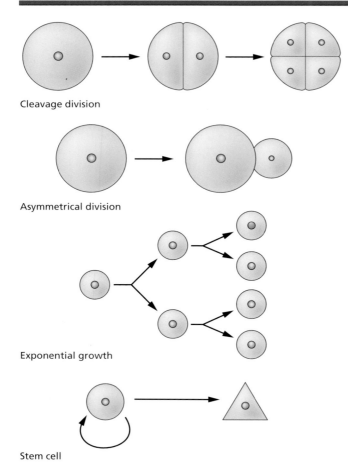

Cleavage division

Asymmetrical division

Exponential growth

Stem cell

Fig. 2.22 Types of cell division.

thereby activates the various components required for mitosis (nuclear breakdown, spindle formation, chromosome condensation). Exit from M phase requires the inactivation of MPF, via the destruction of cyclin, so by the end of the M phase it has disappeared.

Passage of the G1 checkpoint depends on a similar process operated by a different set of cyclins and Cdks, whose active complexes phosphorylate and activate the enzymes of DNA replication. This is also the point at which the cell size is assessed. The cell cycle of G1, S, G2, and M phases is universal although there are some modifications in special circumstances. The rapid cleavage cycles of early development have short or absent G1 and G2 phases and there is no size check, the cells halving in volume with each division. The meiotic cycles require the same active MPF complex to get through the two nuclear divisions, but there is no S phase in between.

The early embryo cleavage divisions occur in the absence of extracellular growth factors. At later stages of development, and particularly in the mature organism, most cells are quiescent. In the absence of growth factors cells enter a state called G0, in which the Cdks and cyclins are absent. Restitution of

growth factors induces the resynthesis of these proteins and the resumption of the cycle starting from the G1 checkpoint. One factor maintaining the G0 state is a protein called Rb (retinoblastoma protein). This becomes phosphorylated, and hence deactivated, in the presence of growth factors. In the absence of Rb, a transcription factor called E2F becomes active and initiates a cascade of gene expression culminating in the resynthesis of cyclins, Cdks, and other components needed to initiate S phase.

True growth, meaning increase in size, is not quite the same as cell division. An obvious example of the dissociation between the two is the embryonic cleavage divisions where cell division occurs without growth (Fig. 2.22a). In later development growth does require an increase in cell number, with doubling of size between divisions, but often also involves an increase in cell size or an increase in the amount of the extracellular matrix. Extracellular material is not extensive in early embryos, but later on a substantial proportion of skeletal tissues such as bone and cartilage are composed of matrix materials. In fact there is no real growth at all in free-living embryo types such as *Xenopus*, zebrafish, or sea urchins. They may expand to some extent because of uptake of water but they do not increase in actual content of matter until they are able to feed themselves. By contrast those embryo types that do have an external nutrient supply, such as the placenta in mammals, or a large extraembryonic yolk mass, as in birds and reptiles, do grow extensively throughout development. This has implications for experiments, because materials injected into cells of a free-living embryo do not become not diluted by growth and may remain active for a considerable time.

In embryos some cells can be seen to undergo visibly **asymmetric divisions** (Fig. 2.22b). This is often associated with the formation of a cytoplasmic asymmetry prior to their division such that cytoplasmic **determinants** inherited by the two daughter cells evoke different patterns of gene expression in their nuclei and thus bring about different pathways of development. The nature of determinants is now quite well understood and include localized mRNAs and self-organizing protein complexes such as the PAR complex (see Chapters 4 and 12).

Cells often have the capability for exponential growth in tissue culture (Fig. 2.22c) but this is very rarely found in animals. Although some differentiated cell types can go on dividing, there is a general tendency for differentiation to be accompanied by a slow down or cessation of division. In postembryonic life most cell division is found among **stem cells** and their immediate progeny called **transit amplifying cells** (see Chapter 18). Stem cells are cells that can both reproduce themselves and generate differentiated progeny for their particular tissue type (Fig. 2.22d). This does not necessarily mean that every division of a stem cell has to be an asymmetrical one, but over a period of time half the progeny will go to renewal and half to differentiation. The term "stem cell" is also used for **embryonic stem cells (ES cells)** derived from early mammalian embryos. These can be expanded without limit in tissue culture and are capable of repopulating embryos and contributing to all tissue types (see Chapter 10 and 21).

Programmed cell death or **apoptosis** is also important in development. Apoptosis is a method of removing cells without spilling their bioactive contents into the surroundings. It involves a molecular pathway that culminates in the activation of proteases called caspases (see Chapter 12). These cause the nucleus to condense, the cell to shrink, and to display on its surface a signal for engulfment by other, phagocytic, cells. Apoptosis is often initiated by a withdrawal of growth factors from the cell, but can sometimes be an active response to a signal. In vertebrate development, apoptosis is particularly important in terms of the reduction of motor-neuron pools in response to the availability of growth factors from their target organ (see Chapter 14), and in limb development where digits form because the tissue in between is removed by programmed cell death (see Chapter 15).

Key points to remember

- The main processes in animal development are regional specification, cell differentiation, morphogenesis, growth, and temporal control.
- Gametes arise from cells of the germ line by meiosis.
- Events at the earliest stages of development involve components preformed in the egg and so depend on the genome of the mother.
- Spontaneous symmetry breaking processes often cause localization of cytoplasmic determinants, which then control the first steps of regional specification.
- Later regional specification involves the action of inducing factors, often in the form of morphogen gradients that can upregulate different developmental control genes at different concentrations.
- Each territory of cells in the embryo has its developmental commitment specified by the activity of a specific combination of developmental control genes. If mutated, these genes may cause homeotic transformations in which one body part is converted into another.
- Early development involves a period of cleavage, leading to the formation of a blastula or blastoderm. This is followed by a phase of morphogenetic movements called gastrulation, during which the three germ layers, ectoderm, mesoderm, and endoderm, are formed.
- Morphogenetic processes in early development include cavitation, involution, invagination, epiboly, and cell condensation. In later development are also found transitions between epithelium and mesenchyme and branching processes.
- Growth normally depends on cell division. The cell cycle of G1, S, G2, and M phases is universal but is modified for specialist developmental processes such as meiosis and cleavage divisions.

Further reading

Germ cells and soma

Matova, N. & Cooley, L. (2001) Comparative aspects of animal oogenesis. *Developmental Biology* **231**, 291–320.

Extavour, C.G. & Akam, M. (2003) Mechanisms of germ cell specification across the metazoans: epigenesis and preformation. *Development* **130**, 5869–5884.

Seydoux, G. & Braun, R.E. (2006) Pathway to totipotency: lessons from germ cells. *Cell* **127**, 891–904.

Strome, S. & Lehmann, R. (2007) Germ versus soma decisions: lessons from flies and worms. *Science* **316**, 392–393.

Fertilization

Strickler, S.A. (1999) Comparative biology of calcium signalling during fertilization and egg activation in animals. *Developmental Biology* **211**, 157–176.

Jungnickel, M.K., Sutton, K.A. & Florman, H.M. (2003) In the beginning: lessons from fertilization in mice and worms. *Cell* **114**, 401–404.

Symmetry breaking, self-organization, asymmetry

Wood, W.B. (1997) Left-right asymmetry in animal development. *Annual Reviews in Cell and Developmental Biology* **13**, 53–82.

Levin, M. & Palmer, A.R. (2007) Left-right patterning from the inside out: Widespread evidence for intracellular control. *BioEssays* **29**, 271–287.

Gönczy, P. (2008) Mechanisms of asymmetric cell division: flies and worms pave the way. *Nature Reviews Molecular Cell Biology* **9**, 355–366.

Karsenti, E. (2008) Self-organization in cell biology: a brief history. *Nature Reviews Molecular Cell Biology* **9**, 255–262.

Oates, A.C., Gorfinkiel, N., Gonzalez-Gaitan, M. & Heisenberg, C.-P. (2009) Quantitative approaches in developmental biology. *Nature Reviews Genetics* **10**, 517–530.

Gradients, thresholds, homeosis

Slack, J.M.W. (1987) Morphogenetic gradients – past and present. *Trends in Biochemical Sciences* **12**, 200–204.

Lewis, E.B. (1994) Homeosis: the first 100 years. *Trends in Genetics* **10**, 341–343.

Xiong, W. & Ferrell, J.E. (2003) A positive-feedback-based bistable 'memory module' that governs a cell fate decision. *Nature* **426**, 460–465.

Tabata, T. & Takei, Y. (2004) Morphogens, their identification and regulation. *Development* **131**, 703–712.

Ashe, H.L. & Briscoe, J. (2006) The interpretation of morphogen gradients. *Development* **133**, 385–394.

Dahmann, C., Oates, A.C. & Brand, M. (2011) Boundary formation and maintenance in tissue development. *Nature Reviews Genetics* **12**, 43–55.

Rogers, K.W. & Schier, A.F. (2011) Morphogen gradients: from generation to interpretation. *Annual Review of Cell and Developmental Biology* **27**, 377–407.

Morphogenesis

Bard, J.B.L. (1992) *Morphogenesis. The Cellular and Molecular Processes of Developmental Anatomy.* Cambridge: Cambridge University Press.

Irvine, K.D. & Rauskolb, C. (2001) Boundaries in development: formation and function. *Annual Reviews in Cell and Developmental Biology* **17**, 189–214.

Schöck, F. & Perrimon, N. (2002) Molecular mechanisms of epithelial morphogenesis. *Annual Reviews in Cell and Developmental Biology* **18**, 463–493.

Thiery, J.P. (2003) Cell adhesion in development: a complex signaling network. *Current Opinion in Genetics and Development* **13**, 365–371.

Hardin, J. & Walston, T. (2004) Models of morphogenesis: the mechanisms and mechanics of cell rearrangement. *Current Opinion in Genetics and Development* **14**, 399–406.

Stern, C.D., ed. (2004) *Gastrulation: From Cells to Embryo.* New York: Cold Spring Harbor Laboratory Press.

Klein, T.J. & Mlodzik, M. (2005) Planar cell polarization: an emerging model points in the right direction. *Annual Review of Cell and Developmental Biology* **21**, 155–176.

Steinberg, M.S. (2007) Differential adhesion in morphogenesis: a modern view. *Current Opinion in Genetics and Development* **17**, 281–286.

Bryant, D.M. & Mostov, K.E. (2008) From cells to organs: building polarized tissue. *Nature Reviews Molecular Cell Biology* **9**, 887–901.

Hammerschmidt, M. & Wedlich, D. (2008) Regulated adhesion as a driving force of gastrulation movements. *Development* **135**, 3625–3641.

Goodrich, L.V. & Strutt, D. (2011) Principles of planar polarity in animal development. *Development* **138**, 1877–1892.

Takeichi, M. (2011) Self-organization of animal tissues: cadherin-mediated processes. *Developmental Cell* **21**, 24–26.

Cell cycle and cell death

Murray, A. & Hunt, T. (1993) *The Cell Cycle: An Introduction.* Oxford: Oxford University Press.

Day, S.J. & Lawrence, P.A. (2000) Measuring dimensions: the reputation of size and shape. *Development* **127**, 2977–2987.

Lawen, A. (2003) Apoptosis – an introduction. *Bioessays* **25**, 888–896.

O'Farrell, P.H. (2003) How metazoans reach their full size: the natural history of bigness. In *Cell Growth: Control of Cell Size*, Hall, M.N., Raff, M., & Thomas, G., eds, Cold Spring Harbor Laboratory Press, pp. 1–21.

Domingos, P.M. & Steller, H. (2007) Pathways regulating apoptosis during patterning and development. *Current Opinion in Genetics and Development* **17**, 294–299.

Lecuit, T. & Le Goff, L. (2007) Orchestrating size and shape during morphogenesis. *Nature* **450**, 189–192.

Time

Moss, E.G. (2007) Heterochronic genes and the nature of developmental time. *Current Biology* **17**, R425–R434.

Tennessen, J.M. & Thummel, C.S. (2008) Developmental timing: let-7 function conserved through evolution. *Current Biology* **18**, R707–R708.

This chapter contains the following animation:

Animation 1 Morphogenetic processes.

 For additional resources for this book visit www.essentialdevelopmentalbiology.com

Chapter 3

Approaches to development: developmental genetics

Developmental mutants

Of all the genes in the genome, approximately 1–2% have functions that are specifically concerned with development. Many more are needed in order that normal development should take place, but their primary functions lie in the central areas of cell biology or metabolism rather than development. Considerable use is made of **mutants** in developmental biology. The **mutations** in specific genes carried by a mutant organism may be spontaneous, or induced by mutagenic treatments such as chemical mutagens or radiation, or, particularly in the mouse, may be specifically designed "targeted" mutations. Mutations fall into several classes in terms of the molecular basis of the change. Chemical mutagenesis usually results in the creation of point mutations in which a single DNA base is changed to another. These may cause the substitution of one amino acid for another, or, if a new termination codon is created, cause a premature chain termination. Addition or deletion of a single nucleotide will produce a frameshift mutation, changing the entire downstream sequence of amino acids in the protein. X-irradiation often induces deletions of a whole stretch of DNA, which may include more than one gene. Spontaneous mutations can be of any of these types, and can, in addition, include the insertion of **transposable elements**.

Homeotic mutations cause the conversion of one part of the body into the likeness of another. The homeotic genes in which such mutations occur code for transcription factors, and are developmental control genes whose normal function is to encode the state of commitment of a cell. However, many developmental control genes do not display homeotic mutations.

The different versions of a gene are called **alleles.** There will usually be just one normal allele, called the **wild type**, but there are an almost infinite number of different possible mutant alleles (Fig. 3.1). The totality of nuclear DNA in an organism is known as the **genome**, and the specific combination of alleles carried is the **genotype**. In developmental biology, the term genotype is usually used in a specific context to refer to the constitution of just one or a few loci. The totality of characteristics of an organism is known as its **phenotype**, and in developmental biology this usually relates to its visible appearance. The normal, or wild-type, phenotype arises from a wild-type genome, and a mutant phenotype arises from the consequences of one or more mutations carried in the genome. It is not possible to deduce the complete function of a gene simply by looking at a mutant phenotype, although when combined with the primary sequence of the gene and the normal expression pattern, a mutant phenotype can be very informative.

It is possible to deduce quite a lot about normal gene function by looking at the effects of several mutant alleles of the same gene, but with different phenotypes. The most common type of mutation is a **loss of function**, meaning that the protein product of the mutant gene is less active than the wild type. A complete loss of function is called a **null** mutation and corresponds to a complete lack of active gene product. Sometimes there exists a set of alleles having different degrees of loss of function which can be arranged into an **allelic series**, ordered by the severity of the abnormal phenotype (Fig. 3.2). The set of phenotypes displayed by an allelic series may make the function of the wild-type gene much more apparent than the phenotype of a single mutant allele. A weak loss-of-function mutant is more likely to be viable and so survive into adult life than a null. Loss-of-function mutations are usually genetically **recessive**, because their effect will be masked by the presence of a wild-type allele on the other chromosome which produces the normal gene product. Sometimes they are genetically **dominant** because the

Essential Developmental Biology, Third Edition. Jonathan M.W. Slack.
© 2013 John Wiley & Sons, Ltd. Published 2013 by John Wiley & Sons, Ltd.

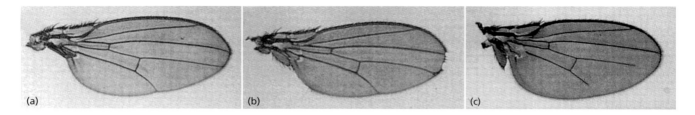

Fig. 3.1 Different alleles of one gene produce different phenotypes. (a) Wild-type *Drosophila* wing. (b) Dominant haploinsufficient mutant of *Notch* showing notches in the wing margin. (c) Recessive allele of *Notch* showing vein shortening. Reproduced from Artavanis-Tsakonas *et al. Science* 1995; 268: 225–232, with permission from AAAS.

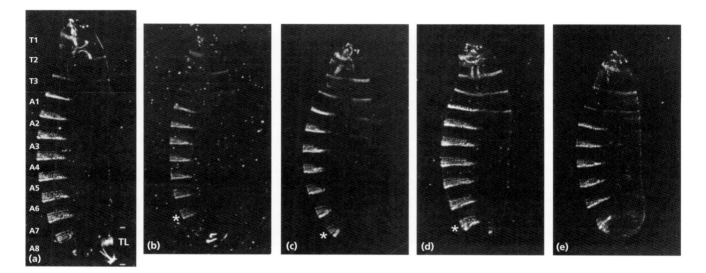

Fig. 3.2 Example of an allelic series of the *Drosophila* gene *tailless*. The figures show cuticular preparations of whole *Drosophila* embryos with the anterior end up. (a) Wild type. (b–e) Alleles showing successively greater loss of function. The more severe mutants progressively lose structures from both the anterior and posterior ends of the body. * Indicates the seventh abdominal segment. Reproduced from Strecker *et al.* (1988), with permission from Company of Biologists Ltd.

loss of 50% of the product is sufficient to cause an abnormal phenotype. This type of dominance is called **haploinsufficiency**. It has the property that the homozygous phenotype, with 100% loss of function, is much more severe than the heterozygous phenotype with only 50% loss.

There are also dominant mutants showing **gain-of-function** phenotypes. For example a cell-surface receptor may normally become activated when the ligand is bound to the extracellular domain. But a gain-of-function mutant may signal all the time regardless of whether the ligand is present or not. Similarly, a transcription factor may normally turn on a target gene in response to some regulatory event, but a gain-of-function mutant may be active all the time and not respond to the regulation. Gene products that are active all the time are called **constitutive**, and gain-of-function mutants often express constitutive versions of the normal gene product. Such mutants are usually genetically dominant because the gain of function persists even in the presence of wild-type gene product made by the allele on the other chromosome. Another type of mutation that is also genetically dominant, but is not constitutive, is the **dominant negative**. Here the mutant form of the gene product itself has no function, but it interferes with the function of the wild-type form. This may arise, for example, where molecules need to form dimers in order to exert their activity. If the dimer of the dominant negative and the normal form is inactive then the overall activity will be well below the 50% characteristic of a recessive mutation. Recessivity of a mutation is generally indicative of loss of function, but dominance may be due either to a haploinsufficient, or constitutive, or dominant negative effect, and further investigation is required to find which is the case. In principle, dominant mutant types can be distinguished by gene dosage analysis since introducing extra copies of a haploinsufficient allele will have little effect, while introducing extra copies of a gain-of-function allele will increase the effect. But such studies are not necessarily easy to perform in all the model organisms.

Genes with many functions are called **pleiotropic**. This is frequently the case for developmental control genes, and they usually have complex 5′ regulatory regions to control expression at several different times and places during development.

It should be remembered that genes are often named because of the first-discovered mutant phenotype and this is often a source of confusion. If it is a loss-of-function phenotype then the function of the wild-type gene may be opposite to that indicated by its name. For example, the *dorsal* gene in *Drosophila* is actually responsible for initiating *ventral* development, and the *white* gene is responsible for making a *red* eye pigment. In the absence of the *dorsal* gene, ventral structures cannot develop and the whole embryo follows the default pathway and becomes dorsal-type all over. In the absent of the *white* gene, the eyes are white because the red pigment is not produced. Another source of confusion is that homologous genes in different organisms often have different names. This is because they will have been named depending on their mutant phenotypes before the actual gene carrying the mutation was identified. In addition, it is often the case that a gene product named on the basis of biochemical studies in vertebrates is known by the names of its mutations in the model invertebrates used in genetics. For example the molecule known as beta-catenin in vertebrates is coded by the *armadillo* gene in *Drosophila* and the *wrm-1* gene in *Caenorhabditis elegans*.

Sex chromosomes

The mechanism of sex determination differs quite substantially between animal groups but usually depends on dimorphic sex chromosomes. In mammals, females have two X chromosomes, while males have one X and one Y. In birds it is the female that has two different chromosome types: the male having two Z chromosomes and the female having one W and one Z. In *Drosophila* the female is XX and the male has just a single X (XO).

Sex-linked mutations are present on a sex chromosome and so their effect depends on the sex of the individual. For example a single-copy recessive mutation on the mammalian X chromosome will be masked by the wild-type copy in females (XX), but will produce a phenotype in males (XY) because the Y chromosome does not carry the corresponding locus. Chromosomes other than sex chromosomes are called **autosomes**, and any gene or mutation not on a sex chromosome is said to be autosomal.

Maternal and zygotic

Normally, in genetics we think of the phenotype as corresponding to the genotype of the same individual. But this is often not the case for the very earliest events of embryonic development. This is because some early developmental events depend on the situation in the *mother* rather than the situation in the embryo itself. For example if a cytoplasmic determinant is placed in a particular region of the oocyte during oogenesis, then all the genes involved in this process will be those of the mother (Fig. 3.3). If the determinant fails to be formed because of a mutation in the mother's genome, then it will be of no avail to receive a good copy of the gene from the father's sperm because by then it is too late to perform the function. A **maternal-effect** gene is one for which the phenotype of an individual depends not on its own genotype but on that of the mother. A good example is the gene stella in the mouse. This encodes a chromosomal protein expressed in germ cells and early embryo blastomeres. If the gene is knocked out, the female −/− mouse is normal but her embryos are defective and die very early.

The period of maternal control of development does not end at fertilization because for most animal types the embryo's own genome, called the **zygotic** genome, remains inactive during the early cleavage stages. Once the zygotic genome has been activated, the normal situation is re-established and the embryo phenotype will thereafter correspond to the embryo genotype. The zygotic genome may be activated at different stages in different animal groups, ranging from early cleavage to late blastula.

Genetic pathways

If a set of genes are involved in a single pathway or process it is often possible to deduce the sequence in which they act from genetic data. There are two common situations, one where a group of different mutations have a similar phenotype and the other where a group of mutations have two phenotypes that are in some sense opposites of one another.

Where several genes have similar mutant phenotypes, the sequence of action can sometimes be established by a rescue protocol. This is illustrated in Fig. 3.4. Imagine that there are three genes in a pathway leading to the formation of the third segment of an animal and that each gene is turned on by the action of the previous one. Loss-of-function mutations in all the genes give the same phenotype, namely loss of the third segment. Now suppose that each of the normal gene products a, b, and c, are injected into embryos that are mutant for gene *b*. Clearly, product a will have no effect because the pathway is blocked at step *b*. Product b will rescue a mutation in its own gene and give a normal phenotype. But so will c rescue a mutation in *b* because *c* lies downstream in the pathway. We could conclude from these data that the pathway goes $a \rightarrow (b, c)$, but to order *b* and *c* we should need to do a similar experiment to find whether a loss-of-function mutant of *c* could be rescued by injection of the product b or c. This type of analysis was used to investigate the posterior group mutants in *Drosophila* and to show that *nanos* was the last-acting member of the pathway.

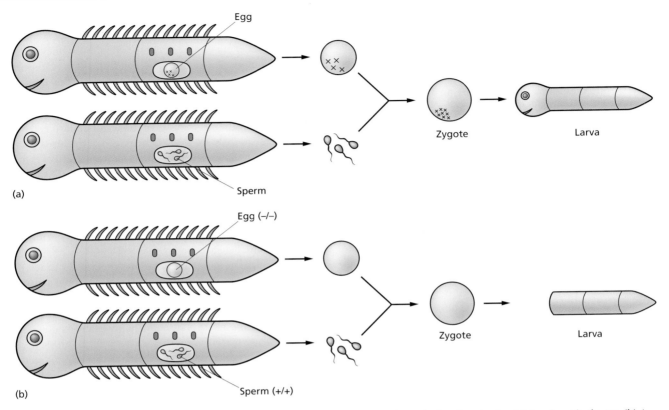

(a)

(b)

Fig. 3.3 Maternal effect gene. (a) Normal development, in which a maternal effect gene is required to deposit a head determinant in the egg. (b) A mutant mother produces eggs lacking the determinant, and consequently the offspring are headless.

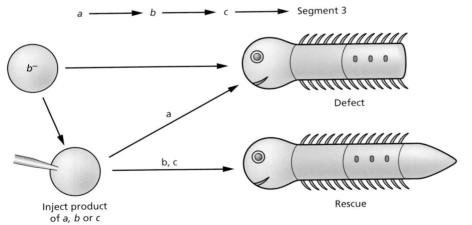

Fig. 3.4 Elucidation of a genetic pathway by rescue experiments. *a, b, c* are a set of genes that upregulate each other in a linear pathway and are required for formation of segment 3. If each of the gene products of *a, b, c* are injected into embryos mutant for *b*, then b and c will rescue a wild-type phenotype but a will not.

If members of a set of mutants have one of two opposite phenotypes then the genes may again code for the successive steps in a pathway but it is likely that some or all the steps will be repressive events rather than upregulations. Figure 3.5 represents this type of situation. Pigment is made in the spots of segment 2 following the operation of a pathway of three genes in which *a* represses *b* which represses *c* which represses pigment formation. Normally, gene *a* is active everywhere except the spots, so only the spots become pigmented. Two types of loss-of-function mutant can be recovered: b^-, which are unpigmented all over, and a^- or c^-, which are pigmented all over. It is possible to deduce the sequence of action of the genes by examining the phenotype of the double mutants. In each case the phenotype of the double mutant is the same as that produced

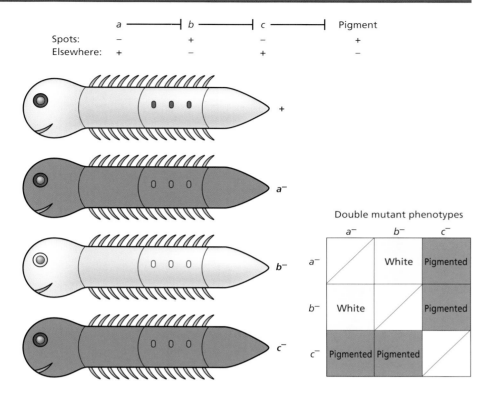

Fig. 3.5 Analysis of a repressive pathway. Normally, a gene *a* is inactivated in three spots of segment 2, resulting in the formation of green pigment. If *a* or *c* is inactive, the whole organism is pigmented, if *b* is inactive there is no pigment. The phenotypes of the double mutants show that *b* must act after *a* and before *c*.

by the mutant of the *later acting* of the two genes. For example, *b⁻c⁻* is pigmented all over because *c* acts after *b*. By looking at the phenotype of each double mutant combination, the genes can be arranged into a pathway. Among many other examples, this type of analysis was used to order the *dorsal* group genes in *Drosophila*. Repressive pathways are remarkably common and they can cause much confusion. The best way to understand them is, as in Fig. 3.5, to write out the state of each gene in the pathway under the two possible conditions: that in which the first step is activated, and that in which the first step is repressed. In developmental biology the two conditions often refer to regions within the embryo where the same pathway is under different regulation, for example the dorsal and ventral sides. They may also relate to the presence or absence of an inducing factor. These methods are examples of **epistasis** analysis, because in genetics if one gene prevents the expression of another it is said to be epistatic to it. Fig. 3.6 shows an actual example of epistasis in the mouse coat color.

Another method of ordering gene action in development depends on the use of **temperature-sensitive** mutations. In contrast to the pathway situations, this does not depend on any particular relationship between different gene products and can be used to order events in time which are mechanistically quite independent of each other. Temperature-sensitive mutants are those that display the phenotype at a **nonpermissive** (usually high) temperature, and do not show a phenotype at the **permissive** (usually low) temperature. They are frequently weak

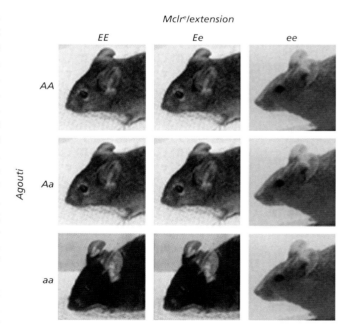

Fig. 3.6 An example of epistasis. In a cross between heterozygotes for the two mouse coat color mutants *agouti* and *extension*, the *extension* phenotype prevails in the double homozygote offspring, yielding a 9:3:4 ratio instead of the usual 9:3:3:1. Reproduced from Phillips (2008), with permission from Nature Publishing Group.

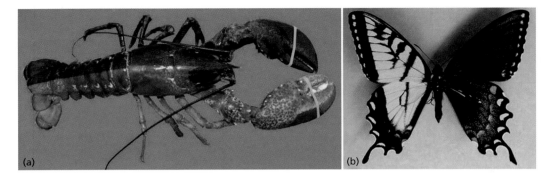

Fig. 3.7 Gynandromorphs. (a) Probable lobster gynandromorph. Reproduced from Levin and Palmer. *BioEssays* 2007; 29: 271–287, with permission from John Wiley & Sons Ltd. (b) Swallowtail butterfly gynandromorph. Reproduced from Agate *et al. PNAS* 2003; 100: 4873–4878, with permission from National Academy of Sciences.

loss-of-function alleles. They arise from changes in the conformation of the protein product, which are sensitive to changes in temperature in the range normally compatible with embryonic survival. The time of action of a gene may be deduced by subjecting groups of temperature-sensitive mutant embryos to the nonpermissive temperature at different stages of development. If the organism ultimately displays the mutant phenotype, this means that the gene product was inactivated at the time of its normal function, in other words that the gene was required during the period of the high-temperature exposure. An example would be the time of action of the gene *cyclops*, which is needed for the induction of the floor plate in zebrafish. To form a floor plate the embryos need to be kept at the permissive temperature between the stages of 60 and 90% epiboly, showing that this is the time at which gene function is needed. Temperature-sensitive mutants are more use in poikilothermic organisms such as *C. elegans*, *Drosophila*, and the zebrafish, rather than in homeotherms such as the mouse, because the range of temperatures to which the embryos can be subjected is much wider.

Genetic mosaics

It is sometimes possible to make organisms that consist of mixtures of cells of different genotypes. These are called **genetic mosaics** and can be useful as they provide information about where in the embryo a particular gene is required. A well-known type of mosaic is the **gynandromorph**, which is an animal composed of half male and half female tissue (Fig. 3.7), and arises from loss of a sex chromosome at the first cleavage.

For developmental analysis the most useful types of mosaic are variable between individuals and have irregular boundaries relative to body structures. Consider the example shown in Fig. 3.8 in which a mutation causes loss of the spots normally present on the second body segment. Fig. 3.8b and c show

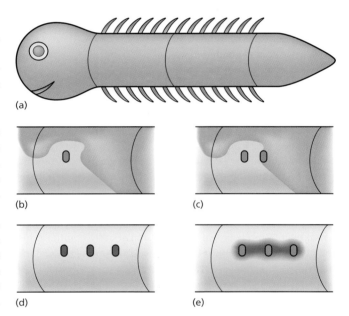

Fig. 3.8 Use of genetic mosaic analysis. (a) A mutant of a gene in which the three spots in segment 2 are missing. In (b) and (c) genetic mosaics are made in which the red tissue is null mutant for *x* and the green tissue is wild type. In case (b) the spot appears in the wild-type zone so the gene must have an autonomous function corresponding, for example, to the wild-type expression pattern in (d). In case (c) the spot appears in the zone of mutant tissue so the gene must have a nonautonomous function, corresponding, for example, to the expression pattern in (e).

genetic mosaics that have part of the body composed of mutant cells and part of wild-type cells. Fig. 3.8b shows the situation in which spots are lost wherever they themselves are of mutant genotype. In this case the mutant is **autonomous**; it affects just the region in which the gene is active. Fig. 3.8c shows the situation in which spots can also form in mutant tissue but only if there is wild-type tissue directly adjacent. This is **nonautonomous** because the wild-type gene is affecting a structure outside its domain of expression. Nonautonomy means that there must be an inductive signaling event that is affected by the mutation. However, it does not necessarily mean that the mutant gene itself codes for a signaling factor, as failure of the signaling event can be a downstream consequence of the mutation.

Genetic mosaics have been widely used in *Drosophila*. A very useful type is made by pole-cell transplantation and consists of germ cells of one genotype in a host of a different genotype. Such mosaics have enabled the understanding of factors controlling the patterning of the oocyte as a result of interactions with the somatically derived follicle cells of the egg chamber. Mosaics have also been used in *C. elegans*, where they arise spontaneously by loss of small duplicated chromosome fragments. In zebrafish, mosaics can be created by grafting as there is quite a lot of early cell mixing to disperse the labeled cells. In mammals, the term **mosaic** is usually reserved for a naturally occurring organism composed of two genetically dissimilar cells (e.g. X-inactivation mosaic, see Chapter 10) and the term **chimera** is used for embryos made experimentally by cell injection or aggregation of blastocysts.

Genetic mosaics should not be confused with embryos said to show **mosaic** behavior (see Chapter 4). This means that surgical removal of parts causes a defect in the final anatomy corresponding exactly to the fate map. Mosaic behavior is contrasted to **regulative** behavior and has nothing to do with genetic mosaics.

Screening for mutants

The term **forward genetics** is sometimes used to describe investigations that start with the discovery of an interesting mutant phenotype. **Reverse genetics** by contrast refers to functional investigations on a known gene. Many interesting mutants have arisen spontaneously, but even more have been recovered in large-scale screens on *Drosophila*, *C. elegans*, zebrafish, and mouse. The details of these screens can be very complex, particularly for *Drosophila* in which there are many selective tricks for reducing the total number of individuals to be dealt with. But the principle is simple as it relies on basic Mendelian genetics. A group of males is mutagenized, for example by treatment with a chemical mutagen. Numerous mutations are induced in the spermatogonial stem cells, so the animals will thereafter produce sperm containing mutations, potentially in any gene in the genome. The mutagenized males are mated to normal females, producing an F1 offspring generation. Each of the F1 individuals is likely to carry a mutation, and each is likely to carry a different mutation from all the others, as the sperm that produced each individual has probably come from a different mutant spermatogonium. Any dominant mutations will be immediately apparent as a change in phenotype of an F1 individual, but the more common recessive mutations will not be

visible since the F1 animals are all heterozygous. So each F1 individual is put in a separate container for further mating to a wild-type animal. This produces a family at the F2 generation. If the F1 individual did carry a mutation, then half the resulting F2 individuals will be heterozygous for it. A set of test matings is carried out between pairs of individuals within each F2 family. If there is a mutation present then 1 in 4 matings will be between heterozygotes and yield an F3 generation that is 25% homozygous mutant. The F3 generation is examined and scored at the embryo stage. By definition developmental mutations are those that perturb the anatomy of the organism, so the homozygous recessive mutants should be visibly abnormal. They may be detected simply by examination of the embryos under the dissecting microscope, or, if the screen is more focused, the mutant phenotype may be visualized following immunostaining or *in situ* hybridization to display a particular structure or cell type.

Any mutation that disables a gene essential for early development is quite likely to be lethal and to prevent development after the time of normal gene function. So the homozygous mutant F3 embryos may well die at an early stage, and they need to be examined early on before they degenerate. It is obviously not possible to maintain a line of a homozygous lethal mutation, but the mutants can be maintained by keeping the heterozygous parents. These will go on producing batches of embryos in which 25% are homozygous mutant, and when the original F2 animals are too old to go on reproducing they can be replaced by their heterozygous offspring. Further descriptions of recessive screens will be found in Chapter 8 (zebrafish) and Chapter 11 (*Drosophila*).

In *Drosophila* there are some very sophisticated methods for reducing the labor involved in screens. The most important is the use of **balancer chromosomes**. These have multiple inversions which mean that there is no recombination between the balancer chromosome and its wild-type homolog. They also carry a recessive lethal mutation, so flies with homozygous balancers are not viable. They also carry some marker gene that will enable all flies carrying the balancer in one copy to be easily identified. The uses of balancers are numerous, but one of the most important is in the simple maintenance of a recessive lethal mutant line. This is shown in Fig. 3.9. The line carries one

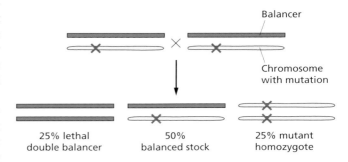

Fig. 3.9 Maintenance of a mutant line by means of a balancer chromosome.

copy of the balancer chromosome and one copy of the homologous chromosome bearing the mutation. In each generation a 1:2:1 ratio is produced of homozygous balancer, heterzygous, and homozygous mutant. The heterozygotes are the only viable offspring and serve to maintain the line. The homozygous mutant embryos are available for experiments. This means that a line can be maintained by repeated mating with no need to test individuals to see whether they are heterozygotes.

Cloning of genes

Developmental genetics existed long before molecular cloning was introduced, but it is now regarded as essential to clone any gene of interest identified by mutagenesis. This caused considerable difficulty in the past but is much easier today with the availability of high-resolution genome maps and genome sequences. A gene is regarded as cloned when the complete coding sequence is incorporated into a bacterial plasmid, or other cloning vector, so that it can be amplified and purified in a quantity suitable for use in any of the types of investigation now described as **reverse genetics**.

Most of the developmentally important genes in *Drosophila* were cloned by inducing mutations with a transposable element called the **P-element**. Once it has been shown that a P-element has integrated into the locus of interest then it can be used as a probe to isolate DNA clones from a genomic library. Nowadays, in most experimental organisms as well as for human genetics, genes are cloned by **positional cloning**. Here a mutation is mapped to very high resolution using microsatellite polymorphisms or restriction fragment length polymorphisms. There are many of these scattered through the genome and so long as the genome has been sequenced all their positions should be known. It is possible to obtain sets of **PCR** primers that enable each polymorphism to be detected in DNA samples by the presence of a specific band visible on a DNA gel. So long as enough offspring can be produced it is now possible to map a mutation to the specific locus using a single cross. The procedure is shown in principle in Fig. 3.10. In general, the test parent will be heterozygous for a recessive lethal mutation of gene *g*, hence designated g^+g^-. It is crossed to an individual of another strain in which most of the polymorphic loci are different. The F1 individuals should contain 50% heterozygotes for the mutant allele of *g*. F1 individuals are crossed together to yield many families of which about one-quarter should contain 25% homozygous lethal g^-/g^- individuals. DNA is isolated from each of the mutants, and also from many of the wild-type siblings, which may be g^+g^+ or g^+g^-. These are then all individually typed for a selection of the polymorphic markers. As shown in Fig. 3.10, a marker that is closely linked to the mutant locus should segregate with it, while others will segregate away by independent assortment of chromosomes or by meiotic crossing over within a chromosome. It is usually necessary to go through several cycles of mapping, using the same DNA samples with markers that are more and more closely spaced around the mutant locus. Eventually, a small chromosome region will be identified which is known from the genome sequence only to contain a few genes. Each of these is then evaluated as a candidate. One consideration is the putative nature of the protein deduced from the sequence, for example if the mutation has a cell autonomous action it is unlikely to code for a signaling molecule. Another is the expression pattern of the candidate gene relative to the domain of action of the mutation. A gene that is not expressed in the relevant region is unlikely to be a good candidate. The expression pattern can be established using *in situ* hybridization (see Chapter 5) with probes designed from the known sequence. Once a good candidate has been found, the mutant DNA can be sequenced at that locus to see if it does, in fact, contain a mutation. Final proof may be obtained by introducing a good copy of the gene by transgenesis and finding whether this will rescue the mutant phenotype towards the wild type.

Whole genome sequencing has enabled reasonably accurate estimates to be made of gene numbers, and hence indirectly of the complexity of living organisms. In terms of protein-coding genes, free-living bacteria have about 2000–4000 genes, unicellular eukaryotes like yeast about 6000, *Drosophila* about 13,000, *C. elegans* about 19,000, and mammals have 20,000–25,000. The gene number in vertebrates may seem low compared to their anatomical and behavioral complexity, but this complexity is also underpinned by extensive alternative splicing and post-translational modifications, as well as the presence of a number of nontranslated RNAs. Furthermore, the control of gene expression can be very complex, especially for developmental control genes.

The power of the current sequencing technologies, so-called "next generation sequencing," is awesome. Samples of a few micrograms of DNA fragments can now be analyzed in a few days yielding gigabases of sequence. The data require extensive bioinformatic analysis, usually based on known genome sequences, and can, for example, now identify all the variant sites in the genome of an individual.

Gain- and loss-of-function experiments

The scope of genetics has been considerably expanded in the molecular era such that many of the genetic manipulations used in experiments no longer arise from mutagenesis, but from the use of more sophisticated and directed methods for altering gene expression. There are two very general types of experiment that may be called the **gain-of-function** and **loss-of-function** approaches. As the names suggest, they refer to adding or removing a specific gene product from the system under study. For experimental purposes, it is often not necessary that the gene should be added or removed from the germ line. Its function can be analyzed if gene activity is altered just in the par-

Positional cloning

Microsatellite markers

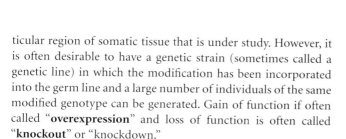

Fig. 3.10 Positional cloning. This is a very simplified presentation of the principle of positional cloning. The mutation to be cloned is in a gene called *g*. Three markers are considered, of which the parental strains have alleles *A, B, C* or *a, b, c*. Mutant or wild-type F2 individuals are analyzed using PCR primers for each of the polymorphic loci. The reaction mixtures are separated by gel electrophoresis and each specific allele is indicated by a DNA fragment of a particular size. It may be seen that the *g⁻* mutant allele is linked to *A* and the *g⁺* wild-type allele is linked to *a*. This indicates that the mutant locus lies near to the *Aa* locus.

ticular region of somatic tissue that is under study. However, it is often desirable to have a genetic strain (sometimes called a genetic line) in which the modification has been incorporated into the germ line and a large number of individuals of the same modified genotype can be generated. Gain of function if often called "**overexpression**" and loss of function is often called "**knockout**" or "knockdown."

Transgenesis

A transgenic animal has incorporated an extra gene introduced by the experimenter. Accordingly, a **transgene** is simply a gene

that has been introduced into an organism by transgenesis. Methods for transgenesis now exist for all the standard model organisms, and are described in Section 2, although the methods for the chick are not yet widely used. All methods of transgenesis other than the targeted **knock-in** (see below) share the property that the integration site in the genome is random, or at least not controlled. Transgenes are usually designed so that their expression is regulated by a promoter within the insert and they are as far as possible immune from the effects of the position within the genome at which they have integrated. This may be done simply by incorporating a large amount of flanking DNA in the construct, or by including specific "insulator" sequences.

However, transgenes are sometimes designed specifically to probe the local environment in the form of an **enhancer trap.** This is a transgene carrying just a minimal promoter, which provides a basal RNA polymerase II binding site but is not sufficient for detectable transcription. If it enters the genome within range of an endogenous promoter or enhancer then this will complement the minimal promoter and activate significant transcription. Enhancer traps are usually not mutagenic as they may be activated anywhere within effective range of an endogenous enhancer. Their main use is providing genetic lines in which particular tissues or cell types are highlighted by expression of markers such as **β-galactosidase** or **green fluorescent protein** (GFP) and are therefore easy to visualize.

Gal4 and Tet systems

A variation of the enhancer trap method is used to drive ectopic expression of any gene of interest. This is based on the Gal4 regulatory system from yeast. Gal4 is a transcriptional activator of the zinc-finger class and can drive expression of any gene sequence cloned downstream of the "upstream activating sequence" (*UAS*) to which it binds. Enhancer trap lines, called "driver lines," are made in which the gene introduced is not a reporter but *Gal4*. Another transgenic line is made in which the gene of interest is linked to *UAS*. When individuals of the two lines are crossed together, the Gal4 protein is expressed under the control of its enhancer and it will upregulate the target gene via the *UAS*. The spatial pattern of expression of the target gene will therefore depend on the enhancer used to drive the Gal4 (Fig. 3.11). This system was developed in *Drosophila* and has been widely used also in zebrafish.

It is often desirable to confer inducibility on an otherwise noninducible transgene. A popular system uses elements of the tetracycline system from *Escherichia coli*. This comes in two forms, allowing expression to be repressed or upregulated by the addition of a tetracycline analog, doxycycline (see Chapter 13). The **Tet** systems were introduced for mice and have been used also for zebrafish and *Xenopus*.

Other gain-of-function techniques

A very useful method of transient overexpression is to inject single cells with mRNA for the gene of interest, synthesized *in vitro*. This is a technique that is suitable for organisms with large eggs, like *Xenopus* and the zebrafish, because they are easy to inject and the injected mRNA can exert its effect on early developmental events before it is degraded or diluted by growth. The RNA persists for at least 1 day which is long enough to exert effects on processes in early development.

Electroporation of DNA is also employed, particularly for transient introduction of DNA into chick embryos.

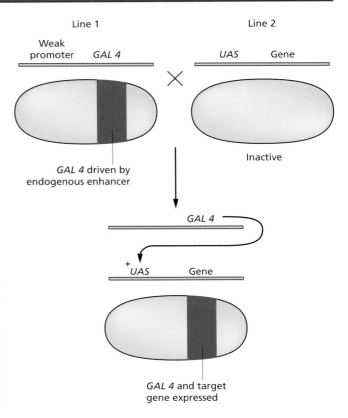

Fig. 3.11 GAL4 method for ectopic overexpression of a transgene. When the two transgenic lines are mated, the offspring will show tissue-specific expression of the target gene.

Targeted mutagenesis

The introduction of specifically engineered mutations at a desired site in the genome is a technology that has been brought to a high level in the mouse. It has mainly been used to produce **knockouts**, or loss-of-function mutations in particular genes selected because they were thought likely to be of developmental or medical importance (see Chapter 10). The same method can achieve **knock-ins**, in which a transgene is inserted at a particular location, usually to take advantage of the regulatory features of the genomic environment. Targeted mutagenesis depends on **homologous recombination**, which is the direct replacement of a gene by a modified version made *in vitro* (see Animation 2: Homologous recombination). Usually, if DNA is added to cells or embryos, most of the integration events will occur in the wrong place in the genome and so there needs to be a selection step to isolate the few homologous recombinant cells in which the input DNA has replaced the endogenous gene. This has in the past been done in tissue culture cells, so the applicability of targeted mutagenesis has depended on the availability of cells that can be grown in culture and can then be reintroduced into an embryo. For mouse embryos, **embryonic stem cells (ES** cells) have played this role and the new technology for making **induced pluripotent stem cells** (**iPS cells**), which are very similar to ES cells, is extending this ability to other mammals. Using the Cre-lox system (see Chapter 13) it is possible to ablate genes in specific tissues by using a tissue-specific promoter to drive the *Cre* gene.

Especially for those model organisms for which pluripotent stem cells are not available, targeted mutagenesis can now be achieved based on the use of **zinc-finger nucleases**. These are artificial proteins constructed using a judicious selection of zinc-finger-type DNA binding domains combined with an endonuclease. In a zinc-finger transcription factor, each of the fingers recognizes three nucleotides of DNA and concatenated zinc fingers can recognize a longer sequence of predetermined specificity. If this domain is combined with an endonuclease then it should cut at the precise sequence recognized by the fingers. The usual type of nuclease used is active as a dimer, so two proteins are expressed, each with three fingers, such that a nine-nucleotide sequence is targeted on either side of the cleavage site. An 18-nucleotide sequence is likely to be unique within an animal genome and to provide enough specificity for targeting. Experiments on embryos of several of the main model organisms have shown that the frequency of specific mutation events is high enough that it is possible to inject a set of fertilized eggs and then screen the resulting embryos for presence of the desired gene ablation.

Other loss-of-function systems

For the species in which targeted mutagenesis is not possible, other methods for producing specific inhibition of known genes have been developed. They are inherently less specific and when using them it is important to ensure that experimental tests of specificity have been carried out.

Dominant negative inhibitors

One method involves the overexpression of **dominant negative** reagents. For example a transcription factor lacking its DNA binding domain may often act as a dominant negative because when overexpressed it will sequester all the normal cofactors needed by the wild-type factor, or it may form inactive dimers with the wild-type factor. A dominant negative version of a gene product can "knock out" the normal gene if it is simply introduced as a transgene anywhere in the genome. It does not need to be a homologous recombination. Alternatively, the mRNA for the dominant negative protein can be produced *in vitro* and injected into the fertilized egg.

Secondly, there is the **domain swap** method, used extensively for transcription factors. Because transcription factors have a modular design (see Appendix) it is possible to replace an acti-

vating region with a repressing domain or vice versa. The domain-swapped factor will still bind to the same site in the DNA but instead of upregulating its target genes it will repress them (or vice versa). This is not quite the same as a loss-of-function mutation, since there will be an active repression of any gene to which the target factor binds, and these genes would not necessarily be inactive following a simple ablation of the transcription factor. Again, this can be introduced either by transgenesis or by injection of mRNA into the fertilized egg. The usual inhibitory domain used in this type of experiment is that from the *Drosophila* gene *engrailed*, and the normal activating domain is that from the herpesvirus gene *VP16*.

Antisense reagents

The third strategy involves the use of **antisense** reagents. If an RNA transcript is made from the noncoding strand of the DNA then it will be an antisense version of the normal messenger RNA. When introduced into the embryo this will form hybrids with the normal mRNA, which are inactive as translation substrates and are often rapidly degraded. There have been fashions for introduction of full length antisense mRNA, and for the external application of antisense oligodeoxynucleotides, but the currently favored methods fall into two groups: the use of morpholinos and the use of RNA interference (RNAi).

Morpholinos are analogues of oligonucleotides, in which the sugar–phosphate backbone is replaced by one incorporating morpholine rings. Unlike normal oligonucleotides they are resistant to degradation by the nucleases that are present in all cells and extracellular fluids, but because the usual four bases are linked to the resistant backbone with the correct spacing, they can still undergo hybridization with normal nucleic acids. Morpholinos are usually synthesized to be about 20 residues long and are designed to be complementary to a region of the mRNA likely to be accessible, such as the translation start region. The hybrid of morpholino and mRNA is usually not degraded but remains inactive for protein synthesis. Because there is no degradation of the mRNA it is necessary to show that specific protein synthesis has been blocked, which requires the availability of an antibody to the protein that can be used for western blotting or immunoprecipitation or *in situ* immunostaining. Morpholinos cannot generally penetrate cell membranes and so their main application has been in early embryos of free living embryos where they can easily be administered by intracellular injection, namely *Xenopus*, zebrafish, sea urchins, or ascidians.

The **RNA interference** (**RNAi**) method depends on normal host defenses against RNA viruses. In some embryo types, introduction of double-stranded RNA (**dsRNA**) of the same sequence as the normal mRNA can be a very effective method for specific mRNA destruction. DsRNA is cleaved by an enzyme called dicer into short (21–23 bp (base pair)) length fragments. These enter a silencing complex that can bind to, unwind, and sometimes cleave mRNAs that contain complementary sequence. In mammalian cells the long dsRNA causes a nonspecific inhibition of translation of all genes, but the short (21–23 bp) length processed fragments are specific, and can be used directly to bring about translational inhibition or destruction of specific mRNAs. DsRNA does not enter cells readily but can be introduced using the same type of lipid transfection reagent that is used to introduce plasmids into cells. Because it is possible to make large libraries of dsRNA this method is now being used instead of chemical mutagenesis to conduct screens. It works particularly well for *C. elegans, Drosophila,* and planarian worms (Fig. 3.12).

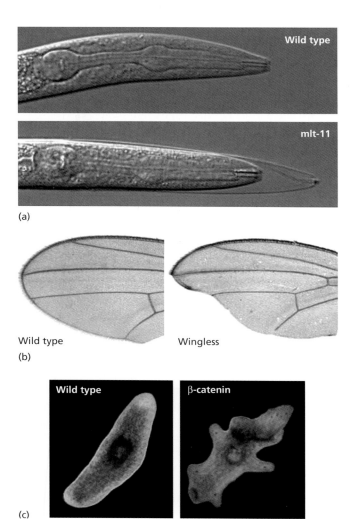

(a)

Wild type   Wingless
(b)

(c)

Fig. 3.12 Results from RNAi screens. (a) *C. elegans* treated with RNAi against a gene whose loss-of-function mutant prevents molting. (b) A *Drosophila* wing blade resulting from expression of an RNAi against the *wingless* gene (driven by *engrailed-Gal4*). (c) A planarian treated with RNAi against β-catenin. Reproduced from Boutros and Ahringer (2008), with permisison from Nature Publishing Group.

Antibodies and inhibitors

A final method of specific inhibition is treatment of the embryo with a specific **neutralizing antibody** directed against the protein product of the gene of interest. Antibodies will not penetrate intact embryos and so they must be injected at the site of interest. A common problem with this method is that most antibodies that bind to a particular protein will not neutralize its activity, so it is necessary to have some independent test to show that the antibody does, in fact, neutralize the target protein.

Numerous small molecule inhibitors are commercially available that have relative specific inhibitory effects on the signaling pathways activated by embryonic inducing factors. They are very often used as transient, nongenetic procedures. It is easy to treat free living embryos or *in vitro* organ cultures with these inhibitors, or if necessary inject substances into *Xenopus* or zebrafish embryos. But no inhibitor is really absolutely specific so it is very important to carry out control experiments to exclude off-target effects.

Gene duplication

Gene duplication is one of the main sources of evolutionary novelty. If a gene becomes duplicated then the constraints on changes to its sequence become relaxed. At one extreme, one copy could continue to be the functional gene and the other copy could accumulate mutations such that it acquired a novel and advantageous function. Alternatively, the second copy could accumulate deleterious mutations until it became nonfunctional, and maybe eventually not even expressed (a pseudogene). More usually, both copies will accumulate some sequence divergence such that they carry out subsets of the original function. Soon after the duplication the overlap in function will be considerable, while after millions of years the functions will diverge. For example the *cyclops* and *squint* genes of the zebrafish arise from duplication of the *Nodal* gene which encodes a critically important mesoderm inducing factor in vertebrate development, but they have diverged in function such that they now act at different stages.

The extreme case of gene duplication occurs when the entire genome becomes duplicated, with a doubling of chromosome number. This is called tetraploidization, as the resulting organisms are **tetraploid** instead of diploid. Tetraploidization can produce a vast array of new genes instantaneously and so enormously enlarge the adaptive possibilities for the evolutionary lines of descent. The pattern of multigene families in vertebrates suggests that two tetraploidization events may have occurred at the time of the origin of vertebrates, temporarily boosting their gene number from about 20,000 to about 80,000. This may account for their subsequent adaptive radiation and evolutionary success, although the count of protein-coding genes in extant vertebrates suggests that the number has been much reduced in subsequent evolutionary time. It also seems that further tetraploidizations have occurred in various lineages. For example *Xenopus laevis* looks as though it underwent a tetraploidization about 30 million years ago, as most genes are recovered in two copies differing in sequence by about 10%. These are known as **pseudoalleles**. They look like alleles, and generally have the similar expression patterns and functions, but they are not alleles because they occupy distinct genetic loci. Bony fish, including the zebrafish, seem to have undergone a tetraploidization at a much more remote time, about 420 million years ago when they first arose as a lineage. Because of the long time interval since this event, each pair of genes is now substantially diverged in sequence and function so bony fish are considered to be diploid organisms today.

Limitations of developmental genetics

Despite its successes there are certain limitations to the standard protocols of developmental genetics. One problem arises directly from gene duplication because this has resulted in the extensive presence of **redundancy** between genes, meaning that there are two or more genes with a significant overlap in function. It is a particular problem in vertebrates because of the repeated gene duplications and divergence. It is unlikely that two genes ever have an exactly identical function unless they are the result of a very recent duplication event. However, there are many examples of substantial overlap when gene function is examined at the laboratory level, where the organisms are not subject to all the challenges of life in the wild. The consequence is that mutation of a single gene to inactivity may produce no abnormal phenotype, or a minimally abnormal phenotype not consonant with the true function of the gene. Mutagenesis screens are usually designed to look at single mutations and so the presence of widespread redundancy in the genome greatly limits the number of informative phenotypes that can be recovered.

Where targeted mutagenesis is possible a mutant can be made and characterized without the need for a specific phenotype, and in fact numerous mouse knockouts show no abnormal phenotype or a very minimally abnormal phenotype. However, an abnormal phenotype is often found when mutations in several members of a multigene family are combined by breeding. For example the knockouts of the myogenic genes *MyoD* and *myf5* have limited effects individually, but in combination show an almost complete inhibition of myogenesis. Likewise, the knockouts of individual Hox genes often show little effect, but if all members of a homologous group (a **paralog** group see Chapter 22) are knocked out, then the abnormality does become significant.

A distinct type of problem for genetic analysis arises where a developmental gene has several functions at different stages of development. In a null mutant the embryos may die because the first of these functions cannot be carried out and this means that the phenotype will not be informative about any of the later functions. An example is the knockout of the gene for Fibroblast Growth Factor 4 (*Fgf4*). Although FGF4 has important functions in gastrulation, brain development, and limb development, the null phenotype is an early lethal because it is needed in preimplantation stages for cell division in the extraembryonic supporting tissues. In this case the problem can be circumvented because of the availability of some sophisticated experimental strategies such as tetraploid complementation (see Chapter 10), but it illustrates how the phenotype of a null mutant does not necessarily reveal much about gene function.

The task of developmental genetics is to analyze development in terms of the functions of individual genes. There is an implicit assumption here that any process under study will be largely explicable in terms of the actions of a few genes, each of which can be studied by gain and loss-of-function experiments. Fortunately, this has often proved to be a correct assumption in developmental biology. However, some processes may defy analysis because they arise from the activity of hundreds or thousands of genes, each contributing a small effect to the overall activity. This is found for example when the genetic predisposition to some common human diseases is investigated by genome-wide association analysis. An approach to such situations may be found in mathematical modeling. There are numerous mathematical models of genetic systems. They range from highly quantitative models of simples systems such as phage lambda to more qualitative models of complex systems, which try to capture some of the dynamical properties without having to know all the molecular detail of every gene product. This topic is outside the scope of a basic developmental biology textbook but promises to become increasingly important in the future.

Key points to remember

- Mutants have been very important for identifying developmental genes and unraveling developmental mechanisms.
- In general, mutations are genetically recessive if they lead to loss of function and genetically dominant if they lead to gain of function.
- Genes may be named after their loss-of-function mutant phenotype, and hence the name may seem opposite to the actual function. The same gene in different organisms may have different names.
- Mutations affecting early developmental processes are often maternal-effect, those affecting later processes are zygotic.
- Regulatory or biochemical pathways can be deduced from genetic experiments, especially from epistasis experiments in which the combined effect is determined of two mutations with opposite phenotypes.

- Screens for developmental mutants can be conducted by mutagenesis followed by breeding to homozygosity.
- Once a mutation has been identified the gene can be cloned by positional cloning.
- Transgenic organisms are those with an extra gene inserted in the genome, and can be made in all of the laboratory model species.
- Targeted mutagenesis based on homologous recombination has so far been applied mostly to the mouse.
- Many other experimental methods exist for inhibiting specific gene activity, including the introduction of dominant negative reagents, antisense oligonucleotides, and RNAi.
- The existence of widespread gene duplication in evolution means that many genes have overlapping functions (redundancy) so the loss-of-function phenotype may not reveal the full activity of a gene.

Further reading

General

Hartwell, L.H., Hood, L., Goldberg, M.L., Reynolds, A.E., Silver, L.M. & Veres, R.C. (2004) *Genetics: From Genes to Genomes*, 4th edn. New York: McGrawHill.

Mardis, E.R. (2008) The impact of next-generation sequencing technology on genetics. *Trends in Genetics* **24**, 133–141.

Karlebach, G. & Shamir, R. (2008) Modelling and analysis of gene regulatory networks. *Nature Reviews Molecular Cell Biology* **9**, 770–780.

Hartl, D.L. & Jones, E.W. (2009) *Genetics: Analysis of Genes and Genomes*, 7th edn. Sudbury, MA: Jones and Bartlett.

Classic mutagenesis screens

Patton, E.E. & Zon, L.I. (2001) The art and design of genetic screens: zebrafish. *Nature Reviews Genetics* **2**, 956–966.

St Johnston, D. (2002) The art and design of genetic screens: *Drosophila melanogaster*. *Nature Reviews Genetics* **3**, 176–188.

Boutros, M. & Ahringer, J. (2008) The art and design of genetic screens: RNA interference. *Nature Reviews Genetics* **9**, 554–566.

Examples of developmental genetic analysis

Anderson, K.V., Jurgens, G. & Nüsslein-Volhard, C. (1985) Establishment of dorso-ventral polarity in the *Drosophila* embryo: genetic studies on the role of the Toll gene product. *Cell* **42**, 779–789.

Schupbach, T. & Wieschaus, E. (1986) Maternal-effect mutations altering the anterior-posterior pattern of the *Drosophila* embryo. *Wilhelm Roux's Archives of Developmental Biology* **195**, 302–317.

Strecker, T.R., Merriam, J.R. & Lengyel, J.A. (1988) Graded requirement for the zygotic terminal gene, tailless, in the brain and tail region of the *Drosophila* embryo. *Development* **102**, 721–734.

Tian, J., Yam, C., Balasundaram, G., Wang, H., Gore, A. & Sampath, K. (2003) A temperature-sensitive mutation in the nodal-related gene cyclops reveals that the floor plate is induced during gastrulation in zebrafish. *Development* **130**, 3331–3342.

Other topics

Villee, C.A. (1942) The phenomenon of homoeosis. *American Naturalist* **76**, 494–506.

Nagy, A. & Rossant, J. (2001) Chimaeras and mosaics for dissecting complex mutant phenotypes. *International Journal of Developmental Biology* **45**, 577–582.

Otto, S.P. (2007) The evolutionary consequences of polyploidy. *Cell* **131**, 452–462.

Phillips, P.C. (2008) Epistasis – the essential role of gene interactions in the structure and evolution of genetic systems. *Nature Reviews Genetics* **9**, 855–867.

Tadros, W. & Lipshitz, H.D. (2009) The maternal-to-zygotic transition: a play in two acts. *Development* **136**, 3033–3042.

Gain and loss-of-function procedures

Rubin, G.M. & Spradling, A.C. (1982) Genetic transformation of *Drosophila* with transposable element vectors. *Science* **218**, 348–353.

Palmiter, R.D. & Brinster, R.L. (1985) Transgenic mice. *Cell* **41**, 343–345.

Thomas, K.R. & Capecchi, M.R. (1987) Site-directed mutagenesis by gene targeting in mouse embryo-derived stem-cells. *Cell* **51**, 503–512.

Kroll, K.L. & Amaya, E. (1996) Transgenic *Xenopus* embryos from sperm nuclear transplantations reveal FGF signalling requirements during gastrulation. *Development* **122**, 3173–3183.

Lagna, G. & Hemmati-Brivanlou, A. (1998) Use of dominant negative constructs to modulate gene expression. *Current Topics in Developmental Biology* **36**, 75–98.

Bishop, J. (1999) *Transgenic Mammals*. Harlow: Longmans.

Hannon, G.J. (2002) RNA interference. *Nature* **418**, 244–251.

Heasman, J. (2002) Morpholino oligos – making sense of antisense. *Developmental Biology* **243**, 209–214.

Nakamura, H., Katahira, T., Sato, T., Watanabe, Y. & Funahashi, J.I. (2004) Gain- and loss-of-function in chick embryos by electroporation. *Mechanisms of Development* **121**, 1137–1143.

Verma, I.M. & Weitzman, M.D. (2005) Gene therapy: twenty-first century medicine. *Annual Review of Biochemistry* **74**, 711–738.

Carroll, D. (2008) Progress and prospects: zinc-finger nucleases as gene therapy agents. *Gene Therapy* **15**, 1463–1468.

This chapter contains the following animation:

Animation 2 Homologous recombination.

 For additional resources for this book visit www.essentialdevelopmentalbiology.com

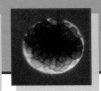

Chapter 4

Approaches to development: experimental embryology

Although the techniques of molecular biology and genetics are now essential in the investigation of development, historically it was the experimental embryologists who gave most thought to developmental mechanisms and who formulated a basic conceptual framework which is still used today.

Normal development

Normal development means the course of development that a typical embryo follows in standard laboratory conditions when it is free from experimental disturbance. A sound knowledge of normal development is necessary in order to understand the effects of experimental manipulations. In order to describe embryos, a number of standard descriptive terms are in use (Fig. 4.1). The front end of an animal is known as the **anterior** or **cranial** end. The rear end is known as the **posterior** or **caudal** end. The upper surface is **dorsal**, the lower surface is **ventral**. For microscope sections, those taken across the long axis of the animal are called **transverse**. Those parallel to the long axis are **longitudinal**. A vertical longitudinal section is known as **sagittal** if it is in the midline and **parasagittal** if it is to one side of the midline. A horizontal longitudinal section, separating dorsal and ventral sides, is called **frontal** or **coronal**.

All of the model species used for laboratory work have published tables of **stage series,** which describe the course of development as a number of standard stages which can be identified by external features under the dissecting microscope. Embryonic development is predictable, so if an embryo has reached stage 10 at a particular time then it is possible to be confident that it will reach stage 20 a particular number of hours later. For mammals and birds, development will always occur at the standard physiological temperature, but for free-living embryos such as *Xenopus* or zebrafish, the rate of development will vary depending on the temperature. The existence of these tables enables investigators to standardize their procedures by using embryos of the same stage, regardless of the temperature in the lab that day.

Features of development are referred to as **maternal** if they are due to components which exist in the egg, having been accumulated during oogenesis. They are said to be **zygotic** if they are due to components newly synthesized by the embryo itself after fertilization.

The fate map

A **fate map** is a diagram that shows what will become of each region of the embryo in the course of normal development: where it will move, how it will change shape, and what structures it will turn into. The fate map will change from stage to stage because of morphogenetic movements and growth, and so a series of fate maps will depict the trajectory of each part from the fertilized egg to the adult. The precision of a fate map depends on how much random cell mixing occurs in development. If there is none, as in the nematode, *Caenorhabditis elegans*, then the fate map can be precise down to the cellular level. For most embryo types there is some local mixing of similar cells and therefore the fate maps cannot be quite this precise. Nonetheless, the fate map is a fundamental concept in embryology and the interpretation of nearly all experiments concerned with early developmental decisions depends on knowledge of the fate map.

Fate maps are constructed by labeling single cells or regions of embryos and locating the position and shape of the labeled patch at a later stage of development (Fig. 4.2). The labeling

Essential Developmental Biology, Third Edition. Jonathan M.W. Slack.
© 2013 John Wiley & Sons, Ltd. Published 2013 by John Wiley & Sons, Ltd.

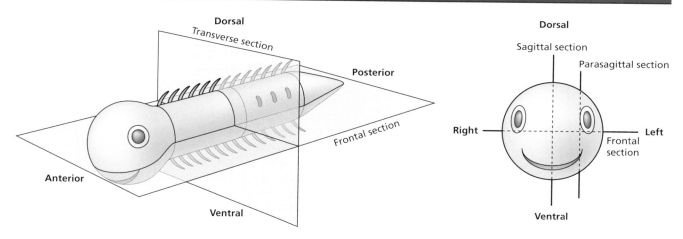

Fig. 4.1 Axes and planes of section used for describing embryos.

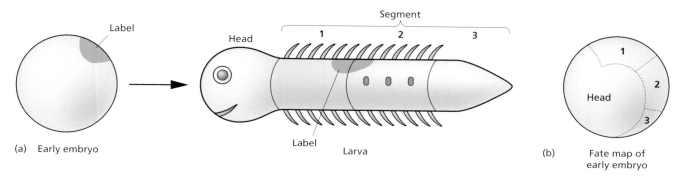

Fig. 4.2 Fate mapping. (a) A label placed at a particular position in the early embryo ends up in a reproducible position on the animal. (b) A possible fate map of the early embryo.

methods used are those described in Chapter 5 below, and may be either application of an extracellular label to a patch of cells, injection of an intracellular label to one cell, or grafting of labeled tissue to replace an exactly equivalent piece in the host embryo. Fig. 4.3 shows a fate mapping experiment conducted by injecting two adjacent early blastomeres of a *Xenopus* embryo with mRNA for green fluorescent protein (GFP), and locating the labeled domains at the neurula stage. The results of many individual labeling experiments can be combined to form the fate map for one particular stage. It is essential to note that a fate map does not indicate anything about developmental **commitment**. All parts of the embryo have a **fate** throughout development, but commitment to form particular structures or cell types is usually acquired through a series of intercellular interactions.

In older embryological literature an embryo is referred to as a **mosaic** if experimentally isolated parts develop strictly according to the fate map, and as **regulative** if an isolated part forms more structures than expected from the fate map. In reality, all types of embryo show some aspects of mosaic and of regulative

behavior depending on the region of the embryo and the developmental stage concerned. The explanation for regulative behavior depends on double gradient systems which are capable of adjusting to an alteration of size of the embryo, for example the ADMP–Chordin system found in *Xenopus*. Note that "mosaic" in this sense is nothing to do with **genetic mosaics**, which are organisms consisting of cells of different genotype (see Chapter 3), and embryonic **regulation** in this sense also differs from gene regulation or regulation at a physiological level.

Clonal analysis

Clonal analysis is a form of fate mapping in which a single cell is labeled and the position and cell types of its progeny identified at a later stage. The labeling may be carried out by injection of one cell with a lineage label. This is a simple method where the cells are large and very suitable for organisms that do not grow significantly such as early stage *Xenopus* or zebrafish or sea

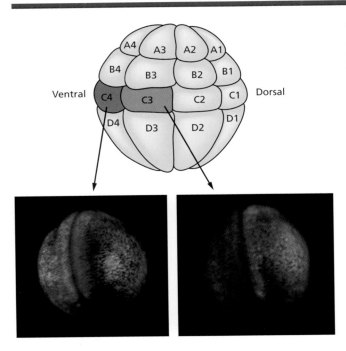

Fig. 4.3 A fate mapping experiment in *Xenopus*. *GFP* mRNA was injected either into the C3 or C4 blastomere at the 32-cell stage.

urchins. For organisms that do grow, such as chick or mouse embryos, it is preferable to introduce into a single cell a genetic label that will persist without dilution. This might be insertion by a replication-incompetent retrovirus, or a genetic recombination event that yields a visible marker. The most useful aspect of clonal analysis is to decide whether a cell is committed to form a particular structure or cell type at the time of labeling. If it is committed to form a particular structure called A then its descendants will only be able to populate structure A and nothing else. It follows that if a clone labels both structures A and B then the cell cannot have been exclusively committed to develop into either A or B at the time of labeling (Fig. 4.4). Sometimes it is found that a label applied early will span A and B while a label applied later will populate only A or B. This may be because the cells have become committed in the time between the two labels. However, it may simply be because the later-induced clones are smaller and so have less chance of populating more than one structure. Thus, a clonal analysis can prove *lack* of commitment but cannot prove the *presence* of commitment.

Clonal analysis has been extensively used particularly in the analysis of *Drosophila* segmentation and of vertebrate hindbrain patterning. The term **compartment** is sometimes used to

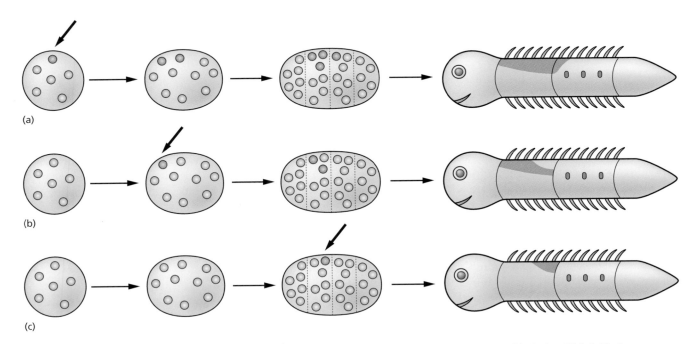

Fig. 4.4 Clonal analysis. No clonal restriction means no determination, but the converse is not necessarily true. (a) Single cell labeled before determination, progeny span later boundary. (b) Cell labeled before determination but progeny fail to span boundary because clone is too small. (c) Cell labeled after determination and cannot cross boundary.

indicate a region in an embryo whose boundaries are boundaries of clonal restriction. Once a compartment is established, no cells may enter and none may leave. In other words, all the cells within a compartment are the descendants of the founder cells. A compartment usually corresponds to a visible structure such as a segment or an organ rudiment and is maintained either by physical boundaries to cell migration, such as basement membranes, or by a differential adhesion of the cells of the compartment compared to those outside, such that all cells of the compartment stick together and cannot mix with their neighbors. In a few cases, notably the anterior–posterior compartment boundaries of *Drosophila* imaginal discs, the boundary of clonal restriction does not correspond to a visible anatomical boundary.

Developmental commitment

As development proceeds, formerly uncommitted cells become committed to form particular body parts or cell types. We now regard commitment as being encoded as a combination of transcription factors present in the cell and so it can be visualized directly by observing the expression of the relevant genes using *in situ* hybridization. But historically commitment was investigated by embryological experiments. This led to two operational definitions, usually called **specification** and **determination**, which are still useful today (see Animation 3: Tests of commitment).

A cell or tissue explant is said to be **specified** to become a particular structure if it will develop autonomously into that structure after isolation from the embryo (Fig. 4.5a,b). If a large number of such experiments are performed, and the results combined, it is possible to construct a **specification map** of the embryo. This shows what the cells have been programmed to do by that particular developmental stage. The specification of a region need not be the same as its fate in normal development. For example the prospective neural plate of a *Xenopus* blastula will differentiate not into neuroepithelium but into epidermis when cultured in isolation. In order to form neuroepithelium it needs to receive an inductive signal from the mesoderm.

A **determined** region of tissue will also develop autonomously in isolation but differs in that its commitment is irreversible

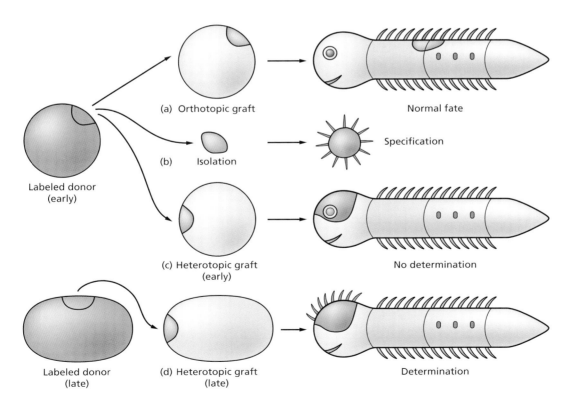

Fig. 4.5 Tests for fate, specification, and determination. (a) The labeled region will normally contribute to the spiny dorsal part of the animal (fate). (b) When isolated this tissue still forms dorsal spines (specification). (c) When grafted at an early stage to another region it differentiates according to the new position (not determined). (d) When grafted at a later stage to another region it differentiates according to its original position (determined).

with respect to the range of environments present in the embryo. In other words, it will continue to develop autonomously even after it is moved to any other region of the embryo (Fig. 4.5c,d). A very large number of embryological experiments consist of grafting a piece of tissue from one place to another and asking whether it develops in accordance with its new position or its old position. Grafts will usually be labeled by one of the methods described in Chapter 5, so that the tissues of graft and host can be distinguished. With practice they can be performed with considerable precision. Figure 4.6 shows a graft of GFP-labeled tissue to the posterior neural fold region of a *Xenopus* closed neurula embryo. A graft to the same position of another embryo is called an **orthotopic** graft, and is one of the usual methods of fate mapping. A graft to a different position in the host is called a **heterotopic** graft, and represents the test for **determination**. If the pathway of development is unaltered by such a graft then the tissue is defined as determined. If a heterotopic graft develops according to its new position then it follows that it was not determined, although it may have been specified, at the time of grafting. A series of such grafts performed at different stages usually show a time at which the tissue becomes determined. For example

the prospective neural plate of a *Xenopus* embryo is not determined at the blastula stage because it will form epidermis or mesoderm if grafted elsewhere in the embryo. It becomes determined to form neural plate during gastrulation, as after this stage grafts to other regions of the embryo will always differentiate into neuroepithelium. In a molecular sense, determination means that the cells have lost their responsiveness, or **competence**, to the signals that originally turned on the relevant combination of transcription factors. This may be because the cells have lost receptors or other components of the signaling machinery, or because the transcription factor combination is maintained by factors other than those responsible for turning it on in the first place.

In the development of any embryo there will be a hierarchy of commitment associated with progressive physical subdivision (Fig. 4.7). For example there will first be formed the three germ layers, ectoderm, mesoderm, and endoderm, then each germ layer will be subdivided, for example the ectoderm into epidermis, neuroepithelium, and neural crest. Later, the neuroepithelium will be subdivided on a smaller scale into subregions such as individual rhombomeres of the hindbrain. This means that any region of cells in the embryos will pass

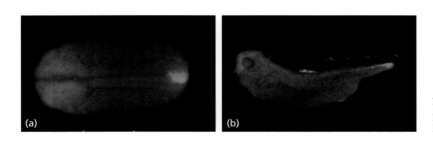

Fig. 4.6 (a) Graft of transgenic *GFP*-labeled tissue to posterior neural fold of a *Xenopus* neurula. (b) The graft later contributes to spinal cord and neural crest.

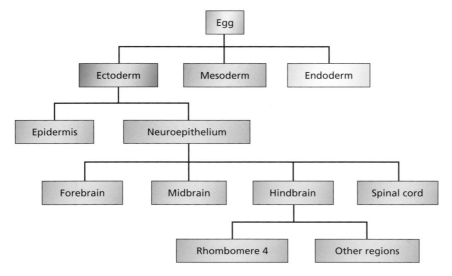

Fig. 4.7 A hierarchy of regional specification in development. Cells that become rhombomere 4 of the hindbrain first need to "decide" to be ectoderm, then to be neuroepithelium, then to be hindbrain, and finally to be rhombomere 4.

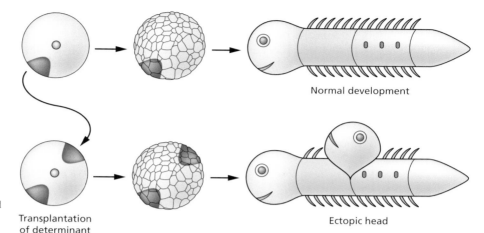

Fig. 4.8 Operation of a cytoplasmic determinant coding for the head of the embryo. In normal development it ensures that the head forms from the cells that inherit it. If the determinant is transplanted elsewhere than it causes formation of an ectopic head.

Transplantation of determinant

Normal development

Ectopic head

through several states of commitment, each defined by a different combination of transcription factors. The embryological methods for defining specification and determination can be applied at any level of this hierarchy.

Because the early steps of the hierarchy of commitment do not correspond to named body parts they are often referred to by position (e.g. dorsal/ventral, anterior/posterior). This can be very confusing to students beginning the study of developmental biology. Experimental manipulation can alter states of commitment so one might encounter a statement such as "overexpression of X makes dorsal cell population Y ventral." This means that cells in a dorsal position in the embryo have been caused to acquire a state of commitment the same as that normally found in the ventral region. So it is important to be very clear when reading publications about whether positional terms like "dorsal" literally refer to position, or whether they refer to a state of commitment associated with that position in normal development.

The term **potency** is sometimes used to mean the range of possible cell types or structures into which a particular cell population can develop. This is similar to competence, but may include also pathways of development that can be provoked *in vitro* by environments not normally found within the embryo. It is now normal to refer to **embryonic stem cells**, which can form any cell type in the body, as being **pluripotent**, and **tissue-specific stem cells**, which can form the cell types characteristic of their own tissue, as **multipotent**.

Cytoplasmic determinants

A cytoplasmic **determinant** is a substance or substances, located in part of an egg or blastomere, which guarantees the assumption of a particular state of commitment by the cells that inherit

it during cleavage (see Animation 4: Cytoplasmic determinant action). By definition, the cell division will be asymmetrical and the two daughters will follow different pathways of development. If cytoplasm containing a determinant is grafted to a different part of the egg, then it will cause formation of the appropriate structure from the new cells that now contain it (Fig. 4.8).

Cytoplasmic determinants are of considerable importance for the very earliest stages of embryonic development because they are often responsible for the establishment of the first two or three distinctly specified regions in the embryo. The subsequent complexity of the pattern develops as a result of interactions between these initial domains. Determinants are sometimes mRNAs that are localized to a part of the cell in association with the cytoskeleton, for example the *bicoid* and *nanos* mRNAs in the *Drosophila* egg. They may also be proteins. In the early stages of *C. elegans* development a complex containing PAR3, PAR6, and aPKC becomes localized in the anterior and controls the fate of the first two blastomeres. A similar system is involved during neurogenesis and epithelial polarization in other types of animal.

Induction

Most regional specification in development arises from the operation of extracellular signals called **inducing factors** (see Animation 5: Induction). The families of signaling molecules involved are briefly described in the Appendix. Many of them are also known as growth factors, cytokines, or hormones in other contexts. The ability to respond to an inductive signal is called **competence** and requires not just the presence of specific receptors, but also a functioning signal transduction pathway coupled to the regulation of transcription factors.

To give a concrete example, in the *Xenopus* embryo the meso-derm is induced from the animal hemisphere tissue in response to signaling molecules called Nodals emitted from the vegetal region (Fig. 4.9 and see Chapter 7). The signals upregulate the expression of various transcription factors that define the meso-dermal state, such as the T box protein Brachyury. The remain-der of the animal hemisphere becomes ectoderm, as does the

whole animal hemisphere in isolation. This interaction can occur between small pieces of the blastula cultured in combina-tion, and so by using pieces taken from embryos of different stages it has been possible to show that the interaction occurs during the blastula stages. This type of interaction is called an **instructive induction** because the responding tissue has a choice before it (either mesoderm or ectoderm), and in normal development the interaction results in an increase in complexity of the embryo.

There are two different types of instructive induction which have somewhat different consequences in terms of regional specification (Fig. 4.10). It may be that the signaling center lies at one end of a cell sheet and is the source of a concentration gradient of the signal substance. The competence of the sur-rounding tissue embodies different **threshold responses** to dif-ferent concentrations and hence a series of territories are formed in response to the gradient. It has become common usage to refer to the signal substance in a case of this sort, where there is more than one positive outcome, as a **morphogen** (see description in Chapter 2). Well established examples of mor-phogen gradients are the gradients of sonic hedgehog protein in the neural tube and the limb bud, of active BMP (bone mor-phogenetic protein) in the early *Xenopus* embryo, or of Dpp (Decapentaplegic) protein in the *Drosophila* imaginal discs.

The other possibility is that the signaling centers lie in one cell sheet and the responding cells in another. When they are brought together, the appropriate structures are induced as a result of a single threshold response in those parts of the responding tissue immediately adjacent to the signaling centers. This happens, for example, in the induction of nasal, lens, and otic **placodes** from the head epidermis of vertebrate neurulae under the influence of the underlying tissues. In the presence

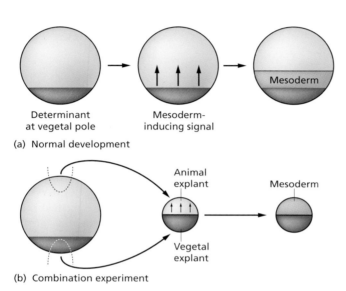

(a) Normal development

(b) Combination experiment

Fig. 4.9 Mesoderm induction in *Xenopus*. (a) As it occurs in normal development. (b) As it occurs in an animal–vegetal combination experiment.

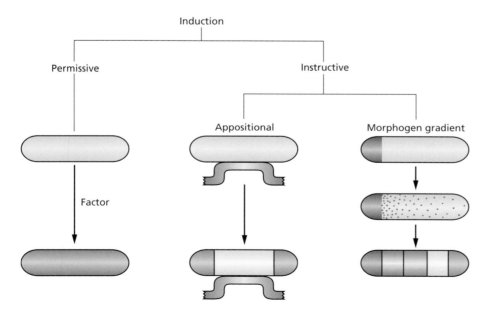

Fig. 4.10 Types of induction: permissive and instructive. Instructive inductions may be appositional or morphogen type.

of the signal the epidermis forms a placode, in its absence it differentiates as normal epidermis. This is called **appositional induction**. Typically, only one threshold response would be made by the responding tissue and the inducing factor for this reason would not be called a morphogen, even though the same substance might function as a morphogen in another context.

There is a further kind of inductive interaction, which is called **permissive**. Here the signal is necessary for the successful self-differentiation of the responding tissue but cannot influence the developmental pathway selected (Fig. 4.10). Permissive interactions are very important in late development. For example in the development of the kidney, the mesenchyme will form tubules on receipt of a permissive signal from the ureteric bud. In the absence of the signal it simply fails to develop, and does not form any alternative tissue. The essential difference between instructive and permissive is that instructive inductions lead to a subdivision of the competent tissue while permissive inductions do not.

It has generally been supposed that inductive signals diffuse through the extracellular space. However, many of the factors concerned are growth factors that are tightly bound by extracellular glycosaminoglycans. For this and other reasons, alternative routes for diffusion have been considered, and one possibility is transport through the fine intercellular processes called **cytonemes** that are found in many developing tissues. Even where inducing factors do travel through the intercellular space their distribution in space may depend on cellular processes of endo- and exocytosis. The transport of inducing factors is best understood in the *Drosophila* imaginal discs where more than one mechanism has been uncovered.

Lateral inhibition

Another important class of cell communication that could be called induction, but is usually considered separately, is **lateral inhibition** (see Animation 6: Lateral inhibition). This is best known in terms of the behavior of the Notch–Delta system that operates in many situations where individual cells from a uniform population follow one pathway of development while the surrounding cells follow another (Fig. 4.11). Examples include **neurogenesis** in the early neural plate, or formation of endocrine cells in the epithelia of the gut. In principle, lateral inhibition systems work because multiple, small signaling centers become established that suppress the formation of the signal in the surrounding cells. This is shown in Fig. 4.12, which

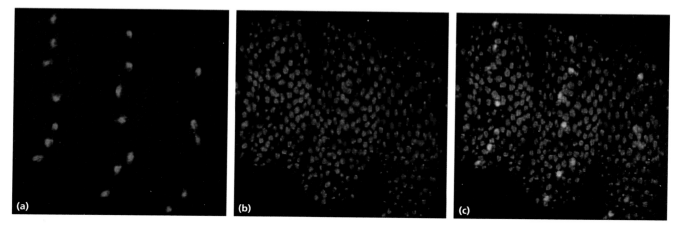

Fig. 4.11 Sensory progenitors arise as individual cells in *Drosophila* imaginal discs following a lateral inhibition process. (a) Early sensory cells visualized by immunostaining for SuH protein (green). (b) A general nuclear stain (propidium iodide). (c) A merged view. Reproduced from Gho *et al. Development* 1996; 122: 1673–1682, with permission from Company of Biologists Ltd.

Fig. 4.12 Lateral inhibition. Cell type A produce both the activator and the inhibitor. Where the activator prevails cell type A is stabilized, where the inhibitor prevails cell type A is suppressed.

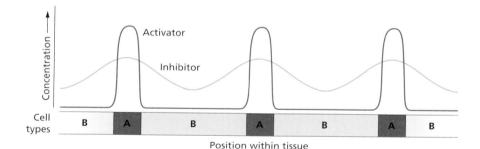

depicts a field of cells that are committed to become cell type B but are spontaneously progressing toward commitment to form cell type A. The first few cells to become type A produce an "activator" substance, which promotes development of cell type A, and an "inhibitor" substance, which antagonizes the action of the activator such that inhibited cells are unable to continue the progress toward type A and revert to becoming cell type B instead. The reason that the system generates pattern is that the activator is short range, perhaps only active through intercellular contacts, while the inhibitor is long range, moving freely by diffusion. This means that near the source of activator, the activator will prevail over the inhibitor and guarantee formation and maintenance of the cell type A, which produces both substances. In the surrounding region the inhibitor will prevail over the activator and so suppress formation of cell type A. Beyond a certain range the action of the inhibitor will be insufficient to prevent the formation of further signaling centers of type A cells and so the final result will be the formation of many signaling centers spaced out across the field of cells in an approximately uniform way. How regular the final pattern is depends on the details of the mechanism.

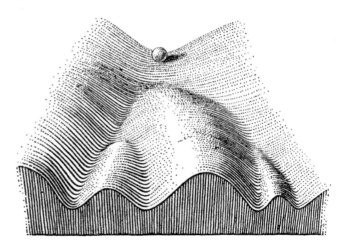

Fig. 4.13 Waddington's "epigenetic landscape." The ball represents a cell, and the valleys represent different developmental choices that it can make. Reproduced from Slack (2002), with permission from Nature Publishing Group.

Stochasticity in development

Development often seems to involve the creation of pattern from homogeneity. This occurs when a lateral inhibition system gets going. It occurs when a cytoplasmic determinant becomes localized at one end of the cell rather than another. It also occurs in any case where multiple cell types appear to differentiate from a single type of progenitor, even when they are cultured *in vitro* in a uniform environment. All of these processes involve **symmetry breaking**. Very small, naturally occurring fluctuations in the concentration of particular substances, or the activity of specific genes, become amplified at particular locations as a result of positive feedback, and become repressed in adjacent regions as a consequence of the amplification. What is the original source of these fluctuations? Ultimately, it is due to the small number of regulatory molecules in a cell. There may be a few hundred copies of a particular transcription factor within each cell and they have to find their binding sites in the DNA by searching a huge amount of DNA sequence. There are only two copies of each gene, and, at any particular time, each copy of a particular regulatory sequence will either have a transcription factor attached or it will not have one attached. The residence times of the individual molecules is quite long, measured in minutes, so a population of apparently identical cells will actually be heterogeneous in terms of the instantaneous occupation of regulatory sites. Given the existence of positive feedback systems, such fluctuations can easily be amplified to become macroscopic and irreversible differences in composition.

The "epigenetic landscape"

Developmental commitment is the subject of a remarkable concept first introduced by the embryologist and geneticist C.H. Waddington in the 1940s. This is the "epigenetic landscape" (Fig. 4.13). His diagrams have recently revived in popularity and are once again being reproduced frequently. A typical diagram depicts a landscape in which a series of valleys arise by branching on the way down the side of a hill, rather as a natural landscape would be sculpted by water. A ball placed at the top will roll down the valleys. At each bifurcation it will move left or right, depending on minor perturbations that it experiences, so each such choice of pathway is a symmetry breaking event. Eventually, it will come to rest in one of the lowest tier of valleys. The ball is supposed to represent a cell in an embryo, the valleys represent states of developmental commitment, and each bifurcation represents a developmental choice. Of course the entire embryo would be represented by many balls, and some balls would adopt each of the possible trajectories down the landscape to generate the complete set of stable cell types. Thus far the diagram captures some important ideas that have been verified in the molecular era. However, it also has some problems. One problem is that the diagram relates only to cell states, and does not capture the structure of the embryo in any way. This means that interactions between parts of the embryo, which are so important because of the inductive signals exchanged, are not represented. In reality, most developmental choices are not symmetry breaking events, they are deterministic events controlled by inductive signals. Another rather more abstract problem is to decide exactly what the surface represents. In fact, it is supposed to be a "potential energy surface" embedded within a multidimensional state space representing all of the levels of gene products and other substances found in cells. Whether such surfaces really exist, or can be computed, remains unclear.

Waddington himself introduced the term **epigenetics** as effectively a synonym of development, recognizing that a large element of development consists of the regulation of gene expression. However, the term is used in a different sense today. **Epigenetic** now refers mostly to the understanding of gene expression in terms of chromatin structure rather than of DNA alone. This recognizes the importance of DNA methylation as a controller of gene expression, and also of the numerous chemical modifications of histones and other chromosomal proteins, which collectively enable or prevent the activity of transcription (see Appendix). The term is also used to relate to the rare examples of inheritance that are not due to DNA sequence, but these are of little importance in animal development.

Criteria for proof

Much research work in developmental biology is concerned with the identification of inducing factors, or of determinants, or of the transcription factors that define the states of developmental commitment. Rigorous proof that a particular molecule really does perform a particular function during normal development requires at least three independent lines of evidence, concerning *expression, activity,* and *inhibition.*

Expression. The molecule in question must be present. There must be evidence that it is expressed in the right place, at the right stage, and in a biologically active form. This is usually obtained from *in situ* **hybridization** or **immunostaining**, but it should be remembered that the presence of mRNA does not guarantee the translation of the polypeptide, nor does presence of protein guarantee its post-translational processing to an active form. Additional evidence can sometimes be obtained from suitable reporter constructs, for example the presence of biologically active levels of retinoic acid can be detected by introducing a gene consisting of a retinoic acid response element (RARE) coupled to a *lacZ* reporter gene. Regions exposed to retinoic acid, and capable of responding to it, will then express β-galactosidase, which can be detected by the X-Gal reaction.

Activity. The molecule in question must have the appropriate biological activity in a suitable test system. For example a candidate inducing factor should be able to evoke the correct responses from its target tissue. For a candidate cytoplasmic determinant, it must be possible to inject it into another part of the cell or a different blastomere, and cause the injected region to develop along the pathway caused by the determinant.

Inhibition. If the molecule is inhibited *in vivo* then the process for which it is thought responsible should fail to occur. Where several similar molecules are responsible for a process (**redundancy**, see Chapter 3) then it may be necessary to inhibit all of them to obtain a result. Inhibition may be achieved at the DNA level by mutation of a gene to inactivity, at the RNA level using antisense morpholinos or RNAi, or at the protein level by introduction of a specific inhibitor of the normal gene product. Extracellular substances may sometimes be successfully inhibited by specific neutralizing antibodies. Although well-characterized mutations are necessarily specific to a single gene, inhibition experiments involving the other methods may not be so specific and need careful evaluation.

Key points to remember

- A fate map shows where each part of the embryo will move, and what it will become. It does not, however, indicate the state of commitment of parts of the embryo at the time of labeling.
- Developmental commitment can be labile (specification) or stable (determination). Specification indicates that a particular structure or cell type will be formed by development in isolation. Determination indicates that a particular structure or cell type will be formed following grafting to other regions of the embryo.
- Clonal analysis comprises the deductions that can be made about developmental mechanisms by labeling a single cell. If the progeny of one cell span the boundary between two structures, it shows that the cell was not determined to become either structure at the time of labeling.
- A cytoplasmic determinant will cause commitment to a particular developmental pathway by the cells that inherit it.
- An inducing factor is an extracellular signal substance that can alter the developmental pathway of cells exposed to it. Many inducing factors are known as growth factors or hormones in other contexts. If the factor is simply necessary for continued development of the target cells, it is said to be permissive. If a different developmental pathway is followed in the absence and the presence of the factor, then it is said to be instructive. If there is more than one positive response at different concentrations, the factor is described as a morphogen.
- Lateral inhibition systems generate two mixed cell populations from a single cell sheet. This occurs by amplification of small initial differences between the cells such that differentiating cells inhibit the differentiation of those around them.
- Proof that a particular gene product is responsible for a particular process requires evidence of appropriate expression pattern, of appropriate biological activity, and of appropriate consequences of ablation.

Further reading

General

Meinhardt, H. (1982) *Models of Biological Pattern Formation.* NY: Academic Press.

Slack, J.M.W. (1991) *From Egg to Embryo. Regional Specification in Early Development*, 2nd edn. Cambridge: Cambridge University Press.

Lewis, J. (2008) From signals to patterns: space, time, and mathematics in developmental biology. *Science* **322**, 399–403.

Tadros, W. & Lipshitz, H.D. (2009) The maternal-to-zygotic transition: a play in two acts. *Development* **136**, 3033–3042.

Kondo, S. & Miura, T. (2010) Reaction-diffusion model as a framework for understanding biological pattern formation. *Science* **329**, 1616–1620.

Bhattacharya, S., Zhang, Q. & Andersen, M.E. (2011) A deterministic map of Waddington's epigenetic landscape for cell fate specification. *BMC Systems Biology* **5**, 85.

Examples of fate mapping and clonal analysis

Garcia-Bellido, A., Lawrence, P.A. & Morata, G. (1979) Compartments in animal development. *Scientific American* **241**, 90–98; 102–111.

Hartenstein, V., Technau, G.M., & Campos-Ortega, J.A. (1985) Fate-mapping in wild-type *Drosophila melanogaster* 3. A fate map of the blastoderm. *Wilhelm Roux's Archives of Developmental Biology* **194**, 213–216.

Kimmel, C.B. & Warga, R.M. (1986) Tissue specific cell lineages originate in the gastrula of the zebrafish. *Science* **231**, 365–368.

Dale, L. & Slack, J.M.W. (1987) Fate map for the 32 cell stage of *Xenopus laevis. Development* **99**, 527–551.

Kimmel, C.B., Warga R.M. & Schilling, T.F. (1990) Origin and organization of the zebrafish fate map. *Development* **108**, 581–594.

Hatada, Y. & Stern, C.D. (1994) A fate map of the epiblast of the early chick-embryo. *Development* **120**, 2879–2889.

Cepko, C., Ryder, E.F., Austin, C.P., Walsh, C. & Fekete, D.M. (1995) Lineage analysis using retroviral vectors. *Methods in Enzymology* **254**, 387–419.

Ruffins, S.W. & Ettensohn, C.A. (1996) A fate map of the vegetal plate of the sea urchin (Lytechinus variegatus) mesenchyme blastula. *Development* **122**, 253–263.

Hejnol, A., Martindale, M.Q. & Henry, J.Q. (2007) High-resolution fate map of the snail Crepidula fornicata: The origins of ciliary bands, nervous system, and muscular elements. *Developmental Biology* **305**, 63–76.

Cytoplasmic determinants

Bowerman, B. (1995) Determinants of blastomere identity in the early *C. elegans* embryo. *BioEssays* **17**, 405–414.

Nishida, H. & Sawada, K. (2001) macho-1 encodes a localized mRNA in ascidian eggs that specifies muscle fate during embryogenesis. *Nature* **409**, 724–729.

Henrique, D. & Schweisguth, F. (2003) Cell polarity: the ups and downs of the par6/aPKC complex. *Current Topics in Genetics and Development* **13**, 341–350.

Nance, J. (2005) PAR proteins and the establishment of cell polarity during *C. elegans* development. *BioEssays* **27**, 126–135.

Goldstein, B. & Macara, I.G. (2007) The PAR proteins: fundamental players in animal cell polarization. *Developmental Cell* **13**, 609–622.

Strome, S. & Lehmann, R. (2007) Germ versus soma decisions: lessons from flies and worms. *Science* **316**, 392–393.

Embryonic induction

Jessell, T.M. & Melton, D.A. (1992) Diffusible factors in vertebrate embryonic induction. *Cell* **68**, 257–270.

Kessler, D.S. & Melton, D.A. (1994) Vertebrate embryonic induction – mesodermal and neural patterning. *Science* **266**, 596–604.

Okada, T.S. (2004) From embryonic induction to cell lineages: revisiting old problems for modern study. *International Journal of Developmental Biology* **48**, 739–742.

Tabata, T. & Takei, Y. (2004) Morphogens, their identification and regulation. *Development* **131**, 703–712.

Zhu, A.J. & Scott, M.P. (2004) Incredible journey: how do developmental signals travel through tissue? *Genes and Development* **18**, 2985–2997.

Ashe, H.L. & Briscoe, J. (2006) The interpretation of morphogen gradients. *Development* **133**, 385–394.

Rogers, K.W. & Schier, A.F. (2011) Morphogen gradients: from generation to interpretation. *Annual Review of Cell and Developmental Biology* **27**, 377–407.

Thresholds, stochasticity, epigenetics

Lewis, J., Slack, J.M.W. & Wolpert, L. (1977) Thresholds in development. *Journal of Theoretical Biology* **65**, 579–590.

McAdams, H.H. & Arkin, A. (1999) It's a noisy business! Genetic regulation at the nanomolar scale. *Trends in Genetics* **15**, 65–69.

Elowitz, M.B., Levine, A.J., Siggia, E.D. & Swain, P.S. (2002) Stochastic gene expression in a single cell. *Science* **297**, 1183–1186

Slack, J.M.W. (2002) Timeline – Conrad Hal Waddington: the last renaissance biologist? *Nature Reviews Genetics* **3**, 889–895.

Goldberg, A.D., Allis, D. & Bernstein, E. (2007) Epigenetics: a landscape takes shape. *Cell* **128**, 635–638.

This chapter contains the following animations:

Animation 3   Tests of commitment.
Animation 4   Cytoplasmic determinant action.
Animation 5   Induction.
Animation 6   Lateral inhibition.

 For additional resources for this book visit www.essentialdevelopmentalbiology.com

Chapter 5

Approaches to development: cell and molecular biology techniques

In Chapters 3 and 4 we have examined the approach to studying development by the use of genetics and experimental embryology. Here we shall consider a set of techniques that are derived from cell and molecular biology but have particular relevance to the study of development. Certain techniques with particular relevance to the study of organogenesis are discussed in Chapter 13.

Microscopy

Embryos are small and their study inevitably requires the use of microscopes. Experiments often involve manual interventions using a **dissecting microscope**. This is a binocular microscope with a magnification in the range of about ×10 to ×50 and a good working distance between the objective lens and the specimen (Fig. 5.1). Dissecting microscopes provide a three-dimensional image, which allows accurate perception of depth by the observer and assists in the performance of manipulations such as microsurgery or microinjection. Unlike most compound microscopes, a dissecting microscope does not invert the image. If specimens are opaque, such as *Xenopus* or chick embryos, incident lighting is used. This means that the beam is shone down from the light source onto the specimen. For living specimens it is important not to overheat them in a powerful incident light beam and so a fiberoptic light guide is used, providing a cold but bright illumination. If specimens are transpar-

ent, such as embryos of zebrafish, sea urchin, or mouse, then a transmitted light base would usually be used.

The **compound microscope** (Fig. 5.2) is used for the examination of sections, or for whole specimens that are small enough to be transparent, such as embryos of *Drosophila, Caenorhabditis elegans*, or zebrafish. Specimens that are not sections are referred to as **wholemounts**. The compound microscope has a magnification range from about ×40 to ×1000, and the upper limit of magnification is set by the wave nature of light, which prevents resolution of points closer together than about 0.2 μm. Most compound microscopes invert the image. This is a natural consequence of the optical system and is not normally corrected because to do so would require extra lenses. Under most circumstances the inversion is no problem.

Optical techniques

There are several different optical techniques commonly used with the compound microscope. Ordinary transmitted light is used to examine stained sections (Fig. 5.3a) or wholemounts. The stains may be traditional histological stains or may be colored histochemical reaction products arising from detection of reporter proteins or specific antigens (see below). **Wholemounts** viewed with transmitted light need to be very thin, as thick specimens tend to be too opaque and detail cannot be seen.

Essential Developmental Biology, Third Edition. Jonathan M.W. Slack.
© 2013 John Wiley & Sons, Ltd. Published 2013 by John Wiley & Sons, Ltd.

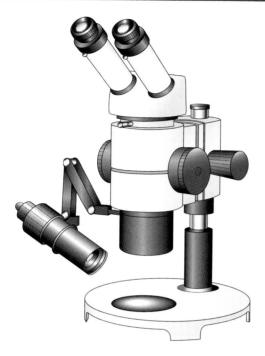

Fig. 5.1 Dissecting microscope.

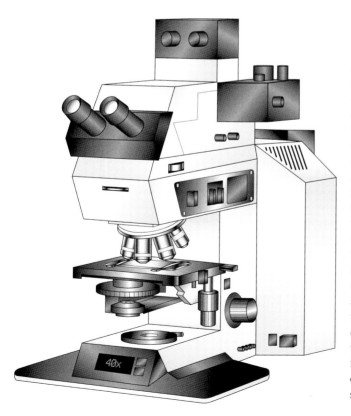

Fig. 5.2 Compound microscope.

Phase contrast microscopy is a technique that converts small differences of refractive index within the specimen into large differences of light intensity (Fig. 5.3b). It is the favored method for examining live cells cultured *in vitro*, or isolated gametes, and is used for small transparent embryos such as the mammalian preimplantation stages.

Differential interference contrast, otherwise known as **Nomarski** optics, converts small differences of refractive index into an apparent difference in height when perceived by the observer. It also provides a sharp resolution of a particular optical section within the specimen, so as one focuses the microscope up and down through the specimen, different optical sections come into view. Nomarski optics provides very clear visualization of single cells within small transparent specimens, and it was this technique that enabled the elucidation of the entire cell lineage of the worm *C. elegans*. It is used for preimplantation mammalian embryos and those of *C. elegans*, zebrafish, and *Drosophila*.

Fluorescence microscopy is used for a variety of purposes but they all depend on visualizing the location of a fluorescent substance, or **fluorochrome**, within a specimen (Fig. 5.3c,d). They include fluorescent antibody staining, fluorescent *in situ* hybridization (FISH), and the visualization of fluorochromes introduced into the specimen to label subsets of cells. The principle of fluorescence is that the fluorochrome absorbs light of a particular energy and emits light of a lower energy. Lower energy means longer wavelength or a color shifted toward the red end of the spectrum. A particular fluorochrome will have an **excitation** spectrum showing how the intensity of fluorescence varies with the excitation wavelength. It will also have an **emission** spectrum showing how the intensity of emitted fluorescence is distributed across the wavelength spectrum (Fig. 5.4). The excitation and the emission spectra are characteristics of the substance. A fluorescence microscope has an attachment that shines the excitation beam down onto the specimen. This consists of a powerful lamp, formerly a mercury arc lamp but now probably a light emitting diode (LED) type lamp. A filter selects a narrow excitation wavelength band suitable for the fluorochrome in use, then a dichroic mirror reflects the excitation beam down onto the specimen A dichroic mirror reflects wavelengths below a certain cutoff and transmits them above this cutoff. Because the emission from the specimen is of longer wavelength, it will be transmitted through the dichroic mirror and can then be visualized by the observer (Fig. 5.4). A fluorescence microscope will usually contain several alternative filter sets, one for each fluorochrome in use. If a specimen contains two or three fluorochromes it should be possible to visualize each one separately using the appropriate filter set. Because of the need to examine fluorescence in whole specimens arising from the various applications of green fluorescent protein and other similar fluorescent proteins (see below), fluorescent dissecting microscopes are now also in common use.

Other techniques common in cell biology do not find so much application in the study of development. Dark-field

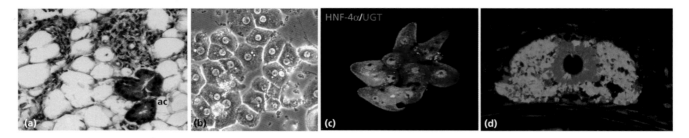

Fig. 5.3 Compound microscope images. (a) Section of a regenerating pancreas stained with hematoxylin and eosin. Two regenerating acini (ac) are clearly visible. (b) Phase contrast view of rat hepatocytes in culture. (c) Immunofluorescence of hepatocytes; green is the transcription factor HNF-4α, red is the enzyme UGT. (d) Section of a regenerated *Xenopus* tadpole spinal cord. Blue is DAPI, a fluorescent stain for DNA; and green is immunostain for neurofilament protein.

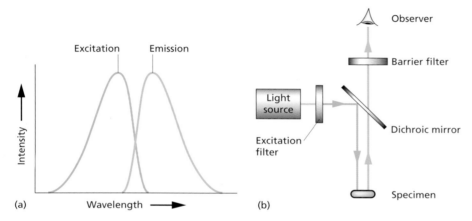

Fig. 5.4 Fluorescence microscopy. (a) Typical spectra for excitation and for emission of a fluorochrome. (b) Arrangement of components in a fluorescent microscope.

microscopy depends on illumination from a very oblique angle so that only points within the specimen that scatter light extensively are visible. They appear as bright points and the remainder of the specimen is dark. In developmental biology it is only really used for visualizing radioactive *in situ* hybridizations, in which the signal consists of an accumulation of silver grains in a photographic emulsion coating the section (see below). However, this technique is obsolescent due to improvements in nonradioactive detection.

Polarization microscopy places the specimen between crossed Polaroid filters. If the specimen contains components, such as muscle fibers, that can alter the plane of polarization of the light beam, then these appear as bright patches, and are said to show birefringence. Again this technique is now obsolescent due to improvements in other methods of detection of specific molecules *in situ*.

Confocal and multiphoton microscopes

Conventional fluorescence microscopy is limited to use on sections or very thin wholemounts because fluorescence from cells

above and below the plane of focus would otherwise swamp the image. However, fluorescence from thicker wholemounts can readily be visualized with the **confocal scanning microscope**. This is a device that uses a laser for illumination so excitation is achieved with just the single wavelength characteristic of that laser. Instead of illuminating the whole specimen, it illuminates just one point at a time. All the points in one particular optical section are scanned in turn and the fluorescence from each point is recorded by a detector and then used to reconstruct the image of the section. Because only one point is viewed at a time, the quality of the optical section is very high. Light from points above or below the plane that is in focus will be dispersed and contribute very little to the signal. The image from a confocal microscope is necessarily a digital image stored on a computer.

The multiphoton microscope can provide even better optical sectioning of fluorescent specimens than the confocal microscope. Here the light source is of less than the energy required to excite the fluorochrome. Thus, an excitation can only occur if two or more photons strike one fluorochrome molecule simultaneously, and this can only happen at the point of focus of the excitation beam where the photons are very dense.

Image capture

Until the 1990s, microscopic images were recorded by photography but all image collection is now done using **charge coupled devices (CCDs)**. The detection chip of a CCD camera contains an array of pixels each of which can be "filled up" with electrons, and the charge each pixel accumulates over a given time is proportional to the light intensity falling on it. An exposure is taken and then the charge on each pixel is read out in sequence by a detector and converted to a digital image. For an 8-bit image, the intensity of each pixel would be represented by a number from 0 to 255. A 24-bit color image would have one 8-bit number for each of three colors: red, green, and blue. There are different types of CCD cameras for different purposes. The most sensitive function only in black and white and can collect long exposures from very faint specimens, in the limiting case being able to count single photons. Less-sensitive ones can collect images fast enough to display on a video screen, and enable the recording of movies in real time.

Once the image is recorded it is transmitted to a computer and displayed on the screen, and then various types of computation can be carried out. For example the contrast and brightness can be altered, the image can be smoothed or sharpened, densitometric measurements can be taken, or multiple fluorescent images obtained through different monochrome channels can be recombined so that several colors are visualized on the same image. One of the many advantages of digital image collection is that the signal/noise ratio of the image can be improved by taking several exposures and then averaging them. The signal will be the same in each exposure and therefore will be unaffected by averaging, whereas the noise will be different in each exposure and therefore will become low and uniform.

Time-lapse imaging is especially useful for the study of morphogenetic movements of cells, organ cultures, or embryos. Individual images are taken at intervals and are then replayed at shorter intervals. For example if an image is recorded every minute for 4 hours and then replayed at 30 frames per second, the 4 hour process will be displayed in 8 seconds. The usual problem is that the specimen moves away from the field of view during the process. Modern microscopes equipped for time-lapse recording will have programmable stage controls such that a series of specimens can be imaged at preset intervals and the microscope will return each to the light path in turn. More sophisticated models will also undertake corrections for small deviations of position or plane of focus. For filming mammalian or avian embryos or organ cultures an environmental chamber is required, similar to a tissue culture incubator, such that the embryos can be kept at 37°C and 5% CO_2, or other such environment as may be appropriate.

Histological methods

Although wholemounts are used extensively in developmental biology, even with Nomarski or confocal microscopy there is a limit to the degree of resolution with which the structure of a specimen can be resolved. Thus, the need to prepare and examine **histological sections** will always remain (Fig. 5.3a,d). The first step in preparing a specimen for any form of microscopical examination is to fix it. **Fixation** means that it is killed and that it is made mechanically robust enough to withstand osmotic shocks or a certain amount of handling. Fixatives work in various ways. The most commonly used is formalin, which is a solution of the gas formaldehyde. This reacts with amino or sulfhydryl groups of macromolecules producing some denaturation and some cross linking. Glutaraldehyde has two aldehyde groups and is a very effective intermolecular cross-linking agent. Other common components of fixatives are acids and organic solvents. These act as denaturing and precipitating agents.

Once the specimen has been fixed it must be embedded in a solid supporting material that will permit it to be cut into very thin sections. The most common material is paraffin wax. In order to infiltrate the specimen with wax it needs to be **dehydrated**, and this is achieved by passing it through a series of baths of ethanol of progressively increasing concentration. This series of concentrations is necessary because a direct transfer from water to pure ethanol causes tissue damage arising from the mixing forces exerted by the solvents. Once the water is removed, the specimen is equilibrated in a solvent miscible with wax, such as xylene, then it is placed in molten wax at about 60°C and left until the wax has thoroughly penetrated every part. The wax is allowed to solidify and the resulting block can be stored permanently at room temperature. In order to make sections, the block is mounted on a **microtome**, which passes it repeatedly across a very sharp knife, each time advancing the block by a few microns. This results in the formation of a connected set, or ribbon, of sections. These are mounted on microscope slides and can then be stained or processed for immunostaining or *in situ* hybridization.

In embryology the orientation of sections is very important. Because it is often desirable to analyze the disposition of structures in the entire specimen it is necessary to have a complete set of connected, or serial, sections. For this reason paraffin wax is a very useful embedding material as the sections naturally form a ribbon as they are cut. Its disadvantage is that the specimen needs to be dehydrated with organic solvents and needs to be heated to 60°C for the embedding period. This can lead to damage to proteins or nucleic acids in the specimen, which may compromise immunostaining or *in situ* hybridization.

For these techniques it is quite common to use frozen sections. Here the specimen may not even need to be fixed but is frozen rapidly in a medium containing a high concentration of sucrose. This is cut into a block and mounted on a **cryostat**, which is simply a microtome operating in a cooled chamber.

The quality of frozen sections is usually not as good as paraffin wax, and it is very difficult to collect serial sections as they do not form a ribbon. But it is a very useful technique if only a few representative sections are required.

For some purposes paraffin wax does not provide sufficient quality of sections, as it is difficult to cut them thinner than about 5 μm. There are other embedding materials based on various plastics that can be cut at 1 μm, or even at fractions of a micron for the electron microscope. But these materials are often not compatible with immunostaining and *in situ* hybridization, and cannot provide serial sections. **Electron microscopy** of sections is only rarely used in developmental biology as it provides more magnification than is required to identify patterns of cell types in tissues or embryos. But the scanning electron microscope can often provide vivid three-dimensional views of wholemounts and is often used for visualizing cell arrangements in the course of morphogenetic movements.

Study of gene expression by molecular biology methods

When studying development it is very important to know the normal expression patterns of the genes under investigation. We need to know at what developmental stages they are active, in which parts of the embryo, and to what level of activity. There are two main classes of method for determining expression patterns: biochemical methods, which give a reasonably accurate quantitative measure but no anatomical information, and *in situ* methods, which give accurate anatomical information but limited quantification. In both cases, there are separate methods for studying mRNA and protein. Where possible, it is desirable to do both as the transcription of a gene does not guarantee its later translation into protein, and the presence of a protein in a particular place does not necessarily mean that it has been synthesized there, as it may have been transported from some other site of synthesis. Recently, the advent of the **laser capture** method (see Chapter 13), has made it possible to do some molecular analysis on specific cells or regions of cells taken from microscope sections.

With the biochemical methods, a limited degree of regional information can be achieved by dissection of the specimen, but they are all intrinsically techniques for making an estimate of the amount of a specific messenger RNA or protein in the whole specimen. They are often used to obtain a **stage series** for expression of a gene, by examining groups of embryos of different developmental stages (Fig. 5.5).

Only a brief description of the biochemical methods is given here, as they are described in detail in textbooks of molecular biology. In all cases, nonradioactive labeling and detection methods are now used instead of radioactivity, although ^{32}P is still occasionally used as it provides the highest level of sensitivity.

Methods for messenger RNA

Reverse transcription polymerase chain reaction (RT-PCR) is the standard method for detecting and measuring mRNA. The total RNA is extracted from the sample and reverse-transcribed into complementary DNA (cDNA) using reverse transcriptase. Then two oligonucleotides are added, chosen to correspond to

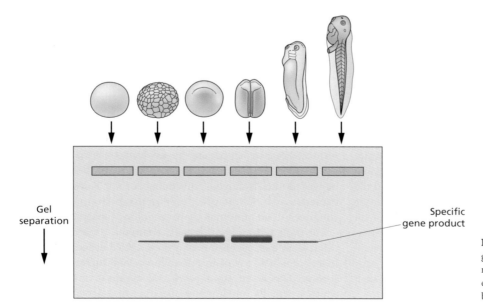

Gel separation

Specific gene product

Fig. 5.5 A "stage series" for a particular gene product. This could be a specific mRNA or a protein extracted from whole embryos at the different stages and separated by a suitable technique.

sequences a certain distance apart in the target cDNA. The DNA sequence between the primers is amplified by repeated cycles of synthesis, melting, and hybridization, which are carried out in a programmable heating block that changes the temperature in such a way as to enable each reaction step to occur in turn. After a suitable number of cycles, the reaction mixture is run on a gel and the DNA bands are visualized by ethidium bromide staining. If the reaction has worked then there should be a band of a size corresponding to the region of cDNA between the two primers. If there is no band then it means that there was no specific target cDNA in the sample. Because what is examined is the accumulated concentration of product at the end of the reaction, when the exponential phase is over, and a plateau has been reached, the method is at best only semiquantitative. Moreover, because of the exponential character of the reaction, RT-PCR is very prone to artifacts and careful controls need to be done to ensure specificity. One of these is the loading control to guarantee the presence of a similar amount of RNA in each sample. This is usually a mRNA that is considered ubiquitous, such as the mRNA for β-actin or glyceraldehyde-3-phosphate dehydrogenase. Another will be a bona fide positive sample, another will be a negative sample and, in addition, it is necessary to include a "no reverse transcription" control in case contaminating DNA generates a signal. Also, it is unwise to use too many cycles; more than 35 usually indicates trace levels of mRNA that have no biological significance.

More accurate quantitative measurement of specific mRNA levels can be achieved with a **real-time PCR** method. Here the rate of formation of the amplified product is monitored in real time. There are various methods of doing this but a commonly used one is to add the dye SYBR Green, which fluoresces brightly when bound to double-stranded DNA but not when bound to single-stranded DNA. The increase of fluorescence thus parallels the buildup of the amplified product. This occurs in an exponential manner and the cycle number is recorded when the double-stranded product exceeds a critical value (Fig. 5.6). Because the early exponential phase of amplification is much more directly related to starting concentration than is the final plateau level, this means that the starting concentration can be accurately measured relative to a control mRNA. The control will normally be a ubiquitous mRNA as is used for loading controls in qualitative RT-PCR.

There are some other methods of mRNA detection that will be encountered in the literature. Ribonuclease protection analysis was the most sensitive method for the detection of specific mRNA before the adoption of RT-PCR but is rarely used today. **Northern blotting** is the oldest and least sensitive technique. It is still sometimes used because it can show the number and size of mRNAs where there are several splice variants formed from the same gene. It involves extracting total mRNA from the specimen and separating it by gel electrophoresis on a denaturing agarose gel. After a good separation has been achieved, the contents of the gel are transferred ("blotted") onto a hybridization membrane. The membrane is hybridized with the specific

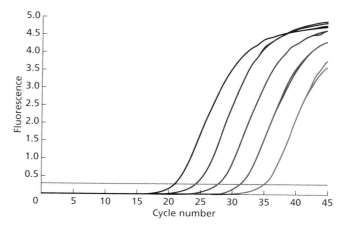

Fig. 5.6 Measurement of relative mRNA abundance by real time PCR. The instrument measures the cycle number at which the critical concentration of double stranded DNA is synthesized (red line).

probe, which is labeled with a chemical tag or with ^{32}P. The chemical tag is detected by an enhanced chemiluminescence reaction and then visualized by autoradiography or with a phosphorimager. Radioactivity may be visualized directly by the same methods.

Microarrays

Microarrays enable examination of large numbers of gene products simultaneously (see Animation 7: Use of microarray). Where the genome sequence is known it is possible to examine the full inventory of cDNAs corresponding to the entire genome, although microarrays often only contain a subset of the complete genome. Microarrays are typically used in the early stage of a research project to identify the genes likely to be involved in a particular system or process. They are made using machines derived from the computer industry and so are sometimes referred to as chips. They come in two basic types: the cDNA array and the Affymetrix oligonucleotide array. cDNA arrays consist of a regular arrangement of closely spaced spots, each of a different cDNA, arranged in a rectangular array on a glass slide. Sometimes, instead of cDNAs, long synthetic oligonucleotides are used. In the Affymetrix system, which is a proprietary product, a larger number of short oligonucleotides are synthesized directly on the sides. Here, several oligonucleotides will represent sequences from one gene. In both cases the array is used for nucleic acid hybridization. A probe is prepared by extracting RNA from the embryo or tissue sample and reverse transcribing the mRNA into cDNA. This is labeled with a fluorescent dye and hybridized to the microarray under suitable conditions. The array is then scanned by a chip reader, which is a fluorimeter that can measure the fluorescence from the dye bound to each spot. For a given cDNA on the array, the intensity of fluorescence should represent the amount of the

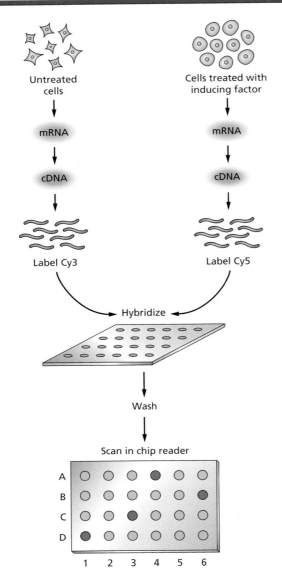

Fig. 5.7 Use of a microarray to compare gene expression in two cell populations. The results show that genes B6 and D1 have been upregulated, and genes A4 and C3 have been downregulated, by the inducing factor treatment.

complementary cDNA in the probe, and thus the amount of that specific mRNA in the tissue sample. In developmental biology, microarrays are often used to make comparisons, for example between two embryonic stages, or between cells treated and untreated with an inducing factor (Fig. 5.7). In this case there is a probe from each of the two different samples to be compared. These are labeled with different fluorochromes, usually the dyes Cy3 (green) and Cy5 (red). The probes are mixed before the hybridization is carried out and then the ratio of green to red fluorescence is measured. Genes whose expres-

sion does not change will give a yellow signal because both probes will bind, while those whose expression goes up or down will give a red or green signal respectively. The analysis of both types of microarray depend on very sophisticated software that can match the position of each spot with the identity of the cDNA.

Deep sequencing methods

The extraordinary capacity of modern ("next generation") sequencing equipment to generate DNA sequence does not only affect genomics but has also extended to expression analysis. This has generated several new techniques, which are likely to find increased application in developmental biology, particularly if the cost can be brought down. **RNA-Seq** refers to the ability to measure the composition of the entire transcriptome in a cDNA sample. This is done by sequencing a very large number of cDNAs, equivalent to at least a fourfold coverage of the whole genome. Each known exon that is encountered is identified by comparison with the known genome of the organism, and then the frequency of each gene product is counted relative to the others. So long as the original RNA preparation, and the reverse transcription reaction, are unbiased, this should give a quantitative profile of the whole transcriptome.

Gene expression analysis has increasingly come to require knowledge of what is happening at the chromatin level, not just the relative abundance of the mRNA product. Deep sequencing can now be combined with DNA methylation analysis to give a genome-wide description of DNA methylation, or with chromatin immunoprecipitation (ChIP, see below) to give a genome-wide view of binding of specific proteins of interest.

Methods for protein

It is possible to identify individual unknown proteins from a complex mixture and the set of techniques employed to do this is referred to as "**proteomics.**" As with microarrays, they are most often employed at the beginning of an investigation where it is required to find which proteins change in expression level during a particular developmental event. Total protein extracts are separated by two-dimensional gel electrophoresis, in which the first dimension is an isoelectric focusing gel separating by isoelectric point, and the second dimension is an sodium dodecyl sulfate (SDS)-polyacrylamide slab gel separating by molecular weight. This gives a pattern of spots, each spot representing one protein. Comparison of preparations from the two samples under investigation will hopefully yield a small number of differences in the spot pattern. Then, individual proteins are identified by mass spectrometry. A mass spectrometer works by ionizing and volatilizing the substance then determining its time of flight to the detector after acceleration in an electric field. This enables measurement of the molecular weight to very high precision. Where the genome has been sequenced the measurement of protein molecular weight may itself allow its identification, since this can be calculated from the known amino acid composition of each polypeptide represented in the genome sequence. Otherwise, the individual protein can be sequenced by tandem mass spectrometry. The protein is digested with trypsin into a number of peptides. These are separated in the first cycle of mass spectrometry, then each peptide is broken up by ion bombardment and the fragments separated in a second cycle of mass spectrometry. Using sophisticated software it is possible to identify the amino acid sequence of the peptide from the characteristic fragmentation pattern.

Immunochemical methods

The methods for measuring a particular protein in a sample usually depend on having a specific **antibody** directed against the protein. This may be a monoclonal or a polyclonal antibody and will have been prepared by immunizing an animal with the pure protein or portion of a protein.

A **western blot** is a method that shows the content of a specific protein. The total proteins of the sample are extracted and separated on an acrylamide gel. The content of the gel is transferred ("blotted") onto a membrane. The membrane is then incubated with the specific antibody, which should bind only to the band of the specific target protein and not to other components separated on the gel. The bound antibody is then visualized with a second antibody, which is directed against the constant region of the first antibody. The second antibody is likely to be purchased commercially and it will be modified for easy detection, nowadays usually by being conjugated with **horseradish peroxidase (HRP)**, which has available very sensitive chemiluminescent substrates. A chemiluminescent substrate will produce a phosphorescent reaction product that decays spontaneously and emits light as it does so. After incubation in the substrate mixture, the blot is exposed to X-ray film or placed in a phosphorimager, and the phosphorescence of the reaction product is recorded in the position where the antibody is bound, corresponding to the position and quantity of the target protein in the original sample.

An **immunoprecipitation** is not a true precipitation but is a method of isolating the protein recognized by a specific antibody. It is often used to find whether a particular protein is being newly synthesized in an embryo or tissue sample. The new synthesis rate may be quite different from the steady state concentration shown by a western blot. For example the embryo may inherit a large maternal store of a particular protein from the egg, but may not be making it any more. For this purpose the live specimen is incubated in radioactive amino acids, usually ^{35}S-methionine or ^{35}S-cysteine, so all of the proteins made during the labeling period become radioactive. The total proteins are extracted and are incubated with the specific antibody. This will bind to its target protein and form an immune complex. The immune complexes are isolated by incubating the mixture with protein A bound to agarose beads. Protein A is a bacterial protein that binds tightly to the constant region of

IgG-type antibodies. It will capture the immune complexes and immobilize them on the beads. The beads are then washed to remove contaminating proteins, and are boiled in a highly denaturing sample buffer to release the bound antibody and target protein. This is run on a SDS protein gel, which separates by molecular weight; the gel is dried, and then the radioactive band of target protein is visualized by autoradiography or phosphorimager. Immunoprecipitations are also often used in nonradioactive mode to establish the presence of a particular protein in a sample. This is usually followed by a western blot and detection of the precipitated protein using a specific antibody and chemiluminescent detection.

Chromatin immunoprecipitation (ChIP) involves shearing chromatin to fragments of about 500 nucleotides in length, then immunoprecipitating with an antibody to a protein of interest, which may be a chromosomal protein such as a histone bearing a particular chemical modification. Then the sample is deproteinized and the DNA is analyzed by RT-PCR to find how much of the region of interest (for example the promoter region of a developmental control gene) has been pulled down. This technique has been combined with microarray analysis so that thousands of loci can be examined simultaneously (ChIP-Chip). It has also been combined with deep sequencing to enable the whole genome to be examined (ChIP-Seq).

Study of gene expression by *in situ* methods

In situ methods are designed to reveal the spatial domains of gene expression in a specimen (see Animation 8: *In situ* and immunostaining). If the specimens are small enough and transparent enough, *in situ* procedures can be performed on wholemounts, which has the advantage of rapid and clear three-dimensional visualization. If specimens are too large or opaque then *in situ* methods can be used on sections, which for embryological work will usually be serial sections.

In situ hybridization

***In situ* hybridization** reveals the regions of a specimen where a specific mRNA is present. The chemistry is the same as a northern blot, as an antisense probe is synthesized *in vitro* complementary to the mRNA to be detected. This is hybridized to the specimen and then visualized (Figs 5.8, 5.9). The probe includes an extra chemical group recognizable by a commercially available specific antibody. The group is attached to one of the nucleoside triphosphates used for synthesis. It is incorporated during synthesis and does not affect the hybridization capacity of the probe. Favorite groups for this purpose are digoxigenin (DIG, a plant sterol) and fluorescein.

For wholemount *in situ* hybridizations, the specimen will usually need to be permeabilized by a short treatment with protease or detergent to enable the large probe molecules to enter the cells. For either wholemounts or sections, the hybridization reaction is conducted overnight, the specimen is washed extensively, then an enzyme-linked antiprobe antibody is added, for example an anti-DIG antibody conjugated to alkaline phosphatase. The location of this can be revealed by placing in a suitable substrate mixture, which yields an intensely colored insoluble precipitate at the site of the reaction. Thus, the color forms where the enzyme is located and this shows where the probe is bound and therefore where the specific mRNA was present.

Some substrates are fluorescent and produce a signal that can be seen under the fluorescence microscope. Fluorescence *in situ* hybridization is often known by its acronym "FISH." This is very convenient when more than one probe has been used and they are visualized through different fluorescence channels to detect regions of separation or overlap in the expression domains. In order to visualize more than one probe at a time it is necessary to use different chemical tags on the probes, which can be detected with different antibodies and substrate mixtures. Note that although fluorescein is often used in probes, the number of fluorescein groups present in the probe is not sufficient on its own to provide a fluorescent signal. Visualization of fluorescein in this context still requires antibody binding and amplification by an enzyme–substrate reaction.

In situ hybridizations were originally performed on sections using radiolabeled probes. This had the advantage of high sensitivity, but the main disadvantage was that they have to be developed at a particular time, which requires prior knowledge of the likely optimal exposure time. A color reaction, by contrast, can be watched as it proceeds and be stopped when the color has reached the desired intensity.

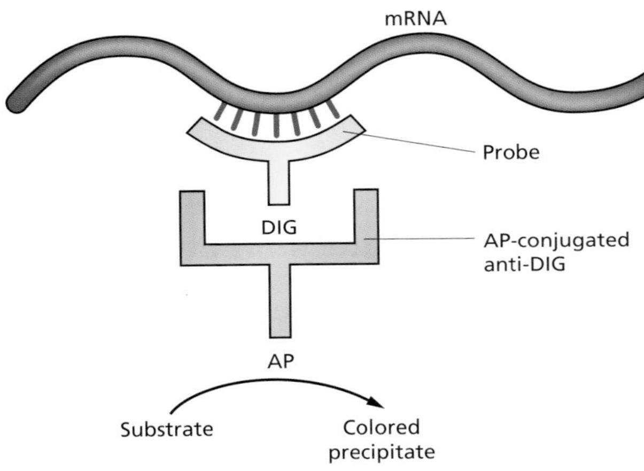

Fig. 5.8 *In situ* hybridization for detection of mRNA.

Immunostaining

Staining of specimens with specific antibodies is also very important in developmental biology (Figs 5.10, 5.11). It is possible to make an antibody against almost any protein or carbohydrate molecule by injecting it into an animal with a suitable adjuvant. The target of a specific antibody is referred to as its **antigen**, regardless of its chemical nature. The particular parts of an antigen molecule that are recognized by the antibody are called **epitopes**. **Polyclonal** antibodies are often made in rabbits. After repeated immunizations there should be a high concentration, or titer, of specific antibody in the serum. Although the antibody is specific for the particular antigen, it will consist of the products of several clones of B lymphocytes, each making antibody from a different antibody gene. In addition to the specific antibody, the serum will contain thousands of other

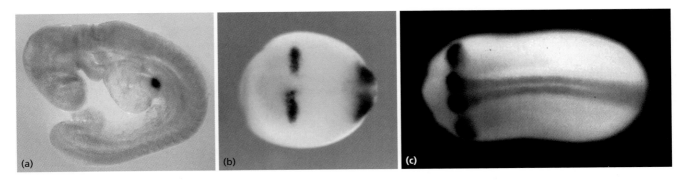

Fig. 5.9 Examples of wholemount *in situ* hybridization. (a) mRNA of the transcription factor C/EBPα in the early liver bud of a 9.5 day mouse embryo. (b) Two domains of expression of the same gene: mRNA of the inducing factor FGF3 in a *Xenopus* neurula. (c) Expression of two genes shown in different colors in a *Xenopus* closed neurula. Dark blue shows mRNA for the transcription factor HoxA4; cyan shows mRNA for the transcription factor Xcad3.

antibodies recognizing unrelated antigens, and thousands of nonantibody proteins. These other components may cause background or nonspecific staining, and therefore the antibody will usually need to be partially purified from the serum before use. Antibody molecules consist of a variable region, containing the part that recognizes the antigen, and a constant region, which is characteristic of the immunoglobulin class and the animal species. A frequently used partial purification method is a column of protein A. As mentioned above, this is a bacterial protein binding to the constant region of IgG-type antibody

molecules. Thus, the protein A column will isolate all the IgG, including the specific antibody, and leave all the other serum components behind. A better method, if enough of the antigen is available, is to make an affinity column carrying the antigen, and select out just the specific antibody from the serum, removing all other antibodies as well as other proteins.

Monoclonal antibodies are usually made by immunizing mice, then fusing their spleen cells with a human tumor cell line called a myeloma. The hybrid cells, called **hybridomas**, are capable both of antibody production and also of growth without limit *in vitro*. Numerous multiwell plates containing clones of the fused cells are screened in order to find the one required. Once the desired cell clone has been isolated, it can be grown and its culture medium harvested as a source of just one particular monoclonal antibody. The advantage of the monoclonal method is that it does not require a pure antigen to start with; in fact, wholly uncharacterized tissue extracts can be used for immunization. The antibodies made by individual hybridomas can then be screened for interesting expression patterns and cloned in the molecular sense by screening an expression library with the same antibody. On the other hand, more work and skill is required to make monoclonal antibodies rather than polyclonal ones, and individual monoclonal antibodies often do not have as high an affinity for their antigen as do polyclonals.

As for *in situ* hybridization, immunostaining can be performed either on **wholemounts** or on **sections**. Wholemounts have the advantage of providing a single three-dimensional view of the location of the antigen in the specimen, while sections provide an intrinsically higher resolution. In both cases, the specimen is incubated with the specific antibody for a suitable period, then washed and incubated with a second antibody. This

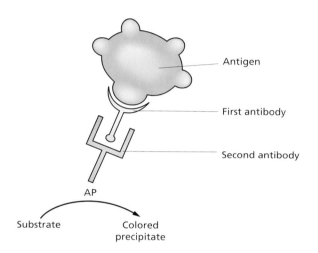

Fig. 5.10 Immunostaining for detection of protein.

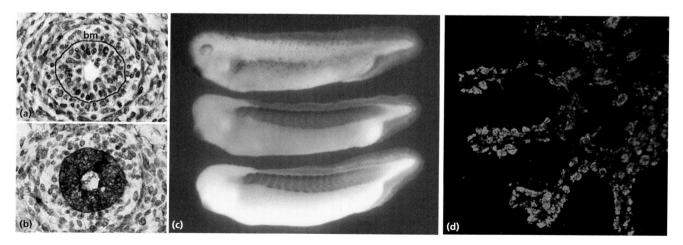

Fig. 5.11 Examples of immunostaining. (a,b) Section through a 13.5 day mouse embryo esophagus stained in hematoxylin–eosin (a) and immunostain for cytokeratin 8 with diaminobenzidine reaction (brown), plus hematoxylin as a nuclear stain (b); bm, position of the basement membrane. (c) Wholemount immunostaining (middle and below) of a muscle antigen in *Xenopus* tailbud-stage embryos. The antibody is the same in both but detection is with alkaline phosphatase and Fast Red (middle) and horseradish peroxidase with diaminobenzidine (below). (d) Wholemount stain of a mouse embryo pancreas organ culture. Three fluorescent antibodies were used: amylase is green, insulin red, and glucagon blue.

is a commercially available antibody directed against the species-specific constant region of the first antibody. It will carry a group suitable for detection, either a fluorescent group like one of the Alexa series of dyes, or an enzyme such as alkaline phosphatase or horseradish peroxidase. If a fluorescent second antibody was used, the specimen can be examined directly in the fluorescence microscope. If an enzyme-linked second antibody was used, then the specimen is incubated in the substrate to allow the colored precipitate to develop, and can then be mounted and visualized in transmitted light.

Enzyme-linked methods are usually more sensitive than fluorescence, because the enzyme–substrate reaction provides an additional amplification step. Also, once the precipitate is formed it is possible to dehydrate the specimen and use a nonaqueous mounting medium. Nonaqueous media are usually preferable to aqueous ones because they have a higher refractive index and render the specimen more nearly transparent. For fluorescence immunostaining it is necessary that the immune complex remains intact, and this makes it impossible to dehydrate the specimen. For the same reason, fluorescent specimens may not be permanent because the immune complexes will eventually dissociate. On the other hand, fluorescence methods are simpler because there are fewer steps, and they provide the best way of looking at more than one antigen in overlapping domains because each fluorochrome can be examined separately in its own channel of the fluorescence microscope.

Reporter genes

There are numerous circumstances in developmental biology where it is convenient to use **reporter genes**, which encode some easily detectable product, to monitor some particular aspect of events in the organism (Fig. 5.12). For example reporters are used to indicate the presence or activity of a transgene, or to label a particular cell type so it becomes readily visible, or to indicate the activity of a specific intracellular signaling pathway. Reporters are also extensively used to analyze the regulatory domains of genes. Each short sequence from a putative regulatory region is attached to a reporter gene and introduced into embryos as a transgene. If the sequence is active it will drive expression of the reporter and the domain of activity can be visualized. Reporter expression is usually monitored by *in situ* methods that have good spatial resolution but are not quantitative. For some purposes it is preferable to use biochemical methods, which are quantitative but lack spatial resolution.

The most popular reporter gene overall is *Escherichia coli lacZ*, coding for **β-galactosidase**. The name arises from the fact that it is the "*Z*" gene of the *lac* operon, subject of classical studies on gene regulation in the 1950s. β-galactosidase is a large, tetrameric enzyme capable of hydrolyzing a whole range of β-galactosides, which are substances consisting of a chemical group joined by a β linkage to the 1-carbon of galactose. Biochemical measurement of β-galactosidase activity is possible with a variety of colorimetric or fluorescent substrates, but in developmental biology this reporter is usually used in *in situ* mode. For this the substrate is 5-bromo-4-chloro-3-indolyl-β-D-galactoside, or X-Gal for short. When hydrolyzed away from the galactose, the X part of the molecule immediately forms a green–blue insoluble precipitate. The sensitivity of the reaction means that very low levels of expression can be detected. β-galactosidase is a very useful reporter both because of its sensitivity and because most animal tissues do not contain cross-reacting enzymes, which means that background staining is usually low. It will work following aldehyde fixation and therefore can be combined with other techniques such as conventional staining or immunostaining.

It should be remembered that the β-galactosidase protein is very stable, so its presence may indicate past as well as present activity of the gene. It is therefore possible to find regions of an embryo in which β-galactosidase enzyme is present but the mRNA encoding it is absent. This phenomenon is known as **perdurance** and may be particularly significant in embryos that do not show much growth. For types that grow rapidly, the

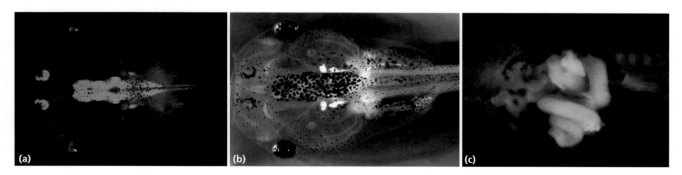

Fig. 5.12 Reporters. (a) A *Xenopus* tadpole carrying an *N-tubulin* reporter driving GFP in the central nervous system. (b) A muscle *actin* promoter driving tdTomato (a red fluorescent protein) in myotomal and jaw muscles. (c) An *elastase* promoter driving GFP, expressed only in the pancreas (two green buds).

protein level will quickly fall by dilution once the gene has been turned off.

β-galactosidase will often remain active as a **fusion protein** when the polypeptide has been fused to some other protein sequence. This is important for some applications including the use of a construct called **β-geo** in mouse knockout technology. β-geo is a fusion of the β-galactosidase enzyme with the product of the **neomycin** resistance gene and possesses both of the biological activities in the one molecule.

A second very important group of reporter genes is based on **green fluorescent protein (GFP)**, which was introduced during the late 1990s. As its name suggests this is a protein showing intense green fluorescent emission. It originates from the jellyfish *Aequorea victoria*. Like β-galactosidase it often remains active when fused to other proteins. Various alterations have been introduced into the coding sequence of GFP to increase the intensity or alter the color of the fluorescence, and so there are now available several derivative proteins emitting red, yellow, or blue fluorescence. One important advantage of the fluorescent proteins is that they are easy to visualize in living specimens and so allow real-time examination of the labeled cells as well as detection in fixed specimens. After fixation, some fluorescence is lost but the molecules may still be detected using specific antibodies.

Another commonly used reporter is firefly **luciferase**. This is an enzyme that catalyzes breakdown of a substrate, luciferin, in the presence of ATP. The reaction leads to emission of light (**luminescence**). The emission spectrum peaks in the green but extends to longer wavelengths, which penetrate tissues easily. The phosphorescence is often measured biochemically using a luminometer to measure the signal from cells or tissue explants. It is also possible to do *in situ* detection using imaging devices containing a highly cooled CCD camera capable to detecting single photons. These are usually used on larger specimens, such as whole mice, to localize luciferase-labeled implants, but can also be used on small specimens. A different luciferase, from the sea pansy *Renilla*, uses another substrate, coelenterazine, and has a blue emission. This is usually used as a normalization control for firefly luciferase.

Microinjection

There are various reasons why it is often desirable to introduce a substance into a single cell of an embryo. Most of the experiments on *Xenopus* embryos involving overexpression of genes are carried out by making synthetic mRNA *in vitro* and injecting it into the fertilized egg. The creation of transgenic mice relies on the ability to inject DNA into a pronucleus of a fertilized egg, and to create transgenic *Drosophila* the DNA must be injected at the posterior end of the egg, where the germ cells are about to form. Microinjection methods may also be used to introduce inhibitors such as specific antibodies or antisense oligonucleotides into cells. Microinjection is also essential for a whole variety of cell-labeling experiments where a particular cell lineage, or a graft needs to be identified by the presence of a visible substance called a **lineage label**.

The equipment required for microinjection depends on the size of the target cell. However, it will always be mounted on a microscope to allow visual control of the injection (Fig. 5.13).

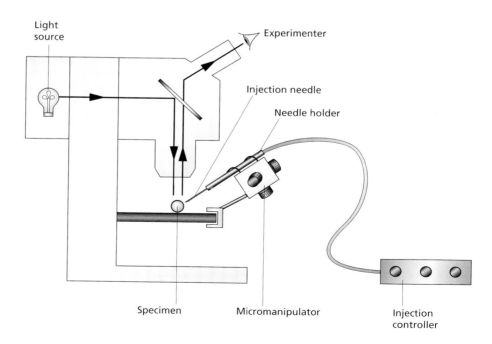

Fig. 5.13 Setup for microinjection under the fluorescence microscope.

It will require some form of micromanipulator to hold the injection needle and reduce the manual movements of the experimenter to a scale commensurate with the target cell. The injection needle itself will be made of glass tubing. This is drawn out into a fine-tipped injection needle using a needle-pulling machine that heats the glass to near melting point and then applies an appropriate pull to draw it out. The substance to be injected is introduced into the needle, either sucked into the tip by capillarity or injected with a syringe into the rear end. Then the needle is connected to an injection controller by a flexible tube. The controller may be a pressure device that applies sharp pulses of pressure to the needle and forces a small volume out of the sharp end. It may be an iontophoretic device which applies an electric field across the needle and causes a migration of appropriately charged molecules out of the tip. In either event, the needle is filled, is connected to the controller and attached to the micromanipulator. Watching down the microscope the experimenter impales the cell required and operates the controller to drive a pulse of the substance out of the tip. If the substance is fluorescent, as is often the case for a lineage label, then the microscope will be fitted with a fluorescent attachment to enable immediate visualization of the effectiveness of the injection. *Xenopus* embryos can be injected under a dissecting microscope, but smaller specimens will require compound microscope magnification. Some setups will use an upright and others an inverted microscope, but in either case the optics of the microscope will usually be arranged so that the image is not inverted, and its mechanics will be such that the stage holding the manipulators is fixed while the lenses move. These are special features, not standard on the normal type of upright compound microscope.

Cell-labeling methods

Cell-labeling methods are used for a whole variety of different reasons. They are used for **fate mapping** to show the normal destiny of embryo regions in the course of development. They are used to label single cells in order to do **clonal analysis**. They are used to label cells into which other substances such as mRNA or antibodies have been introduced. They are used to label whole embryos, which can then be used as donors for labeled grafts. The labels used in developmental biology are normally intended to be **lineage labels**, meaning that they label all progeny of the originally labeled cells and nothing else, although sometimes this ideal is not achieved in practice. The importance of cell labeling lies in the fact that cells in all types of animal embryos can move around considerably and it is impossible to keep track of individual cells by observation alone.

Extracellular labels

The oldest type of extracellular label are the **vital dyes** that were introduced in the early years of the twentieth century for fate mapping. The most commonly used are Nile blue and neutral red. They are called vital dyes because a reasonable intensity of color can be taken up by living cells and does not produce toxic effects. They can be used to label whole specimens by application in the medium, or to label specific regions of an embryo by local application of a small block of agar impregnated with the dye (Fig. 5.14b). They are still sometimes used because they are quick, simple, and cheap. However, they are not really lineage labels. It is difficult to get the dye deep into a specimen, and the dyes do spread and fade, and therefore only produce an approximate indication of the original site of labeling.

More recently, the carbocyanine dyes, **DiI** and **DiO**, have been favored for applications involving small patches of extracellular label. They are applied using an extracellular variant of the microinjection device discussed above. Being very hydrophobic substances they dissolve in the lipid membranes of the labeled cells and are well retained in the progeny. They are intensely fluorescent, DiI producing red and DiO producing

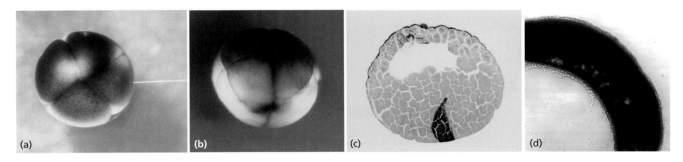

Fig. 5.14 Labeling techniques. (a) Microinjection of a four-cell stage *Xenopus* embryo. (b) A patch of the vital dye Nile blue applied to the dorsal side of a *Xenopus* embryo. (c) A single vegetal blastomere injected with horseradish peroxidase, which is then visualized on a section with the benzidine reaction. (d) A patch of DiI was applied to the early endoderm of a *Xenopus* embryo. The cells are later dispersed in the gut epithelium, indicating cellular intercalation movements.

green emission, so they are visualized using the fluorescence microscope. Although DiI and DiO themselves are removed from the specimen by organic solvents, there are available chemical derivatives that are retained during histological processing and can be examined in paraffin wax sections.

Intracellular labels

The most commonly used intracellular labels are **fluorescent dextrans**. Dextran is a polymer of glucose, freely soluble in water and metabolically inert. Water-soluble molecules cannot enter or leave cells by diffusion through the lipid plasma membrane, and although small molecules can move from cell to cell through gap junctions, these are not able to carry molecules of molecular weight above about 1000. Therefore a dextran of about 10,000 molecular weight will remain confined to the cell into which it has been injected. To make the dextran visible it is conjugated to a fluorochrome, such as fluorescein or rhodamine, and to make it fixable it is also conjugated to the amino acid lysine, which bears a free amino group which readily reacts with aldehyde fixatives. Commonly used substances of this group are FDA (fluorescein dextran amine) and RDA (rhodamine dextran amine).

Another commonly used intracellular label is the enzyme horseradish peroxidase (HRP) which is also retained within cells after injection (Fig. 5.14c). Peroxidases catalyze the oxidation of various substrates by hydrogen peroxide, and the one normally used is diaminobenzidine, which is converted to a brown insoluble material, stable to paraffin wax histology. HRP can be visualized in formalin-fixed wholemounts, but the enzyme itself is not stable to paraffin wax histology and therefore frozen sections need to be cut if the specimen is too big for wholemount staining.

Genetic labels

Substances such as those described above are at their most useful in embryos that show little or no growth during development, that is free-living species like *Xenopus* and zebrafish, although they have also been successfully used for limited periods of mammalian or chick development. Without growth the substances do not become diluted and can remain clearly visible for days. However, with growth they rapidly become diluted and cease to be visible. Under these circumstances the best labels are those incorporated into the genome of the cell such that they become replicated every cell cycle and are not diluted. Much of the early experimentation on mouse and *Drosophila* utilized genetic labels of this kind, although they all had severe disadvantages. For example the mouse glucose phosphate isomerase isozyme system depended on the different electrophoretic mobility of the enzyme present in different mouse strains. But this could not be visualized *in situ*, only by biochemical analysis of dissected tissue pieces, and so had very low spatial resolution. The *yellow* and *multiple wing hairs* mutants of *Drosophila* were widely used but were only expressed in the adult fly and only in the cuticle. They were not suitable for the analysis of larvae or of the internal tissues of the adult.

Nowadays much better genetic labels are available, which are expressed in all tissues, at all developmental stages, and can be visualized by *in situ* methods. For example there are transgenic mouse lines expressing *E. coli* β-galactosidase (encoded by the *lacZ* gene) or human alkaline phosphatase (hPLAP) in all tissues (Fig. 5.15). These can be detected by sensitive histochemical methods, and hPLAP has the advantage that the enzyme activity is stable to paraffin wax embedding so it can be visualized histochemically in sections. In any experiment involving a combination of cells from a labeled and unlabeled source, it is possible to stain the specimen and find which cells derived from each of the original components.

It is also possible to introduce a genetic label by means of a replication-incompetent retrovirus. Retroviruses are RNA viruses that carry a reverse transcriptase enzyme. On infection of a cell the reverse transcriptase makes a DNA copy of the viral genome that can then integrate into the chromosome of the cell, and will remain present in all its progeny. A replication-incompetent virus lacks genes that are necessary for assembly of virus particles and therefore the viral genome simply stays in the chromosome and is unable to produce further virus or lyse

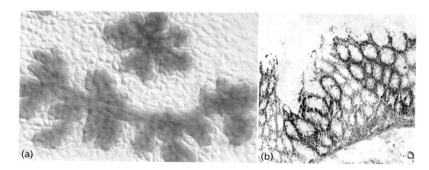

Fig. 5.15 Use of *E. coli lacZ* as a genetic label. Detection by XGal staining. (a) An organ culture of mouse embryo pancreas, similar to Fig. 5.11d, but this one is formed by recombining the epithelium from a ROSA26 mouse, expressing *E. coli* β-galactosidase in all tissues, and the mesenchyme from a wild-type mouse. Only the epithelium is stained. (b) Intestinal crypts from an H253 mouse. This transgenic mouse strain has the *lacZ* gene on its X chromosome. Adult females inactivate the transgene in 50% of cells, leading to the mosaic appearance shown here.

(a)   (b)

the cell. Replication-incompetent viruses are themselves propagated in special cell lines called packaging cell lines, which contain the missing viral genes and are able to complement the functions missing from the virus. A retrovirus used for cell labeling will usually contain a *lacZ* gene driven by the viral **long terminal repeat** which is a strong promoter sequence. They are mainly used for **clonal analysis** (see Chapter 4), therefore infection is instigated at low multiplicity such that each focus of infection represents the progeny of a single labeled cell.

Key points to remember

- Dissecting microscopes are used to handle and manipulate specimens or to examine large wholemounts. Compound microscopes are used to examine sections or small wholemounts.
- Preparation of standard histological sections involves fixation, dehydration, embedding in paraffin wax, sectioning on a microtome, and removing wax from the sections before staining them.
- Gene expression may be studied by biochemical methods that give good quantification but poor spatial resolution. Methods for specific mRNAs include qualitative and real time RT-PCR. Methods for specific proteins include immunoprecipitation and western blotting.

- Conversely, *in situ* hybridization for RNA or immunostaining for protein can give good spatial resolution but limited quantitative information.
- Reporter genes, most often *lacZ* or genes encoding fluorescent proteins, are used for the analysis of promoters in transgenics and for many other purposes.
- Extracellular labels used for fate mapping comprise the traditional vital dyes and the newer carbocyanine dyes. Fluorescent dextrans are often used as intracellular labels.
- Genetic labels are not diluted by growth of the embryo and so are very useful for experiments on mammalian or avian embryos.

Further reading

Teaching-level practical manuals

Gibbs, M.A. (2003) *A Practical Guide to Developmental Biology*. Oxford: Oxford University Press.

Marí-Beffa, M. & Knight, J., eds (2005) *Key Experiments in Practical Developmental Biology*. Cambridge: Cambridge University Press.

Tyler, M.S. & Kozlowski, R.N. (2010) *DevBio Laboratory: Vade Mecum³*. Sinauer Associates http://labs.devbio.com/.

Research-level practical manuals

Stern, C.D. & Holland, P.W.H., eds. (1993) *Essential Developmental Biology. A Practical Approach*. New York: Oxford University Press.

De Pablo, F., Ferrus, A. & Stern C.D., eds (1999) *Cellular and Molecular Procedures in Developmental Biology* (*Current Topics in Developmental Biology*, Vol. 36). New York: John Wiley and Sons.

Tuan, R.S. & Lo, C. (2000) *Developmental Biology Protocols*, 3 vols. Totowa, NJ: Humana Press.

Sharpe, P.T. & Mason, I. (2008) *Molecular Embryology: Methods and Protocols*, 2nd edn. Totowa, NJ: Humana Press.

Histological methods, *in situ* hybridization, and immunostaining

Harland, R.M. (1991) In-situ hybridization – an improved wholemount method for Xenopus embryos. *Methods in Cell Biology* **36**, 685–695.

Polak, J.M. & McGee, J.O'D. (1999) *In Situ Hybridization: Principles and Practice*. Oxford: Oxford University Press.

Wilkinson, D.G., ed. (1999) *In Situ Hybridization–A Practical Approach*, 2nd edn. New York: Oxford University Press.

Polak, J.M. (2003) *Introduction to Immunocytochemistry*, 3rd edn. Oxford: BIOS Scientific Publishers.

Bancroft, J.D. & Gamble, M. (2007) *Theory and Practice of Histological Techniques*, 6th edn. Philadelphia, PA: Churchill Livingstone.

Kiernan, J.A. (2008) *Histological and Histochemical Methods: Theory and Practice*, 4th edn. Bloxham, UK: Scion Publishing.

Organism-specific manuals

Nagy, A., Gertsenstein, M., Vintersten, K. & Behringer, R. (2003) *Manipulating the Mouse Embryo. A Laboratory Manual*, 3rd edn. Cold Spring Harbor Laboratory Press.

Ashburner, M., Hawley, S. & Golic, K. (2005) *Drosophila: A Laboratory Handbook*, 2nd edn. Cold Spring Harbor Laboratory Press.

Strange, K., ed. (2006) *C. elegans. Methods and Applications*. Totowa, NJ: Humana Press.

Bronner-Fraser, M., ed. (2008) *Avian Embryology*, 2nd edn. London: Elsevier.

Westerfield, M., Zon, L.I. & Detrich, H.W., ed. (2009) *Essential Zebrafish Methods: Cell and Developmental Biology (Reliable Lab Solutions)*. Oxford: Academic Press.

Sive, H.L., Grainger, R.M. & Harland, R.M., eds (2010) *Early Development of Xenopus Laevis: A Laboratory Manual.* Cold Spring Harbor Laboratory Press.

Other techniques

Bradbury, H. (1998; repr. 2005) *Introduction to Light Microscopy.* Oxford: Bios Scientific Publishers.

Murphy, D.B. (2001) *Fundamentals of Light Microscopy and Electronic Imaging.* New York: Wiley-Liss.

Avison, M.V. (2007) *Measuring Gene Expression.* Abingdon: Taylor and Francis.

Carroll, D.J., ed. (2008) *Microinjection: Methods and Applications.* Totowa, NJ: Humana Press.

Buckingham, M.E. & Meilhac, S.M. (2011) Tracing cells for tracking cell lineage and clonal behavior. *Developmental Cell* **21**, 394–409.

Supatto, W., Truong, T.V., Débarre, D. & Beaurepaire, E. (2011) Advances in multiphoton microscopy for imaging embryos. *Current Opinion in Genetics and Development* **21**, 538–548.

Kretzschmar, K. & Watt, F.M. (2012) Lineage tracing. *Cell* **148**, 33–45.

This chapter contains the following animations:

Animation 7   Use of microarray.

Animation 8   *In situ* and immunostaining.

 For additional resources for this book visit www.essentialdevelopmentalbiology.com

Section 2

 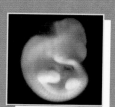

Major model organisms

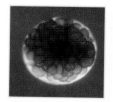 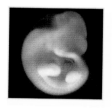 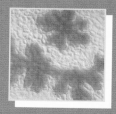 

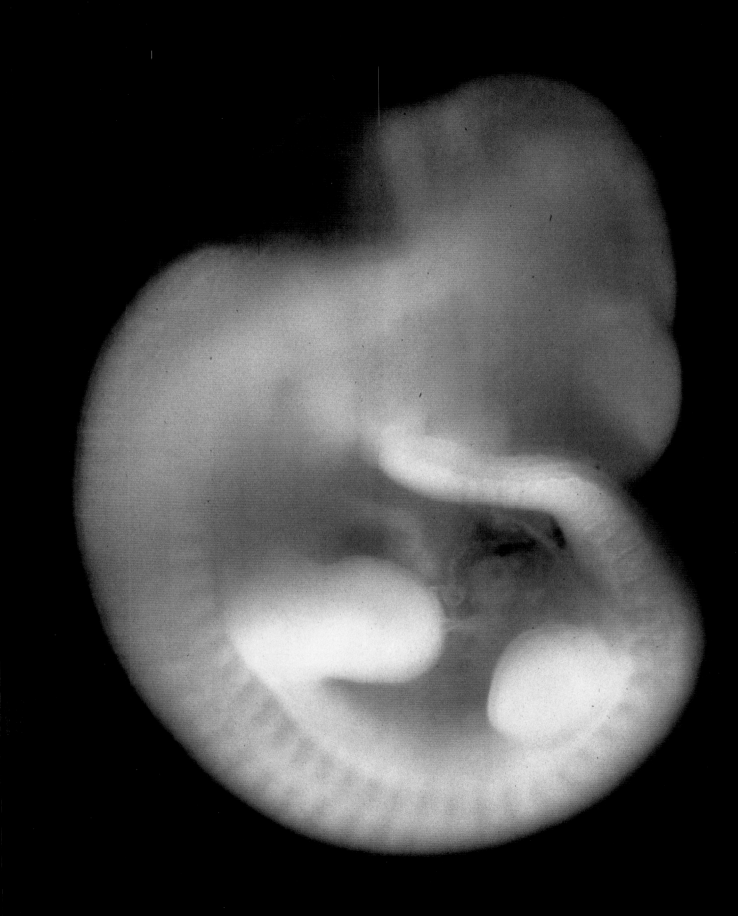

Chapter 6

Model organisms

Out of over 1 million animal species, modern developmental biology has focused on a very small number, which are often described as "**model organisms**." This is because the motivation for their study is not simply to understand how that particular animal develops, but to use it as an example of how all animals develop. Much developmental biology research is supported by medical research funding bodies and their ultimate goal is to understand how the human body develops, even if this is not the immediate goal of the investigators themselves. For this reason, there is often an attraction to studying a process that also occurs in humans. In this chapter it will be explained why research activity has focused down onto a small number of species, with some comparative indication of their strengths and weaknesses. In a text book a line has to be drawn somewhere and so just six of the most important models organisms are covered in detail: the mouse, the chick, the frog *Xenopus*, the zebrafish, the fruit fly *Drosophila*, and the nematode *Caenorhabditis elegans*. Other species have also been extensively studied, the next most popular being various species of sea urchin, but worldwide the largest number of active researchers are working with the big six.

The big six

The organisms discussed in detail in this book are listed in Table 6.1 and their positions in the phylogenetic tree of animals are shown in Fig. 6.1. It will be noted that they do not provide a very good coverage of the animal kingdom, as there are just two invertebrates and four vertebrates. However, the evolutionary distance between them is sufficiently large to make it likely that any developmental features possessed by all six is shared by the entire animal kingdom.

These six model species have each been selected because they have some particular experimental advantages for developmental biology research. It is true that sometimes individual scientists choose their organisms just because they like looking at them and working with them. But there are also some more objective practical considerations that govern the selection of organisms for particular purposes, which are summarized in Table 6.2.

All the six species are available all year round. Without this ready availability they would not have been selected as model organisms at all. It is worth remembering that it can be very difficult to "domesticate" an organism for laboratory life, and to attempt this for a "new" species is a major undertaking. Although some marine invertebrates have been used in developmental research (see below) none of the big six is a marine organism, probably because of the extra difficulties involved in keeping them in the lab. For example it is difficult to breed and rear sea urchins throughout their life cycle in the lab.

In terms of numbers, it is relatively easy to obtain thousands of eggs from *Xenopus*, zebrafish, *Drosophila*, and *C. elegans*. Chick eggs are normally not generated from birds in the lab but are purchased from commercial hatcheries. A large incubator will accommodate a few hundred eggs. Mice are less prolific than the other species but a mated female is likely to produce a litter of, say, 12 embryos, so it is not too difficult to produce moderate numbers.

Cost is an important consideration because the pressure of modern research demands a weekly or even daily supply of embryos. *C. elegans* are very cheap, as they can be grown on agar plates coated with bacteria, and genetic stocks can be stored

Essential Developmental Biology, Third Edition. Jonathan M.W. Slack.
© 2013 John Wiley & Sons, Ltd. Published 2013 by John Wiley & Sons, Ltd.

Species	Common name	Phylum, subphylum, class
Caenorhabditis elegans	Worm	Nematode, phasmida
Drosophila melanogaster	Fruit fly	Arthropod, uniramian, insect
Brachydanio rerio	Zebrafish	Chordate, vertebrate, fish
Xenopus laevis	African clawed frog	Chordate, vertebrate, amphibian
Gallus domestica	Chicken	Chordate, vertebrate, bird
Mus musculus	Mouse	Chordate, vertebrate, mammal

Table 6.1 Organisms discussed in detail in this book.

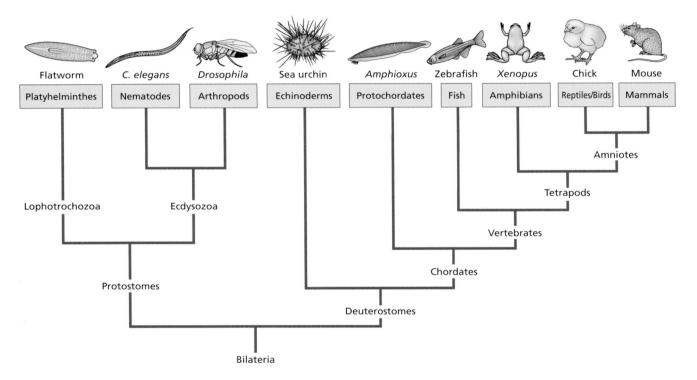

Fig. 6.1 Phylogenetic tree showing the positions of the big six model organisms used in developmental biology.

	C. elegans	*Drosophila*	Zebrafish	*Xenopus*	Chick	Mouse
Number of embryos	high	high	high	high	low	low
Cost	low	medium	medium	medium	low	high
Access	good	good	good	good	good	poor
Micromanipulation	limited	limited	fair	good	good	limited
Genetics	good	good	fair	none	none	good
Gene inventory	known	known	known	known*	known	known

*X. tropicalis.

Table 6.2 Experimental advantages and disadvantages of six model organisms.

frozen. *Drosophila* are potentially cheap to keep but may need temperature- and humidity-controlled fly rooms and significant technician time to maintain all the different genetic stocks. Cryopreservation of *Drosophila* is not yet routine. Zebrafish need an expensive aquarium facility and large-scale stock maintenance. The capital cost of setting up a lab. is quite high and the running cost is probably similar to *Drosophila*. *Xenopus* also need an aquarium although these have traditionally been of a less sophisticated type than those used for zebrafish. The care is relatively cheap since they have "no genetics" and therefore there are few stocks to maintain. Chicks are very cheap to keep as they are purchased from chick hatcheries and just require an egg incubator in the lab. However, if a lab needs to do genetic crosses or have access to very early stages, it will need a chicken-breeding facility, and since the animals are quite large this will be costly. Mice may seem small and cheap but are actually by far the most expensive organisms on the list. The logistics of mammalian breeding and the regulations about space and standards in laboratory animal facilities mean that the technician time and space required for a significant mouse operation is very high. Mouse sperm and embryos can, however, be frozen, reducing the long-term cost of stock maintenance.

Access and micromanipulation

"Access" refers to how easy it is to get at embryos at all stages of development. Just about any embryo can be obtained by dissection, but many experiments require that something should be done to the embryo and it should then be maintained alive until a later stage to observe the consequences. From this point of view, the free-living organisms with external fertilization are the most favorable. This means *Xenopus* and zebrafish. *Drosophila* eggs are also laid soon after fertilization. *C. elegans* need to be dissected from the mother during cleavage stages, although they can survive perfectly well on their own. Chicks undergo their cleavages in the reproductive tract of the hen and are already a double-layered structure containing about 60,000 cells when the eggs are laid. After this stage they are easy to get at because it just requires a hole to be cut in the egg shell. They will survive perfectly well if the hole is resealed with adhesive tape and they are kept in a humidified incubator. Mice are the least good from the access point of view. For the first 4 days the embryos can be flushed from the reproductive tract and cultured *in vitro*. After this, they become implanted into the uterus of the mother and depend on the placenta for their nutrition. It is very difficult to grow preimplantation embryos into postimplantation stages *in vitro*, and if an early postimplantation embryo is removed from the mother's uterus it can only be cultured *in vitro* for another 1–2 days. However, individual organ rudiments from mouse and chick embryos can usually be cultured for long periods and show good differentiation *in vitro*. This means that they are often favored for studies of organogenesis.

The other requirement for embryonic experimentation is the ease of microsurgical manipulation. This may mean removing a single cell, or a small piece of tissue, grafting an explant to another position in a second embryo, or injecting individual cells with substances. All this is relatively easy in *Xenopus*. Because of the large size of amphibian eggs, microsurgery can be done freehand under the dissecting microscope, and microinjection requires relatively cheap and simple equipment. A good level of micromanipulation can also be achieved in the chick, particularly at later stages. The other organisms are somewhat less favorable. With zebrafish it is possible to remove or inject cells during the early stages. *C. elegans* and *Drosophila* eggs are small and surrounded by a tough outer coat. Mouse embryos are small at preimplantation stages and hard to culture at postimplantation stages.

On the other hand, the *C. elegans* and fish embryos have the particular advantage of being transparent and so it is easier to follow cell movements *in vivo* than it is for the other species.

Genetics and genome maps

All the organisms under consideration have genes, but they do not all have "genetics" in the sense of a technology for doing experiments in the lab. involving breeding. This is still important although for the less favorable genetic models, such as *Xenopus*, it is possible to do sophisticated gain and loss-of-function experiments by other means.

Drosophila genetics is very sophisticated as it was practiced for many decades before *Drosophila* was adopted as a model for the study of development. The short life cycle of 2 weeks, and the ease of keeping large numbers of animals are both decisive. Also, the existence of balancer chromosomes (see Chapter 11) simplifies the keeping of stocks of mutants that are lethal in the homozygous form. *C. elegans* is also very favorable for genetics, because of a short life cycle and the ease of keeping large numbers of animals. Because it is a self-fertilized hermaphrodite, new mutations will segregate to the homozygous state automatically without the need to set up any crosses. Mouse genetics has also been practiced for many decades but it still falls below that of *Drosophila* in sophistication, partly because of the difficulty of handling lethal mutations, and also because of the huge cost of keeping the large numbers of animals required for mutagenesis screens. However, the very sophisticated methods now available for targeted mutagenesis enable a large range of experiments to be done. The zebrafish has a shorter history as a laboratory organism, and so lacks some of the sophisticated genetic technology that exists for the more longstanding models. In terms of numbers and life-cycle duration (4 months), it is worse than the invertebrates but good for a vertebrate. *Xenopus* has never been taken seriously for genetics because of the long life cycle; it takes at least 9 months to rear an animal to sexual maturity. However, a related species, *Xenopus tropicalis*, will grow to maturity in 4 months and has been adopted as a genetically tractable model by some labs.

The large-scale genome sequencing activity of recent years means that all the model organisms now have more or less complete inventories of genes and high-resolution genome maps. The main importance of this for developmental biology is that it takes most of the labor out of cloning a new gene. In the past, to obtain a homolog of a known gene in your organism you had to clone it yourself and this can take considerable time and effort. Even worse, the positional cloning of a gene known only as a point mutation could take several years. When complete gene inventories and maps are available any particular gene can be obtained, at least to the level of a large piece of genomic DNA, from a central depository. Furthermore, the complete gene inventory means that all the members of a gene family are known in advance. This is very helpful because there is frequently considerable redundancy of function between members of a gene family and it is necessary to know all the members to be able to interpret the results of both overexpression and loss-of-function experiments.

Unfortunately, it has turned out that *Xenopus laevis* is **pseudotetraploid**. This means that it doubled the number of chromosomes about 30 million years ago, since when mutations have accumulated that make the gene copies on the duplicated chromosomes about 10% different from each other in DNA sequence. But the gene pairs generally retain similar expression patterns and functions. They are sometimes called "**pseudoalleles**" because the sequences make them look like alternative alleles at the same genetic locus but they are really not alleles at all. Pseudotetraploidy is unfavorable to the experimentalist because it means more genome to sequence and more redundancy of function. So the genome sequence of *Xenopus laevis* is still incomplete and attention has focused on *Xenopus tropicalis*, which is a true diploid. Fortunately, the level of sequence divergence between the two *Xenopus* species is low enough that probes from one species will normally hybridize with the homologous gene from the other.

Bony fish, including the zebrafish, also underwent an extra genome duplication, but this is much more ancient (about 420 million years ago) and so the gene pairs have diverged considerably. They have acquired rather different functions, and many have also been lost. Because of this the zebrafish is effectively diploid, although it does tend to have extra copies of many genes important in development.

Relevance and tempo

Table 6.2 summarizes the advantages and disadvantages of the six model organisms. A quick assessment of this table will show that they all have their strengths and weaknesses and that none of them is ideal in all respects. On balance, the mouse and chick score lower than the others, and indeed some of their basic processes of early development were elucidated more recently

because they are technically a little less favorable for experimental work. However, they are among the six favorites because of another important consideration, perceived relevance. Both mouse and chick are **amniotes**, and the mouse is a mammal. This means that they appear much more similar to humans than the other models. This consideration has guaranteed the mouse and chick a good share of medical research funding over the years. At the molecular level, the other organisms are actually much more similar than previously thought to the human, but since we ourselves are mammals we shall always have a special interest in mammalian development. Moreover, the more detailed developmental biology becomes, the more attention becomes focused on the differences between species. All animals have Hox genes which are involved with anteroposterior patterning. But the precise number and expression domains in mouse will be closer to the human than those in lower vertebrates, and much closer than those in invertebrate models.

In today's competitive world it is not possible to work on a system where each experiment will take a very long time. In this context, the rates of development of the big six are shown in Fig. 6.2. This shows that the difference of time span to maturity is enormous, ranging from *C. elegans* at 3 days to *Xenopus laevis* at 9 months. Of course, only the fast models are used for genetic research, but the very short generation time of *C. elegans* and *Drosophila* is a distinct advantage compared to the mouse or zebrafish in terms of getting experiments done at moderate cost in time and personnel. The top part of the diagram covers the embryonic rather than postembryonic period and shows that all the models do enable experiments to be conducted in a few days. This is because the "endpoint" is not usually the end of development but rather it is the stage at which the developmental process under study has been completed. So, for example, most *Xenopus* experiments have concerned early development and are completed in 2–3 days from fertilization, while many mouse experiments are scored in mid-gestation.

Other organisms

A very wide variety of organisms have been used at one time or another for developmental research but have not ended up among the big six. A striking feature of the less-popular models is that particular research communities tend to work on many different similar species within a taxon rather than just one species. This is an important reason why they have not become major model organisms. In the molecular age, research moves faster when probes can be readily exchanged and a total genomic sequence is a very valuable resource, enabling the full set of genes in a particular gene family to be known and the rapid identification of mutants. For multiple species within a taxon the primary sequences of genes and their genomic organization is inevitably somewhat different, and without worldwide agreement on a single species none of this can be done. For example the various planarian and urodele amphibian species used for

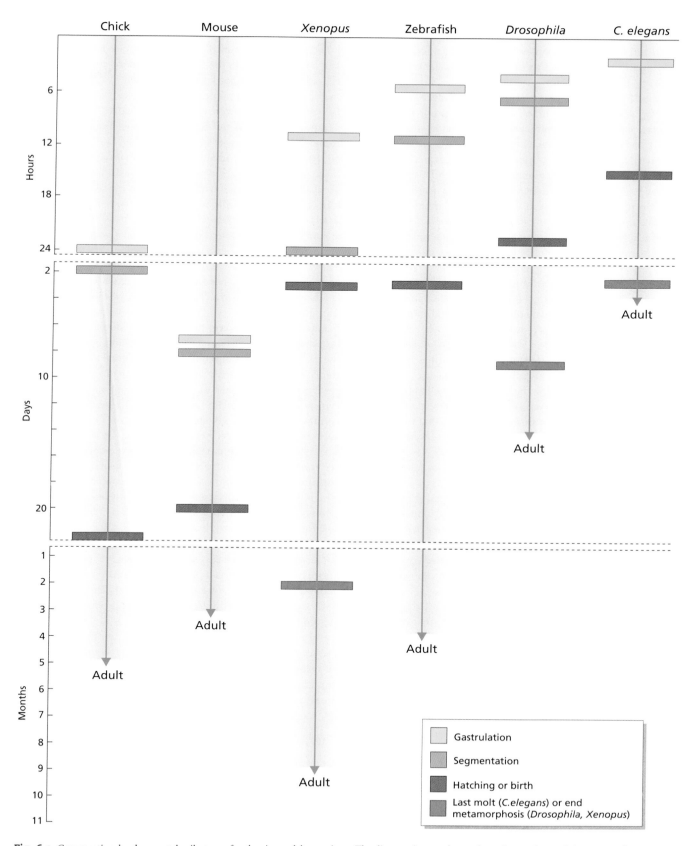

Fig. 6.2 Comparative developmental milestones for the six model organisms. The diagram is spread over three time scales, and times are only approximate since they may depend considerably on temperature for the free-living forms. "Gastrulation" refers to the start of germ layer formation. "Segmentation" refers to the start of visible segmentation. "Adult" refers to reproductive maturity. All times are measured postfertilization, including the chick where this adds approximately 1 day to the usual time scale of days of egg incubation.

regeneration research (Chapter 20) are too far apart for cross-hybridization of probes to be possible and this is a significant handicap to progress.

Probably the most important of the other models is the sea urchin. This is a "senior citizen" of embryological research and did have a much more prominent position in the premolecular era. It was used for the first experiments in the late nineteenth century that demonstrated embryonic regulation (formation of a whole larva from one blastomere), and in the 1930s for experiments demonstrating the existence of a vegetal to animal gradient controlling the body plan. More recently, it has been used to build models of developmental genetic networks. It is easy to obtain large quantities of eggs and to fertilize them *in vitro*. Sea urchin embryos are usually transparent and so morphogenetic processes can be observed *in vivo*. But the life cycle is very long, it is hard to rear the animals through metamorphosis, and they are not suitable for experimental genetics. Although microsurgical experiments have been conducted, the embryos are small (<100 μm diameter) so this type of work is very demanding. Sea urchins had their heyday in the 1960–80 period when it was advantageous to be able to obtain gram quantities of embryos for biochemical studies. This culminated in the discovery of the cyclins, proteins controlling the cell cycle (see Chapter 2). Also sea urchins are very convenient for studying fertilization, but the successful outcome of this work has shown that the molecular mechanisms actually have rather little in common with mammalian fertilization.

Ascidians are lower chordates that have had a modest following. Again they have a distinguished history, having been used for some of the classic cell lineage studies around the beginning of the twentieth century. Isolation and transplantation experiments demonstrated the existence of cytoplasmic determinants, some of which have now been identified. Various gastropods such as *Patella* and *Ilyanassa* have also been used for studies of cell lineage and cytoplasmic determinants. But, in the end, the concentration of attention on *C. elegans* has achieved most of what might have been expected to come from these marine invertebrates in terms of understanding the molecular nature and function of cytoplasmic determinants.

The **planarian** worms display some of the most dramatic regeneration behavior in the animal kingdom and might perhaps one day provide us with a seventh model organism. At present, however, there are rather few labs engaged on the work and several different species are still being studied.

Certain lower eukaryotes, including the cellular slime mould *Dictyostelium*, the acellular slime mould *Physarum*, and the alga *Volvox*, have been studied in relation to particular characteristics that they display clearly, respectively chemotactic cell aggregation, synchronous nuclear division and morphogenetic movements. But all of them are a big evolutionary distance away from the human, substantially more so than even the invertebrate animals, and they do not have the basic developmental mechanisms characteristic of animals.

Then there are the ancillary organisms that lost out historically to the front running models but have some particular technical advantage that keeps them in use in a niche market. The rat, a preeminent model for physiological research, is better than the mouse for whole embryo culture over the early post-implantation period, and is also often used for teratogenicity testing. The quail is easier to keep and breed in the lab than the chicken and so can be used for purposes that require treatment of the mother, such as vitamin A depletion studies. The quail is also used in conjunction with the chick for the labeling of grafts with quail-specific antibody. As mentioned above, *Xenopus tropicalis* offers a simpler genome and the possibility of experimental genetics not shared by *Xenopus laevis*. Also the **urodele** amphibians, such as the axolotl and various newts, show regenerative behavior surpassing that of *Xenopus*. The medaka is another fish that has been domesticated to laboratory life in Japan and was for a period superior to the zebrafish for making transgenics.

Finally, there are some that have been studied not because they are thought to be models for the human but rather from the perspective of trying to understand animal evolution. Here the key issue is not experimental convenience but rather the position in the phylogenetic tree. For example amphioxus is a cephalochordate thought to resemble to some extent the common ancestor of vertebrates. The features that it displays, such as a single Hox gene cluster, are felt to be primitive in the vertebrate lineage. **Cnidarians**, including the familiar freshwater *Hydra*, have a comparable position insofar as they may resemble the common ancestors of all animals. They also display a high degree of regeneration behavior. Many insects other than *Drosophila* have been used for developmental research, including the dragonfly *Platycnemis*, the cricket *Acheta*, the beetle *Tenebrio*, and the silk moth *Bombyx*. Some of these display a mode of development very different from *Drosophila*, in which the blastoderm grows as it produces posterior parts. Their main interest today is also for evolutionary comparison because of the exquisite level of molecular detail with which *Drosophila* development is understood.

All of these organisms have something to offer, but the big six account for most contemporary work and the convenience of using a standard model organism is such that the research funding agencies expect to see good reasons presented for working on something different.

Further reading

General
Bard, J.B.L. (1994) *Embryos. A Colour Atlas of Development*. London: Wolfe Publishing.

Stage series
Hamburger, V. & Hamilton, H.L. (1951) A series of normal stages in the development of the chick embryo. *Journal of Morphology* **88**, 49–92. Reprinted in *Developmental Dynamics* **195**, 231–272 (1992).

Nieuwkoop, P. D. & Faber, J. (1967) *Normal Table of Xenopus laevis*. Amsterdam: N. Holland. Reprinted by Garland Publishing Inc., London (1994).

Eyal-Giladi, H. & Kochav, S. (1976) From cleavage to primitive streak formation: a complementary normal table and a new look at the first stages of the development of the chick. *Developmental Biology* **49**, 321–337.

Theiler, K. (1989) *The House Mouse. Development and Normal Stages from Fertilization to Four Weeks of Age*, 2nd edn. Berlin: Springer-Verlag.

Hausen, P. & Riebesell, M. (1991) *The Early Development of Xenopus laevis*. Berlin: Springer-Verlag.

Kaufman, M.H. (1992) *The Atlas of Mouse Development*. London: Academic Press.

Hartenstein, V. (1993) *Atlas of Drosophila Development*. Cold Spring Harbor, NY: Cold Spring Harbor Laboratory Press.

Kimmel, C.B., Ballard, W.W., Kimmel, S.R., Ullmann, B. & Schilling, T.F. (1995) Stages of embryonic development of the zebrafish. *Developmental Dynamics* **203**, 253–310.

Bellairs, R. & Osmund, M. (1997) *The Atlas of Chick Development*. London: Academic Press.

Campos-Ortega, J.A. & Hartenstein, V. (1997) *The Embryonic Development of Drosophila melanogaster*, 2nd edn. Berlin: Springer-Verlag.

Some other organisms used in developmental biology

Sea urchin

Ohlendieck, K. & Lennarz, W. J. (1996) Molecular mechanisms of gamete recognition in sea urchin fertilization. *Current Topics in Developmental Biology* **32**, 39–58.

Davidson, E.H., Cameron, R.A. & Ransick, A. (1998) Specification of cell fate in the sea urchin embryo: summary and some proposed mechanisms. *Development* **125**, 3269–3290.

Angerer, L.M. & Angerer, R.C. (2000) Animal-vegetal axis patterning mechanisms in the early sea urchin embryo. *Developmental Biology* **218**, 1–12.

Ettensohn, C.A. & Sweet, H.C. (2000) Patterning the early sea urchin embryo. *Current Topics in Developmental Biology* **50**, 1–44.

McDougall, A., Shearer, J. & Whitaker, M. (2000) The initiation and propagation of the fertilization wave in sea urchin eggs. *Biology of the Cell* **92**, 205–214.

Oliveri, P. & Davidson, E.H. (2004) Gene regulatory network controlling embryonic specification in the sea urchin. *Current Opinion in Genetics and Development* **14**, 351–360.

McClay, D.R. (2011) Evolutionary crossroads in developmental biology: sea urchins. *Development* **138**, 2639–2648.

Other

Nishida, H. (1997) Cell fate specification by localized cytoplasmic determinants and cell interactions in ascidian embryos. *International Review of Cytology – a Survey of Cell Biology* **176**, 245–306.

Aubry, L. & Firtel, R. (1999) Integration of signaling networks that regulate *Dictyostelium* differentiation. *Annual Review of Cell and Developmental Biology* **15**, 469–517.

Wittbrodt, J., Shima, A. & Schartl, M. (2002) Medaka – A model organism from the Far East. *Nature Reviews Genetics* **3**, 53–64.

Satoh, N. (2003) The ascidian tadpole larva: Comparative molecular development and genomics. *Nature Reviews Genetics* **4**, 285–295.

Manahan, C.L., Iglesias, P.A., Long, Y. & Devreotes, P.N. (2004) Chemoattractant signaling in *Dictyostelium* discoideum. *Annual Review of Cell and Developmental Biology* **20**, 223–253.

Halpern, M.E., ed. (2004) Special issue medaka. *Mechanisms of Development* **121**, Issues 7–8.

da Fonseca, R. N., Lynch, J. A. & Roth, S. (2009) Evolution of axis formation: mRNA localization, regulatory circuits and posterior specification in non-model arthropods. *Current Opinion in Genetics and Development* **19**, 404–411.

Lemaire, P. (2011) Evolutionary crossroads in developmental biology: the tunicates. *Development* **138**, 2143–2152.

For planarians and urodele amphibians, see Chapter 20. For amphioxus, see Chapter 22.

Chapter 7

Xenopus

Although other amphibian species have been used for experimental work in the past, the African clawed frog *Xenopus laevis* has been the world standard for many years. This is because of its ease of maintenance, ease of induced spawning (pronounced "spore-ning"), and robustness of the embryos. The experimental production of *Xenopus* embryos is very simple as they can be caused to spawn following injection of both male and female with chorionic gonadotrophin. Nowadays, it has become more common to perform *in vitro* fertilization, which generates smaller numbers of embryos but ones whose time of fertilization is known precisely and whose development shows a high degree of synchrony. *Xenopus* early embryos are about 1.4 mm in diameter, which is large enough for quite discriminating microsurgery. Because the egg contains a high content of yolk granules and other reserve food materials it is possible for small multicellular explants to survive and differentiate for several days in very simple media, and the use of such explants is the basis of many experiments. The introduction of cell lineage labels, such as fluorescent dextrans, has enabled the construction of accurate fate maps. Methods have been introduced for overexpressing genes by making synthetic mRNA *in vitro* and injecting it into the fertilized egg, and for inhibiting gene action by injection of antisense **morpholinos**. These techniques have been heavily used for the study of early development, which is now very well understood. More recently, various techniques of **transgenesis** have enabled *Xenopus* also to be used for molecular genetic studies of organogenesis, regeneration, and metamorphosis.

Recently, a convention for *Xenopus* gene and protein names has been published, which requires both to be written in lower case. Although this convention is by no means fully established, it will be followed here. *Xenopus* gene and protein names are sometimes prefixed with an "X" or "x" to indicate their species of origin, for example *Xbra* for the *Xenopus brachyury* gene, or *Xgsc* for the *Xenopus goosecoid* gene. However, the convention is not uniformly adhered to, and in this book these "Xs" are omitted.

Oogenesis, maturation, and fertilization

The frog ovary consists of large numbers of **oocytes** surrounded by layers of follicle cells and blood vessels. The **oogonia** in the frog ovary persist into adulthood and continue to divide throughout life. Oogonia become primary oocytes following their last mitotic division. The growth of the oocyte then takes several months, during which time it acquires the food reserves needed to support the embryo over the early days of development before larval feeding can commence (Fig. 7.1). The oocyte nucleus is very large and is known as the **germinal vesicle**. The chromosomes are the four-stranded **bivalents** characteristic of meiotic prophase, but they remain transcriptionally active during oocyte growth and display numerous protruding loops of chromatin. Because of this appearance they are called **lampbrush chromosomes**.

To support protein synthesis in early development, the oocytes need to accumulate a large store of ribosomes and transfer RNA. For this reason the gene cluster coding for ribosomal RNA becomes amplified at an early stage into about 1000 additional extrachromosomal copies, all of which are active as templates for ribosomal RNA synthesis. During the growth phase the oocyte also acquires a large amount of **yolk** proteins. These are not made by the oocytes themselves, but by the liver of the mother, and are absorbed by the oocytes from the bloodstream. The early previtellogenic (= preyolky) oocyte is transparent, but it becomes opaque as the yolk granules begin to

Essential Developmental Biology, Third Edition. Jonathan M.W. Slack.
© 2013 John Wiley & Sons, Ltd. Published 2013 by John Wiley & Sons, Ltd.

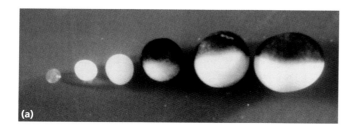

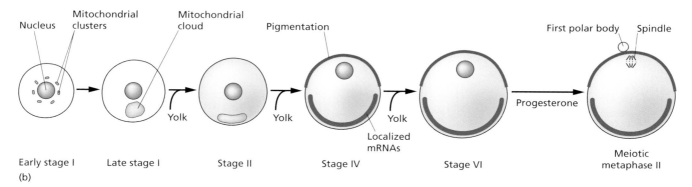

Fig. 7.1 Oogenesis in *Xenopus*. (a) Photographs of oocytes at stage I through VI (Roman numerals are used for staging). (b) Diagram of events occurring at each stage. Part (a) reproduced with permission from Smith et al. Oogenesis and oocyte isolation. In *Xenopus Laevis Practical Uses in Cell and Molecular Biology*, B.K.Kay and H.B.Peng (eds). Methods in *Cell Biology*, Vol. 36, San Diego Academic Press, 1991.

accumulate. When the oocyte is fully grown transcription ceases but protein synthesis and degradation continues. This is significant for experimental purposes because it means that if a specific mRNA is depleted, it will not be replaced, and the protein product will disappear over 1–2 days.

During growth the **animal–vegetal polarity** of the oocyte arises. In the previtellogenic stage a special cytoplasmic region becomes assembled. This is rich in mitochondria and called the mitochondrial cloud, or Balbiani body. It is the precursor of the **germ plasm**, and its location determines the future vegetal pole of the egg. A number of mRNAs, including *vegT* (see below), are associated with the cloud, and they become localized to the vegetal **cortex** in early vitellogenesis. The hemisphere containing the germinal vesicle is called the **animal hemisphere**. In mid-oogenesis this becomes dark due to an accumulation of pigment granules, while the vegetal hemisphere remains light colored. At about the stage that the pigmentation difference appears, a second group of mRNAs, including *vg1* (see below), also become localized to the vegetal cortex. The organization of the RNA localization depends on the *vegT* RNA itself, as if it is degraded following antisense oligonucleotide treatment then the other mRNAs become delocalized.

The fully grown primary oocyte is about 1.2 mm in diameter. Maturation is achieved in response to gonadotrophins secreted by the pituitary gland of the mother. These travel through the bloodstream and provoke release of progesterone from the ovarian follicle cells. This binds to steroid receptors in the oocyte and activates translation of an oncoprotein, c-Mos, which activates the phosphatase Cdc25 and hence activates **maturation promoting factor** (**MPF**, also referred to as M phase promoting factor, the complex of Cdk and cyclin required to initiate M phase, see Chapter 2). The germinal vesicle breaks down and the first meiotic division takes place, resulting in the formation of the first polar body. The result is often called an **unfertilized egg**, although it is strictly a secondary oocyte arrested in second meiotic metaphase. The metaphase arrest is due to the fact that cyclin breakdown is inhibited by a complex of c-Mos with Cdk2, known as cytostatic factor. The eggs are shed into the body cavity, enter the oviducts through the fimbriae (funnels) at the anterior end, and travel down the oviducts where they become wrapped in jelly.

In a normal mating the male clasps the female and fertilizes the eggs as they emerge from the cloaca, although in the laboratory eggs are usually fertilized *in vitro* simply by adding sperm. On fertilization the secondary oocyte/egg becomes a **fertilized egg** or **zygote**. The rise in intracellular calcium caused by sperm entry brings about the destruction of the cytostatic factor, leading to the breakdown of cyclin and progression into the second meiotic division, with release of the second **polar body**. The calcium also causes exocytosis of **cortical granules** near the egg surface whose protein and carbohydrate contents lift the **vitelline membrane** off the egg surface and allows the egg to rotate freely under the influence of gravity to bring the animal hemisphere uppermost.

Although our concern here is with the *Xenopus* oocyte as the precursor cell for the egg and embryo, it may be noted that it has also been an important tool in cell biology research. This is because it has all the properties of a normal cell and is also large enough to permit microinjection of mRNA or other substances, and to allow for sophisticated physiological studies.

Normal development

Development up to the general body plan stage may be subdivided into **cleavage, gastrulation**, and **neurulation**, and in *Xenopus* these stages are completed within 24 hours at 24°C. A numerical stage series was devised by Nieuwkoop and Faber, according to which stage 8 is the mid-blastula, stage 10 the early gastrula, stage 13 the early neurula, and stage 20 the end of neurulation (Fig. 7.2).

The sperm enters the animal hemisphere and initiates a cytoplasmic rearrangement called the **cortical rotation** (Fig. 7.3). This is a rotation of the egg cortex relative to the interior, which is associated with the transient appearance of an orientated array of microtubules in the vegetal hemisphere. It leads to a reduction in the pigmentation of the animal hemisphere on the prospective dorsal side, opposite the sperm entry point.

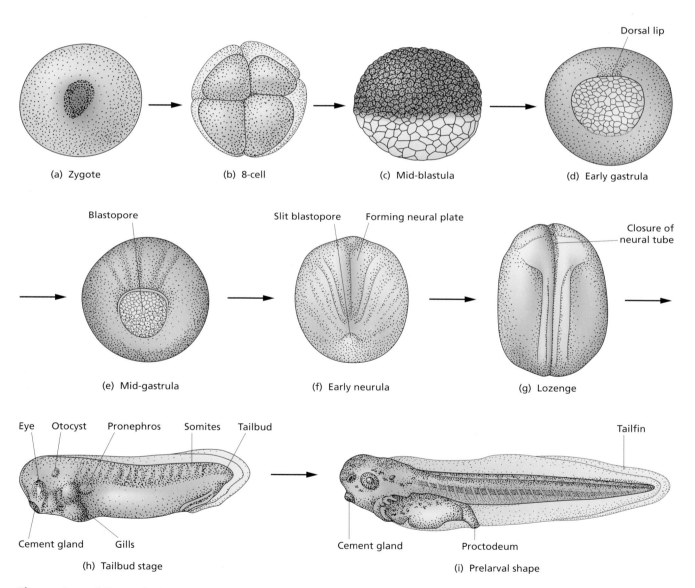

Fig. 7.2 Stages of *Xenopus* development. Here the numbers of the Nieuwkoop–Faber series are used for staging. (a) Zygote, from animal pole; (b) eight cells, stage 4, from animal pole; (c) mid-blastula, stage 8, from side; (d) early gastrula, stage 10, from vegetal pole; (e) mid-gastrula, stage 11, from vegetal pole; (f) early neurula, stage 14, from dorsal side; (g) postneurula "lozenge," stage 22, about 1 day old, from dorsal side; (h) tailbud, stage 30, about 2 days old; (i) prelarva, stage 40, about 3 days old.

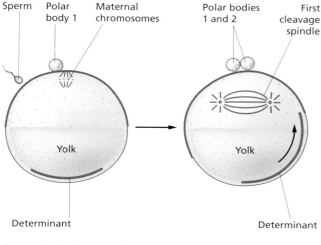

Fig. 7.3 Cortical rotation. The egg cortex moves about 30° relative to the internal cytoplasm, in the direction of the sperm entry point.

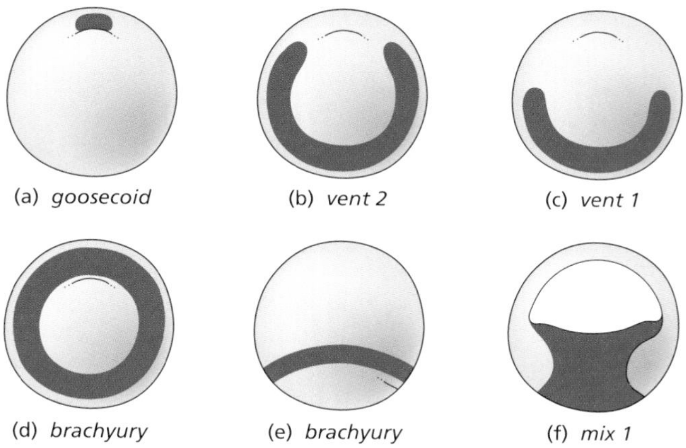

Fig. 7.4 Early domains of zygotic genes. (a–d) View from vegetal pole; (e) from side; (f) section from side.

Internally a dorsal **determinant** (see below) is moved from the vegetal pole to the dorsal side, ensuring that the dorsal structures will develop opposite the point of sperm entry. In some other frog species a similar pigmentation change gives rise to a surface feature known as the **grey** (or **gray**) **crescent**.

Cleavage

The first cleavage is vertical and separates the egg into right and left halves. The second cleavage is also vertical, and at right angles to the first, separating prospective dorsal from ventral halves. The third cleavage is equatorial, separating animal from vegetal halves. Subsequent cleavages vary to some extent between individuals, but it is usually possible to obtain some embryos showing four tiers of eight cells at the 32-cell stage. As in other species, the large cells resulting from early cleavage divisions are called **blastomeres**. As cleavage takes place a cavity called the **blastocoel** forms in the center of the animal hemisphere and the embryo is referred to as a **blastula**. The outer surface of the blastula consists of the original oocyte plasma membrane. A complete network of tight junctions around the exterior cell margins seals the blastocoel from the exterior and renders the penetration of almost all substances highly inefficient. This is why radiochemicals and other substances need to be introduced into the embryo by microinjection. Oocytes, on the other hand, are permeable up to the stage of maturation to substances such as amino acids and sugars. Internal cells of the blastula are connected by cadherins and are readily dissociated by removal of calcium ions from the medium. Desmosomes are not found until the neurula stages, but all the cells of early cleavage stages are connected by **gap junctions**.

Cleavage continues rapidly for 12 divisions after which an important transition occurs, known as the **mid-blastula transi-**

tion or **MBT**, although it is actually at the *late* blastula stage. The rate of cleavage slows down, the synchrony of cell divisions is lost, and the strength of intercellular adhesion increases so the blastula appears as a smooth sphere instead of a knobbly one. The MBT is the time at which significant transcription of the zygotic genome commences although it is possible to detect low-level transcription of some genes during the cleavage stages.

The onset of significant zygotic transcription makes it possible to visualize the early domains of commitment by *in situ* hybridization for specific transcription factors (Fig. 7.4). Only a few examples can be given here of the large number of genes that have been studied. The entire mesoderm can be visualized by the expression of the T-box gene *brachyury*, which is needed to upregulate later mesodermal genes and to control gastrulation movements. The dorsal sector, which will form the **organizer** region (= Spemann's organizer), is characterized by the expression of a variety of transcription factor genes including *siamois, goosecoid, not*, and *lim1*. All these are required for axial differentiation and, in addition, *siamois* is involved in the original formation of the organizer, and *goosecoid* is needed for gastrulation movements. The ventral mesoderm expresses the homeobox genes *vent1* and *vent2* in nested pattern. These act in a combinatorial way to specify the lateral plate (*vent1+2*) and somitic (*vent2* only) regions. The endoderm can be visualized as the domain of expression of several transcription factor genes including *mix1* and *sox17*; these are later required for upregulation of various genes of particular endoderm-derived tissues.

Gastrulation

Gastrulation is a phase of morphogenetic movements in the course of which the belt of tissue around the equator, called

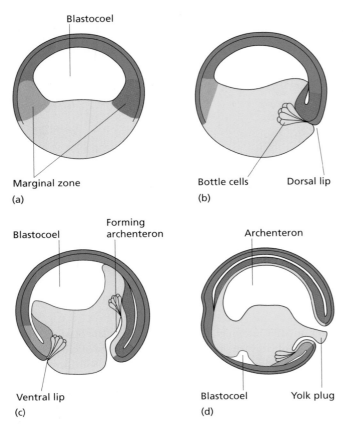

Fig. 7.5 Gastrulation. (a) Blastula; (b) early gastrula; (c) mid-gastrula; (d) late gastrula. All views medial section.

Labels in figure: Blastocoel; Marginal zone; (a); Bottle cells; Dorsal lip; (b); Blastocoel; Forming archenteron; Ventral lip; (c); Archenteron; Blastocoel; Yolk plug; (d).

the **marginal zone**, becomes internalized through an opening called the **blastopore** (Fig. 7.5). This establishes the typical three-layered structure of an animal body, with outer **ectoderm**, middle **mesoderm**, and inner **endoderm**. The start of gastrulation is marked by the appearance of a pigmented depression in the dorsal vegetal quadrant. This is the **dorsal lip** of the blastopore. The blastopore becomes elongated laterally and soon becomes a complete circle. When it is circular the part referred to as the dorsal lip is the dorsal segment of the complete circle and the part referred to as the ventral lip is the ventral part of the complete circle. Elongated cells called bottle cells are present around the blastopore, although their precise mechanical role is not understood. Tissue invaginates all round the circular blastopore but the invagination is much more extensive on the dorsal side, where it proceeds until the leading edge of invaginating tissue is well past the animal pole. In the lateral and ventral parts of the blastopore there is only a small extent of invagination.

Several components of the gastrulation movements may be distinguished, which are to some extent independent of each other:

1 Active expansion of the animal hemisphere (**epiboly**) such that it eventually covers the whole embryo surface.

2 Invagination of the marginal zone. This starts on the dorsal side and spreads to the lateral and ventral side until the blastopore is circular. The cavity formed by the invagination is called the **archenteron** and expands at the expense of the blastocoel as gastrulation proceeds. It has the form of a cylinder, which is very much longer on the dorsal than the ventral side. When the yolk plug becomes internalized at the end of gastrulation it becomes part of the archenteron floor. In the course of the invagination the leading edge of the endoderm literally crawls up the inside of the blastocoel, and requires a layer of fibronectin on the blastocoelic surface to do so.

3 The prospective mesoderm is internal from the start of gastrulation. As the invagination proceeds, the mesoderm separates from the endoderm and **involutes** as a separate tissue layer between the ectoderm and endoderm.

4 Elongation of the dorsal axial mesoderm in the anteroposterior direction. This occurs by an active process of cellular intercalation in all three germ layers, called **convergent extension**, which helps drive the internalization of the marginal zone and the closure of the blastocoel.

5 Movement of the ventrolateral mesoderm towards the dorsal midline.

The cellular mechanisms of gastrulation are still incompletely understood. Most attention has been devoted to convergent extension but is must be remembered that this is only one of several cell behaviors required to generate the coordinated gastrulation process. Convergent extension involves cells becoming polar and actively moving in between their neighbors to generate a shape change at the tissue level, which is an elongation at right angles to the direction of cell movement and a corresponding narrowing in the direction of cell movement. Convergent extension requires the small GTP exchange proteins rho and rac, as introduction of dominant negative versions of these proteins will block the process. They are activated by the Wnt planar polarity pathway (see Appendix), with wnt5a probably being the dominant ligand. A cadherin-type adhesion molecule, called paraxial protocadherin, is expressed in the paraxial mesoderm and helps to generate activation of rho and rac though its cytoplasmic domain. In addition, there is evidence that the Wnt Ca pathway regulates cell adhesion during gastrulation, with the dorsal cells undergoing convergent extension showing an increase in intracellular calcium.

By the end of gastrulation the former animal cap ectoderm covers the whole external surface of the embryo and the yolky vegetal tissues has become a mass of endoderm in the interior. The former marginal zone has generated a cylindrical layer of mesoderm extending on the dorsal side from the slit-shaped blastopore right the way to the anterior end, and on the ventral side just a limited distance from the blastopore. The mesodermal layer remains incomplete in the anteroventral region for some time.

New directions for research

Xenopus development is now quite well understood but there are still opportunities to discover new things.

An important opportunity in early *Xenopus* is the study of *morphogenetic movements*, which remain only partly understood. The advantage is the ability to set up simple *in vitro* systems such as the animal cap or the dorsal marginal zone, which can be observed in real time and which can easily be modified by injection of mRNA into the fertilized egg, or by treatment with biologically active substances.

Later-stage *Xenopus* can now be used for research into *organogenesis, metamorphosis,* and *regeneration,* because transgenic methods can be used to modify gene expression, and can be combined with microsurgical procedures such as isolation or grafting experiments.

A genetic dimension to *Xenopus* experiments has been added by the availability of the related species *Xenopus tropicalis*, which has a much shorter breeding cycle and is a diploid, making genomic analysis simpler than for the pseudotetraploid *X. laevis*.

The fates of the three germ layers in terms of tissue type are as follows:

ectoderm becomes epidermis, nervous system, lens and ear, cement gland;

mesoderm becomes head mesoderm, notochord, somites, kidney, lateral plate, blood, blood vessels, heart, limbs, gonads;

endoderm becomes epithelial lining of gut, lungs, liver, pancreas, bladder.

In the course of the invagination of the marginal zone, the archenteron becomes the principal cavity at the expense of the blastocoel, and the embryo rotates so that the dorsal side is uppermost. It now has a true **anteroposterior** axis, which runs from the leading edge of the mesoderm at the **anterior** to the residual blastopore at the **posterior**.

If embryos are placed in a salt solution of osmolarity similar to the blastocoel, instead of the more usual hypotonic solution, the gastrulation movements are seriously deranged. Instead of invaginating into the interior, the endomesoderm evaginates from the ectoderm to form a dumbbell-like structure (Fig. 7.6). This is known as **exogastrulation**. Both the ectoderm and the endomesoderm of the exogastrula is remarkably normally patterned, although the central nervous system is substantially defective and there is no tail.

Although superficially similar, the gastrulation movements of *Xenopus* differ somewhat from the urodele species that were used for classical studies on amphibian embryology in the 1920s to 1940s. This means that older textbook accounts may differ somewhat from this one.

Neurulation and later stages

The next stage of development is called the **neurula**, in which the ectoderm on the dorsal side becomes the central nervous system. The **neural plate** becomes visible as a keyhole-shaped region covering much of the dorsal surface of the embryo and delimited by raised neural folds. Genes expressed in the whole neural plate include those coding for sox2, also associated with neural stem cells in mammals, and the neural cell adhesion molecule N-CAM. Quite rapidly the folds rise and move together to form the **neural tube**, which, after closure, becomes covered by the ectoderm from beyond the folds, now known as the **epidermis** (Fig. 7.7). During and after neurulation there is a striking elongation of the body, which means that the whole trunk and tail region are derived from the posterior quarter of the neurula. This is driven by a continuing process of convergent extension both in the notochord and in other tissues. The neural tube, notochord, and somites are collectively known as the **axis**, not to be confused with the geometrical axes used for anatomical description. Here the term "dorsal axis" will be used for these structures to avoid confusion.

By the **tailbud** stage all major body parts are in their final positions (Fig. 7.8a,b). The **notochord** forms from the dorsal midline of the mesoderm, rows of segmented **somites** appear on either side, the lateral plate mesoderm later gives rise to limb buds, kidney, and coelomic mesothelium, and in the ventroposterior region of the mesoderm is a string of blood islands, which provide the early tadpole with its erythrocytes. Another population of blood cells later arise from the dorsal aorta. A complex region at the posterior end, comprising the posterior neural plate and the mesoderm beneath it, becomes the **tailbud** (Fig. 7.8c), which generates the notochord, neural tube, and somites of the tail over the next 1–2 days. Because the neural folds close over the blastopore there is a connection created between the neural tube lumen and the gut, called the **neuroenteric canal**. This persists for about a day and is then blocked. The epithelia of the gut, comprising pharynx, lungs, stomach, liver, pancreas, and intestine, develop from the endoderm, although differentiation occurs much later than for the ectodermal and

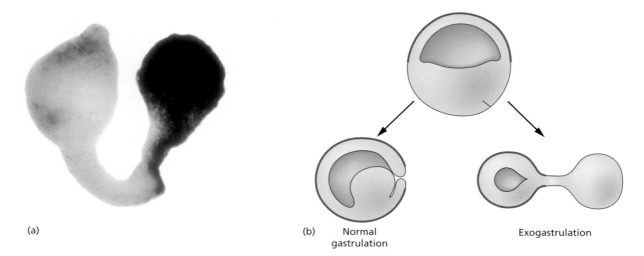

Fig. 7.6 Exogastrulation. (a) Exogastrulation occurs when embryos are placed in isotonic salt solutions. (b) A *Xenopus* exogastrula, with mesoendoderm on the left and ectoderm on the right.

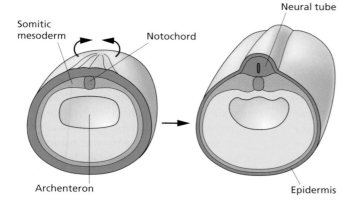

Fig. 7.7 Neurulation. Section through mid-body region.

mesodermal tissues. The **proctodeum**, terminating in the anus, forms from the channel between neuroenteric canal and exterior formed by the process of neural tube closure. The mouth develops somewhat later from a new aperture formed at the anterior end.

In the head the anterior neural tube forms three vesicles that become the **forebrain, midbrain**, and **hindbrain**. The epidermis forms various columnar thickenings called **placodes**. These include: the nasal placode, consisting of sensory cells from which originate the axons of the olfactory nerve connecting to the telencephalon; the lens placodes forming the lenses of the eyes; and the otic placodes that form the ears. The eyes develop as outgrowths of the forebrain (**optic lobes**) that invaginate into a cup shape such that the **pigment epithelium** forms from the outer layer and the retina forms from the inner layer. The lens invaginates from the surface epidermal lens placode and

becomes surrounded by the optic cup. The optic nerve grows back from the retina down the optic stalk and projects to the **optic tecta** in the midbrain (see also Chapter 14).

Some cells from the folds of the neural plate which comes to lie on the dorsal side of the neural tube later become the **neural crest**. This is a migratory tissue that forms a variety of tissue types. In the head the neural crest forms most of the skeletal tissues of the skull. In the trunk it forms the dorsal root ganglia, the sympathetic ganglia, and the **melanocytes** (pigment cells). In the vagal and sacral regions of the trunk it also forms the parasympathetic ganglia of the gut. The neural crest is discussed further in Chapter 14.

Anteriorly, the head mesoderm forms part of the jaw muscles and branchial arches. A structure known as the cement gland develops from the anterior epidermis, ventral to the future mouth. This is a prominent external feature of the head from the late neurula. The heart is formed from the ventral edges of the anterior lateral mesoderm that move down and fuse in the ventral midline around the end of neurulation.

Fate maps

Numerous **fate maps** have been published for amphibian embryos at stages from the fertilized egg to the end of gastrulation. Until 1983, all studies were by localized **vital staining** with the dyes neutral red or Nile blue, applied to the embryo surface from a small fragment of impregnated agar.

More recently, injectable **lineage labels** have been preferred, such as horseradish peroxidase (HRP) or fluorescein-dextran-amine (FDA) (see Chapter 5). These can fill whole cells and do not diffuse. For surface marking **DiI** is now preferred. Because there is little increase in size of the early embryo, passive labels

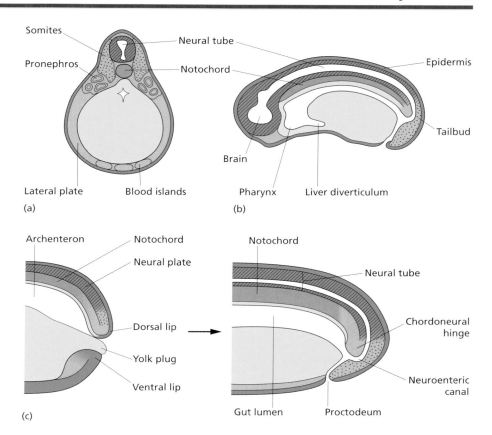

Fig. 7.8 Tailbud stage. (a) Transverse section through trunk; (b) median section; (c) formation of the tailbud by closure of the neural plate over the residual blastopore.

of these sorts do not become diluted and so remain clearly visible for several days. The lineage labels have revealed a certain degree of local cell mixing and this means that the fate map cannot be quite precise since the mixing causes some overlap between prospective regions when results are combined from different individual embryos. As this feature is not so apparent with vital dyes, maps shown in older textbooks imply a spurious degree of precision, which does not exist in reality. A modern fate map is shown in Fig. 7.9b, and Fig. 7.9c shows results of filling a dorsal blastomere (C1) and a ventral blastomere (C4) with a lineage label.

Important features of the fate map are:

1 The neural plate arises from the dorsal half of the animal hemisphere and the epidermis from the ventral half.

2 The mesoderm arises from a broad belt around the equator of the blastula, much from the animal hemisphere.

3 The endoderm arises from the vegetal hemisphere.

4 The somitic muscle arises from most of the marginal zone circumference, much from the ventral half of the blastula.

5 Both dorsal and ventral structures at the anterior end of the body come from the dorsal side of the blastula.

The *Xenopus* early embryo fate map shows that the side conventionally called "dorsal" projects largely to the head end of the body, although it does also populate the dorsal midline along the whole anteroposterior axis. Conversely, the side conventionally called "ventral" projects largely to the posterior of the later body. This fact has caused some confusion but need not do so as long as the essentials of the gastrulation movements are borne in mind. It is important to note that the animal–vegetal axis of the egg is not the future anteroposterior axis, which arises as a result of the elongation of the dorsal tissues during gastrulation.

Experimental methods

As discussed in Chapter 4, when attempting to establish the function of any gene product in development it is necessary to know at least the expression pattern, the biological activity, and the effect of specific inhibition *in vivo*. The expression pattern is normally determined by *in situ* **hybridization**. The biological activity and inhibition experiments are particularly easy in *Xenopus* because of the ease of injecting materials into embryos and the use of ancillary techniques such as microsurgical isolation of explants or ultraviolet irradiation.

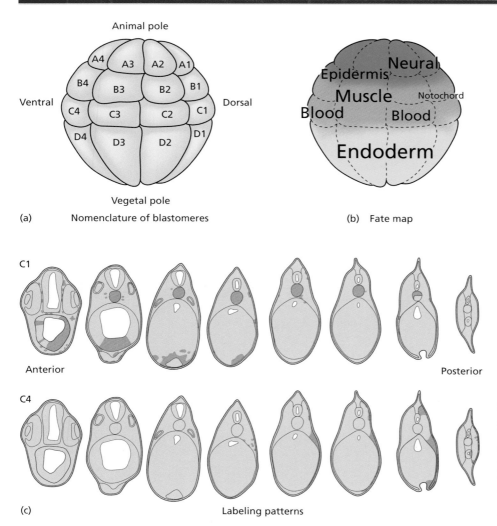

Animal pole

A4 A3 A2 A1
B4 B3 B2 B1
Ventral C4 C3 C2 C1 Dorsal
D4 D3 D2 D1

Vegetal pole

(a)   Nomenclature of blastomeres

Neural
Epidermis
Muscle   Notochord
Blood   Blood
Endoderm

(b)   Fate map

C1

Anterior   Posterior

C4

(c)   Labeling patterns

Fig. 7.9 A *Xenopus* fate map for the 32-cell stage. (a) Nomenclature of blastomeres at the 32-cell stage; (b) fate map of tissue types projected onto 32-cell stage; (c) reconstruction of labeling pattern from blastomeres C1 and C4. The figure shows transverse sections taken at equal intervals along the body from anterior to posterior.

Gain of function

For biological activity measurement the material to be injected will usually be an mRNA made *in vitro* (Fig. 7.10a,b). The RNA synthesis is performed using plasmids carrying promoters for RNA polymerases of bacteriophage such as Sp6, T3, or T7, and a poly(A) addition site to ensure *in vivo* addition of poly(A) to stabilize the message. The mRNA can be injected into a whole fertilized egg, or into a specific blastomere during cleavage, which gives control over its location. The mRNA is translated by the protein synthesis machinery of the egg and is likely to persist and remain active throughout early development, although it will eventually be degraded. Usually, the same plasmid can be used for preparation of the *in situ* probes required for expression studies, by transcription from the oppositely oriented promoter.

As many developmentally active molecules have different effects at different times, it may be desirable to induce activity of the introduced gene product at a particular stage. A useful method for doing this for transcription factors involves adding the hormone-binding domain from the glucocorticoid receptor to the protein of interest. This then causes it to be sequestered in the cytoplasm by binding to the cytoplasmic protein hsp90, until such time as a glucocorticoid, usually dexamethasone, is added. As a lipid-soluble substance, dexamethasone can penetrate into the embryo and will bind to the receptor, liberating it from the hsp90, and allowing it to move to the nucleus where the transcription factor part of the molecule can exert its biological activity (Fig. 7.10c).

It is also possible to introduce genes by **transgenesis**, particularly important for studying events in late development by which time injected RNA may have been degraded. There are now several ways for making *Xenopus* transgenics. Most often the plasmid DNA is incorporated into swollen sperm heads, which are then injected into unfertilized eggs to create transgenic zygotes (Fig. 7.10d). It is also possible to inject

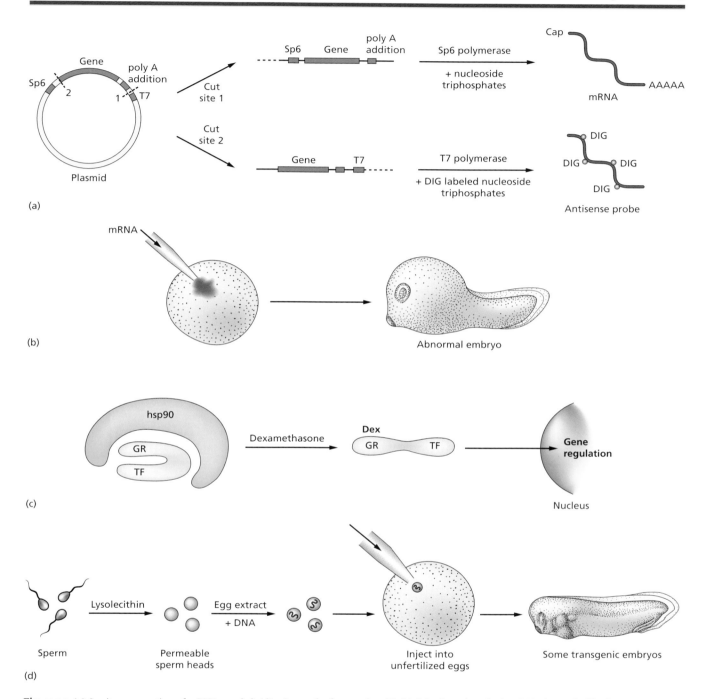

Fig. 7.10 (a) *In vitro* preparation of mRNA or a hybridization probe from a plasmid. (b) Injection of synthetic mRNA into a fertilized egg. (c) Activation of a glucocorticoid receptor fusion protein by dexamethasone. (d) Introduction of genes by transgenesis.

DNA into fertilized eggs in the presence of a transposase or a rare-cutting restriction endonuclease. It is usual to incorporate a GFP coding sequence into the transgene so that the transgenic embryos can be identified by their green fluorescence. Because the long generation time precludes much breeding, most transgenics are used as founders. This means that each individual transgenic embryo will have a different insertion site and copy number, and this may cause some variability of biological behavior compared with a pure-bred transgenic line.

The effects of overexpression of a specific RNA on the development of the whole embryo are often not very informative because of the nonspecific nature of the defects observed. But two procedures combining overexpression with another tech-

nique have been particularly useful. One is the animal cap autoinduction assay (Fig. 7.11a). If an embryo is injected with a component of the mesoderm-inducing or neural-inducing systems and then the animal pole region of the blastula, called the **animal cap**, is explanted, it will autonomously undergo the induction. This is because some or all of the cells in the cap make and secrete the factor and all the cells are competent to respond to it. Untreated animal caps develop into spherical balls of epidermis. Those induced to form axial tissues undergo a convergent extension process and become very elongated. Those induced to form ventral-type tissues will swell to form translucent vesicles. As these changes can be observed under the dissecting microscope, this offers a very simple and convenient assay method.

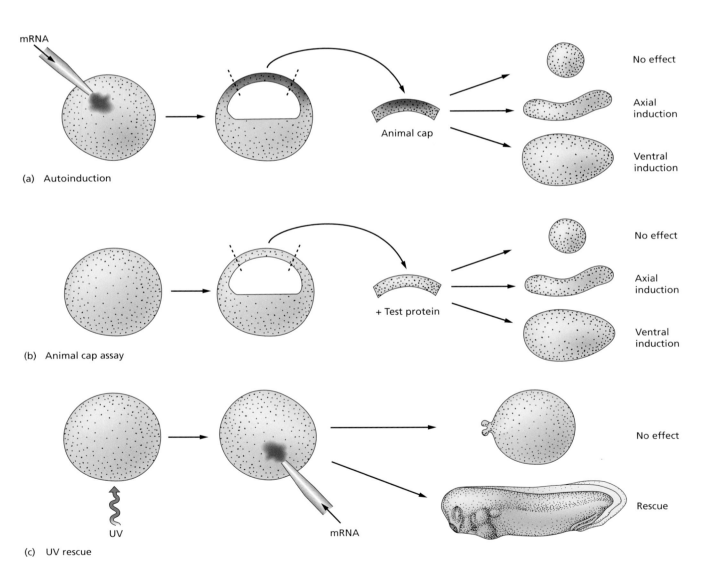

Fig. 7.11 Use of animal caps and UV embryos.

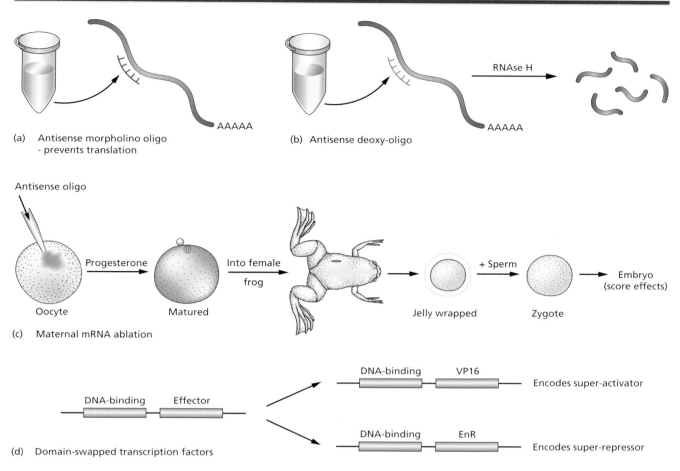

Fig. 7.12 Methods for inhibition of specific gene activity. (a) Antisense morpholino oligo. (b) Antisense deoxy-oligo. (c) Making an oocyte into an embryo after ablation of a specific maternal mRNA. (d) Domain swapped transcription factors.

For the study of extracellular factors it is possible simply to apply proteins to explants from embryos and observe the change in developmental pathway (Fig. 7.11b). Since the outer surface of the embryo is impermeable, these treatments must be applied to explants before they round up and become re-sealed. Whether the assay depends on mRNA injection or on treatment with a protein, the visual scoring of the results can be supplemented by conventional histology, *in situ* hybridization or immunostaining for specific markers, or by biochemical methods for specific mRNAs (see Chapter 5).

Another important combined protocol is the ultraviolet radiation (UV) rescue method (Fig. 7.11c). As we shall see, it is possible to create embryos lacking all axial structures by UV radiation of the zygote. Injection of an mRNA coding for any component of the axial induction system can restore part or all of the dorsal axis to such embryos, and this is easily scored by visual inspection. Because the head is formed mainly from the dorsal part of the blastula, the degree of head formation provides a semiquantitative measure of the degree of dorsal rescue following the injection. These analytical tech-

niques are also depicted in Animation 9: Animal cap and UV assays.

Loss of function

There are three standard protocols for inhibition experiments, which are very effective. First, the most commonly used is the injection of **antisense morpholino** oligos into fertilized eggs (Fig. 7.12a). Morpholinos hybridize to their complementary mRNA and block translation. It is necessary to have an antibody to the target protein in order to show that the morpholino has been effective and really prevented synthesis of the protein. Second, if it is desired to inhibit a maternally acting gene, then antisense deoxyoligonucleotides are also effective (Fig. 7.12b). These are injected into the oocyte since it is in the oocyte that various maternal components are laid down that are essential to later, zygotic, development. The antisense oligos, like the morpholinos, hybridizes to the target message. But in this case the resulting RNA–DNA hybrid is a target for the nuclease

RNAseH, which destroys the message. It is necessary to confirm destruction of the specific mRNA by RT-PCR. In order to establish later developmental effects of these experiments, it is then necessary to make the treated oocytes into embryos. This is a difficult manipulation but can be achieved by maturing the oocytes *in vitro* with progesterone, coloring them with a vital dye, and reimplanting them into the abdomen of an ovulating frog so that they pass down the oviduct and become wrapped in jelly. They will then be laid by the female, can be fertilized, and the course of development of the colored embryos observed (Fig. 7.12c).

A third inhibitory protocol involves the design of **dominant negative** versions of gene products, which can specifically inhibit the normal protein on overexpression. There are many possible types of dominant negative reagent. In *Xenopus*, particularly extensive use has been made of **domain swapped** versions of transcription factors (Fig. 7.12d). Here, the effector domain of the factor is replaced by a strong activator (e.g. VP16 activation domain) or a strong repressor (e.g. engrailed repression domain, EnR), as required. mRNA is made from each of these constructs and is injected into fertilized eggs. If the effect of, say, the VP16 fusion is similar to the normal factor this shows that the normal factor must be an activator. The phenotype of the EnR fusion, in which the normal targets of the transcription factor are repressed, gives an indication of the functions for which the normal factor is needed. Conversely, if the EnR version gives a normal phenotype then the transcription factor is likely to be a repressor, and the phenotype obtained with the VP16 fusion will indicate what functions require the normal gene product.

Xenopus has no embryonic stem cells, so there is no equivalent to the method for making knockouts in the mouse by targeted mutagenesis (see Chapter 10). However, a new method of gene ablation is the use of **zinc-finger nucleases** to cause double-stranded breaks in target loci. The DNA repair process generates small insertions or deletions that are often loss-of-function mutations. The fertilized egg is injected with mRNA encoding the nuclease and most cells show specific gene ablation with very limited side effects.

Processes of regional specification

Summary of processes

The fertilized egg contains two determinants, a **vegetal** and a **dorsal** determinant (Fig. 7.13). The vegetal determinant becomes established during oogenesis as a result of mRNA localization to the vegetal cortex and causes formation of the **endoderm**. This is the source of **mesoderm-inducing** signals that induce a ring of tissue around the equator to become the **mesoderm**. The dorsal determinant causes formation of the **organizer** in the region that will become the **dorsal lip** of the blastopore. It is initially localized at the vegetal pole, and

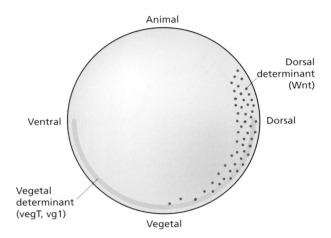

Fig. 7.13 Dorsal and vegetal determinants in the fertilized egg of *Xenopus*.

shifts to the dorsal side during the cortical rotation. The organizer is the source of later dorsalizing signals that both induce the neural plate and pattern the mesoderm into zones forming different tissues.

Determinants

Evidence for the existence of the two determinants comes from the effects of early ablation and isolation experiments, which show that formation of a complete body pattern requires the presence of both the vegetal and the dorsal regions. Complete twins can be produced by separation of the first two blastomeres, or the equivalent lateral subdivision of a blastula into right and left halves. In this case, some vegetal and some dorsal material is present in each half. However, if early cleavage stages or blastulae are divided **frontally** (separating prospective dorsal and ventral halves) then the dorsal half forms a slightly hyperdorsal whole embryo while the ventral half forms a "belly piece" of extreme ventral character. In the frontal separation, the dorsal half has both vegetal and dorsal tissue, but the ventral half does not. Direct evidence for the existence of the dorsal determinant has been obtained by grafting cytoplasm from the vegetal pole of the oocyte into UV-ventralized embryos, which results in some rescue of the dorsal axis. The dorsal determinant is now known to consist of components of the Wnt signaling pathway, and the vegetal determinant to consist of mRNA encoding the transcription factor vegT. (For a summary of the various signaling pathways and transcription factor classes important in development, see the Appendix.)

Another determinant believed to exist in the *Xenopus* zygote is the **germ plasm**. This is visible in the electron microscope as a patch of fibrillar granular material not bounded by any membrane. Similar in appearance to the germ plasm of other animal types, it is sometimes referred to as "nuage." It appears in the mitochondrial cloud (Balbiani body) of stage II oocytes and ends up in the vegetal cortical region of mature oocytes and early embryos. The germ plasm contains *cat2* mRNA (usually called *xcat2*, a homolog of *nanos*, see Chapter 11). Although it is not completely certain that the presence of the germ plasm actually does specify primordial germ cells, the plasm is inherited by cells that express primordial germ cell markers at later stages.

Early dorsoventral patterning

The dorsal determinant is moved from the vegetal pole to the dorsal side by the cortical rotation that occurs following fertilization (Fig. 7.3 above). This depends on an array of parallel microtubules, which form transiently in the sense of the rotation, and on their associated motor proteins, kinesin and dynein (see Appendix). The microtubules elongate from the sperm aster and when they meet the vegetal cortex they become anchored to the cell surface by kinesin-related proteins such that the tubules are aligned with their + ends in the direction of the future cortical rotation. The rotation movement is probably due to the relative displacement of the surface-tethered microtubules relative to internal dynein protein.

The cortical rotation can be prevented by microtubule-depolymerizing drugs such as nocodazole or by injection of neutralizing antibodies directed against the kinesin-related proteins. It can also be prevented if the vegetal hemisphere is irradiated with UV light before the rotation starts. The embryos that subsequently develop following any of these treatments are radially symmetrical and extreme ventral in character, with the mesoderm mainly consisting of blood islands of the type normally arising in the ventral midline (Figs 7.14, 7.15a–c). This shows the importance of the cortical rotation for the development of dorsal structures. The fact that the dorsal side forms opposite the site of sperm entry is explained by the relationship of the microtubule array to the sperm aster.

The dorsal determinant itself consists of components of the canonical Wnt signal transduction pathway, which cause the stabilization of β-catenin. Injection of *wnt* mRNAs will rescue

formation of a dorsal axis in UV-ventralized embryos, or induce a second dorsal axis in normal embryos. Evidence for the essential role of β-catenin has been obtained by antisense oligonucleotide-mediated ablation of the *β-catenin* mRNA from oocytes. When such oocytes are matured and fertilized they develop as ventralized embryos. The dorsal determinant probably comprises several of the components of the Wnt pathway, not just the extracellular ligand. Although other Wnt family members are more active in overexpression assays, the ones that are present in the oocyte are wnt5a and wnt11. The *wnt11* mRNA is localized to the vegetal hemisphere and shifts toward the dorsal side during the cortical rotation. Antisense ablation of *wnt11* in the oocyte, followed by maturation of embryos, leads to ventralization (Fig. 7.16), which can be rescued towards a normal pattern by co-injection of *β-catenin* mRNA. This suggests that wnt11 is acting via β-catenin even though it, like wnt5a, is generally considered to signal through the alternative Wnt pathways. Similar results have been obtained with other components of the canonical Wnt pathway, including dishevelled and gsk3. Suppression of Wnt pathway activation in ventrolateral regions of the embryo is maintained by the presence of maternally encoded inhibitors including dickkopf, which interacts with the Wnt co-receptor lrp.

In normal development β-catenin can be seen by immunostaining to enter the nuclei on the dorsal side. This is seen in the dorsovegetal region and also extends to some extent above the equator where it is important in the initial formation of the neural plate.

If Wnt pathway components are overexpressed in early embryos, and the injection is close enough to the original dorsal side not to produce a second dorsal axis, then the result is a hyperdorsal embryo, which has too much head and not enough trunk and tail. Hyperdorsal embryos may also be produced simply by treating early blastulae with lithium salts. They have a structure that is the opposite of the UV embryo and in the extreme case resemble radially symmetrical heads (Figs 7.14, 7.15c). They arise because the entire mesoderm has been caused to develop as organizer tissue. UV-ventralized embryos can be rescued back to a normal pattern by a localized injection of lithium (Fig. 7.15d), and a localized injection of lithium on the ventral side of a normal embryo will induce a secondary dorsal

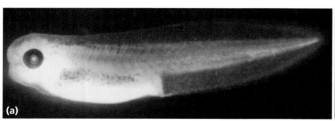

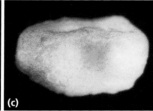

Fig. 7.14 Appearance of lithium- and UV-treated embryos at 3 days of development. (a) Normal; (b) lithium-treated at morula stage; (c) UV irradiated before first cleavage.

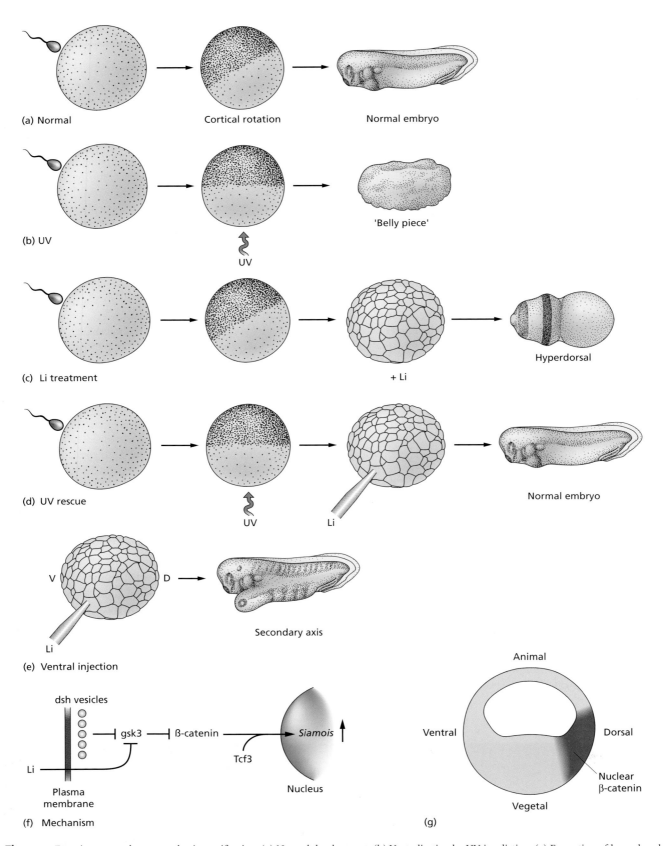

Fig. 7.15 Experiments on dorsoventral axis specification. (a) Normal development. (b) Ventralization by UV irradiation. (c) Formation of hyperdorsal embryo by lithium treatment. (d) Rescue of UV embryo by localized injection of lithium ion. (e) Induction of secondary axis in normal embryo by localized injection of lithium ion. (f) The Wnt pathway in the *Xenopus* egg. Gsk3 will normally be repressed by the determinant on the dorsal side, but will be active elsewhere. Lithium can also inhibit gsk3. (g) Location of nuclear β-catenin in the late blastula.

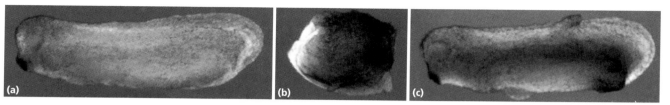

Fig. 7.16 Effect of ablating *wnt11* mRNA from oocyte and converting the oocytes into embryos. (a) Control; (b) *wnt11* ablated; (c) *wnt11* ablated, with injection of *wnt11* mRNA. Reproduced from Tao *et al. Cell* 2005; 120: 857–871, with permission from Elsevier.

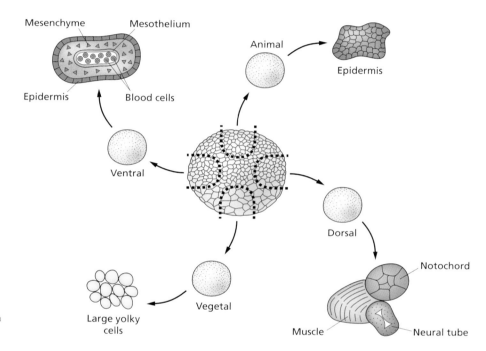

Fig. 7.17 Differentiation of explants from different parts of the blastula.

axis (Fig. 7.15e). Lithium has these effects because it is an inhibitor of gsk3, so mimics the effects of Wnt signaling. It may also exert additional effects through inhibition of the inositol phosphate pathway. The fact that these lithium effects can be exerted as late as the early blastula, even though the normal process occurs in the first cell cycle, is because there is little zygotic transcription before MBT, so the target genes of β-catenin do not become significantly expressed until this stage. The immediate targets of β-catenin are the Tcf transcription factors. Tcf1 becomes converted from an activator to a repressor by association with β-catenin, while tcf3 has its repressive activity neutralized. The net result is the upregulation in the dorsal sector of organizer genes, including those encoding many transcription factors, such as goosecoid, and many signaling molecules, including the bone morphogenetic protein (BMP) inhibitors responsible for organizer activity.

Inductive interactions

The fate map shows that the three germ layers arise from different animal–vegetal levels: the endoderm from the vegetal region, the mesoderm from the equatorial region, and the ectoderm from the animal region. However, explantation experiments on small pieces taken from blastulae (Fig. 7.17) show that the region forming mesoderm extends less far toward the animal pole than is shown by the fate map. The fate map also shows that the dorsal axial structures, particularly the somites, arise from a wide area extending around most of the dorsal to ventral circumference. However, explant studies show that only a restricted dorsovegetal region, about 60° of circumference, will form dorsal axial structures. The differentiation of explants represents the specification of the tissues at the time of isolation. So the mismatch between fate map and explant behavior indicates that inductive interactions must be necessary for achieving the final fates, particularly with regard to the formation of the mesoderm and the dorsal axial structures.

Germ-layer formation

The primary cause for this pattern is the localization of maternal mRNAs to the vegetal hemisphere during oogenesis. Among these is the mRNA for a T-box type transcription factor called vegT. This mRNA appears to act as a **determinant** for the **endoderm**. The evidence is: firstly, that *vegT* mRNA is localized to the prospective endoderm; secondly, that the mRNA will induce endodermal markers if injected into other parts of the embryo; and, thirdly, that antisense oligonucleotide-ablation of maternal *vegT* mRNA in oocytes, followed by maturation and fertilization, will produce embryos lacking endoderm.

The endoderm is defined by the expression of a group of transcription factors, all of which depend directly or indirectly on vegT. The transcription of some, including the genes for the sry-related factor sox17a, and the homeodomain factor mix1, are induced directly by vegT. These genes become upregulated from MBT even if the cells have been dissociated so that no intercellular signaling is possible. Expression of other endodermal transcription factors, including the homeodomain factor mixer and the zinc finger factor gata4, depend on the signals emitted from the vegT-containing cells and are not induced if these signals are blocked. Later on, the endoderm becomes regionalized into zones that will form the different tissue types of the gut and respiratory system, and this involves the upregulation of transcription factor genes of the Parahox cluster: *pdx1* (= *Xlhbox8*) in the future foregut and *cdx* genes in the future intestine.

 The **mesoderm** arises from the equatorial region of the blastula and is characterized by the expression of various transcription factors, including the T-box transcription factor brachyury. The part on the vegetal side of the equator may be formed by the endodermal determinants acting at lower concentration than is required to form endoderm. But at least that part of the mesoderm arising from the animal hemisphere is formed by **induction** (Fig. 7.18; and see Animation 5: Induction). Animal caps (explants from the animal hemisphere) develop into **epidermis** after isolation. If an animal cap is combined with endoderm, then substantial amounts of mesodermal tissues are induced in the cap (Fig. 7.18b).

The factors responsible for mesoderm induction are members of the transforming growth factor (TGF)-β superfamily, especially the nodal-related factors (see Chapter 10 for the original *nodal*). This subset of inducing factors activates cell-surface receptors that phosphorylate smad2 and smad3 proteins in the cytoplasm, and the activated smads can then move into the nucleus and upregulate target genes (see Appendix). The evidence for the importance of nodal-related factors is very good. Firstly, several such factors are expressed in the vegetal hemi-

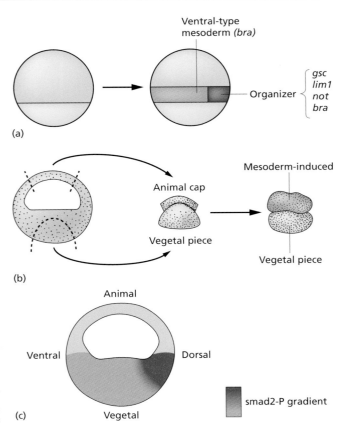

Fig. 7.18 Mesoderm induction. (a) In normal development; (b) mesoderm induced in animal cap combined with vegetal explant; (c) location of phosphorylated smad 2 in late blastula.

sphere (mostly upregulated by vegT). Secondly, they will induce mesoderm if applied to animal caps. Thirdly, inhibition of their action, especially by a specific inhibitor called cerberus-short (see cerberus below) will prevent mesoderm induction. The signal transduction associated with mesoderm induction can also be visualized in the late blastula. Antibodies specific for the phosphorylated forms of smad2 and 3 show them present in the vegetal and equatorial region. This shows that the signal *in vivo* affects the equatorial region but does not reach to the animal pole. It also shows that mesoderm induction mostly occurs following MBT, after the *nodal-related* genes have become upregulated. The competence of the ectoderm to respond to the mesoderm-inducing signals rises in the early blastula and falls in the early gastrula. One aspect of competence is the presence of active fibroblast growth factor (FGF) signaling within the animal hemisphere, and for this reason antagonists of FGF signaling will block mesoderm induction.

Transcription of some of the nodal-related factors is upregulated directly by vegT, and of others indirectly via the initial signaling process. The nodal-related factors acting by themselves upregulate pan-mesodermal genes such as *brachyury*, and ventral genes such as the *vents*. In the dorsal quadrant where β-catenin is also activated, there is a synergistic response to both pathways, resulting in the upregulation of genes encoding goosecoid, not, lim1, and other transcription factors whose activity defines the organizer region. The region of the endoderm with organizer-inducing ability is sometimes called the **Nieuwkoop center**, after the great Dutch embryologist of that name. There is also a difference of **competence** within the animal hemisphere such that it is much easier to obtain dorsal axial-type inductions from the dorsal region. This is because the β-catenin activation caused by the dorsal determinant extends into the animal hemisphere.

There is also a contribution to mesoderm induction from an inducing factor whose mRNA is localized to the vegetal hemisphere of the oocyte, called vg1 (pronounced "veg1"). Despite the similarity of name, vg1 is unrelated to vegT, and is an inducing factor of the Nodal class, not a transcription factor. Although believed for some years to be biologically inactive, it is now clear that it does contribute to mesoderm induction. Ablation of *vg1* mRNA from oocytes using antisense oligonucleotides, followed by their conversion to embryos, reduces the elevation of smad1 phosphorylated smads and subsequent mesoderm formation.

Limits to the spread of mesoderm induction depends partly on the range of the inducing factors, but also on inhibitors whose mRNAs are localized to the animal hemisphere of the oocyte. One of these is ectodermin whose mRNA encodes a RING family ubiquitin ligase. This ubiquitinates and thereby reduces the level of smad4, which is an essential part of both the nodal and BMP signaling pathways.

The cellular properties of the mesoderm germ layer are exhibited in the form of convergent extension movements during gastrulation. This depends on the activity of the Wnt-PCP pathway which is inhibited by the intracellular protein sprouty. Convergent extension occurs in the mesoderm because of the actions of the transcription factor brachyury, which represses expression of the *sprouty* gene, and especially in the dorsal axial mesoderm because of the presence of the cell adhesion molecule paraxial protocadherin, which antagonizes sprouty action.

Dorsalization and neural induction

Mesoderm induction leads to the creation of a belt of mesoderm around the equator of the blastula with a large ventrolateral region of extreme ventral character, and a small organizer region at the dorsal side. The organizer region is often called **Spemann's organizer** after the great German embryologist who originally discovered its properties. It is a key signaling center for the subsequent stages of development, and it acts by secretion of factors that *inhibit* the action of BMPs (see Appendix).

During gastrulation the ventrolateral mesoderm becomes partitioned into zones forming, respectively, somites, kidney, lateral plate, and blood islands, as a function of distance from the organizer (Fig. 7.19a). Tissue explants from the ventrolateral mesoderm retain an extreme ventral character. But if they are experimentally combined with organizer tissue, they will be dorsalized and form large muscle blocks and pronephric tubules (Fig. 7.19b).

Also, during gastrulation, the neural plate becomes induced from the ectoderm under the influence of the organizer. When gastrula ectoderm is brought into contact with the organizer it will form **neuroepithelium**, characterized by expression of various transcription factors including sox2, and the cell adhesion molecule N-CAM. There are various ways of achieving this experimentally. The most straightforward is a combination of competent ectoderm with organizer tissue (Fig. 7.20b). Another is the Einsteckung procedure, where bits of organizer tissue are inserted into the blastocoel of an early gastrula and become pressed against the ventral ectoderm by the gastrulation movements (Fig. 7.20c). Finally, in the **organizer graft** (see below) a secondary neural plate is formed from the ectoderm overlying the secondary dorsal mesodermal axis.

If blastulae are placed in a salt solution isotonic to the embryo rather than the usual very dilute solution, they will **exogastrulate** such that the endomesoderm becomes extruded from the animal hemisphere instead of invaginating into it (see Fig. 7.6 above). The elongation of the embryo proceeds normally but the mesoderm is joined to the ectoderm end-to-end rather than being apposed as layers. Another situation in which mesoderm and ectoderm are adjacent, but not apposed as layers, is the **Keller explant**. This is an explant of dorsal tissue extending from the dorsal lip to the animal pole and displays the convergent extension movement without the other features of gastrulation (Fig. 7.20d). In both these situations where the axial mesoderm does not underlie the ectoderm there is still some neural induction, resulting in a correct anteroposterior patterning of structures. This shows that the signals can propagate in the plane of the tissue (tangential induction) as well as from one layer to the other (**appositional induction**).

In normal development, the neuralization of ectoderm and the dorsalization of mesoderm are both caused by the same group of inducing factors secreted from the organizer. These are chordin, noggin, and follistatin, which all function as inhibitors of BMPs. BMP4 is produced by the embryo in all regions except those dorsal regions with high nuclear β-catenin. The intracellular signal transduction pathway for BMPs is similar to that for the nodal-related factors, but uses different receptors and smads (smad1 and 5; see Appendix). There is abundant evidence that BMP inhibitors are the agents of neural induction and mesoderm dorsalization. Firstly, the organizer region secretes the BMP inhibitors, while the prospective epidermis makes BMPs. After mid-blastula transition, smad1 and 5 becomes phosphorylated over the whole ventrolateral sector of the embryo, indicating active BMP signaling in this region. In terms of biological activity, treatment of isolated animal caps with BMP inhibitors causes neuralization while treatment of isolated ventral mesoderm explants will dorsalize them. Finally,

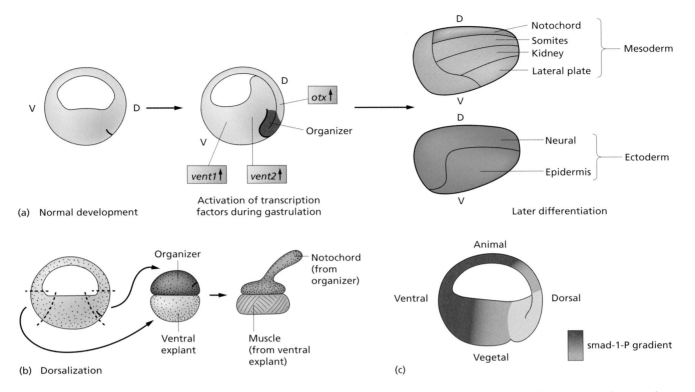

(a) Normal development

(b) Dorsalization

(c)

Fig. 7.19 Dorsalization of mesoderm. (a) In normal development; (b) muscle induced in ventral explant combined with organizer; (c) location of phosphorylated smad1 in early gastrula.

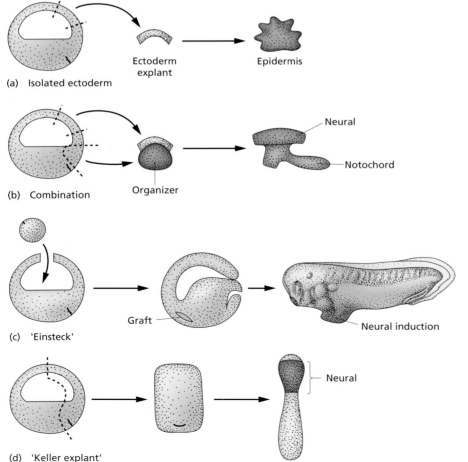

(a) Isolated ectoderm

(b) Combination

(c) 'Einsteck'

(d) 'Keller explant'

Fig. 7.20 Neural induction. (a) Isolated ectoderm from prospective brain turns into epidermis; (b) ectoderm is neuralized in combination with the organizer; (c) ventral ectoderm can be neuralized in the Einsteckung procedure; (d) Keller explant.

Classic experiments

Discovery of inducing factors

The inducing factors that control embryonic development were discovered in the 1980s using *Xenopus*. These are the factors that are now used to control the differentiation of pluripotent stem cells (see Chapter 21).

The key step in their discovery was the ability to apply purified protein solutions to isolated animal caps and to observe the consequences down the dissecting microscope by changes in shape of the explant. This assay led to discovery of the inducing properties of the FGFs and the activins (close relatives of the nodal-related factors).

The other vital technique was the injection of messenger RNA, prepared *in vitro*, into fertilized eggs, or into UV-ventralized eggs. This led to the discovery of the role of the Wnt pathway in the formation of the dorsal axis. Further, it allowed the introduction of expression cloning protocols whereby a whole library of expression clones could be divided into subsets, transcribed, and the mixed mRNA assayed by injection. This led to the discovery of noggin and eventually to the mode of action of the organizer.

First inducing factors

Slack, J.M.W., Darlington, B.G., Heath, J.K. & Godsave, S.F. (1987) Mesoderm induction in early Xenopus embryos by heparin-binding growth-factors. *Nature* **326**, 197–200.

Kimelman, D. & Kirschner, M. (1987) Synergistic induction of mesoderm by FGF and TGF-beta and the identification of an messenger-RNA coding for FGF in the early Xenopus embryo. *Cell* **51**, 869–877.

Smith, J.C., Price, B.M.J., VanNimmen, K. & Huylebroeck, D. (1990) Identification of a potent Xenopus mesoderm-inducing factor as a homolog of activin-A. *Nature* **345**, 729–731.

Axis duplication from Wnt

Smith, W.C. & Harland, R.M. (1991) Injected XWnt-8 RNA acts early in Xenopus embryos to promote formation of a vegetal dorsalizing center. *Cell* **67**, 753–765.

Sokol, S., Christian, J.L., Moon, R.T. & Melton, D.A. (1991) Injected Wnt RNA induces a complete body axis in Xenopus embryos. *Cell* **67**, 741–752.

Discovery of noggin

Smith, W.C. & Harland, R.M. (1992) Expression cloning of noggin, a new dorsalizing factor localized to the Spemann organizer in Xenopus embryos. *Cell* **70**, 829–840.

inhibition experiments give the predicted result. Morpholinos directed against the BMP inhibitors, especially chordin, will suppress the dorsal pattern. mRNA encoding dominant negative BMP receptors has a stronger effect and will induce secondary dorsal axis formation following injection on the ventral side.

Chordin, noggin, and follistatin act by direct binding to and inhibition of BMP protein and thereby bring about a gradient of BMP activity from ventral to dorsal. There are many target genes of BMP signaling, although just one example may illustrate how the pattern of the embryo is built up. A high level of BMP activity induces expression of the homeobox gene *vent1* while a lower level upregulates the gene *vent2* (see Figs 7.4, 7.19a). The region in which both *vent* genes are on later becomes the lateral plate. In the mid-ventral region this forms the blood islands, with BMP upregulating expression of the gene encoding SCL (stem cell leukemia factor), a bHLH-type transcription factor which later activates hematopoietic differentiation. The region in which *vent2* but not *vent1* is on later becomes the somites, and soon shows upregulation of genes for **myogenic** factors such as *myf5*. The proteins coded by the *vent* genes are themselves transcriptional repressors as may be shown by the fact that VP16 fusions have a dorsalizing effect (i.e. the opposite effect to the native factors).

Experimentally, neuralization can be provoked by disaggregation of gastrula ectoderm. This is because when cells are disaggregated the extracellular signal regulated kinase (ERK) signaling pathway becomes activated as a general "wounding" response. This leads to inhibition of the action of smads 1 and 5 by phosphorylation at a second site (the linker region; see Appendix).

In addition to the inhibition of BMP signaling, an element of Wnt inhibition is also involved in the normal dorsalization process. After MBT, the *wnt8* gene becomes activated in the nonorganizer part of the mesoderm and the Wnt pathway now has a *ventralizing* effect. Although the Wnt pathway has a dorsalizing function following cortical rotation in the egg, after MBT the transcription factor tcf3 becomes replaced by lef1. Both of these factors cooperate with β-catenin, but they show a different spectrum of specificity in relation to their target genes such that tcf3 is dorsalizing while lef1 is ventralizing. This can be demonstrated because dominant negative constructs for these two transcription factors show the opposite effects when overexpressed in embryos: dominant negative tcf3 is ventralizing and dominant negative lef1 is dorsalizing. This change of competence explains the hitherto puzzling fact that treatment with lithium after MBT has ventralizing rather than dorsalizing effects. The dorsalizing signals from the organizer include at least one Wnt inhibitor called frzb (pronounced "frizzbee"), which resembles the extracellular part of the Wnt receptors, and which inhibits the action of wnt8.

Proportion regulation

An aspect of regional specification in early development that has long attracted attention is the capacity of many types of early embryo to exhibit **proportion regulation**. In other words, if material is removed from the embryo, the pattern that eventually forms does not have a gap but rather consists of a reduced-scale version of the normal pattern (see also Chapter 4). In reality, the patterns usually are not perfectly proportioned but they certainly can tend toward the normal. The simple diffusion gradient of a morphogen does not explain proportion regulation; in fact, it can be shown that it exhibits antiregulative behavior because of the morphogen accumulating at the cut surface (see Fig. 2.14c). Models that can show regulative behavior always have two gradients with different diffusion characteristics, and involve a comparison of concentrations between them.

Only in *Xenopus* has a mechanism been uncovered for regulation. The key is the expression of a BMP-type inducing factor called antidorsalizing morphogenetic protein (ADMP) in the *dorsal* region. Despite being expressed in the organizer this protein has ventralizing properties and titrating its activity will change the proportion of structures formed (Fig. 7.21). Its transcription is repressed by BMP signaling. So if the amount of ventral tissue is experimentally reduced, and the BMP gradient thereby lowered, the domain of *admp* expression expands (Fig. 7.22). The ADMP is more diffusible than the BMP inhibitors so this means that the ADMP level in what is now the ventral region exceeds that of the BMP inhibitors, so ADMP upregulates *bmp* expression in the new ventral region. This explains why dorsal half embryos, although they do not recover perfectly

normal proportions, do develop a complete dorsal–ventral pattern.

The ability to regulate proportions is lost if these molecular components are removed. So embryos injected with morpholinos to ablate all four factors: BMP2, 4, and 7, and ADMP, become fully neuralized and show no proportion regulation (Fig. 7.23).

Anteroposterior patterning

During gastrulation not only is the dorsal to ventral pattern of territories specified, but also the anterior to posterior pattern. Neural-inducing activity is shown both by the organizer and by the axial mesoderm into which the organizer develops. In both cases the inductions show regional specificity. Anterior organizer or anterior axial mesoderm induces brain structures, while posterior organizer or posterior axial mesoderm induces both brain and spinal cord. Since a complete neural axis can be induced by a posterior inducer it has long been thought that a posterior signal controls anteroposterior patterning during the formation of the central nervous system (CNS).

The initial anteroposterior pattern exists in the form of two domains within the organizer prior to gastrulation, which are derived from the different ratio of β-catenin and nodal-related signaling arising from cortical rotation and mesoderm induction, respectively. The anterior part will later become the ante-

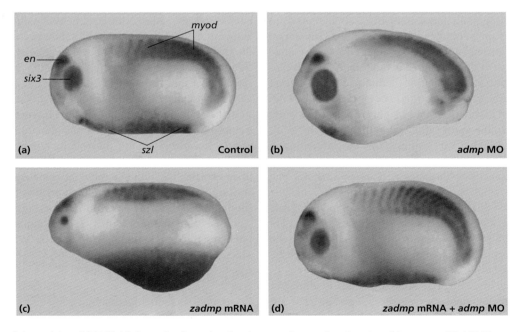

Fig. 7.21 Ventralizing activity of ADMP. (a) Control embryo showing the normal expression domains of four genes. (b) ADMP morpholino causes a dorsalization of pattern. (c) Overexpression of zebrafish ADMP mRNA causes ventralization. (d) As the zebrafish *admp* is not recognized by the morpholino, it can neutralize its effect. Reproduced from Reversade and De Robertis (2005), with permission from Elsevier.

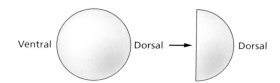

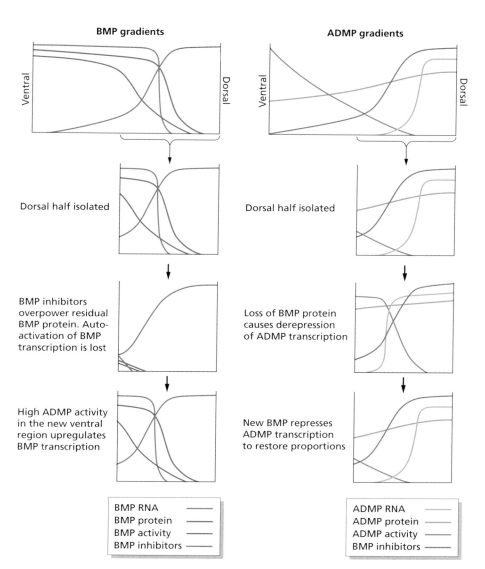

Fig. 7.22 Proportion regulation in isolated dorsal half embryo. Isolation of the dorsal half reduces the domain of BMP activity and hence leads to loss of *bmp* expression. This causes an increase in the region of *admp* expression, and the increased ADMP activity re-establishes *bmp* expression in the new ventral region. This is a very oversimplified diagram. In reality there are other components at work and additional effects on transcription of BMP inhibitors.

BMP gradients

Dorsal half isolated

BMP inhibitors overpower residual BMP protein. Auto-activation of BMP transcription is lost

High ADMP activity in the new ventral region upregulates BMP transcription

BMP RNA	———
BMP protein	———
BMP activity	———
BMP inhibitors	———

ADMP gradients

Dorsal half isolated

Loss of BMP protein causes derepression of ADMP transcription

New BMP represses ADMP transcription to restore proportions

ADMP RNA	———
ADMP protein	———
ADMP activity	———
BMP inhibitors	———

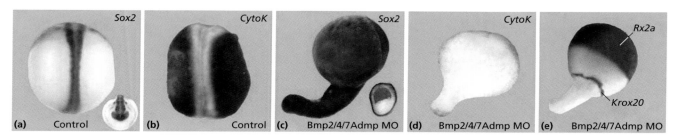

Fig. 7.23 Effects of removing all three BMPs and ADMP from the embryo with morpholinos. (a,b) Controls show expression of *sox2*, encoding a neuroepithelial transcription factor, and an mRNA for an epidermal cytokeratin. (c,d) Total neuralization in the absence of the four factors. (e) Anteroposterior pattern is still retained as shown by expression of a forebrain (*rx2a*) and hindbrain (*krox20*) marker. Reproduced from Reversade and De Robertis (2005), with permission from Elsevier.

rior endoderm and the prechordal mesoderm. During gastrulation this moves towards the animal pole by crawling up the blastocoelic surface of the ectoderm, adhering to the fibronectin substratum. It is characterized by expression of certain transcription factor genes, including *goosecoid*. It will induce expression of anterior-type genes from ectoderm, such as *otx2* (fore/ midbrain) and *ag1* (cement gland); and its anterior inductive activity is due to secreted factors including cerberus (an inhibitor of Wnts, BMPs, and nodal-related factors) and dickkopf (a Wnt inhibitor).

The posterior part will later become the notochord and somites and, during gastrulation, it elongates considerably by convergent extension. It is characterized by expression of a different group of transcription factor genes including the homeobox gene *not* and the T-box gene *brachyury*. It will induce both anterior- and posterior-type genes from the ectoderm (e.g. both *otx2* and **Hox genes**), and its posteriorizing activity due to secretion of FGFs and Wnts. Together these upregulate a group of homeodomain transcription factors encoded by the *cdx* genes and these in turn upregulate the posterior Hox genes of paralog groups 6–13 (Fig. 7.24). This subset of Hox genes specifies the trunk–tail part of the pattern and the genes are normally turned on sequentially during gastrulation.

Evidence that FGF and Wnt signaling is required to induce the trunk–tail region is very good. Several FGFs and Wnts are expressed in the posterior region, that is around the blastopore, during gastrulation. Their activity can be visualized by immunostaining for the activated form of ERK (FGF) or for nuclear β-catenin (Wnt). If animal caps are treated with the BMP inhibitor noggin, then only anterior-type neural genes are induced,

but addition of Wnt or FGF will also induce posterior neural genes. Overexpression in embryos of a dominant negative FGF receptor that inhibits endogenous FGF signaling, or of the dickkopf Wnt inhibitor, will prevent formation of the posterior. The FGF effects are mostly felt in the trunk–tail region while the Wnt effects reach also into the hindbrain, controlling expression of genes such as *krox20* (a Zn finger transcription factor expressed in the hindbrain) and *engrailed* (a homeodomain transcription factor expressed at the midbrain–hindbrain boundary). Another component often believed to have a role in anterior–posterior patterning is retinoic acid. Overexpression or inhibition of retinoic acid does have effects on the pattern but in *Xenopus* they are much less than those of FGFs and Wnts and are largely confined to the hindbrain.

Repression of Wnt in the anterior is also necessary for formation of the foregut. The endoderm is initially partitioned into an anterior and posterior domain by the expression of the transcription factor genes *hex* and *pdx1* in the anterior and *cdx* in the posterior, although these domains remain labile for about 24 hours. Experiments involving overexpression of Wnts in the anterior will repress formation of the stomach, liver, and pancreas, while inhibition of Wnt in the posterior will induce ectopic structures of these tissue types.

The organizer graft

All three of the processes, dorsalization, neural induction, and anteroposterior patterning, are shown in the **organizer graft** first performed by Spemann and Mangold in 1924. This is the

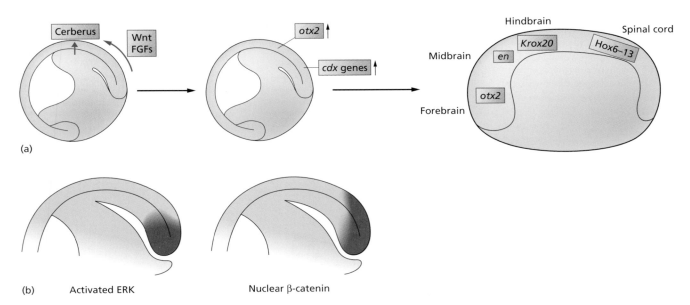

Fig. 7.24 Anteroposterior patterning of the CNS. (a) Action of cerberus from the anterior part of the organizer, and of FGFs and Wnts from the posterior part. (b) Location of phosphorylated ERK (FGF target) and nuclear β-catenin (Wnt target) in late gastrula.

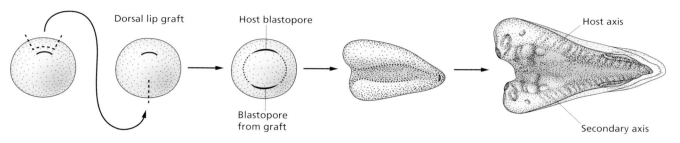

Fig. 7.25 The organizer graft. The graft forms the notochord and head mesoderm of the secondary axis and induces the other parts from the host.

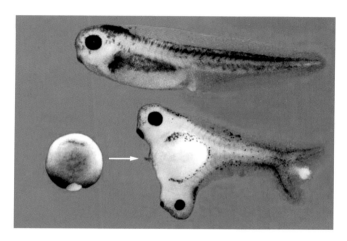

Fig. 7.26 The organizer graft. The graft may be seen as a white patch on the ventral side of the gastrula. The resulting double embryo, and the control embryo, are about 3 days old. Reproduced from De Robertis and Kuroda. *Ann Rev Cell Devel Biol* 2004; 20: 285–308, with permission from Annual Reviews.

most famous experiment in embryology, although its true significance has only recently become clear now that the various component processes have become understood. In the graft, a piece of tissue from above the dorsal blastopore lip is implanted into the ventral marginal zone (Fig. 7.25). It leads to the formation of a double dorsal embryo in which the notochord of the

secondary embryo is derived from the graft and the remainder from the host (Fig. 7.26). The ectoderm above the graft becomes induced to form a second neural tube. As gastrulation proceeds, both host and graft axes form progressively more posterior parts. In favorable cases, the final result is a symmetrical pair of embryos joined belly to belly.

The movements of the grafted organizer are autonomous and preprogrammed. Therefore it invaginates and undergoes active elongation by convergent extension. The graft emits dorsalizing signals (BMP and Wnt inhibitors). These diffuse to the neighboring mesoderm, and the reduction in BMP signaling suppresses upregulation of *vent1* and so causes upregulation of myogenic genes to form a file of somites on either side of the graft. The same substances diffuse to the neighboring ectoderm and upregulate pan-neural genes such as *sox2* and *n-cam*. During gastrulation the graft emits FGFs and Wnts, which induce *cdx* and Hox genes needed for development of the trunk–tail region.

The secondary axes arising from organizer grafts often lack a head, but if the graft includes the deep anterior region of the organizer then this will emit cerberus and dickkopf and induce a head in the overlying ectoderm. The net result is the formation of the secondary embryo. Thus the organizer graft nicely illustrates the coordinated working of the whole sequence of inductive interactions that operate between early cleavage and the end of gastrulation, and result in the formation of three germ layers, and a dorsoventral and anteroposterior pattern in each of those germ layers.

Key points to remember

- The descriptive embryology of *Xenopus* is typical of animals generally although some details are specific. Cleavages are rapid and synchronous, leading to blastula formation. Gastrulation movements lead to the formation of the three germ layers with simultaneous development of regional pattern in the dorsoventral and anteroposterior axes.
- Accurate fate maps have been constructed by injection of fluorescent dextrans into individual cells.
- Overexpression of gene products is usually carried out by injection of mRNA into fertilized eggs. Inhibition of specific gene products is usually carried out by injection of antisense morpholinos or of mRNA for dominant negative constructs.
- The pattern of the embryo is specified in relation to two determinants in the egg. The vegetal determinant includes the mRNA for the T-box transcription factor vegT which becomes localized to the vegetal cortex during oogenesis. The dorsal determinant comprises Wnt pathway components which become moved from the vegetal to the dorsal side during the sperm-induced cortical rotation.
- VegT protein upregulates expression of genes encoding endoderm-specific transcription factors (e.g.

sox17) and also the genes encoding nodal-type inducing factors. The nodal factors are secreted and induce the expression of mesoderm-specific transcription factors (e.g. brachyury) in an equatorial belt of neighboring cells.
- On the dorsal side, the Wnt pathway upregulates expression of genes encoding dorsal-specific transcription factors (e.g. *siamois*). The region containing both mesodermal and dorsal transcription factors becomes the organizer. Here genes for BMP inhibitors (noggin, chordin) are upregulated. These are secreted and create a gradient of BMP activity from ventral to dorsal. Low BMP activity enables expression of transcription factors such as sox2 in the animal (ectodermal) region, leading to formation of the neural plate. It also induces transcription factors such as myf5 in the equatorial (mesodermal) region, leading to the formation of the myotomes.
- Anteroposterior patterning depends on a posteriorizing signal emitted from the blastopore region during gastrulation. This consists of FGFs and Wnts, and causes upregulation of Hox genes in a nested manner such that each gene is upregulated at a particular anteroposterior level and remains on posterior to this.

Further reading

Website

http://www.xenbase.org/

General

Nieuwkoop, P.D. & Faber, J. (1967) *Normal Table of Xenopus laevis.* Amsterdam: N. Holland. Reprinted Garland Publishing Inc. (1994).

Hausen, P. & Riebesell, M. (1991) *The Early Development of Xenopus laevis.* Berlin: Springer-Verlag.

Brown, D.D. (2004) A tribute to the *Xenopus laevis* oocyte and egg. *Journal of Biological Chemistry* **279**, 45,291–45,299.

Carruthers, S. & Stemple, D.L. (2006) Genetic and genomic prospects for *Xenopus tropicalis* research. *Seminars in Cell and Developmental Biology* **17**, 146–153.

Brown, D.D. & Cai, L. (2007) Amphibian metamorphosis. *Developmental Biology* **306**, 20–33.

Green, S.L. (2010) *The Laboratory Xenopus sp.* Boca Raton FL: CRC Press.

Harland, R.M. & Grainger, R.M. (2011) Xenopus research: metamorphosed by genetics and genomics. *Trends in Genetics* **27**, 507–515.

Oocyte and maternal factors

Tunquist, B.J. & Maller, J.L. (2003) Under arrest: cytostatic factor (CSF)-mediated metaphase arrest in vertebrate eggs. *Genes and Development* **17**, 683–710.

Heasman, J. (2006) Maternal determinants of embryonic cell fate. *Seminars in Cell and Developmental Biology* **17**, 93–98.

Fate maps

Keller, R.E. (1975) Vital dye mapping of the gastrula and neurula of *Xenopus laevis* I. Prospective areas and morphogenetic movements of the superficial layer. *Developmental Biology* **42**, 222–241.

Keller, R.E. (1976) Vital dye mapping of the gastrula and neurula of *Xenopus laevis* II. Prospective areas and morphogenetic movements of the deep layer. *Developmental Biology* **51**, 118–137.

Dale, L. & Slack, J.M.W. (1987) Fate map for the 32 cell stage of *Xenopus laevis*. *Development* **99**, 527–551.

Bauer, D.V., Huang, S. & Moody, S.A. (1994) The cleavage stage origin of Spemann's organiser: analysis of the movements of blastomere clones before and during gastrulation in *Xenopus*. *Development* **120**, 1179–1189.

Gastrulation

Keller, R.E., Danilchik, M., Gimlich, R. & Shih, J. (1985) The function and mechanism of convergent extension during gastrulation of *Xenopus laevis*. *Journal of Embryology and Experimental Morphology* **89** (Suppl.), 185–209.

Keller, R., Shih, J. & Domingo, C. (1992) The patterning and functioning of protrusive activity during convergence and extension of the *Xenopus* organiser. *Development* (Suppl.) **116**, 81–91.

Beetschen, J.C. (2001) Amphibian gastrulation: history and evolution of a 125 year old concept. *International Journal of Developmental Biology* **45**, 771–795.

Winklbauer, R., Medina, A., Swain, R.K. & Steinbeisser, H. (2001) Frizzled-7 signalling controls tissue separation during Xenopus gastrulation. *Nature* **413**, 856–860.

Early dorsoventral polarity

Gerhart, J., Danilchik, M., Doniach, T., & Roberts, S. (1989) Cortical rotation of the Xenopus egg: consequences for the anteroposterior pattern of embryonic dorsal development. *Development* (Suppl.) **107**, 37–51.

Weaver, C. & Kimelman, D. (2004) Move it or lose it: axis specification in *Xenopus*. *Development* **131**, 3491–3499.

Endoderm

Dale, L. (1999) Vertebrate development: multiple phases to endoderm formation. *Current Biology* **9**, R812–R815.

Woodland, H.R. & Zorn, A.M. (2008) The core endodermal gene network of vertebrates: combining developmental precision with evolutionary flexibility. *BioEssays* **30**, 757–765.

Mesoderm induction

Heasman, J. (2006) Patterning the early *Xenopus* embryo. *Development* **133**, 1205–1217.

Kimelman, D. (2006) Mesoderm induction: from caps to chips. *Nature Reviews Genetics* **7**, 360–372.

Dorsoventral signaling and the organizer

Dale, L. & Wardle, F. (1999) A gradient of BMP activity specifies dorsal-ventral fates in early Xenopus embryos. *Seminars in Cell and Developmental Biology* **10**, 319–326.

Niehrs, C. (2004) Regionally specific induction by the Spemann-Mangold organizer. *Nature Reviews Genetics* **5**, 425–434.

Reversade, B. & De Robertis, E.M. (2005) Regulation of ADMP and BMP2/4/7 at opposite embryonic poles generates a self-regulating morphogenetic field. *Cell* **123**, 1147–1160.

Simeonia, I. & Gurdon, J.B. (2007) Interpretation of BMP signaling in early Xenopus development. *Developmental Biology* **308**, 82–92.

De Robertis, E.M. (2009) Spemann's organizer and the self-regulation of embryonic fields. *Mechanisms of Development* **126**, 925–941.

Anteroposterior pattern

Doniach, T. (1992) Induction of anteroposterior neural pattern in *Xenopus* by planar signals. *Development* (Suppl.) **116**, 183–193.

Piccolo, S., Agius, E., Leyns, L., Bhattacharyya, S., Grunz, H., Bouwmeester, T. & De Robertis, E.M. (1999) The head inducer Cerberus is a multifunctional antagonist of Nodal, BMP and Wnt signals. *Nature* **397**, 707–710.

Kiecker, C. & Niehrs, C. (2001) A morphogen gradient of Wnt/beta-catenin signalling regulates anteroposterior neural patterning in Xenopus. *Development* **128**, 4189–4201.

Niehrs, C. (2004) Regionally specific induction by the Spemann–Mangold organizer. *Nature Reviews Genetics* **5**, 425–434.

This chapter contains the following animations:

Animation 5 Induction.
Animation 9 Animal cap and UV assays.

 For additional resources for this book visit www.essentialdevelopmentalbiology.com

Chapter 8

The zebrafish

The zebrafish was introduced as a model organism for developmental biology largely because it was considered a vertebrate favorable for genetic experimentation. The adults are small and large numbers can be kept in recirculating aquaria, which are typically subdivided into numerous small tanks to enable many genetic lines to be kept. Animals can be bred at about 3–4 months, and can be re-mated at 2-week intervals.

In contrast to *Xenopus*, the original approach to research with the zebrafish was through mutagenesis, and this means that loss-of-function mutations are available for many of the developmentally important genes. A short list of some key developmental genes, which are those mentioned in the text, is provided in Table 8.1. Most are homologs of the corresponding genes in *Xenopus* and the mouse, but because most were originally discovered by mutation, and named before the identity of the gene was known, their names are specific to the zebrafish. Because they are named after recessive alleles, these gene names are written with an initial lower case letter.

In addition to the study of early development described here, the zebrafish has become increasingly important in organogenesis research and will feature again in later chapters. It is also becoming important outside developmental biology, especially as a test bed for drug screening, because of the ability easily to visualize the morphological consequences of disturbing particular biochemical pathways.

Normal development

Oogenesis

As in other lower vertebrates, oocytes arise from oogonia throughout adult life. The primary oocyte is in meiotic prophase so chromosomes are arranged as bivalent pairs, each chromosome consisting of two chromatids arising from the last DNA replication cycle. Growth of the primary oocyte is divided into three stages: designated as I, II, and III. During growth, primary oocytes expand from about 7 to 700 µm in diameter. They are transcriptionally active and, especially in the early stages, display lampbrush chromosomes. Over 1000 nucleoli appear in stage II, indicating active ribosomal RNA synthesis. In stage I–II, a Balbiani body (mitochondrial cloud) is visible, and, as in *Xenopus*, this is associated with localization of maternal RNAs. During these stages, the follicle cells that invest the oocyte change from a columnar to a flat form and secrete the vitelline envelope (the future chorion). One follicle cell, called the micropylar cell, maintains a contact with the oocyte such that a small hole, the micropyle, is left in the vitelline membrane, and this later serves as the route for sperm entry. During stage III, the oocyte becomes opaque due to formation of yolk granules. The vitellogenin protein is taken up from the circulation in a process mediated by the follicle cells. Ovulation is caused by a rise in gonadotrophin, which triggers the follicle cells to produce a steroid: 17α,20β-dihydroxy-4-pregnene-3-one (17α,20β-DP). This stimulates nuclear hormone receptors in the oocyte and activates a wide range of signaling cascades, resulting in the events of maturation. These include the migration of the nucleus from the center of the oocyte to the animal pole, the subsequent dissolution of the nuclear membrane, the elevation of maturation promoting factor (MPF = complex of M-phase cyclin and cyclin-dependent kinase), and the resumption of meiosis, with first division and expulsion of the first polar body. The mature egg (strictly a secondary oocyte arrested in second meiotic metaphase) has a diameter of about 0.74 mm.

Fertilization and cleavage

Most zebrafish embryos in the laboratory are obtained by natural matings, although *in vitro* fertilization can also be

Essential Developmental Biology, Third Edition. Jonathan M.W. Slack.
© 2013 John Wiley & Sons, Ltd. Published 2013 by John Wiley & Sons, Ltd.

Table 8.1 Some developmental genes in the zebrafish.

Gene	Homolog	Developmental function	Gene product
ichabod	*β-catenin*	dorsal axis formation	co-TF
radar	*Gdf6a*	activates *Bmp* expression	IF
cyclops	*Nodal*	mesendoderm induction	IF
squint	*Nodal*	mesendoderm induction	IF
one-eyed pinhead	*Cripto*	needed for nodal action	IF
bozozok (dharma, nieuwkoid)		defines organizer	paired homeo TF
notail	*brachyury/T*	defines posterior mesoderm	T-box TF
spadetail	*Tbx6/VegT*	defines trunk mesoderm	T-box TF
floating head	*Not*	suppresses *spadetail*	homeodomain TF
chordino (dino)	*chordin*	inhibits BMP protein	IF
swirl	*Bmp2b*	ventralizing	IF
snailhouse	*Bmp7*	ventralizing	IF
vent (vega2)	*Vent1,2*	defines ventral	homeodomain TF
vox (vega1)	*Vent1,2*	defines ventral	homeodomain TF
mini fin	*Tolloid*	tail morphology	protease
casanova	*Sox32*	defines endoderm	Sox type TF
spiel-ohne-grenzen	*Oct4*	needed for endoderm	POU type TF
bonnie-and-clyde	*Mixer*	defines endoderm	T-box TF
faust	*Gata5*	heart/ endoderm	Zn finger TF
silberblick	*Wnt11*	needed for convergent extension	IF
acerebellar	*Fgf8*	posteriorizing	IF

Note that in conformity with genetic custom these genes have been given names based on the loss-of-function phenotype. The wild-type function is typically the opposite of that implied by the name, for example the wild-type allele of *notail* is needed to make the tail, and of *acerebellar* to make the cerebellum.

TF, transcription factor; IF, inducing factor.

carried out (especially from "high clutch size, easy to squeeze" lines). To obtain eggs, the male and female are placed together overnight in a tank with a perforated floor. This allows the fertilized eggs to settle without risk of being eaten by the parents. The eggs and sperm are shed into the water and fertilization is external. Unlike a mammalian sperm, the zebrafish sperm has no acrosome. It enters through the micropylar orifice in the chorion, binds to microvilli on the egg surface, and fuses with the egg. Within 10 minutes, the second meiotic division is completed with expulsion of the second polar body. Cortical granules beneath the egg surface exocytose and release their contents, causing the chorion to lift off the egg surface. There is a pronounced cytoplasmic streaming process such that nonyolk cytoplasm moves to the animal pole and forms a clear island called the cytoplasmic cap. Pronuclear fusion occurs about 17 minutes, and first cleavage about 40 minutes, postfertilization.

The course of development is shown in Fig. 8.1. The standard developmental temperature is 28.5°C, so developmental times relate to this temperature. Cleavage is **meroblastic** as it involves only the animal pole region. Cleavages occur about every 15 minutes and the first five are usually vertical, generating a 32-cell stage composed of one layer of blastomeres. At this stage the outer blastomeres remain connected to the main acellular yolk mass (the yolk cell) by cytoplasmic bridges. The sixth cleavage is usually horizontal, generating an outer layer of blastomeres without connections to the yolk cell. Cleavages continue

in a synchronous manner until the **mid-blastula transition (MBT)** occurs after nine divisions (during the tenth cell cycle). This is similar to *Xenopus* in that the synchrony of divisions breaks down, the average cell cycle duration increases, cells become more motile, and the transcription of the zygotic genome commences. At MBT, the outer blastomeres that have cytoplasmic connections to the yolk cell (sometime called Wilson cells) sink into the yolk cell so that it becomes a multinuclear syncytium. This is called the **yolk syncytial layer (YSL)**. The YSL nuclei divide rapidly three to four times and then stop. The YSL is thicker round the periphery of the embryo but also extends under the blastoderm. The remainder of the yolk cell is surrounded by a thin, anuclear layer of cytoplasm. The main blastoderm generates an external layer of epithelial cells called the **enveloping layer**. This will later form the periderm, or outer layer, of the larval epidermis. The major part of the blastoderm consists of a cap of cells called **deep cells**, about six to eight thick, occupying the animal part of the embryo. The deep cells are quite motile, making it difficult to construct fate maps for pregastrulation stages.

Gastrulation

The overall course of the morphogenetic movements is quite similar to that in *Xenopus* although the mechanics may be

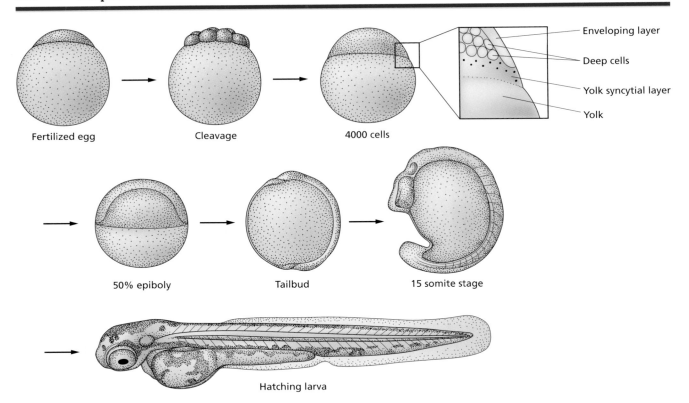

Fig. 8.1 Normal development of the zebrafish.

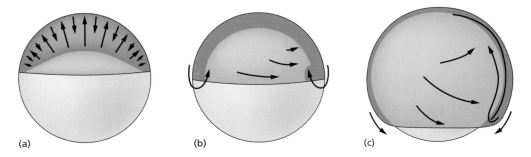

Fig. 8.2 Cell movements during gastrulation of the zebrafish. (a) Epiboly; cells intercalate radially, contributing to the expansion. (b) involution and dorsal convergence; (c) continued epiboly, dorsal convergence and involution, with convergent extension. Reproduced from Schier and Talbot (2005), with permission from Annual Reviews.

somewhat different. Three distinct processes can be distinguished: **epiboly**, **involution**, and **convergent extension** (Fig. 8.2).

Starting shortly after MBT, the blastoderm commences an active expansion called **epiboly** such that the margin moves down progressively to cover the yolk cell. Stages of zebrafish development are identified as percentage of epiboly, depending on how much of the yolk cell has been surrounded. This movement is driven by the yolk syncytial layer and continues even if most of the rest of the yolk cell is ablated. Epiboly depends on the activity of microtubules within the YSL and can be dis-

rupted by treatment with microtubule-disrupting agents such as nocodazole. At 50% epiboly (approximately 5–6 hours) there is a pause as the margin of the blastoderm begins to involute to form a lower layer called the **hypoblast** or **mesendoderm**. In the zebrafish the term **gastrulation** is normally reserved specifically for the involution movement of the mesoderm as distinct from the epibolic spreading process. Involution takes place all around the blastoderm margin, but, as in *Xenopus*, the dorsal involution is much more pronounced than the ventral. As a result of involution the blastoderm becomes thicker around the margin than elsewhere and this thickening is called the **germ**

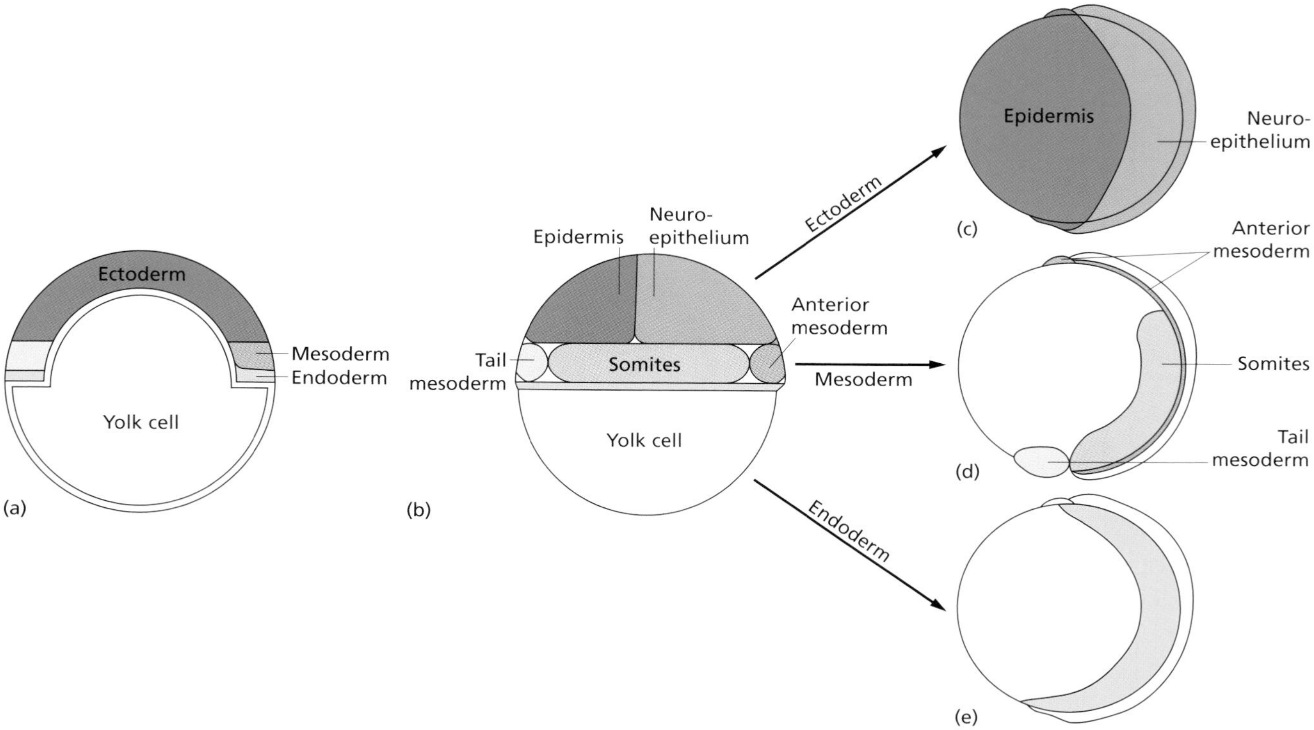

Fig. 8.3 Formation of each germ layer in the course of zebrafish gastrulation: (a) median section and (b) surface view at 50% epiboly; (c–e) the germ layers at the end of gastrulation. Reproduced from Solnica-Krezel (2002), with permisson from Springer.

ring. The involution movement is carried out only by the deep cells and the outer enveloping layer cells do not participate. It was formerly thought that some cells ingressed as individuals but filming of individual cells during gastrulation indicates that this only occurs very close to the blastoderm margin, so the process is correctly described as an involution. Simultaneous with the involution, the dorsal region thickens to form the **embryonic shield**, a structure comparable to the **dorsal lip** of *Xenopus* or the **node** of amniote embryos. As involution proceeds, the shield elongates in an anteroposterior sense by **convergent extension**, drawing in cells from more lateral regions. The dorsal midline elongates substantially to become the midline of the anteroposterior body axis. The lateral tissue converges substantially to the midline but elongates less. The midventral sector does not participate in dorsal convergence but moves to the vegetal pole and later forms the tail bud. Collectively, these processes leave the anteroventral region of the yolk cell depleted of mesoderm (Fig. 8.3). The YSL also undergoes convergent extension and forms a strip below the notochord. As in *Xenopus*, the driving force for convergent extension is an active process of cell intercalation in a mediolateral direction. The Wnt-PCP pathway is necessary for convergent extension to proceed, and the process is inhibited in loss-of-function mutants of *wnt11* (*silberblick*). A small population of cells called forerunner cells originate from the enveloping layer and migrate ahead

of the blastoderm. They become subducted around 90% epiboly and end up in a structure called Kupffer's vesicle, which is lined with monocilia and is involved in the generation of left–right asymmetry in the heart and digestive tract. Epiboly and involution reach completion at about 10 hours when the yolk cell is completely covered by the blastoderm, and the dorsal axis has been formed.

Neurulation and subsequent stages

In the hypoblast the anterior region moves past the animal pole to what is now the anterior of the embryo. The central part forms the prechordal plate and the lateral part forms the hatching gland and pharyngeal endoderm. More posteriorly, the notochord appears in the midline in the later stages of gastrulation. The outer layer of the shield, the dorsal ectoderm, becomes the **neural plate**. This appears as a thickening of the ectoderm at the end of gastrulation. The lateral regions thicken further and fuse in the midline to form a solid neural keel at about 13 hours, sinking into the interior as a solid neural rod by 16 hours. Unlike in *Xenopus*, the neural tube lumen does not arise from the closure process but from a secondary cavitation of the neural rod to create a neural tube. The first neurons arise by 10 hours with a second phase of neurogenesis at 14 hours.

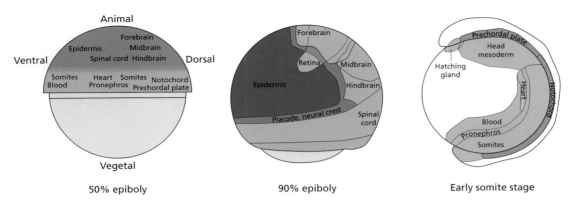

Fig. 8.4 Fate maps of the zebrafish at different stages. Reproduced from Schier and Talbot (2005), with permission from Annual Reviews.

The axons of these neurons develop over the succeeding 10 hours.

In the mesoderm, the notochord continues to differentiate in the midline. Somites start to appear in the paraxial mesoderm shortly after the end of gastrulation. As in all vertebrates, somitogenesis proceeds from anterior to posterior. The first five to six somites form about every 20 minutes, and succeeding ones every 30 minutes.

The basic body plan structures of the embryo become visible by about 14 hours, and the axis straightens by 24 hours, when it may be called a **pharyngula**. Hatching from the chorion occurs in the period 48–72 hours. The swim bladder, an important structure in teleost fishes for adjusting buoyant density, inflates and becomes clearly visible on the fourth day, and feeding commences about 5 days after fertilization.

Fate map

The fate map for the zebrafish has been constructed by injection of individual blastomeres with fluorescent dextrans. As there is considerable mixing of the deep cells, a cell in a particular position will move differently in different individual embryos and so the fate maps from the cleavage stage are only statistical. This means that in an aggregated set of data from many embryos it will appear that each individual blastomere contributes to a very wide region of the later embryo. By the beginning of gastrulation the range of random cell movement is much reduced and a fate map can be produced with resolution approaching that of *Xenopus* (Fig. 8.4). The fate map shown is for the beginning of gastrulation and concerns just the deep cells. Individual cells may populate more than one tissue type or even more than one germ layer if labeled in the early stages of epiboly. But by the start of involution all clones have become restricted to tissue types such as neuroepithelium or somite.

As in *Xenopus*, the process of convergent extension and the corresponding dorsal convergence means that cells originating at the dorsal midline of the blastoderm project mostly to the anterior, although also the whole dorsal midline, while cells from the ventral side of the blastoderm project mostly to the posterior.

Genetics

Mutagenesis screens

The zebrafish has been the subject of several mass mutagenesis screens to isolate mutations in developmental genes. The most common protocol involves the identification of recessive mutations, sometimes lethal ones, by breeding for three generations (Fig. 8.5). Mutations are induced in a small number of founder males by treatment with the chemical **mutagen** ethyl nitrosourea (ENU). This induces point mutations at high frequency in dividing cells, including the spermatogonia, up to about one mutation in a particular gene per 500 gametes. The males are allowed to recover for 2 weeks and during this period the sperm already undergoing spermatogenesis are lost and new ones are produced from the mutated spermatogonia. ENU introduces adducts onto individual DNA bases and these become converted into single-strand mutations after one DNA replication cycle, and double-strand mutations after two cycles. ENU adducts onto the DNA of mature sperm will therefore generate genetically mosaic embryos. But if the spermatogonia are allowed two division cycles after ENU treatment, then the mutations in the sperm will be double stranded and will give rise to fully transgenic embryos.

The mutagenized males are mated to normal females, producing an F1 offspring generation. Each of the F1 individuals may carry a mutation, and each is likely to differ from the other F1 individuals because very many spermatogonia, mutated at many different loci, are producing the mutant sperm. Any dominant mutations affecting development should be immediately apparent as a change in phenotype of an F1 individual, but the

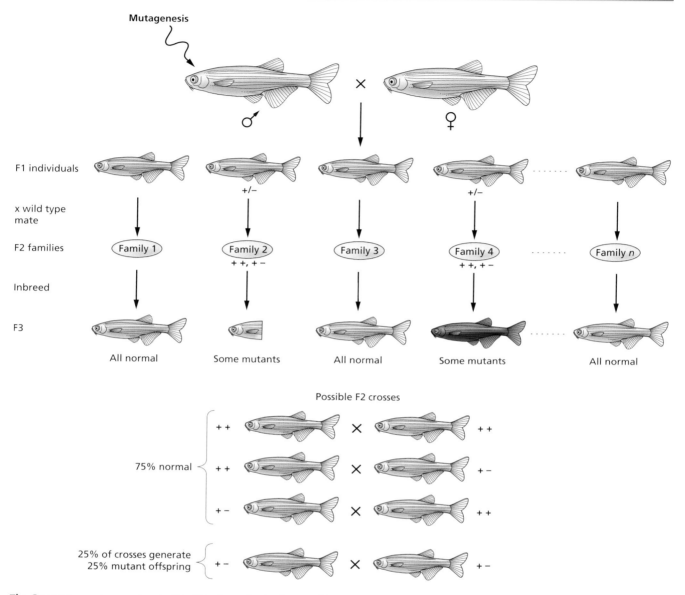

Fig. 8.5 Mutagenesis screen in zebrafish. This shows the simplest possible type of screen for zygotic recessives. Each F1 individual is outcrossed to generate a family at the F2 generation. Pairs of F2 animals are mated and if the family contains a mutation then about one mating out of four will be between two +/− individuals, and such matings will yield 25% −/− progeny, which should show an abnormal phenotype.

Classic experiments

Because of the simplicity of the genetic methods employed, the original mutagenesis screen performed in the 1990s represents a paradigm for a zygotic mutagenesis screen for developmental mutants. This was complemented by adoption of the morpholino approach for loss-of-function studies of known genes, described in the last reference.

Haffter, P., Granato, M., Brand, M., Mullins, M.C., Hammerschmidt, M., Kane, D.A., Odenthal, J., vanEeden, F.J.M., Jiang, Y.J., Heisenberg, C.P.,

Kelsh, R.N., Furutani-Seiki, M., Vogelsang, E., Beuchle, D., Schach, U., Fabian, C. & NüssleinVolhard, C. (1996) The identification of genes with unique and essential functions in the development of the zebrafish, Danio rerio. *Development* **123**, 1–36.

Driever, W., SolnicaKrezel, L., Schier, A.F., Neuhauss, S.C.F., Malicki, J., Stemple, D.L., Stainier, D.Y.R., Zwartkruis, F., Abdelilah, S., Rangini, Z., Belak, J. & Boggs, C. (1996) A genetic screen for mutations affecting embryogenesis in zebrafish. *Development* **123**, 37–46.

Nasevicius, A. & Ekker, S.C. (2000) Effective targeted gene "knockdown" in zebrafish. *Nature Genetics* **26**, 216–220.

more common recessive mutations will not be visible since the F1 animals are all heterozygous. For the next generation each F1 individual is put in a separate container and mated to a wild-type animal. This produces a set of F2 families, each family derived from one founder fish, which may carry one or more recessive mutations. If the F1 individual did carry a mutation, then half the resulting F2 individuals will be heterozygous for it. A set of test matings is carried out between pairs of individuals within each F2 family. If there is a mutation present then 1 in 4 matings will be between heterozygotes and yield an F3 generation that is 25% homozygous mutant. The F3 generation is examined and scored at the embryo stage. Homozygous recessive mutants that affect development should be visibly abnormal. They may be detected simply by examination of the embryos under the dissecting microscope, or, if the screen is more focused, by examination of a specific structure or cell population.

Any mutation that disables a gene essential for early development is quite likely to be lethal and to prevent development after the time of normal gene function. So the homozygous mutant F3 embryos may die at an early stage and they need to be examined and scored before they degenerate.

The overall screening procedure contains various elements of randomness. It is quite likely that F2 families will contain no mutation, or none giving an abnormal phenotype, in which case they are discarded. It is also possible for an F3 phenotype to be caused by more than one mutation in the original sperm. If there are two mutations showing independent segregation then the F3 offspring of double heterozygous parents will actually be a 9:3:3:1 mix of wild type, single homozygotes of each type, and double homozygotes. This may not be apparent immediately, but can be resolved by further breeding. When an interesting-looking phenotype is found among the F3 generation, the F2 parents are outcrossed to wild-type fish to reduce the burden of other, nondevelopmental mutations and to set up a permanent line. The majority of mutations in developmentally important genes are lethal in the homozygous state, therefore the line has to be maintained by identifying the heterozygotes and mating them together when mutant embryos are required.

Figure 8.6 shows some mutants arising from a zebrafish screen. Although in principle inbred lines (i.e. lines with a homozygous genetic background) are good because of limited variability, in practice inbreeding leads to low fertility and viability, so useful mutant lines are often outcrossed again to maintain some variation in the genetic background. It is also possible to store mutant lines in the form of sperm or testis frozen in liquid nitrogen, and this can reduce the labor of keeping large numbers of lines as breeding colonies.

The initial screens were carried out by simple dissecting microscope examination of the F3 embryos with the intention of isolating all possible mutations in all developmentally significant genes. But it has become clear that more mutations can be identified if the screen is more focused, so more recent screens have tended to examine one particular organ system in detail.

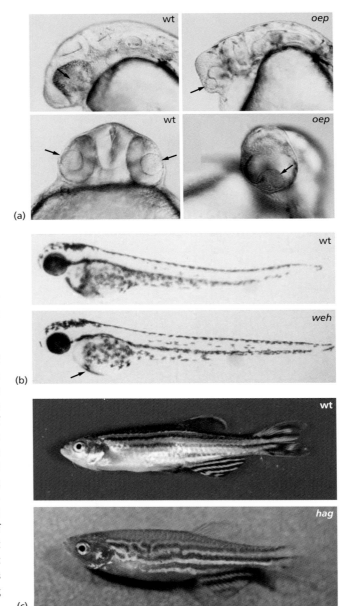

Fig. 8.6 Examples of mutants identified in zebrafish mutagenesis screens. (a) *one-eyed pinhead* (*oep*), causes cyclopia (arrows show lenses, note single midline eye). Top, lateral views; bottom, anterior views. wt, wild type. (b) *weissherbst* (*weh*) decreases blood formation. The effect is shown by O-dianisidine stain for hemoglobin in 2-day embryos. wt, wild type. (c) *hagoramo* (*hag*) is a dominant mutation which results in a disrupted stripe pattern in adult fish and was generated by insertional mutagenesis. wt, wild type. Reproduced from Patton and Zon (2001), with permission from Nature Publishing Group.

Some of these have used immunostaining or *in situ* hybridization to highlight the structures of interest, and in other cases the parent line is a transgenic line in which the structures of interest express GFP, or another fluorescent protein, and can easily be visualized under the fluorescence dissecting microscope.

The results of a screen are initially classified according to the phenotype and then **complementation analysis** is carried out for mutations with similar phenotypes to find whether they are different alleles of the same gene. This is done by mating the heterozygotes of the two lines. If the mutations are indeed in the same gene then there will be 25% offspring with the phenotype, whereas if they are in different genes the offspring will not show the phenotype at all. Then the mutation will be mapped by conventional genetic mapping. The map position will be the starting point for **positional cloning** of the gene. Positional cloning used to be a slow and arduous procedure, but nowadays, with a near complete genome sequence and a large number of molecular polymorphisms available to use in mapping, it has become relatively easy (see Chapter 3, Fig. 3.10). However, many zebrafish mutants have also been identified without positional cloning by the testing of likely candidates based on the similarity of the phenotype to loss-of-function effects found in *Xenopus* or other model organisms.

Genome duplication

Evidence from the results of genome sequencing suggests that the zebrafish, like all teleosts (the largest group of bony fishes), often contain two similar genes for every one found in higher vertebrates. These related copies have a high level of sequence divergence, suggesting a rather ancient duplication of the entire genome. It is estimated that this would have occurred in the ancestor of all teleosts shortly after it separated from the vertebrate main line of descent about 420 million years ago. The effect is that a gene in mouse that has more than one expression domain or developmental function may be represented by two genes in the fish, and these often share out the domains and functions between them such that their combined expression and function resembles that of the single mouse gene. This fact has advantages and disadvantages to the experimenter. On the one hand there are somewhat more genes to sift through and analyze. On the other, a gene whose null mutant is an early lethal in mouse may appear as two genes each of whose null phenotype is milder, allowing development to a more advanced stage or even long-term viability.

Other screening methods

Haploid screens
Apart from the normal type of recessive screen, the zebrafish allows two other techniques for mutational screening, which

have advantages in particular circumstances. One is the generation of **haploid** embryos (Fig. 8.7). Female F1 fish are taken and squeezed gently to obtain unfertilized eggs. These are fertilized *in vitro* with sperm heavily irradiated with ultraviolet light to render them inviable. Such sperm can activate development but the sperm nucleus does not participate and so the embryo is a haploid, arising from the maternal pronucleus alone. In haploid embryos the effects of recessive mutations will be immediately apparent and so an interesting mutation can be identified without the need for the F2 family. However, the haploid embryos are not viable in the long term. They die after a few days and so a mutation needs to be maintained by conventional breeding starting from the mother. Moreover, haploid embryos themselves are not normal even in the absence of mutations, and therefore screens are only feasible for characteristics that would not be masked by the typical haploid syndrome.

The haploid syndrome can be avoided if the eggs are converted into **gynogenetic diploids**. Again, the eggs are squeezed from the female fish and fertilized with inviable sperm. Then

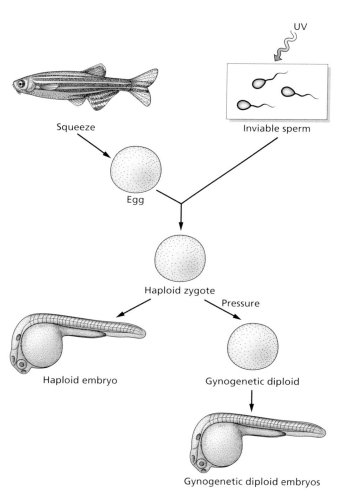

Fig. 8.7 Production of haploid embryos and gynogenetic diploids.

they are subjected to a pressure pulse to drive the second polar body back into the egg and cause it to fuse with the maternal pronucleus. This reversal of the second meiotic segregation means that the egg pronucleus is diploid but composed entirely of maternal genetic material. However, loci will only be homozygous if there has been no meiotic recombination. If recombination has occurred at the bivalent stage then one mutant copy can be lost into the first polar body and the gynogenetic diploid will potentially be heterozygous for all loci distal to the crossover point. Thus, this technique can be quite useful for genes located near the centromeres, which are unlikely to be lost by recombination, but it does not have a general validity.

Insertional mutagenesis

Before the existence of high-resolution genomic maps and sets of genetic markers for positional cloning, the cloning of genes was rendered much easier if mutations were induced by a procedure that results in the insertion of some molecular probe. This led to methods for **insertional mutagenesis** based on the introduction of replication-defective **retroviruses**. The method involves production of Moloney leukemia virus, which is provided with a new coat protein to enable infection of nonrodent species, and injection of high-titer virus into blastulae. When these are grown to adults the majority have germ-line insertions and produce some offspring carrying them. Insertion of the retrovirus into a gene will often abolish its function, generating a null allele, so most mutations will be recessive. To make these homozygous, the offspring of the founders are bred in the same way as the F1 generation from an ENU screen. Retroviral insertion has also been used as a method to create many **enhancer trap** lines, which have particular cell types or structures expressing a fluorescent protein. These are particularly useful as starting material for other genetic screens, or for chemical screens for useful small molecules affecting the development of the structure that is highlighted.

Embryological and molecular techniques

Although the zebrafish embryo is not quite as favorable as *Xenopus* for micromanipulation, it is still possible to carry out cell labeling and transplantation experiments. It is usual to use an upright compound microscope with a fixed stage and a mobile optical tube. This enables micromanipulators to be clamped to the immobile stage and provides stability. Fluorescent dextrans are usually used for cell labeling. Because of the persistence of cytoplasmic bridges to the yolk cell in early cleavage, very high molecular weight dextrans are favored as diffusion is very slow and they stay confined to the injected blastomere while the cytoplasmic bridges persist. Transplantation is generally carried out by using microinjection equipment to suck cells out of the donor and inject them into the host. Transplantation

is easiest at the early epiboly stages (30–50%) and again during somitogenesis. Because of the significant randomness of cell movement during gastrulation, an injected cell group will generally become quite widely dispersed.

Reverse genetic methods

As with *Xenopus*, **overexpression** of gene products may be achieved by injecting synthetic mRNA into embryos. Because of the early cytoplasmic streaming to the animal pole, RNA can be injected into the yolk mass of newly laid eggs and will be carried into the blastomeres. Although the injected RNA may fill the whole embryo, it does not necessarily do so and it is usual to accompany it with *Gfp* RNA so that the domain of action can be visualized later on. The RNA becomes translated right away and the protein builds up to active levels before MBT. The RNA, and the protein product, will probably be degraded by 24 hours, so this is a technique most useful for modifying early developmental stages.

Transgenic zebrafish were initially prepared by injection of DNA into fertilized eggs. This produces mosaic expression and quite a low proportion of founder fish will transmit the gene through the germline. More recently, other methods have been used. One involves flanking the DNA with sites for the rare cutting restriction enzyme I-SceI (meganuclease), and injecting the enzyme along with the DNA. This enzyme cuts at specific 18-mer sequences and the effect is to catalyze integration events with greater efficiency and with less formation of concatamers than simple DNA injection.

A similar method utilizes the transposon Tol2 as the DNA vector, with coinjection of mRNA encoding the transposase enzyme. This also gives a high proportion of germ-line transmission, mostly single insertions, and has become the favored method.

For conditional transgenesis, the Gal4 system (see Chapter 3) has found favor and many driver lines have been made. These are mostly enhancer trap lines in which an endogenous enhancer drives expression of the activated form of Gal4, Gal4VP16, in a particular cell type or organ. Also the heat shock promoter *Hsp70* works well and is often used to induce activity of a transgene by heating the fish to 37°C for 1 hour.

For ablation of specific gene products, **morpholinos** have been found successful and are widely used. Like RNAs, they need to be injected into fertilized eggs or early blastomeres. The effects persist for up to 48 hours. As always, it is necessary to ensure specificity by doing the right controls. Lack of effect of an unrelated morpholino is not necessarily sufficient. Ideally, it should be demonstrated the translation of the protein target is inhibited, usually by western blotting, and that the phenotype can be rescued by overexpression of an RNA encoding the same protein but rendered insensitive to the mopholino. Other types of antisense method, such as RNAi, have not been found generally reliable.

As discussed in Chapter 4, the association of a particular gene product with a particular developmental function requires assessment of the expression pattern, the biological activity, and the effects of ablation. Once a zebrafish gene is cloned, the expression pattern is established by *in situ* hybridization or immunostaining. Some idea about biological activity may be established by overexpression of mRNA in wild-type embryos, and more specific information by overexpression in various mutant hosts. Because the starting point for the study of a gene is often a null mutant, the effect of ablation may already be known. If it is not, then this would be established with a morpholino experiment.

Regional specification

Maternal determinants

Germ plasm
The mitochondrial cloud (Balbiani body) includes mRNAs encoding Nanos, Vasa, and Dazl, all associated with primordial germ cell formation in other model organisms. After the cloud disassembles in the stage II oocyte, *dazl* migrates to the vegetal cortex, *vasa* to the whole cortex, and *nanos* becomes delocalized. During the first cleavage, these RNAs relocalize to two points on the cleavage membrane, then four points on the second cleavage membranes, in a microtubule-dependent process. These four patches of RNA then remain coherent and become inherited by four small groups of blastomeres within the prospective mesoderm ring of the blastoderm. These are the primordial germ cells (PGCs). Unlike in *Drosophila* and *C. elegans*, the PGCs remain transcriptionally active. Because the first cleavages are random with respect to the later dorsoventral axis, the PGC groups are not aligned with the forming body axis, but during gastrulation they migrate to form one patch on each side of the embryo at the level of somite 1, and subsequently move posteriorly to become incorporated into the gonad around the level of somite 8. Removal of the germ plasm at the cleavage stage will inhibit subsequent germ-cell formation.

Dorsoventral polarity
Unlike *Xenopus*, where the sperm entry point breaks the radial symmetry of the egg and specifies the dorsoventral axis, in the zebrafish the sperm must enter at the animal pole through the micropyle so the immediate cause of dorsoventral symmetry breaking is still uncertain. Yolk ablation experiments suggest that there is a migration of a dorsal determinant, probably mRNA for *wnt8*, from the vegetal pole to the dorsal side, which is completed by early cleavage stages. There is no obvious cortical rotation in the zebrafish, but treatment with the microtubule depolymerizing drug nocodazole does suppress axis formation, suggesting a role for microtubules in the establishment of the dorsal center. Formation of the dorsal axis is also suppressed in

embryos laid by *ichabod⁻* females. *ichabod* is a loss-of-function mutant of one of two *β-catenin* genes. In normal development there is an elevation of nuclear β-catenin on the dorsal side at about the 128-cell stage, visible by immunostaining.

As in *Xenopus*, formation of the dorsal axis can be suppressed by overexpression of *gsk3*, and a dominant negative *gsk3* can induce a secondary axis. Similar results are obtained with other products that activate or inhibit the Wnt–β-catenin pathway, including lithium, which has a dorsalizing effect when applied before MBT. Overexpression of β-catenin upregulates a gene of the paired-homeobox class called *bozozok* (alias *dharma* or *nieuwkoid*). *bozozok* is necessary for formation of the organizer, and the loss-of-function mutant lacks notochord, prechordal plate, and neural tube. This evidence suggests that initial dorsoventral polarity determination events are comparable to *Xenopus*, with the Wnt–β-catenin pathway upregulating genes that specify the organizer on the future dorsal side.

Oct4
oct4 (spiel-ohne-grenzen), a gene very important for control of pluripotency in mammalian embryonic stem cells, has maternal functions in the zebrafish. Maternal and zygotic ablation of *oct4* reduces *bmp* expression and has dorsalizing effects. It is also needed together with the *casanova* product (see below) for formation of the endoderm.

Zygotic events

The sequence of inductive events that build up the body pattern is very similar to those in *Xenopus* with a few variations (Fig. 8.8). There is an initial mesoderm induction, involving the

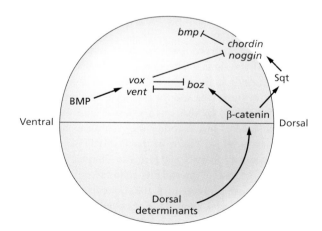

Fig. 8.8 Sequence of inductions in the zebrafish. Reproduced from Kimelman and Schier, Mesoderm Induction and Patterning. In Solnica-Krezel, L. (2002). Pattern Formation in Zebrafish (Results and Problems in Cell Differentiation Series): Springer Verlag, Berlin and Heidelberg, with permission.

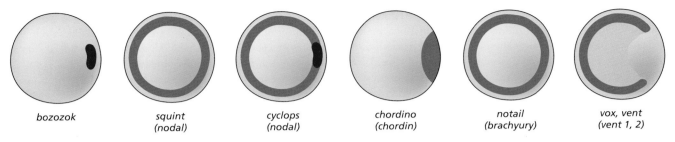

| bozozok | squint (nodal) | cyclops (nodal) | chordino (chordin) | notail (brachyury) | vox, vent (vent 1, 2) |

Fig. 8.9 Some key gene expression patterns in the early embryo. General vertebrate gene names are given below the zebrafish-specific names.

formation of a ring of mesoderm, with an organizer region (the embryonic shield) on the dorsal side. The organizer region then emits signals leading to neural induction of the ectoderm and to dorsalization of the lateral mesoderm. Some relevant gene expression domains are shown in Fig. 8.9.

Mesendoderm induction

The yolk cell, which is the zebrafish equivalent of the *Xenopus* vegetal hemisphere, emits a mesendodermal inducing signal. If the yolk cell, with its associated yolk syncytial layer, is recombined with an animal cap, then mesoderm is induced around the junction, as shown by expression of markers like *notail* (the zebrafish homolog of *brachyury*) in the responding tissue. If the yolk cell is removed from the embryo before the 16-cell stage, then formation of the mesoderm is prevented. Induction is also inhibited by injection of RNAse enzyme into the yolk cell at an early stage, suggesting that at least some maternal RNA must be needed for generation of the mesoderm-inducing signal.

However, the role of VegT is very different from that in *Xenopus*. In zebrafish this is encoded by the gene *spadetail*, which is expressed zygotically in the mesoderm and whose mutant phenotype is a defect in the trunk rather than loss of all mesoderm and endoderm. *spadetail* is not expressed maternally.

There is good evidence that mesoderm induction is carried out by Nodal-type signals. There are two zebrafish homologues of *Nodal*: *cyclops* and *squint*. The double loss-of-function mutant has no germ ring and forms little mesoderm. A similar phenotype is shown by loss-of-function mutants of the gene *one-eyed-pinhead*. This encodes the zebrafish homolog of Cripto, an extracellular factor of the EGF-CFC class required for receptor activation by Nodal (see also Chapter 10). Furthermore, a similar loss of mesoderm follows overexpression of Cerberus-short, the fragment of the *Xenopus* Cerberus factor that antagonizes Nodal. Conversely, the overexpression of *Nodal* mRNA, or of a mRNA for a constitutive Nodal receptor, will induce mesoderm in zebrafish **animal caps**. In terms of expression patterns, *squint* is expressed in the YSL and later in the mesendoderm, *cyclops* just in the mesendoderm, so there must be some relay element to the process with initiation of the signal in the YSL and amplification in the responding tissue.

The mesoderm-inducing signal operates all round the blastoderm margin and the cell layers closest to the yolk, which receive the highest signal, become endoderm, with more distant cell layers becoming mesoderm. The overall signal range is short, about four cell layers at the 1000- to 2000-cell stage. Inhibitors that block the action of Nodal-specific receptors do not prevent mesoderm induction if administered and then removed before the MBT, so the main spread of the induction must occur after zygotic gene upregulation at MBT.

Transcription factor genes characteristic of the early endoderm include *bonnie and clyde* (= *mixer*), *faust* (= *gata5*), and *casanova* (= *sox32*), which are direct targets of Nodal signaling. Casanova has a key position as it will induce endoderm on overexpression, along with expression of the key endoderm transcription factor Sox17. Interestingly, for its action, Casanova requires the cooperation of Oct4 (encoded by *spiel-ohne-grenzen*), which is active maternally and zygotically. In the mesoderm, *brachyury* (*notail*) is expressed around the entire ring, while *goosecoid* is found just in the forming organizer (shield) region. Nodal will upregulate both genes, with *goosecoid* at higher and *notail* at lower concentration, so it has been suggested that a dorsoventral difference of intensity controls the regional pattern in the mesoderm. But *in vivo* the formation of the organizer also depends heavily on the maternally initiated activation of β-catenin on the dorsal side.

The organizer

The elevation of β-catenin on the dorsal side causes upregulation of genes that define the properties of the organizer. These include *bozozok* (= *dharma*, *nieuwkoid*), which encodes a paired-homeodomain transcription factor. The loss-of-function mutant of *bozozok* is ventralized, and this can be rescued by injection of mRNA for the EnR domain swap version of the protein, showing that the factor normally acts as a transcriptional repressor (see Chapter 3 for domain swaps). Bozozok upregulates a number of inducing factor genes including those for Chordin (*chordino*, = *dino*), and Noggin; presumably, it does so indirectly since the Bozozok protein itself is a transcriptional repressor.

The organizer region in zebrafish is called the embryonic shield although the *chordino* domain extends further than the shield, to about 90° of circumference. Grafting of the shield to a ventral position can induce a secondary axis containing somites and a neural tube derived from the host. Extirpation of the shield does not ablate the whole dorsal axis, but removal of the whole *chordino* domain does so. The tissues of the shield will later form the hatching gland, head mesoderm, notochord, somites, dorsal endoderm, and the ventral neural tube. Expression of ADMP in the organizer suggests a similar mechanism for regulating proportion to that found in *Xenopus* (see Chapter 7).

As in other vertebrates, the dorsoventral patterning of the early embryo depends on a ventral-to-dorsal gradient of BMP. Two BMP genes, *swirl* (*bmp2b*) and *snailhouse* (*bmp7*), are expressed in the ventral part of the gastrula. Their loss-of-function mutants are dorsalized, showing that BMP signaling is needed for ventral development. The transcription factor Bozozok, expressed in the organizer region, normally suppresses expression of *swirl*. The loss-of-function mutant of the *chordin* homolog, *chordino*, causes ventralization. As mentioned earlier, because of the dorsal convergence during gastrulation, dorsali-

zation causes increased representation of the head as well as enlargement of notochord, somites, and neural tube, while ventralization causes reduction of the head as well as enlargement of posterior structures and blood islands (Fig. 8.10).

Initial upregulation of *swirl* and *snailhouse* is due to maternal BMPs including Gdf6a (encoded by *radar*). Subsequent stabilization of the pattern depends on inhibitory interactions at both the inducing factor and transcription factor level. At the inducing factor level, Chordin and Noggin bind to BMPs and reduce their activity. At the transcription factor level there is mutual inhibition between the transcription factors defining the organizer and the ventrolateral mesoderm. The ventral state is defined by expression of two homeodomain transcription factors, Vent and Vox, which are the homologues of vent1 and -2 in *Xenopus* and are transcriptional repressors. These have a redundant action, but loss-of-function mutants of both genes together produce a dorsalized phenotype. Their expression is repressed by Bozozok, and they themselves repress expression of *bozozok*, ensuring that any one cell adopts the organizer or ventrolateral state, but not some intermediate. As in *Xenopus*, *wnt8* becomes

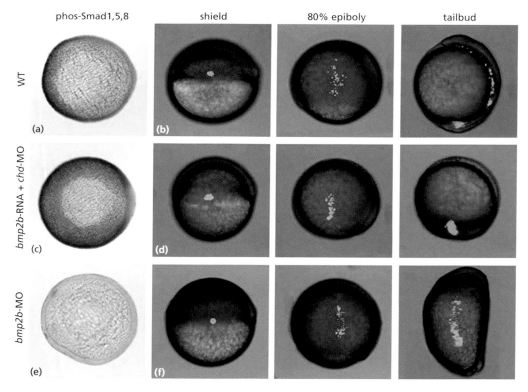

Fig. 8.10 Effects of the BMP gradient on cell movements. Here a small patch of cells in the lateral mesoderm has been labeled by photoactivation of a caged fluorescein. Normally, this will converge dorsally to flank the main body axis (b). In (d) the embryo is heavily ventralized by overexpression of BMP and inhibition of chordin, and the labeled cell group migrates to the tail. In (f) the embryo is dorsalized and the labeled cells remain lateral, failing to converge. (a,c,e) Vegetal views of embryos at 85% epiboly showing immunostaining for antiphospho Smad1,5,8; dorsal is to the right. (b,d,f) Live embryos with labeled cell group in the lateral mesoderm. Reproduced from Hardt *et al. Current Biology* 2007; 17: 475–487, with permission from Elsevier.

upregulated ventrolaterally at about 70% epiboly and has a ventralizing function, notwithstanding the earlier dorsalizing effects of the Wnt pathway.

In the ectoderm, BMP signaling upregulates expression of ΔNp63, a key transcription factor for development of stratified epithelia, and this initiates differentiation of the epidermis.

In the tailbud, Tolloid is produced, which is a metalloprotease that degrades Chordin. It is encoded by the gene *minifin*, whose loss-of-function phenotype is loss of the ventral fin. This additional mechanism to reduce Chordin activity is likely to be needed in the tail because of the short dorsoventral distance compared to that in the main body.

In summary, the dorsalizing action of the organizer arises from inhibition of BMPs by direct transcriptional inhibition within the organizer itself and by secreted BMP inhibitors in the surrounding regions.

Anteroposterior patterning

In the mesoderm, the anteroposterior pattern consists of the prechordal plate and the notochord, which extends through the hindbrain, trunk, and tail regions. The notochord is flanked by the somites, which in the trunk form rib-bearing vertebrae, and in the tail form vertebrae without ribs. The notochord expresses *brachyury* (*notail*) and its transcriptional target *not* (*floating head*), which encodes a homeodomain transcription factor.

The earliest involuting part of the shield expresses *goosecoid* and populates the prechordal plate. The initially more animally located region, which becomes more posterior in the course of involution, forms the notochord. This initial subdivision is probably due to the activity of the maternally initiated β-catenin activation combined with the Nodal signal from the yolk cell. Once involution is underway the anteroposterior pattern becomes stabilized and refined through the action of FGF and Wnt factors emanating from the posterior.

Overexpression of FGFs, or Wnts, or, to an extent, retinoic acid, will posteriorize the whole embryo (Fig. 8.11). *dickkopf1* (*dkk1*) is expressed in the blastoderm dorsal margin and in the dorsal YSL after MBT. Dkk1 inhibits Wnt signaling by binding to the coreceptor Lrp6. It will induce head formation on overexpression and helps to sustain a posterior-to-anterior gradient of Wnt activity. The ventral marginal zone has been described as a "tail organizer," as it will form a tail if combined with an animal cap, much of the tail arising from the animal cap cells. This effect can be mimicked by injection of a combination of mRNAs encoding Nodal + BMP + Wnt and, since the combination of Nodal + BMP might be expected to generate ventral mesoderm, this is further evidence for the posteriorizing activity of Wnt.

Three FGF genes, *fgf3*, *-8*, and *-24* are expressed in the mesoderm, initially at the dorsal margin and later around the germ ring. Overexpression of dominant negative FGF receptor causes loss of the trunk and tail, associated with lack of upregulation of *spadetail* and *notail*. Normally, *spadetail* is repressed by *float-*

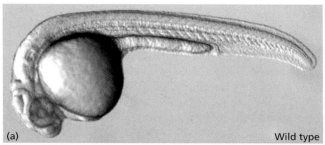

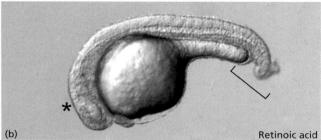

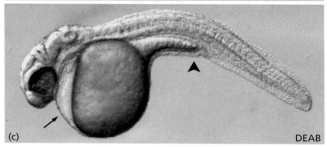

Fig. 8.11 Effects of retinoic acid on the anteroposterior pattern. (a) Normal; (b) treatment with retinoic acid produces defects in the tail (bracket) and brain (asterisk); (c) depletion of retinoic acid by treatment with an inhibitor of retinaldehyde dehydrogenase (diethylaminobenzaldehyde, DEAB) produces some heart edema (arrow) and tail bending (arrow head). Reproduced from Skromne and Prince. *Developmental Dynamics* 2008; 237: 861–882, with permission from John Wiley & Sons, Ltd.

ing head, leading to the formation of the presomite plate. *spadetail* in the presomite plate upregulates expression of Paraxial protocadherin, a cell adhesion molecule that is essential for convergence extension movements.

Fgf8 is encoded by the gene *acerebellar*, but its loss-of-function phenotype, loss of the cerebellum, results from absence of Fgf8 at the midbrain–hindbrain boundary (see Chapter 14). However, when the *acerebellar* mutant is also treated with morpholino to ablate Fgf24 activity, then a posterior defect is produced. So Fgf8 and -24 together probably make up the majority of the posteriorizing FGF activity.

Other uses for the zebrafish

This is a developmental biology textbook and the current section focuses on the early development of the main model organisms. However, the zebrafish is now being used for a much wider range of biomedical research functions than developmental biology. Because the molecular genetics of development is generally similar to other vertebrates, zebrafish mutants have been adopted as disease models, where they are homologous to a human disease mutation or where they produce similar physiological consequences to those found in a human disease condition. The general similarity of the cardiovascular, digestive, and central nervous system to those of mammals is an advantage. Because of the rapid development of the embryos and their transparency, it is possible to test hundreds or thousands of compounds on zebrafish and to observe the effects *in vivo*, often with transparent embryos displaying the structures of interest highlighted by expression of a fluorescent protein. This means that the zebrafish can be useful for screening of compound libraries for biological activity against particular targets and can also be used for preliminary toxicity studies of candidate drugs.

By combining mutants in genes required for pigment cell development, a fully transparent adult zebrafish has recently been prepared, which enables visualization of the cells of hematopoietic cell transplants (see Chapter 18) in the intact host in real time. This example illustrates that the full opportunities offered by a genetically tractable small vertebrate are by no means confined to developmental biology.

New directions for research

The zebrafish will probably make most contribution in the following three areas:
1 For research into the organogenesis stages of development there will be further mutagenesis screens that are targeted to particular organs. This may reveal previously unknown genes with key roles in the formation of specific organs.
2 For research into morphogenetic movements, and other applications where cells need to be visualized, the transparency of the embryo will enable high-resolution studies in real time. Individual cell populations can be labeled by transgenic markers.
3 For high-throughput applications on a whole animal model, which will include screening for small molecule compounds active against particular molecular pathways, for drugs effective against particular disease models, and toxicity screens of drugs or other substances.

Key points to remember

• The zebrafish is a vertebrate well suited to genetic experimentation. Embryonic development is rapid, the generation time is short, and large numbers of fish can be kept in a facility.
• Many developmental mutants have been isolated from mutagenesis screens. The genes within which the mutations lie have been identified either by positional cloning or by testing candidates.
• Transparency of the embryos enables visualization of cell behavior *in vivo*.
• In general, the fate map and sequence of inductive steps in early development is similar to *Xenopus*. The Wnt signaling pathway is required for early dorsoventral axis formation, Nodal factors are required for mesoderm induction, and BMP inhibitors for the dorsalizing effects of the organizer. Anteroposterior patterning involves FGFs, Wnts, and retinoic acid.
• The zebrafish has an important role in provision of experimentally convenient
models for aspects of human physiology or disease, and for drug screening and toxicity testing.

Further reading

General

Beis, D. & Stainier, D.Y.R. (2006) In vivo cell biology: following the zebrafish trend. *Trends in Cell Biology* **16**, 105–112.

Skromne, I. & Prince, V.E. (2008) Current perspective in zebrafish reverse genetics: Moving forward. *Developmental Dynamics* **237**, 861–882.

White, R.M., Sessa, A., Burke, C., Bowman, T., LeBlanc, J., Ceol, C., Bourque, C., Dovey, M., Goessling, W., Burns, C.E. & Zon, L.I. (2008) Transparent adult zebrafish as a tool for in vivo transplantation analysis. *Cell Stem Cell* **2**, 183–189.

Lieschke, G.J., Oates, A.C. & Kawakami, K., eds (2009) *Zebrafish: Methods and Protocols. Methods in Molecular Biology*, vol. **546**. New York: Humana Press.

Embryology

Kimmel, C.B., Warga, R.M. & Schilling, T.F. (1990) Origin and organization of the zebrafish fate map. *Development* **108**, 581–594.

Warga, R.M. & Kimmel, C.B. (1990) Cell movements during epiboly and gastrulation in zebrafish. *Development* **108**, 569–580.

Kimmel, C.B., Ballard, W.W., Kimmel, S.R., Ullmann, B. & Schilling, T.F. (1995) Stages of embryonic development of the zebrafish. *Developmental Dynamics* **203**, 253–310.

Solnica-Krezel, L., ed. (2002) *Pattern Formation in Zebrafish.* Berlin/Heidelberg: Springer-Verlag.

Genetics

Patton, E.E. & Zon, L.I. (2001) The art and design of genetic screens: zebrafish. *Nature Reviews Genetics* **2**, 956–966.

Golling, G., Amsterdam, A., Sun, Z.X., Antonelli, M., Maldonado, E., Chen, W.B., Burgess, S., Haldi, M., Artzt, K., Farrington, S., Lin, S.Y., Nissen, R.M. & Hopkins, N. (2002) Insertional mutagenesis in zebrafish rapidly identifies genes essential for early vertebrate development. *Nature Genetics* **31**, 135–140.

Kawakami, K. (2005) Transposon tools and methods in zebrafish. *Developmental Dynamics* **234**, 244–254.

Asakawa, K. & Kawakami, K. (2008) Targeted gene expression by the Gal4-UAS system in zebrafish. *Development Growth and Differentiation* **50**, 391–399.

Halpern, M.E., Rhee, J., Goll, M.G., Akitake, C.M., Parsons, M. & Leach, S.D. (2008) Gal4/UAS transgenic tools and their application to zebrafish. *Zebrafish* **5**, 97–110.

Shoji, W. & Sato-Maeda, M. (2008) Application of heat shock promoter in transgenic zebrafish. *Development Growth and Differentiation* **50**, 401–406.

Abrams, E.W. & Mullins, M.C. (2009) Early zebrafish development: It's in the maternal genes. *Current Opinion in Genetics and Development* **19**, 396–403.

Lawson, N.D. & Wolfe, S.A. (2011) Forward and reverse genetic approaches for the analysis of vertebrate development in the zebrafish. *Developmental Cell* **21**, 48–64.

Inductive interactions

Agathon, A., Thisse, C. & Thisse, B. (2003) The molecular nature of the zebrafish tail organizer. *Nature* **424**, 448–452.

Dougan, S.T., Warga, R.M., Kane, D.A., Schier, A.F. & Talbot, W.S. (2003) The role of the zebrafish nodal-related genes squint and cyclops in patterning of mesendoderm. *Development* **130**, 1837–1851.

Leung, T.C., Bischof, J., Soll, I., Niessing, D., Zhang, D.Y., Ma, J., Jackle, H. & Driever, W. (2003) Bozozok directly represses bmp2b transcription and mediates the earliest dorsoventral asymmetry of bmp2b expression in zebrafish. *Development* **130**, 3639–3649.

Schier, A.F. & Talbot, W.S. (2005) Molecular genetics of axis formation in zebrafish. *Annual Review of Genetics* **39**, 561–613.

Ota, S., Tonou-Fujimori, N. & Yamasu, K. (2009) The roles of the FGF signal in zebrafish embryos analyzed using constitutive activation and dominant-negative suppression of different FGF receptors. *Mechanisms of Development* **126**, 1–17.

Modeling human disease for drug discovery

Lieschke, G.J. & Currie, P.D. (2007) Animal models of human disease: zebrafish swim into view. *Nature Reviews Genetics* **8**, 353–367.

Chakraborty, C., Hsu, C.H., Wen, Z.H., Lin, C.S. & Agoramoorthy, G. (2009) Zebrafish: a complete animal model for in vivo drug discovery and development. *Current Drug Metabolism* **10**, 116–124.

Chapter 9

The chick

The visible course of development of the chick is superficially very different from the lower vertebrates and is much closer to the mammalian type. But even though the visible morphogenetic movements in the early embryo can appear quite different, at a molecular level it is clear that essentially the same basic processes are taking place in all vertebrates. Because the chick has been extensively used for work on later stages of development, a general outline of vertebrate organogenesis is given in this chapter. In Section 3 this will be revisited in greater detail and with incorporation of relevant evidence about mechanisms from other vertebrate species.

Fertilized eggs are usually obtained from commercial hatcheries, and, at the stage of laying, the embryo is a flat **blastoderm** of 20,000–60,000 cells, depending on the exact stage of development, lying on top of the yolk. Development is arrested at low temperatures and therefore eggs can be stored for some days at 10°C, during which time embryonic development remains suspended. It recommences when the eggs are incubated at 37.5°C. Unlike *Xenopus* and the zebrafish, the chick undergoes extensive growth during embryonic life because it has at its disposal the food reserves of the egg.

For the experimentalist the chick has the great advantage over the mouse that the embryo is accessible at all stages following laying of the egg (Fig. 9.1). Early blastoderms can be cultured *in vitro* for long enough to form a recognizable primary body plan. Alternatively, a hole can be cut in the egg shell and the embryo can be manipulated *in ovo*, then the hole is resealed with adhesive tape and the whole egg incubated until the embryo has reached a later stage of development. Furthermore, it is possible to explant small pieces of tissue onto the **chorioallantoic membrane (CAM)** of advanced embryos, where they become vascularized and will grow and differentiate in effective isolation. Culture of some organ rudiments is also possible *in vitro*. For the labeling of grafts, extensive use has been made of interspecies combination between chick and quail. Quail embryos are anatomically very similar to chick although they are slightly smaller and develop a little faster. Originally, they were used because all quail cells possess a condensation of **heterochromatin** associated with their nucleolus, and this is easily visualized as a dark blob by staining for DNA using the Feulgen histochemical reaction. Nowadays, the quail grafts are visualized by staining with a species-specific antibody.

The chick is not well suited to genetic work. The life cycle is long, the existing mutants are limited in number, and there is no routine protocol for transgenesis or targeted mutagenesis. However, there are several useful methods for overexpression of genes in chick embryos. Much use has been made of **retroviruses** carrying the gene in question. These are often called **RCAS viruses** for **r**eplication-**c**ompetent, **a**vian-**s**pecific. The virus can be injected locally in the region to be modified, and the effect will spread as new virus particles are produced by the infected cells. If the range of infection is to be limited, this can be achieved by making a graft of tissue from a sensitive strain of chick into an embryo of a resistant strain. Then just the tissue of the graft will become infected and express the virus-encoded gene. **Electroporation** is also widely used. This involves injecting DNA into the region of interest, usually a cavity such as the neural tube lumen, and then subjecting the embryo to a series of low-voltage electric pulses, which can drive DNA into the cells without doing too much damage. DNA is negatively charged and so moves towards the anode, entering cells as it goes. The tissue on the cathode side of the injection will be untransformed and can serve as an internal control for the effect of the gene. For investigation of embryonic induction, localized treatment of embryos with **growth factors** is done by absorbing the pure factor onto affinity chromatography beads, which are implanted into the embryo in the desired position. Such beads can bind a large amount of the growth factor and then release

Essential Developmental Biology, Third Edition. Jonathan M.W. Slack.
© 2013 John Wiley & Sons, Ltd. Published 2013 by John Wiley & Sons, Ltd.

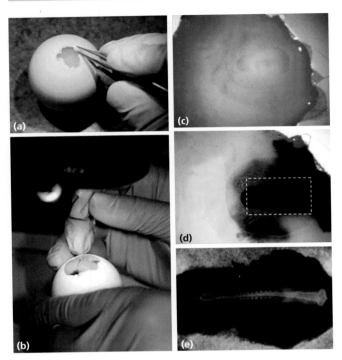

Fig. 9.1 Preparation of the egg for microsurgery. (a) Windowing the egg; (b) injecting ink under the blastoderm to improve contrast; (c) a blastoderm under incident light; (d,e) following ink injection. Reproduced from Hassan Rashidi (2009), with permission from John Wiley & Sons, Ltd.

it slowly for a few days, mimicking to some extent the normal release of an inducing factor from a signaling center. Many experiments of this type have also been conducted by implantation of pellets of mammalian cells expressing the factor in question.

Loss-of-function studies are usually performed by overexpression of dominant negative reagents using the above methods. Antisense morpholinos are less useful than for *Xenopus* or zebrafish, because there are no large blastomeres to inject. They can be introduced by electroporation but results are often not satisfactory.

Normal development

The hen's egg is a familiar object, with its shell, albumen layer ("white"), and yolk. But the true egg consists of just the yolk, and, in a fertilized egg, an inconspicuous **blastoderm** of cells, which is surrounded by a vitelline membrane. The yolk corresponds to the mature **oocyte**. It is generally believed that avian, like mammalian, oogenesis occurs only during fetal life and that the complement of oocytes present in the hatchling represented a lifetime's supply. Up to about 1–2 weeks before ovulation, the hen's oocyte remains quite small, but it then puts on a tremendous growth spurt and over a few days acquires a weight of

about 55 g. On ovulation it is released from the ovary and enters the **oviduct** where, if the hen has recently mated, it will be fertilized. Passage down the oviduct takes about 24 hours, in the course of which the egg is invested successively with the albumen layer, the shell membranes, and the shell itself. It then passes into the vagina and is laid.

Cleavage is highly **meroblastic** and involves just the patch of cytoplasm 2–3 mm in diameter, which is present in the zygote (and oocyte) at the edge of the yolk mass (Fig. 9.2a). The early cleavages take place in the oviduct, producing a circular blastoderm, initially one cell thick and later several cells thick. The cleavage pattern is very variable from one embryo to the next and the blastomeres at the ventral and lateral faces of the sheet remain connected to the yolk for some time by large cytoplasmic bridges. The egg spends about 20 hours in the lower part of the oviduct, called the uterus, undergoing slow rotations driven by uterine peristalsis while the calcareous shell forms around it. When the blastoderm consists of a few hundred cells, a space called the subgerminal cavity opens beneath it (Fig. 9.2b). Cells are shed from the lower surface of the blastoderm into this cavity and probably die so that by the end of the uterine period the central region of the blastoderm has thinned to an organized epithelium one or a few cells thick. Because of its translucent appearance this is known as the **area pellucida**. The outer, more opaque, part of the blastoderm is called the **area opaca** and the junctional region the **marginal zone**. Note that the region called the "marginal zone" of the chick embryo is not a homologous structure to the marginal zone of an amphibian embryo.

A lower layer of cells, the **hypoblast**, then develops, partly by ingression of small groups of cells all over the area pellucida (the primary hypoblast) and partly by spreading of cells from the deep part of the posterior marginal zone (the secondary hypoblast). The two cell populations are now known to be different in gene expression, with the primary hypoblast expressing Goosecoid, Hex, Cerberus, Otx2, and Crescent (a Wnt inhibitor). For this reason the secondary hypoblast is also known as the **endoblast**. The upper layer of cells now becomes known as the **epiblast**. A crescent-shaped cluster of cells beneath the epiblast at the posterior margin is known as **Kollar's sickle** (Fig. 9.2c,d). The primary hypoblast may be homologous to the anterior visceral endoderm of the mouse egg cylinder (see Chapter 10), but both components of the hypoblast contribute only to extraembryonic structures and are not homologous to the hypoblast of the zebrafish, which is part of the embryo itself. At the time the egg is laid it will usually just have commenced endoblast formation, which may be regarded as the beginning of gastrulation. The early developmental stages are described by the stage series of Eyal-Giladi and Kochav, which uses Roman numerals. In this series, I represents the fertilized egg, X represents the single-layered blastoderm stage, and XIII represents the complete two-layered blastoderm stage. The stage of egg laying, after which the embryos become accessible for experimentation, is about stage X–XI. Subsequent development is

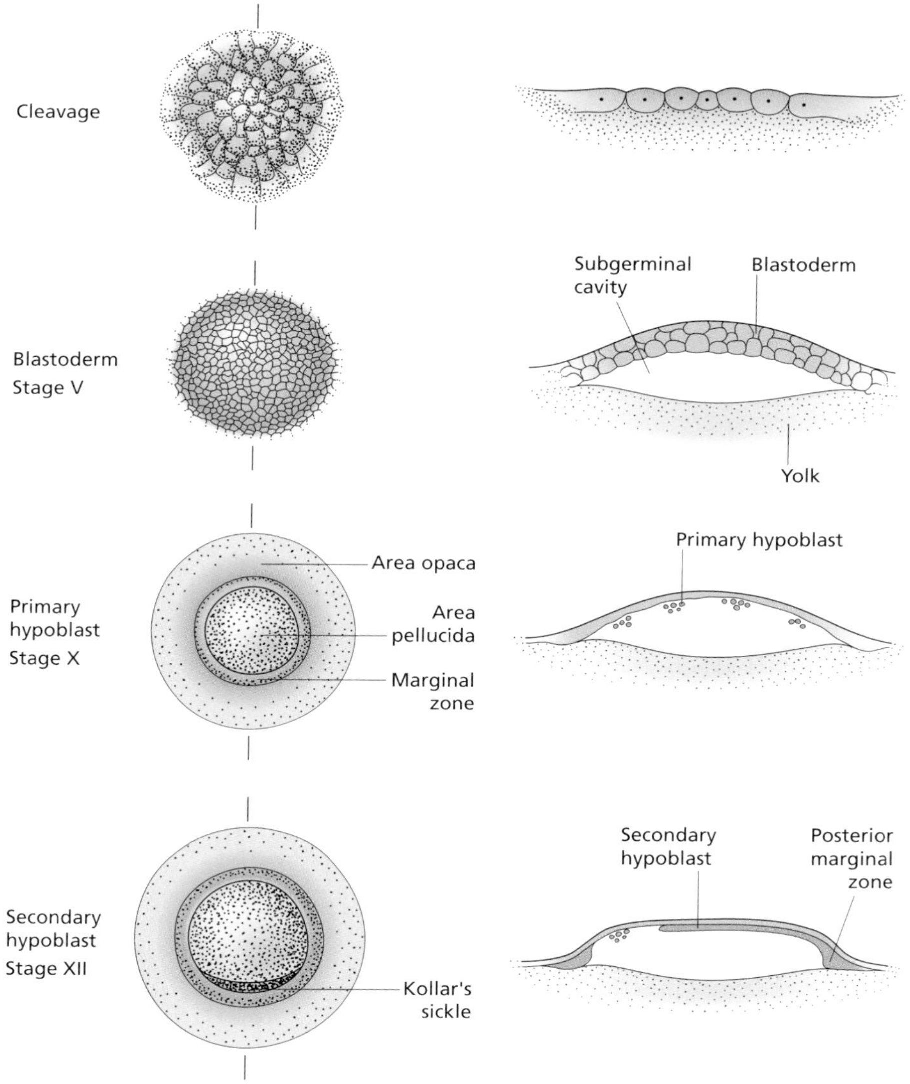

Fig. 9.2 Development of the chick blastoderm up until the time of egg laying. Vertical lines indicate planes of section of figures on right side.

described by the stage series of Hamburger and Hamilton (H & H), which uses Arabic numbers and is indicated in what follows.

The stages of body plan formation are depicted in Fig. 9.3 showing top views, and Fig. 9.4 showing transverse sections. A condensation of cells called the **primitive streak** arises at the posterior edge of the area pellucida (stage 2) and elongates until it reaches the center (Fig 9.3a). Prior to its formation, cells of the epiblast undergo extensive circular movements, tending to bring them to the posterior midline and then anteriorwards. They are known as "polonaise" movements (Fig. 9.5). These movements depend on the Wnt-PCP pathway and can be inhibited by electroporation of a dominant negative form of the dishevelled (*DVL*) gene into the cells. The process of hypoblast formation and gastrulation are shown in Animation 10: Chick gastrulation.

The streak expresses the T-box transcription factor gene brachyury (*T*). It extends to the center of the area pellucida by active stretching, as may be shown by the fact that it will push a bead placed at its anterior end. Although some cells remain resident in the streak, it consists mainly of cells in the process of moving through it, as there is a migration of epiblast cells from both sides which enter and move through to form both the mesoderm and endoderm (Fig. 9.4a). As a result of the entrance of new cells into the midline of the lower layer, the original hypoblast and endoblast cells become pushed to the outer rim. Cells from the epiblast that pass through the streak become the **mesoderm** and the **definitive endoderm** part of the lower layer. This process is called **gastrulation** in the chick, although the future gut lumen, or **archenteron**, is not a new cavity but is the pre-existing space below the endoderm. The

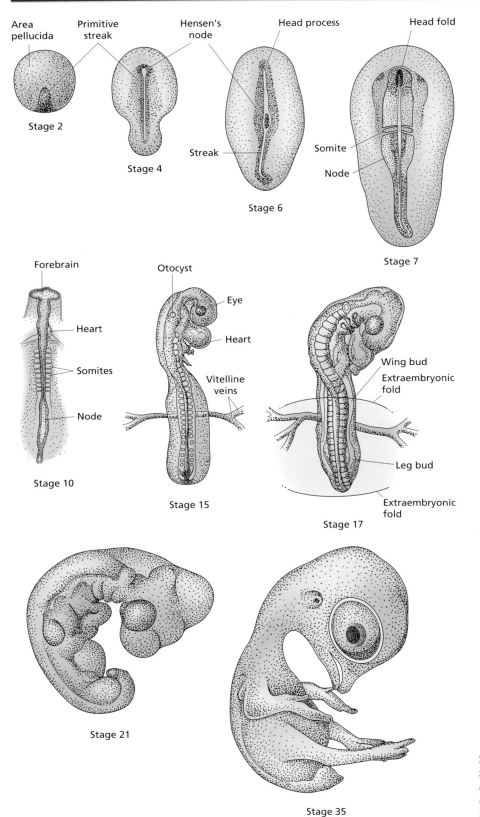

Fig. 9.3 Normal development of the chick. Stage 7 is reached after about 1 day, stage 12 after 2 days, stage 17 after 3 days, stage 24 after 4 days, and stage 35 after 9 days of incubation.

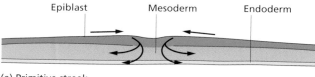

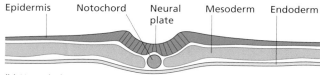

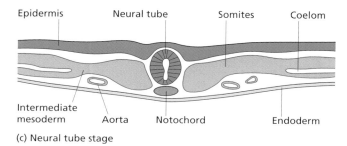

Fig. 9.4 Normal development of the chick. Transverse sections during formation of main axial structures. (a) Stage 4, posterior to node, arrows indicate invagination of cells; (b) mid-body stage 8; (c) mid-body stage 10.

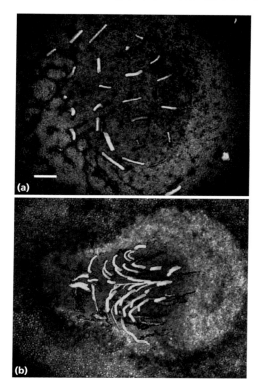

Fig. 9.5 "Polonaise" movements of epiblast cells during primitive streak formation. Posterior is to the left. Small groups of cells were labeled by DiI injection and filmed by time lapse. The colored lines indicate 160 minutes of movement of individual marks, with the green portion being the last 40 minutes. (a) Labeled at stage 1; (b) labeled at stage 3. Reproduced from Cuia *et al. Dev Biol* 2005; 284: 37–47, with permission from Elsevier.

area pellucida gradually changes from a disc to a pear shape and a further condensation, called **Hensen's node**, appears at the anterior end of the primitive streak (stage 4; Fig. 9.3b). This expresses homologs of various transcription factor genes characteristic of the organizer region in the lower vertebrates, including goosecoid (*GSC*), *NOT*, and *FOXA2* (= *HNF3β*). The node contains the presumptive notochord cells, some of which migrate anteriorly to form the head process, or that part of the notochord lying within the head (stage 6; Fig. 9.3c). The remainder of the node moves posteriorly and as it does so the principal structures of the body plan appear in its wake: the **notochord** in the midline, the **somites** on either side of it, and the **neural plate** in the epiblast (Fig. 9.4b). The **primordial germ cells** appear at the extreme anterior edge of the area pellucida, outside the embryo proper.

By about 1 day of incubation the anterior end of the embryo is marked by an uplifting of the blastoderm called the head fold, and one somite and the anterior neural folds have appeared in the track of the regressing node (stage 7; Figs 9.3d, 9.6). From this stage the embryo proper becomes progressively separated from the surrounding extraembryonic tissue. This is achieved by the appearance of folds involving all three germ layers that appear around the embryo and undercut it such that initially

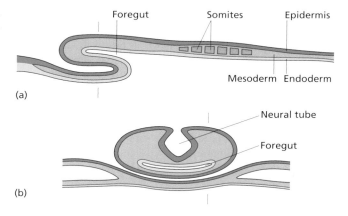

Fig. 9.6 Head fold at stage 8. (a) Parasagittal section; lines indicate plane of section of (b). (b) Transverse section; lines indicate plane of section of (a).

the head, and later the tail and trunk, project above the surface of the extraembryonic tissue (Fig. 9.6).

Early on the second day, blood islands appear in the outer extraembryonic part of the blastoderm, and the heart primordium forms by fusion of the rudiments on the right and left side

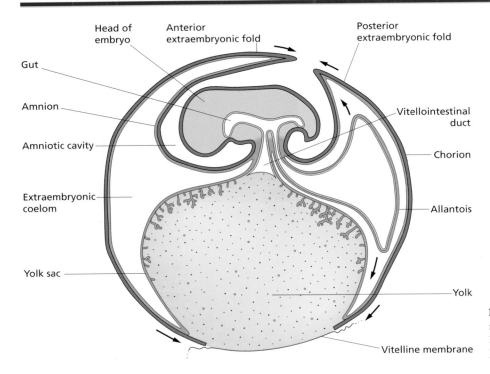

Fig. 9.7 Formation of the extraembryonic membranes in the chick. Reproduced from Hildebrand (1995), with permission from John Wiley & Sons, Ltd.

of the anterior mesoderm (Fig. 9.3e). The heart is able to form in this anterior mid-ventral position because the formation of the anterior body fold has now lifted the head above the level of the surroundings. The formation of the body fold has also caused the foregut to become enclosed as a pocket, while the rest of the presumptive gut endoderm is still a lower layer of cells facing the yolk.

The **neural tube** closes first over the midbrain and then progressively in both directions. **Somites** continue to arise in antero-posterior sequence from the segmental plates of mesoderm which flank the notochord and neural tube. By about 36 hours there are 10 somites and the neural tube has closed to form forebrain, midbrain, and hindbrain vesicles (stage 10). The **lateral plate mesoderm** becomes divided into a **somatic** layer, adhering to the epidermis, and a **splanchnic** layer, adhering to the endoderm. The space in between is the **coelom** (pronounced "see-loam": Fig. 9.4c). A further subdivision of mesoderm appears as a longitudinal strip in between the presomitic mesoderm and the lateral plate. This is the **intermediate mesoderm** that later forms the kidney, adrenals, and gonads. Although node regression and the formation of the posterior part of the body continues for some time, this stage marks approximately the junction between early and late development since the general body plan has been laid down and the formation of individual organs is about to begin.

Extraembryonic membranes

In embryos that have an external food supply the formation and arrangement of the **extraembryonic membranes** is essential to

their survival. Amniotes (reptiles, birds, and mammals) are so called because they all have an **amnion**, and in fact many of the extraembryonic structures are obviously homologous. The arrangement in the chick embryo is shown diagrammatically in Fig. 9.7.

From the time of gastrulation onwards, the outer **area opaca** expands over the surface of the yolk as a membrane consisting of extraembryonic ectoderm and endoderm. The mesoderm has a more restricted spread and is coincident at any one time with the region of extraembryonic vasculature (see below). Initially, the coelom is continuous between the embryonic and extraembryonic regions. The inner extraembryonic layer, which is composed of splanchnic mesoderm, blood-forming tissue, and endoderm, is called the **yolk sac**. This gradually surrounds the entire yolk mass and serves as a digestive organ, the products of digestion of the yolk being absorbed into the blood vessels and conveyed to the embryo. The outer extraembryonic layer consists of somatic mesoderm and ectoderm and is called the **chorion**. During the third day, a fold of the chorion starts to cover the anterior end of the embryo, and a corresponding fold also grows from the posterior. These folds are shown as lines on Fig. 9.3g. The folds fuse in the middle to form two complete membranes covering the embryo, the outer still being called the **chorion** and the inner being called the **amnion**.

The **allantois** consists of a layer of endoderm covered by splanchnic mesoderm. It grows out from the hindgut into the extraembryonic coelom and expands to fuse with the undersurface of the chorion. The allantois serves two functions, one being as an excretory receptacle in which uric acid accumulates, the other as the main respiratory organ of the embryo. The **chorioallantoic membrane**, as it is called after fusion, is located just under the shell and is richly supplied with blood vessels for gaseous exchange. It is often used by experimentalists as a site for *in ovo* culture of isolated organ rudiments taken from the embryo.

Fate map

Only a small proportion of the blastoderm cells become incorporated into the embryo itself, the rest forming extraembryonic structures. The **fate map** of the early embryo has been extensively studied, most recently using small localized marks of **DiI** and **DiO**, which can be visualized at later stages. These studies show that the primitive streak and the node originate from the most posterior part of the area pellucida and from the posterior marginal zone. Like the area pellucida itself, the marginal zone contains two cell layers; the upper layer making a substantial contribution to the ectoderm of the streak, and the lower layer to the endoblast.

Fate maps of early stages are not very precise because of the degree of random cell movement between different individual embryos. However, the prospective region for all of the axial structures lies in the posterior. It should also be noted that the chick blastoderm is conventionally described such that the node is "anterior" and the remainder of the streak "posterior." However, as in *Xenopus* and zebrafish, the projection from early to late stages is not simple and it must be realized that cells that had an anterior position according to the blastoderm description may not be anterior in the later, general body plan, anatomy.

Initially, the prospective head lies in the extreme posterior and the prospective posterior and lateral regions lie more anterior in the blastoderm. But the polonaise movements followed by the lateral invagination of cells through the streak inverts this arrangement until by stage 3, the elongated streak, the prospective head lies anterior in the blastoderm and the prospective posterior and lateral tissues lie posterior (Fig. 9.8).

Labeling experiments conducted at stage 3–6 show that the node, like the rest of the streak, has a continuous flux of cells moving through it. During the phase of node regression, the node itself is the prospective region for the notochord along the entire length of the body (Fig. 9.9a). The neural plate arises from epiblast around the node, with a conservation of anteroposterior levels (Fig. 9.9b), the somites arise from the region just around the node and the lateral plate from more posterior parts of the streak. The origin of the endodermal layer closely follows the mesoderm, with medial parts from the anterior streak and lateral parts from the posterior streak (Fig. 9.9c). It is important to note that the primitive streak does not, as is often thought, map in a one-to-one manner onto the later anteroposterior body plan. Instead, the anteroposterior axis of the streak maps to the mediolateral axis of the later embryo and this means that the posterior half of the streak is destined to become extraembryonic tissues.

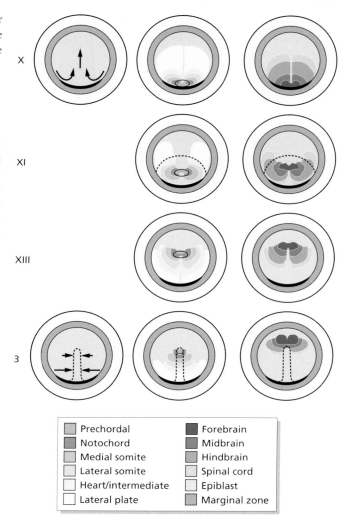

Prechordal	Forebrain
Notochord	Midbrain
Medial somite	Hindbrain
Lateral somite	Spinal cord
Heart/intermediate	Epiblast
Lateral plate	Marginal zone

Fig. 9.8 Fate map of the early chick blastoderm deduced from localized vital dye marks. Mesodermal and neural structures are shown separately. The predominant cell movements are the polonaise movement at the beginning of this period and the ingression through the streak at the end. Reproduced from Stern (2004), with permission from Cold Spring Harbor Laboratory Press.

Regional specification of the early embryo

As long ago as 1828, von Baer propounded a rule that enabled the anteroposterior axis to be predicted in the majority of eggs. This states that if the egg is horizontal with the pointed end to the right then the tail of the embryo should be towards the observer. The rule arises because the egg undergoes a continuous rotation when it is in the uterus and this usually in the same direction relative to the sharp and blunt ends of the egg (Fig. 9.10). The embryo and yolk do not rotate along with the outer surface of the egg but are nevertheless tipped in the direction of rotation and, in fact, a simple tipping of the blastoderm

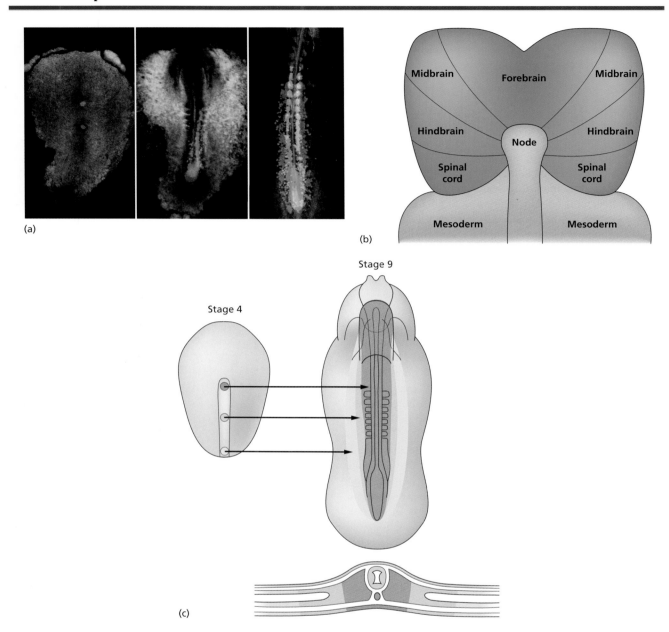

Fig. 9.9 Fate map of the node-stage chick embryo. (a) Marks of DiI (red) and DiO (green) at different streak levels, show a projection from the node to dorsal midline structures and from more posterior streak to lateral mesoderm. Reproduced from Iimura *et al. PNAS* 2007; 104: 2744–2749, with permission from National Academy of Sciences of the United States of America. (b) The prospective region for the central nervous system about stage 4. (c) The origin of mesodermal and endodermal levels from the primitive streak.

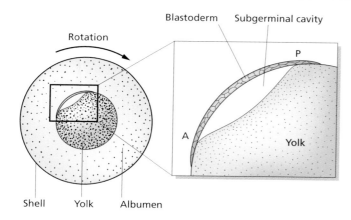

Rotation

Blastoderm   Subgerminal cavity

P

A

Yolk

Shell   Yolk   Albumen

Fig. 9.10 Acquisition of anteroposterior polarity by the chick blastoderm, as a result of intrauterine rotation of the egg. A, anterior; P, posterior.

at the critical period of 14–16 hours of uterine life is enough to establish a polarity such that the posterior, streak-forming, region forms at the uppermost end of the blastoderm. The molecular or cellular mechanisms that bring about this symmetry breaking are not known.

The early blastoderm can be cut into two or three parts, each of which will form a complete embryonic axis. In isolated anterior halves the streak always forms from the posterior edge, and with an 8-hour delay after formation of the streak from the corresponding posterior half. This shows that there is no requirement for a localized cytoplasmic determinant to form the streak, and that some inhibitory signal must suppress formation of multiple streaks in normal development.

Early inductive interactions

The same processes of mesoderm induction, dorsalization, and anteroposterior patterning can be identified in the chick as in *Xenopus* and zebrafish. The genes and gene products involved are generally similar although there are some differences of detail. Because the chick blastoderm is flat, it is more appropriate to speak of "mediolateral" patterning rather than "dorsoventral," and the term "dorsoventral" tends to be used to indicate the arrangement of germ layers after gastrulation (ectoderm dorsal and endoderm ventral).

Mesoderm induction
The primitive streak expresses the T-box transcription factor brachyury (*T*), and its formation corresponds to **mesoderm induction**. Normally, the streak arises from the extreme posterior part of the area pellucida epiblast. If an anterior piece of epiblast is combined with a piece of posterior marginal zone (PMZ), then a new streak can be induced with polarity such

that its posterior end abuts the PMZ (Fig. 9.11). Use of quail grafts into chick hosts demonstrates that this is a true induction with the new streak formed from host tissue. Competence for streak induction by the PMZ is lost by stage XI, as formation of the endoblast is commencing. As in *Xenopus*, the inducing factors responsible are a combination of Vg1, Wnt, Nodal, and FGF. Vg1 is present as a maternal determinant in the *Xenopus* egg, and is partially responsible for mesoderm induction (Chapter 7). In the chick it is expressed in the PMZ region that shows streak-inducing activity, and subsequently in the streak itself (Fig. 9.12). Wnt factors are expressed in the marginal zone, and in particular *WNT8C* RNA is found expressed as a gradient from posterior to anterior. A combination of Vg1 and Wnt1, applied in a cell pellet or adsorbed onto affinity beads, will mimic the action of the PMZ and induce a streak when implanted in the anterior area pellucida. One of the earliest consequences of the action of a PMZ explant, or of Vg1 + Wnt, is to induce expression of *NODAL* in the adjacent area pellucida. This probably represents a relay mechanism that extends the range and intensity of the streak-inducing signal.

The requirement for Wnt and Nodal in normal development is supported by the fact that administration of the Wnt inhibitor FzN8, or the Nodal inhibitor Cerberus short, will both suppress streak formation when given before stage XI. As in *Xenopus*, the role of FGFs in mesoderm induction appears to be permissive. Fgf8 is expressed in the hypoblast. FGF beads can induce a streak, and application of the FGF inhibitor SU5402 will suppress streak formation.

In normal development it is thought that multistreak formation is prevented initially by an inhibition from the hypoblast, and subsequently by an inhibition from the first-formed streak. Removal of the hypoblast often leads to multiple streaks arising from the epiblast, and application of Nodal will only induce a streak if the hypoblast has been removed. In normal development, the displacement of the primary hypoblast, which secretes Wnt and Nodal inhibitors, by the endoblast, which does not, creates the conditions for Nodal to induce a streak adjacent to the Vg1 domain.

Evidence that the first-formed streak inhibits the appearance of further streaks is obtained from Vg1 pellet implantation experiments. An isolated anterior half blastoderm will usually form one streak at the posterior margin. If a pellet releasing Vg1 is grafted laterally then it will induce a streak locally and suppress the host streak. If two pellets are grafted then they will both induce streaks unless the time interval between them exceeds 5 hours, in which case only the first induces a streak. This indicates that an existing streak emits a signal inhibiting the formation of additional streaks in its vicinity.

Organizer effect and anteroposterior patterning
The next inductive interaction is the regionalization of the ectoderm and mesoderm under the influence of Hensen's node, described as dorsalization/neural induction in lower

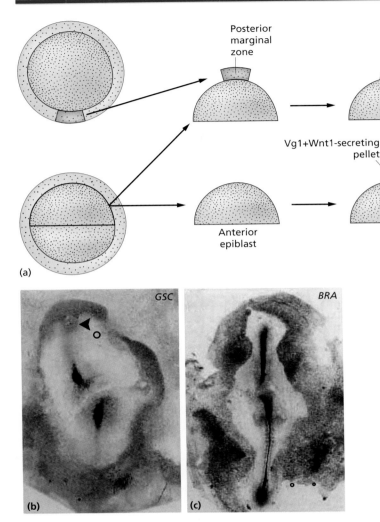

(a)

Posterior marginal zone

Vg1+Wnt1-secreting pellet   Induced streak

Anterior epiblast

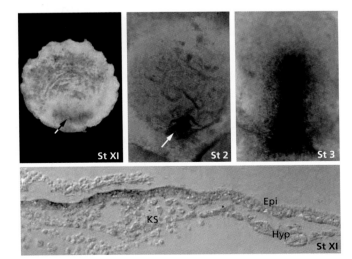

GSC

BRA

(b)   (c)

Fig. 9.11 (a) Induction of a primitive streak by a posterior marginal zone graft, or an implant secreting Vg1 and Wnt. (b,c) Induction of streaks by Vg1 pellet in the marginal zone, which already expresses Wnt. In (b) the graft site is marked by a black arrowhead, the original posterior with carbon particles (white arrowhead), and the streaks are visualized at 24 hours by *in situ* hybridization for goosecoid (*GSC*). The circle near the black arrowhead is a small air bubble. In (c) streaks are visualized at 48 hours by *in situ* hybridization for brachyury (*T*). Parts (b) and (c) reproduced from Shah *et al. Development* 1997; 124: 5127–5138, with permission from Company of Biologists Ltd.

St XI   St 2   St 3

Epi

KS

Hyp

St XI

Fig. 9.12 Expression of *VG1* in the chick blastoderm. Expression commences in the epiblast of the posterior marginal zone (arrows) and later extends into the streak. KS, Kollar's sickle; epi, epiblast; hyp, hypoblast. Reproduced from Shah et al. Development 1997; 124: 5127–5138, with permission from Company of Biologists Ltd.

vertebrates. As mentioned above, it is not appropriate to refer to the organizer effect in the chick as "dorsalization." Because the chick blastoderm is flat, the axis homologous to the dorsoventral axis of *Xenopus* and zebrafish runs, at this stage, from medial to lateral. In the early chick embryo the "dorsoventral axis" is often referred to as the axis running through the blastoderm from the dorsal surface to the subgerminal cavity. This is homologous to the animal–vegetal axis and not the dorsoventral axis of *Xenopus*. After ventral closure, the notochord, somites, and lateral plate do indeed run from dorsal to ventral and the nomenclature becomes the same for the chick and the amphibian.

A node grafted into the area pellucida not too far from the host streak can induce a secondary axis in which the notochord is derived from the graft but the neural tube and somites are derived from the host (Fig. 9.13). This closely resembles the behavior of the *Xenopus* organizer and shows that there is an inductive signal emitted by the node. The activity of the node persists until about H & H stage 8 (four somites), with younger nodes inducing anterior neural structures and older nodes (stage 4–8) inducing posterior structures. The competence of

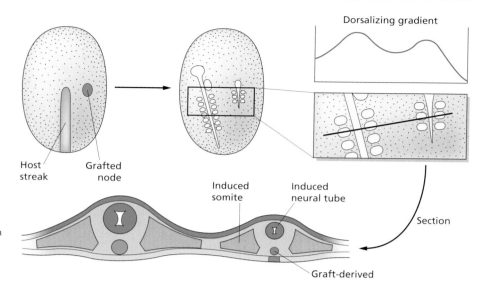

Fig. 9.13 A node grafted to the area pellucida will induce a partial second axis in which the notochord is graft-derived and the neural tube and somites are host-derived.

the epiblast disappears before this, after H & H stage 4 (definitive streak). Several "preneural" markers, including Sox3 and Otx2, appear widely expressed at early stages but this does not constitute a specification of neuroepithelium as epiblast explants do not progress to actual neuroepithelium expressing Sox2 without the signals from the node.

It is tempting to suppose that the organizer effect is due to BMP inhibition, as several BMPs are expressed in the peripheral part of the blastoderm and the BMP inhibitor Chordin is expressed in the node. There is also a reduction of Smad1-phosphate detectable by immunostaining in the prospective neural plate region (Fig. 9.14). However, there has been some controversy over this because, in general, the BMP inhibitors do not have the expected neuralizing activity when applied to explants of epiblast. Probably more than one factor is involved, as it is also known that FGFs have some neuralizing activity, FGF inhibitors can suppress neural induction, and the Wnts are also antagonistic. As far as effects on the mesoderm are concerned, applied BMPs can suppress somite formation in favor of the lateral plate, and application of Noggin will induce ectopic somites, suggesting that mediolateral mesodermal patterning does depend on a graded inhibition of BMP activity, as in *Xenopus*.

Formation of the trunk–tail parts of the anteroposterior pattern is probably due to FGF, as in *Xenopus* and zebrafish. *FGFs* are expressed in the primitive steak, and an FGF-impregnated bead will induce posterior neural tube from the epiblast. FGFs will induce expression of *CDX* genes and these will in turn induce expression of the posterior Hox genes (*HOX6–13*).

A complete axis can still be produced following extirpation of Hensen's node at an early stage. This is a good example of embryonic **regulation** (see Chapter 4). The node produces

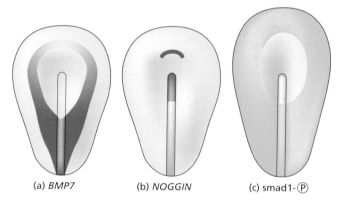

(a) *BMP7* (b) *NOGGIN* (c) smad1-Ⓟ

Fig. 9.14 Expression domains of: (a) a BMP; (b) a BMP inhibitor; (c) visualization of BMP signaling by detection of Smad1-phosphate.

ADMP, which has BMP-like activity and is thought to be responsible for regulation of proportions in *Xenopus* (Chapter 7). The node reforms from the posterior margin of the defect and the process requires the presence of the mid-primitive streak. Once a node has been formed it suppresses formation of other nodes in the vicinity.

Left–right asymmetry

As mentioned in Chapter 2, vertebrates are not exactly bilaterally symmetrical. In the chick, the deviation arises at an early stage with a slight tilt of Hensen's node to the left. This is soon followed by the S shape of the heart tube and the flexion of the whole embryo, with the head lying to the right when viewed from above. When the stomach and intestine develop, they are markedly asymmetrical.

Classic experiments

Left–right asymmetry

Vertebrate embryos are more or less bilaterally symmetrical, and so for many years the nature of the mechanism producing asymmetry was mysterious. In 1995 it was shown that some key genes (*NODAL*, sonic hedgehog (*SHH*), and activin receptor IIa (*ACVR2A*)) have asymmetrical expression patterns in the early chick, and regulated each other to form a pathway linking an initial symmetry-breaking event to the final morphological asymmetry of the heart and viscera.

Levin, M., Johnson, R.L., Stern, C.D., Kuehn, M., & Tabin, C. (1995) A molecular pathway determining left–right asymmetry in chick embryogenesis. *Cell* **82**, 803–814.

Subsequently, it was shown in the mouse that the node bears cilia which, because of their inherent molecular asymmetry, drives fluid preferentially to the left.

Nonaka, S., Tanaka, Y., Okada, Y., Takeda, S., Harada, A., Kanai, Y., Kido, M. & Hirokawa, N. (1998) Randomization of left–right asymmetry due to loss of nodal cilia generating leftward flow of extraembryonic fluid in mice lacking KIF3B motor protein. *Cell* **95**, 829–837.

While primary cilia are believed to cause left–right asymmetry in mouse, zebrafish, and *Xenopus*, it is not yet proved that they do so in the chick.

Several of the gene products involved in this process have been identified in the chick because of their asymmetrical expression patterns, and the sequence of events has been worked out from experiments in which these factors are applied locally, either as RCAS virus or as protein on affinity beads, and the effects on expression of other factors is observed. The process involves four steps. Firstly, a breakage of the basic bilateral symmetry of the embryo in the node or midline structures. Secondly, an amplification of the initial asymmetry to create different regimes of gene expression on either side of the midline. Thirdly, a spread of the information out to the lateral mesoderm, which is the tissue layer most involved in the formation of the asymmetrical organs; and, finally, the control of the events of cell adhesion and movement that actually bring about the asymmetrical morphology.

In the chick the original symmetry-breaking event is still unclear, although in the mouse, *Xenopus*, and zebrafish there is good evidence that it depends on the intrinsically asymmetrical structure of primary cilia. These cilia are found in the node of the mouse (see Chapter 10), Kupffer's vesicle of the zebrafish, and the roof plate of the archenteron in *Xenopus*. In each case the cilia generate a leftwards flow of fluid that initiates the cascade of events, and mutations or inhibitors that prevent ciliary action prevent the development of left–right asymmetry. In the chick there is currently no evidence for an involvement of cilia and the initial event seems to be a depolarization of membrane potential on the left side of the streak, mediated by an asymmetrical activity of the H+/K+-ATPase. This leads to a movement of cells such that sonic hedgehog (*SHH*)-expressing cells become more abundant to the left of the node and *FGF8*-expressing cells on the right.

The key player among the asymmetrically expressed genes is believed to be *NODAL*, since it is preferentially expressed on the left side in all types of vertebrate, and application of Nodal to the right side will randomize the asymmetry of multiple organ systems. The components upstream and downstream of *NODAL* do, however, vary considerably between vertebrate species. In the chick, Nodal appears in a small domain on the left of the regressing node at about H&H stage 6 and then spreads to a much wider domain in the left lateral plate mesoderm. Its expression on the left side is preceded by *SHH*, and on the right by *ACTIVINβB*, *FGF4*, and *FGF8* (Fig. 9.15). Sonic hedgehog will upregulate *NODAL* expression when applied to the right side and Activin or FGFs will repress *NODAL* if applied to the left side, indicating that they are upstream regulatory components.

Spread of the *NODAL* zone to the lateral plate mesoderm seems to involve partly an autocatalytic loop whereby Nodal signaling upregulates *NODAL* transcription, and partly a Cerberus-like BMP inhibitor called Caronte whose transcription is upregulated by Nodal. BMPs are expressed on both sides of the embryo and can suppress *NODAL* expression. But Caronte suppresses BMP signaling on the left and thereby allows the spread of *NODAL* expression on this side. The limitation of spread is controlled by another TGFβ-like factor called Lefty, which becomes expressed on either side of the *NODAL* domain. The Lefty protein is an inhibitor of Nodal and reduces its signaling activity. The end product of the gene cascade is expression of the homeodomain transcription factor Pitx2 on the left side. This is controlled by Nodal signaling through transcriptional repression of Snail-related, a zinc finger transcription factor, which itself represses *PITX2* transcription. The sequence of events is shown in Fig. 9.15 and in Animation 11: Left–right asymmetry.

Although some steps of this mechanism are not be shared by other vertebrates, *NODAL* and *PITX2* are always expressed on the left and seem, respectively, to represent the principal signaling step and final controller of cell differentiation.

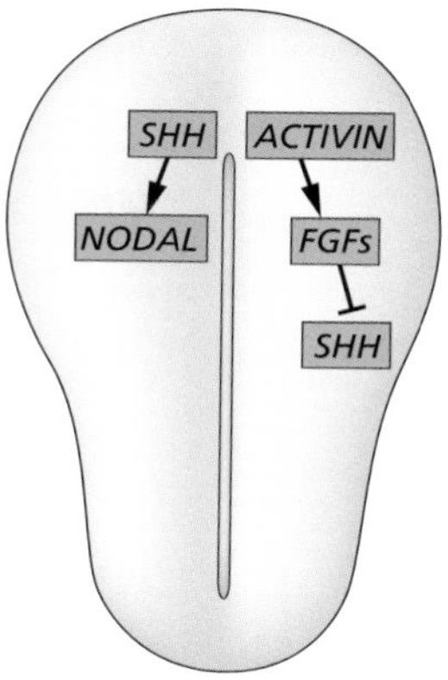

Fig. 9.15 Development of left–right asymmetry in the chick embryo.

	L	R
Activin βB	–	+
FGFs	–	+
Shh	+	–
Nodal	+	–

Early events

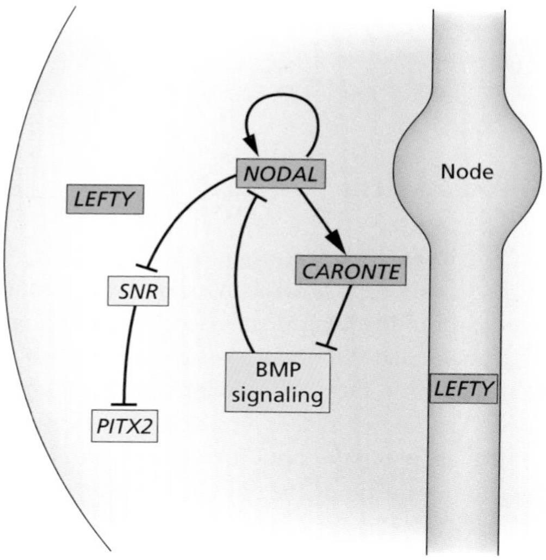

Later events

Description of organogenesis in the chick

The later stages of chick embryo development have been very well known for a long time and for this reason serve as the basic resource for our knowledge of vertebrate organogenesis in general. As the experimental work on several organ systems is described in more detail in later chapters, this section contains just a brief summary of some of the major morphological events.

Whole embryo

By the second day the heart is bent to the right and the **optic vesicles** appear from the forebrain (see Fig. 9.3e). The head turns to face the right and at this time (H&H stage 13) an anterior extraembryonic fold rises to cover the head. This fold moves progressively posteriorly and will later become the chorion and amnion (see Fig. 9.7). The head becomes sharply flexed between the region of the forebrain and the hindbrain. The first three **pharyngeal pouches** appear, the optic vesicles invaginate and the lenses of the eyes appear in the adjacent epidermis. On day 3 the limb buds appear in the lateral plate mesoderm (stage 17) and after a further day they have become as long as they are broad. A posterior extraembryonic fold appears and moves anteriorly, eventually meeting the anterior fold. Shortly after the appearance of the limb buds these folds

fuse to enclose the embryo within an amniotic cavity (see Fig. 9.7). Eye pigmentation appears from about 3.5 days.

By the third day the original head fold has deepened into an anterior body fold, such that the anterior half of the body has become elevated above the surroundings (see Figs 9.6, 9.7). This results in the formation of a closed tube of **foregut** running anteriorly from an **anterior intestinal portal** connecting it to the subendodermal space. Over the fourth day a corresponding posterior body fold lifts the rear part of the embryo and results in the formation of a **hindgut**. The residual ventral opening of the gut becomes progressively smaller and eventually becomes narrowed to a vitellointestinal duct joining the yolk mass to the midgut (see Fig. 9.7). The **allantois** arises from the hindgut and expands rapidly into the space between chorion and amnion.

A mouth is formed at the anterior end of the foregut. The face is formed by the fusion of a set of paired processes: above the mouth are the frontonasal and maxillary processes and below the mouth the mandibular processes. Each pair fuses in the midline to make up the face. A beak appears from about 5.5 days, the upper part from the maxillary, and the lower part from the mandibular process. The outer **epidermis** of the embryo is often still called **ectoderm** for several days because of its undifferentiated appearance, but this is a misnomer as it is no longer able to form neural tissues after the primitive streak stage. In the chick the feather germs start to appear in the epidermis from about 6 days.

From 3 to 4 days the posterior extremity of the embryo consists of a **tailbud**. This consists of a juxtaposition of the various axial tissue types: the notochord, neural tube, somites, and hindgut. In the chick it is only responsible for producing the most posterior four to five somites, although in other vertebrates it can produce many more.

Hatching of the chick occurs on the 20th or 21st day.

Nervous system

The early brain is shown in Fig. 9.16. The three primary brain vesicles, visible from the second day, are the **forebrain**, **midbrain**, and **hindbrain**. The anterior part of the forebrain is the telencephalon, later forming the cerebral hemispheres, and the posterior part is the diencephalon, which produces the optic vesicles. The midbrain later forms the **optic tecta**, which are the receptive areas for the optic nerves. The hindbrain forms the cerebellum, controlling the body's movements, and the medulla oblongata, site of control centers for various vital functions. The remainder of the neural tube forms the spinal cord. Ten pairs of cranial nerves leave the brain to innervate various muscles and sense organs, and the spinal cord produces pairs of spinal nerves in between the vertebrae.

The dorsal part of the neural tube gives rise to a migratory population of cells called the **neural crest**. In the head this contributes to the cranial nerve ganglia and a large proportion of the skeleton of the skull. In the trunk it forms the spinal ganglia, autonomic ganglia, adrenal medulla, and pigment cells.

Further details of nervous system and neural crest development will be found in Chapter 14.

Pharyngeal arch region

At the level of the hindbrain the body has an obviously segmental arrangement (Fig. 9.17). This is made up of elements from different germ layers. In the endoderm of the pharynx, paired lateral **pharyngeal pouches** develop. These are the famous "gill slits," or branchial clefts, which all vertebrate embryo possess in rudimentary form, although in the chick they do not become fully patent. The hindbrain itself is divided into seven **rhombomeres**. Rhombomere 1 becomes the cerebellum. Each pair of rhombomeres 2–7 produce the neural crest cells that migrate to form one cartilaginous **branchial arch** surrounding the pharynx and separating the clefts. The first of these, lying anterior to the first cleft, is the mandibular arch, which subsequently becomes the mandibular and maxillary processes which form the lower half of the face. The second arch is called the hyoid arch. Each pair of rhombomeres also produces one cranial nerve running into its associated arch. The cranial ganglia are composed partly of cells from the neural crest and partly of the corresponding epibranchial placodes, which form in the adjacent epidermis. Each pharyngeal arch is associated with a vascular aortic arch connecting the ventral aorta from the heart with the paired dorsal aortas (see below). In the endoderm the thyroid forms

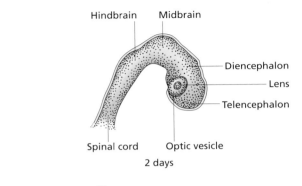

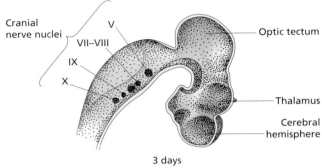

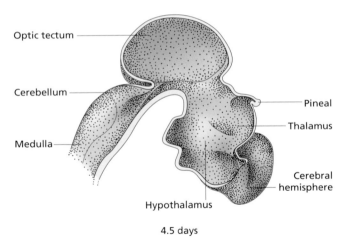

Fig. 9.16 Change in shape of the chick embryo brain from 2 to 4.5 days of development.

in the ventral midline from tissue of the second pouch, and paired thymus and parathyroid rudiments form from the third and fourth pouches. This whole segmental arrangement is transitory, so the clefts are only open for a short time and the aortic arches are not all functional simultaneously.

Heart and circulation

The heart originates at the hindbrain level from paired endothelial condensations between the splanchnic mesoderm and the

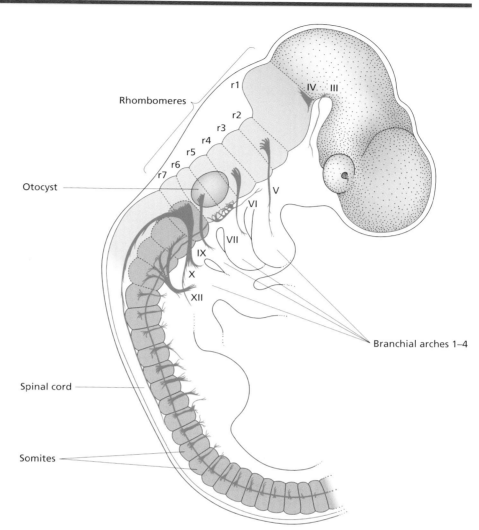

Fig. 9.17 Pharyngeal arch region of a 3-day chick embryo. The rhombomeres are labeled r1–r7 and the associated cranial nerves V–XII. Each pair of rhombomeres innervate one arch. The otocyst is at the level of r5–6. Reproduced from Lumsden. *Phil Trans R Soc* (Lond.) B 1991; 331: 281–286, with permission from The Royal Society.

gut. These initially form separate tubes and then early on the second day, from about seven somites, they fuse in the ventral midline to form a single tube of **endocardium**. The **myocardium**, or muscular wall of the heart, originates from the adjacent splanchnic mesoderm. As the fusion is taking place, the heart moves posteriorly, following the progress of the anterior body fold. By 48 hours the heart is a coiled tube consisting of, from posterior to anterior, sinus venosus, atrium, ventricle, and outflow tract.

During the early stages of heart formation a system of blood islands and blood vessels appears in the area opaca. The vessels grow into the area pellucida and join up with vessels arising from the mesoderm within the embryo. The heart starts to beat about the middle of the second day, and by the third day establishes a blood circulation between the embryo and the yolk mass. From the heart, blood flows into the short ventral aorta, through the aortic arches, initially one for each of the first three

pharyngeal arches, into the dorsal aortas (Fig. 9.18). These are initially paired but become progressively united, from anterior to posterior, into a single aorta. From these, blood flows to all parts of the embryo and also out of the embryo through paired vitelline arteries located at the level of the trunk. It then becomes oxygenated in the extraembryonic capillary bed and returns via the vitelline veins, which approach the embryo from both anterior and posterior directions, and, with the embryonic venous system, join up at the sinus venosus. As the body folds reduce the connection between the embryo and the yolk mass, the vitelline arteries and veins are moved together into the umbilical cord. Later on, from the sixth day, most of the blood flow becomes directed through the **allantois** (see Fig. 9.7), as this becomes the principal respiratory organ.

Further details about the heart, blood, and vascular system will be found in Chapter 15.

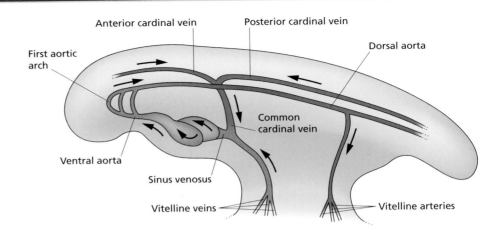

Fig. 9.18 The basic circulation of an amniote embryo.

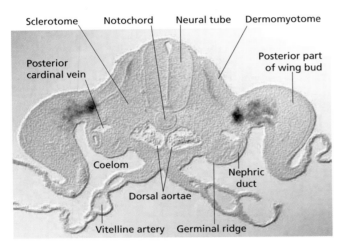

Fig. 9.19 Transverse section through trunk region of 3-day chick embryo. The blue color is the product of *in situ* hybridization for mRNA of the transcription factor *LBX1*, and indicates myoblasts migrating into the limb buds. Reproduced from Bryson-Richardson and Currie. *Nat Rev Genet* 2008; 9: 632–646, with permission from Nature Publishing Group.

Trunk mesoderm

The disposition of mesodermal structures is shown in the section of Fig. 9.19, taken through the forelimb buds at about 3 days of incubation. **Somites** arise from the mesoderm on either side of the notochord, called the **paraxial mesoderm**. From the end of day 1 the somites appear in anterior-to-posterior sequence. The visible event of segmentation corresponds to a transition of each somite from a mesenchymal morphology to epithelial spheres of tightly apposed cells (Fig. 9.20a,b). Somites continue to form until there are 45 by 4 days. They will later form three types of structure (Fig. 9.20c). The inner part, flanking the notochord, is called the **sclerotome** and later becomes the vertebrae. The formation of vertebrae is out of phase with the somites, so each vertebra is formed from the *posterior* sclerotome of one somite and the *anterior* sclerotome of the next, in a process known as "resegmentation." The outer part of the somite is the dermomyotome, which contributes cells both to the dermis of the skin and to the segmental muscles of the body (**myotomes**). In the chick, the most anterior two somites disperse shortly after their formation and the next four are occipital somites, which contribute to the occipital part of the skull rather than forming vertebrae.

The kidney develops from the intermediate mesoderm (Fig. 9.21). First the **pronephros** develops on day 2–3 at the level of the seventh to 15th somite. A nephric duct grows posteriorly from the pronephric area down to the cloaca. In fish and amphibians the pronephros is functional, but in amniotes it soon degenerates as a **mesonephros** appears from intermediate mesoderm at the level of the 16th to 27th somite. In the chick this develops on the third and fourth day and consists of glomeruli and tubules that attach to the nephric duct. The mesonephros is the functional kidney in birds during embryonic life. It is superseded after hatching by the **metanephros**. This develops from about 5–8 days from the posterior end of the intermediate mesoderm. A branch from the nephric duct, the ureteric bud, grows into the neighboring mesoderm and provokes the formation of the metanephric tubules in a mesenchymal to epithelial transition.

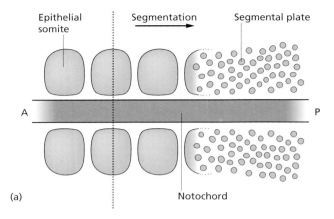

Epithelial somite   Segmentation →   Segmental plate

A   P

(a)   Notochord

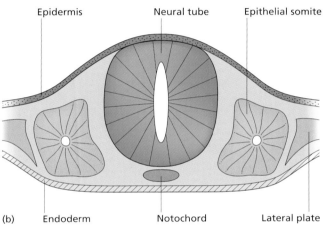

Epidermis   Neural tube   Epithelial somite

(b)   Endoderm   Notochord   Lateral plate

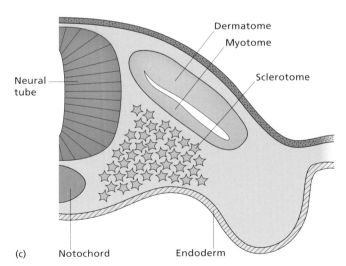

Dermatome

Myotome

Sclerotome

Neural tube

(c)   Notochord   Endoderm

Fig. 9.20 Somitogenesis. (a) Somites are formed sequentially from anterior (A) to posterior (P). (b) Transverse section through recently formed somite of (a), showing epithelial form. (c) Later formation of dermatome, myotome, and sclerotome from the somite.

New directions for research

The early chick offers a good opportunity to study the still somewhat mysterious property of embryonic regulation. In most embryo types, regulation requires the persistence of the signaling centers but in the chick the main signaling center, Hensen's node, can itself be replaced following extirpation.

The later chick embryo will continue to provide material for research on organogenesis because of the ease of micromanipulation. Research programs now often combine microsurgical experiments on the chick with the use of tissues from knockout mouse lines.

The strip of intermediate mesoderm ventromedial to the kidney forms the adrenal gland and the **gonads**. The adrenal gradually becomes a compact body over days 4–8, with the outer cortex formed from intermediate mesoderm and the inner medulla formed from the neural crest. The gonads arise both from the mesenchyme and from the overlying coelomic epithelium, which together form a protrusion into the coelom called the **genital ridge**. The **primordial germ cells** arise from an anterior extraembryonic position (Fig. 9.22) and enter the germinal ridge after a long migration. On the fourth day the gonads are still of similar appearance in males and females, but subsequently they differentiate as testes or ovaries, respectively.

The limbs appear on day 3 as buds formed from the lateral plate mesoderm and the overlying epidermis. They elongate, and from the fourth day start to differentiate in proximal to distal sequence. In birds the forelimb buds become the wings and the hindlimb buds the legs.

Further details about development of the principal mesodermal organs will be found in Chapter 15.

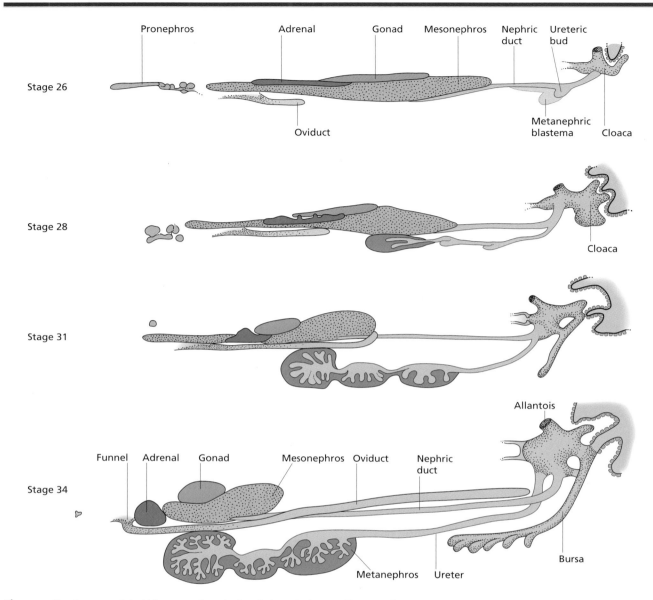

Fig. 9.21 Development of the kidney, gonads, and adrenals from the intermediate mesoderm. Reproduced from Witschi (1956) *Development of Vertebrates*. Philadelphia: Saunders, with permission from Elsevier.

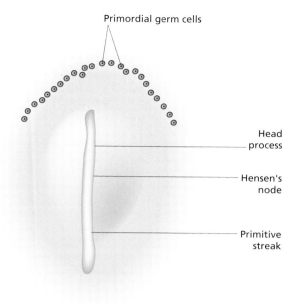

Primordial germ cells

Head process

Hensen's node

Primitive streak

Fig. 9.22 Extraembryonic position of the primordial germ cells.

Key points to remember

- The chick is an amniote with a generally similar morphology to mammalian embryos.
- Early development occurs as a flat blastoderm. The primitive streak is induced at the posterior margin and elongates to the anterior. During gastrulation cells pass through the streak and become the major body parts of the embryo. Hensen's node forms at the anterior tip of the streak and regresses to the posterior with the body pattern appearing behind it.
- The chick possesses a set of extraembryonic membranes needed to support the embryo and to transfer nutrients and oxygen to the embryo. These are the yolk sac, the chorion, the amnion, and the allantois.
- The fate map of the primitive streak shows that the anteroposterior axis of the streak becomes the mediolateral axis of the primary body plan.
- The sequence of inductive interactions is comparable to *Xenopus*. The induction of the primitive streak corresponds to mesoderm induction in *Xenopus* and the behavior of Hensen's node is similar to the *Xenopus* organizer. However, some of the molecular components underlying these processes may not be identical.
- The formation of left–right asymmetry of the embryo depends on the expression of Nodal on the left side of the axis.
- The chick is very important for research on organogenesis (see later chapters).

Further reading

General

Hamburger, V. & Hamilton, H.L. (1951) A series of normal stages in the development of the chick embryo. *Journal of Morphology* **88**, 49–92. *Reprinted Developmental Dynamics* **195**, 231–272 (1992).

Eyal-Giladi, H. & Kochav, S. (1976) From cleavage to primitive streak formation: a complementary normal table and a new look at the first stages of the development of the chick. *Developmental Biology* **49**, 321–337.

Bellairs, R. & Osmund, M. (1997) *The Atlas of Chick Development.* London: Academic Press.

Stern, C., ed. (2004) The chick in developmental biology. *Mechanisms of Development* **121** (special issue 9).

Claudio, D.S. (2005) The chick: a great model system becomes even greater. *Developmental Cell* **8**, 9–17.

Hassan Rashidi, V.S. (2009) The chick embryo: hatching a model for contemporary biomedical research. *BioEssays* **31**, 459–465.

Fate maps

Psychoyos, D. & Stern, C.D. (1996) Fates and migratory routes of primitive streak cells in the chick embryo. *Development* **122**, 1523–1534.

Fernández-Garre, P., Rodríguez-Gallardo, L., Gallego-Diaz, V., Alvarez, I.S. & Puelles, L. (2002) Fate map of the chicken neural plate at stage 4. *Development* **129**, 2807–2822.

Lawson, A. & Schoenwolf, G.C. (2003) Epiblast and primitive streak origins of the endoderm in the gastrulating chick embryo. *Development* **130**, 3491–3501.

Early development

Lemaire, L. & Kessel, M. (1997) Gastrulation and homeobox genes in chick embryos. *Mechanisms of Development* **67**, 3–16.

Bachvarova, R.F., Skromne, I. & Stern, C.D. (1998) Induction of primitive streak and Hensen's node by the posterior marginal zone in the early chick embryo. *Development* **125**, 3521–3534.

Skromne, I. & Stern, C.D. (2001) Interactions between Wnt and Vg1 signalling pathways initiate primitive streak formation in the chick embryo. *Development* **128**, 2915–2927.

Bertocchini, F. & Stern, C.D. (2002) The hypoblast of the chick embryo positions the primitive streak by antagonizing nodal signaling. *Developmental Cell* **3**, 735–744.

Faure, S., de Santa Barbera, P., Roberts, D.J. & Whitman, M. (2002) Endogenous patterns of BMP signaling during early chick development. *Developmental Biology* **244**, 44–65.

Stern, C.D. (2004) Gastrulation in the chick. In: *Gastrulation: From Cells to Embryo.* New York: Cold Spring Harbor Press.

Stern, C.D. (2005) Neural induction: old problem, new findings, yet more questions. *Development* **132**, 2007–2021.

Stavridis, M.P., Lunn, J.S., Collins, B.J. & Storey, K.G. (2007) A discrete period of FGF-induced Erk1/2 signalling is required for vertebrate neural specification. *Development* **134**, 2889–2894.

Left-right asymmetry

Levin, M. (1998) Left–right asymmetry and the chick embryo. *Seminars in Cell and Developmental Biology* **9**, 67–76.

Capdevila, J., Vogan, K.J., Tabin, C.J. & Izpisua-Belmonte, J.C. (2000) Mechanisms of left-right determination in vertebrates. *Cell* **101**, 9–21.

Mercola, M. & Levin, M. (2001) Left-right asymmetry determination in vertebrates. *Annual Reviews in Cell and Developmental Biology* **17**, 779–805.

Gros, J., Feistel, K., Viebahn, C., Blum, M. & Tabin, C.J. (2009) Cell movements at Hensen's node establish left/right asymmetric gene expression in the chick. *Science* **324**, 941–944.

Organogenesis, general

Witschi, E. (1956) *Development of Vertebrates.* Philadelphia: W.B. Saunders.

Balinsky, B.I. & Fabian, B.C. (1981) *An Introduction to Embryology*, 5th edn. Philadelphia: Saunders.

Hildebrand, M. (2001) *Analysis of Vertebrate Structure*, 5th edn. New York: John Wiley & Sons.

Kardong, K.V. (2002) *Vertebrates: Comparative Anatomy, Function, Evolution*, 3rd edn. New York: McGraw-Hill.

Also see references in Chapters 1, 14, 15, and 16.

This chapter contains the following animations:

Animation 10 Chick gastrulation.
Animation 11 Left–right asymmetry.

 For additional resources for this book visit www.essentialdevelopmentalbiology.com

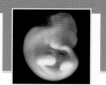
The mouse

Unlike the other model organisms considered here, mammals are **viviparous** and the resulting inaccessibility of the postimplantation stages of development means that microsurgical procedures are used less than in *Xenopus*, zebrafish, or chick. For this reason, the developmental biology of the mouse has depended to a greater extent on genetic manipulation. There are many laboratory strains of mice that have each been inbred to homozygosity at all loci and each have their own advantages and disadvantages for different types of experiment. Each strain has a characteristic coat color (e.g. black, albino, agouti), and in experiments involving embryos of more than one strain these differences can serve as a quick visual indication of the genetic constitution. Gene symbols have an initial small letter if they relate to recessive mutants. Unlike the lower vertebrates, but similar to the human, it is conventional to spell the protein products of mouse genes in capitals, for example "NODAL."

Mice mate in the night and so the age of the embryos is often expressed as days and a half, for example a 7.5-day embryo, designated E7.5, is recovered on the eighth day after the mice mated. A solid white deposit, or "plug," that is formed in the female vagina after mating, allows determination of the mating night. Although "E" designations are the usual way to indicate stage there is also a numerical stage series by Theiler, in which stages 1–5 are preimplantation, 6–14 are early postimplantation, comprising body plan formation and turning, and 15–27 comprise organogenesis and fetal growth up to birth, which occurs at about 20 days after fertilization.

Ovulation occurs a few hours after mating and fertilization takes place at the upper end of the oviduct. For the early, preimplantation, stages the embryos are located first in the oviduct and then the uterus of the mother. During this time they can be collected and kept *in vitro* in simple media. It is possible to do microsurgical manipulation of these early stages, including the microinjections required to make transgenics and knock-outs. In order to turn a modified preimplantation embryo into a late-stage embryo or into an adult mouse, it must be transplanted into the uterus of a **foster mother** that has been made **pseudopregnant**, and thus receptive to the embryos, by previous mating with a vasectomized, and therefore sterile, male. So long as the transplanted embryos implant, they should continue to develop and be born in the normal way.

Because of their viviparity, postimplantation mammalian embryos have a considerable external nutrient supply and undergo extensive growth during development. In this respect the mouse resembles the chick and differs from *Xenopus* and the zebrafish. At late stages it is possible, as in the chick, to explant individual organ or tissue rudiments from embryos into *in vitro* culture, where they are accessible to manual intervention. The scope of these experiments is greatly increased in the mouse by the ability to use tissues from transgenic and knockout strains.

Mammalian fertilization

From a biological standpoint, the fertilization mechanism needs to bring the male and female gametes together in a productive union, while avoiding both cross-species fertilization (**hybridization**) and fertilization of the egg by multiple sperm (**polyspermy**). Fertilization is also a topic of great practical importance in human reproduction. The mechanisms have been studied in a variety of animal models, including several marine invertebrates such as the sea urchin. However, the molecular mechanisms of fertilization are more diverse than those of many other developmental processes, and it has turned out that there are rather few features in common between the sea urchin and mammals. So in this case the value of invertebrate models is somewhat reduced. To maintain coherence, the

Essential Developmental Biology, Third Edition. Jonathan M.W. Slack.
© 2013 John Wiley & Sons, Ltd. Published 2013 by John Wiley & Sons, Ltd.

Classic experiments

The main contribution of the mouse to developmental biology has been through transgenic and knockout technology. These techniques are now fundamental to virtually every investigation of early development or organogenesis. They have also been used to create many mouse models for human diseases, which are used to investigate both the molecular mechanisms of the disease and to test potential therapies.

First transgenesis

Gordon, J.W., Scangos, G.A., Plotkin, D.J., Barbosa, J.A. & Ruddle, F.H. (1981) Genetic-transformation of mouse embryos by micro-injection of purified DNA. *Proceeding of the National Academy of Sciences USA* **77**, 7380–7384.

Discovery of ES cells

Evans, M.J. & Kaufman, M.H. (1981) Establishment in culture of pluripotential cells from mouse embryos. *Nature* **292**, 154–156.

Martin, G.R. (1981) Isolation of a pluripotent cell line from early mouse embryos cultured in a medium conditioned by teratocarcinoma cells. *Proceeding of the National Academy of Sciences USA* **78**, 7634–7638.

Invention of basic knockout procedure

Thomas, K.R. & Capecchi, M.R. (1987) Site directed mutagenesis by gene targeting in mouse embryo-derived stem cells. *Cell* **51**, 503–512.

Mansour, S.L., Thomas, K.R. & Capecchi, M.R. (1988) Disruption of the protooncogene int-2 in mouse embryo-derived stem cells: a general strategy for targeting mutations to non-selectable genes. *Nature* **336**, 348–352.

following account relates just to the mouse (Fig. 10.1; and also see Animation 12: Mammalian fertilization) although most of the features of mouse fertilization are found also in other mammals.

The sperm is a highly specialized cell (Fig. 10.2). The nucleus is haploid and the DNA is highly condensed, with very basic proteins called protamines making up much of the protein content of the chromatin. In front of the nucleus lies a large Golgi-like body called the **acrosome**. Behind the nucleus lies a **centriole**, a mid-piece rich in mitochondria, and the tail, which is a reinforced flagellum having the usual 9 + 2 arrangement of microtubules. The swimming movements of the sperm are driven in an ATP-dependent process by dynein arms, which are attached to the microtubules.

In most mammals the sperm are not capable of fertilization immediately after release. They need to spend a period in the female reproductive tract during which they become competent to fertilize, a process known as **capacitation**. This can be brought about *in vitro* in simple synthetic media containing albumin, calcium, and bicarbonate. One element of capacitation is the loss of cholesterol, and a medium rich in cholesterol will inhibit the process. It is thought that the loss of cholesterol makes the membrane permeable to the Ca^{2+} and HCO_3^-, which can directly activate an adenylyl cyclase, resulting in the production of cAMP and activation of protein kinase A. This has various consequences including protein tyrosine phosphorylation, increase of membrane potential from about −30 to −50 mV, increase in intracellular pH and Ca^{2+}, and increased motility. Capacitation also involves the loss of glycoproteins that prevent the sperm–zona interaction, and the display of some acrosomal proteins at the cell surface.

The "**egg**" of most mammals, including the mouse and the human, is strictly speaking an **oocyte** arrested in second meiotic

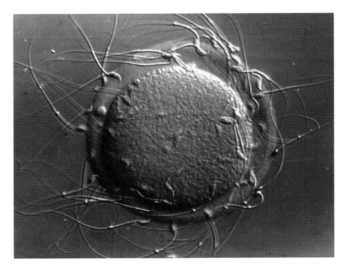

Fig. 10.1 Mouse sperm binding to the zona pellucida of an unfertilized mouse egg. Reproduced from Wassarman *et al.* (2001), with permisison from Nature Publishing Group.

metaphase. It is released from the ovary as a complex with ovarian follicle cells (**cumulus** cells). The oocyte itself is surrounded by a transparent layer of extracellular material called the **zona pellucida**, secreted by the follicle cells. Outside this lie some cumulus cells, which are embedded in an extracellular matrix rich in hyaluronic acid. The oocyte–cumulus complex is picked up by the funnel (infundibulum) at the entrance to the **oviduct**, a process that depends on adhesion of the cilia of the infundibulum to the extracellular matrix of the complex. It is then "churned" to compress the matrix and allow it through the narrow neck (ostium) into the oviduct itself.

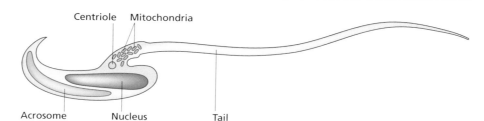

Fig. 10.2 Diagram of a mouse sperm.

Although chemotaxis of sperm towards eggs occurs in some other types of animal, the evidence for this in mammals is not clear cut. The transport of the sperm depends to some extent on muscular movements of the female reproductive tract, which assist its passage up the vagina, through the uterus, and into the oviducts. The main steps of sperm–egg interaction are shown in Fig. 10.3. The sperm carry a membrane-bound hyaluronidase that assists their passage through the extracellular matrix of the oocyte–cumulus complex. The next stage is the binding of sperm to the zona pellucida and this is the stage at which species specificity is controlled. If the zona is removed, then cross-species fertilization is possible, and this is the basis for the routine assay of the effectiveness of human sperm using hamster eggs. The zona is composed of three glycoproteins called ZP1, ZP2, and ZP3, which share a common "ZP" peptide sequence motif at the C-terminus. Of these, ZP3, with its specific O-linked oligosaccharide, is the specific sperm receptor. Low concentrations of ZP3, or the oligosaccharide alone, will prevent sperm–zona binding. Female mice in which the gene for ZP3 has been knocked out will form normal oocytes but they do not have a zona and the mice are infertile (Fig. 10.4). If the human ZP3 gene is "knocked in" to replace the mouse gene, then zona formation and fertility is restored. But, interestingly, eggs from these mice do not acquire a competence to be fertilized by human sperm. This is because the species-specificity resides in the carbohydrate attached to the ZP3 polypeptide, and this still has the murine structure because it is assembled by the murine glycosyl transferases.

The recognition protein for ZP3 on the sperm is a cell surface β-1,4-galactosyl transferase (GalT) and binding of ZP3 to this protein provokes the acrosome reaction. The evidence for this is that GalT will bind tightly to the ZP3 oligosaccharide, and that sperm from mice in which the GalT gene has been knocked out show reduced binding to the zona and are not provoked into an acrosome reaction. The acrosome reaction is a rapid exocytosis of the acrosomal vesicle and the released products are needed for the later events of fertilization. The coupling between GalT and the exocytosis is thought to proceed by the sequence: G protein activation, less negative membrane potential, opening of voltage-gated Ca^{2+} channels, which cause a rise in intracellular Ca^{2+}. There is also a rise in intracellular pH. GalT itself is a single-pass membrane protein with a G-protein activa-

tion domain in the cytoplasm. If GalT is expressed in *Xenopus* oocytes, then the oocytes will bind ZP3, and ZP3 will stimulate G protein activation and activation events such as cortical granule exocytosis. Here the *Xenopus* oocyte is being used simply as an experimental test system, because it is a large cell into which it is easy to inject mRNA. The fact that it is an oocyte is not relevant since the normal location of GalT is the sperm and not the oocyte.

The materials released by the acrosome reaction include hydrolytic enzymes such as the serine protease, acrosin, that help digest a path through the zona and enables the sperm to reach the egg surface. At this stage there is a second recognition process. The sperm–egg recognition is carried out by ADAM proteins on the sperm, which bind to integrins on the egg surface. ADAM stands for **a d**isintegrin **a**nd **m**etalloprotease domain, and proteins of this class will bind tightly to integrins. The sperm contains three ADAM proteins: fertilin α, fertilin β, and cyritestin. All probably play some role in the interaction. Peptides from fertilin β will block sperm–egg binding, and sperm from mice in which the genes for fertilin β or cyritestin have been knocked out show greatly reduced fertility. The best candidate for the target integrin on the egg is integrin α6 because binding is prevented by a monoclonal antibody to this protein. However, as the knockout mouse for this integrin has normal fertility there must be some other components also involved, which have a redundant function.

The binding of sperm to egg is the first stage in fusion of the plasma membranes, which occurs at a domain on the side of the sperm head. Fusion requires a four-pass membrane protein on the egg called tetraspanin or CD9. Evidence for this is that female knockout mice for CD9 show no fusion and are infertile; but the fusion ability of the eggs can be restored by injection of CD9 mRNA. The fusion process also requires one or more glycosylphosphatidylinositol (GPI)-anchored cell surface proteins on the egg, as their removal prevents fusion even when all other processes are occurring normally. In the course of fusion the whole sperm, including the tail, enters the egg.

Cell fusion causes a rise of intracellular Ca^{2+} concentration, which is responsible for all the subsequent events of fertilization. This calcium rise is perhaps the only aspect of fertilization that does seem to be universal across the animal kingdom. In

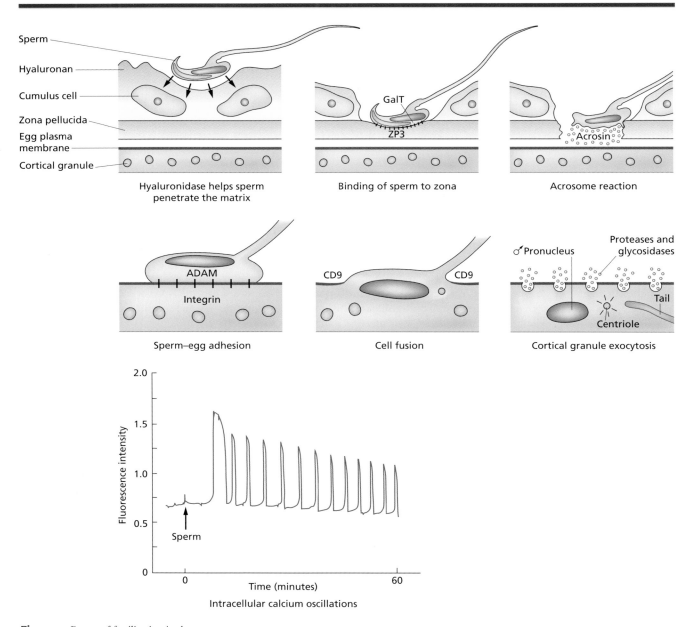

Fig. 10.3 Events of fertilization in the mouse.

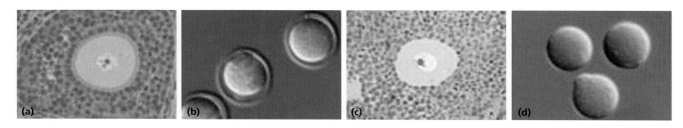

Fig. 10.4 Loss-of-function mutant of ZP3 produces eggs with no zona pellucida: (a,b) normal; (c,d) ZP3 knockout. Reproduced from Dean (2004), with permission from John Wiley & Sons, Ltd.

many mammals, including the mouse, the initial calcium spike is followed by a series of others making up an oscillatory pattern over several hours (Fig. 10.3g). Calcium transients of this sort are observed by loading the eggs with calcium-sensitive reagents such as the phosphorescent protein aequorin or the fluorescent dye fura2, and measuring the resulting light emission or fluorescence. Ca^{2+} release is caused by activation of the inositol trisphosphate (IP_3) pathway by a specific phospholipase C introduced by the sperm. The evidence for this is as follows:

1 Injection of IP_3 will provoke Ca^{2+} release.

2 Inhibitors of phospholipase C or of IP_3 receptor will inhibit Ca^{2+} release.

3 Injection of whole sperm, or demembranated sperm heads, or sperm extracts, will provoke Ca^{2+} release.

4 Sperm contains a specific phospholipase C (PLCζ), which itself will cause Ca^{2+} release.

5 Immunodepletion of this PLCζ from sperm extract will abolish activity.

Injection of Ca^{2+}, or treatment with calcium ionophore, which allows entry of Ca^{2+} from the medium, will both cause the same events of egg activation as fertilization by sperm. Such eggs are **parthenogenetic** (i.e. contain no paternal nucleus) and because of this they cannot develop far (see imprinting below).

The events dependent on the Ca^{2+} entry comprise the exocytosis of **cortical granules**, the completion of the second meiotic division, the resumption of DNA synthesis, the recruitment of maternal mRNA into polysomes, and a general metabolic activation. These events are mediated by the γ isoform of calcium/calmodulin-dependent protein kinase II (CaMKII). The cortical granules are found just under the plasma membrane and their contents include glycosidases and proteases that modify the zona pellucida receptors so that they can no longer bind sperm. In the mouse this process is the main factor preventing polyspermy.

The completion of the second meiotic division results in expulsion of the second **polar body** containing the surplus chromosomes. The sperm nucleus itself decondenses, assisted by reduction of protamine disulfide bonds by the peptide glutathione, which is present in the egg. The protamines become replaced by histones and the sperm DNA becomes actively demethylated, although without affecting imprinted loci (see below). The two pronuclei migrate slowly towards each other and undergo DNA replication. In mammals they do not fuse to form a true zygote nucleus, instead the pronuclear envelopes break down as they meet, and the chromosomes become aligned on the mitotic spindle ready for the first cleavage. Both pronuclei form a nucleolus, and this depends on components present in the oocyte nucleolus, as the sperm does not have one.

In addition to the nucleus, the sperm also introduces some mitochondria although these degenerate and do not participate in later development. In most mammals, but not in the mouse, a **centriole** is contributed by the sperm and becomes the **microtubule organizing center** for the sperm **aster**, later dividing to form the first mitotic spindle. But in the mouse both centrioles are maternal in origin, and this is a reason why parthenogenetic development can occur to some extent following artificial egg activation.

Normal development

Preimplantation stages

The course of preimplantation development is shown in Fig. 10.5. Following fertilization the first few cleavages are very slow and, in contrast to *Xenopus* and the zebrafish, expression of the zygotic genome commences early, at the two-cell stage. The first cleavage occurs after about 24 hours and the second and third cleavages, which are not entirely synchronous, follow at intervals of about 12 hours. This slow tempo of early development may be an adaptation to the time required for the uterus to prepare for implantation. In the early eight-cell stage the shapes of individual cells are still clearly visible but they cease to be visible when the whole embryo acquires a more nearly spherical shape in a process called **compaction** (Fig. 10.5d). This consists of a flattening of blastomeres to maximize intercellular contacts

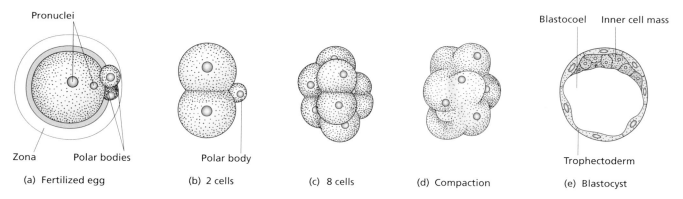

Fig. 10.5 Preimplantation development. The zona remains present but is not shown in (b) to (e).

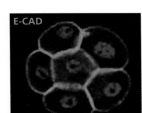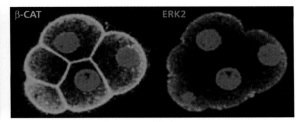

Fig. 10.6 Polarization of cells beginning at the eight-cell stage. β-catenin and ERK are concentrated in the outer part of each cell. DAPI is a blue fluorescent DNA-binding dye that marks cell nuclei. Reproduced from Lu *et al. Nat Genet* 2008; 40: 921–926, with permission from Nature Publishing Group.

and is mediated by the calcium-dependent adhesion molecule E-cadherin (also called leucocyte cell adhesion molecule (L-CAM) or uvomorulin). At this stage the cells become polarized in a radial direction (Fig. 10.6). This is apparent from the appearance of **microvilli** on the outer surfaces but it also involves a variety of changes in the cell interior. **Gap junctions** are also formed at this stage and allow diffusion of low molecular weight substances throughout the embryo.

The embryo is called a **morula** from compaction until about the 32-cell stage. During this period desmosomes and tight junctions appear, creating a permeability seal between the inside and outside of the embryo, and a fluid-filled **blastocoel** begins to form in the interior. This is about 3 days after fertilization and around the time that the embryo moves from oviduct to uterus. The cavity expands the embryo into a **blastocyst**, which consists of an outer cell layer of epithelial morphology called the **trophectoderm** and a clump of cells attached to its interior, the **inner cell mass (ICM)** (Fig. 10.5e). At the 60-cell stage about one-quarter of the cells are found in the ICM and three-quarters in the trophectoderm. The ICM expresses a set of transcription factors associated with pluripotent behavior (see below), including OCT4 (POU domain), SOX2 (SRY type), and NANOG (homeodomain), and it secretes FGF4. The trophectoderm expresses an FGF receptor, FGFR2, and the transcription factors TEAD4 (TEF family) and CDX2 (homeodomain), both of which are necessary for trophectoderm differentiation.

From E3.5 to E4.5 both the ICM and the trophectoderm diversify into two tissue types (Fig. 10.7a). Within the ICM, some cells lose expression of NANOG and acquire GATA4 and -6. These cells constitute the **primitive endoderm** or **hypoblast**, and they sort out from the others to form a layer on the blastocoelic surface of the ICM. The primitive endoderm contributes to extraembryonic tissues but not to the **definitive endoderm** of the embryo itself. The trophectoderm becomes divided into a polar component, which overlies the ICM, and a mural component, which makes up the remainder. While the polar trophectoderm continues to proliferate, the mural trophectoderm becomes transformed into **polytene** giant cells in which the DNA continues to be replicated but without mitosis. In these cells the DNA content is increased between 64 and 512 times.

Early postimplantation stages

At about this stage the embryo hatches from the zona and becomes implanted in a uterine crypt. The uterus is only competent to receive embryos during a short period about 4 days from mating. The uterus is attached to the body wall by a membrane called the mesometrium, which carries the uterine blood vessels. The mesometrial side of the uterus is the side where the placentas will form. When the embryos implant, they are orientated such that the ICM lies away from the mesometrium.

The course of early postimplantation development is shown in Fig. 10.7. After implantation the trophectoderm becomes known as the **trophoblast** and it stimulates proliferation of the connective tissue of the uterine mucosa to form a **deciduum** (plural decidua). From this stage onward the embryo receives a nutrient supply from the mother and can begin to grow in size and weight. As in the chick, the zygote does not just form an embryo but also a substantial complex of extraembryonic membranes and the entire product of fertilization is referred to as the **conceptus**.

The next stage of development is known as the **egg cylinder** (Fig. 10.7b). The cylinder itself can be regarded as homologous to the area pellucida of the chick embryo and consists of an "upper" layer of **primitive ectoderm** or **epiblast**, homologous to the chick epiblast, and a "lower" layer of **primitive endoderm**, whose affinities are less clear but may resemble the chick endoblast. The terms "upper" and "lower" are in quotes because the embryo is actually the shape of a deep cup with the epiblast on the inside and the endoderm on the outside. It appears U-shaped in a sagittal section or O-shaped in a transverse section. Cells derived from the primitive endoderm move out to cover the whole inner surface of the mural trophectoderm and start to secrete an extracellular basement membrane, known as Reichert's membrane, containing laminin, entactin, and type IV collagen. These cells are called the **parietal endoderm**. The remainder of the primitive endoderm remains epithelial and forms a layer of **visceral endoderm** around the epiblast. The cells of this layer somewhat resemble the later fetal liver, being characterized by the synthesis of α-fetoprotein, transferrin, and other secreted proteins. In the distal region the inner layer of

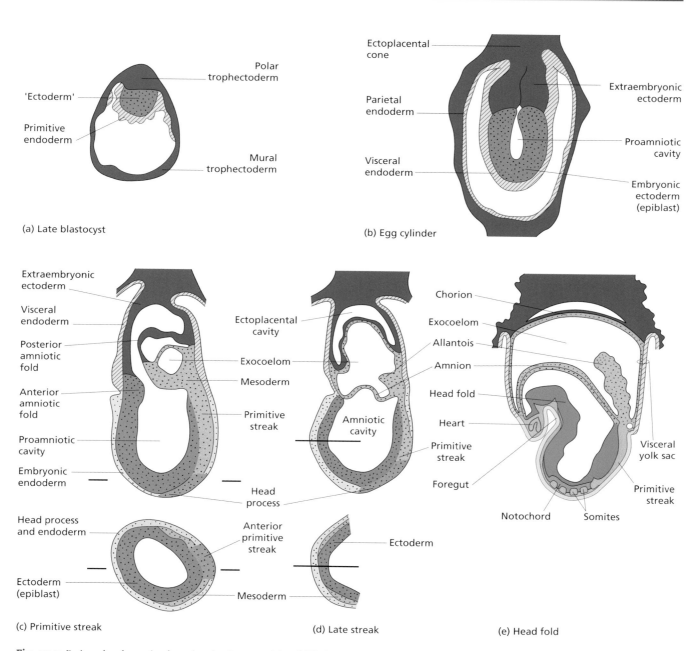

(a) Late blastocyst

Polar trophectoderm

'Ectoderm'

Primitive endoderm

Mural trophectoderm

(b) Egg cylinder

Ectoplacental cone

Parietal endoderm

Visceral endoderm

Extraembryonic ectoderm

Proamniotic cavity

Embryonic ectoderm (epiblast)

(c) Primitive streak

Extraembryonic ectoderm

Visceral endoderm

Posterior amniotic fold

Anterior amniotic fold

Proamniotic cavity

Embryonic endoderm

Head process and endoderm

Ectoderm (epiblast)

Ectoplacental cavity

Exocoelom

Mesoderm

Primitive streak

Head process

Anterior primitive streak

Mesoderm

(d) Late streak

Amniotic cavity

Primitive streak

Ectoderm

(e) Head fold

Chorion

Exocoelom

Allantois

Amnion

Head fold

Heart

Foregut

Notochord

Somites

Visceral yolk sac

Primitive streak

Fig. 10.7 Peri- and early postimplantation development. (c) and (d) show both sagittal and transverse sections at levels indicated by the lines. Due to growth, the headfold stage embryo is about 8 times longer than the early egg cylinder.

the egg cyliner is the epiblast and in the proximal region it is extraembryonic ectoderm derived from the polar trophectoderm. The polar trophectoderm also forms the ectoplacental cone and as this proliferates it produces further layers of giant cells, which move around and reinforce the trophoblast. In contrast to the various extraembryonic tissues, the epiblast from which the entire embryo will be derived remains visibly undifferentiated at this stage.

At about E6.5 the anteroposterior axis of the future embryo becomes apparent with the formation of the **primitive streak** at one edge of the epiblast. The streak marks the posterior end of the future embryonic axis, which will extend across the ectoderm toward the distal tip of the cup. It is a region of cell movements similar to that found in the chick embryo and these movements result in the formation of the **definitive endoderm** and the mesoderm (Fig. 10.8). By this stage the egg cylinder has become somewhat laterally compressed. The streak elongates until it has reached the distal tip of the egg cylinder and the **node** appears at its anterior end. This is homologous to Hensen's node in the chick and consists of two cell layers while the remainder of the streak has three. By E7.5 a **head process** appears anterior to the node, consisting of a forming **notochord** flanked by definitive endoderm in the lower layer and **neural plate** in the upper layer. As in the chick, the node then moves posteriorly and the axial body structures appear in anteroposterior sequence in its track, convergent extension movements being involved in the formation of the trunk notochord. By E8.5 the embryo has elongated somewhat in length and a massive head fold has formed at the anterior end, mainly composed of the anterior neural tube. The **somites** begin to form from E8 in anteroposterior sequence, at the rate of about one somite per 1.5 hours.

Although at first sight the morphology of the mouse embryo and the lower vertebrates seems very different, the homology of parts is clearly displayed by the gene expression patterns during gastrulation: for example *Otx2* in the anterior neural plate, *brachyury* (*T*) and *Mesp1* in the mesoderm, and *Foxa2* in the definitive endoderm (Fig. 10.9). The transcription factor gene *brachyury* is usually called *T* in the mouse, and in fact the original mouse mutant gave the name "T box" to this transcription factor family.

The amniotic fold forms at about 7 days as an outpushing of the ectoderm and mesoderm at the junction of posterior primitive streak and extraembryonic ectoderm (Fig. 10.7c,d). The side of this fold nearer the embryo becomes the **amnion** and the side nearer the ectoplacental cone becomes the **chorion**. The fold pushes across the proamniotic space and divides it into three: an amniotic cavity above the embryo, an exocoelom separating amnion and chorion, and an ectoplacental cavity lined with extraembryonic ectoderm. Into the exocoelom grows the

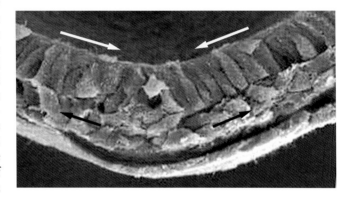

Fig. 10.8 Transverse scanning electron micrograph through a mouse embryo primitive streak. The arrows represent the direction of cell movements. Reproduced from Arnold and Robertson (2009), with permission from Nature Publishing Group.

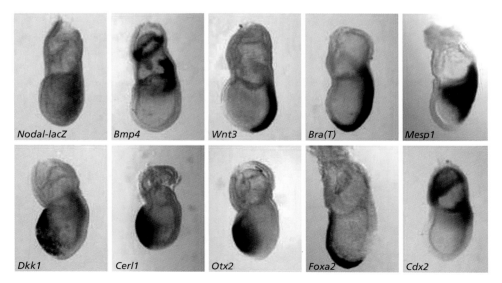

Fig. 10.9 Expression of various inducing factor and transcription factor genes at the primitive streak stage. The anterior of the embryo is to the left and the axis extends around the lower part of the egg cylinder. Reproduced from Arnold and Robertson (2009), with permission from Nature Publishing Group.

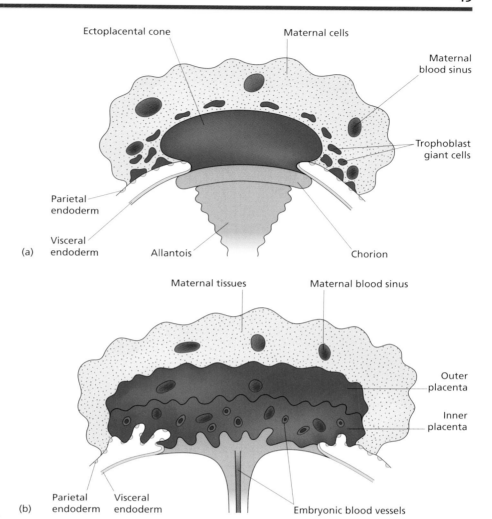

Ectoplacental cone

Maternal cells

Maternal blood sinus

Trophoblast giant cells

Parietal endoderm

Visceral endoderm

Allantois

Chorion

(a)

Maternal tissues

Maternal blood sinus

Outer placenta

Inner placenta

Parietal endoderm

Visceral endoderm

Embryonic blood vessels

(b)

Fig. 10.10 Schematic views of the mouse placenta: (a) E8.5; (b) E14.5. Reproduced from Hogan *et al.* (1994), *Manipulating the Mouse Embryo*, 2nd edn, with permission from Cold Spring Harbor Laboratory Press.

allantois (Fig. 10.7e), consisting of extraembryonic mesoderm from the posterior end of the primitive streak. This grows to contact the chorion and later forms the embryonic blood vessels of the placenta. Unlike the allantois of the chick (Chapter 9), or the human (see below), the mouse allantois contains no endodermal layer.

The placenta itself forms from the ectoplacental cone region, which is directed toward the mesometrial side of the uterus. Trophoblast giant cells invade the decidual tissue, accompanied by polar trophectoderm, and mesoderm and blood vessels from the chorion and allantois. From the maternal side the mature placenta consists of the following layers: maternal decidual tissues; giant trophoblast cells with maternal blood sinuses; diploid trophoblast with fetal blood vessels; and, finally, crypts lined with extraembryonic endoderm (Fig. 10.10). The placenta is not only important as the organ supplying nutrients to the fetus, but also has endocrine functions, producing progesterone and estrogen to maintain the pregnancy as well as a number of other important hormones.

Around 8.5 days a most remarkable process takes place known as **turning**, which brings the germ layers into the proper orientation within the embryo. This is best described as a rotation of the embryo around its own long axis. It starts as a U-shaped structure with the dorsal side concave and the rotation converts it into an inverted U with the dorsal side convex (Fig. 10.11). This change of orientation has drastic consequences for the arrangement of the extraembryonic membranes. The amnion becomes enlarged so that it surrounds the whole embryo instead of just covering the dorsal surface. The membrane lining the exocoelom, which consists of visceral endoderm and mesoderm, also becomes stretched to cover the entire embryo, and becomes known as the visceral yolk sac. The final arrangement of membranes after turning has the amnion on the inside, the visceral yolk sac next and the parietal yolk sac, formed from the trophectoderm lined with parietal endoderm, on the outside. Turning leads to a rapid closure of the midgut, which starts as a large area of endoderm exposed on the ventral side and becomes constricted to a small umbilical

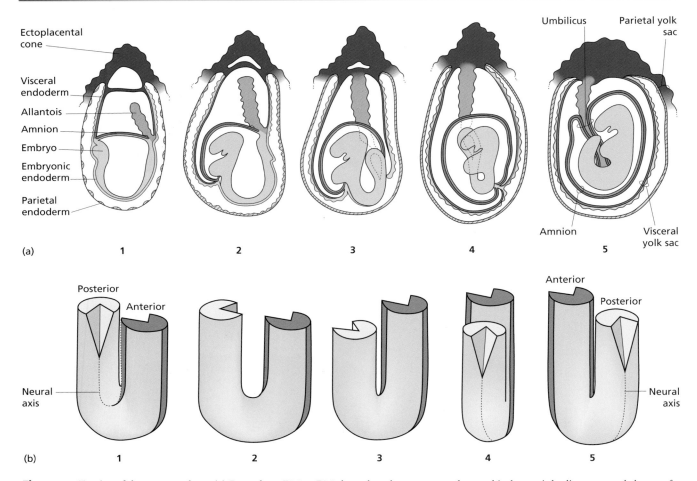

Fig. 10.11 Turning of the mouse embryo. (a) From about E7.5 to E9.5 the embryo becomes rotated around its long axis leading to ventral closure of the gut; (b) diagrammatic representation of turning. Modified from Kaufman, in *Postimplantation Mammalian Embryos*, Copp and Cockroft (eds) 1990. IRL Press, with permission from Oxford University Press.

tube containing the vitellointestinal duct, the vitelline vessels, and the allantois. Turning is a process found only in rodents, which have the egg cylinder type of embryo. It does not occur in the human embryo where the epiblast is flat (see below).

Organogenesis stages

By about E9.5, when the axis has formed, the embryo has turned, and the gut has closed, the embryo has reached the junction between the early phase of body plan formation and the later phase of **organogenesis**. Organogenesis is in most respects very similar to that in the chick so only a brief sketch is given here, with Fig. 10.12 showing exterior views of the embryo. Further details of mouse organogenesis will be found in Chapters 14–16.

Neural tube closure takes place simultaneously with turning. It commences in the hindbrain and proceeds both anteriorly

and posteriorly, with the anterior neuropore closing at E9 and the posterior neuropore at E10 to E10.5. The **optic vesicles** form at E9.5 and the lens has been incorporated into the eye by E11.5. The **neural crest** emerges from the neural tube over the period E8.5 to E10.5. As in the chick, it forms the skeletal structures of the head, the dorsal root ganglia, Schwann cells, sympathetic ganglia, pigment cells, adrenal medulla, and enteric ganglia. **Somites** continue to form until E14 by which time there are about 65, many being in the tail, which is much longer than that of the chick. As in other vertebrates, the somites form the vertebrae and myotomes and contribute to the dermis. The paired heart primordia fuse at E8.5 and the heartbeat begins at E9. The left and right atria become separated at E11.5 and the ventricles at E14. The gut originates from fore- and hindgut pockets as in the chick, although the closure of the midgut is faster because it is driven by the turning process. It is now thought that there is some contribution to the gut epithelium from the visceral endoderm as well as the definitive endoderm that passes through

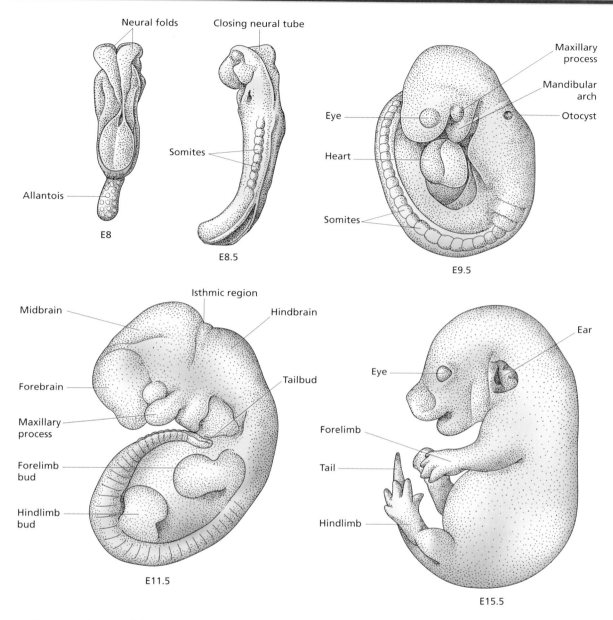

Fig. 10.12 Organogenesis stages of the mouse embryo. During these stages there is considerable growth. The E8 embryo is about 1 mm long, whereas the E15.5 fetus is about 13 mm from crown to rump.

the primitive streak. There are six **pharyngeal arches** forming from E9, although the fifth arch is vestigial. The mouth forms at E9.

The mesoderm lateral to the somites is the **intermediate mesoderm**, which becomes the kidney and gonads. The pronephric primordium arises about E9, but has no function. The **genital ridges** become visible from about E10 from mid-trunk to mid-tail. The lateral part of these ridges forms the mesonephros, although in mammals this too is vestigial and without function. The nephric duct produces the **ureteric bud** at E11.5 and this grows into the posterior nephrogenic mesoderm to

produce the **metanephros**, which is the functional kidney. The medial part of the urogenital ridges produces the **gonads**. The **germ cells** originate from the extraembryonic mesoderm in the posterior amniotic fold. These enter the hindgut at E10 and migrate up the mesentery, reaching the gonads between E11 and E13. Lateral to the intermediate mesoderm is the **lateral plate**, which is divided by the coelom into the outer **somatopleure** (epidermis and somatic mesoderm) and the inner **splanchnopleure** (endoderm and splanchnic mesoderm). The limb buds arise from the somatopleure at E9.5 to E10 with the forelimb bud at the level of somites 8–12 and the hindlimb buds

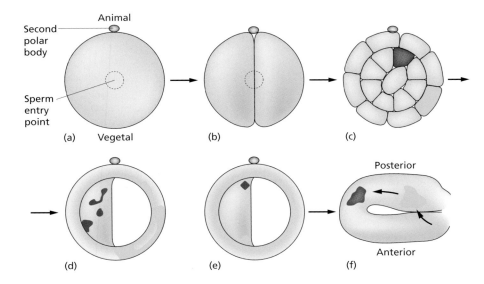

Animal
Second
polar
body

Sperm
entry
point

(a)   Vegetal   (b)   (c)

(d)   (e)   (f)

Posterior

Anterior

Fig. 10.13 Fate mapping of early stages. (a) The second polar body marks the animal pole of the zygote. (b) The first cleavage is meridional. (c,d) In the compacted blastocyst, labeling of inner cells shows that they enter the inner cell mass, while the outer cells enter the trophectoderm. (e,f) In the inner cell mass, labeled animal cells become distal and labeled vegetal cells become proximal in the egg cylinder.

at the level of somites 23–28. The outer epidermis of the mouse, like other mammals, is covered with hair follicles, arising from E14 onwards.

Fate map

It used to be believed that mammalian embryos, with their high regulative capacity (described below), had no definable fate map at the stage of the fertilized egg. However, it is now recognized that the fertilized egg does possess a polarity, which is retained through the early embryonic stages (Fig. 10.13a–d). This has been established by a combination of careful observation and the labeling of portions of the egg surface with fluorescent lectin-coated beads and other markers. The original animal pole of the egg can be identified by the position of the second polar body. The first cleavage is approximately meridional and separates blastomeres that will preferentially become the embryonic pole (i.e. the end with the inner cell mass) and the abembryonic pole (the end of the mural trophectoderm) of the blastocyst. There is also some evidence for a coincidence of the first cleavage plane with the site of sperm entry.

Labeling of individual blastomeres by injection of horseradish peroxidase (HRP) or fluorescent dextrans shows that the trophectoderm arises from the polar cells on the exterior of the morula, while the inner cell mass arises from the apolar cells in the interior. By the time the blastocyst expands, there is no further interchange of cells between these two populations. In the early postimplantation embryo, single labeled cells of the polar trophectoderm give rise to clones of postmitotic cells in the mural trophectoderm, showing that the polar region is a proliferating zone feeding the mural trophectoderm.

There is also a predictable relationship between the axes of the blastocyst and the later egg cylinder stage (Fig. 10.13e,f). If the cells on the animal pole end of the blastocyst are labeled by injection of mRNA for GFP it is found that they end up distally in the visceral endoderm of the egg cylinder. Conversely, if ICM cells at the vegetal pole end are labeled, they end up proximally in the egg cylinder. This distal–proximal distribution of cells arises from the early stages of the morphogenetic movements, which form the primitive streak of the embryo. It follows from this that the original animal pole of the zygote is likely to end up as the posterior side of the embryo, that is the side of the forming primitive streak, when it becomes visible in the egg cylinder stage.

Fate mapping of postimplantation stages was undertaken by labeling the embryo by injection of cells with HRP, or extracellularly with DiI, then culturing it *in vitro* for up to 48 hours (Fig. 10.14). These studies show that the primitive streak and the mesoderm of the amnion and allantois all arise from the posterior edge of the epiblast. The streak pushes distally and the extraembryonic tissue expands into the proamniotic space, mainly driven by growth but also by some lateral movement of cells around the cup of epiblast. The anterior part of the streak becomes the node, which forms the notochord and part of the somites along the entire length of the body. Cells also enter the somites and the neural plate from positions lateral to the midline. The middle part of the streak populates mainly lateral plate mesoderm in the posterior half of the body. The posterior part of the streak populates mainly mesoderm of the amnion, visceral yolk sac, and allantois. It seems that the primitive streak in the mouse works in a similar way to that of the chick, with cells from the surface layer moving laterally towards and then through the streak to become definitive endoderm and mesoderm. As in the chick, the posterior streak does not become the posterior part of the midline but instead populates the lateral parts of the body. All the studies show a substantial degree of mixing between labeled and unlabeled cells, showing that the fate map of the postimplantation mouse is a statistical construction and that the boundaries between prospective regions are not precise.

More recently, the Cre labeling technique (see Chapters 3 and 13) has been used for fate mapping. This has shown that the *Foxa2* domain of the anterior primitive streak later populates the definitive endoderm and notochord. The *Mesp1* domain of the early streak later mainly populates the heart.

Regional specification

Formation of extraembryonic structures

The early blastomeres of a mouse embryo are known to be **totipotent**. It is possible to obtain formation of a complete

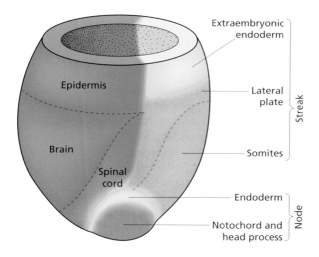

Fig. 10.14 Fate map of the late primitive streak stage. The boundaries are actually much fuzzier than shown, because of cell mixing.

blastocyst from each single blastomere isolated from the two- or four-cell stage. These blastocysts tend to have a higher proportion than normal of trophectoderm but they will form complete normal embryos after reimplantation. From the eight-cell stage it is no longer possible to obtain a complete embryo from one blastomere, although blastomeres to the 32-cell stage can still form a complete fetus when introduced into a tetraploid host. Tetraploid embryos, made by electrofusion of the two-cell stage, are able to generate the extraembryonic membranes and to support development of a fetus.

In normal development the formation of ICM and trophectoderm depends on the cell polarization that occurs at the eight-cell stage (Fig. 10.15). Exactly how this converts into the two cell types is still not completely clear, but it is likely that the PAR proteins are involved. These were first discovered in *C. elegans* (see Chapter 12) and are associated with unequal cell division in many other systems. There is some segregation of PAR proteins at the eight-cell stage: EMK1 (= PAR1) concentrates internally and PAR6b externally. Administration of dsRNA or dominant negative versions of certain PAR proteins will increase the proportion of cells becoming ICM and reduce the proportion becoming trophectoderm. ERK2, a form of MAP kinase involved in FGF signaling, is also concentrated apically. Activation of FGF signaling increases the upregulation of *Cdx2* and the proportion of trophectoderm cells produced, while inhibition of the FGF pathway has the opposite effect. The TEF family transcription factor TEAD4 is needed for expression of *Cdx2*, and CDX2 is needed for trophectoderm formation, and suppresses the expression of *Oct4* and *Nanog* in the trophectoderm. CDX2 may also help to promote the original polarization event. Although the normal specification commences at the eight-cell stage, it is still possible for a period to force polar cells to become ICM, or apolar cells to become trophectoderm, by putting them, respectively, on the inside or the outside of a cell aggregate. By the 64-cell stage the two cell types have stabilized and are no longer interconvertible.

After implantation the ICM becomes divided into an outer layer of primitive endoderm and an inner core of epiblast. In another process requiring FGF signaling, some cells of the ICM

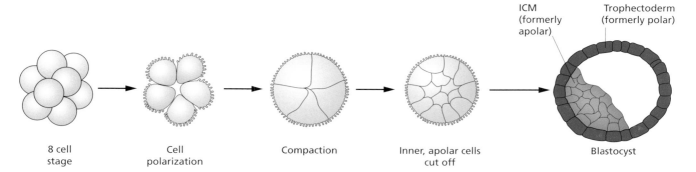

Fig. 10.15 Origin of ICM and trophectoderm from cell polarization at the eight-cell stage.

randomly upregulate *Gata6* and lose *Nanog*, and then these cells sort themselves into a layer of primitive endoderm at the blastocoel surface (Fig. 10.16).

The trophectoderm becomes divided into the polar trophectoderm, neighboring the ICM, and the mural trophectoderm surrounding the rest of the blastocyst. The maintenance of cell division in the polar region depends on the proximity of the ICM, and requires the transcription factor ELF5 (ETS family) whose expression is upregulated by FGF4 and NODAL.

The primitive endoderm becomes divided into visceral endoderm, in contact with the epiblast, and parietal endoderm, in contact with the mural trophectoderm. The parietal endoderm arises as a result of an epithelial to mesenchymal transition and

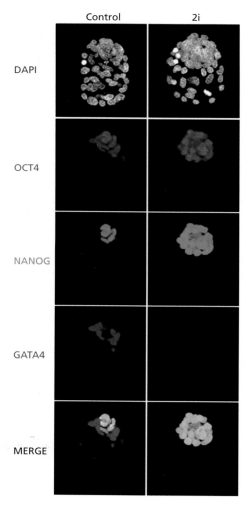

Fig. 10.16 Formation of the primitive endoderm (hypoblast). Here the ICM is shown by immunostaining for OCT4, the epiblast with NANOG, and the primitive endoderm with GATA4. Formation of the primitive endoderm is suppressed by inhibitors of GSK3 and of FGF signaling (2i). Reproduced from Nichols *et al. Development* 2009; 136: 3215–3222, with permission from Company of Biologists Ltd.

secretes the Reichert's membrane on the inner surface of the trophectoderm. The formation of parietal endoderm is stimulated by fibronectin and parathyroid hormone related peptide (PTHrP) and reduced by laminin.

Embryonic body plan

Following the formation of the various extraembryonic structures, the whole embryo derives from the epiblast of the egg cylinder. Up to the late primitive streak stage, grafts of tissue within the epiblast develop in accordance with their surroundings. In addition, it is possible at these stages to induce twinning by treatment with cytotoxic drugs, which kills large numbers of cells and leads to regeneration from small nests of survivors. These data indicate that regional determination has not become irreversible before the late streak stage, although the pattern of gene expression indicates a progressive specification of parts of the body axis throughout the course of gastrulation. Unlike lower vertebrates, this is a stage of very rapid growth. The epiblast increases from about 660 cells to about 15,000 from egg cylinder to late streak stages.

As in *Xenopus*, a critical role in the sequence of inductions is held by NODAL. *Nodal* was originally discovered by retroviral mutagenesis of the mouse, and owes its name to the fact that it is expressed in the node. The homozygous null embryos are unable to form any embryonic pattern, and many of the developmentally important genes active during formation of the primitive streak and the node are not expressed. The knockout of the so-called activin receptor IIA and B (*Acvr2a* and *Acvr2a*), which are receptors for NODAL, or of *Smad2*, which is required for signal transduction of NODAL-like factors, or *Foxh1*, which is a partner of SMAD2 in transcriptional regulation, all abolish the anteroposterior polarity of the embryo, such that the mesoderm that forms is entirely extraembryonic. *Nodal* is initially expressed throughout the epiblast and later is downregulated in the anterior and becomes concentrated in the streak and node (Fig. 10.9). The formation of the primitive streak requires an epithelial–mesenchymal transition. This is effected by loss of E-cadherin, which depends on repression by the T-box transcription factor Eomesodermin. Eomesodermin has a number of other important target genes including *Lmx1* and *Mesp1*.

The anteroposterior pattern of the embryo arises from an egg cylinder that is initially radially symmetrical but polarized along the proximodistal axis. Initially, the transcription factor gene *Hex* and the inducing factor genes cerberus-like1 (*Cerl1*), dickkopf 1(*Dkk1*), and *Lefty1* are all expressed at the distal tip of the visceral endoderm, and a number of genes including *T* are expressed at the proximal end in the epiblast (Fig. 10.17). The distal group of inducing factors are all antagonists of other factors. LEFTY belongs to the TGFbeta superfamily but is antagonistic to NODAL action, DKK1 is a WNT antagonist and CERBERUS has inhibitory domains for NODAL, BMP, and WNT. The establishment of this proximodistal pattern is due to signals

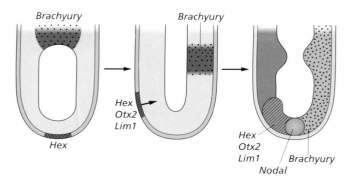

Fig. 10.17 Anteroposterior patterning of epiblast in the egg cylinder to primitive streak stage.

emitted from the extraembryonic ectoderm comprising WNT3 and various BMPs *Wnt3* is expressed in the proximal epiblast, later becoming active in the primitive streak. *Bmp2,-4*, and *-8b* are expressed in the distal extraembryonic ectoderm, abutting the epiblast. Knockouts of these genes are early lethals, the *Wnt3* knockout having no primitive streak or mesoderm, and the individual *Bmp* knockouts being defective in embryonic mesoderm and primordial germ cells. Furthermore, physical removal of the extraembryonic ectoderm causes the distal gene expression domains to spread throughout the rest of the visceral endoderm, indicating that the signals originating from the proximal region act to limit their extent.

As the cup-shaped egg cylinder expands and the proamniotic cavity forms, the radial symmetry of the egg cylinder is broken. Morphogenetic movements shift the *Hex* domain to one side and the *T* domain to the other. These, respectively, become **anterior** and **posterior** ends of the embryo. As indicated above, there is a statistical association between the position of the original animal pole of the zygote and the posterior side of the early embryo, so it is possible that the breakage of the radial symmetry of the egg cylinder arises as it does because of small asymmetries derived from the fertilized egg. The former distal visceral endoderm becomes the **anterior visceral endoderm (AVE)** and expresses a group of genes for transcription factors (including *Otx2*, *Foxa2*, and *Lim*, = *Lhx1*) and genes for secreted factors (including *Cerl*, *Dkk1*, and *Lefty1*) that are associated with anterior development. Expression of *Nodal* is needed for formation of the AVE as well as of the primitive streak. Knockouts of the inducing factor genes expressed in the AVE lead to anterior truncations, although the AVE itself will not induce neural structures from other parts of the epiblast. Loss of *Cerl* or *Lefty1* can cause multiple primitive streaks to form. This effect is prevented by removing one copy of the *Nodal* gene, and indicating a role of NODAL inhibitors in suppressing ectopic streak development.

As in lower vertebrates, the anterior neural plate can be regarded as a default state of development. SOX2 is expressed all over from the ICM stage onwards, and OTX2 is expressed in the

whole epiblast, being suppressed in the posterior half as the streak forms. The posterior epiblast, then the anterior streak, then the node itself, all express genes associated with the organizer such as *Foxa2*, *Lim1*, and goosecoid (*Gsc*). But this is not a single cell population, as cells are passing through these zones throughout gastrulation. The node behaves like the organizer in *Xenopus*, as it can induce a second axis containing host-derived neural tube and somites if transplanted to another part of the epiblast. Mouse nodes will also induce second axes in Chick blastoderms. Activity is maximal at the mid-streak stage. Chordin (*Chrd*) and noggin (*Nog*) are both expressed in the anterior streak and the node epiblast, and these proteins will themselves induce anterior neural tissue from the epiblast. Moreover, the double knockout of *Chrd* and *Nog* prevents formation of the forebrain and gives defects in notochord and sclerotome, indicating a role in neural induction and mesodermal dorsalization. However, the *Chrd* and *Nog* knockouts do not prevent formation of the trunk and tail nervous system, indicating the likelihood that additional BMP inhibitors operate in concert with the usual caudalizing factors of FGF and Wnt.

As in *Xenopus*, the FGF–CDX–HOX pathway is thought to control patterning of the posterior region. *Fgf4* and *Fgf8* genes are expressed in the streak, activated ERK can be detected by immunostaining, and knockouts of *Fgf8*, or *Fgf receptor 1*, or *Cdx* genes, show posterior defects associated with reduction of HOX activity. Because *Cdx2* is also required for trophectoderm development, investigation of the null phenotype requires the use of tetraploid complementation (see below). Overexpression of trunk Hox genes under a *Cdx2* promoter will rescue posterior pattern in *Cdx* mutants with trunk defects, confirming that *Cdx* genes operate to upregulate Hox genes *in vivo*.

A gradient of FGF8 protein can be seen by immunostaining in all three germ layers at tailbud stages. This arises by transcription of *Fgf8* in the tail bud followed by gradual decay of the mRNA after the cells have left the tail bud (Fig. 10.18). This mechanism depends on the active growth of the embryo and would not work in lower vertebrates where there is little growth at early stages. However, the end result of a gradient of FGF activity is similar. Some *Wnt* genes are also important in posterior body formation in the mouse. *Wnt3A* in particular is expressed in the posterior and the knockout for *Wnt3A* lacks some posterior parts. Furthermore, retinoic acid is probably also involved as loss-of-function mutants of retinaldehyde dehydrogenase, the enzyme responsible for the production of endogenous retinoic acid, lead to posterior defects.

The available information suggests that the mechanisms whereby regional specification is achieved in the mouse is comparable to *Xenopus*, zebrafish, and chick although not identical. The initial regionalization is into just two parts, a head and a trunk region, the head pattern depending on a group of genes upregulated initially in the anterior visceral endoderm, with an inductive signal required to induce a corresponding anterior territory in the epiblast. The trunk pattern depends on the signaling center associated with the tip of the primitive streak

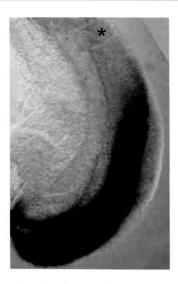

Fig. 10.18 Expression of gradient of *Fgf8* mRNA in the posterior of E9.5 mouse embryo. Asterisk, last-formed somite. Reproduced from Dubrulle and Pourquie. *Nature* 2004; 427: 419–422, with permission from Nature Publishing Group.

and later the node, which is responsible both for neural induction and for dorsalization of the mesoderm. In all the vertebrate species the NODAL type factors have a key role in mesoderm formation, the BMP inhibitors are required to some extent for neural induction, and the FGF and WNT systems are needed for formation of posterior parts.

Germ cell formation

Unlike most animal species, the primordial germ cells (PGCs) in the mouse do not arise from a cytoplasmic determinant in the egg. They are instead formed by induction. About 30–40 PGCs become visible at E7.5 located at the base of the allantois and can be recognized using a histochemical stain for alkaline phosphatase. They are induced by BMP4 from the extraembryonic ectoderm. The evidence is that the *Bmp4* knockout forms no PGCs, and that BMP4 can induce the formation of PGCs when applied to competent epiblast. PGCs lose DNA methylation and reactivate expression of *Oct4* and *Sox2*, transcription factors associated with pluripotency of the epiblast and of embryonic stem cells. Although initially formed in the posterior mesoderm, the PGCs then enter the hindgut and subsequently migrate to the gonads, as described in Chapter 15.

Left–right asymmetry

As in the chick, *Nodal* becomes expressed preferentially on the left side of the node from the two- to three-somite stage, along with *Lefty2*. Expression then spreads to the left lateral plate

mesoderm. The sequence of subsequent gene upregulations is somewhat different from the chick but eventually these factors bring about the asymmetries of organogenesis by differential upregulation of downstream targets on the two sides. For example there are two bHLH-type transcription factors whose activity contributes to the asymmetry of the early heart. The cascade initiated by left-sided NODAL causes dHAND to be expressed mainly on the right side and eHAND mainly on the left. Similarly, the counterclockwise coil of the intestines arises from the presence of denser mesenchyme on the left side. This is caused by the presence of N-cadherin on the left, whose synthesis is controlled by the transcription factors PITX2 and ISL1, which are downstream members of the cascade.

The original cause of the asymmetry, also called "situs" (pronounced "see tuss"), is thought to arise from the action of the cilia found in the node. Each node cell bears a single cilium and they are motile in the period just before the onset of asymmetric gene expression, driving a flow of fluid from right to left. This can be visualized by addition of fluorescent dyes to embryos cultured *in vitro*. The fluid flow is causally related to the later gene expression, as may be shown by experimentally reversing its sense, which brings about *Nodal* expression on the right instead of the left. The exact mechanism remains debated but it is likely that the fluid flow stimulates sensory cilia on the left side, provoking an increase of intracellular calcium, which affects gene expression.

Cilia are intrinsically asymmetric structures built around the standard 9+2 arrangement of microtubules and containing motor proteins to generate their movement. In the mouse there are numerous mutants that perturb the normal left–right asymmetry of the body and a high proportion of these cause defects in cilia. Three distinct classes of mutant phenotype can be distinguished (Fig. 10.19). First are mutants that cause randomization of asymmetry. This means that half the embryos have the normal *situs solitus*, and half have *situs inversus*. Such an outcome suggests that the breakage of bilateral symmetry still occurs but that the bias that normally guarantees the usual *situs solitus* has been removed. In such mutants the expression of *Nodal* and *Lefty2* may be on either side, but not both, indicating that the mutant genes are upstream of these factors. These mutations are in genes encoding **motor proteins**, for example the kinesin components KIF3A and KIF3B, and a dynein, LRD, encoded by the *iv* gene. The motor proteins are needed either for the assembly of the cilia, in the case of kinesin, or for ciliary motion, in the case of dynein.

The second class of mutant leaves the embryo in a state of bilateral symmetry. This is known as an **isomerism** (not, of course, to be confused with chemical isomerism of molecules). An example is the *TG737* mutant. Here, the gene product, called POLARIS, is also involved in the assembly of cilia. The mutant, which is loss of function, has no cilia and shows symmetrical *Nodal* expression.

The third class of mutant has a reversed asymmetry (*situs inversus*). An example is *inv*, which arises from a transgene

Wild type

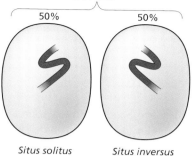
iv⁻; KIF3B⁻
50%   50%

inv⁻

Fig. 10.19 Asymmetry mutants in the mouse.

Situs solitus   Situs solitus   Situs inversus   Situs inversus

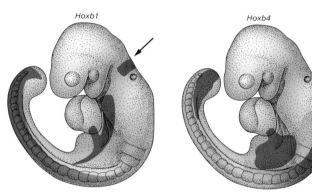
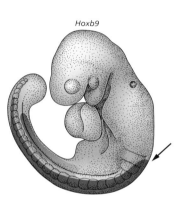

Hoxb1   Hoxb4   Hoxb9

Fig. 10.20 Expression of three Hox genes, showing the different anterior boundaries.

insertion but, like *iv* and *Kif3b⁻*, is genetically recessive. Its homozygotes all have *situs inversus*. In *inv⁻* embryos, *Nodal* and *Lefty2* are expressed on the right side. The *inv* product, INVERSIN, is a cytosolic protein with ankryn repeats. It is expressed early and affects the direction of the ciliary beat. In the mutant the fluid flow is reduced and rightwards instead of leftwards and expression of *Nodal* is on the right side.

The human genetic disease Kartagener's syndrome involves a randomization of asymmetry. Genetically, it is somewhat heterogeneous but many cases involve dynein mutants. Individuals also suffer from male infertility, because their sperm are immotile, and from lung diseases, because the cilia of their bronchial epithelia are immotile and cannot remove the accumulated mucus from the lungs.

Hox genes

In the mouse the four **Hox gene** clusters contain a total of 39 genes. Most of the paralog groups have two or three members. As in other vertebrates, the genes are maximally expressed at the **phylotypic** stage, they tend to have sharp anterior expression boundaries in the central nervous system (CNS) and mesoderm

and to fade out in the posterior, and members of the same paralog group have similar anterior boundaries (Fig. 10.20). Extensive work has been done on Hox function by making knockouts of the Hox genes, or by expressing them ectopically by making transgenics in which a Hox gene is driven by a foreign promoter.

In general, a knockout results in an anterior transformation. The effects are usually strongest at the anterior expression boundary of the gene in question, because it is there that this particular gene distinguishes the overall Hox combination from that in more anterior positions. Usually, the knockout of just one Hox gene has only a modest effect, but when the whole paralog group is knocked out then the effect becomes substantial. For example knockout of the paralog group 10 causes all lumbar vertebrae to convert to thoracic character, and knockout of the paralog group 11 causes all sacral vertebrae to convert to lumbar. The substantial redundancy arises because the members of a paralog group are usually quite similar to each other and are expressed in similar domains.

Ectopic expression of Hox genes generally results in a posterior transformation. For example if *Hoxa7*, normally with its anterior boundary in the thoracic region, is expressed in the head, then the basal occipital bone of the skull is transformed into a pro-atlas type vertebra. Upregulation of Hox genes can also be provoked by treatment of the mothers with retinoic acid, and this produces multiple posteriorization events due to simultaneous ectopic expression of several Hox genes.

Human early development

The mouse is usually considered to be a good model for the human. Indeed, the preimplantation stages of human development (Fig. 10.21) do resemble those of the mouse quite closely, but there are some significant differences. Fertilization occurs in the oviduct and the embryo takes about 4 days to move into the uterus. Cleavage is slow, with the first division at about 30 hours, the second about 40 hours, accelerating a little to produce an early blastocyst by the fourth day, and a mature blastocyst, which hatches from the zona, by 5 days. The blastocyst looks very similar to a mouse blastocyst, with an inner cell mass surrounded by the trophectoderm, normally called **trophoblast** in the human. Implantation occurs on day 6–7. The equivalent of the murine mural trophectoderm forms a syncytium called the **syncytiotrophoblast**, and this burrows into the endometrium of the mother. But the blastocyst remains completely surrounded by the **cytotrophoblast**, which is similar to the murine polar trophectoderm, and remains as proliferating cells.

The **primitive endoderm**, or **hypoblast**, becomes visible about day 7, on the blastocoelic surface of the ICM, and cells then migrate across the internal surface of the trophoblast. But from day 8 the course of development differs from the mouse in three respects (Fig. 10.22). Firstly, an **amniotic cavity** arises by cavity formation within the ICM. Secondly, unlike the rodent embryo, the epiblast does not form an egg cylinder. Instead the epiblast and primitive endoderm, also called hypoblast, make up a flat **blastodisc** situated between the amniotic cavity and the yolk sac cavity. Thirdly, a large volume of mesenchyme appears between the primary endoderm and the trophoblast. The origin of this extraembryonic mesoderm has been controversial. It appears some time before the primitive streak so is not likely to be a product of gastrulation. Examination of Rhesus monkey embryos suggests an origin from the primary endoderm, and this is also consistent with analysis of chromosome defects in human chorionic villi samples. Although most chromosome abnormalities are common to the fetus and the extraembryonic membranes, sometimes there are discrepancies, and a proportion of defects in the mesoderm-derived tissues are discordant both from the trophoblast and from the fetus. This suggests a likely origin from an early nondisjunction event in another extraembryonic tissue such as the primary endoderm.

The placenta begins to form with the appearance of lacunae in the syncytiotrophoblast. These become continuous with maternal blood sinusoids. **Chorionic villi** consist of projections of cytotrophoblast containing blood vessels derived from the extraembryonic mesoderm. These grow into the blood-filled lacunae and provide for an exchange between maternal and the future fetal circulations. Chorionic villus sampling is one method of prenatal diagnosis for genetic diseases. The component cells are part of the conceptus and so of fetal genotype. Germ-line mutations can be looked for, and also chromosomal nondisjunctions arising at meiosis. However, as noted above, occasionally the chromosomal nondisjunctions which arise in

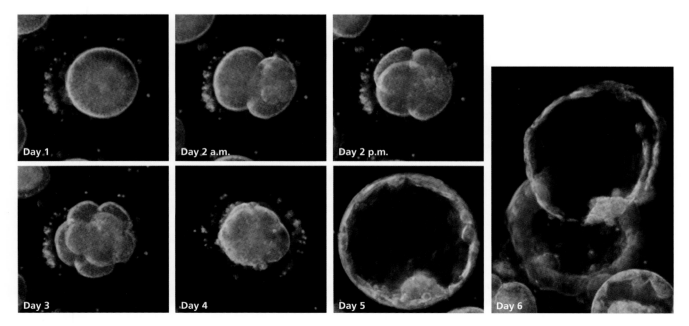

Fig. 10.21 Early human development: fertilized egg, two-, four- and eight-cell stages, compacted morula, blastocyst, and blastocyst hatching from zona. Courtesy of Dr K. Loewke, Stanford University.

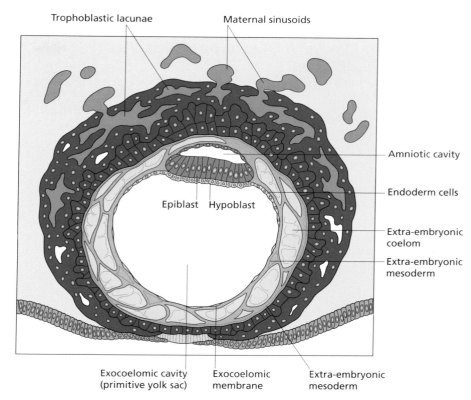

Fig. 10.22 Implanted human embryo at 12 days postfertilization. Unlike the mouse it has an amniotic cavity formed in the inner cell mass, a flat blastodisc, and a mass of extraembryonic mesoderm. Reproduced from T.W.Sadler, *Langman's Medical Embryology*, 6th edn. Williams and Wilkins: Baltimore 1990, with permission.

early development are not be shared by the fetus and the extraembryonic membranes.

The **primitive streak** arises in the blastodisc on about day 15. Gastrulation is probably similar to the chick and mouse, with cells migrating through the streak to form the mesoderm and a definitive endoderm, which replaces the hypoblast. A node arises at the tip of the primitive streak and the notochord pushes anteriorly from it. Unlike the mouse notochord this is hollow, being formed by an invagination process. It later fuses with the endoderm, causing formation of a transient **neuroenteric canal** joining the amniotic cavity and yolk sac cavity. Other aspects of primary body plan formation are comparable to the mouse, except that the allantois is an evagination from the hindgut and is lined with endoderm, whereas the allantois of the mouse is composed only of mesoderm.

The events of early human development are of some relevance to the ethical debate about human **embryonic stem cells** (see Chapter 21) as well as to the issue of whether any experimentation should be permitted on the early human embryo. In some jurisdictions, such as the United Kingdom, experimentation is permitted, under license, for the first 14 days after fertilization. The rationale for this is a biological one: that there is no identifiable human organism until the formation of the primitive streak. Prior to this, all of the cells of the conceptus probably contribute some of their descendants to the placenta, and the occurrence of twinning indicates that more than one

individual can arise from the conceptus as a whole. It is thought that human identical twins usually arise from division of the ICM (70–75%), less often from blastomere separation (25–30%), and most infrequently from division of the primitive streak itself (1%). These figures are arrived at on the basis of whether the twins share a common placenta and amnion. In cases of blastomere separation the placentas will be separate, in the case of ICM division the placenta will be common but the amnions separate, and in the case of primitive streak division, both placenta and amnion will be common.

Mouse developmental genetics

Transgenic mice

The creation of mice with an engineered genetic constitution has become a substantial industry, with applications to many branches of biomedicine in addition to developmental biology. Sometimes the term **transgenic** is used to include all forms of genetic modification; here it will be used more narrowly to indicate introduction of new genes into the germ line. For many years the standard method was to inject the DNA directly into one **pronucleus** of the fertilized egg, but recently the **knock-in** method described below has gained popularity, because this means that the environment of the modified gene is completely

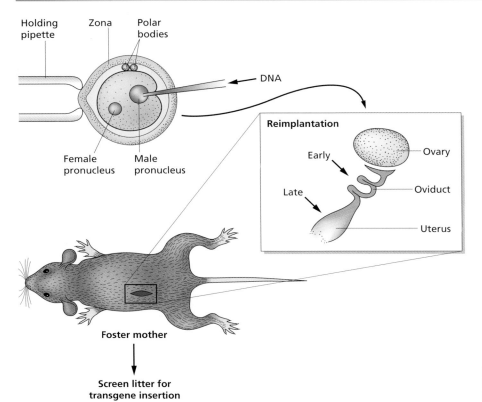

Fig. 10.23 Making transgenic mice by injection of DNA into a pronucleus of the fertilized egg.

normal. There are also methods being developed using integrase enzymes that can insert DNA at a predetermined site, previously introduced by homologous recombination.

The pronuclear injection methods (Fig. 10.23) lead to a reasonable yield of transgenics with a good probability that the germ line will be transgenic. Normally, integration is at a random position in the genome and comprises many copies of the transgene joined head to tail in a tandem array. Better levels of gene expression are achieved if the gene to be injected is purified away from plasmid DNA as the prokaryotic sequences are usually inhibitory to transgene expression. It is also important to use genomic DNA containing introns, as this gives better expression levels than complementary DNA (= cDNA; i.e. DNA copied from mRNA, without introns). Normally, linear rather than circular DNA is preferred for injection. The newborn mice arising from a transgenesis experiment will be screened for incorporation of the gene, by performing a genomic **Southern blot** or **polymerase chain reaction** (PCR) on a DNA sample from the tail tip of each individual. A transgene will usually behave as a simple dominant gene in subsequent breeding. It may be desirable to breed it to homozygosity so that all embryos from the line are known to possess the gene.

Despite the abnormal chromosomal location of pronuclear injection transgenes, they can show correct temporal and regional expression so long as sufficient flanking DNA sequences are included. Bacterial artificial chromosomes (BACs) can accept much larger inserts than plasmids, and a motivation for the use of BACs instead of plasmids for cloning the insert is the desire to include as much flanking DNA as possible. A very large amount of work has been performed in which the flanking sequences of the genes are modified in order to identify the particular regulatory sequences responsible for control of their expression. For this type of work it may be sufficient to use transient transgenics. This means that the injections are done, the embryos reimplanted, and then later recovered and analyzed directly without any breeding to establish a permanent line of animals. Instead of monitoring the normal gene product, such studies are usually done using a **reporter gene** whose product is more easily detected. Commonly used reporter genes are *lacZ* and *Luciferase*. The β-galactosidase encoded by the *lacZ* gene can easily be detected in wholemounts or sections using the X-Gal reaction. Luciferase is an enzyme from an insect that will convert a specific substrate, luciferin, into a luminescent product. It can be measured in tissue samples with very high sensitivity using a luminometer, and can also be visualized in intact mice using sensitive whole-body imaging systems.

By making the appropriate molecular constructs it is possible to express the coding region of one gene under the control of the promoter of another, and this enables ectopic expression of specific genes, and the resulting modification of the course of development. Sometimes uniform expression is required, which can be achieved using promoters for housekeeping genes such as cytoskeletal (β) actin (*Actb*) or histone H4 (*Hist4*).

Embryonic stem cells

Cell lines showing developmental **pluripotency** can be produced by putting mouse blastocysts into tissue culture. After the blastocyst hatches from the zona it will adhere to the substrate and, in a suitable medium, the ICM cells will proliferate. The resulting cells are called **embryonic stem cells (ES cells)**. Although they are derived from the ICM they do not have exactly the same gene expression profile and it is unclear whether they have an exact counterpart in the normal embryo. They can be cultured for many passages and be frozen for later use. The standard method for cultivation has been to suppress differentiation by growing the ES cells on **feeder layers** of irradiated fibroblasts or in the presence of leukemia inhibitory factor (LIF). Recently, it has been found that the pluripotency of ES cells can be maintained in the presence of inhibitors for components of the FGF and Wnt signaling pathways (2i or 3i medium). It used to be the case that ES cells could only be derived from a few strains of mice, but use of the new inhibitor-containing media enables production of ES cells from a wide range of mouse strains and also from the rat. When the feeder cells, or the LIF, or the inhibitors, are removed, the ES cells will differentiate into **embryoid bodies** in which a temporally normal but spatially very abnormal process of differentiation takes place. The reason that mouse ES cells are normally described as **pluripotent** rather than **totipotent** is that they do not normally form any trophectoderm on differentiation.

ES cells are described as **stem cells** because they display the characteristics of indefinite self-renewal and the ability to generate differentiated offspring (see Chapter 18). Their *in vivo* counterparts, ICM or early epiblast cells, are not stem cells because they quickly develop into other cell types under the influence of inductive signals. The stem-cell property of ES cells depends on a mutually upregulating set of transcription factors, including OCT4, NANOG, and SOX2, described as pluripotency factors. These upregulate each other and repress expression of the developmental control genes responsible for germ layer formation and other early events of regional specification. The developmental control genes are maintained in a state of competence by a "bivalent mark," namely the presence of antagonistic pairs of modified histones that can upregulate or repress activity of the associated genes. For example bivalent domains with histone methylation at both H3K27 and H3K4 have been described on many genes. Bivalent chromatin domains are not, however, unique to ES cells.

When implanted into immunologically compatible adult mice, ES cells form tumors called teratomas containing several differentiated tissue types arranged in a somewhat disorganized manner. When ES cells are injected into mouse embryo blastocysts they will integrate into the ICM, giving a high frequency of **chimerism**. In many cases the chimerism extends to the **germ cells**, making it possible to breed intact mice from cells that have been grown in culture (Fig. 10.24). A minority of ES lines are capable of forming a complete embryo on their own, after injection into **tetraploid** blastocysts, where the tetraploid host cells are capable of forming extraembryonic membranes but not a fetus. The existence of ES cells demonstrates the interesting fact that it is possible to disengage growth from developmental commitment, as lines have been passaged as many as 250 times without loss of the ability to repopulate embryos.

Human embryonic stem cells are an important topic in regenerative medicine (see Chapter 21). But so far the main practical importance of mouse ES cells is that they offer a sophisticated route for the reintroduction of genes into mouse embryos. Although genes can be introduced by simple injection of DNA into the zygote, this offers no control over the number of copies introduced or the location of the insertion site. By contrast, with ES cells in culture it is possible to select for rare events, particularly **homologous recombination** of exogenous DNA into the complementary site in the genome (see Animation 2: Homologous recombination). This has been used to **knock out** individual genes by replacing them with an inactive variant, to **knock in** one gene in place of another.

Knockouts and knock-ins

In developmental biology, the main object of knocking out individual genes is an expectation that the null phenotype will reveal the gene's function. Usually, the product of the gene in question is believed to be important for some reason, such as its biochemical activity, or expression pattern, or the existence of a developmental function for the homolog in another organism, such as *Drosophila*. But knockouts are also useful to create mice mutant for particular gene variants responsible for human genetic diseases. This can create an **animal model** of the human disease, so that further information can be gained about the pathology of the disease, and strategies for therapy can be tested. An example would be the creation of mice mutant for the cystic fibrosis transmembrane conductance regulator (CFTR protein) which serve as a model of human cystic fibrosis. Another is a mouse lacking the gene for apolipoprotein B, which develop atherosclerosis over 2–3 months. A further application in medical research is the creation of mice especially susceptible to cancer, for example, by removal of tumor-suppressor genes such as the *p53* gene.

Although knockout technology is now very sophisticated, the basic method usually employs the positive–negative selection procedure for homologous rather than random integrations. A targeting construct is assembled that consists of genomic DNA for the region to be replaced, with an essential functional region

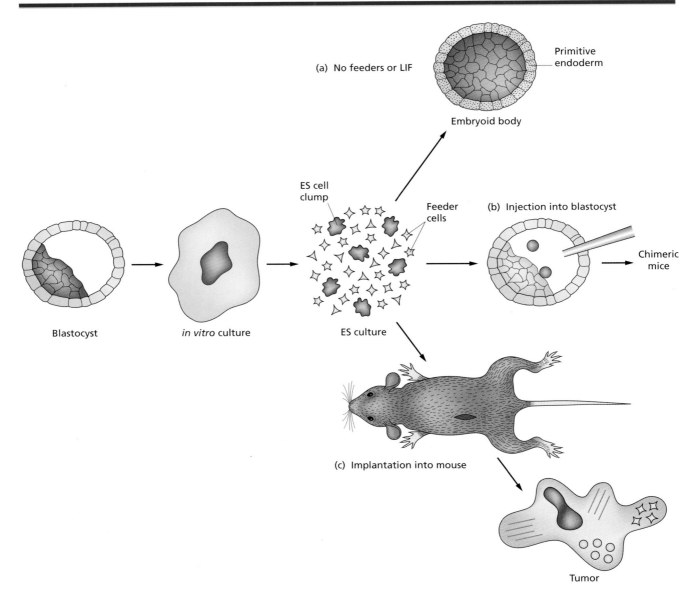

Fig. 10.24 Embryonic stem cells. (a) Removal of feeder cells will cause differentiation *in vitro* into "embryoid bodies." (b) Injection into blastocysts can generate chimeric mice. (c) Implantation under the kidney capsule can produce a teratoma.

of the gene replaced by an antibiotic-resistance gene, usually neomycin resistance (*neo*). Flanking this is a copy of a viral gene coding for thymidine kinase (*TK*). The construct is transfected into the ES cells and if it integrates by homologous recombination, only the *neo* will be incorporated (Fig. 10.25a). If it integrates at random, in the wrong place, both *neo* and *TK* will probably be incorporated (Fig. 10.25b). Then the cells are subject to selection using two drugs. **Neomycin** will kill the cells that have not incorporated the targeting vector at all, as the host ES cells are sensitive to it (the feeder cells are resistant). **Ganciclovir** will kill cells that have incorporated the construct in such a way that the *TK* gene is present. This is because the thymidine kinase converts the drug into a cytotoxic product. The net effect is that the surviving cells should be the ones that have undergone homologous recombination such that the target gene has been exactly replaced by the inactive version. These cells are grown up as individual clones and are screened by Southern blotting or PCR to ensure that the targeting construct has indeed integrated at the predicted position (Fig. 10.26). Then the cells are injected into host blastocysts and the embryos reimplanted into foster mothers. The strain of the host embryos will be chosen such that the coat color differs from that of the parent strain of the ES cells. The chimerism should then be indicated by a patchy coat color, and this is confirmed by analy-

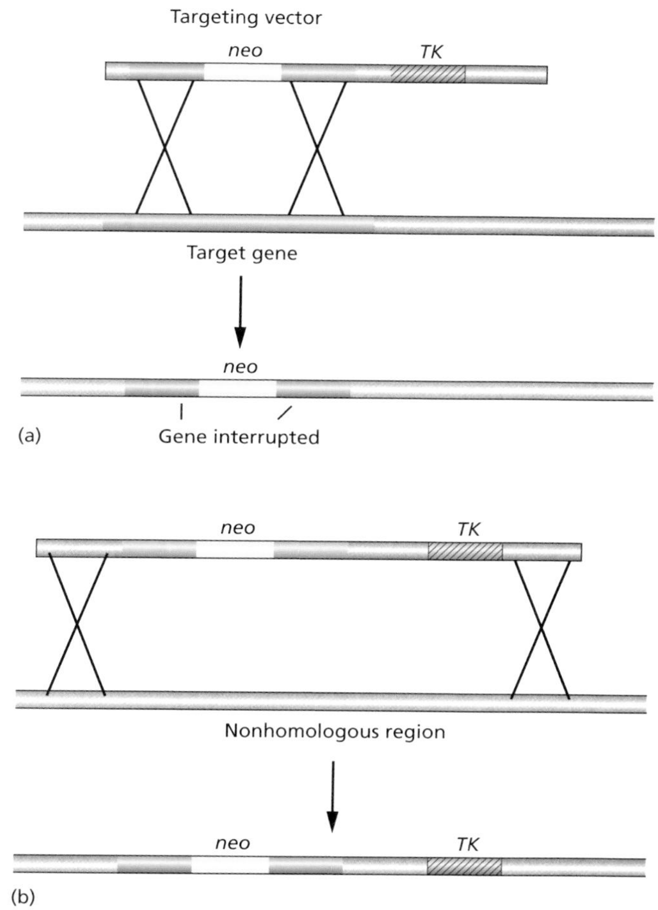

Fig. 10.25 Gene targeting: (a) recombination at the homologous site disrupts the host gene and introduces *neo* but not *TK*; (b) recombination at a nonhomologous site introduces both *neo* and *TK*.

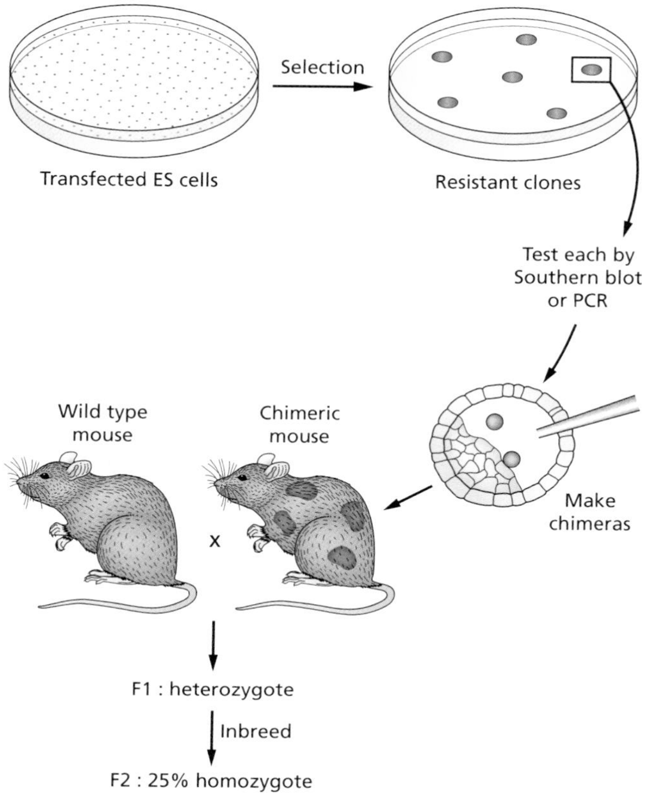

Fig. 10.26 Procedure for making a gene knockout via homologous recombination into ES cells.

sis of DNA from tail tips. Chimeric mice may or may not have any of the donor cells in their germ line, so they are mated and the resulting offspring also subject to tail-tip DNA analysis. If the F1 generation do contain the targeting construct, then they can be mated together to generate homozygotes, and the phenotype of the null mutation can be established.

Exactly the same technology can be used to perform "**knock-ins**." Here instead of replacing the endogenous gene with an inactive version, it is replaced by another gene, maybe a modified version of itself, or is supplemented by a reporter. The knock-in method has an advantage over traditional transgenesis because the insertion site and its environment are defined and so there should be no unpredictable consequences of random insertion.

There are various problems that have been encountered with the use of knockout mice. Firstly, there may be no abnormality of phenotype, or a barely perceptible abnormality. This is usually ascribed to **redundancy**, or the presence of other genes in the genome with overlapping functions, which is very common in vertebrates because of the large number of multi-gene families (see Chapter 3). In order to establish the function of a set of genes showing significant redundancy it is necessary to knock them all out and then assemble a multiple homozygous null by mating of the individual knockout lines together. Considerable work of this sort has been done, for example, on Hox gene paralog groups and on the various retinoic acid receptors. Secondly, the effects of the knockout may vary considerably depending on the genetic background, or strain of mouse into which the mutation is introduced. Although it may complicate the functional analysis this can be a useful property as it can help to identify other interacting genes important for the process under investigation. Finally, the phenotype may be so severe that the embryos die at an early stage, which precludes any examination of later functions of the same gene. There are various solutions to this problem. One is to use a **conditional knockout** strategy such as the Cre-lox system (see Chapters 3 and 13). Another is to make **chimeras** in which the gene is knocked out in only part of the conceptus.

Chimeric knockouts

The usual application of the chimera method is the case where a gene is required at an early stage in the extraembryonic tissues and at a later stage in the embryo itself, so the early death is due to failure of nutritional support from the extraembryonic tissues. As mentioned above, when ES cells are injected into a **tetraploid** blastocyst, for reasons that are not well understood the ES cells mainly form the embryo and the tetraploid cells mainly form the extraembryonic tissues. Tetraploid embryos are made by electrofusion of the blastomeres at the two-cell stage and are not themselves viable in the long term. To establish the postimplantation function of a gene, homozygous null ES cells, or ICM cells from a homozygous null blastocyst, are injected into tetraploid blastocysts, which are implanted into foster mothers. The tetraploid-derived extraembryonic structures support development until the gene is needed in the embryo itself, and from this stage an identifiable defect will arise, which will hopefully reveal something of the normal gene function. An example is the case of Fgf receptor 2 (*Fgfr2*). If knocked out, this causes preimplantation death, due to failure of proliferation of the polar trophectoderm. But a chimeric embryo, composed of *Fgfr2*⁻ cells in a wild-type tetraploid blastocyst, will develop to about E10.5, and at this stage it is clear that the embryos lack both lungs and limbs, showing that FGF signaling through this receptor is needed for the formation of both these organs.

Gene and enhancer traps

The knockout technology is very powerful, but it can only be applied to identified genes. Unknown genes need to be identified by mutagenesis. This may be chemical mutagenesis, as described in Chapters 3 and 8, followed by positional cloning of the genes. It can also be useful to carry out **insertional mutagenesis** with a suitable DNA vector, as this allows easy identification of the modified locus. The relevant constructs are called **gene traps** and contain a splice acceptor site, a reporter gene, and a selectable marker (Fig. 10.27). Often the reporter and selectable marker are combined in a single gene called **β-geo**, which is a fusion of *E. coli lacZ* and the neomycin-resistance gene, having the activities of both. The construct is transfected into ES cells. If it integrates outside a gene it remains inactive because it has no promoter of its own. If it integrates within a gene it should function as the last exon, with the splice acceptor

enabling incorporation into the coding sequence. Because the endogenous gene is disrupted by the insertion, it is likely to be mutated to inactivity or reduced activity. The β-geo will be expressed as a fusion on the C-terminal end of the truncated endogenous protein, and if the splicing is in frame then the cells should be selectable by neomycin resistance, and should also express β-galactosidase.

The gene trap is transfected into ES cells, then introduced into mice by the same procedure as used for knockouts. It is often possible to examine the normal expression pattern of the gene by doing X-Gal staining on the chimeras themselves. Although only the donor cells will carry the construct, it may be enough for this purpose. If the expression pattern looks interesting, the chimeras will be bred to establish an F1 generation heterozygous for the gene trap, and these will be mated with each other to establish an F2 with 25% homozygosity. At this stage it will be clear whether the trap has produced an interesting recessive phenotype. If so the gene will be cloned by making a genomic DNA library from the gene trap line and probing it for the gene trap vector sequence. The 5′ junction sequence should be within an intron of the trapped gene and enable rapid cloning of the remainder.

One important gene trap mouse strain is called **Rosa26**. In this case the trapped gene encodes an untranslated RNA of unknown function. But it has the property of being reliably expressed in a ubiquitous manner, at all stages and in all tissues. For this reason mouse lines designed for constitutive ubiquitous expression of a transgene are now often made by "knocking in" the required gene to the *Rosa26* locus, using the techniques of homologous recombination.

An **enhancer trap** is a gene trap that lacks the splice acceptor and carries its own minimal promoter, which provides a basal RNA polymerase II binding site but is not sufficient for detectable transcription (Fig. 10.27). If it enters the genome within range of an endogenous promoter or enhancer then this will complement the minimal promoter and enable significant transcription of the *lacZ*. Enhancer traps are usually not mutagenic as they may be activate anywhere within effective range of an endogenous enhancer. Because they may be at some distance from any endogenous gene they are also not so useful for cloning purposes. Their main use is providing lines of mice in which particular tissues or cell types are highlighted by expression of β-galactosidase and are therefore very easy to visualize.

Other topics in mouse development

Nuclear transplantation and imprinting

Nuclei transplanted from early blastomeres to enucleated eggs can support development, although the ability to do so falls off rapidly with stage, perhaps associated with the early onset of

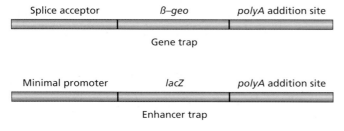

Fig. 10.27 Gene trap and enhancer trap.

zygotic transcription in the mouse. An essential condition for normal development is for the egg to contain one paternally derived and one maternally derived pronucleus. **Parthenogenetic** diploid embryos, made by activation of the egg and suppression of second polar body formation, develop poorly because of inadequate formation of extraembryonic tissues. **Androgenetic** diploids, made by removing the female pronucleus and injecting a second male pronucleus, have abundant extraembryonic tissue but the embryo arrests at an early stage. This suggests that there is some difference between the state of homologous genes on chromosomes from the two parents such that the paternal copy is more readily activated in the trophoblast and the maternal copy in the embryo.

The reason for this is the existence of **imprinted** genes on several chromosomes. These are genes that are only expressed from either the maternal or the paternal chromosome. It has been thought that there are about 100 such genes, mainly concerned with growth control. However, the number with parent-of-origin specific expression in the brain may be tenfold higher. In relation to the growth control functions the conventional explanation for the phenomenon is an evolutionary one, as natural selection will always favor traits that maximize the number of offspring. For the father the maximum offspring are achieved by mating with many females and by the embryos achieving the maximum growth rate relative to any other male's offspring. For females the maximum offspring are achieved by devoting equal resources to each embryo and by surviving the birth so that another reproductive cycle can take place. There is thus a potential evolutionary conflict between the traits that will be favored in males and females. Put crudely, the males "want" fast growth of the embryos while the females "want" uniform and controlled growth. An example supporting this principle is the insulin-like growth factor (IGF)-2 system. IGF2 is a growth factor promoting prenatal growth of the embryo and its gene is active only on the paternal chromosome. There is an inhibitor of IGF2 misleadingly called the "IGF2 receptor," and its gene is active only on the maternal chromosome. In such cases the effects of genetic heterozygosity depend on which chromosome (maternal or paternal origin) a mutation lies on. If an inactive allele is present on the nonexpressed chromosome then there will be no effect, while if it is present on the expressed chromosome the effect will be the same as that of a homozygous loss-of-function mutant. Conversely, it is possible to have mutations that lead to loss of imprinting on the normally nonexpressing chromosome and this may have effects by doubling the normal gene dosage. For example in humans the Beckwith–Widemann syndrome is a condition involving embryonic overgrowth and predisposition to cancer. It involves the disruption of the gene for a voltage-gated K^+ channel, but a high proportion of the cases also show loss of imprinting of IGF2. Another example is loss of imprinting of IGF2 in most cases of Wilm's tumor, a pediatric cancer of the kidney.

Although there are many other imprinted genes, the *Igf2* system itself seems particularly significant in terms of prevention of parthenogenetic development. If an oocyte is reconstructed to contain two haploid female nuclei, one of which is modified to enable expression of *Igf2* like a paternal nucleus, then the parthenogenetic embryo can develop to term.

Imprints become established in the germ line during gametogenesis. This process can be followed by nuclear transplantation into eggs, followed by analysis of whether both maternal and paternal alleles at an imprinted locus are expressed (Fig. 10.28). Whole embryo clones made with nuclei from primordial germ cells show an erasure of the original imprints from E11.5, probably associated with a genome-wide demethylation of DNA in these cells. The imprints become re-established late in female gametogenesis, during the growth phase of the oocyte, and earlier than this in male gametogenesis. It remains unclear how they are initially established, although their maintenance generally correlates with the DNA methylation state of the genes (see Appendix). Several experimentally introduced transgenes have been found to become imprinted, and are differentially methylated in sperm and eggs. In the early stages of embryonic development there is another general demethylation of the genome but this does not affect the imprints, which must be protected in some way.

New directions for research

Mice will continue to be used to make transgenic and knockout lines to address a huge range of biological issues, by no means confined only to mechanisms of development. For example these include problems in cell biology, immunity, and neurobiology as well as the study of disease mechanisms.

Increased sophistication in methods for labeling and modifying specific tissues will continue to provide an impetus for research in mechanisms of organogenesis and the properties of tissue-specific stem cells.

The ability to make ES cells from rats promises to create a new range of genetically modified rats which, because of their size, are more suitable for physiological and pharmacological investigations than are mice.

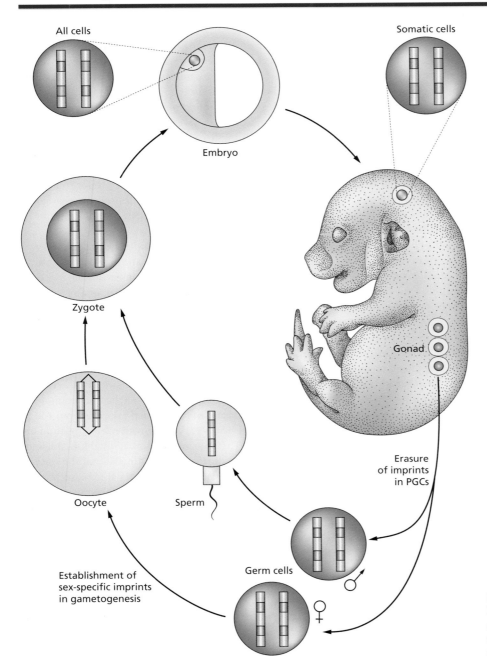

All cells

Embryo

Somatic cells

Zygote

Oocyte

Sperm

Germ cells

Gonad

Erasure
of imprints
in PGCs

Establishment of
sex-specific imprints
in gametogenesis

Fig. 10.28 "Life cycle" of imprinting. Two loci are shown, with red for an expressed gene and blue for an imprinted, inactive, gene. Imprints are erased in the primordial germ cells and reset during gametogenesis.

X-inactivation

In all female mammals one copy of the **X-chromosome** becomes heterochromatic and inactive, and is visible as a Barr body within the nucleus. This occurs at random at the egg cylinder stage such that either the maternally derived or the paternally derived X becomes inactive. The inactivation of one copy of the X-chromosome produces a **mosaic** with respect to any heterozygous locus on the chromosome, since some cells will express the gene in question and others will not and this ensures that all female mammals are naturally mosaic for all X-linked heterozygous loci. X inactivation is accompanied by late replication of the DNA relative to the rest of the genome, and by hypoacetylation of the histones on the inactive chromosome. In the mouse X-inactivation occurs earlier in the extraembryonic tissues than in the epiblast itself, and in these tissues the paternal X-chromosome is specifically inactivated (Fig. 10.29a). This is known as imprinted X-inactivation and shows some differences from the random X-inactivation found in the embryonic tissues. X-inactivation can be easily visualized in the H253 strain of

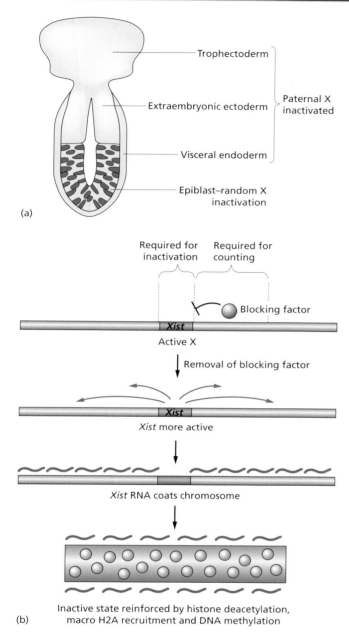

(a)

(b)

Fig. 10.29 Inactivation of the second X-chromosome in females. (a) Expression of *lacZ* in a heterozygous H253 mouse embryo which has *lacZ* on the paternal X-chromosome. (b) Events of X-inactivation.

mice, which carry a number of copies of bacterial *lacZ* on the X-chromosome. In female heterozygotes the chromosome carrying the *lacZ* is inactivated in 50% of cells so when a specimen is stained with X-Gal individual cells are either blue or white. This enables some types of **clonal analysis**, as any structure formed from a single cell must become either all blue or all white but cannot be of mixed composition.

X-inactivation depends on the presence in the chromosome of an X-inactivation center (Xic). This contains a gene called

Xist (pronounced "exist") which encodes a nontranslated RNA. *Xist* is critically important for X-inactivation and works only on its own chromosome, not on others within the same nucleus. It is active at a low level in all X chromosomes of the early embryo. In the chromosomes that become inactivated the activity increases and the *Xist* RNA coats the whole of the chromosome. If the *Xist* locus is deleted then the chromosome cannot become inactivated. Conversely, a transgene for *Xist* can cause inactivation of the chromosome where it lies, even if this is an autosome. Experiments with an inducible *Xist* transgene in ES cells have shown that it will cause inactivation of its own chromosome when induced and that its presence is required for 48 hours for maintenance of inactivation. After this it is no longer required, as the inactivation becomes stabilized by other mechanisms, such as recruitment of a special histone, macroH2A, and histone hypoacetylation (Fig. 10.29b). The promoter for *Xist* is differentially methylated during gametogenesis, such that it is demethylated in sperm and methylated in oocytes. This pattern is maintained in the extraembryonic tissues, correlating with the preferential inactivation of the paternal X. In the epiblast, the imprint is erased and replaced by random methylation of the promoter. There is an antisense gene called *Tsix* that overlaps the *Xist* gene on the other DNA strand. This is active in maternally derived X-chromosomes and helps make the maternal X resistant to early inactivation.

X-inactivation must involve a counting mechanism that senses the ratio between the number of X chromosomes and the number of autosomes. The evidence for this is that individuals who possess multiple X chromosomes inactivate all but one. A model consistent with these data involves a blocking factor, produced by autosomes, which is produced in limiting amount so that it can only block *Xist* expression on one X chromosome (Fig. 10.29b).

Teratocarcinoma

Most of the thoughts about the nature of ES cells and the uses to which they might be put were anticipated by earlier work on **teratocarcinomas**. These are malignant and transplantable tumors, which consist of several types of tissue and also contain undifferentiated cells. The undifferentiated cells, which will grow in tissue culture, are considered to be the stem cells for the tumor and are often called embryonal carcinoma (EC) cells. Three types of teratocarcinoma can be distinguished: spontaneous testicular, spontaneous ovarian, and embryo derived. Spontaneous testicular teratocarcinomas arise in the testes of fetal male mice of strain 129. They are thought to arise from primordial germ cells and the well-known F9 cell line is of this type. Spontaneous ovarian teratocarcinomas arise in females of LT mice at about 3 months of age. They are derived from oocytes that have completed the first meiotic division and undergo approximately normal embryonic development as far as the egg cylinder stage. Embryo-derived teratocarcinomas can be produced by grafting early mouse embryos to extrauterine sites in immunologically compatible hosts, usually the kidney or testis. The embryos become disorganized and produce various adult tissue types together with proliferating EC cells.

Different teratocarcinoma cell lines differ greatly in their properties: some need feeder cells to grow and others do not. Some will grow as dispersed cells in the peritoneal cavity (called ascites tumors) and others will not. Some will differentiate *in vivo* or *in vitro* while others will not. Most lines have an abnormal karyotype although a few are normal or nearly normal. Most of the differences have probably arisen by selection of variants during establishment of the tumor and during culture, although it may to some extent also reflect a heterogeneity of the parent cell type. Chimeric mice can be produced by injection of some types of teratocarcinoma cells into blastocysts, but this does not work nearly as well as with ES cells. It does, however, have some theoretical interest since it shows that at least one sort of tumor can be made to revert to normal behavior if it is placed in the appropriate biological environment.

Key points to remember

- Fertilization comprises several distinct steps. Firstly, recognition of the zona pellucida by the sperm. Secondly, the acrosome reaction leading to release of hydrolytic enzymes that digest a path through the zona. Thirdly, the fusion of the sperm and oocyte plasma membranes with introduction of a phospholipase C into the oocyte cytoplasm. Fourthly, the elevation of intracellular calcium, which provokes the release of cortical granules, the completion of the second meiotic division of the oocyte, and the resumption of DNA synthesis ready for the first cleavage division.
- During the first 4 days of development the mouse embryo is in the preimplantation phase. It lies free in the oviduct or uterus and develops up to the blastocyst stage. Preimplantation embryos can be cultured *in vitro* and manipulated. Subsequent postimplantation development occurs within the uterus with nutrition provided via the placenta.
- The overall course of development resembles the chick, but the arrangement of extraembryonic membranes is different and there are distinct features associated with the egg cylinder arrangement found in rodents. The embryo body plan is established in the egg cylinder epiblast by inductive signals from the extraembryonic ectoderm and the anterior visceral endoderm. NODAL is essential for formation of the mesoderm and the node region has similar properties to the organizer in *Xenopus*. FGF and WNT signaling are required for formation of posterior body parts.
- The breakage of symmetry that leads to left–right differences of body structure depends on the intrinsic asymmetry of the nodal cilia.
- Embryonic stem (ES) cells are cells cultured from the inner cell mass of blastocysts. They can renew themselves indefinitely *in vitro*, differentiate to embryoid bodies, form teratomas if implanted into immunocompetent hosts, and can contribute to all of the tissue types in the body if reintroduced into a blastocyst.
- Transgenic mice are those containing an extra gene. This can be introduced by injection of the DNA into a pronucleus of a fertilized egg, or by "knock-in" using ES cells. The injected eggs are implanted into the reproductive tract of a "foster mother" and reared to term. Then they are bred to create a transgenic line of mice. The specificity of transgene expression can be controlled by the promoter to which it is attached.
- Knockout mice are those in which a gene has been inactivated. Knockouts and knock-ins are made by homologous recombination of a targeting construct containing the modified gene into ES cells. The cells are injected into mouse blastocysts to create chimeric embryos. They are reimplanted into foster mothers and reared to term. Offspring that transmit the mutation through their gametes are used to breed a line of genetically modified mice.
- Some genes are expressed only from the maternal or the paternal chromosome (imprinting).
- In early embryos of females one copy of the X chromosome becomes inactivated.

Further reading

Websites

The Mouse Genome Informatics database (Jackson Laboratory) contains gene lists, expression patterns, phenotypes, and much other data: http://www.informatics.jax.org/
The Edinburgh databases of mouse embryo anatomy and gene expression: http://www.emouseatlas.org/emap/home.html

General

Theiler, K. (1989) *The House Mouse. Development and Normal Stages from Fertilization to Four Weeks of Age*, 2nd edn. Berlin: Springer-Verlag.
Kaufman, M.H. (1992) *The Atlas of Mouse Development*. London: Academic Press.
Alexandre, H. (2001) A history of mammalian embryological research. *International Journal of Developmental Biology* **45**, 457–467.
Nagy, M., Gertsenstein, M., Vintersten, K. & Behringer, R. (2003) *Manipulating the Mouse Embryo*, 3rd edn. Cold Spring Harbor, NY: Cold Spring Harbor Laboratory Press.
Niakan, K.K., Han, J., Pedersen, R.A., Simon, C. & Pera, R.A.R. (2012) Human pre-implantation embryo development. *Development* **139**, 829–841.

Fertilization

Wasserman, P.M., Jovine, L. & Litscher, E.S. (2001) A profile of fertilization in mammals. *Nature Cell Biology* **3**, E59–E64.
Runft, L.L., Jaffe, L.A. & Mehlmann, L.M. (2002) Egg activation at fertilization: where it all begins. *Developmental Biology* **245**, 237–254.
Talbot, P., Shur, B.D. & Myles, D.G. (2003) Cell adhesion and fertilization: steps in oocyte transport, sperm-zona pellucida interactions, and sperm-egg fusion. *Biology of Reproduction* **68**, 1–9.
Dean, J. (2004) Reassessing the molecular biology of sperm-egg recognition with mouse genetics. *Bioessays* **26**, 29–38.
Jones, K.T. (2005) Mammalian egg activation: from Ca2+ spiking to cell cycle progression. *Reproduction* **130**, 813–823.
Ogushi, S., Palmieri, C., Fulka, H., Saitou, M., Miyano, T. & Fulka, J., Jr. (2008) The maternal nucleolus is essential for early embryonic development in mammals. *Science* **319**, 613–616.
Schatten, H. & Sun, Q.Y. (2009) The role of centrosomes in mammalian fertilization and its significance for ICSI. *Molecular Human Reproduction* **15**, 531–538.

Morphology and fate map

Rossant, J. (1995) Development of extraembryonic lineages. *Seminars in Developmental Biology* **6**, 237–246.
Brune, R.M., Bard, J.B.L., Dubreuil, C., Guest, E., Hill, W., Kaufman, M., Stark, M., Davidson, D. & Baldock, R.A. (1999) A three dimensional model of the mouse at embryonic day 9. *Developmental Biology* **216**, 457–468.
Lawson, K.A. (1999) Fate mapping the mouse embryo. *International Journal of Developmental Biology* **43**, 773–775.
Gardner, R.L. (2001) Specification of embryonic axes begins before cleavage in normal mouse development. *Development* **128**, 839–847.
Piotrowska, K. & Zernicka-Goetz, M. (2001) Role for sperm in spatial patterning of the early mouse embryo. *Nature* **409**, 517–521.

ES cells and gene targeting

Gossen, M. & Bujard, H. (2002) Studying gene function in eukaryotes by conditional gene inactivation. *Annual Reviews of Genetics* **36**, 153–173.

Tam, P.P.L. & Rossant, J. (2003) Mouse embryonic chimeras: tools for studying mammalian development. *Development* **130**, 6155–6163.
Capecchi, M.R. (2005) Gene targeting in mice: functional analysis of the mammalian genome for the twenty-first century. *Nature Review Genetics* **6**, 507–512.
Niwa, H. (2007) How is pluripotency determined and maintained? *Development* **134**, 635–646.
Ying, Q.-L., Wray, J., Nichols, J., Batlle-Morera, L., Doble, B., Woodgett, J., Cohen, P. & Smith, A. (2008) The ground state of embryonic stem cell self-renewal. *Nature* **453**, 519–523.
Evans, M. (2011) Discovering pluripotency: 30 years of mouse embryonic stem cells. *Nature Reviews Molecular Cell Biology* **12**, 680–686.
Nichols, J. & Smith, A. (2011) The origin and identity of embryonic stem cells. *Development* **138**, 3–8.

Early developmental mechanisms

Liu, P.T., Wakamiya, M., Shea, M.J., Albrecht, U., Behringer, R.R. & Bradley, A. (1999) Requirement for Wnt3 in vertebrate axis formation. *Nature Genetics* **22**, 361–365.
Bachiller, D., Klingensmith, J., Kemp, C., Belo, J.A., Anderson, R.M., May, S.R., McMahon, J.A., McMahon, A.P., Harland, R.M. & Rossant, J. (2000) The organizer factors Chordin and Noggin are required for mouse forebrain development. *Nature* **403**, 658–661.
Brennan, J., Lu, C.C., Norris, D.P., Rodriguez, T.A., Beddington, R.S.P. & Robertson, E.J. (2001) Nodal signalling in the epiblast patterns the early mouse embryo. *Nature* **411**, 965–969.
Lohnes, D. (2003) The Cdx1 homeodomain protein: an integrator of posterior signaling in the mouse. *Bioessays* **25**, 971–980.
Sutherland, A. (2003) Mechanisms of implantation in the mouse: differentiation and functional importance of trophoblast giant cell behaviour. *Developmental Biology* **258**, 241–251.
Arnold, S.J. & Robertson, E.J. (2009) Making a commitment: cell lineage allocation and axis patterning in the early mouse embryo. *Nature Reviews Molecular Cell Biology* **10**, 91–103.
Johnson, M.H. (2009) From mouse egg to mouse embryo: polarities, axes, and tissues. *Annual Review of Cell and Developmental Biology* **25**, 483–512.
Rossant, J. & Tam, P.P.L. (2009) Blastocyst lineage formation, early embryonic asymmetries and axis patterning in the mouse. *Development* **136**, 701–713.
Takaoka, K. & Hamada, H. (2012) Cell fate decisions and axis determination in the early mouse embryo. *Development* **139**, 3–14.

Left-right asymmetry

Mercola, M. & Levin, M. (2001) Left-right asymmetry determination in vertebrates. *Annual Reviews of Cell and Developmental Biology* **17**, 779–805.
Hamada, H., Meno, C., Watanabe, D. & Saijoh, Y. (2002) Establishment of vertebrate left-right asymmetry. *Nature Reviews Genetics* **3**, 103–113.
McGrath, J. & Brueckner, M. (2003) Cilia are at the heart of vertebrate left-right asymmetry. *Current Opinion in Genetics and Development* **13**, 385–392.
Levin, M. & Palmer, A.R. (2007) Left-right patterning from the inside out: Widespread evidence for intracellular control. *BioEssays* **29**, 271–287.

Hox genes

Hunt, P. & Krumlauf, R. (1992) Hox codes and positional specification in vertebrate embryonic axes. *Annual Review of Cell and Developmental Biology* **8**, 227–256.

van den Akker, E., Fromental-Ramain, C., de Graaff, W., Le Mouellic, H., Brulet, P., Chambon, P. & Deschamps, J. (2001) Axial skeletal patterning in mice lacking all paralogous group 8 Hox genes. *Development* **128**, 1911–1921.

Wellik, D.M. & Capecchi, M.R. (2003) Hox10 and Hox11 genes are required to globally pattern the mammalian skeleton. *Science* **301**, 363–367.

Deschamps, J. & van Nes, J. (2005) Developmental regulation of the Hox genes during axial morphogenesis in the mouse. *Development* **132**, 2931–2942.

Imprinting and X-inactivation

Moore, T. & Haig, D. (1991) Genomic imprinting in mammalian development: a parental tug of war. *Trends in Genetics* **7**, 45–49.

Avner, P. & Heard, E. (2001) X-chromosome inactivation: counting, choice, and initiation. *Nature Reviews Genetics* **2**, 59–67.

Wilkins, J.F. (2005) Genomic imprinting and methylation: epigenetic canalization and conflict. *Trends in Genetics* **21**, 356–365.

Reik, W. (2007) Stability and flexibility of epigenetic gene regulation in mammalian development. *Nature* **447**, 425–432.

Augui, S., Nora, E.P. & Heard, E. (2011) Regulation of X-chromosome inactivation by the X-inactivation centre. *Nature Reviews Genetics* **12**, 429–442.

Ferguson-Smith, A.C. (2011) Genomic imprinting: the emergence of an epigenetic paradigm. *Nature Reviews Genetics* **12**, 565–575.

This chapter contains the following animations:

Animation 2   Homologous recombination.
Animation 12   Mammalian fertilization.

 For additional resources for this book visit
www.essentialdevelopmentalbiology.com

Chapter 11

Drosophila

The first organism whose development was understood in molecular detail was the fruit fly *Drosophila melanogaster*. *Drosophila* is highly suited to genetic experimentation because of its small size and its short life cycle of 2 weeks. A number of very sophisticated techniques have been developed for constructing stocks and carrying out mutagenesis screens. Historically, developmental biology research on *Drosophila* started with mutagenesis to produce mutants with interesting-looking phenotypes. Then the genes were cloned and the expression patterns studied by *in situ* hybridization. Then the interactions between genes were deduced by examining the expression of one gene in embryos in which there is loss of expression, or ectopic expression, of another. Finally, more details about the molecular biology were obtained by *in vitro* studies of interactions between transcription factors and regulatory regions in the DNA.

More recently, the toolkit of techniques has been applied to a wider range of cell and developmental biology issues, especially the behavior of morphogen gradients and the control of growth. Because *Drosophila* genes were mostly discovered as mutations, their names are written with the first letter lower case if the first allele to be discovered was recessive. A table of genes mentioned in this chapter is presented as Table 11.1.

Insects

All adult and larval insects are built up of an anteroposterior sequence of segments which fall into the three principal body regions of head, thorax, and abdomen. The prototype body plan is most clearly seen at the embryonic stage, which is called the **extended germ band** and is the **phylotypic** stage at which all insect species display their maximum morphological similarity. The head may consist of as many as six segments: three pro-

cephalic and three gnathal (pronounced "naythal"), the gnathal segments bearing leg bud-like appendages, which later become the mouthparts. The middle part of the body is the thorax, which always consists of three segments: the **prothorax** (**T1**), **mesothorax** (**T2**), and **metathorax** (**T3**). All of these develop a pair of legs on the ventral side and the meso- and metathorax also produce a pair of wings on the dorsal side. The number of abdominal segments varies with species but is usually in the range 8–11.

Drosophila, like other Diptera (two-winged flies), follows the general insect pattern except that it does not display the procephalic head segments even in the embryo, the three gnathal segments appear only transiently, and only the mesothorax bears wings, the metathoracic wings being represented by small balancing structures called **halteres**. *Drosophila* has the usual three thoracic segments (T1–3) and has eight abdominal segments (A1–8). Although the conventional segments are dominant in the larval and adult body plan, during early development the most important repeating units are the **parasegments**, which have the same period but are out of phase with the later segments (see below).

Inside the insect body the principal nerve cord is on the ventral side. The heart is on the dorsal side, and its action moves the hemolymph around the body cavity, there being no specialized vascular system as in vertebrates. Oxygen is brought to the tissues by diffusion through tracheae, which are long, branched ingrowths of the epidermis.

Drosophila is a **holometabolous** insect, meaning that it undergoes an abrupt and complete metamorphosis, so the egg hatches into a larva which is quite different in structure from the adult. The larva grows and passes through two molts before becoming a resting stage called a **pupa** in which the body is remolded to form the adult. Much of the adult body is formed from the **imaginal discs** and the abdominal histoblasts, which

Essential Developmental Biology, Third Edition. Jonathan M.W. Slack.
© 2013 John Wiley & Sons, Ltd. Published 2013 by John Wiley & Sons, Ltd.

Gene	Nature of protein	Vertebrate homolog
Dorsoventral systems		
K10	DNA binding	
gurken	Inducing factor, ligand for EGFR	TGFα
Egfr (=torpedo)	Cell surface receptor	EGF receptor
spätzle	Inducing factor, ligand for Toll	
easter	Extracellular protease	
pipe	HSPG sulfotransferase	
Toll	Cell surface receptor	Toll
cactus	Signal transduction	IκB
dorsal	Signal transduction	NFκB
twist	Transcription factor	Twist
snail	Transcription factor	Snail
zerknüllt	Transcription factor	Hox 3
rhomboid	Transmembrane protein	
decapentaplegic	Inducing factor	BMPs
screw	Inducing factor	BMPs
short gastrulation	Inhibitor of Dpp	Chordin
tinman	Transcription factor	Nkx2.5
Maternal anteroposterior systems		
Delta	Inducing factor, ligand for Notch	Delta
Notch	Cell surface receptor	Notch
unpaired (=outstretched)	Inducing factor	
gurken	Inducing factor, ligand for EGFR	TGFα
bazooka	Cytoplasmic protein	Par3
bicoid	Transcription factor	
caudal	Transcription factor	Cdx
oskar	Cytoplasmic protein	
nanos	RNA binding protein	
pumilio	RNA binding protein	
vasa	RNA helicase	Vasa
torso	Cell surface receptor	
torsolike	Secreted protein	
trunk	Inducing factor, ligand for Torso	
Gap genes		
orthodenticle	Transcription factor	Otx
hunchback	Transcription factor	
Krüppel	Transcription factor	
knirps	Transcription factor	
giant	Transcription factor	
tailless	Transcription factor	Tlx
huckebein	Transcription factor	
Pair-rule genes		
hairy	Transcription factor	Hes, Her
runt	Transcription factor	Runx
even skipped	Transcription factor	Evx
odd skipped	Transcription factor	
fushi tarazu	Transcription factor	
paired	Transcription factor	
odd paired	Transcription factor	
sloppy paired	Transcription factor	
Segment polarity genes		
hedgehog	Inducing factor, ligand for Patched	Hh
wingless	Inducing factor, ligand for Frizzled	Wnt
engrailed	Transcription factor	En
frizzled	Cell surface receptor	Frizzled
zeste-white 3	Signal transduction	Glycogen synthase kinase 3

Table 11.1 Key genes involved in *Drosophila* early development.

Table 11.1 (*Continued*)

Gene	Nature of protein	Vertebrate homolog
armadillo	Signal transduction	β-catenin
pangolin	Transcription factor	Lef, Tcf
patched	Cell surface receptor	Patched
smoothened	Signal transduction	Smoothened
cubitus interruptus	Transcription factor	Gli
Hox genes		
labial	Transcription factor	Hox 1
Deformed	Transcription factor	Hox 4
Sex combs reduced	Transcription factor	Hox 5
Antennapedia	Transcription factor	Hox 6
Ultrabithorax	Transcription factor	Hox 7
abdominal A	Transcription factor	Hox 8
Abdominal B	Transcription factor	Hox 9–13

Note that many of the genes featured have additional functions at later stages of development (e.g. the Delta–Notch system is very important in neurogenesis). The vertebrate homologs are usually gene families arising from duplication events. Some vertebrate genes are named after the *Drosophila* genes because they were discovered first in *Drosophila*.

Classic experiments

The breakthrough from mutants to gene function

Homeotic mutants had been known in *Drosophila* since the 1920s but few investigators were interested in them. One was Ed Lewis who published a classic study of the Bithorax complex in 1978 and proposed a model for body-plan development. The full range of developmental genes was revealed by the mutagenesis screens of Christiane Nüsslein-Volhard and Eric Wieschaus, which provided the raw material for the work of most labs in the 1980s. These three workers were awarded the Nobel Prize for Physiology in 1995.

The newly invented techniques of molecular biology made it possible to clone the genes that gave rise to developmental mutants and to study their expression patterns. This showed that the homeotic genes were indeed expressed in spatial domains, and that genes with periodic phenotypes, like the pair-rule genes, were expressed in periodic patterns.

Lewis, E.B. (1978) A gene complex controlling segmentation in Drosophila. *Nature* **276**, 565–570.
Nüsslein-Volhard, C. & Wieschaus, E. (1980) Mutations affecting segment number and polarity in Drosophila. *Nature* **287**, 795–801.
Akam, M.E. (1983) The location of Ultrabithorax transcripts in Drosophila tissue sections. *EMBO Journal* **2**, 2075–2084.
Hafen, E., Kuroiwa, A. & Gehring, W.J. (1984) Spatial distribution of transcripts from the segmentation gene fushi-tarazu during Drosophila embryonic development. *Cell* **37**, 833–841.

are present as undifferentiated buds in the larva. Imaginal disc development is described in Chapter 17. Some other insect orders are **hemimetabolous**, meaning that the larva resembles a small-scale version of the adult and acquires the final adult form via a series of nymphal stages separated by molts.

Normal development

Oogenesis

Events during **oogenesis** are very important for regional specification in the *Drosophila* embryo, and the egg is laid with considerable amount of its pattern already specified. At the start of oogenesis, one germ cell divides four times to produce 16 cells, one of which becomes the **oocyte** and the other 15 all become **nurse cells**. Interestingly, the *par1* gene, important for cytoplasmic asymmetry in *C. elegans* (see Chapter 12) is also essential for oocyte formation in *Drosophila*, and without it all 16 cells become nurse cells. The whole cluster of oocyte and nurse cells is surrounded by ovarian follicle cells to form the **egg chamber** (Fig. 11.1). The follicle cells are derived from the gonads and are thus of **somatic** rather than **germ-line** origin. As the egg chamber enlarges, the follicle cells become divided into three populations. Those over the nurse cells are **squamous** in form and those over the oocyte are columnar. At both ends of the oocyte there is a special group of follicle cells called **border cells**, which are important in the determination of anteroposterior pattern. The nurse cells become polyploid and export large amounts of RNA and protein into the oocyte, contributing to its increase in size. In the later stages the oocyte becomes visibly polarized in both dorsoventral and

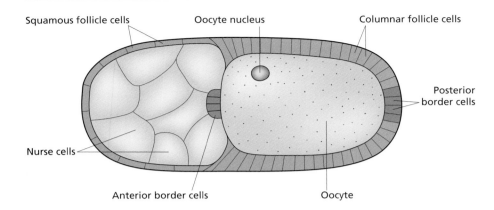

Squamous follicle cells   Oocyte nucleus   Columnar follicle cells

Posterior border cells

Nurse cells

Anterior border cells   Oocyte

Fig. 11.1 The *Drosophila* egg chamber.

anteroposterior axes and a granular **pole plasm** forms at the posterior end. The follicle cells secrete both the **vitelline membrane** and the **chorion** which is a tough outer coat surrounding the egg. The major production site for the **yolk** of the egg is the fat body of the female fly, the yolk proteins being carried to the ovary through the hemolymph.

Embryogenesis

Fertilization occurs in the uterus, the sperm entering at the anterior end through a hole in the chorion called the micropyle. Development of *Drosophila* is very fast compared with most other insects and the larvae hatch after less than 24 hours at normal laboratory temperatures. The early embryonic stages are depicted in Fig. 11.2. The initial period is called "cleavage" but as in most insects it is actually a period of rapid synchronous nuclear divisions without cellular cleavage. This is sometimes called **superficial cleavage** (see Chapter 2) and at this early stage the whole embryo forms a **syncytium**, in which all the nuclei lie in a common cytoplasm. After the first eight divisions the **pole cells** are formed at the posterior end, incorporating those nuclei lying within the pole plasm (Fig. 11.3). These later become the **germ cells**. After nine divisions most of the nuclei migrate to the periphery to form the syncytial **blastoderm**, those remaining internally are later incorporated into vitellophages, which end up in the gut lumen. After four more nuclear divisions, during the third hour, cell membranes grow inwards from the plasma membrane to separate the nuclei and the **cellular blastoderm** is formed. At this stage there are about 5000 surface cells, 1000 yolk nuclei, and 16–32 pole cells. The division rate of the blastoderm cells slows dramatically, and the pole cells divide just once more before gastrulation. In insect embryology positions along the anteroposterior axis of the early stages are often expressed as percentage egg length (%EL) from the posterior pole. So, for example, the region designated 100–50%EL is the anterior half, and the region 10–0%EL is the posterior tenth of the embryo.

The columnar epithelium of the blastoderm is quite thick and most of it is destined to become part of the embryo, only a thin dorsal strip becoming the extraembryonic amnioserosa. Gastrulation commences at about 3 hours with the formation of a deep **ventral furrow** along much of the embryo length (Figs 11.2e–g, 11.4). This consists of a mesodermal invagination along the ventral midline, joined shortly later by invaginations of **anterior** and **posterior midgut** at the respective ends. The cephalic furrow appears laterally at 65%EL. Concurrent with gastrulation the germ band begins to elongate, driving the posterior end with the pole cells round to the dorsal side of the egg. By about 4 hours the first neuroblasts appear in the neurogenic ectoderm, which is now mid-ventral, having closed over the ventral furrow.

Segmentation initially appears at the extended germ-band stage. The initial repeating pattern is of **parasegments**. The definitive segments each form from the posterior two-thirds of one parasegment combined with the anterior third of the next. Although their morphological appearance is transient, about 5–7 hours postfertilization, parasegments are important because they are fundamental units for the construction of the body plan (see below).

At about 7.5 hours the germ band retracts and as it does so the epidermal grooves rearrange themselves so that they separate the definitive segments. The anterior and posterior midgut fuse in the middle and the ventral nerve cord becomes segregated. **Dorsal closure** of the epidermis occurs at 10–11 hours, displacing the amnioserosa into the interior. At about this time the head "involutes" into the interior and is therefore scarcely represented on the outer surface of the larva. This happens also in other Diptera but is unusual for insects in general. In later stages the Malpighian tubules (excretory organs) are formed at the junction of posterior midgut and hindgut; muscles, the fat body, and the gonads arise from the mesoderm, and the central nervous system is formed by the ganglia of the ventral nerve cord.

Larval stages

The *Drosophila* larva has no legs, its head is tucked away in the interior, and it has three thoracic and eight visible abdominal segments. The specializations of the epidermal cuticle, which

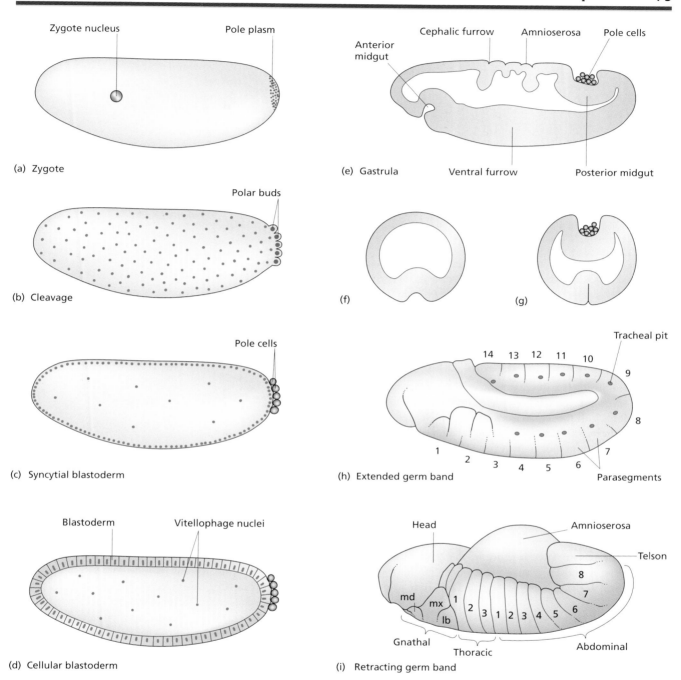

Fig. 11.2 Early development of *Drosophila*. (f) and (g) are, respectively, middle and posterior transverse sections of the gastrula. In (i) md, mandible; mx, maxilla; lb, labium.

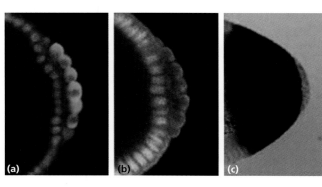

Fig. 11.3 Pole cells. (a) Immunostain for RNA polymerase II (red) and Vasa protein (green) distinguishes somatic and pole cells. (b) Immunostain for RNA polymerase II (red) and histone 3 K4Me (green) (somatic nuclei have both and are yellow). (c) *In situ* hybridization for *tailless* mRNA (dark blue). This is expressed by somatic but not pole cells. Reproduced from Strome & Lehmann, Germ Versus Soma Decisions: Lessons from Flies and Worms. *Science* 2007: 316; 392, with permission from American Association for the Advancement of Science.

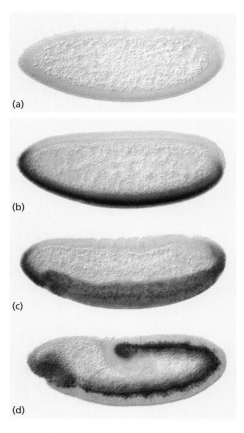

Fig. 11.4 *Drosophila* gastrulation visualized by expression of the Twist protein (brown): (a) syncytial blastoderm; (b) cellular blastoderm; (c) start of gastrulation; (d) extended germ band. Reproduced from Leptin, M. (2004) Gastrulation in Drosophila. In *Gastrulation: From Cells to Embryo*, C.D.Stern (ed.), with permission from Cold Spring Harbor Laboratory Press.

are formed during late development, are important as they are used to assess the phenotypes of late embryo lethal mutations. On the dorsal side the region from T2 to A8 is covered with fine hairs. On the ventral side are **denticle** belts on each of the thoracic and abdominal segments (Fig. 11.5). Each belt occupies mainly the anterior part of a segment but it also straddles the segment boundary and extends slightly into the posterior of the next segment. The thoracic and abdominal segments can be distinguished by the shapes and sizes of the denticle belts. Segment A8 bears the posterior spiracles, which are openings of the tracheal system. The extreme posterior is unsegmented and is called the **telson**. The structure of the larval head is very complex, but its most prominent component is a horny cephalopharyngeal skeleton secreted by the stomodeal part of the alimentary tract.

The **imaginal discs** are present as small nests of cells in the first-instar larva. They expand slightly during embryonic development, but most of their growth occurs during the larval stages. During pupation they differentiate and form the main epidermal structures of the adult body. The abdominal segments are formed from abdominal histoblasts, which grow only in the pupal stage.

Fate map

Fate maps for the cellular blastoderm-stage *Drosophila* embryo have been constructed by localized UV irradiation to produce small defects, and by injection of cells labeled with **horseradish peroxidase** (Fig. 11.6). The principal features of the fate map are as follows: prospective regions exist for all the larval seg-

ments, there being no regions of indeterminacy representing later cell mixing or later growth from a small bud. The prospective regions for the recognizable gnathal, thoracic, and abdominal segments are arranged in anteroposterior sequence from 75 to 15% of the egg length. The prospective procephalic head structures occupy the anterior 25%. The prospective anterior midgut also maps to the anterior, while the posterior midgut, Malpighian tubules, proctodeum, and germ cells map to the posterior, as one would expect from the descriptive embryology.

Pole plasm

The pole plasm is formed during oogenesis and contains a **determinant** for germ cells, important components of which are the products of the genes *oskar* and *vasa*. Direct evidence that the pole plasm can program the nuclei to become germ cells was obtained from transplants of pole plasm to the anterior

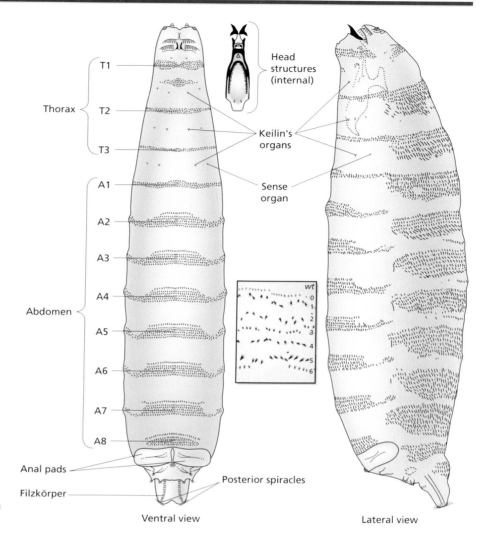

Fig. 11.5 *Drosophila* larva, ventral and lateral views. Inset: denticle belts in a third instar larva, the rows 0–1 belong to the posterior compartment and 2–6 to the adjoining anterior compartment. Reproduced from Repiso *et al. Development* 2010; 137: 3411–3415, with permission from Company of Biologists Ltd.

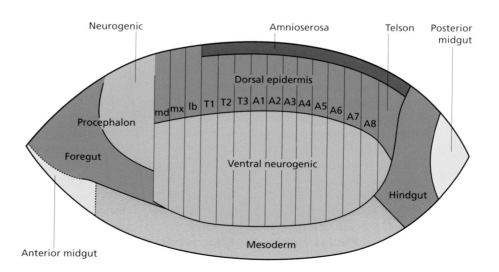

Fig. 11.6 Fate map of the cellular blastoderm stage. Reproduced from Hartenstein (1993) *Atlas of Drosophila Development*. Cold Spring Harbor Laboratory Press.

end of host eggs at cleavage stages. Ectopic pole cells arise from those nuclei that are surrounded by the grafted pole plasm. If these ectopic pole cells are then grafted back to the posterior of a second host, they can be incorporated into the gonad and form functional germ cells. A number of maternal-effect mutations prevent the formation of the pole plasm and result in sterility of the resulting offspring.

Drosophila developmental genetics

The understanding of *Drosophila* development depends largely on the sophistication of the genetic methods available. The genome size is small in comparison with vertebrates, both in gene number and in DNA content. The generation time is short and the animals are small, so experiments involving complicated breeding protocols and large numbers of individuals can be completed in a few weeks. The genetic maps are very detailed as are the cytological maps of the giant polytene chromosomes from the larval salivary glands.

Transgenesis

A transposable element known as the **P-element** is important for the creation of transgenic lines. The gene of interest is cloned into a P-element vector containing a marker gene, often the wild-type allele of the *white* gene which produces the normal red pigment in the eyes. The recombinant P-element is injected into the posterior pole of early embryos of the *white* genotype together with plasmid or mRNA encoding the transposase enzyme. The resulting flies are crossed to *white* flies. If the P-element has integrated into some germ cells the flies resulting from the injected embryos will be mosaics and will produce some uniformly transgenic offspring, which can be identified by their possession of red eyes. From these a stable line can be bred.

P-element transgenesis has been used for many purposes: to introduce enhancer traps (see also Chapters 3 and 10), to create strains carrying reporter constructs, to rescue endogenous mutations, and to express genes in an ectopic manner. It does have some limitations. The maximum size of insert is 20–25 kb, and the element tends to integrate near the 5′ end of genes, thereby causing insertional mutagenesis. Recently, methods for *Drosophila* transgenesis have been introduced using phage ΦC31 integrase, which catalyzes recombination between special *attB* sites in the injected plasmid and *attP* sites in the host. The *attP* sites need to be previously introduced with a P-element, but this means that the position of integration can be chosen to avoid position effects, and larger DNA sequences can be introduced, enabling the use of genomic sequences which give normal expression levels of the transgene.

Because of its propensity to introduce insertional mutations, the P-element has been specifically used for insertional muta-

genesis. For this purpose, a line carrying a P-element is crossed to another line carrying the gene for the transposase enzyme. In the offspring the transposase can move the P-elements around in the genome, so the gametes contain a range of new insertions, each tagged by the presence of the P-element.

Identification of developmentally important genes

The *Drosophila* genome contains about 13,000 genes altogether, including about 5000 that can be mutated to lethality. By mutagenesis screening it was possible to assemble a fairly complete collection of genes that, when mutated, give rise to pattern alterations in the embryo. There are several types of protocol for doing this, of which a simple example is shown in Fig. 11.7. Although the basic principle is the same as the zebrafish screen outlined in Chapter 3, the special tricks available for *Drosophila* select against the unwanted classes of recombinant and greatly reduce the labor involved. The screen illustrated in Fig. 11.7 is for autosomal recessive mutations and like most screens depends on the use of special **balancer chromosomes**, which have three important features: they suppress recombination, they carry some visible marker gene, and they are lethal in the homozygous condition. Males carrying a visible marker (*a* – distinct from the balancer marker) are mutagenized and are then mated to females carrying the balancer. Each *individual* offspring fly represents one mutagenized gamete, so individual males from the F1 generation are isolated and crossed again. For creation of the F2, females are used that carry a dominant temperature-sensitive lethal mutation on the chromosome opposite the balancer and this enables selection against offspring not carrying the mutant chromosome. In the F2 generation each *tube* of flies represents one of the original mutagenized gametes. The viable F2 flies are mated with each other to produce the F3 generation. As the F2s are heterozygous for the mutagenized chromosome and the balancer, 25% of the F3 will have a homozygosed mutant chromosome. If these homozygous mutants are viable, they will display the recessive marker phenotype (*a/a*) from the mutagenized males. If this marker is not visible, then the dead embryos or larvae are examined to see whether they have any significant pattern abnormality. In the case where the mutants are lethal the stock is maintained by breeding from the heterozygotes that have one mutant chromosome and one balancer. These will breed true because both the possible homozygous chromosome combinations that can arise from mating are lethal. The original mutagenesis will induce mutations on all the chromosomes, but those that do not lie on the balanced chromosome are mostly lost in the outcrosses, so the method enables the screen to be carried out for one chromosome at a time. Once obtained, mutations are sorted into groups of alleles of the same gene by **complementation tests**, and are genetically mapped.

In *Drosophila* much pattern information is laid down during oogenesis and mutations in genes required during oogenesis

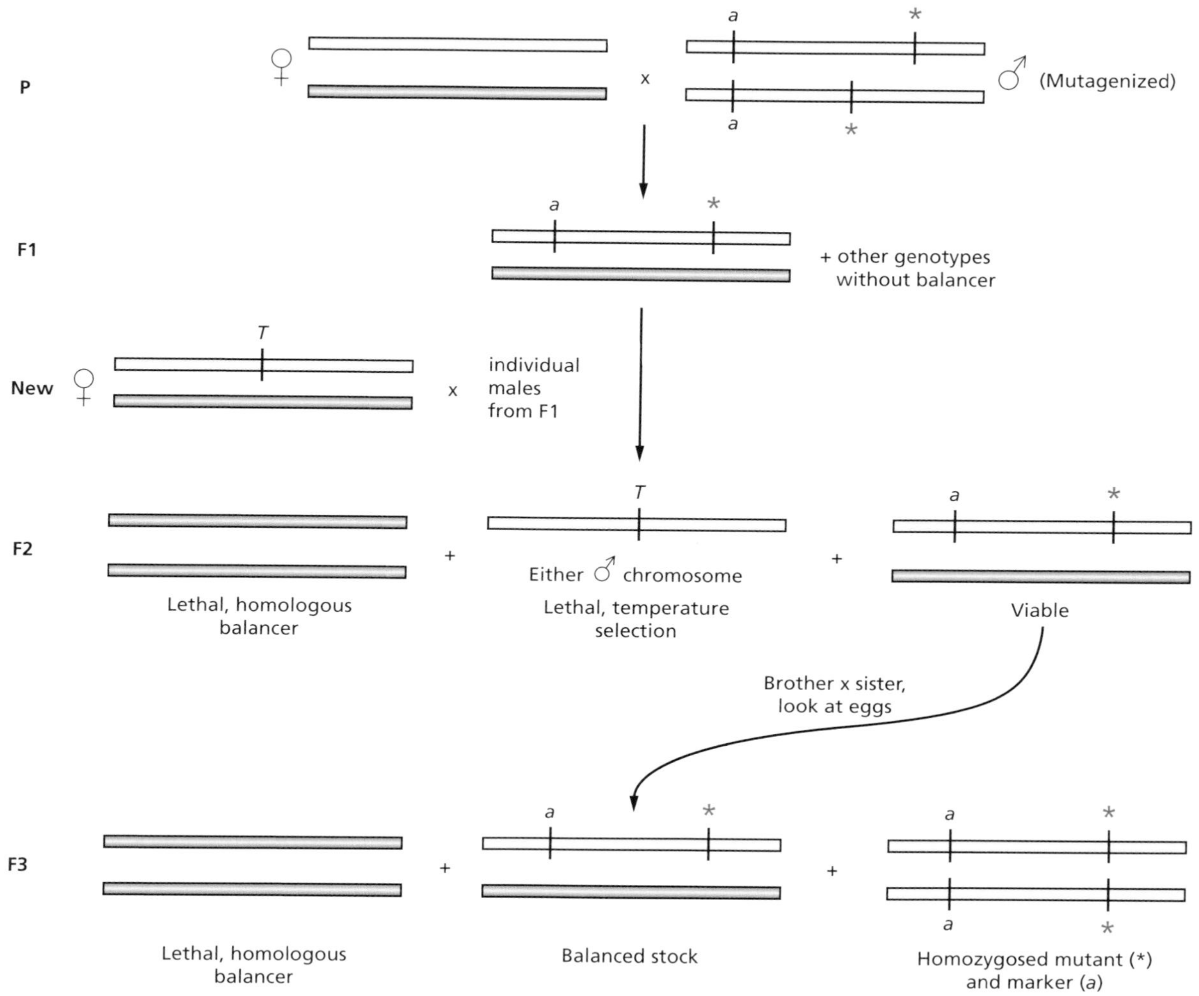

Fig. 11.7 A mutagenesis screen for zygotic autosomal recessive mutants affecting early development. The dark blue chromosome is a **balancer**, which suppresses recombination, carries a visible dominant marker gene, and is lethal in homozygous form. "*a*" is a different visible marker gene, which is recessive. "*T*" is a dominant temperature-sensitive lethal. * represents a new mutation. The matings are carried out as shown and the F3 eggs are examined for abnormalities.

manifest themselves as **maternal effects**, meaning that the structure of the embryo corresponds not to its own genotype but to the genotype of the mother (see also Chapter 3). Strictly speaking, when discussing maternal effect mutations one should always say "eggs from mutant mothers" rather than "mutant eggs" but this convention is rarely adhered to. Maternal-effect genes are often identified as female sterile mutations in screens similar in principle to those for zygotic lethals. But some important maternal-effect mutations are also zygotic lethals. In other words, the gene is needed for embryonic development as well as for oogenesis. In such a case the homozygous individuals do not grow up to become flies whose fertility can be tested, so maternal screens are usually incomplete. It is, however, possible to test individual zygotic lethals for maternal effects by making

mosaic embryos in which the germ line is mutant but the somatic tissues are wild-type (see below).

Types of mutation

Most mutations are reduced function or loss of function, representing the production of a smaller amount of gene product, or a gene product of reduced effectiveness. These are usually recessive. The alleles are called **hypomorphs** and the phenotypes are called **hypomorphic**. In *Drosophila* genetics it is usual to make several alleles for each locus to obtain at least one that has lost all function, the so-called **null** alleles giving amorphic phenotypes. Frequently, the hypomorphic alleles can be arranged in a series of increasing severity with the amorphic phenotype as the limit form. Such **allelic series** can often give useful information about function, particularly where the weaker alleles give something recognizable and the stronger ones do not.

Some alleles may be **temperature sensitive**, usually because the mutation affects the thermal stability of the protein product. In general, the protein is active at a low temperature (the **permissive** temperature) but inactive at a high temperature (the **nonpermissive** temperature). Temperature-sensitive mutants are useful because they can help to establish the developmental stages at which a gene product is required. This is done by shifting between the temperatures, and if the shift to the nonpermissive temperature is made before the time of gene function then most cases will be mutant, while if it is made afterwards, most cases will be normal.

Dominant mutations are sometimes loss-of-function, in those cases where the locus is **haploinsufficient**. This means that a reduction in the level of the product to 50% of the wild-type level is sufficient to cause a mutant phenotype. In such cases the homozygous phenotype will be more severe than the heterozygous one. More often dominant mutations are **gain of function**, resulting in the production of active gene product in positions or at times when it is not normally found, or **dominant negative** (=antimorphic), where the mutant version of the gene product interferes with the function of the wild-type version.

The names of *Drosophila* developmental genes usually derive from the appearance of the mutant phenotype. This means that they may indicate the opposite to the normal function. For example the *dorsal* gene is so called because null mutants are dorsalized, but this is because the normal function of the gene is to form ventral parts. Some genes are named after the adult phenotypes of viable alleles that were discovered before the big screens were carried out for embryonic lethal mutations. For example *Antennapedia* is so named because of a gain-of-function allele that converts antenna to leg, although the null phenotype of *Antennapedia* is a conversion of parasegments 4 and 5 to parasegment 3 and the homozygous null is lethal at embryonic stages.

Cloning of genes

Historically, most of the *Drosophila* developmental genes were cloned by **P-element mutagenesis**. When a P-element integrates into or near a gene it may mutate it to inactivity and it is possible to screen for a particular mutation by methods similar to those of Fig. 11.7. A mutation caused by P-element insertion can then be used as a starting point for cloning the gene. A genomic library from this strain is screened with the P-element probe and the resulting clones are tested by *in situ* hybridization to polytene chromosomes to find which particular P-element they represent. The one lying nearest to the genetic map posi-

tion of the mutation is then used as the starting point for a chromosomal "walk" to obtain the whole of the required gene. Proof that the cloned candidate really *is* the required gene is obtained by three criteria:

1 that several known mutants have identifiable sequence changes in the candidate gene;
2 that the candidate gene is not expressed in null mutants (although sometimes null mutants can produce an inactive protein product);
3 by rescuing the phenotype of the null mutant by transgenesis with the candidate gene.

Once the gene is cloned and sequenced the next step is to determine the normal expression pattern using *in situ* hybridization to different stage embryos. The next step is to make antibodies to the protein product, usually of a fusion protein expressed in bacteria, and use this to determine the protein expression pattern.

Gene function is studied by finding how the mutation of one gene affects the expression of others. If A is normally needed to upregulate B, then removal of A will cause loss of expression of B, and overexpression of A will cause corresponding ectopic expression of B. On the other hand, if A normally represses B, then removal of A will cause ectopic expression of B, and overexpression of A will cause repression of B in its normal domains. But such results do not reveal whether the effects are direct, meaning that the protein product of A is actually interacting with gene B to regulate it. It could equally well be indirect, with gene A turning on something else that regulates gene B. If gene A codes for an inducing factor or receptor then the effect is necessarily indirect. If gene A codes for a transcription factor, then there are three methods of establishing directness:

1 it may be possible to demonstrate interactions between gene-A-product and gene B-regulatory region, using **band shift** assays;
2 genes A and B can be transfected together into *Drosophila* tissue culture cells to see whether B is turned on in the absence of other developmental machinery;
3 it may be possible to do a domain swap such that a new DNA recognition domain is put onto A and the corresponding DNA sequence is inserted into the regulatory region of B. If the effects of A on B are maintained under these circumstances then it must be due to a direct molecular interaction.

When the effect of A upon B is examined, it is quite common to look not at the endogenous B gene product but at a construct composed of the regulatory region of B fused to a reporter gene, usually *lacZ*. The reasons for this may be simply improved sensitivity if the endogenous product is hard to detect. Also, in many constructs the β-galactosidase protein is quite stable and so its concentration effectively "integrates" the cumulative gene activity up to the time of examination. Most often it is because the regulatory region of B has been dissected into a number of parts to find which DNA sequences are responsible for each component of the expression pattern. When a reporter construct is used it may include just a single enhancer from the original regulatory region of gene B.

The developmental program

The detail of *Drosophila* early development can be understood if it is appreciated that the overall program can be regarded as a set of subprograms. One subprogram operates in the **dorsoventral axis** and is responsible for the formation of the mesoderm, the neurogenic region, and the epidermis from ventral to dorsal. As a result of a series of events in the egg chamber of the mother, the product of a maternal gene *dorsal* becomes distributed in the nuclei of the blastoderm in a ventral–dorsal gradient (Fig. 11.8). It regulates a set of zygotic genes including *twist*, *rhomboid*, and *zerknüllt*, which control formation of the various bands of tissue from ventral to dorsal. A more-or-less independent set of subprograms operates in the **anteroposterior axis** and is more complex. The first phase of specification occurs in the egg chamber with the establishment of three maternal systems. The anterior system is concerned with the production of a gradient of the Bicoid protein from *bicoid* mRNA localized in the anterior. The posterior system, of which the pole plasm is an integral part, deposits the mRNA of a gene called *nanos* in the posterior. The **terminal system** involves activation of the receptor encoded by *torso* at both extremities, leading to activation of the ERK signaling pathway (Fig. 11.8).

The products of these three maternal systems divide the embryo into several anteroposterior zones depending on their relative concentrations. The later regionalization of the anteroposterior axis is shown in Fig. 11.9, with the expression patterns of a few of the key genes. Before cellularization one nucleus can affect the gene activity of nearby nuclei simply by producing a transcription factor, no receptors or signal transduction mechanisms being required. Because of the overlaps between the domains of activity, and because different concentrations of the same substance can have different effects, the maternal systems can upregulate a spatial pattern of zygotic gene activity which is more complex than their own. The genes upregulated at this stage belong to the **gap** class, for example *Krüppel*, and to the pair-rule class, for example *even skipped*. The gap genes are expressed in one or a few domains while the pair-rule genes are expressed in stripy patterns with a periodicity of two segment widths. Their periodicity arises because many different combinations of maternal and gap genes can upregulate the same pair-rule gene. The overlapping periodic patterns of pair-rule genes lead to the upregulation of a repeating pattern of segment polarity genes, for example *engrailed*, which have single segment periodicity and cause the subdivision of the axis into parasegments. Simultaneously, the combined maternal, gap, and pair-rule gene product combinations upregulate the Hox genes, for example *Ubx*, which control the character of each parasegment and thus its subsequent pathway of differentiation.

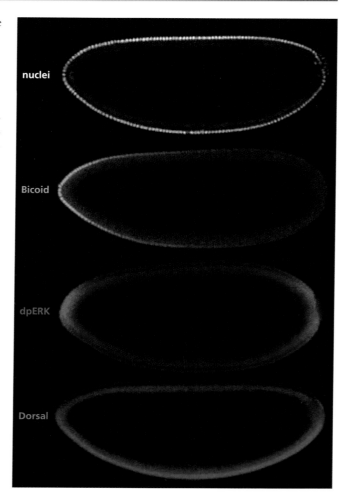

Fig. 11.8 Immunostaining of three components of the maternal systems. The nuclei of the cellular blastoderm are shown at the top. The gradient of Bicoid protein (green), mostly intranuclear, runs from anterior to posterior. The phosphorylation of ERK (red) is seen at both termini, resulting from activation of Torso. The gradient of Dorsal protein (blue) runs from ventral to dorsal, again mostly intranuclear. Reproduced from Shvartsman *et al. Curr Opin Genet Dev* 2008; 18: 342–347, with permssion from Elsevier.

The dorsoventral pattern

The pattern along the dorsoventral axis is relatively simple. It consists at the cellular blastoderm stage of four strips committed to become, from ventral to dorsal, the mesoderm, the ventral neurogenic region, the dorsal epidermis, and the amnioserosa (Fig. 11.10). The disposition of these four territories is controlled by a ventral–dorsal nuclear gradient of the *dorsal* gene product, which is a transcription factor (see Animation 13: DV patterning system). To avoid confusion between dorsal position and the presence of the product of the *dorsal* gene, the gene product will be referred to here as "Dorsal protein." The high point of the Dorsal protein gradient depends on signaling

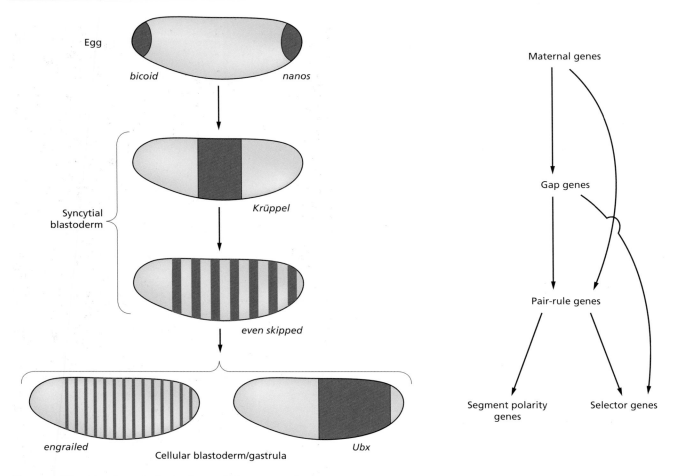

Fig. 11.9 Hierarchy of steps in the development of anteroposterior pattern.

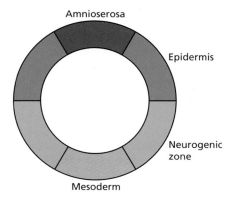

Fig. 11.10 Territories formed during dorsoventral specification of the early *Drosophila* embryo.

from the ventral follicle cells (Spätzle protein), and this signal is only active ventrally because it is repressed on the dorsal side by a previous inhibitory signal from the oocyte (Gurken protein). The Dorsal protein gradient itself works by regulating various zygotic genes such that each ventral to dorsal strip

of cells has a different combination of transcription factors active.

Maternal control of dorsoventral patterning

There is a group of maternal-effect genes affecting dorsoventral pattern. For most of them the loss-of-function phenotype consists of a folded tube of larval cuticle bearing the fine hairs typical of the dorsal epidermis, but lacking structures normally derived from the lateral or ventral territories of the blastoderm. The prototype mutation of this dorsalizing class was called *dorsal*. In addition, there are three genes, *gurken*, *Egfr* (=*torpedo*), and *cactus*, whose loss-of-function phenotype is ventralizing, with denticle bands extending all round the embryo. One gene, called *Toll*, has loss-of-function dorsalizing alleles and also has a gain-of-function allele with a ventralizing effect.

Several types of genetic and embryological experiment were important in understanding how these genes worked. First, several of the genes have dorsalizing alleles with different degrees of function and these can be arranged in allelic series in which structures are progressively lost from the ventral side until, in the amorphic embryos, only the symmetrical tubes of dorsal cuticle remain. This shows that the normal function of these genes is to promote ventral development.

Next, the genes can be arranged in a temporal series by making double mutants of which one has a dorsalizing and the other a ventralizing effect. Whichever predominates is assumed to act later in the developmental program (**epistasis,** see also Chapter 3). Some idea of the time of action can also be gained by using temperature-sensitive mutants and shifting between the permissive and nonpermissive temperature at different times.

Next, it can also be asked whether a particular gene is required in the **germ line** (i.e. the oocyte itself and the nurse cells) or in the **soma** (i.e. the ovarian follicle cells). To answer this, embryos were created by grafting mutant pole cells into normal embryos. When these are grown up the females will produce offspring that are mutant if the gene was required in the germ line, or normal if it was required in the soma. Studies of this sort showed that some genes are required in the germ line (e.g. *gurken, Toll, dorsal*) and some in the soma (e.g. *Egfr, pipe*). Nowadays, this type of experiment is usually done not by transplantation but using the **FLP system** (see Chapter 17).

Finally, even before the genes were cloned and the gene products were identified, it was possible to find out something about their function by performing cytoplasmic **transplantations**. Most of the dorsal group mutants can be "rescued" towards a normal appearance by the injection of small amounts of cytoplasm from wild-type embryos before the pole cell stage.

These various types of experiment led to the following account, in which the term "mutant" will indicate loss of function unless otherwise stated. The first known gene to act is called *K10*. The mutant has a dorsalizing effect and the function of its product is to sequester the mRNA for *gurken* in the vicinity of the oocyte nucleus. *gurken* and *Egfr* are two genes with similar ventralizing mutant phenotypes (Fig. 11.11a). *gurken* encodes a growth factor related to vertebrate TGFα, and is required in the oocyte. *Egfr* codes for an EGF/TGFα type receptor and is required in the follicle cells. *gurken* mRNA is present just in the vicinity of the nucleus, which is positioned on the dorsal side of the oocyte, while *Egfr* is expressed all over the follicle cells. The protein product of *gurken* is secreted only on the dorsal side, it stimulates the *Egfr* product in the dorsal follicle cells thereby activating the ERK signaling pathway and causing changes of gene expression, in particular the repression of *pipe*, which would otherwise be upregulated. Mutations in *gurken* and *Egfr* are ventralizing because if these genes are inactive, then *pipe* becomes upregulated all over.

pipe is one of a group of genes whose function is to create an active extracellular ligand localized on the ventral side of the oocyte (Fig. 11.11a). *pipe* itself codes for an enzyme responsible for adding sulfate groups to extracellular glycosaminoglycans. In normal development *pipe* is expressed just in the ventral follicle cells, since expression on the dorsal side is repressed by the *gurken–Egfr* system. The local sulfation of glycosaminoglycans sequesters a group of proteins on the ventral side including

proteases produced by the genes *snake* and *easter*, which can activate the actual ventral signal. This signal is the protein product of the *spätzle* gene, active in the oocyte, and is synthesized as an inactive precursor requiring proteolytic cleavage for activation. Spätzle has some homology to vertebrate nerve growth factor and its function is to activate a receptor encoded by the *Toll* gene, which is present all over the surface of the oocyte. *Toll* has two types of mutant, recessive dorsalizing mutants in which receptor function is lost, and dominant ventralizing mutants in which the receptor is signaling continuously even in the absence of ligand.

The activation of *Toll* on the ventral side commences after fertilization and the subsequent events occur during the embryonic cleavage stages (Fig. 11.11b). *dorsal* is the final gene in the maternal dorsoventral pathway. The Dorsal protein is a transcription factor homologous to the vertebrate factor NFκB. Its mRNA is uniformly distributed in the oocyte and the protein is synthesized after fertilization. The distribution is initially uniform but during the syncytial blastoderm stage it enters the nuclei preferentially on the ventral side, forming a ventral–dorsal gradient of nuclear protein. The entry to the nuclei depends on the proximity of activated Toll, which causes dissociation of dorsal protein from a complex formed with another protein, IκB. This releases dorsal protein and allows it to enter the nuclei and regulate its target genes. *Drosophila* IκB is encoded by the gene *cactus*. *cactus* mutants are ventralizing because the normal role of Cactus is to inhibit the action of Dorsal protein, so in the absence of Cactus, Dorsal protein can enter the nuclei all over the embryo and make it ventral in character all over.

Zygotic control of dorsoventral patterning

The gradient of Dorsal protein works by upregulating or repressing the expression of various transcription factors that are encoded by zygotic genes (Figs 11.11c, 11.12).

The **mesoderm** is defined by two transcription factors encoded by the genes *twist* and *snail*. These both code for transcription factors and are upregulated by dorsal protein at high nuclear concentration. *twist* encodes a bHLH protein and is required for correct mesodermal differentiation. *snail* encodes a zinc-finger transcription factor and is needed for invagination of the mesoderm.

Other genes are turned on at lower concentrations of Dorsal protein so they become expressed laterally as well as ventrally. Some of these are repressed by Snail, leading to a lateral stripe in the prospective neuroectodermal region, for example *rhomboid*, coding for a transmembrane protein involved in EGF signaling, is upregulated by Dorsal protein and repressed by Snail, such that it is expressed as a stripe in the neurogenic region (Fig. 11.12).

The **dorsal ectoderm** is defined by a homeoprotein encoded by *zerknüllt*, normally expressed in about 40% of the embryo circumference. It is repressed by Dorsal protein, and therefore

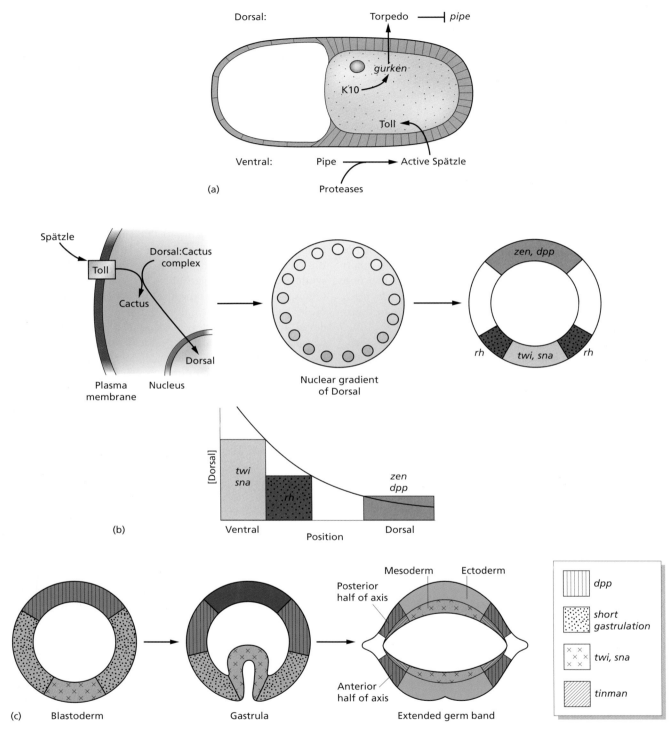

Fig. 11.11 Operation of the dorsoventral system. (a) Dorsally, the Gurken signal causes repression of *pipe*. Ventrally, *pipe* is active and enables the activation of Spätzle (actually produced by the oocyte). (b) Spätzle activates Toll, which causes nuclear translocation of Dorsal, and Dorsal regulates zygotic genes. (c) The territories are later refined by means of the gradient of Dpp. *tinman* is maintained in the lateral mesoderm. Abbreviations: *twi*, twist; *sna*, snail; *rh*, rhomboid; *zen*, zerknüllt; *dpp*, decapentaplegic; *sog*, short gastrulation.

wildtype   hypomorphic serpin   null serpin

Dorsal

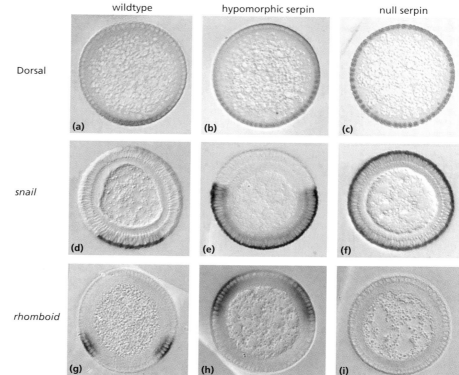

snail

rhomboid

Fig. 11.12 Disruption of the normal dorsoventral system by loss of function of a gene encoding a protease inhibitor (*serpin*). (a,d,g) Normal gradient of Dorsal protein and upregulation of two zygotic genes, *snail* and *rhomboid*. (b,e,h) Partial loss of function of *serpin*, giving a stronger Dorsal gradient and pushing the response thresholds to more dorsal levels. (c,f,i) Complete loss of *serpin* function resulting in complete ventralization of the pattern. Reproduced from Ligoxygakis *et al. Curr Biol* 2003; 13: 2097–2102, with permission from Elsevier.

becomes expressed all round in dorsalizing mutants. Another gene repressed by Dorsal protein and normally expressed in the dorsal 40% zone is *decapentaplegic* (*dpp*). This encodes a signaling molecule homologous to vertebrate BMPs, which brings about the patterning of the dorsal half of the embryo circumference. Injection of *dpp* mRNA can induce amnioserosa at high dose and dorsal hairs at lower dose. Although this suggests gradient-like behavior, there is not a gradient of production of the Dpp protein itself. Instead, the induction of dorsal genes is conducted by Dpp acting together with another BMP homolog called Screw, which is expressed ubiquitously. The graded effect of these factors arises because of the action of an inhibitor encoded by the *short gastrulation* (*sog*) gene. This is expressed in a lateral belt because it is repressed by Snail, and it inhibits Screw, leading to a dorsal–ventral gradient of BMP activity. Sog is the homolog of the vertebrate Chordin, also an inhibitor of BMP4 (see Chapter 7).

The patterning of the mesoderm also depends on Dpp, whose expression later resolves into a pair of broad longitudinal stripes on the dorsal side. *tinman* encodes a homeodomain transcription factor necessary for heart formation. It is initially upregulated in the whole mesoderm by *twist* but then is turned off except in the lateral region that is adjacent to the *dpp*-expressing epidermis. This lateral region forms the heart while the ventral part forms the body wall muscles.

Thus, the full dorsoventral pattern arises from a maternal gradient of dorsal protein controlling the ventral half, and the

zygotic gradient of Dpp controlling the dorsal half. Interestingly, the dorsal Dpp–ventral Sog pattern is homologous to the vertebrate ventral BMP4–dorsal chordin pattern, providing evidence that one of the groups must have "turned upside down" during evolution (see Chapter 22). Dpp and BMP4 have similar biological activity, as do Chordin and Sog.

The dorsoventral system provides examples of several key developmental processes. The Gurken and Spätzle proteins are both **inducing factors** that cause a regional patterning of their competent tissue. The localized region of activation of Toll is an example of a **cytoplasmic determinant** in the egg, in this case a determinant that is not made up of localized mRNA. The nuclear gradient of Dorsal protein, although intracellular, exemplifies the conversion of a simple pattern into a more complex one through the formation of a **gradient**. The various dorsoventral tissue types arise because each is encoded by a combination of transcription factor genes upregulated, directly or indirectly, by the Dorsal protein gradient.

The anteroposterior system

Events in the egg chamber

The specification of structures along the anteroposterior axis is controlled by maternal systems, which are largely, although not entirely, independent from that controlling the dorsoventral pattern. The basic mechanism is summarized in Fig. 11.13 and

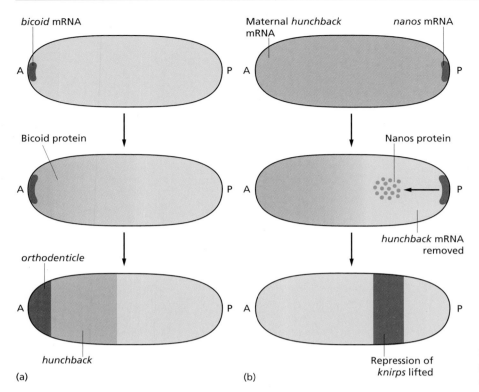

Fig. 11.13 Maternal anteroposterior systems. (a) Anterior system; (b) posterior system.

Classic experiments

The first real gradient

Developmental biologists had speculated for some years about the existence of morphogen gradients that might control the pattern of gene expression. This was an element in the theoretical paper by Wolpert, which revived the problems of experimental embryology and restated them in modern language. But the first gradient to be detected experimentally was that of the Bicoid protein within the early *Drosophila* embryo. This was actually an intracellular gradient since the early embryo is a syncytium. But it did control gene expression and it did follow the predicted behavior, so people finally started to believe in gradients.

Wolpert, L. (1969) Positional information and the spatial pattern of cellular differentiation. *Journal of Theoretical Biology* **25**, 1–47.
Driever, W. & Nüsslein-Volhard, C. (1988) A gradient of bicoid protein in *Drosophila* embryos. *Cell* **54**, 83–93.

in Animation 14: AP polarity of oocyte. In the anterior, mRNA for *bicoid* is deposited in the egg and a gradient of Bicoid protein upregulates various genes to generate regional subdivisions (Fig. 11.13a). In the posterior, mRNA for *nanos* is deposited, and its protein product lifts the inhibition on gene activity in the future abdomen (Fig. 11.13b).

As in the dorsoventral system, the basic regionalization of the embryo occurs in the developing oocyte of the mother and depends on interactions between the oocyte and the follicle cells. Right from the beginning a small group of terminal follicle cells at each end of the egg chamber are specialized, and will later become the anterior and posterior **border cells**. Early in oogenesis the follicle cells are still dividing and the oocyte is a similar size to the nurse cells. At this stage the germ cells emit a

Delta signal, which stimulates the receptor Notch on the follicle cells. This brings about the transition from mitotic division to endoreduplication of DNA in the follicle cells and it also confers upon them the competence to respond to the ligand Unpaired, which is secreted from the terminal follicle cells and causes subdivision of follicle cells into their various subpopulations. The oocyte then starts to produce Gurken. Until this stage both sets of terminal follicle cells have a default anterior character. But the Gurken signal causes the population adjacent to the oocyte to acquire a posterior character (Fig. 11.14a). The posterior follicle cells then emit another signal that causes polarization of the oocyte itself (Fig. 11.14b). The nature of this signal is still unclear but its production is dependent on activation of the Hippo pathway (see Chapter 19 and Appendix) within the

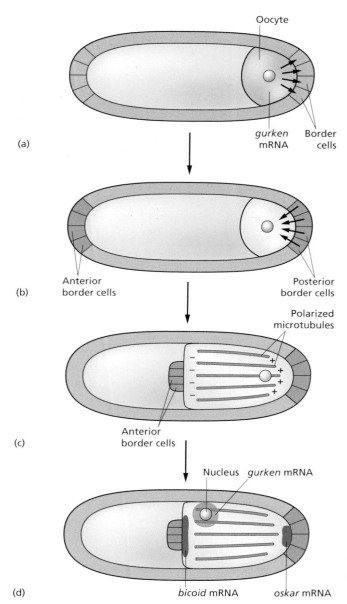

(a)

Oocyte

gurken mRNA Border cells

(b)

Anterior border cells

Posterior border cells

(c)

Polarized microtubules

Anterior border cells

(d)

Nucleus *gurken* mRNA

bicoid mRNA *oskar* mRNA

Fig. 11.14 Origin of anteroposterior polarity. (a) Posteriorization of border cells by Gurken. (b) Reverse signal from follicle cells to oocyte. (c) Formation of oriented microtubule array. (d) Movement of the nucleus, and localization of *oskar* and *bicoid* mRNAs by transport along orientated microtubules.

sion experiments show that they each repel the other. The other main type of evidence for these events is that loss-of-function mutants of components of the *par* system, the Hippo pathway, Gurken, or Delta/Notch all prevent formation of the anteroposterior polarity of the egg chamber. In *Drosophila* the *par3* gene is known as *bazooka*.

The orientation of the tubules means that plus or minus-directed **motor proteins** can, respectively, localize *oskar* mRNA to the posterior and *bicoid* mRNA to the anterior (Fig. 11.14d). This process can be demonstrated in flies transgenic for a gene for a fusion protein of kinesin with β-galactosidase. In the egg chambers of these flies the fusion protein can be located by staining for β-galactosidase, and is found at the posterior end of the oocyte, along with *oskar* mRNA. Kinesin normally migrates to the plus end of microtubules and this experiment shows that the tubule array can serve as a polarized substrate for intracellular localization.

As the oocyte grows larger, the nucleus moves back towards the anterior along the microtubule array. The tubules are arranged around the exterior of the oocyte, so the nucleus has to follow a track around the exterior rather than travel straight down the central axis. This movement breaks the former radial symmetry of the oocyte and causes the nucleus to lie closer on one side than the other. Now the *gurken* mRNA, still near the nucleus, causes Gurken protein to be secreted just from the side on which the nucleus is present. The adjacent follicle cells become dorsal in character because the Gurken signal stimulates Egfr and thereby represses expression of *pipe*, as described above. Thus, some of the same genes that are responsible for the dorsoventral patterning of the follicle cells are also, at an earlier stage of oogenesis, responsible for the anteroposterior patterning of the oocyte and border cells. The reason that the same signal can be involved in polarization along two anatomically orthogonal axes is threefold: it acts at different times; the responding populations of follicle cells have different competence; and the intervening growth of the oocyte has changed the effective location of the signal.

bicoid

bicoid is a maternal-effect gene coding for a homeodomain transcription factor. Loss-of-function mutations cause a deletion of the head and thorax. Transcription of *bicoid* occurs during oogenesis in both oocyte and nurse cells, and the mRNA becomes localized at the anterior of the oocyte. Study of the Bicoid protein by antibody staining and by observation of a Bicoid–GFP fusion protein shows that it is synthesized during the syncytial stage at 1–3 hours of embryonic development. By 90 minutes postfertilization it has formed an exponential concentration gradient from anterior to posterior. Originally, it was believed that this gradient was maintained by the presence of the localized RNA forming the source, and by degradation elsewhere in the embryo acting as the sink. However, studies of the diffusion behavior by photobleaching of Bicoid–GFP *in vivo*

follicle cells (Fig. 11.15). The consequence is the formation of a gradient of microtubules from anterior to posterior and a polarization of the microtubules such that minus ends are anterior and plus ends posterior directed (Fig. 11.14c). This process of symmetry breaking also depends on the Par system. Rather similarly to the *C. elegans* zygote, Par1 moves to the posterior and Par3/Par6/aPKC to the anterior in association with the rearrangement of the microtubules. GFP fusions of the Par1 and 3 proteins can be observed moving in this way, and overexpres-

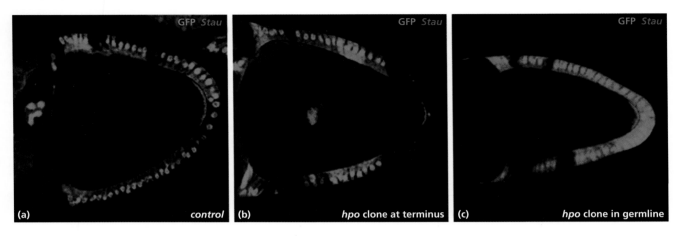

Fig. 11.15 Dependence of the signal from follicle cells on an active Hippo pathway. Here, loss-of-function clones of *hippo* (*hpo*) are induced by the FRT method (see Chapter 17). The clones are visualized by loss of GFP. (a) Normal situation, the posterior group mRNA *staufen* (*stau*, red) is correctly localized in the posterior. (b) If the loss-of-function clone lies in the follicle cells the localization of *staufen* is deranged. (c) If the clone lies in the germ cells, localization of *staufen* is normal. Reproduced from Polesello and Tapon. *Curr Biol* 2007; 17: 1864–1870, with permission from Elsevier.

indicates that the mechanism of gradient maintenance must be more complex than this.

The function of the Bicoid protein gradient is to regulate the zygotic expression of several of the gap genes, such as *orthodenticle*, *hunchback*, and *Krüppel*. *orthodenticle* and *hunchback* both encode homeodomain transcription factors, and *Krüppel* encodes a zinc-finger transcription factor. Each of these genes has a promoter of different sensitivity to Bicoid, so they become upregulated at different anteroposterior levels.

The system does show gradient-like behavior in terms of the formation of pattern. The level of protein can be manipulated by changing the number of active copies of the gene in the female and this displaces in the expected directions features such as the cephalic furrow or stripes of pair-rule gene expression. Furthermore, eggs lacking *bicoid* can be rescued toward a normal phenotype by injection of *bicoid* mRNA. The position at which the injection is made determines the position of the induced anterior end, so, for example, injection of mRNA to a central position of a *bicoid⁻* egg produces a head at the injection site, flanked by two thoraxes in mirror symmetrical arrangement.

Posterior system

The posterior system depends on the deposition of mRNA for *nanos* in the posterior and the action of its protein product in lifting an inhibition of the transcription of various gap genes in the future abdomen (Fig. 11.13b). The mechanism of this system has been deduced from a combination of genetic and embryological experiments.

Pricking of the posterior pole and extrusion of a small volume of cytoplasm causes defects in the larva; not in the telson as might be expected, but in the abdomen whose prospective region lies around 50–20%EL. This indicates that something

present at the posterior pole is needed for development of the abdomen. There are also a number of mutations in maternal-effect genes that cause loss of the abdomen. Among these are *nanos*, *oskar*, and *pumilio*. Mutant embryos for all these genes can be rescued toward normality by injection of cytoplasm taken from the pole plasm region of a wild-type embryo and injected into the abdominal region of the mutant, confirming the localization of an "abdomen-forming substance" at the posterior pole of normal embryos. This was eventually identified as the *nanos* product, which is an RNA-binding protein.

nanos mRNA is normally localized in the pole plasm in a process depending on *oskar*, but the protein is found in the prospective abdomen. Its function is to allow the transcription of the zygotic gap gene *knirps*, in concert with the product of *pumilio*. The mechanism of action involves a double repression with *knirps* being repressed by Hunchback, and translation of *hunchback* being inhibited by Nanos. *hunchback* is another zygotic gap gene and is upregulated by Bicoid in the anterior. But *hunchback* is also active during oogenesis such that the egg is normally filled with a uniform concentration of maternally derived *hunchback* mRNA. Translation of this mRNA commences in early cleavage and is normally inhibited in the posterior half of the embryo by the Nanos protein. *knirps* becomes upregulated in the posterior but not in the anterior because Hunchback protein is absent from the posterior. If Nanos is missing then the Hunchback protein is made all over and *knirps* cannot be expressed anywhere. Nanos is required again at a later stage for germ cell development, and for this it is produced from the zygotic gene.

Terminal system

A third maternal system is concerned with the formation of the embryo termini (Fig. 11.16). It involves a signal from the follicle

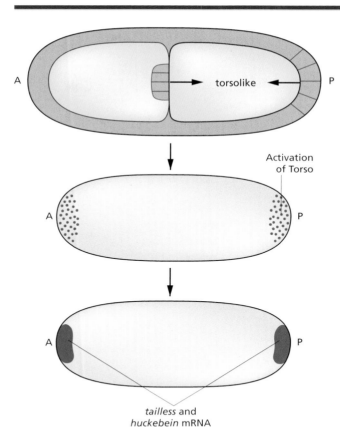

A → torsolike ← P

Activation
of Torso

A → P

A → P

tailless and
huckebein mRNA

Fig. 11.16 The terminal system.

cells activating a receptor at both termini and thereby turning on zygotic genes.

Embryos resulting from loss-of-function mutations of the terminal system show defects at both ends with a normal pattern in between. The key gene in the pathway is *torso*, which also has a gain-of-function allele causing substantial suppression of segmentation in the thorax and abdomen. *torso* encodes a cell-surface receptor of the tyrosine kinase class, which stimulates the ERK signal transduction pathway, and the gain-of-function phenotype arises from a constitutively active form of the receptor. As with *Toll*, expression is uniform in the oocyte and in normal development the receptor becomes activated at the termini. The ligand is encoded by *trunk*, which is active in the oocyte and secreted into the perivitelline fluid. The Trunk protein is activated at the termini by the product of *torsolike*, encoding a novel protein, which is present in the anterior and posterior border cells.

Activation of Torso leads, via the ERK pathway, to upregulation of two zygotic gap genes: *tailless* and *huckebein*. *tailless* encodes a transcription factor of the nuclear receptor family and *huckebein* encodes a zinc-finger transcription factor.

Although different molecules are involved, the terminal system has a mechanism remarkably similar to the dorsoventral system, and once again the activated receptor protein can be regarded as a type of **cytoplasmic determinant**.

Gap genes

The system of morphogens and determinants bequeathed by the mother becomes elaborated into increasingly complex patterns of gene activity by successive levels of the developmental hierarchy. The first zygotic level is made up by the **gap genes**, so called because mutant embryos have patterns bearing gaps of up to eight contiguous segments. All the gap genes code for transcription factors, and because the early embryo is a syncytium, these can diffuse from one nucleus to another and exert their effects directly, with no need for cell–cell signaling. Some important members of this group are *orthodenticle*, *hunchback*, *Krüppel*, *knirps*, and *giant*. The regulatory relationships have been deduced mainly by two types of experiment:

1 By examining the expression pattern of the gene of interest in the absence of another. Expansion of the domain indicates that the absent gene normally has a repressive function, whereas reduction of the domain indicates that the absent gene is responsible for positive regulation in normal development.

2 By examining the expression pattern of the gene of interest following uniform overexpression of another gene.

The type of result predicted from these two protocols is shown in Fig. 11.17.

Along with Bicoid protein, another important early regulator of gap gene expression is the product of the *caudal* gene. This encodes a homeodomain transcription factor and is homologous to the *Cdx* gene family in vertebrates. It is expressed maternally to produce a uniform distribution of mRNA in the oocyte. At the syncytial blastoderm stage, mRNA is differentially lost, resulting in a posterior-to-anterior gradient of mRNA and protein. Embryos lacking *caudal* have severe posterior defects. Some gap genes are upregulated by both Bicoid and Caudal proteins, and since these two factors form inverse gradients in the early embryo, it means that the gene regulation appears to be autonomous and not spatially regulated. *caudal* also has a posterior zygotic domain in the prospective proctodeum.

orthodenticle mutants have defects in the head. The gene encodes a homeodomain transcription factor. Expression is in the head and is upregulated by high levels of Bicoid and by Torso. An important vertebrate homolog of *orthodenticle* is *Otx2*, which is required for the formation of the forebrain and midbrain.

Embryos homozygous for null alleles of *hunchback* have a large anterior gap, which removes the labium and thorax. The gene codes for a zinc-finger transcription factor. Transcription during oogenesis leaves mRNA uniformly distributed in the egg but, as we have seen, its stability and translation is antagonized in the posterior half by the Nanos protein and so an anterior to posterior gradient of Hunchback protein arises during cleavage. Zygotic transcription commences in the syncytial blastoderm in the anterior half of the embryo, and at cellular blastoderm also in a posterior stripe. The anterior, but not the posterior, zygotic

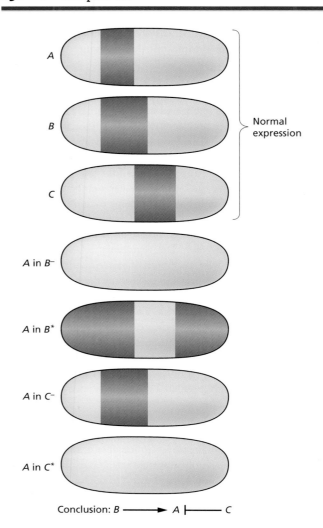

Fig. 11.17 How to work out regulatory relationships between genes. The top three drawings show the normal expression of three genes. The bottom four show expression of gene *A* in embryos mutant (−) or with ubiquitous overexpression (*) of *B* or *C*. The results show that *B* upregulates *A* and that *C* represses *A*.

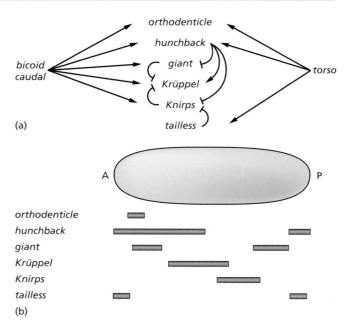

Fig. 11.18 Gap genes. (a) Some regulatory relationships; (b) simplified expression domains of six gap genes.

domain is upregulated directly by the Bicoid protein at a lower concentration than that required to upregulate *orthodenticle*. The posterior zygotic domain is upregulated by Torso.

Krüppel is an entirely zygotic gene whose null mutants have a large deletion of the central part of the body, comprising the thorax and abdominal segments 1–5. The gene codes for a zinc-finger transcription factor. Expression commences in the syncytial blastoderm as a central band. *Krüppel* is upregulated by Bicoid and Hunchback, and repressed by Knirps and Giant. This ensures its expression as a broad band from about 60–50%EL.

Null mutants of *knirps* are similar to the maternal posterior group, having abdominal segments 1–7 replaced by a single large abdominal segment of uncertain identity. The gene codes for a transcription factor belonging to the nuclear hormone receptor family. The expression pattern shows a band between

about 45 and 30%EL, appearing at the syncytial blastoderm stage. Gene activity is constitutive and the position of the main band in normal development is regulated by repression due to Hunchback and Tailless.

giant mutants have defects in the anterior thorax and in the abdomen at A5–A7. *giant* encodes a leucine-zipper transcription factor and expression starts in the syncytial blastoderm in two zones, an anterior zone about 80–60%EL and a posterior zone about 33–0%EL. By the cellular blastoderm the posterior band is fading and the anterior one has resolved into three stripes. *giant* is upregulated by Bicoid and Caudal, and repressed in the anterior by Hunchback.

tailless and *huckebein* are the zygotic genes upregulated by Torso. The mutants each have terminal defects, which together add up to the phenotype produced by mutants of the maternal terminal genes.

The main regulatory relationships that have been mentioned are summarized in Fig. 11.18. An experiment showing the alteration of some gap gene domains in the absence of Bicoid and the presence of uniform Bicoid is shown in Fig. 11.19.

Pair-rule system

The **pair-rule genes** function as a layer in the developmental hierarchy between the gap genes and the segment polarity genes. They also have a role, along with the gap genes, in controlling the expression of Hox genes and aligning their domains with the segmental repeating pattern. The upregulation of the pair-rule genes represents the first formation of a reiterated pattern in the embryo.

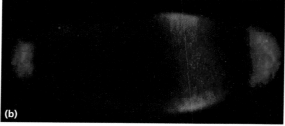

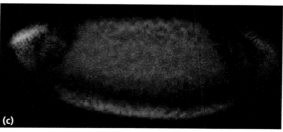

Fig. 11.19 Expression of three gap genes. (a) Normal pattern of expression of *cap-n-collar* (green), *giant* (red), and *tailless* (blue). (b) Removal of *bicoid* causes loss of the *cap-n-collar* domain and shift of the *giant* domain to the posterior. (c) Ubiquitous expression of *bicoid*, in a *bicoid⁻* background, causes a mirror symmetrical pattern, with the terminal system providing most of the patterning information. Reproduced from Löhr *et al. PNAS* 2009; 106: 21695–21700, with permission from National Academy of Sciences.

All the pair-rule genes code for transcription factors and the expression patterns consist of seven stripes, corresponding to two segment-wide bands of the syncytial blastoderm (Fig. 11.20). The mutant phenotypes typically show deletions with a periodicity of two segments, although more complex and more severe phenotypes can be found. For example *even-skipped* gets its name from the fact that in the original mutant the even numbered segments were reduced or lost. But this allele is now known to be a hypomorph, and in the null mutant all the segments are lost.

Those pair-rule genes designated primary are regulated mainly by the maternal and gap genes, while those designated secondary are regulated mainly by the primary genes. Primary pair-rule genes are *hairy, even-skipped (eve)*, and *runt*, and perhaps also *fushi-tarazu (ftz)* and *odd-paired (odd)*. They have complex regulatory regions containing several enhancers.

In general, the rule is "one enhancer, one stripe." Each enhancer contains overlapping binding sites for activators and repressors such that it will be on if the activators prevail and off if the repressors prevail. To form a stripe it is necessary that the gene be turned on at one anteroposterior level and turned off again at a slightly different level. Consider for example the *even-skipped* gene (Fig. 11.21). This encodes a homeodomain protein that acts as a transcriptional repressor and the gene has a large regulatory region containing 12 enhancers. Among these, the stripe 2 element is a positive regulator in the presence of Bicoid and Hunchback and is repressed by Giant and Krüppel, so in normal development stripe 2 is formed in the thin strip in between the *giant* and *Krüppel* domains. Two of the enhancers control pairs of stripes. The 4 + 6 and the 3 + 7 enhancers are both active in the presence of ubiquitous components and repressed by Hunchback and Knirps. Hunchback and Knirps also repress each other, ensuring that the *knirps* domain lies in

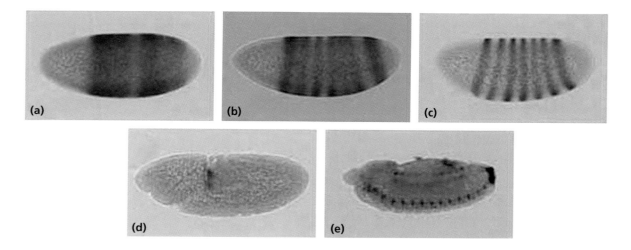

Fig. 11.20 Expression of *even-skipped* (*eve*) (*in situ* hybridization). (a–c) Development of the pair-rule pattern in the syncytial blastoderm. (d) Most *eve* expression is lost by extended germ-band stage. (e) *eve* expression later returns in the ventral ganglia. Reproduced from Peterson *et al. Big Genomes Facilitate the Comparative Identification of Regulatory Elements PLoS ONE* 4, e4688, with permission from Public Library of Science.

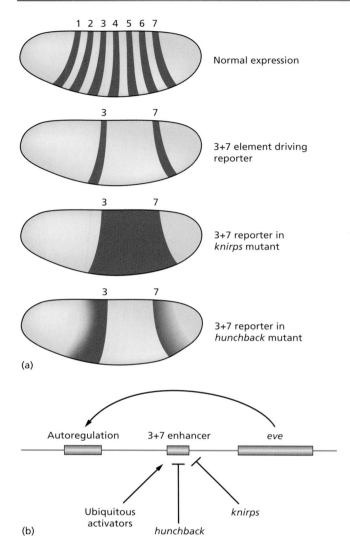

(a)

(b)

Fig. 11.21 Regulation of *even-skipped*. (a) Normal expression pattern, reporter driven by 3 + 7 enhancer, effects of mutations on pattern of reporter expression; (b) two enhancers controlling *even-skipped* expression.

ments are upregulated by *paired*, and *eve* enables this to occur by repressing the expression of *sloppy-paired*, such that there is a domain in which *paired* predominates over *sloppy-paired*. The *engrailed* stripes of the even-numbered parasegments are upregulated by *ftz* in the region where there is a slight overlap with *eve* (Fig. 11.22). This is because here *eve* represses expression of *odd-paired* and enables *ftz* to prevail. The fact that the odd- and even-engrailed stripes are regulated by different mechanisms explains the origin of the "pair-rule" phenotypes when expression of one pair-rule gene is altered. Note that the gene names *even-skipped* and *odd-skipped* relate to visible segment defects. But because the initially formed parasegments are out of phase with the final segments the *eve* mutants give odd parasegmental defects and *odd* mutants give even parasegmental defects!

Segment polarity system

The **segment polarity genes** function to create the **parasegment** boundaries of the early embryo (Fig. 11.23). Once upregulated, they maintain their repeating pattern through a positive-feedback loop between the cells on either side of each boundary, defined by activity of the transcription factor genes *engrailed* and *cubitus interruptus* (*ci*). The *engrailed* cells emit an inducing factor called Hedgehog, which maintains activity of *ci* in the neighboring cells, and the *ci* cells emit a factor called Wingless which maintains activity of *engrailed* in the neighboring cells (Fig. 11.24; and see Animation 15: Ci-engrailed loop establishment). These factors are very familiar in vertebrate development: Hedgehog is homologous to a series of important ligands including Sonic hedgehog and Indian hedgehog, while Wingless is homologous to the Wnt family of factors.

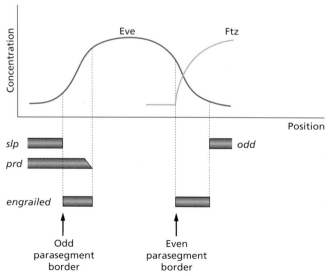

Fig. 11.22 Initial establishment of *engrailed* stripes. Low levels of Even-skipped (Eve) repress *sloppy-paired* (*slp*) and *odd-paired* (*odd*), thus enabling *engrailed* to be upregulated by Paired (Prd) and Fushi-tarazu (Ftz).

between the two zygotic *hunchback* domains. These two *even-skipped* enhancers show different levels of sensitivity to repression so each enhancer controls two expression domains and they are nested such that stripes 3 and 7 form outside stripes 4 and 6. In addition to the stripe enhancers, there is an autocatalytic element which stabilizes the seven-stripe pattern once it is formed, and there are other elements active at later stages driving expression in the mesoderm or in subsets of neurons.

The way that the pair-rule transcription factors upregulate the 14 stripes of the segment polarity genes is very complex indeed. But in principle it is a matter of creating 14 narrow bands in which activators predominate over inhibitors. For example the *engrailed* stripes of the odd-numbered paraseg-

After cellularization of the blastoderm, it is no longer possible for one nucleus to influence another simply by producing a transcription factor and allowing it to diffuse into the surroundings. In a multicellular embryo communication necessarily involves the secretion of inducing factors and the activation of cell-surface receptors. Whereas all the gap and pair-rule genes code for transcription factors, the segment polarity genes, which start to function after cellularization, may code either for transcription factors or for components of the signaling machinery. Most of the segment polarity genes have mutant phenotypes in which the segmental pattern of denticle bands is replaced by a continuous lawn of denticles. This spiky appearance is underlined by some of the gene names, such as *hedgehog, armadillo,* or *pangolin*.

engrailed codes for a homeodomain transcription factor and comes on at the cellular blastoderm stage, forming a 14-stripe pattern by the extended germ-band stage (Fig. 11.23). Each band characterizes the anterior quarter of a parasegment. Expression of *engrailed* is initially upregulated by the pair-rule

genes as described above (Fig. 11.22). *ci* encodes a Gli-type zinc-finger transcription factor and is repressed by Engrailed such that its expression pattern consists of 14 stripes in between the *engrailed* stripes.

The maintenance of the pattern depends on mutual interactions. This may be shown by the fact that if a gene required for one of the states is absent, the pattern is initially set up correctly, but it then rapidly decays. The key genes for maintenance make up the components of the two intercellular signaling systems. In the *ci* cells, Wingless is produced. This stimulates a Wingless receptor coded by two *frizzled* genes. The signal inhibits a kinase coded by the *zeste-white-3* (=*gsk3*) gene, which in turn inhibits a protein coded by the *armadillo* (=β-*catenin*) gene. As two inhibitions equal one activation, this means that Wingless activates Armadillo and causes it to move into the nucleus, together with the product of the *pangolin* gene (=Lef/Tcf), to upregulate target genes. Among the targets is *hedgehog*, so Hedgehog protein is secreted by the *engrailed* cells and binds to the receptor encoded by *patched*. Patched is a constitutively active repressor of Smoothened, another cell-surface protein, which activates the Ci protein. Hedgehog inhibits Patched, thus lifts the inhibition of Smoothened and thereby enables Ci protein to enter the nucleus and to upregulate the *wingless* gene. The system is shown in Fig. 11.24.

There is an essential polarity to this system, because the signals Hedgehog and Wingless only activate their targets on one side and not on both. This polarity arises from the action of other pair-rule genes that restrict the competence to express *wingless* to the posterior half of the parasegment.

The segment polarity genes all have vertebrate homologs that are important in development and have been mentioned in

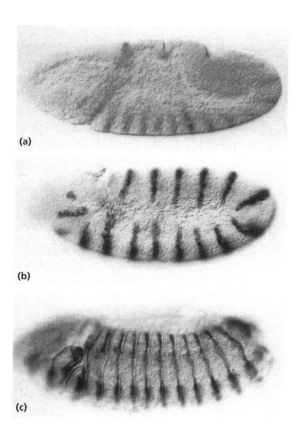

(a)

(b)

(c)

Fig. 11.23 Expression of Engrailed protein (immunostaining). The stripes are just beginning to appear at gastrulation and are fully developed in the extended germ band. (a) Gastrula; (b) extended germ band; (c) retracted germ band. Reproduced from Hama *et al. Genes Dev* 1990; 4: 1079–1093, with permission from Cold Spring Harbor Laboratory Press.

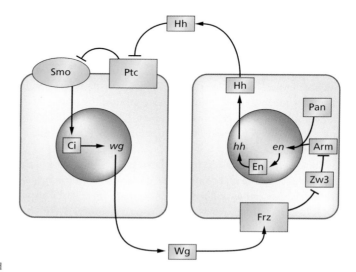

Fig. 11.24 Maintenance of the pattern by the action of Hedgehog and Wingless systems. En, engrailed; Hh, hedgehog; Ptc patched; Smo, smoothened; Ci, cubitus interruptus; Wg, wingless; Frz, frizzled, Zw3, zeste white-3; Arm, armadillo; Pan, pangolin.

Discovery of the homeobox

The homeobox was discovered as a DNA sequence that was present in all of the genes of the Bithorax and Antennapedia complexes. It was called the homeobox because of its location in homeotic genes. It was soon found to exist also in the DNA of other animals, including vertebrates. Excitement grew that it might label all homeotic genes, or perhaps all genes concerned with segmentation, and be a real "Rosetta stone" for developmental genetics. Things are not quite so simple, as it turned out that transcription factors containing the homeodomain DNA binding region are not confined to animals and do not have any single biological function. However, within the animal kingdom a high proportion of them are indeed concerned with some aspect of development, and the Hox clusters homologous to Antennapedia/ Bithorax are homeotic genes concerned with anteroposterior patterning.

McGinnis, W., Levine, M.S., Hafen, E., Kuroiwa, A. & Gehring, W.J. (1984). A conserved DNA sequence in homeotic genes of the *Drosophila* Antennapedia and Bithorax complexes. *Nature* **308**, 428–433.

Scott, M.P. & Weiner, A.J. (1984). Structural relationships among genes that control development – sequence homology between the *Antennapedia, Ultrabithorax* and *fushi tarazu* loci of *Drosophila. Proceedings of the National Academy of Sciences USA* **81**, 4115–4119.

McGinnis, W., Garber, R.L., Wirz, J., Kuroiwa, A. & Gehring, W.J. (1984). A homologous protein-coding sequence in Drosophila homeotic genes and its conservation in other metazoans. *Cell* **37**, 403–408.

Carrasco, A.E., McGinnis, W., Gehring, W.J. & Derobertis, E.M. (1984). Cloning of an *X. laevis* gene expressed during early embryogenesis coding for a peptide region homologous to Drosophila homeotic genes. *Cell* **37**, 409–414.

previous chapters (Table 11.1). Although these homologs exist and the biochemical pathways are usually similar, this does not mean that the developmental functions are necessarily the same. The *engrailed–cubitus* loop is the motor of segmentation in insects and other arthropods, but it is probably not involved in vertebrate segmentation. Likewise, the Wnt pathway is an essential feature of dorsoventral polarity determination in *Xenopus*, but has no comparable function in *Drosophila*.

Hox genes

The **homeobox** and the **Hox genes** were first discovered in *Drosophila* but we have already met them in previous chapters. Like other animals, *Drosophila* contains a Hox cluster as well as many other homeobox genes that are involved in development, but are not part of the Hox cluster. The Hox cluster is on a single chromosome, but is split into two gene groups: the Antennapedia complex and the Bithorax complex. This split is probably quite recent in evolutionary history, as some other insects maintain a single cluster.

The Antennapedia complex corresponds to vertebrate paralog groups 1–6 and contains the genes *labial, proboscipedia, Deformed, Sex combs reduced,* and *Antennapedia*. The Bithorax complex corresponds to vertebrate paralog groups 7–10 and contains the genes *Ultrabithorax (Ubx), abdominal-A,* and *Abdominal-B*. The Antennapedia complex also contains *bicoid* and *zerknüllt*, which do not function as Hox genes in *Drosophila*, but have Hox-like homologs in other insects. Except for *Deformed* where a substantial protein concentration is achieved by the cellular blastoderm stage, the Hox genes are turned on slightly later than the gap and pair-rule classes, and their protein

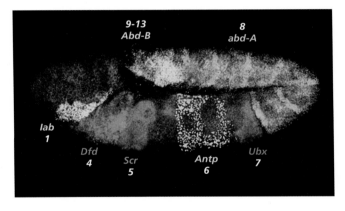

Fig. 11.25 Expression of the Hox genes visualized by seven-color *in situ* hybridization and confocal microscopy. Reproduced from Lemons and McGinnis. *Science* 2006; 313: 1918–1922, with permission from the American Association for the Advancement of Science.

products have generally built up to an effective level by the extended germ-band stage (Fig. 11.25). As in other animals, the order in which the genes are expressed in the anteroposterior axis is also the order of their arrangement on the chromosome.

The approximate expression domains of the Hox genes in the parasegments of the extended germ-band stage are shown in Fig. 11.26 (*proboscipedia* is expressed only in the larva). The function of these genes is to impart different characters to the different segments. In general, if the expression domains are altered by mutation or by overexpression experiments, then the identity of the segments is altered in a predictable way. Loss-of-function mutations generally produce anteriorizations and gain-of-function mutations generally produce posteriorizations. However, their action does not depend simply on the combination of gene products present in a region, because individual structures within a segment may also be specified by a peak of expression of one of these genes. For example expression of *Antennapedia* in parasegment 5 excludes the tracheal pits, which are the only cells of this parasegment to express *Ubx*.

The properties of the system are illustrated by a consideration of the *Ubx* gene, which is the homolog of the Hox7 paralog group in vertebrates. The expression of *Ubx* starts around cellularization and shows an initial peak in parasegment 6 and subsequent lesser expression from parasegment 5–13. It is also expressed in the mesoderm. Later expression is prominent in the ventral ganglia of these segments and in the metathoracic (T3) imaginal discs of the larva. Null mutants of *Ubx* show a transformation of parasegments 5 and 6 to parasegment 4.

Initial control of *Ubx* expression is by Hunchback, which acts as a repressor. There is an upregulation by Fushi tarazu, which gives a transient pair-rule character to the *Ubx* pattern. This pair-rule control is important for aligning the parasegmental register of segment polarity and homeotic genes. Once it is established, the posterior boundary of *Ubx* expression is maintained by repression from Abd-A protein. Ubx itself represses *Antp*, so maintaining its posterior boundary. Maintenance, at least in the visceral mesoderm, also depends on a positive-feedback loop involving Wingless and Decapentaplegic. More permanent regulation in the long term is achieved by regulation of chromatin structure and the demarcation of active and inactive domains within the genome. *Polycomb* and *extra sex combs* encode components of key regulatory complexes which are ubiquitously present and generally repressive (see Appendix). Loss-of-function mutants of these chromatin regulators lead to ectopic Hox activity (e.g. the *extra sex combs* mutant derepresses *Sex combs reduced* in T2–3, leading to the formation of male sex combs on the second and third legs as well as on the first leg where they are normally found. Hence the gene name: *extra sex combs*).

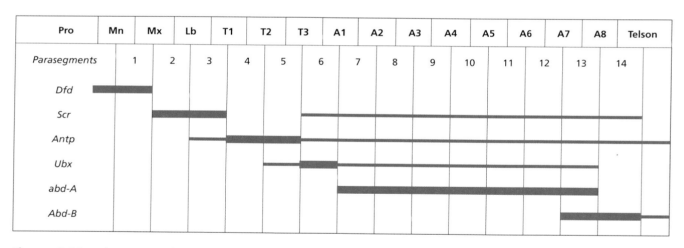

Fig. 11.26 Schematic expression of Hox genes at the extended germ-band stage.

New directions for research

As the development of *Drosophila* is rather well understood, its main role in the future is likely to be in the area of cell biology. The genetic tools for ablating genes in specific regions or labeling specific cell types are extremely powerful. These are increasingly enabling the investigation of phenomena such as cell polarity, cell movement, and intracellular trafficking of materials, in a way which is complementary to the continued use of mammalian tissue culture cells. Also, *Drosophila* imaginal discs (Chapter 17) continue to offer exciting opportunities for work on growth control and regeneration.

The anteroposterior body pattern

The long and complex series of interactions described above leads to a fully specified body plan by the extended germ-band stage. This specification has two essential components. There is the repeating pattern composed of parasegments, whose boundaries are defined by the juxtaposition of bands of cells in which the *engrailed* (anterior) and *ci* (posterior) systems are active. There is also the nonrepeating sequence of Hox gene expression zones. Although the Hox domains overlap, there is a clear sequence from anterior to posterior in which expression of a single gene predominates.

Each element of this pattern appears to be initiated locally by combinations of concentrations of the products of the maternal systems, the gap genes, and the pair-rule genes. The pair-rule genes have a particularly important role in that they must control the register between the segment polarity and the Hox genes so that each segment acquires the correct identity. By the extended germ-band stage the products of the controlling systems have decayed or are decaying and so the maintenance of the pattern is ensured by separate means: such as positive-feedback loops, for example of *Ubx*; mutual reinforcement of neighboring states, for example of *engrailed* and *ci*; inhibition between neighboring states, for example of *Antp* by *Ubx*; or long-term changes in chromatin configuration.

Up to the formation of the general body plan the regional specification of *Drosophila* is quite well understood. But the ultimate role of the developmental genes is to upregulate appropriate combinations of genes to carry out the differentiated functions of the relevant cells. How this is done is still only partly understood, mainly because the cell biology, histology, and general biochemistry of *Drosophila* is not as well known as that of vertebrates.

Key points to remember

- The mutagenesis screens for genes affecting early *Drosophila* development led to the discovery of most of the classes of gene that control development not just in *Drosophila* but in all animals.
- *Drosophila* and other insect embryos initially develop as a syncytium. This means that during the first 3 hours transcription factors can diffuse from one nucleus to another to control gene expression.
- The dorsoventral and anteroposterior patterns are specified largely independently. The dorsoventral pattern is set up by a series of interactions between oocyte and follicle cells, starting with the eccentric position of the oocyte nucleus and culminating in a gradient of dorsal protein in the nuclei of the syncytial blastoderm.
- The cytoplasmic determinants that control anteroposterior pattern are also laid down in the oocyte before fertilization. They are: *bicoid* mRNA in the anterior, *nanos* and *oskar* mRNA in the posterior, and activation of Torso at the termini.

- Development proceeds in a stepwise fashion with each domain of cells being defined by the expression of a combination of transcription factors encoded by the gap and pair-rule genes. These regulate the expression of the segmentation genes, defining the 14 repeating units of the embryo; and the Hox genes that define the anteroposterior character of each parasegment.
- Segmentation is controlled by a cross activation between stripes of *engrailed*-expressing cells that secrete Hedgehog, and *cubitus interruptus*-expressing cells that secrete Wingless.
- Anteroposterior pattern is controlled by transcription factors encoded by Hox genes which are upregulated in a nested pattern such that all are on at the posterior end with each gene having a specific anterior expression limit. Loss-of-function mutants generally cause anteriorization, while gain-of-function mutants generally cause posteriorization.

Further reading

See also the general textbooks referenced in Chapter 2.

Websites

The Interactive Fly: http://www.sdbonline.org/fly/aimain/1aahome.htm
www.virtual library *Drosophila*: http://ceolas.org/fly/
Volker Hartenstein's Atlas: http://www.sdbonline.org/fly/atlas/00atlas.htm

General

Lawrence, P.A. (1992) *The Making of a Fly*. Oxford: Backwell Science.

Bate, M. & Martinez Arias, A., eds. (1993) *The Development of Drosophila melanogaster*, vols 1 and 2. Cold Spring Harbor, NY: Cold Spring Harbor Laboratory Press. Reprinted 2009.

Campos-Ortega, J.A. & Hartenstein, V. (1997) *The Embryonic Development of Drosophila melanogaster*. Berlin/Heidelberg: Springer-Verlag.

Gehring, W.H. (1998) *Master Control Genes in Development and Evolution. The Homeobox Story*. New Haven: Yale University Press.

Genetics

Nüsslein-Volhard, C. & Wieschaus, E. (1980) Mutations affecting segment number and polarity in *Drosophila*. *Nature* **287**, 795–801.

St Johnston, D. (2002) The art and design of genetic screens: *Drosophila melanogaster*. *Nature Reviews Genetics* **3**, 176–188.

Venken, K.J.T. & Bellen, H.J. (2007) Transgenesis upgrades for *Drosophila melanogaster*. *Development* **134**, 3571–3584.

Dorsoventral patterning

Morisato, D. & Anderson, K.V. (1995) Signalling pathways that establish the dorso-ventral pattern of the *Drosophila* embryo. *Annual Review of Genetics* **29**, 371–399.

Leptin, M. (1995) *Drosophila* gastrulation: from pattern formation to morphogenesis. *Annual Review of Cell and Developmental Biology* **11**, 189–212.

Anderson, K.V. (1998) Pinning down positional information: dorsoventral polarity in the *Drosophila* embryo. *Cell* **95**, 439–442.

Stathopoulos, A. & Levine, M. (2002) Dorsal gradient networks in the *Drosophila* embryo. *Developmental Biology* **246**, 57–67.

Moussian, B. & Roth, S. (2005) Dorsoventral axis formation in the *Drosophila* embryo – shaping and transducing a morphogen gradient. *Current Biology* **15**, R887–R899.

Maternal anteroposterior systems

St Johnston, D. & Nüsslein-Volhard, C. (1992) The origin of pattern and polarity in the *Drosophila* embryo. *Cell* **68**, 201–219.

Munn, K. & Steward, R. (1995) The anteroposterior and dorsoventral axes have a common origin in *Drosophila melanogaster*. *Bioessays* **17**, 920–922.

López-Schier, H. (2003) The polarisation of the anteroposterior axis in *Drosophila*. *Bioessays* **25**, 781–791.

Montell, D.J. (2003) Border-cell migration: the race is on. *Nature Reviews Molecular Cell Biology* **4**, 13–24.

Poulton, J.S. & Deng, W.-M. (2007) Cell–cell communication and axis specification in the *Drosophila* oocyte. *Developmental Biology* **311**, 1–10.

Shvartsman, S.Y., Coppey, M. & Berezhkovskii, A.M. (2008) Dynamics of maternal morphogen gradients in *Drosophila*. *Current Opinion in Genetics and Development* **18**, 342–347.

Grimm, O., Coppey, M. & Wieschaus, E. (2010) Modelling the Bicoid gradient. *Development* **137**, 2253–2264.

Zygotic anteroposterior systems

Pankratz, M.J. & Jäckle, H. (1990) Making stripes in the *Drosophila* embryo. *Trends in Genetics* **6**, 287–292.

Forbes, A.J., Nakano, Y., Taylor, A.M. & Ingham, P.W. (1993) Genetic analysis of hedgehog signalling in the *Drosophila* embryo. *Development* (Suppl.) 115–124.

Rivera-Pomar, R. & Jäckle, H. (1996) From gradients to stripes in *Drosophila* embryogenesis: filling in the gaps. *Trends in Genetics* **12**, 478–483.

González-Gaitán, M. (2003) Endocytic trafficking during *Drosophila* development. *Mechanisms of Development* **120**, 1265–1282.

Dmitri, P. (2009) Stripe formation in the early fly embryo: principles, models, and networks. *BioEssays* **31**, 1172–1180.

Schroeder, M.D., Greer, C. & Gaul, U. (2011) How to make stripes: deciphering the transition from non-periodic to periodic patterns in Drosophila segmentation. *Development* **138**, 3067–3078.

Modeling

O'Connor, M.B., Umulis, D., Othmer, H.G. & Blair, S.S. (2006) Shaping BMP morphogen gradients in the Drosophila embryo and pupal wing. *Development* **133**, 183–193.

Reeves, G.T., Murator, C.B., Schüpbach, T. & Shvartsman, S.Y. (2006) Quantitative models of developmental pattern formation. *Developmental Cell* **11**, 289–300.

Umulis, D.M., Shimmi, O., O'Connor, M.B. & Othmer, H.G. (2010) Organism-scale modeling of early Drosophila patterning via bone morphogenetic proteins. *Developmental Cell* **18**, 260–274.

Hox genes

Mann, R.S. & Morata, G. (2000) The developmental and molecular biology of genes that subdivide the body of *Drosophila*. *Annual Review of Cell and Developmental Biology* **16**, 243–271.

Maeda, R.K. & Karch, F. (2006) The ABC of the BX-C: the bithorax complex explained. *Development* **133**, 1413–1422.

Hueber, S.D. & Lohmann, I. (2008) Shaping segments: Hox gene function in the genomic age. *BioEssays* **30**, 965–979.

Mallo, M., Wellik, D.M. & Deschamps, J. (2010) Hox genes and regional patterning of the vertebrate body plan. *Developmental Biology* **344**, 7–15.

This chapter contains the following animations:

Animation 13   DV patterning system.
Animation 14   AP polarity of oocyte.
Animation 15   Ci-engrailed loop establishment.

 For additional resources for this book visit
www.essentialdevelopmentalbiology.com

Chapter 12

Caenorhabditis elegans

Caenorhabditis elegans is a small, free-living, soil nematode and has been used for developmental biology research since the 1960s. Among developmental biologists it is usually known as "the worm." In one sense it is the best known animal on Earth since the location and lineage of every cell in embryo, larva, and adult is known. Also, its genome was the first of any animal to be completely sequenced. The genome contains 19,099 genes, of which about 2000 are mutatable to lethality.

C. elegans is kept on Petri plates and feeds on bacteria. Genetic screening is easy because it is possible to examine large numbers of worms, and the generation time is only 3 days. The worms are self-fertilized **hermaphrodites** so recessive mutations will automatically segregate as homozygotes in two generations without the need to set up any crosses (Fig. 12.1). Genetic stocks can be preserved in liquid nitrogen. This ease of genetic analysis means that large numbers of mutants are available, and, historically, the investigation of a biological problem often started with discovery of a mutant affecting the process. As an alternative to the isolation of mutants, it is now also easy to inhibit gene action by **RNA interference** (see Chapter 3). Double-stranded RNA complementary to an endogenous message can be administered by feeding. A convenient method is to express the required dsRNA from a plasmid in *E. coli* and then use these bacteria as the food for the worms: sufficient dsRNA is absorbed intact to exert its biological effect.

It is also easy to make **transgenics** by injection of the required DNA into the gonad, where it is incorporated as an extrachromosomal element into the germ cells. Such transgenics are not stable because the element can be lost at meiosis or mitosis. However, transient transgenesis is often sufficient for experimental purposes. There is now also a method for incorporation of transgenes into the genome by excision of a transposon followed by repair of the resulting double stranded break by DNA synthesis. In the presence of an extrachromosomal transgene flanked by ends homologous to the break point, DNA synthesis during the repair process can copy in the required gene.

Genetic mosaics can be made by the spontaneous loss of small, free chromosome fragments, which are duplications of normal chromosomal regions. If the main chromosome carries a mutant allele and the free chromosome fragment the wild-type allele, then when the fragment is lost from a cell, this cell and its descendants will all be mutant while the remainder of the animal stays wild type. Mosaics can be very useful for establishing in which cells of the embryo the function of a gene is required.

The negative features of *C. elegans* are the small size of the eggs and the tough egg case, both of which make microsurgical experiments difficult.

In *C. elegans* it is conventional to capitalize the names of proteins, and to hyphenate both gene and protein names. So for example the *glp-1* gene encodes the GLP-1 protein. As usual, many mutations affecting early development are maternal-effect and in such cases it is the genotype of the mother and not the zygote that determines the embryo phenotype. In *C. elegans* a gene that is lacking maternally will also usually be lacking in the embryo. This is because the normal mode of propagation from a self–fertilized hermaphrodite means that a −/− parent will produce −/− zygotes, so both will lack the gene. But it is still necessary to remember that the gene products that control early development are deposited in the egg during oogenesis.

Adult anatomy

The adult is highly elongated ("worm shaped"; Fig. 12.2). The outer layer is the **hypodermis**, which is one cell thick, largely syncytial, and secretes a thick cuticle. Beneath the hypodermis are four longitudinal bands of mononucleate muscle cells.

Essential Developmental Biology, Third Edition. Jonathan M.W. Slack.

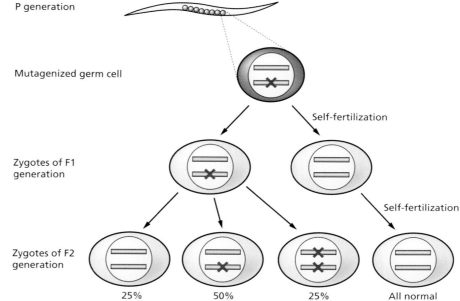

Fig. 12.1 Segregation of homozygotes after mutagenesis and two generations of self-fertilization. The red cross denotes a mutation. In the P generation, a particular mutation will be found in just a few gametes, having occurred in a single mitotic germ cell.Self-fertilization involving these gametes gives some heterozygous mutant worms in the F1 generation. Self-fertilization of these individuals gives 25% homozygous mutants, which are part of the F2 generation.

P generation

Mutagenized germ cell

Self-fertilization

Zygotes of F1 generation

Self-fertilization

Zygotes of F2 generation

25%　　　50%　　　25%　　　All normal

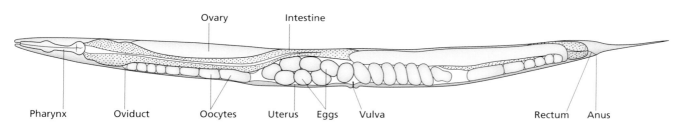

Ovary　　　Intestine

Pharynx　　Oviduct　　Oocytes　　Uterus　　Eggs　　Vulva　　　　Rectum　Anus

Fig. 12.2 Adult anatomy of *C. elegans*. Reproduced from Sulston and Horvitz. *Dev Biol* 1977; 56: 110–156, with permission from Elsevier.

There is a through gut with a muscular pharynx and an intestine. There is a nerve ring surrounding the pharynx, a ventral nerve cord, and tail ganglia. The main body cavity of nematodes is described as a **pseudocoelom** rather than a coelom because it is not lined all round with mesoderm. The **gonad** opens into a mid-ventral vulva. In the hermaphrodite this has two arms which are both bent back on themselves. Within the gonad, the cells nearest the vulva mature as sperm, while the more distant ones divide for a while and then become **oocytes**. These mature into eggs, become fertilized as they encounter the sperm on their way out, and are laid as cleavage-stage embryos.

Although most worms are hermaphrodite, there are also occasional males whose gonad has just one arm and opens posteriorly at the cloaca. Hermaphrodites have an XX-chromosome constitution while males are XO. They arise when an X-chromosome is lost by disjunction during meiosis. If a male and hermaphrodite mate, then the male sperm outcompete those of the hermaphrodite, resulting in an outcross.

Even at the adult stage nematodes have rather few cells, and during embryonic and larval development *C. elegans* shows almost complete invariance of cell lineage, meaning that every individual embryo shows exactly the same sequence and orientation for every cell division. Embryos are laid at about 30 cells and hatch at about 14 hours with 558 cells. The larva feeds and grows, undergoing four molts before reaching the adult stage with 959 somatic cells plus about 2000 germ cells. The second stage larva may enter a dormant **dauer larva** phase if nutrients are in short supply during the first larval stage. After the last molt the adult worm shows no further cell division of somatic tissues and can grow only by cell enlargement.

C. elegans does possess six Hox genes, although they do not form a true cluster as there are some intervening genes. In general, the genes obey the rule of colinearity of chromosomal position and anterior expression limit, although the second gene, *ceh-13*, has the most anterior expression limit and is homologous to the anterior Hox genes in other animals. The

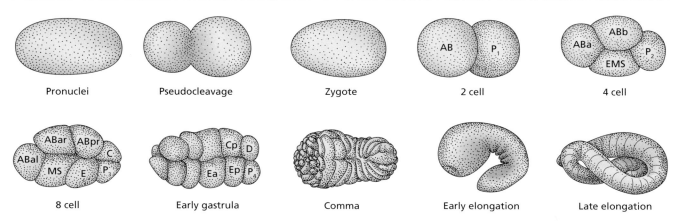

Pronuclei   Pseudocleavage   Zygote   2 cell   4 cell

8 cell   Early gastrula   Comma   Early elongation   Late elongation

Fig. 12.3 Embryonic development of *C. elegans*.

genes are called: *lin-39*, *ceh-13*,[gap], *mab-5*, *egl-5*,[gap], *php-3*, *nob-1*. The last three belong to the *Abdominal B*, or posterior class. Only *ceh-13*, *php-3*, and *nob-1* have embryonic phenotypes in loss-of-function mutants.

Embryonic development

Fertilization in *C. elegans* is somewhat unusual. The sperm are amoeboid, with no flagellum or acrosome. Oocytes are fertilized before the first meiotic division, and the initiation of maturation, with germinal vesicle breakdown, depends on a Major Sperm Protein released from nearby sperm. The sperm can enter the oocyte at any position and the point of sperm entry defines the future **posterior** of the zygote. An early sign of this is the appearance of a smooth posterior cortical region, while the remainder of the egg cortex becomes ruffled. Following completion of both meiotic divisions, there is a cytoplasmic rearrangement whereby internal cytoplasm moves posteriorly and cortical cytoplasm moves anteriorly. This is associated with a "pseudocleavage" or formation of a furrow, which does not progress to a full cleavage. As a result of this polarization, the early cleavages are asymmetrical (Fig. 12.3). The first forms an anterior AB and posterior P_1 cell. AB then forms ABa and ABp while P_1 behaves in a reiterated manner, keeping a P-like daughter (successively called P_2, P_3, P_4) while cutting off the EMS, C, and D blastomeres. The final P cell (P_4) is the **germ cell** precursor, dividing only once more in embryonic life to become Z2 and Z3, which generate all the germ cells of larva and adult. Maternal components are sufficient to direct development up to about 26 cells, as this is the earliest stage that defects are apparent if embryos are raised in α-amanitin, an inhibitor of RNA polymerase II. Zygotic transcription of RNA polymerase II genes in the germ line remains repressed for longer, until about the 100-cell stage.

The egg contains RNA-rich **P-granules**, which are initially randomly dispersed but which concentrate in the posterior

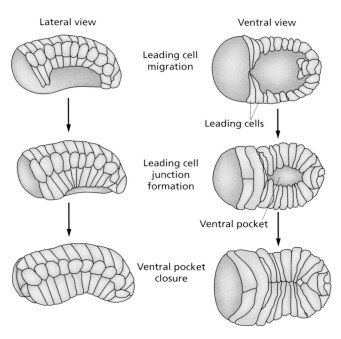

Lateral view   Ventral view

Leading cell migration

Leading cells

Leading cell junction formation

Ventral pocket

Ventral pocket closure

Fig. 12.4 The later stage of *C. elegans* "gastrulation." Modified from Simske and Hardin, *Bioessays* 2001; 23: 2066–2075, with permission from John Wiley & Sons, Ltd.

during the cytoplasmic rearrangement period. During each successive division these granules concentrate in the region that will become the new P cell. Of the original founder cells, AB, MS, and C all produce a variety of cell types while the others generate a single cell type: P_4 becoming the germ line, E becoming the gut, and D becoming muscle. "**Gastrulation**" in *C. elegans* is rather atypical and prolonged but can be considered as starting at the 26-cell stage when the two E cells move into the interior, in an autonomous invagination process. These are followed by the myoblasts derived from C and D, and the pharyngeal cells derived from ABa. The ventral cleft, which resembles a blastopore, closes at about the 300-cell stage (Fig. 12.4).

Because of the relatively small cell number and the invariance such that all individuals undergo exactly the same sequence of cell divisions, the complete **cell lineage** of embryo, larva, and adult has been determined by direct observation, the first part of which is shown in Fig. 12.5. Although in one respect setting a high standard of precision, the lineage falls short of a complete fate map as it shows only the "family tree" of the cells but not their spatial relationships at the different stages. Development was originally thought to be entirely **mosaic** in character, because in almost all cases when a cell is removed by laser microbeam irradiation all of its descendants are lost and there is no consequence for the development of neighboring cells. However, a number of inductive interactions are now known, so *C. elegans* does not really differ greatly from the other model species in this regard.

The precision of the cell lineage makes it less useful to define which parts of the embryo belong to the different germ layers than it is for the other animal types. The "official" germ layers are:

ectoderm: AB, Caa, Cpa;
mesoderm: MS, Cap, Cpp, D;
endoderm: E.

However, AB produces the pharyngeal muscles, which would normally be considered a mesodermal type, and MS produces some pharyngeal neurons, which would normally be considered ectodermal.

Regional specification in the embryo

Asymmetrical cleavages

Cell polarization followed by asymmetrical division is often important in stem cell behavior in higher animals and some of the basic mechanisms were discovered by studying *C. elegans*.

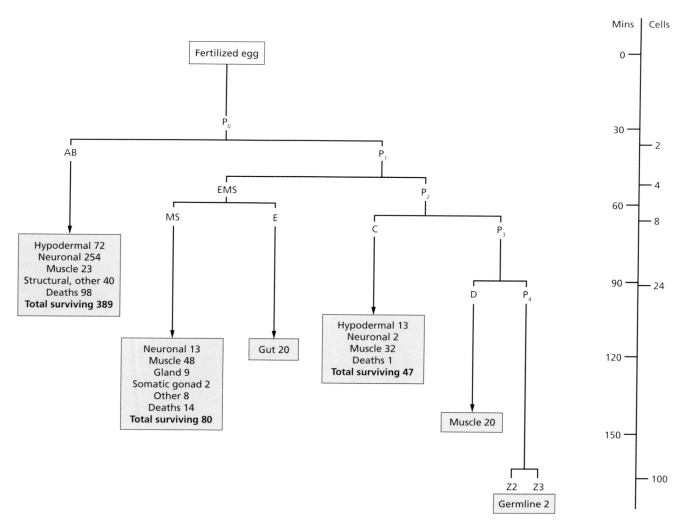

Fig. 12.5 Early cell lineage of *C. elegans*.

Classic experiments

Mechanism of unequal cell division

The breakthrough depended on isolation of mutants of maternal-effect genes in which the normal polarization of the zygote was lost. Homologues of the *par* genes are now known to be involved in asymmetric cell divisions in many other animals.

Kemphues, K.J., Priess, J.R., Morton, D.G. & Cheng, N. (1988) Identification of genes required for cytoplasmic localization in early *C. elegans* embryos. *Cell* **52**, 311–320.

Guo, S. & Kemphues, K.J. (1995) *Par-1*, a gene required for establishing polarity in *C. elegans* embryos, encodes a putative Ser/Thr kinase that is asymmetrically distributed. *Cell* **81**, 611–620.

Asymmetrical cell division requires two processes, the establishment of a cytoplasmic polarity and the correct orientation of the mitotic apparatus (Fig. 12.6). The key players are a series of proteins called the PAR proteins. Their genes were identified by screens for maternal-effect lethal mutations deranging early cleavages ("**par**titioning defective") (Fig. 12.7). It is now know that there are mammalian and *Drosophila* homologs of the *par* genes, and these are also involved in the acquisition of cell polarity and in the control of asymmetrical cell division (Chapter 18). Operation of the system is shown in Animation 16 "PAR system" on the website.

The original screen was carried out by mutagenizing worms that are themselves unable to lay eggs. Such worms can still reproduce, because the larvae arising from self-fertilization simply eat their way out of the body of the hermaphrodite. To do such a screen, one F1 larva is put in each dish. If it carries a mutation on one chromosome, then 25% of its (F2) offspring will be homozygous for that mutation. If the mutation is zygotic lethal, then the affected embryos will simply fail to develop. However, if the mutation is a maternal-effect lethal, then the homozygous mutant F2 generation will develop into worms but they will then fill up with unviable F3 embryos that cannot develop and so do not eat their way out (Fig. 12.8).

- PAR-1 is a SerThr kinase which binds nonmuscle myosin. After the cytoplasmic rearrangement it is found in the posterior cortex of the zygote.
- PAR-2 is a cytoplasmic protein with adenosine triphosphate (ATP)-binding and zinc-binding (RING) domains. It is also localized to the posterior of the zygote.
- PAR-3 and -6 are cytoplasmic proteins containing a PDZ (protein–protein recognition) domains. They form a complex with an atypical protein kinase C (aPKC3), and the complex becomes associated with the plasma membrane in the anterior of the zygote.

In the unfertilized oocyte the PAR proteins are uniformly distributed. Following fertilization the centrosome introduced by the sperm somehow causes the polarization to be initiated. UV laser ablation of the centrosome, but not the sperm pronucleus, prevents polarization. The cortical flow away from the sperm entry point carries some of the PAR-3 complex anteriorly. The PAR-3 complex then repels PAR-1 and -2 such that they become concentrated in the posterior (Fig. 12.9). This mutual repulsion is the key element of the cell polarization and

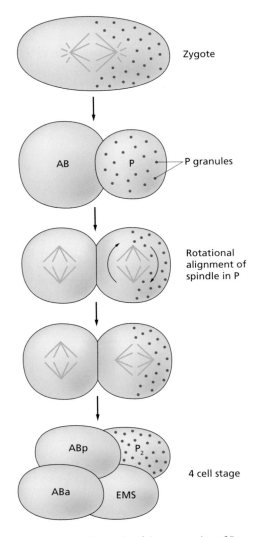

Fig. 12.6 Asymmetrical division involving segregation of P-granules into P cells and rotational alignment of the P$_1$ spindle.

serves to amplify the small change brought about by sperm entry into a big change affecting the overall structure of the zygote. Once the polarization process has started, it can no longer be modified by ablating the sperm centrosome.

Evidence for the mutual repulsion mechanism comes from observing the distribution of one PAR protein in the absence of

Fig. 12.7 Embryo produced by *par-3* defective mother. (a) Wild type showing normal AB and P₁ cells. (b) *par-3* mutant embryo with first two blastomeres of equal size. Reproduced from Goldstein and Macara. *Developmental Cell* 2007; 13: 609–622, with permission from Elsevier.

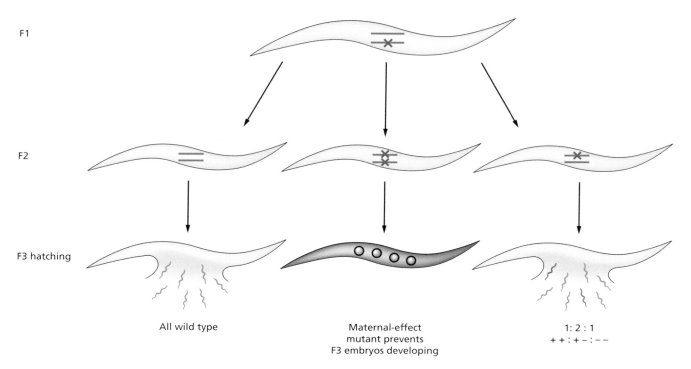

F1

F2

F3 hatching

All wild type

Maternal-effect
mutant prevents
F3 embryos developing

1 : 2 : 1
+ + : + − : − −

Fig. 12.8 Maternal screen for isolation of *par* mutants. The hermaphrodites are vulvaless and cannot lay eggs, so the larvae eat their way out, destroying the parents. But those F2 worms carrying arrested embryos due to a maternal-effect mutation will persist.

another. In the absence of PAR-2 there is no movement of the PAR-3 complex to the anterior, and in the absence of PAR-3 there is no movement of PAR-1 and -2 to the posterior. The localizations are achieved partly by actual movement, partly by differential degradation, and the mechanism is known to involve the phosphorylation of PAR-3 by PAR-1, which destabilizes the PAR-3 complex; and the phosphorylation of PAR-2 by aPKC, which prevents association with the cortex.

Mitotic orientation

Normally, the early blastomeres will divide in a direction at right angles to their last cleavage. The AB cell follows this rule as it divides orthogonal to the first cleavage, but the P₁ cell does not do so, instead dividing parallel to the first cleavage. It does this because of "rotational alignment", which is a 90° rotation of the

centrosomes and nucleus, driven by the positioning of microtubule attachments on the cell cortex.

The PAR proteins help to control the orientation of mitoses by regulating rotational alignment. Embryos lacking PAR-2 show rotational alignment in neither AB nor P, while embryos lacking PAR-3 show rotational alignment of both AB and P cells. The double mutant, *par-2⁻ par-3⁻*, produces embryos showing rotational alignment in both cells, like *par-3⁻*. This means that something other than the *par* genes must be causing the rotational alignment and that the PAR-3 complex normally suppresses it in the AB blastomere and its absence allows it in P₁.

Once the first asymmetrical cleavage has occurred the composition of the cytoplasm in the two daughters is very different. The AB cell contains an excess of the PAR-3 complex and the P₁ cell an excess of PAR-1 and -2. These proteins are cytoplasmic

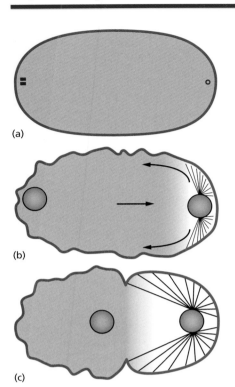
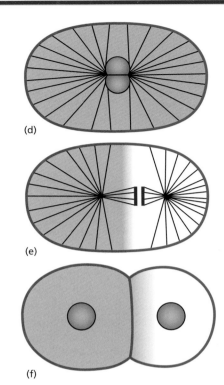

(a)

(b)

(c)

(d)

(e)

(f)

Fig. 12.9 Sequence of events leading to the unequal first division of the *C. elegans* zygote, anterior to left. (a) Chromatin from oocyte (left) and sperm (right) in blue, PAR-3 in red, MEX-5 in pink. (b) Cortex smoothes in posterior, PAR-2 (in green) accumulates there, pronuclei form, cytoplasmic flows commence. (c) PAR-2 zone expands, oocyte pronucleus moves posteriorly, pseudocleavage. (d) Pronuclear fusion, cortical ruffles disappear. (e) First mitosis. (f) Two-cell stage, AB is larger than P_1, PAR3 complex extends into P_1. Reproduced from Nance. *BioEssays* 2005; 27: 126–135, with permission from John Wiley & Sons, Ltd.

determinants because the difference in their representation is the cause of the subsequent events of regional specification.

Determinants

In addition to the PAR proteins, several other maternally encoded proteins responsible for regional specification have been identified (Fig. 12.10; Table 12.1; and see Animation 4: Cytoplasmic determinant action). Their mode of action has been deduced from the maternal effect mutant phenotype, and from the effect on their localization of mutating other genes, including the *par* genes. Localization may be studied by immunostaining for the protein, or by observing the intracellular position of a transgenic GFP fusion protein by fluorescence microscopy.

SKN-1 (pronounced "skin-1"), is a transcription factor of the bZIP type and confers an EMS type of development on its nuclei. The mRNA is present maternally and is not localized, but the protein accumulates only in the P_1 nucleus, and later in the descendant P_2 and EMS nuclei, and then becomes lost after the 12-cell stage. Embryos without SKN-1 lack pharynx and intestine because the E and MS blastomeres develop like the C blastomere (hence too much "skin"). Although in the normal embryo SKN-1 is present both in EMS and in P_2, its transcription factor activity is repressed in P_2 by the PIE-1 protein, which is responsible for repression of all RNA polymerase II-mediated transcription in the early germ line. Embryos lacking PIE-1 still

have a normal distribution of SKN-1 protein, but the P_2 cell now develops like EMS, because SKN-1 is active in both cells.

mex-1 mRNA is initially ubiquitous but becomes lost from cells other than the P lineage. The protein also concentrates in the posterior of the zygote. MEX-1 appears to prevent SKN-1 from entering AB. Embryos lacking MEX-1 have SKN-1 in the nuclei of the two AB daughter cells as well as in P_2 and EMS. As a consequence, they have AB descendants developing like the normal MS descendants, leading to too much muscle. Embryos lacking both MEX-1 and SKN-1 have a similar phenotype to that caused by lack of SKN-1 alone, with AB normal but EMS developing like C. These results confirm that normal AB behavior depends on the absence of SKN-1.

pal-1 is a homolog of the *Drosophila caudal* gene and the vertebrate Cdx gene family. Like them it is needed for posterior development. The mRNA is present all over the early embryo, but is normally only translated in EMS and P_2. Translation is repressed during the early stages, and in AB cells, by MEX-3, which acts on the 3' UTR of the *pal-3* mRNA. The *mex-3* mRNA and MEX-3 protein are initially uniform then become more abundant in AB cells and are lost after the four-cell stage. Embryos lacking MEX-3 express PAL-1 protein all over and are posteriorized in morphology, with the AB descendants resembling the normal descendants of blastomere C.

These results show how the character of each of the early blastomeres is specified by the particular combination of determinants which it inherits. The spatial disposition of the

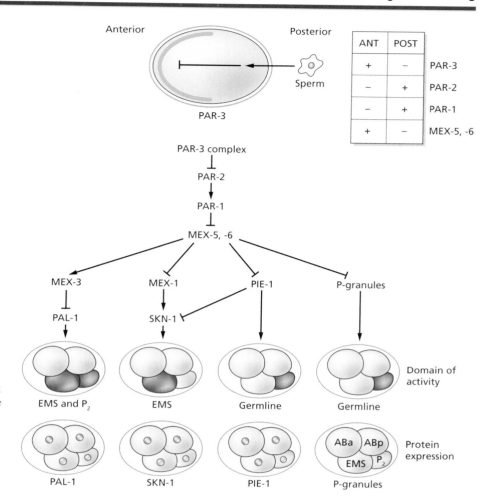

Fig. 12.10 Cytoplasmic determinants in *C. elegans*. Following fertilization the PAR-1 and PAR-2 proteins become localized to the posterior and the PAR-3 complex to the anterior. Localization of SKN-1, PIE-1, and PAL-1 proteins occurs as shown by the four-cell stage. Note that SKN-1 protein is present in P₂ but not active because of the presence of PIE-1.

Table 12.1 Determinants active in *C. elegans*.

Protein	Biochemical nature	Loss-of-function phenotype
SKN-1	bZIP transcription factor	Too much hypodermis
PIE-1	Zn finger transcription factor	Excess pharynx and intestine; no germline
MEX-1	Cytoplasmic Zn finger protein	Muscle excess
MEX-3	RNA/protein-binding protein	Muscle excess
MEX-5,6	Cytoplasmic Zn finger proteins	Muscle excess
PAL-1	Homeodomain transcription factor	Posterior defective

determinants is controlled by the PAR system and PAR-1 is the main effector, acting on the cytoplasmic proteins MEX-5 and MEX-6 such that they become localized to the anterior (Fig. 12.10). This is achieved by an increase in the intracellular diffusion rates of the MEX proteins following phosphorylation by PAR-1. MEX-5 and -6 themselves act on the MEX-1 and PIE-1 proteins, and the P-granules, to localize them all to the posterior where they direct the formation of the P lineage of blastomeres.

The evidence for this is that in the absence of PAR-1, the MEX-1, -5, -6 proteins, the PIE-1 protein, and the P-granules are all uniformly distributed. In the absence of the MEX-5 and -6 proteins, the PAR-1 distribution is normal, but the distribution of MEX-1, PIE-1, and the P-granules are all uniform, showing that PAR-1 normally regulates the localization of MEX-5 and -6 and they in turn control the disposition of the other components. These effects are exerted mostly through differential protein degradation.

Blastomere identity   Tissue/organ identity   Differentiation

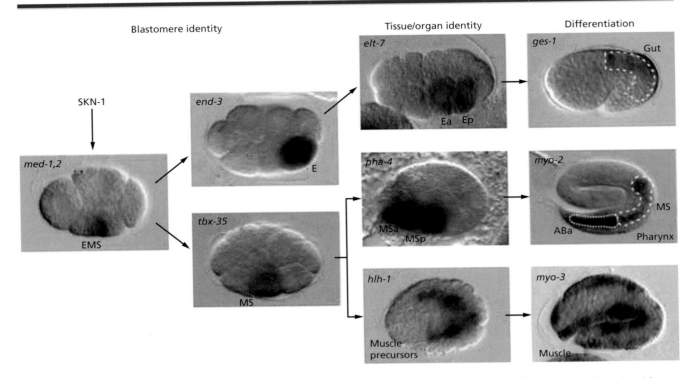

Fig. 12.11 *In situ* hybridizations showing zygotic gene activation consequent on the localization of determinants described above. Reproduced from Maduro and Rothman. *Dev Biol* 2002; 246: 68–85, with permission from Elsevier.

MEX-5 and -6 also concentrate MEX-3 in the anterior. In the absence of PAR-1, MEX-3 is present all over the embryo. Normally, MEX-3 in the anterior inhibits the production of PAL-1 by translational control, confining the activity of PAL-1 to the posterior. But in the absence of PAR-1, leading to uniform MEX-3, there is no expression of PAL-1 and posterior development is defective.

The components dealt with so far are all maternal. But specific zygotic gene activation follows in each of the cell lineages (Fig. 12.11). This depends both on the correct placement of the maternal cytoplasmic determinants, and on the occurrence of inductive signals between adjacent blastomeres. The localization of both the determinants and the components of the signaling systems depend ultimately on the operation of the PAR system.

Induction of the pharynx

The pharynx of the adult worm is a tube in three sections. The corpus joins the buccal cavity and serves to pump food further in. The isthmus moves food on by peristalsis, and the terminal bulb grinds the food finely to prepare it for the intestine. The pharynx is thus a highly muscular organ, but also contains other cell types including glands and neurons.

The pharynx arises from descendants of both ABa and MS blastomeres. The part arising from ABa requires two successive inductive signals in order to form. The first of these is repressive and comes from the P_2 cell, and the second is positive and comes

from the descendants of MS (Fig. 12.12). Both signals operate through the Notch pathway, but utilize different ligands of Notch.

The repressive signal normally prevents formation of pharynx from the sister blastomere of ABa, ABp. If ABa and ABp are interchanged then ABp will form the anterior pharynx instead of ABa, showing that position of the cell rather than its lineage is important. However, if P_2 is prevented from touching ABp then ABp forms pharynx as well as ABa, showing that there must normally be a signal from P_2 to ABp that suppresses pharynx formation.

This repressive signal is encoded by the maternal-effect *apx-1* (anterior pharynx excess) gene. Embryos lacking APX-1 show formation of anterior pharynx from both AB blastomeres, instead of just ABa. *apx-1* codes for a Delta-like ligand. The receptor is encoded by another maternal-effect gene, *glp-1* (the name refers to an effect on **g**erm-**l**ine **p**roliferation) whose product is a Notch-type receptor. Embryos lacking GLP-1 also have ABp developing as ABa in most respects, but unlike *apx-1⁻*, *glp-1⁻* mutants do not actually go on to form pharynx from the two equivalent ABa-like blastomeres. This is because formation of pharynx is not, in fact, a default for ABa, but depends on a subsequent positive inductive interaction. This may be shown by the fact that an isolated AB cell does not produce any pharynx. Also, laser ablation of the MS cell between eight and 12 cells prevents ABa forming pharynx, showing that its presence must be necessary during this time interval. At the 12-cell stage the

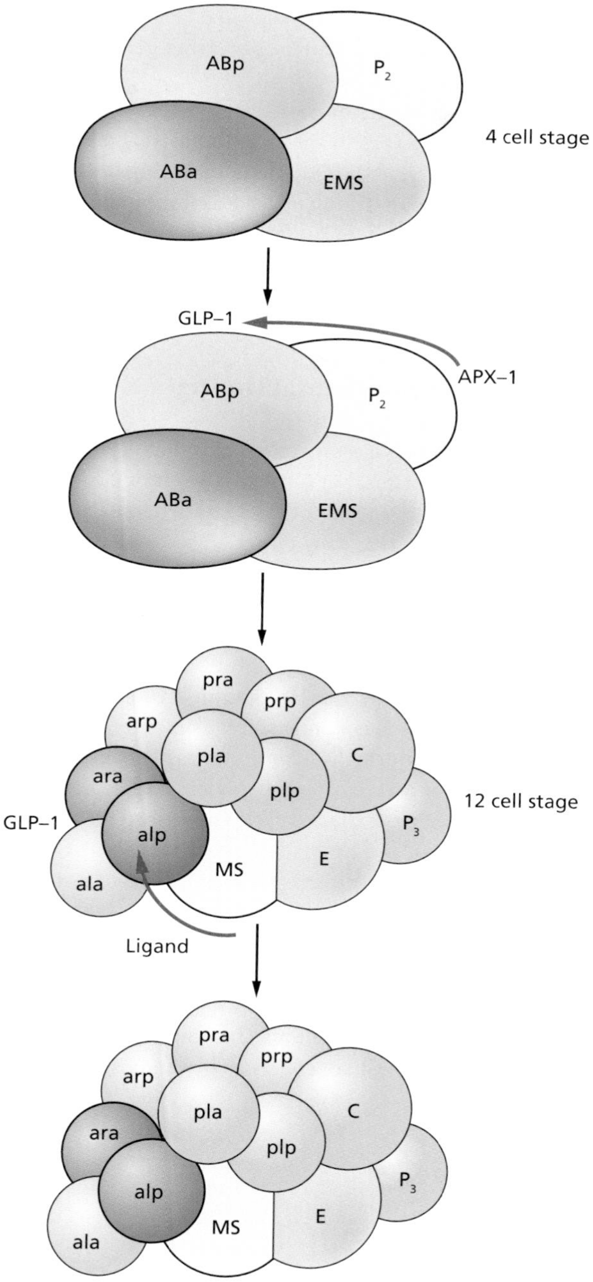

Fig. 12.12 Two inductive interactions leading to formation of the pharynx.

behave the same. The *glp-1* mRNA is uniformly distributed in the embryo up to the eight-cell stage but the protein is found only in the AB descendants. This is because of differential translation regulated by a sequence in the 3′-UTR of the message, with translation in the posterior being repressed by a cascade of factors ultimately controlled by PAR-1. The GLP-1 protein disappears at the 28-cell stage (when there are 16 AB descendants). The ligand for GLP-1 expressed by MS is distinct from APX-1.

Hence there is a double requirement for GLP-1, first as the receptor mediating repression of pharynx formation by the action of P$_2$ on ABp, and then as the receptor mediating positive induction of pharynx formation by the action of MS on ABa. These two separate requirements are clearly shown by the phenotypes of temperature-sensitive mutants of *glp-1*, kept for different time periods at the nonpermissive temperature. If the nonpermissive temperature is given only around the four-cell stage then the phenotype is just like the maternal effect phenotype of *apx-1⁻*, with ABp as well as ABa forming pharynx. If the nonpermissive temperature is maintained until the 12-cell stage then the phenotype is like the maternal effect *glp-1* null mutant, with equivalent cell divisions of ABp and ABa but no subsequent pharynx formation.

Ultimately, formation of the entire pharynx is dependent on the zygotically expressed gene *pha-4*, encoding a winged helix transcription factor homologous to the FoxA genes important in vertebrate gut development. *pha-4* is activated by T-box transcription factors, which become zygotically upregulated in both the ABa- and the MS-derived parts of the pharynx. In the ABa-derived part they are the genes *tbx-37* and *-38*, and in MS the gene *tbx-35*. In MS *tbx-35* is upregulated by the products of the zygotic genes *med-1* and *-2* which encode GATA-type transcription factors and are themselves activated by SKN-1 (Fig. 12.11).

PHA-4 is responsible for controlling expression of pharyngeal genes in all the component cell types of the organ through a "pharyngeal enhancer." Among cell-type-specific genes that are activated are *hlh-6*, encoding a bHLH factor controlling gland differentiation, and *tbx-2*, controlling muscle differentiation. The loss-of-function mutant of *pha-4* lacks the entire pharynx, both the part formed from ABa and the part formed from MS. Use of a temperature-sensitive allele shows that there is a requirement for *pha-4* throughout development, for both early and late differentiation events. The level of expression increases progressively during development. This means that promoters with a high affinity for PHA-4 are turned on early, and those with a lower affinity are turned on later when the concentration of PHA-4 protein has built up to a sufficient level. Apart from its role in the pharynx, PHA-4 also has functions in gonadal development and affects the lifespan of the adult worm.

Intestinal development

The intestine is composed of 20 cells derived from the E blastomere. These cells polarize, intercalate with each other, and become arranged around a gut lumen and joined with

MS blastomere touches the two ABa grand-daughters (ABalp and ABara), and emits the second signal responsible for inducing the pharynx. Moreover, in *apx-1⁻* embryos the descendants of the ABp cell that produce pharynx are ABpra, ABprp, and ABplp, all of which also contact MS at the 12-cell stage. Remarkably, it seems that the receptor for this second signal is also GLP-1, since, as we have seen, the mutant embryos do not form a pharynx, even though in other regards the two AB lineages junctional complexes. The developmental specification of the E blastomere depends on induction from the P₂ cell. This emits a Wnt-like signal that causes the nearer, posterior, part of EMS to become E and the further, anterior, part to follow the default specification of MS. This may be shown by removing the P₂ cell, which causes both daughters of EMS to resemble MS. A series of *mom* (more mesoderm) mutants have a similar effect to loss of P₂. These encode members of a Wnt-like pathway, and it was shown by mosaic analysis that the signaling components including the product of the Wnt homolog itself (*mom-2*) were required in the P₂ cell, while the receptor homolog (*mom-5*) was required in EMS. Loss-of-function, maternal-effect mutations of these genes will convert E into a second MS. The reverse phenotype results from loss of function of *pop-1*, which encodes an HMG domain transcription factor similar to the Tcf and Lef factors in vertebrates. This converts MS into a second E, suggesting that formation of E depends on inhibition of POP-1 activity by the Wnt-like signal. In vertebrates the Wnt-beta catenin pathway would normally activate TCF/LEF. The *C. elegans* Wnt-like pathway is somewhat different and is variously called the "Wnt/MAPK pathway" or "noncanonical Wnt pathway" but these terms are misleading in that they imply close similarity to similarly named vertebrate pathways.

The *C. elegans* Wnt-like pathway operates as follows. The Wnt-like (MOM-2) signals though MOM-5 to activate both a homolog of beta-catenin (WRM-1) and a SerThr kinase (LIT-1, for loss of intestine, a homolog of a *Drosophila* kinase called NEMO). LIT-1 phosphorylates POP-1, and WRM-1 then facilitates its export from the nucleus. Another distant homolog of beta-catenin, SYS-1 (**sym**metrical **s**isters), becomes elevated in the prospective E cell. POP-1 normally represses expression of *end-1* and *-3*, which encode GATA factors needed for intestinal development. In the MS nucleus where POP-1 remains high, these genes are repressed. But in the presence of an excess of SYS-1, POP-1 becomes an activator of the same target genes. So

in the E lineage *end-1* and *-3* are activated (Fig. 12.11) and they bring about activation of a battery of further genes controlling intestinal differentiation.

The same mechanism of Wnt-like signaling affecting the ratio of POP-1 and SYS-1 in the two daughters, operates many more times in *C. elegans* development, including the formation of the distal tip cells that control germ-cell proliferation.

Analysis of postembryonic development

Control of developmental timing

Developmental timing is a critical process that is poorly understood in most organisms, but *C. elegans* has provided a window into the problem by means of its **heterochronic mutations**. These are mutations that may either advance or retard events from the normal developmental sequence, and which typically affect not just one structure but multiple lineages (Fig. 12.13). After hatching *C. elegans* undergoes four larval stages called L1, -2, -3, -4, each separated from the next by a **molt** of the extracellular cuticle secreted by the hypodermis. During each larval stage a precise set of events normally takes place. For example the intestinal cells normally divide during L1 but not subsequently. In L2 a subset of lateral hypodermal seam cells divide and in L4 they exit from the cell cycle and fuse to form cuticular lateral ridges (alae).

The first heterochronic mutant was called *lin-4*. Loss of function causes the L1 cell division patterns to be repeated, with extra molts (Fig. 12.13). Another mutation, *lin-14* (loss of function) causes a skipping of the L1 divisions and occurrence of L2-type divisions during L1. Loss of both shows a *lin-14* like phenotype. But the *lin-14* gene also has gain-of-function mutations that have *lin-4*-like effects. This indicates that *lin-4*

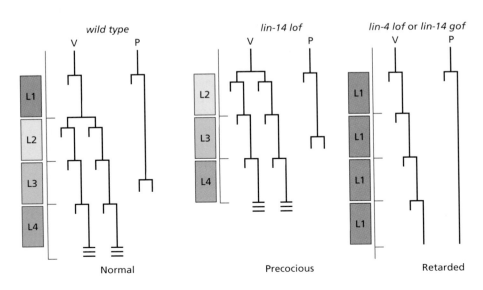

Fig. 12.13 Examples of phenotypes of heterochronic mutations. Two larval cell lineages are shown, called V and P, which undergo very different division programs in normal development. They are coordinately affected in the mutants. The precocious mutant causes skipping of a stage, the retarded ones cause reiteration of the L1 events. lof, loss of function; gof, gain of function. Reproduced from Moss. *Curr Biology* 2007; 17: R425–R434, with permission from Elsevier.

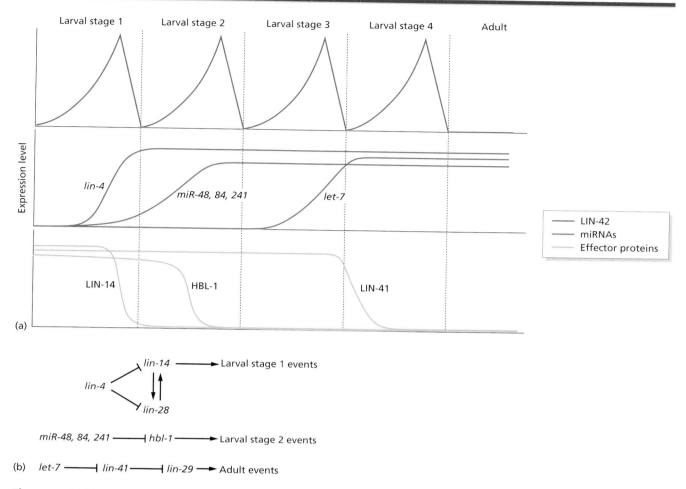

Fig. 12.14 (a) Diagram of relative expression behavior of the *per* homolog *lin-42*, the microRNAs, and the proteins whose translation they inhibit. (b) Some of the regulatory relationships controlling timing of larval stages.

normally suppresses the action of *lin-14*, and *lin-14* is needed for the occurrence of L2 cell divisions. *lin-14* encodes a transcription factor, and its translation into protein is inhibited by the product of *lin-4*, which is a microRNA (miRNA, see Appendix). The gain-of-function alleles of *lin-14* have lost the 3' untranslated region from the mRNA and are therefore resistant to the inhibitory action of *lin-4*.

LIN-14 is normally present during L1, and brings about the L1 division events. *lin-4* miRNA is normally expressed from the end of L1, so by the start of L2, the translation of LIN-14 is being suppressed (Fig. 12.14a).

LIN-14 acts in parallel with LIN-28, a cytoplasmic RNA binding protein. Loss of function of *lin-28* causes skipping of L2 divisions. Like LIN-14, LIN-28 translation is also inhibited by *lin-4* miRNA. LIN-14 and LIN-28 also stimulate each other's translation (Fig. 12.14b). Because they are mutually dependent in this way, removal of either will generate a phenotype and suppress effects of loss of *lin-4*.

A similar switch occurs at the end of L2. This depends on three related miRNAs (*miR-48, -84, -241*) which suppress translation of a transcription factor HBL-1 (a homolog of *Drosophila* Hunchback). HBL-1 controls the divisions of L2. During L2, the three miRNAs are expressed and block its activity. If all three miRNAs are removed, the L2 divisions are reiterated, due to persistence of HBL-1. If just one of them is overexpressed before L2 then the L2 divisions are skipped due to premature removal of HBL-1.

At the end of the larval stages loss of function of *let-7* causes reiteration of the last larval stage events. Conversely, loss of function of *lin-41* causes a precocious adult transition at the end of L3 instead of L4. *let-7* encodes a miRNA that suppresses translation of LIN-41. LIN-41 is a cytoplasmic protein that inhibits translation LIN-29, a Zn finger transcription factor that triggers cell cycle exit and cell fusion characteristic of the adult state. Normally, *let-7* miRNA increases from late L3 and suppresses LIN-41 by the end of L4.

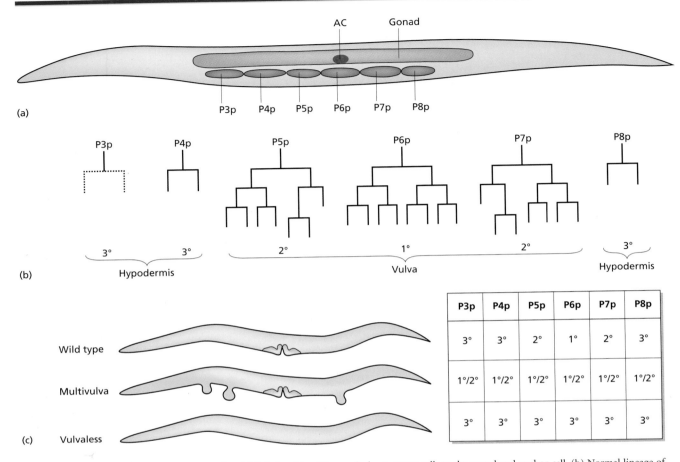

Fig. 12.15 Development of the *C. elegans* vulva. (a) Relationship of the equivalence group cells to the gonad and anchor cell. (b) Normal lineage of each cell. (c) Mutant phenotypes. The fate of each of the cells is indicated for each phenotype.

How does this set of repressive interactions build up into a developmental timing controller? This is not really known, but a clue comes from the properties of the *lin-42* gene. Its mRNA and protein are expressed in a periodic manner, high in intermolts, and low during molts (Fig. 12.14a). The protein is located in nuclei and is homologous to the *Drosophila* Per protein that has a central role in the control of circadian rhythms. Loss of function of *lin-42* gives precocious behavior. It has been suggested that expression of LIN-42 provides a gating mechanism within which the cycles of miRNA expression and action occur.

Is there something about miRNAs that make them uniquely suitable to be regulators of developmental timing? This also is not known. Many of the genes involved in *C. elegans* timing have vertebrate homologs. *let-7* miRNA is expressed during vertebrate development, and has been implicated in the progression of tumors, but a timing role is unclear. *lin-28A* has become famous as one of the genes used to make **induced pluripotent stem cells** (**iPS cells**, see Chapter 21), and in vertebrates is a target for the *lin-4* homolog as it is in *C. elegans*. A possible link with timing has come out of genome-wide association studies,

showing that the presence of mutations in a closely related gene, *lin-28B*, is correlate with an earlier onset of puberty in humans. In *Drosophila* loss of *let-7* causes persistence of some juvenile features into the adult stage, providing another possible link to timing mechanisms.

The vulva

The vulva is an epidermal structure formed in larval life around the mid-ventral opening of the gonad that is used for egg laying and mating (Fig. 12.15). Its formation is controlled by an Epidermal Growth Factor (EGF)-like signal from an internal cell called the **anchor cell**. The vulva arises from the cells called P5p, P6p, and P7p, which are the three posterior daughters of the embryonically generated ectodermal cells P5, P6, and P7:

- P5p makes seven vulval descendants;
- P6p makes eight vulval descendants;
- P7p makes seven vulval descendants.

In the fourth larval stage these 22 cells undergo various movements and fusions to make the vulva itself. In addition, the surrounding cells, called P3p, P4p, and P8p, are competent to make vulva, but in normal development they each divide just once to make two cells that later enter the syncytial hypoderm. In discussions of the vulva the following convention is used:

- formation of eight cells = primary fate 1°; normally followed by P6p;
- formation of seven cells = secondary fate 2°; normally followed by P5p and P7p;
- formation of two cells = tertiary fate 3°; normally followed by P3p, P4p, P8p.

Hermaphrodites are able to reproduce without a vulva, because the larvae can chew their way out of the body. Therefore it is possible to screen worms for vulval defects which prevent egg laying. However, viable mutations are often hypomorphic, with the corresponding null alleles of the same genes being lethal. The main classes of mutant are *vulvaless* and *multivulva*, the latter forming supernumerary vulvas from the same P3p–P8p cell group.

The six P3p–P8p cells are said to make up an **equivalence group**, because they are all competent to form vulva and they can replace each other in various experimental situations. This is clearly shown by their relations with the gonadal **anchor cell**, which lies internally adjacent to P6p. If the anchor cell is ablated by laser microbeam radiation, then all the P3p–P8p cells follow the tertiary fate 3° and no vulva is produced. If one of the P3p–P8p cells is removed by laser microbeam ablation, then in certain cases one of its neighbors will take its place and a normal vulva will result. If the anchor cell is moved relative to the P3p–P8p cells, as in various *displaced gonad* mutants, then whichever three of the P3p–P8p cells that are nearest will produce the vulva.

There are several *vulvaless* mutants, giving a similar phenotype to anchor cell ablation. Some encode components of the familiar EGF signaling pathway (see Appendix). The anchor cell ligand is a homolog of EGF encoded by *lin-3*. The receptor is an EGF receptor homolog encoded by *let-23*. The homolog of the vertebrate Ras protein is encoded by *let-60*, which has opposite phenotypes for loss-of-function and gain-of-function alleles. Loss-of-function mutants give a vulvaless phenotype while gain-of-function (constitutively active) mutants give a multivulva phenotype. Double-mutant combinations work in a predictable way, for example the combination of *let-23-* and *let-60 gof* gives a multivulva phenotype, because the Ras requirement lies downstream of the receptor. It is possible to visualize a gradient of EGF activity centered on P6p in worms that are transgenic for an EGF reporter gene.

Although the gradient of EGF signaling should theoretically be enough to generate the three cell fates, there is also a secondary signal emitted by P6p that activates the Notch pathway in P5p and P7p. In normal development the combination of both these signals serves to control the formation of the vulva.

The germ line

An important aspect of postembryonic development is the maturation of the **germ line**. This derives from the P_4 cell of the early embryo, which inherits various determinants, some associated with the P-granules. As mentioned above, the PIE-1 protein causes a general repression of genes transcribed by RNA polymerase II during the early stages. In the newly hatched larva the germ line consists of just two cells descended from the P_4 cell: Z2 and Z3. These express the *cgh-1* gene, which encodes an RNA helicase homologous to *vasa* in *Drosophila*. The *cgh-1* gene is active in the germ line thereafter, and the CGH-1 protein is one of the components of the P-granules found in the egg and segregated to the germ-line lineage in early development. Treatment of worms with RNAi directed against *cgh-1* causes death of the oocytes and formation of nonfunctional sperm, suggesting a key function in the later stages of germ cell development.

In the larval and adult worm the germ cells lie within the gonad (Fig. 12.16a,b). There are about 60 germ cells by the L2 stage and about 2000 in the adult hermaphrodite. Initially, the whole germ line is mitotic. As development proceeds a meiotic zone appears. Initially, this produces sperm and later it switches to oocyte production. Interestingly, most gene regulation during germ-cell development is carried out by translational rather than transcriptional control. The sequence of events is spatially organized along the adult gonad such that the most mature cells lie near the vulva and the most immature cells, which are still mitotic, at the blind ends of the gonad. Although the mitotic germ cells are joined by cytoplasmic bridges, their cell divisions are not synchronized, so they do not form a true syncytium.

The tip of each branch of the gonad contains an important somatic cell called the **distal tip cell** whose function is to maintain the neighboring germ-cell nuclei in mitosis (Fig. 12.17). These two cells derive from Z1 and Z4, progeny of the MS blastomere. They are specified by the same Wnt-like mechanism that is earlier responsible for the E-MS decision. The two distal progeny of Z1 and Z4 experience a high ratio of SYS-1 to POP-1, and become distal tip cells, while the proximal daughters, with low and therefore repressive POP-1, form the anchor cells that control formation of the vulva.

As the gonads grow during larval life, the mitotic germ-cell zones elongate such that cells progressively leaves the range of influence of the distal tip cell, whereupon they stop mitotic division and enter meiosis. Ablation of the distal tip cell at any stage by laser irradiation causes all the remaining mitotic nuclei to enter meiosis. The composition of the gonad following this is then appropriate to the stage of maturation reached, so an early ablation will create all sperm while a late ablation will result in a nearly normal arrangement of oocytes and sperm (Fig. 12.16c). The distal tip cell acts by expression of *lag-2*, which is a homolog of *Delta*. The receptor, encoded by *glp-1* (the same gene required for pharynx induction) is present on

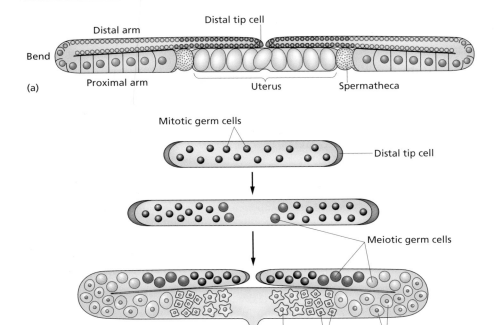

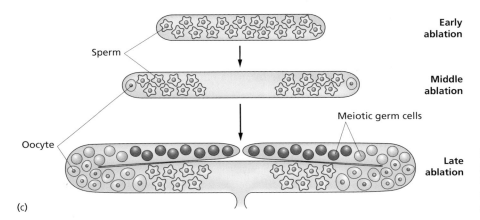

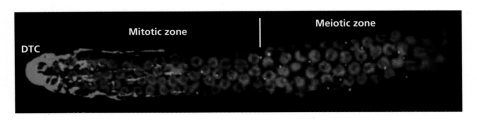

Fig. 12.16 Development of the *C. elegans* gonad. (a) Adult hermaphrodite gonad. (b) Normal development. (c) Effect of removing the distal tip cell at different stages.

Fig. 12.17 The distal tip cell, visualized with a GFP reporter. Reproduced from Kimble and Crittenden. *Ann Rev Cell Dev Biol* 2007; 23: 405–433, with permission from Annual Reviews.

the germ-cell syncytial membrane. Zygotic loss-of-function mutations of *glp-1* or of *lag-2* have the same effect as ablation of the distal tip cell.

The GLP-1 signal inhibits the activity of a pair of proteins GLD-1 and -2, respectively an RNA binding protein and a polyA polymerase, that are present in the germ line and are needed for progression to meiosis. It also upregulates *daz-1*, encoding another RNA binding protein required for oocyte maturation. *daz-1* is a homologue of the human gene *daz* (**d**eleted in **az**oospermia) although this is required for spermatogenesis rather than oogenesis. So, although the development of the *C. elegans* germline is quite different from that of the other model organisms, a number of the key molecular components controlling the process are still found to be conserved.

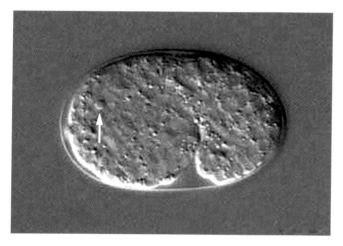

Fig. 12.18 An embryonic cell death (arrow). Reproduced from Lettre and Hengartner. *Nature Rev Mol Cell Biol* 2006; 7: 97–108, with permission from Nature Publishing Group.

Programmed cell death

During normal development of the *C. elegans* hermaphrodite 131 cells die out of 1090 generated, mostly during a death wave from 250–450 minutes of development (Fig. 12.18). They are mainly small cells and collectively represent only 1% of the biomass. Most deaths are autonomous and occur shortly after the cell was born. A few depend on signals from neighboring cells, shown by the fact that certain cells normally destined to die will survive if their neighbor has been ablated by laser irradiation. Although there are some differences, the sequence of events is generally similar to mammalian apoptosis, with a condensation of the nucleus, a shrinkage of the cell to a membrane-bound body, and engulfment by neighboring cells.

Cell-death-defective (*ced*) mutants affect all the cell deaths in the organism, while some other mutants affect the decisions of particular cells to die. Most of the *ced* mutants interfere with the engulfment of the dead cells, but three of them are components of the actual death program itself. In loss-of-function mutants of *ced-3* and *ced-4*, all of the cells normally destined to die now survive. The *ced-9* gene has both loss-of-function and gain-of-function alleles. In loss-of-function mutants there is excessive cell death, while in gain-of-function mutants there is some survival of cells that normally die. Excess cell survival is also shown on overexpression of wild-type *ced-9*. Double mutants of the type *ced-9⁻;ced-3⁻* or *ced-9⁻;ced-4⁻* also show target cell survival, so it follows that the normal function of *ced-9* must be to repress *ced-3/4*, which are themselves downstream of *ced-9* and necessary to execute the death program (Fig. 12.19). Genetic mosaic experiments show that the *ced-3/4* wild-type gene has to be present in the target cell itself in order for it to die.

ced-9 codes for the homolog of the mammalian protein BCL2. Discovered originally as an **oncogene** product, it is a cytoplasmic protein that is an inhibitor of apoptosis. CED-9 and BCL2 are interchangeable, therefore worms transgenic for

Classic experiments

Cell lineage and cell death

The first two references are the papers describing the complete cell lineage of *C. elegans*, derived from painstaking observation by interference microscopy. The third is a detailed anatomical study. In the course of this work a number of programmed cell deaths were described. Later analysis of the cell death (*ced*) mutants showed that the biochemistry of the process was common to higher animals and allowed the elucidation of the pathway. Sydney Brenner, Robert Horvitz, and John Sulston were awarded the Nobel Prize for Physiology for this work in 2002.

Sulston, J.E. & Horvitz, H.R. (1977) Postembryonic cell lineages of the nematode *Caenorhabditis elegans*. *Developmental Biology* **56**, 110–156.

Sulston, J.E., Schierenberg, E., White, J.G. & Thomson, J.N. (1983) The embryonic-cell lineage of the nematode *Caenorhabditis-elegans*. *Developmental Biology* **100**, 64–119

White, J.G., Southgate, E., Thomson, J.N. & Brenner, S. (1986). The structure of the nervous-system of the nematode *Caenorhabditis-elegans*. *Philosophical Transactions of the Royal Society of London Series B-Biological Sciences* **314**, 1–340.

Yuan, J.Y., Shaham, S., Ledoux, S., Ellis, H.M. & Horvitz, H.R. (1993) The *C-elegans* cell-death gene *ced-3* encodes a protein similar to mammalian interleukin-1-beta-converting enzyme. *Cell* **75**, 641–652.

Hengartner, M.O. & Horvitz, H.R. (1994). *C. elegans* cell-survival gene *ced-9* encodes a functional homolog of the mammalian protooncogene *bcl-2*. *Cell* **76**, 665–676.

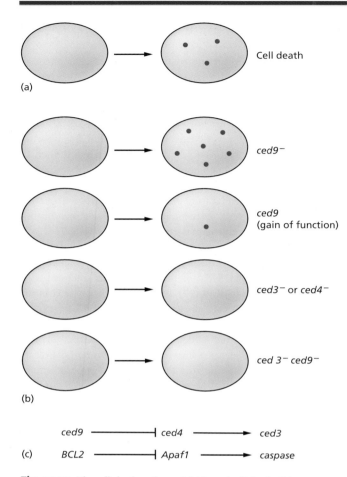

(a)

(b)

Cell death

ced9⁻

ced9
(gain of function)

ced3⁻ or ced4⁻

ced 3⁻ ced9⁻

(c) ced9 ⎯⎯⎯⎯⎯⫞ ced4 ⎯⎯⎯⎯⟶ ced3

BCL2 ⎯⎯⎯⎯⎯⫞ Apaf1 ⎯⎯⎯⎯⟶ caspase

Fig. 12.19 The cell-death pathway. (a) Normal cell death. (b) Phenotypes of mutants. (c) The pathway in *C. elegans* and in mammals.

mammalian BCL2 show inhibition of cell death, and *ced-9⁻* mutants can be rescued by BCL2. *ced-3* codes for a homolog of the interleukin 1β-converting enzyme (ICE). This is a cysteine protease that cleaves at Asp-X sequences, and is the prototype member of the family of caspases, of which many are now known. The caspases are the enzymes that actually bring about cell death, and the targets for the CED-3 protease include poly-ADPR polymerase (involved in DNA repair), lamins (nuclear membrane proteins), and other nuclear proteins. CED-3 itself will cause apoptosis if introduced into mammalian cells. *ced-4* codes for a protein whose mammalian homolog is called Apaf1 and which activates procaspase 9. CED-4 itself similarly activates CED-3 and is inhibited by CED-9.

The final stage of the cell death process is the engulfment of the apoptotic cells by neighboring cells. *C. elegans* does not possess specialized phagocytes and the engulfment is carried out by nonspecialized cells. But as in mammalian phagocytosis a critical step is the recognition of a particular phospholipid, phosphatidyl serine, on the surface of the cell to be engulfed. A number of the *ced* genes are defective in engulfment and their homologs are thought also to be important for the properties of mammalian phagocytes. Examples are *ced-1*, encoding a homolog of the mammalian SREC (scavenger receptor from endothelial cells), and *ced-7*, encoding an ABC (ATP binding cassette) transporter protein, one of a large class of proteins responsible for transport of small molecules and ions across cell membranes.

As with asymmetric cell division, the discovery of the mechanism of programmed cell death is an area in which *C. elegans* genetics has made an important contribution to general cell biology.

New directions for research

In recent years *C. elegans* has been used extensively for research into the mechanisms of aging (see Chapter 19), and other topics such as the mechanisms of pathogenesis. An understanding of basic developmental mechanisms is an important foundation for these types of work.

Key points to remember

- *C. elegans* is very favorable for genetic experimentation. It is a self-fertilized hermaphrodite with a short generation time.
- The precise cell lineage of all cells in the embryo and adult has been described. This serves as a resource for all kinds of experimental work.
- The early regional pattern of the embryo arises through the action of the PAR proteins, which become segregated along the anteroposterior axis after fertilization and control the distribution of other cytoplasmic determinants between the early blastomeres.
- Various steps of *C. elegans* development depend on inductive interactions. These utilize the same molecular pathways as other animals. For example the formation of the pharynx from the AB cells depends on a Delta-like signal from the P_2 and MS cells. The formation of the E lineage depends on a Wnt-like signal from the MS cell.
- Postembryonic development also involves inductive interactions. The EGF/Ras pathway is important for vulval development and the Notch pathway for the maintenance of the mitotic germ line.
- Heterochronic mutations affect multiple cell lineages in a coordinated manner, and may cause advancement or retardation of developmental events relative to the normal schedule.
- *C. elegans* has made important contributions to cell biology by helping to elucidate the mechanisms of cell polarization and of apoptotic cell death.

Further reading

Websites

http://www.wormbook.org/
http://www.wormbase.org/
http://elegans.som.vcu.edu/

General

Riddle, D.L., ed. (1997) *C. elegans* II. Cold Spring Harbor, NY: Cold Spring Harbor Laboratory Press.

Simske, J.S. & Hardin, J. (2001) Getting into shape: epidermal morphogenesis in *Caenorhabditis elegans* embryos. *BioEssays* **23**, 12–23.

Singson, A. (2001) Every sperm is sacred: fertilization in *Caenorhabditis elegans*. *Developmental Biology* **230**, 101–109.

Joshi, P.M. & Rothman, J.H. (2005) Nematode gastrulation: having a BLASTocoel! *Current Biology* **15**, R495–R498.

Genetics

Kuwabara, P.E. & Kimble, J. (1992) Molecular genetics of sex determination in *C. elegans*. *Trends in Genetics* **8**, 164–168.

Salser, S.J. & Kenyon, C. (1994) Patterning in *C. elegans*: homeotic cluster genes, cell fates and cell migrations. *Trends in Genetics* **10**, 159–164.

Hunter, C.P. (1999) A touch of elegance with RNAi. *Current Biology* **9**, R440–R442.

Yochem, J. & Herman, R.K. (2003) Investigating *C. elegans* genetics through mosaic analysis. *Development* **130**, 4761–4768.

Antoshechkin, I. & Sternberg, P.W. (2007) The versatile worm: genetic and genomic resources for *Caenorhabditis elegans* research. *Nature Reviews Genetics* **8**, 518–532.

Frokjaer-Jensen, C., Wayne Davis, M., Hopkins, C.E., Newman, B.J., Thummel, J.M., Olesen, S.-P., Grunnet, M. & Jorgensen, E.M. (2008) Single-copy insertion of transgenes in *Caenorhabditis elegans*. *Nature Genetics* **40**, 1375–1383.

Asymmetric division and determinants

Rose, L.S. & Kemphues, K.J. (1998) Early patterning of the *C. elegans* embryo. *Annual Review of Genetics* **32**, 521–545.

Schneider, S.Q. & Bowerman, B. (2003) Cell polarity and the cytoskeleton in the *Caenorhabditis elegans* zygote. *Annual Review of Genetics* **37**, 221–249.

Macara, I.G. (2004) Parsing the polarity code. *Nature Reviews Molecular Cell Biology* **5**, 220–231.

Nance, J. (2005) PAR proteins and the establishment of cell polarity during *C. elegans* development. *BioEssays* **27**, 126–135.

Goldstein, B. & Macara, I.G. (2007) The PAR proteins: fundamental players in animal cell polarization. *Developmental Cell* **13**, 609–622.

Induction, the germ line, the vulva, cell death

Priess, J.R. & Thomson, J.N. (1987) Cellular interactions in early *C. elegans* embryos. *Cell* **48**, 241–250.

Horvitz, H.R. & Sternberg, P.W. (1991) Multiple intercellular signalling systems control the development of the *C. elegans* vulva. *Nature* **351**, 535–541.

Sundaram, M. & Han, M. (1996) Control and integration of cell signalling pathways during *C. elegans* vulval development. *Bioessays* **18**, 473–480.

Kornfeld, K. (1997) Vulval development in *C. elegans*. *Trends in Genetics* **13**, 55–61.

Metzstein, M.K., Stanfield, G.M. & Horvitz, H.R. (1998) Genetics of programmed cell death in *C. elegans*: past present and future. *Trends in Genetics* **14**, 410–417.

Seydoux, G. & Strome, S. (1999) Launching the germ line in *Caenorhabditis elegans*: regulation of gene expression in early germ cells. *Development* **126**, 3275–3283.

Gaudet, J. & Mango, S.E. (2002) Regulation of organogenesis by the *Caenorhabditis elegans*, FoxA protein PHA-41. *Science* **295**, 821–825.

Lettre, G. & Hengartner, M.O. (2006) Developmental apoptosis in *C. elegans*: a complex CEDnario. *Nature Reviews Molecular Cell Biology* **7**, 97–108.

Maduro, M.F. (2006) Endomesoderm specification in *Caenorhabditis elegans* and other nematodes. *BioEssays* **28**, 1010–1022.

Kimble, J. & Crittenden, S.L. (2007) Controls of germline stem cells, entry into meiosis, and the sperm/oocyte decision in *Caenorhabditis elegans*. *Annual Review of Cell and Developmental Biology* **23**, 405–433.

Mango, S.E. (2009) The molecular basis of organ formation: insights from the *C. elegans* foregut. *Annual Review of Cell and Developmental Biology* **25**, 597–628.

Developmental timing

Jeon, M., Gardner, H. F., Miller, E.A., Deshler, J. & Rougvie, A.E (1999) Similarity of the *C. elegans* developmental timing protein LIN-42 to circadian rhythm proteins. *Science* **286**, 1141–1146.

Rougvie, A.E. (2005) Intrinsic and extrinsic regulators of developmental timing: from miRNAs to nutritional cues. *Development* **132**, 3787–3798.

Moss, E.G. (2007) Heterochronic genes and the nature of developmental time. *Current Biology* **17**, R425–R434.

Resnick, T.D., McCulloch, K.A. & Rougvie, A.E. (2010) miRNAs give worms the time of their lives: Small RNAs and temporal control in *Caenorhabditis elegans*. *Developmental Dynamics* **239**, 1477–1489.

Ambros, V. (2011) MicroRNAs and developmental timing. *Current Opinion in Genetics and Development* **21**, 511–517.

This chapter contains the following animations:

Animation 4   Cytoplasmic determinant action.
Animation 16   PAR system

 For additional resources for this book visit www.essentialdevelopmentalbiology.com

Section 3

Organogenesis

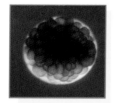

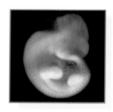

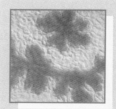

Chapter 13

Techniques for studying organogenesis and postnatal development

Most of the techniques described in Chapter 5 are used for the study of late as well as early development. But there are some that are employed more especially for the late stages, and in particular these include various ramifications of the transgenic techniques used for mice, especially the Cre system. These are described here. In addition, this chapter reviews clonal analysis methods for mammals, methods for tissue and organ culture, and methods for flow cytometry for separation of cell populations. These methods are not only used for the study of organogenesis. They are also very important when analyzing postnatal developmental events relating to tissue organization and stem cells, which will be considered in Section 4 below.

The majority of organogenesis work is done with the mouse or chick, although *Xenopus* and zebrafish can sometimes make useful contributions. In this section the mouse developmental stages are expressed as days of embryo gestation (e.g. E12.5) and chick stages by the Hamburger–Hamilton (H&H) stage series.

Genetics

Conventional overexpression transgenics and knockout mouse strains are widely used in organogenesis research. However, they have limitations and some very sophisticated methods have been added to mouse genetics in recent years to enable a much wider range of experiments to be carried out. Several of these involve using components from other organisms in the mouse, such as bacteria, phage, or yeast. Such components can rarely be used as they come, but need to be modified to function well in a mammalian cells. For example the usage of synonymous codons in mammals and bacteria is different because the relevant transfer RNAs are present in different ratios. So the coding sequences of genes often need to be modified by altering individual codons to those that are more commonly used in mammals, while still specifying the same amino acid.

The Cre system

Cre (pronounced "kree") is a recombinase enzyme from phage P1 which can excise segments of DNA flanked by binding sequences called *loxP* sites. Here we shall consider its uses for two purposes: knockout of a gene in a specific cell type, and fate mapping of specific gene expression domains. It can also be used for clonal analysis, as described below.

Tissue-specific knockouts

This is especially important in organogenesis research since the phenotype of a loss-of-function mutant is dominated by the first function of the gene in development. Quite often the embryo will die at an early stage if it lacks a key gene, and so it is not possible to analyze the effects of loss of the gene on any of its later functions.

In order to knock out a gene in just one tissue, and thereby circumvent the problem of early lethality, two mouse lines need to be created (see Animation 17: Cre-lox recombination). In the first the target gene has *loxP* sites inserted on either side of an essential part of the coding region (the gene is said to be "**floxed**"). The second is a transgenic line in which the Cre recombinase is driven by a tissue-specific promoter. The nomenclature for this type of construct follows the form "*Pdx1-Cre*," meaning the *Cre* coding region driven by the *Pdx1* promoter. When the two strains of mice are mated together, the target gene should be excised in the offspring, but only in the tissue producing the Cre (Fig. 13.1). In practice, the Cre does not give a 100% effective excision, so it is best if the parent with the floxed target gene is a heterozygote with the second copy of the gene inactive. This means only one copy of the gene needs to be excised per cell and also that the protein product starts out at only half the normal level.

Cre-based fate mapping

The Cre-lox system can also be used for the tissue-specific activation of a transgene. This can be used to express one functional gene in the domain of another, which provides more sophistication than ubiquitous overexpression of a gene. Most important, when the gene being regulated is a reporter producing an easily detectable but inactive product, the system can be used for fate mapping. For this purpose the Cre functions to excise an inhibitory DNA sequence that normally prevents expression of the marker gene. For example a transcription termination sequence flanked with *loxP* sites can be inserted between a constitutive promoter and the coding region. This is sometimes written as "stoplox," for example *stoplox-lacZ* or *stoplox-GFP*. When the inhibitory sequence is removed by the Cre, the reporter gene will be expressed. Because the system brings about targeted excision of a length of DNA, the change is heritable on cell division and will persist in all descendants of each modified cell. The nomenclature for the experimental animals will follow the form "*Pdx1-Cre x stoplox-lacZ*" where the first component refers to the tissue-specific promoter driving expression of the Cre enzyme, and the second relates to the reporter used.

This heritable property has made this application of the Cre-lox system very widely used for fate mapping. In fact, it is the only method of fate mapping used in mouse development after the egg cylinder stage. If it is desired to know what will normally be formed by a particular group of cells that activate a particular promoter at a particular stage, then a mouse can be assembled in which the promoter in question drives expression of the Cre, and the Cre activates a reporter gene (Fig. 13.2). An example of this strategy is the proof that all cell types of the pancreas come from endoderm expressing the transcription factor Pdx1 (see Chapter 16). Of course, the fidelity of the promoter is absolutely key to this kind of experiment, and needs to be confirmed carefully. If there is any ectopic expression of the promoter driving the Cre, or any normal domains in which the promoter is not active, then the results will be misleading.

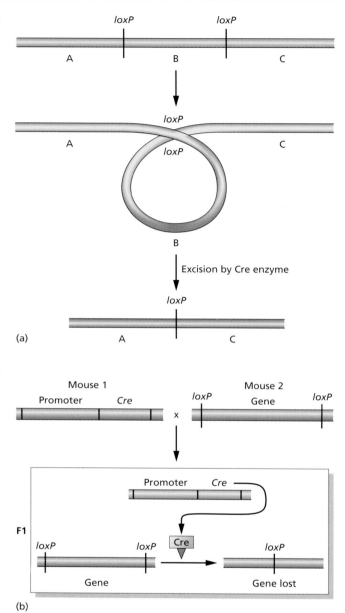

Fig. 13.1 Cre-lox system. (a) The Cre recombinase will excise DNA between two *loxP* sites. (b) A binary mouse. When mouse 1 and 2 are mated, the Cre recombinase is expressed according to the specificity of the promoter to which it is coupled and will drive excision of the gene between the *loxP* sites.

Several reporter strains have been used for this type of fate mapping. Most popular is a mouse called R26R. The transgene is *E. coli β-galactosidase* fused to *neomycin resistance*, also called *β-geo*. This is prefixed by a transcription stop signal flanked by loxP sites and the whole is knocked in to the *Rosa26* locus. *Rosa26* is a locus encoding an untranslated RNA that is expressed in all tissues at all developmental stages. It was originally discovered as a gene trap with ubiquitous expression (see Chapter 10). The R26R strain therefore does not express *lacZ* unless Cre is present, and it will express *lacZ* forever in any cell that expressed the Cre enzyme, even if only transiently. Other reporter lines have been devised in which one marker is expressed prior to Cre action and another afterwards. These include Z/AP which switches from β-galactosidase to alkaline phosphatase, and Z/EG which switches from β-galactosidase to GFP.

Inducible Cre

It is possible to make the Cre system inducible by using a fusion of the Cre enzyme with the ligand-binding domain of the estrogen receptor (ER). In the same way as discussed for the glucocorticoid receptor in Chapter 7, nuclear hormone receptors like ER are normally sequestered in the cytoplasm by binding to heat-shock proteins. When their specific ligand becomes bound it displaces the heat-shock protein and allow the migration of the receptor, or in this case the ER fusion protein, to the nucleus. The expression of *CreER* is still driven by a tissue-specific promoter but the time at which it becomes active is controlled by injecting the mother with a suitable receptor ligand. The ER sequence in use has been modified not to be sensitive to endogenous estrogens, so the ligand used experimentally is not estrogen itself but the synthetic antagonist tamoxifen, or its metabolically active metabolite, 4-hydroxytamoxifen. Following injection of tamoxifen, biologically active 4-hydroxytamoxifen is produced in the liver after a few hours, and it is mostly cleared from the circulation after a further 24 hours. So the system has a time discrimination of about 1 day.

Any experiment that can be done with Cre can also be done with CreER/tamoxifen with the tamoxifen dose controlling the timing of the Cre activity. This has particular value for fate mapping because it enables the descendants of a cell population

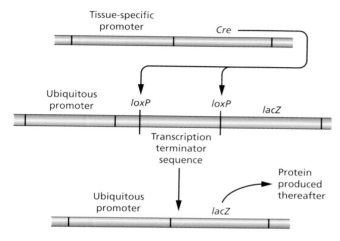

Fig. 13.2 Use of Cre-lox system for fate mapping. A cell population will be labeled permanently following upregulation of the promoter driving the *Cre*.

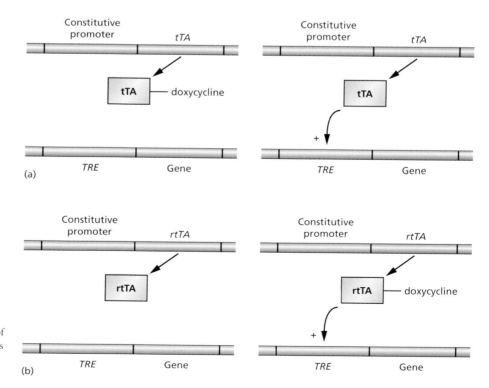

Fig. 13.3 Tet systems. (a) In Tet-off, tTA upregulates the target gene in the *absence* of doxycycline. (b) In Tet-on, rtTA upregulates the target gene in the *presence* of doxycycline.

to be traced depending on the expression of a particular promoter on a particular day of embryonic development.

Tet systems

The CreER system embodies one type of inducibility in which the effect of a transient induction is permanent. It is also possible to use inducible transgene systems that do not bring about heritable changes in the DNA, but are active only as long as the stimulus is provided.

For this the preferred system uses elements of the *tetracycline* operon from *E. coli*. The enzymes that degrade tetracycline are encoded by the genes of the operon. In the absence of tetracycline the operon is repressed, but in the presence of tetracycline it becomes activated. The system works because the tetracycline combines with a Tet repressor protein (TetR), and thereby causes it to dissociate from the *tet* operator (*tetO*) sequence in the DNA, enabling transcription to take place.

For mouse experiments this system has been modified, and thereby improved in specificity, in various ways (Fig. 13.3). The response element used (*TRE*) includes several copies of the *E. coli tetO* together with a minimal promoter. The original TetR has been converted into an activator (TetA) in which its normal repression domain is replaced by the VP16 activation domain, a strong transcription activating sequence from the herpes simplex virus. In a transgenic situation, the *TRE* is combined with the target gene and the TetA is expressed from a tissue-specific promoter elsewhere in the genome. The inducer used is not tetracycline itself but a stable analog called doxycycline. In this variant, called **Tet-off**, the gene is inactive in the presence of doxycycline but, when the drug is withdrawn, the TetA will bind to the *TRE* and activate transcription. Doxycycline is supplied continuously to the pregnant female mice by including it in their drinking water, so the TetA is sequestered and the target gene remains off. When it is desired to activate the target gene the doxycycline is withdrawn. It becomes cleared from maternal and fetal circulation within a few hours, the TetA is then able to bind the *TRE*, and the target gene becomes activated.

The TetA has also been converted by judicious mutagenesis into a "reverse activator," called rtTA. This molecule requires doxycycline before it can bind to the *TRE* and it activates the gene expression when it does so. When doxycycline is withdrawn, the rtTA will disengage from the *TRE* and the gene turns off. This version is called **Tet-on**. Whether Tet-on or Tet-off is more appropriate depends on the particular experimental design. Both systems are also extensively used in tissue-culture-based systems, such as embryonic stem cells, often being used to regulate gene expression from viral gene-delivery vectors.

Late gene introduction

In mice, the majority of overexpression experiments are done using traditional injection transgenesis or the knock-in procedure described in Chapter 10. But despite all the sophistication of inducible systems, sometimes it is preferable just to introduce the gene when and where it is required.

This can be done by **electroporation**, used mostly in the chick. A cavity, such as the spinal cord lumen, is injected with the DNA, then a square-wave electric pulse is applied across the region of tissue. DNA is negatively charged so gets driven into cells towards the anode of the electroporator. In the mouse at organogenesis stages the embryos are located within the uterus of the mother. However, surgical methods have now been perfected for opening the uterus, exposing the embryo or fetus, introducing a gene by electroporation to the required position, and then closing the uterus and allowing the mother to continue the pregnancy and bear the pups. Unlike transgenesis, electroporation is a transient technique only useful for short-term gene overexpression. The DNA rarely becomes integrated into the genome of the host cells, and so it becomes diluted on cell division and is also degraded metabolically.

Viral vectors for gene delivery

Considerable use is now made of various types of vector derived from animal viruses to introduce genes into cultured cells or intact organisms. They fall into several categories, but the types most commonly used in research are **retroviruses**, the related lentiviruses, and adenovirus.

Retroviruses are RNA viruses and, after infection of a cell, the single-stranded RNA genome is reverse transcribed to double-stranded DNA, which can become integrated in the host genome. In retroviral vectors, the gene of interest, usually with its own promoter, replaces all three of the viral coding regions. For viral vectors, the term "transduction" is generally used in place of "infection." Following transduction they can become integrated and the transgene is then expressed under the control of its own promoter. Lentiviruses are also retroviruses, being related to human immunodeficiency virus, but they have the additional ability to transduce and integrate into nondividing cells.

For most overexpression experiments replication-defective vectors are used (Fig. 13.4). These lack most or all of the genes required for replication and so cannot produce new virus particles. In order to produce the vector, a plasmid carrying the vector sequence is transfected into a packaging cell line, together with plasmids encoding the missing components. Since all required components are then present, the packaging cells will secrete virus particles containing the defective genome, which can then transduce host cells, integrate, and be expressed, but not produce new virus particles. It is common to replace the normal virus coat protein with one of broader host range, such as the coat protein of vesicular stomatitis virus (VSV-G). This is known as "pseudotyping." Following transduction not all cells necessarily integrate the viral DNA but most will express it. In tissue culture integration can be selected for using antibiotic resistance genes. In experiments on whole embryos or animals it is likely that expression in many cells will be transient, ceasing

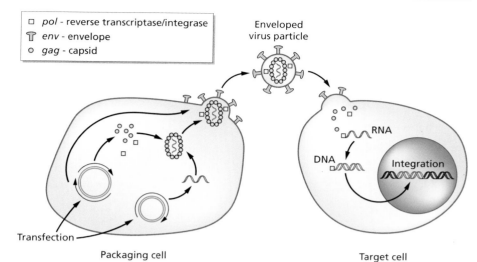

□ *pol* - reverse transcriptase/integrase
⊤ *env* - envelope
○ *gag* - capsid

Enveloped
virus particle

RNA

DNA

Integration

Transfection

Packaging cell

Target cell

Fig. 13.4 A replication-defective lentiviral vector. The vector is produced in a packaging cell line from components expressed from transfected plasmids. It can transduce a target cell, and generate DNA for integration into the genome, but cannot make new virus particles.

when nonintegrated viral genomes are degraded. The chances of obtaining integration in the **germ line** following infection of whole embryos or animals is very low, so the use of retroviral vectors does not usually result in the formation of a transgenic strain. Replication-defective retroviruses are also used for **clonal analysis**. For this purpose transduction is deliberately at very low multiplicity to give occasional labeled cells on a wide unlabeled background.

The other vector commonly used in research is based on adenovirus. This is a double-stranded DNA virus that does not integrate into the host genome. It gives high efficiency of infection for many cell types but expression is also transient, lasting only as long as the viral DNA.

All of the viral vectors in use may incorporate conditional regulation elements such as Gal4, the Tet system, or Cre, so it is often possible to regulate gene expression from them in a sophisticated manner.

For studies of chick embryo development use has also been made of replication-competent retroviruses (**RCAS**). Here, infected cells make and export virus particles, which infect the neighboring cells and therefore in a few days the infection, and the transgene expression, spreads through the embryo. As this usually kills the embryo, the experiments are timed so that just the region of interest has been infected by the stage of interest. There are also available strains of chicken that are resistant to the retroviruses used. So it is possible to graft a piece of tissue, such as a limb bud, from a sensitive to a resistant embryo, and then do the infection. Under these circumstances, the overexpression is confined to the sensitive region.

Cell ablation

There are several ways of killing cells in a developmentally specific manner but probably the best involves the use of diphtheria

toxin, an exotoxin from the pathogen *Corynebacterium diphtheriae*. This molecule has two subunits, the A subunit, which kills the cell, and the B subunit, which interacts with a cell surface molecule normally present on human cells (called diphtheria toxin receptor, DTR) and thereby introduces the A subunit. The A subunit is an enzyme that inactivates the elongation factor 2 essential for protein synthesis. It is extremely toxic and one molecule is enough to kill a cell.

DT-A can be used on its own, controlled by appropriate tissue-specific promoters or inducible systems. But since it is so toxic the slightest degree of leaky expression may kill the mouse. A more sophisticated method involves expressing the DTR under the control of a tissue-specific promoter, a Cre-lox system, or a Tet system. Mouse cells are not susceptible to DT, but they become susceptible if they express the DTR. Following this, the mouse can be injected with the complete toxin (A and B subunits) and just those cells expressing the DTR will bind the B subunit and allow in the A subunit.

Clonal analysis

Clonal analysis was introduced in Chapter 4 in relation to early development. At this stage the issue is usually finding the time of determination, which involves detecting the inability of a single clone to span two structures after they have become determined. For the study of organogenesis and postnatal development the significance is rather the identification of progenitor cells and understanding the proliferative organization of tissues. If a whole structure arises by division of a single cell, then the whole structure should be labeled following its labeling. On the other hand, if the structure arises from several cells then it will always be of mixed composition following labeling of a single progenitor. Here will be described the methods for clonal analysis that have been used with mice. Clonal analysis

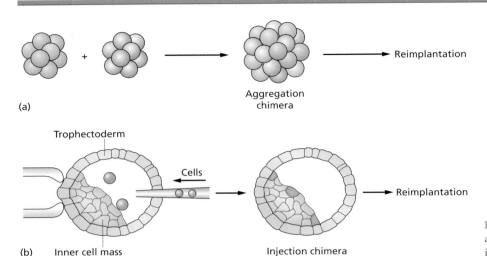

Fig. 13.5 Making chimeras: (a) by aggregation; and (b) by injection of cells into the blastocyst.

has also been very important in *Drosophila*, to investigate the development of the imaginal discs at larval stages. But the techniques for this are somewhat specialized and they are described in Chapter 17.

Mosaicism and chimerism

Among mammals an animal composed of cells that are genetically dissimilar is called a **mosaic** or a **chimera**. Usually the term "mosaic" is used if the animal has arisen from a single zygote, such as the mosaicism of all female mammals arising from X chromosome inactivation, and "chimera" if it has arisen from some experimental or natural mixture of cells taken from different sources. Genetic mosaics should not be confused with mosaic development, an entirely different concept mentioned in Chapter 4.

There are two experimental methods for making mouse chimeras (Fig. 13.5). **Aggregation chimeras** are made by removing the zona pellucida of four- to eight-cell-stage embryos, then gently pressing them together so that the cells adhere. The fused embryo is reimplanted into the uterus of a pseudopregnant recipient. Such embryos show normal development and consist of a mixture of the cell types of the two components. Injection chimeras are made by injecting cells into the blastocoelic cavity of expanded blastocysts, with zona intact. Again they are implanted in foster mothers for further development. It is often possible to detect chimerism by examination of the coat pattern, if the two mouse strains used have different coat colors. Although the pigment cells themselves are derived from the neural crest, the expression of pigment in the hair follicles will depend also on the local genetic composition of the epidermis. Therefore chimeras will often be blotchy in coat pattern, or at least have some admixture of the hair colors of the two contributing strains.

Injection chimeras have enormous practical importance because they represent a key step in the production of **knockout** mice. But both types of chimera, and also X-inactivation mosaics, can also be used for the clonal analysis of structures in the fetal or postnatal mouse. This type of analysis has been carried out for structures such as hair follicles and intestinal crypts, as described in Chapter 18. Injection chimeras can be made multicolored, having more than one population of labeled cells, by injecting ES cells from several different cell lines, each expressing a different colored fluorescent protein. An example of this is shown in Chapter 15, relating to developmental hematopoiesis.

Induction of labeled clones

A method for labeling clones in mouse or chick embryos is by infection with a low dose of replication-incompetent retrovirus expressing *β-galactosidase* or *GFP*. Ideally, the dose is low enough that the infected cells are well separated and each positive cell group can be considered as a single clone. This method has been used a great deal in the central nervous system. But, in practice, it is not necessarily easy to tell clones apart, so often a library of virus is used containing a large number of variant sequences, and tissue taken from each labeled patch is typed by PCR to find whether it belongs to the same or different clones.

There are also some methods for clone induction not requiring access to the embryo. The *CreER x stoplox-reporter* mouse can be used with a low dose of tamoxifen such that only the occasional cell undergoes a recombination event. The time of labeling will be during the day after administration of the tamoxifen, and the frequency of clones can be tuned by adjusting the dose.

A simpler but less controllable method uses a transgene called *laacZ*. This is a version of *lacZ* containing an internal duplica-

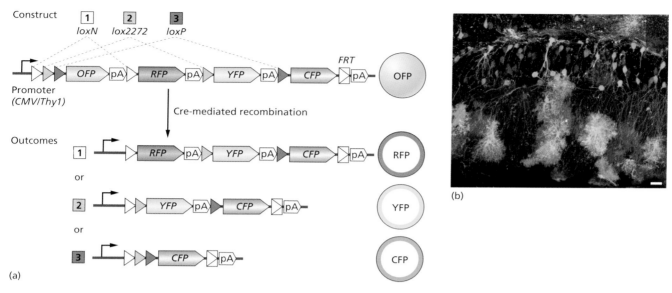

Fig. 13.6 The "Brainbow" technique for generating multiple clones of different colors. (a) One construct used, containing three different variants of the loxP site. In the presence of Cre there are three possible excision events generating different expressed fluorescent proteins. (b) Many labeled clones are present in this section from the dentate gyrus. Scale bar 20 μm. Reproduced from Livet *et al.* (2007), with permission from Nature Publishing Group.

tion with termination codons. At a frequency of about $1/10^6$ per cell division, the defect is spontaneously repaired by an intragenic recombination, which removes the termination codons. This generates a clone in which the normal *lacZ* is expressed and can be visualized by XGal staining. So a transgenic mouse in which *laacZ* is driven by a tissue-specific promoter should always have a low frequency of clones in this tissue.

Sometimes it is useful to be able to visualize more than one lineage at the same time. As mentioned above, this is achieved in chimeras made by injecting more than one differently labeled ES lines into an unlabeled host embryo. There are also transgenic techniques for labeling with more than one color. The "brainbow" method was introduced for study of the nervous system but can be used for any organogenesis or postnatal development problem (Fig. 13.6). This is a Cre-lox system using modified *lox* sites. The normal lox site is called *loxP*, but there are variants, such as *loxN* and *lox2272*, that can also be recognized by the Cre enzyme, but only in identical pairs. In Fig. 13.6a is shown a construct that contains three different pairs of lox sites. When crossed to a CreER mouse and induced with tamoxifen, Cre is produced and can cause one of three mutually exclusive excision events. A cell containing this construct will start off orange, and can become either red, yellow, or cyan depending on the particular excision event experienced. Because the transgenes are present as tandemly repeated arrays, each repeat within a single array may undergo a different excision event, generating a large number of possible hues for different clones.

Finally, a technique called mosaic analysis with double markers (MADM) can enable the generation of labeled clones

that also carry homozygosed mutations. The principle is similar to those used for the study of mutant clones in *Drosophila*, except that the Cre-lox system is brought into play. The double marker loci consist of half the coding region of each of two fluorescent proteins separated by a *loxP* site (Fig. 13.7). For example one chromosome may bear *N-terminal GFP-loxP-C-terminal RFP* and the homologous chromosome carried the reciprocal *N-terminal RFP-loxP-C-terminal GFP*. These constructs have to be knocked in to an identical locus of two strains of mice, and be brought together by mating. The Cre enzyme is capable of catalyzing recombination *between* the two homologous chromosomes bearing *loxP*. If this occurs at the G1 stage of the cell cycle, the outcome is two clones, one making both GFP and RFP and therefore being yellow, and the other being unlabeled. At the G2 stage, each homologous chromosome is now duplicated so the outcome of a single recombination event depends on how the chromosomes segregate at mitosis. In the event that the red and green chromosomes segregate from each other, the result is an adjacent red clone and green clone, or **twin spot**, to use *Drosophila* nomenclature.

The potential value of this technique lies in the fact that a recombination generating red and green clones will also have homozygosed all the other loci distal to the *loxP* site on the same chromosome arm. So if this arm contains a developmental locus of interest, and is heterozygous for a loss-of-function mutation, then one of the labeled clones will be null and the other +/+. This enables the study of the behavior of clones carrying mutations that are lethal if they affect the whole organism, or even all of a particular tissue.

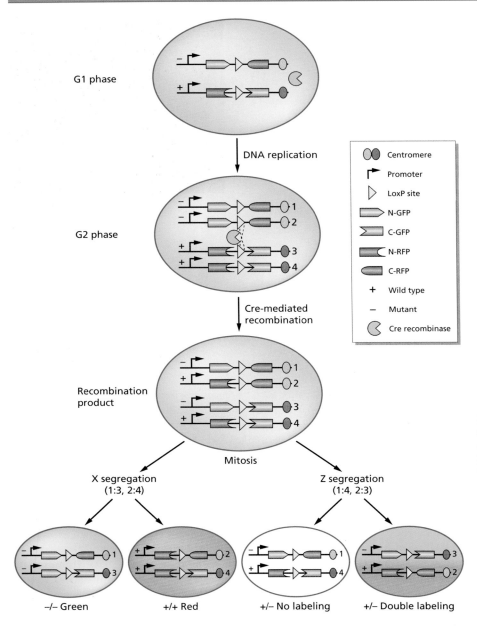

G1 phase

DNA replication

G2 phase

Cre-mediated
recombination

Recombination
product

Mitosis

X segregation
(1:3, 2:4)

Z segregation
(1:4, 2:3)

−/− Green   +/+ Red   +/− No labeling   +/− Double labeling

Centromere
Promoter
LoxP site
N-GFP
C-GFP
N-RFP
C-RFP
+   Wild type
−   Mutant
Cre recombinase

Fig. 13.7 The mosaic analysis with double markers (MADM) technique. The Cre catalyzes a recombination between homologous chromosomes at the *loxP* site. This causes the assembly of complete coding sequences for GFP or RFP. If the recombination occurs in the G2 phase of the cell cycle, and if mitotic chromosome segregation is the "X type," then a twin spot is generated of red and green clones. Any heterozygous locus more distal than the loxP site, indicated by +/−, will be homozygosed by the recombination. Reproduced from Zong *et al.* (2005), with permission from Elsevier.

Tissue and organ culture

Tissue culture

Tissue culture has tended to play a peripheral role in developmental biology, but it must now be considered as a core technique because of the importance of **embryonic stem cells** and **induced pluripotent stem cells** (Fig. 13.8a,b; and see Chapter 21). Tissue culture means growing cells *in vitro*, in a nutrient medium. Usually, cells are grown as a monolayer on a plastic surface, although more complex three-dimensional substrates are increasingly used. Primary cell cultures are those derived directly from tissues within the organism. Usually, primary cell lines will only grow for a certain number of cell generations before senescing. For rodent cells this appears to be mostly because of an accumulation of cyclin-dependent kinase inhibitors (CKIs), and for human cells because of a lack of **telomerase** leading to eventual degradation of chromosome ends (see Chapter 19). Because there is an automatic selection for the fastest growing variants, tissue culture cells are vulnerable to the acquisition of chromosome abnormalities and other mutations, and this must always be borne in mind when relating their behavior to *in vivo* cell properties. Permanent cell lines have always accumulated some of these mutations, which

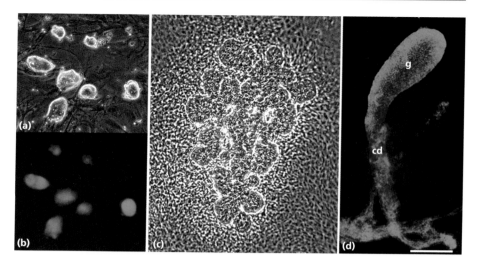

Fig. 13.8 Tissue and organ culture. (a,b) Mouse induced pluripotent stem cells (iPSC) in tissue culture: (a) phase contrast; (b) fluorescence visualizing an endogenous *Oct4-GFP* reporter. (c) Organ culture of mouse embryo pancreas, phase contrast. (d) Organ culture of mouse embryo gall bladder, stained green with a fluorescent lectin (DBA). g, gall bladder; cd, cystic duct; scale bar 200 μm.

enable them to evade the normal process of senescence. In general, tissue culture is well developed for mammals, especially for rodent and human cells, and poorly developed for nonmammals.

Mammalian cell culture **media** (singular **medium**) generally have a total osmolarity about 350 mOsm. The pH needs to be tightly controlled, usually 7.4 is normal. The pH control is usually achieved with bicarbonate-CO_2 buffers (2.2 g/L bicarbonate and 5% CO_2 being a common combination). These give better results than other buffers, perhaps because bicarbonate is also a type of nutrient. The medium must contain a variety of components: salts, amino acids, sugars, plus low levels of specific hormones and growth factors required for the particular cells in question. The requirement for hormones and growth factors is usually met by including some animal serum, often 10% fetal calf serum. This is long standing practice but has some disadvantages. Serum can never be completely characterized and there are often differences between batches of serum that can be critical for experimental results. Serum-free media need to include the necessary hormones and growth factors in pure form.

Assuming that cells are kept in a near-optimal medium they can grow exponentially, with a constant doubling time. In order to keep them growing at maximal rate they need to have their medium renewed regularly and to be subcultured and replated at lower density whenever they approach confluence, that is covering all the available surface. Subculturing is usually carried out by treatment with the enzyme trypsin, which degrades some of the extracellular and cell-surface protein and makes the cells drop off the plastic and become roughly spherical bodies in suspension. Once the trypsin is diluted out, the cells can be transferred at lower density into new flasks. The cells take an hour or two to resynthesize their normal complement of surface molecules and can then adhere to the new substrate and carry on growing.

Although exponential growth is often sought and encountered in tissue culture, cells very rarely grow exponentially in the body. Most cells are quiescent, rarely undergoing any division at all, and thus resemble static confluent tissue cultures more than growing ones. Some tissues in the body do undergo continuous renewal such that the production of new cells is balanced by the death and shedding of old ones. In such cases there are centers of cell proliferation but the overall cell number remains constant.

Organ culture

Tissue culture can be contrasted to **organ culture** where a small explant of a tissue or organ rudiment from an embryo is maintained intact in a nutrient medium (Fig. 13.8c,d). Organ cultures generally maintain their three-dimensional structure and do not grow extensively during the culture period, although they remain viable and may differentiate and undergo morphogenetic movements. Some organ culture is carried out on simple plastic surfaces, some on extracellular matrix materials such as collagen or fibronectin-coated plastic. In some cases the explant is embedded in extracellular matrix such as a collagen gel or Matrigel, which is a commercially available extract of extracellular matrix from a cell line. There are available various designs of membrane for organ culture such that the explant can be kept at an air–liquid interface, which is often advantageous. Other types of membrane allow culture of different tissue components on the two sides, particularly useful for analysis of inductive interactions between them.

In the chick embryo it is possible to grow organ cultures *in vitro* similar to the mouse. In addition, it is possible to explant small pieces of tissue onto the **chorioallantoic membrane (CAM)** of advanced embryos, where they become vascularized and will grow and differentiate in effective isolation.

Cell separation methods

Flow cytometry

Stem cell research in particular has depended on the ability to identify, classify, and sort cells from complex mixed populations such as those found in the mammalian bone marrow. This is known as **flow cytometry**. It can be applied to cells from any tissue source so long as they can be converted into a single cell suspension, for example by trypsinization. The smaller and less complex flow cytometers can analyze the composition of a cell mixture, while the larger and more complex machines can also sort live cells of different types into different tubes. If the machine can sort cells as well as analyze them it is called a **fluorescence-activated cell sorter** or **FACS** machine (see Animation 18: FACS operation).

Cells may be analyzed by size, by "granulosity," or by their specific fluorescence signal, dependent either on the binding of fluorescent antibodies or on the expression of endogenous reporter genes. If live cells are to be sorted, then only antibodies to cell surface molecules can be used. If the aim is just analysis then the cells can be fixed and antibodies used that recognize internal antigens. To carry out an analysis run, the cells are placed in a reservoir in suspension in a suitable medium. They then pass through a nozzle into a stream of a sheath fluid, which controls the flow characteristics. A laser beam is directed at the stream and as each cell passes its light scattering and fluorescence signal is measured by detectors. The forward light scatter provides a measure of cell size and the sideways light scatter a measure of granulosity.

For fluorescence-activated cell sorting, the instrument will vibrate the nozzle, causing the stream containing the cells to break into a series of tiny droplets. The cell density is such that very few droplets contain more than one cell. The experimenter programs the machine to set criteria for fluorescence and size, and the machine will impart a positive or negative charge through a drop charge electrode to each droplet, depending on the measurements of the detectors. The droplets then pass between positively and negatively charged metal plates and those carrying a charge become appropriately deflected and can be collected. This means that pure subsets of cells can be isolated in viable form from a mixed starting population (Fig. 13.9). Machines may have more than one laser and detector, enabling multichannel detection and very sophisticated separation procedures.

Laser capture microdissection

Until recently, it was not possible to do molecular biological analysis on tissue sections. Either the bulk tissue, usually containing many different cell types, was analyzed for gene expression, or it could be sectioned, stained, and examined anatomically. Only where cells could be separated by flow cytometry could both types of analysis be carried out.

Now, it is possible to isolate single cells, or small groups of cells, from microscope sections and analyze them using any technique that will work on small quantities, including gene sequencing or expression measurements of specific genes. This is known as **laser capture microscopy**. A special microscope is required fitted with a UV pulsed laser microbeam, which can cut round a specified area. This may be just a single cell, or some microstructure of interest such as a stem cell domain or a small patch in a tumor. The software enables the guidance of the cutting laser beam by the operator. Then the tissue fragment is isolated, usually by "catapulting" into a sample tube using a defocused laser pulse. Like all techniques involving minute samples and significant amplification of the material by PCR, great care is needed to avoid contamination and to perform adequate positive and negative controls.

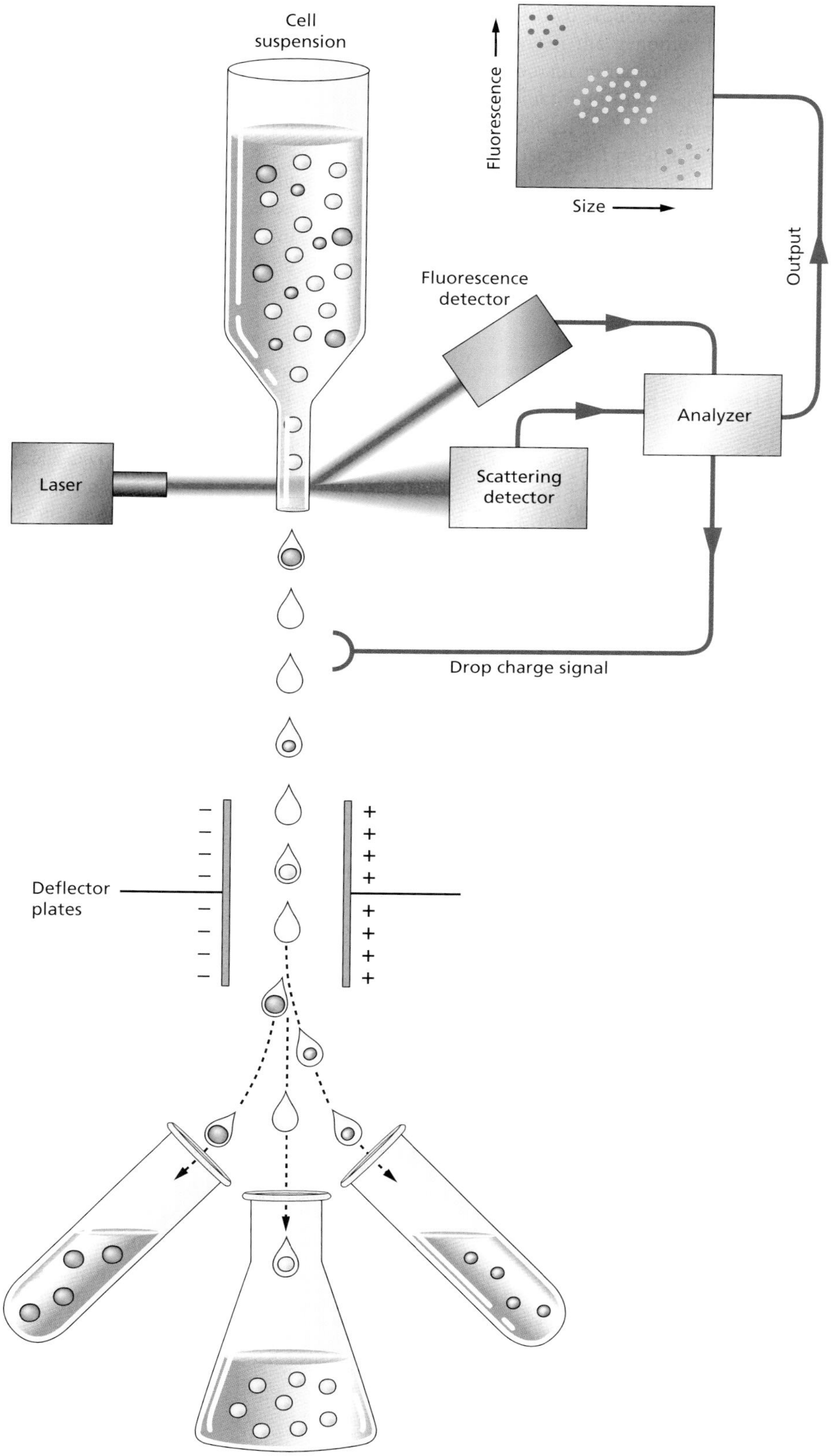

Fig. 13.9 Fluorescence-activated cell sorter. The output shows that the cell population contains minority fractions, which are small, high fluorescence (green) and large, low fluorescence cells (red). These can be sorted into different tubes by appropriate settings of the machine.

Key points to remember

- The Cre-lox system can be used for two distinct types of procedure: (1) tissue-specific knockout of a gene; and (2) fate mapping a domain in which a particular promoter is active. The use of inducible versions of the Cre enzyme enable the timing of its effects to be regulated by the experimenter.
- The Tet-on and Tet-off systems enable the expression of transgenes to be controlled by the administration of tetracycline analogs.
- In addition to transgenesis, genes may be introduced into late mouse embryos by *in utero* electroporation.
- Targeted cell ablation may be achieved using diphtheria toxin.
- Clonal analysis in mouse can be carried out in several ways: (1) analysis of mosaics or chimeras made with early embryos; (2) administration of low multiplicity replication-defective retrovirus carrying a marker gene; (3) induction of low-level Cre labeling; (4) spontaneous reversion of *laacZ*; (5) various more sophisticated Cre-based techniques including Brainbow and MADM.
- Tissue culture involves the growth of cells *in vitro*, usually as a monolayer on plastic. Organ culture involves the maintenance of a viable tissue rudiment such that it can mature *in vitro*.
- Cells that can form a population in suspension can be analyzed by flow cytometry. Different subsets can be separated by fluorescence-activated cell sorting.
- Single cells, or small cell groups, can be isolated from microscope sections by laser capture and subject to analysis.

Further reading

Genetics

Petit, A.-C., Legué, E. & Nicolas, J.-F. (2005) Methods in clonal analysis and applications. *Reproduction Nutrition Development* **45**, 321–339.

Buch, T., Heppner, F.L., Tertilt, C., Heinen, T.J.A.J., Kremer, M., Wunderlich, F.T., Jung, S. & Waisman, A. (2005) A Cre-inducible diphtheria toxin receptor mediates cell lineage ablation after toxin administration. *Nature Methods* **2**, 419–426.

Zong, H., Espinosa, J. S., Su, H.H., Muzumdar, M.D. & Luo, L. (2005) Mosaic analysis with double markers in mice. *Cell* **121**, 479–492.

Legue, E. & Nicolas, J.-F. (2005) Hair follicle renewal: organization of stem cells in the matrix and the role of stereotyped lineages and behaviors. *Development* **132**, 4143–4154.

Joyner, A.L. & Zervas, M. (2006) Genetic inducible fate mapping in mouse: Establishing genetic lineages and defining genetic neuroanatomy in the nervous system. *Developmental Dynamics* **235**, 2376–2385.

Dymecki, S.M. & Kim, J.C. (2007) Molecular neuroanatomy's "three Gs": a primer. *Neuron* **54**, 17–34.

Livet, J., Weissman, T.A., Kang, H., Draft, R.W., Lu, J., Bennis, R.A., Sanes, J.R. & Lichtman, J.W. (2007) Transgenic strategies for combinatorial expression of fluorescent proteins in the nervous system. *Nature* **450**, 56–62.

Buckingham, M.E. & Meilhac, S.M. (2011) Tracing cells for tracking cell lineage and clonal behavior. *Developmental Cell* **21**, 394–409.

Kretzschmar, K. & Watt, F.M. (2012) Lineage tracing. *Cell* **148**, 33–45.

Cells

Fukuda, K., Sakamoto, N., Narita, T., Saitoh, K., Kameda, T., Iba, H. & Yasugi, S. (2000). Application of efficient and specific gene transfer systems and organ culture techniques for the elucidation of mechanisms of epithelial–mesenchymal interaction in the developing gut. *Development, Growth and Differentiation* **42**, 207–211.

Givan, A.L. (2001) *Flow Cytometry: First Principles*. New York: Wiley-Liss.

Freshney, R.I. (2005) *Culture of Animal Cells: A Manual of Basic Technique*, 5th edn. New York: Wiley-Liss.

Espina, V., Wulfkuhle, J.D., Calvert, V.S., VanMeter, A., Zhou, W., Coukos, G., Geho, D.H., Petricoin, E.F. & Liotta, L.A. (2006) Laser-capture microdissection. *Nature Protocols* **1**, 586–603.

This chapter contains the following animations:

Animation 17   Cre-lox recombination.
Animation 18   FACS operation.

 For additional resources for this book visit
www.essentialdevelopmentalbiology.com

Chapter 14

Development of the nervous system

Overall structure and cell types

The vertebrate central nervous system (CNS) consists of the brain and spinal cord and is formed from the **neural plate** of the early embryo. The peripheral nervous system (PNS) consists of the cranial, spinal, and autonomic nerves and their associated **ganglia**. It is formed from the **neural crest**, from epidermal **placodes**, and from axons growing from cell bodies in the CNS. The nerve cells, or neurons, of the CNS generally have numerous dendrites receiving input from other neurons, and one axon emitting signals from the cell body and connecting to the dendritic trees of numerous other neurons (Fig. 14.1a). In the CNS, "grey matter" means mainly cells and "white matter" means mainly nerve axons, the whiteness coming from the high lipid content of the myelin sheaths that enclose the axons. Condensations of neuronal cell bodies in the CNS are often called nuclei, not to be confused with cell nuclei.

The connections between neurons, or between neurons and non-neuronal target cells, are called synapses. They work by the release of a neurotransmitter from the presynaptic terminal and the opening of a ligand-gated ion channel in the postsynaptic terminal, which initiates an action potential in the receiving cell. The overall complexity of the vertebrate nervous system is enormous: the human brain containing around 10^{11} neurons, and each of these forming synapses with around 10^3 others. The capabilities of the system depend largely on the placement and connectivity of the neurons, which is achieved during embryonic development, but the vast complexity of the end product is built up by the sequential operation of a number of quite simple processes.

There are many types of neuron (Fig. 14.1a). They can be classified by the morphology of the axon and the dendritic tree, by the neurotransmitter used, for example acetylcholine, norepinephrine (= noradrenaline), or dopamine, and by whether their target is another neuron, a muscle (motor neurons), a gland, or a sense organ (sensory neurons). In addition to the neurons there are about 10 times as many **glial cells** in the nervous system (Fig. 14.1b). **Astrocytes** are the most numerous cell type in the CNS and perform many structural and metabolic functions, including maintenance of the blood–brain barrier, neurotransmitter uptake, and extracellular matrix production. Type 1 (= fibrous) astrocytes are more numerous in white matter, and type 2 (= protoplasmic) astrocytes are more numerous in grey matter. **Oligodendrocytes** are the cells of the CNS that form the myelin sheaths around axons, and the **Schwann cells** are their equivalents in the PNS. NG2 cells (= polydendrocytes, synantocytes) are a recently defined class of glia that serve as oligodendrocyte precursors and remain present throughout life. **Radial glial cells** are present only during development and are now known not really to be glia but instead constitute a type of neuronal progenitor cell. They have cell bodies near the lumenal surface of the neural tube and extend a fine process to the outer surface, which guides the radial migration of daughter neurons. **Microglia** are not formed from the neural plate at all but are phagocytic cells derived from a primitive population of embryonic macrophages. Finally, the **ependyma** is the simple cuboidal epithelial lining of the ventricles and spinal canal.

Neurons carry out most of their protein synthesis in the cell body. Materials are transported along the axon by slow axonal

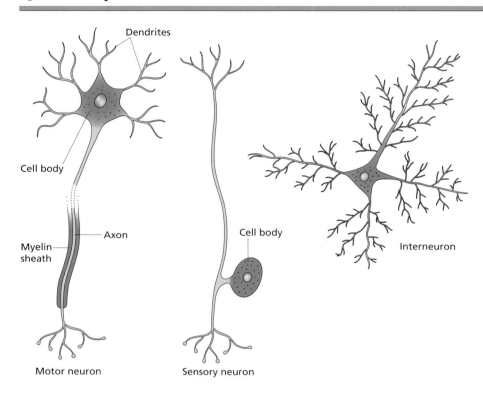

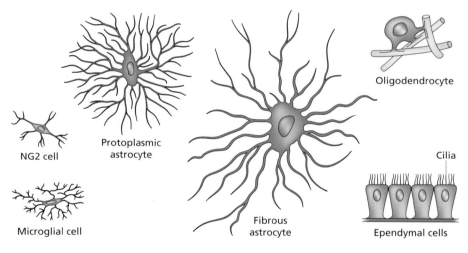

Fig. 14.1 Cell types found in the central nervous system. Note that some cells with astrocytic morphology are now considered to be neural stem cells.

transport, or, in the case of neurotransmitter vesicles, by fast axonal transport. This flow of materials is balanced in a mature neuron by retrograde transport of materials from the axon terminal to the cell body. This feature of neuronal behavior is very useful to the neuroanatomist as it is possible to trace the pathways of axons through the complexity of the CNS by injecting at the periphery a marker such as the enzyme **horseradish peroxidase** (**HRP**). This will be taken up by the axons and transported back to the cell bodies, which can then be located by histochemical staining or other appropriate detection method. For example if HRP is injected into a single muscle,

then it will be transported back into the motor neurons of the spinal cord that are innervating the muscle and enables them to be visualized.

The brain

The main parts of the brain (Fig. 14.2) appear from the neural tube during early development. The anterior forebrain becomes the **telencephalon**, which forms the paired cerebral hemispheres and serves, particularly in lower vertebrates, as the receptive area

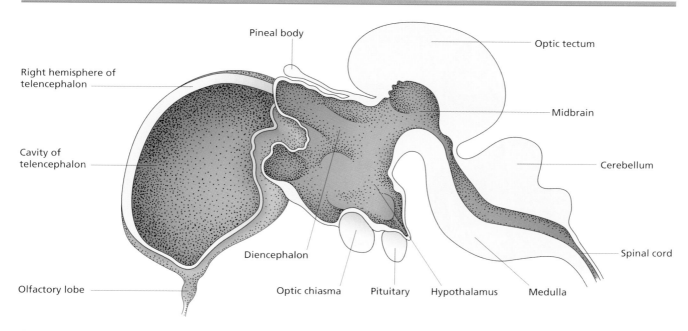

Fig. 14.2 Structure of the brain of an 8-day chick embryo. The mid- and hindbrain are sectioned in the median plane, while the right telencephalon and cerebellar swelling are shown in parasagittal section. Reproduced from Lillie, F.R. (1919) *The Development of the Chick. An Introduction to Embryology.* New York: Henry Holt & Co.

for the olfactory organs. The olfactory sensory cells themselves arise from the nasal placodes, which are thickenings of the epidermis adjacent to the telencephalon. The cerebral hemispheres are very large in higher vertebrates where more complex associative and control behaviors are apparent, and in mammals the multilayered cerebral cortex is often called the **neocortex**. The posterior forebrain becomes the **diencephalon**, which produces the optic vesicles. These grow out and invaginate to form the **optic cups** (Fig. 14.3). The lens of the eye arises from the epidermis where it is contacted by the optic cup. This forms a thickened lens placode which then rounds up and invaginates to be incorporated into the eye. The inner layer of each cup becomes the sensory retina, the outer layer becomes the pigmented retina, and the tube leading from the brain to the eyecup later carries the optic nerve. When it differentiates, the retina forms a multilayered structure containing the sensory photoreceptors, and various other neurons. These send processes back to the brain down the optic stalks. The optic nerve fibers terminate not in the diencephalon but in the **midbrain**, whose dorsal region becomes enlarged into paired optic lobes, called **optic tecta** (singular **tectum**) in nonmammals, which are the primary centers for reception of visual information. The superior colliculus of mammals is a homologous structure to the optic tectum. The midbrain also forms the auditory lobes.

Apart from the eyes, the diencephalon also forms the thalamus, the pineal body (= epiphysis) dorsally, and the pituitary (= hypophysis) ventrally. The thalamus occupies most of the side walls and is an important center of integration. The pineal is thought to be concerned with diurnal rhythms, and is later a source of melatonin. The pituitary, a key endocrine organ secreting a wide variety of hormones, has a dual origin. The anterior part arises from a placode initially just anterior to the neural folds, which invaginates through the oral aperture to become Rathke's pouch in the dorsal pharynx. The posterior part arises from the hypothalamus, the floor of the diencephalon, to which it remains permanently connected.

The hindbrain contains the cerebellum, important in the control of movement, dorsoanteriorly, and the medulla oblongata, containing a series of cranial nerve nuclei, more posteriorly. The early embryonic hindbrain becomes segmented to form seven **rhombomeres**, of which the first forms the cerebellum. Opposite rhombomeres 5 and 6 are the **otic placodes**, which invaginate and develop to form the inner ear, consisting of the cochlea, which is the organ of hearing, and the semicircular canals, which are concerned with perception of motion and control of balance.

The ventricles of the brain and the central canal of the spinal cord represent the persistent original lumen of the neural tube. There is a basic set of 10 **cranial nerves** in all vertebrates: I is the olfactory nerve; II the optic nerve; III, IV, and VI control the eye muscles; V is the trigeminal; VII the facial; VIII the auditory; IX the glossopharyngeal; and X the vagus.

Spinal cord

The structure of the spinal cord consists of an inner ependymal layer, a central region of grey matter, containing most of the cell

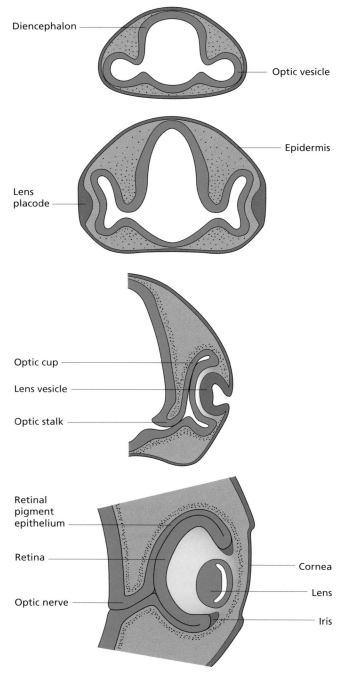

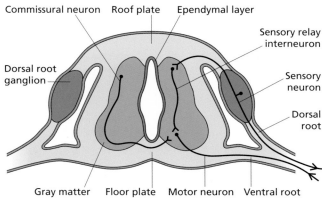

Fig. 14.4 Main structures of the developing spinal cord.

Fig. 14.3 Development of the eye. The optic vesicle grows out of the diencephalon while the lens develops from the epidermis. The inner layer of the cup becomes the retina, while the outer layer becomes a pigmented epithelium. Reproduced from Hildebrand (1995). *Analysis of Vertebrate Structure*, 4th edn, John Wiley and Sons Inc., New York, with permission.

bodies, and an outer layer of white matter containing mainly myelinated fibers (Fig. 14.4). In the grey matter, the motor neurons are located ventrally, while the interneurons, which receive sensory input, and the **commissural neurons**, which connect one side of the spinal cord with the other, lie dorsally.

One pair of **spinal nerves** originates from each trunk segment and is formed by the combination of nerves called the ventral and dorsal roots. The ventral roots consist of motor fibers from the spinal motor neurons, which are destined to innervate the muscles. The dorsal roots consist of sensory fibers, which connect the peripheral sense organs with the interneurons of the dorsal spinal cord. The sensory neurons are located in the **dorsal root ganglia**, which are formed from the neural crest (Fig. 14.4). At the levels of the fore- and hindlimb buds the spinal nerves join to form networks, respectively, called the **brachial plexus** and the **lumbosacral plexus**, from which emerge the nerves supplying the limbs.

The autonomic system with its numerous small ganglia and CNS connections is also formed from the neural crest. It is divided into a distinct sympathetic and parasympathetic system, also producing an extensive system of enteric neurons in the gut wall.

Anteroposterior patterning of the neural plate

The development of the nervous system involves an initial process of regionalization of the neural plate into territories, by mechanisms that are similar to those in other processes of organogenesis. Subsequently, the unique features of the nervous system become manifest in terms of specialized cell divisions leading to neuro- and gliogenesis, and the later formation of specific nerve connections guided by various extracellular and cell-surface factors.

The initial induction and regionalization of the neural plate has already been covered with regard to *Xenopus* in Chapter 7 and the chick in Chapter 9. Neural induction is due partly to the effects of fibroblast growth factors (FGFs), and partly to the inhibition of bone morphogenetic proteins (BMPs) by substances such as Chordin released from the **organizer**. This induces a **neuroepithelium** with an anterior specification such that, in the absence of further signals, it will express transcrip-

Classic experiments

Patterning of the CNS

The "real" neural inducer was eventually established to consist of factors inhibiting BMP signaling. The first paper is the culmination of this work, described in Chapter 7.

Wilson, P.A. & Hemmati Brivanlou, A. (1995) Induction of epidermis and inhibition of neural fate by BMP-4. *Nature* **376**, 331–333.

The next two papers describe the hindbrain in terms of developmental units characterized by expression of particular combinations of transcription factors. This helped to unify the traditional neuroanatomical description with the principles of modern developmental biology.

Lumsden, A. & Keynes, R. (1989) Segmental patterns of neuronal development in the chick hindbrain *Nature* **337**, 424–428.

Wilkinson, D.G., Bhatt, S., Cook, M., Boncinelli, E. & Krumlauf, R. (1989) Segmental expression of Hox-2 homeobox-containing genes in the developing mouse hindbrain. *Nature* **341**, 405–409.

The last describes the signaling activity of the floor plate, which controls the dorsoventral pattern in the neural tube.

Yamada, T., Placzek, M., Tanaka, H., Dodd, J. & Jessell, T.M. (1991) Control of cell pattern in the developing nervous-system – polarizing activity of the floor plate and notochord. *Cell* **64**, 635–647.

tion factors such as Otx and form forebrain structures. The determination of more posterior structures starts with the action of various posteriorizing factors. Important among these are the FGFs and Wnts, which induce expression of Cdx transcription factors, which in turn induce Hox genes of the trunk and tail. In the hindbrain region retinoic acid seems to have an important posteriorizing function. This overall picture is similar in all vertebrates although there are differences of detail.

Forebrain

These early inducing factors produce a CNS that is divided into just a few large anteroposterior domains, and this then becomes refined by a number of local interactions. Patterning of the forebrain works in the same way as other embryological interactions: inducing factors cause activation of different combinations of transcription factors in different regions, and these control the subsequent pattern of differentiation (Fig. 14.5). It was formerly thought that patterning of the cerebral cortex depended on innervation from the thalamus. But mouse knockouts in which this innervation is lost nonetheless show a normal initial pattern of transcription factor domains in the cortex. So now it is considered that the thalamic inputs refine a pattern that is initially established autonomously. One inducing signal required for forebrain regionalization is FGF from the anterior neural ridge. This was first discovered in zebrafish as a region at the extreme anterior of the neural plate that caused expansion of the diencephalon when removed. When grafted posteriorly it could induce ectopic telencephalon. A similar source of FGF8, -17, and -18 exists in the mouse. Overexpression of FGFs anteriorly will shift the boundaries of forming structures to the posterior (Fig. 14.5d). Introduction of an ectopic source of FGF in the posterior can cause duplication of structures. Both of these behaviors indicate response to an FGF gradient.

At the posterior edge of the developing cerebral cortex, in the cortical hem region, a number of Wnts and BMPs are expressed. Mouse knockouts for *Wnt3A* or *Lef1* show defects in the posterior telencephalon, such as lack of the hippocampus, suggesting that a Wnt signal is required there. The transcription factor genes that are expressed in the forebrain in response to the FGF, Wnt, and BMP signals (*Emx2, Pax6, Couptf1, Sp8*) show predictable changes of expression in loss-of-function and gain-of-function mutants affecting the inducing factors. Loss of function shifts the later structures toward the weakened source, while gain of function pushes structures further away from the source. The genes for these transcription factors show graded expression, like the inducing factors themselves, but the subsequent boundaries that form between structures are nonetheless discrete.

Isthmic organizer

An important anteroposterior signaling center is the isthmic organizer. The constriction between midbrain and hindbrain vesicles is known as the **isthmus** (Fig. 14.6). At the open neural plate stage, transcription factor genes such as *Otx2* and *Lim1* are expressed in the prospective fore- and midbrain. At their posterior boundary, just anterior to the future isthmus, a strip of cells starts to produce the inducing factors Fgf8 and Wnt1. This organizing region controls the regional specification of the neural tissue on either side. To the anterior it induces the **midbrain**, and also establishes a gradient of expression of the transcription factors Engrailed 1 and 2. To the posterior it induces rhombomere 1 and thence the **cerebellum**. These effects may be shown by transplantation of the isthmic region of the chick embryo, which can induce midbrain from the prospective diencephalon, and cerebellum from the anterior part of the hindbrain. These activities are shared by pure Fgf8 protein, applied on a bead. Fgf8 will also induce expression of *Engrailed* (*En*) and *Wnt1* and of its own gene, showing that the organizer can maintain itself after its initial establishment.

The mouse knockouts of both *Wnt1* and *En1* lack not just the isthmus region where these genes are expressed, but also large parts of the midbrain and cerebellum, confirming that they are essential to the signaling function of the isthmic organizer. The mouse knockout of *Fgf8* is an early lethal because it has a large

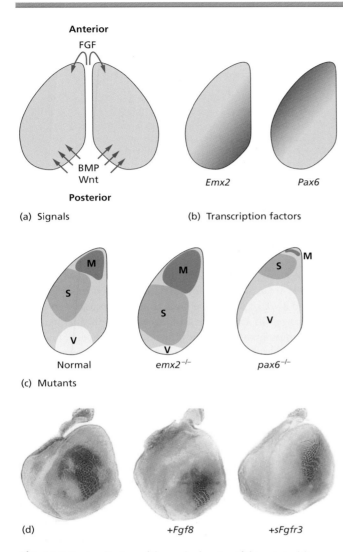

Anterior

FGF

BMP
Wnt

Posterior

(a) Signals

Emx2   *Pax6*

(b) Transcription factors

Normal   *emx2*⁻ᐟ⁻   *pax6*⁻ᐟ⁻

(c) Mutants

(d)   +Fgf8   +sFgfr3

Fig. 14.5 Regionalization of the cerebral cortex of the mouse. (a) Source regions of inducing factors are the anterior rim and the cortical hem. (b) Various transcription factors show early regionally specific expression. (c) The phenotype of loss-of-function mutations suggests that these transcription factors control subsequent development. M motor area, S sensory area, V visual area. (d) Shift in position of the sensory area in layer 4 of the postnatal cortex, following anterior electroporation of FGF8 or an FGF inhibitor into E11.5 embryos. The sensory barrelfield is viewed by histochemistry for cytochrome oxidase. For simplicity the diagrams represent the 11.5-day mouse embryo, but the signals operate at earlier stages. Reproduced from O'Leary *et al.* (2007), with permission from Elsevier.

posterior body defect, but there is a hypomorphic allele of *Fgf8* that has enough residual function to survive early development, and this mutant displays a midbrain–cerebellum defect. A similar phenotype arises in the zebrafish mutant *acerebellar*, which has a loss-of-function mutation of *Fgf8* but is viable for a longer period than the corresponding mouse mutant.

A consequence of the activity of the isthmic organizer is the expression of different combinations of Pax family transcrip-

tion factor genes in different regions. The *Pax6* gene is expressed in forebrain and hindbrain, while *Pax2* and *Pax5* are expressed in the midbrain and rhombomere 1 region. A mouse double knockout of *Pax2* and *-5* lacks the midbrain and cerebellum, showing that these genes are necessary for development of the corresponding structures.

Segmentation of the hindbrain

The pattern of the CNS is more obviously segmental in the hindbrain than in the more anterior parts of the brain. A series of seven bulges becomes visible after neural tube closure, called **rhombomeres** (Figs 14.6, 14.7). The first rhombomere gives rise to the cerebellum. Each subsequent pair of rhombomeres contains a reiterated set of motor nuclei and contributes fibers to one **cranial nerve**. Thus, the trigeminal (V) nerve comes from rhombomere 2–3, the facial (VII) nerve from rhombomere 4–5, and the glossopharyngeal (IX) nerve from rhombomere 6–7. Every other rhombomere also contributes **neural crest** cells to one of the **branchial arches**. Therefore rhombomere 2 contributes cells to the first arch, rhombomere 4 to the second arch, and rhombomere 6 to the third arch. The segmental character of the rhombomeres is also evident from **clonal analysis**. If a single cell is labeled, its progeny will cross between rhombomeres at an early but not at a late stage. The clonal restriction arises because of the aggregation of cells in the individual rhombomeres. This is at least partly due to the expression by

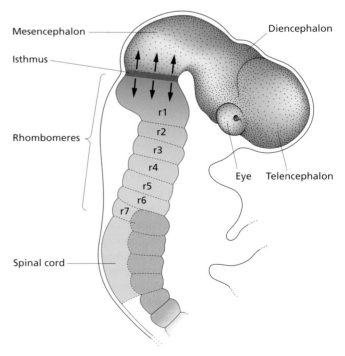

Mesencephalon

Isthmus

Rhombomeres

Spinal cord

Diencephalon

r1
r2
r3
r4
r5
r6
r7

Eye   Telencephalon

Fig. 14.6 Position of the isthmic organizer in relation to main parts of the brain of the chick embryo.

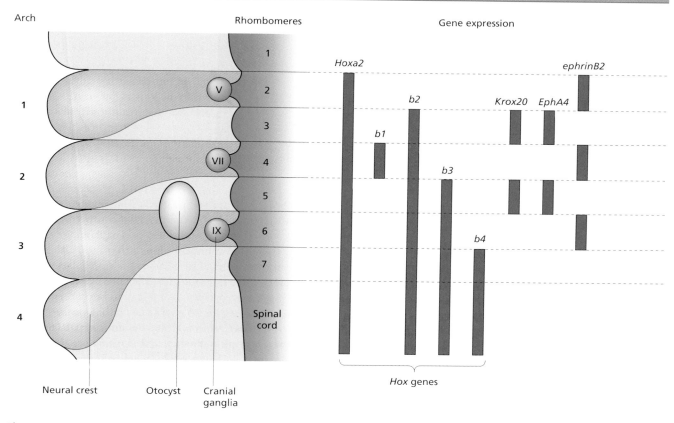

Fig. 14.7 Diagrammatic view of the rhombomeres showing gene expression domains and contributions to the cranial nerves and neural crest.

alternating rhombomeres of ephrins and their Eph-type receptors. Ephrin B-type ligands are expressed on rhombomeres 2, 4, and 6 while Ephs are found on rhombomeres 3 and 5. A repulsive interaction between the cell groups bearing Ephs and ephrins creates the boundaries, and the boundaries are not formed if a dominant negative Eph is overexpressed in the region. In the open neural plate stage the identity of the rhombomeres is still labile and so prospective hindbrain regions can be exchanged in position without affecting the final pattern. However, once the rhombomere boundaries begin to appear, their character can no longer be re-specified by transplantation. Actual formation of the boundary cells depends on the Notch system. In the zebrafish it has been shown that cells overexpressing constitutive Notch are driven into the boundary regions while those overexpressing a dominant negative Su(H), which antagonizes Notch signaling, are excluded from the boundaries.

Hox genes are not expressed in the fore- and midbrain but are important for anteroposterior patterning more posteriorly. The most anterior-type Hox genes have their anterior boundaries in the hindbrain, and the combinations of activity are such that each rhombomere carries a unique code (Fig. 14.7). For example, *Hoxa2* is expressed with its anterior limit at the future rhombomere 1/ 2 boundary, and expression in the emergent

neural crest cells is seen in rhombomere 4 but not 2. In the mouse knockout of *Hoxa2*, there is a homeotic shift of arch 2 towards the arch 1 morphology. Conversely in overexpression experiments, carried out in chick and *Xenopus*, there is a shift from the arch 1 to arch 2 morphology.

The Hox genes are upregulated by other transcription factors, which are in turn upregulated by the early gradients of FGF, Wnt, and retinoic acid. An example of one such intermediate transcription factor is Krox 20 which is a zinc-finger protein expressed in rhombomeres 3 and 5. This is the rhombomere pair that expresses Eph-type receptors and that does not emit a neural crest stream. If *Krox20* is knocked out in the mouse, then rhombomeres 3 and 5 fail to form and 2/4/6 arise all fused together with no boundaries to separate them.

Manipulations of retinoic acid levels show particularly strong effects on the hindbrain. Retinoic acid is produced from dietary vitamin A and the enzyme that makes it (retinaldehyde dehydrogenase, RALDH2) is found in the somites and lateral plate of the trunk region. The enzyme that destroys it (Cyp26, a cytochrome P450) is found in the fore- and midbrain. Since the source lies to the posterior and the sink lies to the anterior, the retinoic acid forms a concentration gradient falling from posterior to anterior across the hindbrain region. Quail embryos can be made almost entirely retinoid-free by depriving the

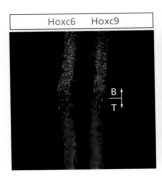
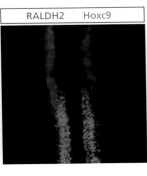
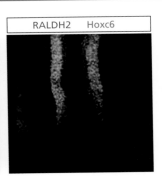

Hoxc6	Hoxc9
RALDH2	Hoxc9
RALDH2	Hoxc6

Fig. 14.8 Immunostaining of Hoxc6 and c9 showing the boundary between brachial and thoracic (B/T) regions of the chick spinal cord at 4 days of development. RALDH2 is retinaldehyde dehydrogenase, an enzyme responsible for production of retinoic acid, also expressed in the brachial region. Reproduced from Dasen *et al.* (2003), with permission from Nature Publishing Group.

mothers of vitamin A. Such embryos show small hindbrains completely lacking rhombomeres 5–7, but the missing parts do develop if the embryos are given vitamin A before the early somite stage. There is also a loss of ventral neuronal types in the anterior spinal cord. Mouse embryos with *Raldh2* knocked out die at mid-gestation and also lack rhombomeres 5–7. Similar effects are shown in *Xenopus* by overexpression of a dominant negative retinoic acid receptor. Mouse embryos can be treated with an excess of retinoic acid by dosing the mothers. Treatment at E7.5 causes *Hoxb1* to be expressed in rhombomere 2 as well as the normal expression in rhombomere 4. This alters the character of the neural crest cells that come out of rhombomere 2 and so converts the first arch into a copy of the second arch. This is a good example of a **teratogenic** effect that is now well understood at the molecular level. Retinoic acid has in the past presented a real teratogenic risk to humans as it is used in oral form for acne therapy. Before the risk was understood, this led to many abortions and to a number of birth defects showing brain damage.

Spinal cord

Within the spinal cord the anteroposterior pattern is not so evident as in the brain. However, there are Hox gene expression boundaries at various levels and there is an anteroposterior arrangement of certain groups of motor neurons. For example the lateral motor neuron columns, which supply the limbs, are present only at limb levels, at the forelimb level corresponding to expression of *Hoxc6*. The column of autonomic motor neurons (called the column of Terni in the chick) lies in between the limb levels, corresponding to expression of *Hoxc9* (Fig. 14.8). Electroporation of *Fgf8* into this region of the chick neural tube causes an anterior shift of the *Hoxc6/ Hoxc9* boundary, with corresponding shrinkage of the lateral motor column.

Dorsoventral patterning of the neural tube

The spinal cord has the same basic dorsoventral arrangement along its length, and in fact many of the features extend through the hindbrain and midbrain as well. Mid-ventrally there is a **floor plate**, composed of columnar cells in close proximity to the notochord. Flanking this is the region containing the motor neurons. There are various concentrations of these, the medial column supplying the axial and body wall muscles, and the lateral columns the limb muscles. The motor neurons in each column express a particular combination of Lim-type transcription factors and this is associated with the group of muscles to which they later project. If the combination of *Lim* genes expressed is experimentally altered, then the motor neuron connectivity pattern is changed accordingly. In the dorsal half of the spinal cord are the dorsal sensory relay interneurons and the **commissural neurons**, which project to the opposite side of the spinal cord. In the dorsal midline is the **roof plate**.

This dorsoventral pattern is controlled by inductive signals, initially from the tissues neighboring the neural plate, and later from the floor plate and the roof plate (Fig. 14.9). The **notochord** emits Sonic hedgehog (Shh) which induces the floor plate and, at lower concentration, the motor neuron-forming territory. This can be shown by recombination experiments in which the notochord is grafted adjacent to different dorsoventral levels of the chick neural tube. An ectopic floor plate will form in close proximity and motorneurons a little further away. Evidence that the signal is Shh is threefold: *Shh* is expressed in the notochord; it will induce floor plate and motor neurons when applied to the neural tube; and the floor plate and motor neurons are lost in the mouse knockout of *Shh*. The gradient can also be visualized in mice transgenic for a fusion protein of Shh with GFP. This shows a punctate pattern of protein concentrated at the apical (ventricular) surface of the responding neuroepithelial cells. This intracellular pattern indicates transcytosis as a possible mechanism for spreading (Fig. 14.10; see also Chapter 17).

The floor plate itself also makes Shh, and once it has been formed it contributes to the inducing role of the notochord. Shh upregulates the expression of one set of transcription factor genes, including *Foxa2*, *Nkx2.2*, and *Nkx 6.1*. It represses the expression of various others such as *Pax6* and *Pax7*. There follows a sharpening of dorsoventral boundaries arising from a mutual repression between pairs of these transcription factors. For example Pax6 and Nkx2.2 repress the expression of each other's gene, leading to the formation of the p3/pMN boundary, as shown in Fig. 14.9c. Loss of either of these genes causes

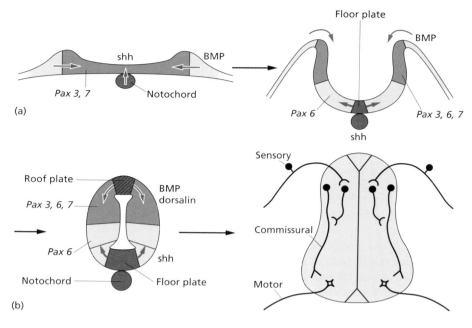

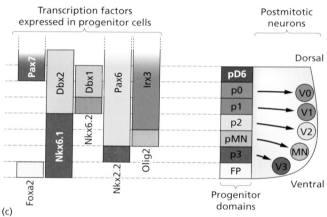

Fig. 14.9 Dorsoventral patterning of the neural tube. (a) Sonic hedgehog from the notochord induces expression of its own gene in the floor plate. BMP from the epidermis induces the roof plate. (b) The signals from the two centers induce a nested expression of genes that later brings about the pattern of neuronal differentiation. (c) Diagram of the initial gene expression domains and the corresponding progenitor domains in the spinal cord. FP, floor plate; MN, motorneurons; p-domains are progenitor domains; V0–V3 are interneuron domains. Reproduced from Dessaud et al. (2008), with permission from Company of Biologists Ltd.

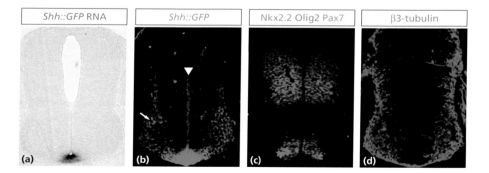

Fig. 14.10 The Sonic hedgehog gradient visualized in an E10.5 mouse transgenic for a Shh–GFP fusion protein. (a) In situ hybridization showing presence of mRNA in the floor plate. (b) GFP fluorescence showing transport of protein up the ventricular zone (arrowhead), as well as to nonreceptive areas of differentiated neurons laterally (arrow). (c) Three of the early gene expression domains visualized by immunostaining of transcription factor products (domains labeled from bottom to top). (d) β3 tubulin immunostain shows positions of neurons. Reproduced from Dessaud et al. (2008), with permission from Company of Biologists Ltd.

expansion of the expression domain of the other. There is also a formation of dorsoventral territories expressing specific transcription factors. For example, Olig2, involved in the later differentiation of motorneurons and oligodendrocytes, is upregulated by Shh but subsequently repressed by Nkx2.2 and ends up defining the pMN region.

The elucidation of this mechanism has led to understanding of the human congenital defect called **holoprosencephaly**. This involves midline defects ranging in severity from the presence of a single midline incisor tooth, to very severe cases in which there is a single midline eye with a proboscis above it. The defect arises from a partial loss of function of Shh, either through mutation in the *SHH* gene itself or through defects in cholesterol metabolism that prevent the addition of cholesterol to the active form of the Shh molecule. Although holoprosencephaly can also arise from mutations in other genes, those attributable to loss of function of Shh also show defects in limbs and vertebrae, consistent with the role of Shh in the patterning of these structures (see Chapter 15).

The dorsal midline of the neural tube arises from the lateral margin of the neural plate, where it contacts the epidermis. The epidermis produces BMP4 and BMP7, and these proteins can elevate expression of *Pax6* and *Pax7* when applied to the neural plate. After neural tube closure, the roof plate expresses *Bmp4* and *-7* and also the gene for another TGF-β superfamily member called Dorsalin. These factors all have the activity of inducing dorsal cell types such as sensory relay neurons. If formation of the roof plate is prevented, as for example in the mouse knockout of the *Lim1a* gene, then these dorsal neuronal types fail to form. The overall dorsoventral pattern of the spinal cord arises from the combined effects of the gradients of Shh from the floor plate and of BMPs and Dorsalin from the roof plate.

Neurogenesis and gliogenesis

Drosophila

Some of the principles of neurogenesis were learned from studies on *Drosophila*. In *Drosophila* the ventral nerve cord arises from a neurogenic region, which originates lateral to the mesoderm and becomes midventral once the mesoderm has invaginated. The neurogenic region contains a segmentally reiterated pattern of **proneural clusters** defined by expression of the bHLH transcription factors produced by the achaete–scute complex (AS-C; pronounced eh-keet-scute). The positions of the clusters depend on the operation of the anteroposterior and dorsoventral patterning systems, described in Chapter 11. Each cluster produces only one neuroblast together with a number of epidermal cells, and the selection of the future neuroblast is controlled by the Notch **lateral inhibition** system (see Animation 19: Neuroblast formation). AS-C factors activate production of Delta and this stimulates Notch on the adjacent cells. The Notch signal represses expression of AS-C and so reduces

Delta production in the adjacent cells (Fig. 14.11). This positive feedback system amplifies initially small differences such that eventually there is just one cell with a high level of AS-C, which becomes the neuroblast, surrounded by several cells with a low level of AS-C, which become epidermal. This mechanism explains why loss of function of *Notch* leads to an overproduction of neuroblasts. The mechanism is an example of a **symmetry-breaking** system. The initial uniform state is unstable because there is slight fluctuation in the levels of AS-C or Delta between cells, and this will automatically become amplified by the feedback inherent in the interaction.

Once formed the neuroblasts detach from the neighboring cells and sink into the interior, remaining temporarily connected by an apical stalk in the plane of the epithelium. This apical specialization contains the Par complex described in Chapter 12 (i.e. the complex of Par3/Par6/aPKC; the *Drosophila* homolog of *par6* being called *bazooka*). The Par complex phosphorylates components of the intracellular trafficking system so as to repel other proteins to the basal region, including Numb which is a tyrosine phosphatase, Prospero, which is a homeodomain transcription factor, and Brat, an RNA-binding protein. Subsequent divisions are asymmetrical with a large neuroblast on the apical side, inheriting the apical stalk, and a smaller ganglion mother cell on the basal side, enriched in Numb, Prospero, and Brat. These components control the properties of the ganglion mother cell, which subsequently divides once to yield a pair of neurons or glia. Evidence for this mechanism is largely derived from the behavior of loss-of-function mutants, for example embryos lacking components of the Par complex do not show localization of Numb and Prospero, and their neuroblasts divide symmetrically. Loss of function of *prospero* and *brat* causes defective ganglion mother cells.

The same themes are seen in vertebrate neurogenesis: lateral inhibition underlying primary neuron formation and asymmetrical division in later neurogenesis, often involving homologous molecular components to those employed by *Drosophila*.

Vertebrate primary neurogenesis

In lower vertebrates the first neurons arise directly from the neuroepithelium of the open neural plate. At this stage the neuroepithelium is just one cell thick but appears stratified because the nuclei are present at many different levels. There are several neurogenic transcription factors that drive neuronal differentiation. Among these are Neurogenin and NeuroD, both bHLH-type factors belonging to the same biochemical family as AS-C in *Drosophila*. *Neurogenin* is expressed in the areas of neural plate that give rise to the first primary neurons and causes upregulation of *NeuroD*, which becomes highly expressed in the primary neurons. Overexpression of either factor in *Xenopus* will increase the proportion of neurons formed from the neuroepithelium. The spacing of the primary neurons is determined by a lateral inhibition mechanism very similar to

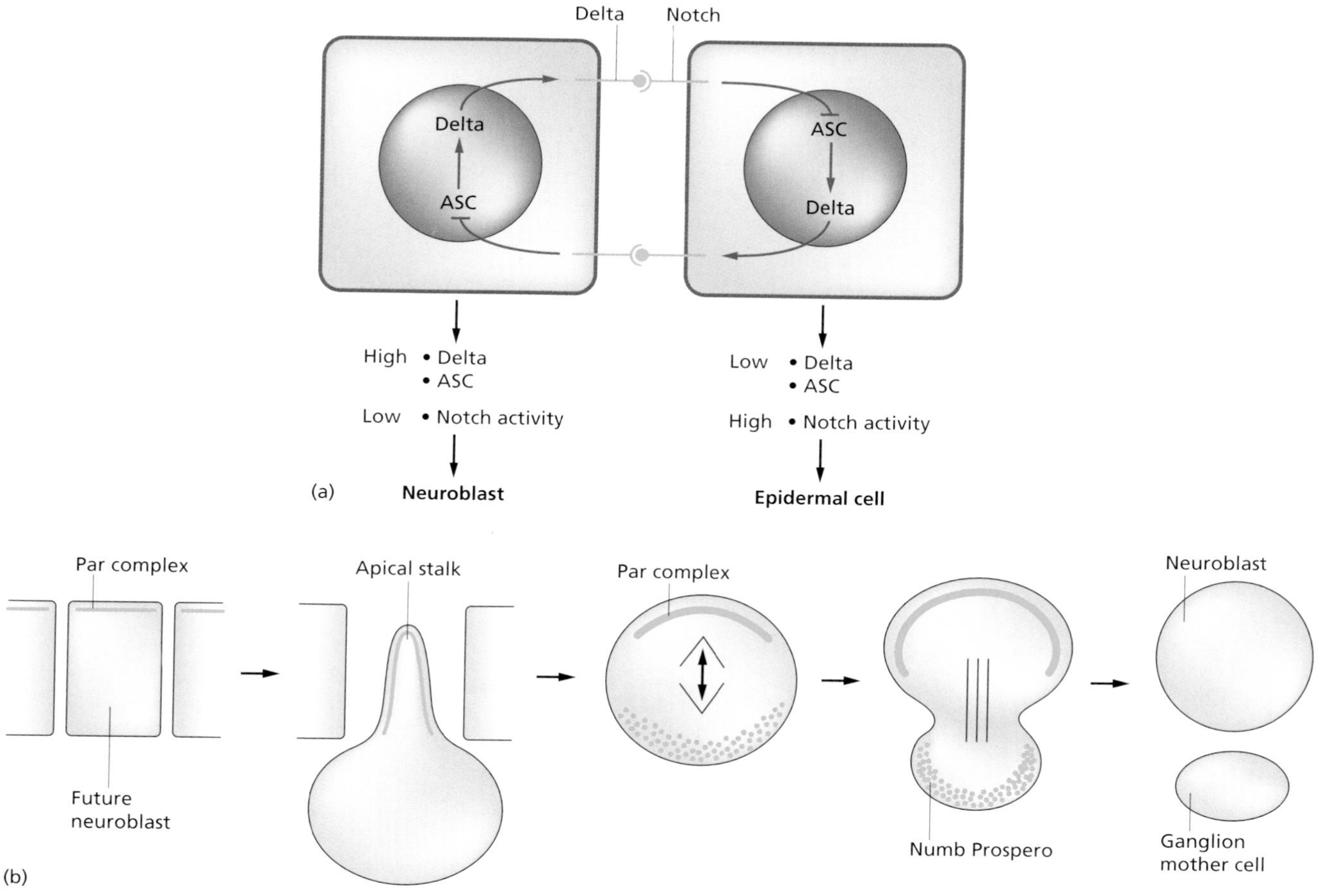

Delta   Notch

Delta

ASC

ASC

Delta

(a)

High • Delta
      • ASC

Low   • Notch activity

Neuroblast

Low   • Delta
      • ASC

High • Notch activity

Epidermal cell

Par complex   Apical stalk   Par complex   Neuroblast

Future
neuroblast

(b)

Numb Prospero   Ganglion
                       mother cell

Fig. 14.11 Neurogenesis in *Drosophila*. (a) The Delta–Notch lateral inhibition system amplifies initial small differences between cells of a proneural cluster ensuring that only one ends up with high ASC and becomes the neuroblast. (b) Asymmetrical division of neuroblast to generate another neuroblast and a ganglion mother cell.

that in *Drosophila* (Fig. 14.12a; and see Animation 6: Lateral inhibition). All of the cells in a *Neurogenin*-expressing patch are competent to become neurons. But this tendency is inhibited by Notch signaling, and Notch on one cell is stimulated by Delta on the neighboring cells. Overexpression of Delta, or of a constitutive form of Notch, in the *Xenopus* neural plate, reduces the formation of primary neurons. Conversely, overexpression of a dominant negative Delta generates a higher density of primary neurons. The lateral inhibition process leads to the formation of a number of individual neurons scattered along six longitudinal neurogenic belts. Although the symmetry breaking event was originally ascribed to molecular-scale statistical fluctuations between cells, it is now known that this biochemical system is also capable of sustained oscillation. Because there is no synchrony between adjacent cells, spontaneous oscillation can also create plenty of heterogeneity to get the process started.

Later neurogenesis

There is a little primary neurogenesis in the mouse but most occurs in the second half of gestation when the neuroepithelium is more than one cell thick. Early (E9–10) labeling experiments with retrovirus show that clones are usually composed of all neurons or all glia, but a sizeable minority are mixed. This shows that at least some of the early progenitors are multipotent. It is often considered that cells giving rise to single-type clones are committed to do so, but this is not the case, as explained in Chapter 4, since the same result could arise from similar microenvironmental effects on all the progeny from multipotent progenitors.

In terms of the conventional terminology for epithelial structure, the **apical** side of the neuroepithelium is that facing the lumen of the neural tube while the **basal** side is that facing the exterior (this means that apical is usually at the *bottom* of a

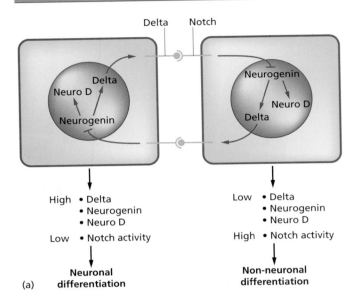

High
• Delta
• Neurogenin
• Neuro D

Low
• Notch activity

(a) **Neuronal differentiation**

Low
• Delta
• Neurogenin
• Neuro D

High
• Notch activity

Non-neuronal differentiation

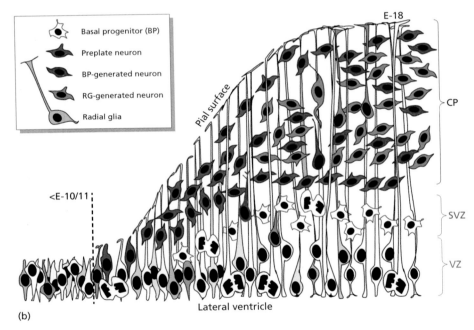

(b)

Fig. 14.12 (a) Lateral inhibition mediated by the Delta–Notch system ensures that only some cells from the neurogenic regions become neurons. (b) Neurogenesis from radial glial cells. The diagram depicts the thickening of the cerebral cortex during the second half of gestation in the mouse. The original neuroepithelium becomes multilayered with a ventricular zone (VZ), subventricular zone (SVZ), and cortical plate (CP). Some early neurons arise from the simple neuroepithelium, but most derive from radial glia, either directly or via the formation of basal progenitors. In terms of epithelial morphology, the outer, pial, surface is basal and the inner, ventricular, surface is apical. The basal progenitors are so called because they lie basal to the ventricular zone. Reproduced from Malatesta *et al.* (2007), with permission from Springer.

diagram and basal at the *top* – opposite to the convention for other epithelia). In neuroanatomy the apical side becomes known as **ventricular**, as it abuts the brain ventricles, while the outer surface becomes known as **pial**, after the enveloping connective tissue layer called the pia mater (Latin for "tender mother").

At early stages all of the cells are mitotic and there is a migration of nuclei such that they approach the exterior surface during S phase and the lumenal surface during mitosis. This is called **interkinetic migration**. As neurogenesis gets going most of the neuroepithelium becomes genuinely stratified and a pro-

portion of neuroepithelial cells become transformed into **radial glia** that continue to span the full ventricular to pial distance (Fig. 14.12b). These cells acquired their name because of the expression of various glial markers, but they are actually progenitor cells for both neurons and glia.

Radial glia retain the apical Par complex from the single-layer neuroepithelium, together with an apical concentration of a membrane protein, Prominin-1 (= CD133), the intermediate filament protein Nestin, and the Brain lipid binding protein (Blbp). In mice transgenic for *Blbp-Cre* crossed to an R26R-type reporter, some neurons, basal neuron progenitors, astrocytes,

Classic experiments

Neural crest and neurogenesis

Le Douarin, N.M. (1986) Cell line segregation during peripheral nervous system ontogeny. *Science* **231**, 1515–1522.
This is actually a review, but nicely summarizes a series of papers published in French in the 1970s. It establishes the fate and potency of regions of the neural crest using quail to chick grafts.
Heitzler, P. & Simpson, P. (1991) The choice of cell fate in the epidermis of Drosophila. *Cell* **64**, 1083–1092.
Analysis of the lateral inhibition system controlling *Drosophila*, and, as it later turned out, vertebrate, neurogenesis.
Lee, J.E., Hollenberg, S.M., Snider, L., Turner, D.L., Lipnick, N. & Weintraub, H. (1995) Conversion of Xenopus ectoderm into neurons by NeuroD, a basic helix-loop-helix protein. *Science* **268**, 836–844.

Evidence that overexpression of a single proneural transcription factor could drive the formation of neurons in a vertebrate.
Eriksson, P.S., Perfilieva, E., Bjork-Eriksson, T., Alborn, A.M., Nordborg, C., Peterson, D.A. & Gage, F.H. (1998) Neurogenesis in the adult human hippocampus. *Nature Medicine* **4**, 1313–1317.
Although evidence had existed for many years that adult neurogenesis occurred in mammals, this paper finally convinced people.
Noctor, S.C., Flint, A.C., Weissman, T.A., Dammerman, R.S. & Kriegstein, A.R. (2001) Neurons derived from radial glial cells establish radial units in neocortex. *Nature* **409**, 714–720.
Establishment of radial glia as the principal source for neurogenesis in development.

oligodendrocyte precursor cells, and ependyma can all be labeled. This shows the importance of radial glia as progenitor cells. The progeny neurons tend to crawl up the parent radial glial cell to their final position in the neuroepithelium and observation of this phenomenon accounts for why radial glia were for some time believed simply to provide guidance cues for radial migration of neurons arising from somewhere else.

Unlike in *Drosophila*, in vertebrate neurogenesis the Par complex seems to attract Numb to the same region of the cell rather than repelling it to the opposite side. After division the proliferative cell retains the Numb protein and the postmitotic neuron is depleted in Numb. In mouse Numb (and a similar protein, Numblike) seem to maintain radial glia in a proliferative mode by sustaining the intercellular contacts that control the structure of the cell. A localized knockout of both genes in the dorsal forebrain, using *Emx1-Cre* crossed to mates floxed for *Numb* and *Numblike*, showed a delay of neurogenesis, resulting in numerous defects.

In lower vertebrates, which show considerable adult neurogenesis, radial glia persist through adult life. In the mouse most radial glia mature into astrocytes during the first postnatal week. This transition depends on cardiotrophin, a growth factor belonging to the IL6 family, that is secreted by newly formed neurons. A few radial glia also become transformed into **neural stem cells** (see below). This is demonstrated by labeling a reporter mouse postnatally with adenovirus encoded Cre, delivered to the striatum, such that only the radial glia of the cortex can take it up. Most radial glia labeled in this way shortly after birth become astrocytes, but a few become subventricular neural stem cells, and these will found clones containing neurons, astrocytes, and oligodendrocytes.

Neuronal birthdays and layers

Once a neuron is formed it ceases to divide and this means that a pulse of BrdU or ^{3}HTdR (see Chapter 18), given during development, can reveal the time of formation of each class of neuron. If the tissue is examined some time after a pulse label, all the cells that have continued to divide will have lost the label by dilution through succeeding rounds of DNA replication, whereas those that ceased dividing just after incorporating the label will still retain it. The time of the last S phase before neuronal differentiation is often called the **birthday** of the neuron.

The layered structure of the CNS emerges through an "inside-out" pattern of morphogenesis. Cells moving inwards from the ventricular layer form a **mantle layer**, later known as the grey matter, containing both neurons and glia. As the neurons produce their axons, these build up to form a cell-poor **marginal zone** under the pial surface. Once the axons are myelinated this becomes the white matter. As the neural tube thickens, the neurons have to migrate radially from the site of their formation in the ventricular zone to reach their final positions. The neurons with early birthdays tend to migrate shorter distances and those with late birthdays migrate further.

The initial three-layered structure of ventricular, mantle, and marginal zones is maintained in the spinal cord, but in some parts of the brain it gets much more complex. In the mammalian cerebral cortex the first formed neurons form a preplate on the pial side. This is composed of an outer layer of cells called Cajal–Retzius neurons, and an inner layer of subplate neurons. New neurons from the mantle zone then invade the marginal zone to establish six further layers of **neocortex** (numbered I–VI from pia to ventricle). Each of these contains a different type of neuron. Most cortical neurons of mouse embryos are generated during the period E12.5–17.5 and each new cohort migrates past the subplate to create new layers of neocortex. The layers are formed from inside out, such that the first-born neurons populate layer VI and the later-born ones migrate through the existing layers to populate higher ones. Some classic experiments on ferret embryos showed that early migrating neurons could populate upper layers when grafted to a later

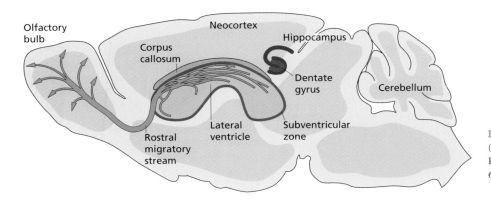

Fig. 14.13 Location of neural stem cells (red areas) in the adult mouse brain. Reproduced from *Zhao et al. Cell* 2008; 132: 645–660, with permission from Elsevier.

stage brain, but later migrating ones cannot be respecified to populate lower layers.

In the cerebellum some mitotic cells migrate to the outer layer to establish an external germinal layer, with subsequent "outside-in" neurogenesis. Also, the mantle zone becomes subdivided into several layers including one for the Purkinje neurons, which are large cells, found only in the cerebellum, and contacting as many as 10^5 other neurons.

In the eye the outer region of the optic cup becomes **pigment epithelium** and the inner layer becomes the retina, consisting of six cell layers with the photoreceptors on the original ventricular surface and the fibers running to the optic nerve on the original pial surface.

Neural stem cells

In lower vertebrates there is cell renewal throughout life in many areas of the CNS, and this also happens in two regions of the brain in mammals: the subventricular zone adjacent to the lateral ventricles and the subgranular zone of the hippocampus, which is a structure on the inside of the temporal lobe, concerned with learning and memory (Fig. 14.13). The new neurons from the lateral ventricles migrate anteriorly to end up as interneurons in the olfactory bulbs. Those from the hippocampus migrate inwards to become neurons of the granular zone.

It is now believed that **neural stem cells** are organized in a similar way to stem cells of other renewal tissues (see Chapter 18). The stem cells exist in a specific niche, and give rise to transit amplifying cells, which are progenitors of finite division potential that in turn give rise to neuroblasts. The stem cells are identified with morphologically identifiable "type B" cells. These express glial fibrillary acidic protein (GFAP), formerly considered characteristic of mature astrocytes, Nestin, Sox2, important in the embryonic neuroepithelium, integrins, and certain carbohydrate epitopes such as Prominin1 (= CD133) and SSEA1. Although occupying the subventricular zone, they do extend processes between the ependymal cells to contact the ventricular surface. The **stem cell niche** (Fig. 14.14) is thought

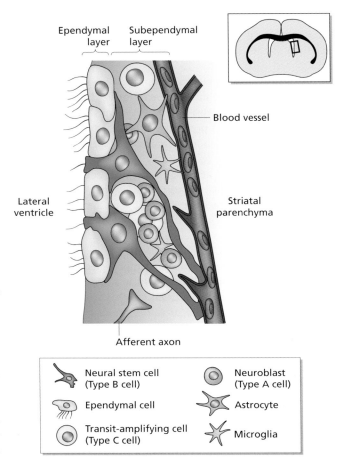

Fig. 14.14 A model for the stem cell niche in the subventricular zone of the mammalian lateral ventricle. Reproduced from Miller and Gauthier-Fisher (2009), with permission from Elsevier.

to be formed by the proximity of blood vessels, with their basal laminae sequestering growth factors. Because the stem cells contact both the vascular system and the cerebrospinal fluid they are potentially able to respond to environmental signals from both sources.

Evidence that the neural stem cells are the astrocyte-like B cells is drawn from several sources. Firstly, these cells are the only ones that show long-term retention of **BrdU**. Second, they survive treatment with the antimitotic drug ara-C, which kills the rapidly dividing transit amplifying cells. Since neurogenesis resumes after the cessation of treatment it must originate from these survivors. Thirdly, if astrocyte-like cells are specifically labeled with GFP, their progeny can be seen to include dividing transit amplifying cells and neurons migrating to the olfactory bulbs. This specific labeling is done by viral infection of transgenic mice that express a receptor for avian leukosis virus under the control of a glial-specific (GFAP) promoter. The virus cannot normally infect mice, but in this transgenic line it will specifically infect and label the GFAP-positive cells, which include both normal astrocytes and the type B cells. Lastly, **neurospheres** (see below) can be grown from this region, but they will not grow if all the GFAP-positive cells have been killed, which can be done using the drug ganciclovir on cells from a transgenic mouse carrying the *Gfap-TK* gene (see Chapter 10 for TK selection). Evidence for the role of blood vessels in maintaining the niche is that endothelial cells can maintain self-renewal of neural stem cells *in vitro*, in a co-culture or a **transfilter** culture situation.

The potency of individual stem cells has been examined by two methods. One is *in vivo* **clonal analysis** by transduction with a replication-incompetent retrovirus expressing **β-galactosidase** or **GFP**. If this is done at low multiplicity, the infected cells are well separated and each positive cell group can be considered as a single clone. This method confirms the predominant radial migration of the progeny and shows that most clones consist of either neurons or glia, indicating transduction of a transit amplifying cell. However, a minority of clones contain both neurons and glia, showing that the labeled cell was **multipotent**. The other method is *in vitro* culture of individual cells. In cultures from the lateral ventricular zone of the rat embryo cerebral cortex, the majority of clones form small numbers of neurons, a few form small numbers of glia, and about 7% form large clones containing neurons, astrocytes, and oligodendrocytes. In both types of experiment the large mixed clones are presumed to arise from the stem cells and the small single-type clones from transit amplifying cells.

In terms of regional specification of stem cells, it has been shown that different subclasses of olfactory neuron arise from stem cells in different regions of the subventricular zone. If the generative regions are grafted to new position in a newborn mouse, then the subclass specificity is retained, indicating a dependence on the initial regionalization of the forebrain.

An intriguing aspect of neurogenesis is that in both embryonic life and postnatally the frequency of retrotransposon mobilization is much higher than in somatic cells. The normal genome contains large numbers of retrovirus-like elements. Of these a small number are active retrotransposons. They are normally kept inactive by DNA methylation but occasionally they may become transcribed into RNA and then, using a reverse transcriptase encoded in their own sequence, to a DNA molecule that can insert elsewhere in the genome. These genetic insertion event may cause a somatic mutation or may modulate differentiation by affecting activity of neighboring genes. Although the case is not proved, it is conceivable that this process is the cause of those aspects of personality and mental ability that are not inherited but also not modifiable by experience; in other words, those individual differences that exist within a litter of inbred animals or between identical human twins.

Neurospheres

Despite the restriction of true neural stem cells to the lateral ventricle linings and the hippocampus of the mammalian CNS, it is possible to grow cells showing a stem-like property in tissue culture from a much wider range of brain regions than this. The stem-like cells grow in a form called **neurospheres** (Fig. 14.15). These are clumps of cells, up to 0.3 mm wide, that grow in suspension culture in a medium containing epidermal growth factor (EGF) and FGF. The cultures can be initiated from any part of the fetal CNS and often from parts of the adult CNS as well, even from regions that do not undergo continuous renewal. It has been shown that mouse spinal cord in which the ependymal cells are Cre-labeled will give rise to labeled neurospheres, indicating an ependymal origin.

Neurospheres are thought each to contain a few neural stem cells, which are capable of self-renewal, plus a certain number of transit amplifying cells, that have finite division potential. When neurospheres are plated on a laminin substrate in the presence of serum, they will differentiate into neurons, astrocytes, and oligodendrocytes. If neurospheres are dissociated into single cells, a few percent of these cells can establish new neurospheres, with similar properties to the original. Repeated cycles of dissociation and growth can provide substantial expansion capacity.

The phenomenon of neurospheres is an example of the fact that cells may behave in a different manner in tissue culture from *in vivo*. Neurospheres have created enormous interest because, unlike hematopoietic stem cells, they are readily expandable *in vitro*, and because there is a hope that they might one day be used for **cell therapy** of the very intractable human neurodegenerative diseases that involve widespread neuronal death.

Gliogenesis

Glial cell formation tends to occur after neurogenesis. Some oligodendrocyte progenitors arise from radial glia, and radial glia themselves turn into astrocytes once neurogenesis is finished, and this switch depends on cardiotrophin secreted by the newly formed neurons. Glial cells often migrate considerable distances before assuming their final position.

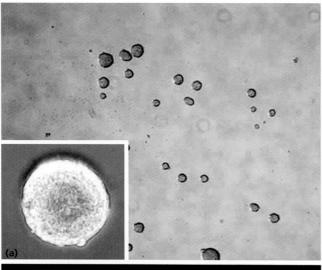

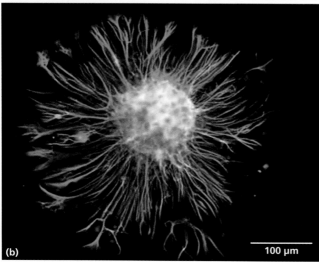

Fig. 14.15 Neurospheres. (a) Neurospheres in suspension; inset shows single sphere. (b) Neurosphere differentiating on an adhesive substrate with serum. Immunostaining shows: GFAP, green;β-tubulin, red; DNA, blue. Reproduced from Marshall *et al. Curr Pharm Biotechnol* 2007; 8: 141–145, with permission from Bentham Science Publishers.

In the spinal cord, astrocytes originate from the p1, p2, and p3 regions, while oligodendrocytes originate from the dorsal and the pMN regions (Fig. 14.9c). The more dorsally originating astrocytes produce the guidance molecule Reelin, which binds low-density lipoprotein receptors, and the more ventrally originating ones produce Slit, which binds receptors of the Robo class. These factors help to control subsequent migration and positioning.

The pMN region, like other spinal cord domains, is specified by a combination of transcription factors expressed in accordance with the Shh gradient from the notochord and floor plate. Initially, the transcription factors Ngn2 and Olig1 and -2 are expressed in this region and it produces motor neurons during the phase of neurogenesis. In late gestation the Ngn2 becomes

downregulated and the Nkx2.2 domain expands dorsally to include this region, which then produces oligodendrocyte precursor cells.

Oligodendrocyte precursor cells have a bipolar morphology. Their normal fate, as shown by retroviral labeling *in vivo*, is to form oligodendrocytes. But in culture they can also form type 2 astrocytes on exposure to ciliary neurotrophic factor (CNTF). The NG2 glia (polydendrocytes) are long-term oligodendrocyte progenitors, which continue to express Olig1 and -2. Cre labeling using the promoter for the protein part of the NG2 molecule, CSPG4, indicates that in white matter NG2 cells can self-renew and produce oligodendrocytes, while in grey matter they can also produce type 2 astrocytes.

The neural crest

The neural crest is a population of cells derived from the neural folds of all vertebrates (Fig. 14.16). They are present in the dorsal part of the neural tube at the time of closure but soon migrate away from the neural tube into the surrounding tissues. They form a variety of different cell types of which the following are the most significant:

- neurons and glia of the sensory and autonomic systems;
- adrenal medulla and calcitonin cells of thyroid;
- pigment cells (excluding those of the pigmented retina);
- skeletal tissues of the head;
- part of the cardiac outflow tract.

The normal fate of the crest has been established by two major techniques. One is the grafting of segments of neural tube from quail to chick embryos, followed by localization of the quail cells at a later stage. The second method is *in vivo* labeling of neural crest by localized injection of dyes. Extracellular injection of **DiI** can be used to label a patch of cells, or intracellular injection of a **fluorescent dextran** can be used to label an individual cell. Following application of the label, the embryo is allowed to develop for a while and then the position and cell types of the labeled cells are identified at a later stage.

Experiments of these types show that the crest is divided into distinct anteroposterior domains as far as normal fate is concerned (Fig. 14.17). The cephalic crest gives rise to a remarkable range of tissues. Initially, it is the source of most of the head mesenchyme. As well as general connective tissue, this later becomes cartilage, bone, cranial ganglia, the pericytes and smooth muscle of blood vessels, and the odontoblasts (dentine-forming cells) of the teeth. The trunk crest shows two distinct pathways of migration (Fig. 14.18). The dorsolateral pathway passes between the epidermis and the somites and give rises to melanocytes. The ventral pathway passes through the sclerotome to become the dorsal root ganglia, the sympathetic ganglia, the adrenal medulla, and also makes a contribution to the outflow tract of the heart (cardiac crest). This migration occurs only through the anterior half of each sclerotome segment, and each dorsal root ganglion includes a contribution

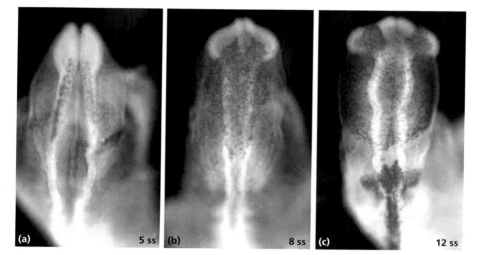

Fig. 14.16 Migration of cranial neural crest from the neural tube of the chick embryo. Cells are visualized by *in situ* hybridization for *Sox10*: (a) 5-somite embryo; (b) 8-somite embryo; (c) 12-somite embryo. Reproduced from Crane and Trainor. *Ann Rev Cell Devel Biol* 2006; 22: 267–286, with permission from Annual Reviews.

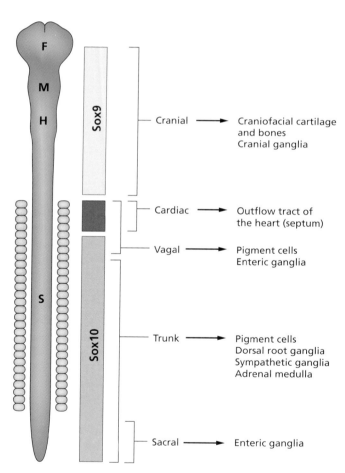

Fig. 14.17 Fate map of neural crest compiled from chick–quail orthotopic grafts. F, forebrain; M, midbrain; H, hindbrain; S, spinal cord.

of cells from the crest adjacent to the posterior sclerotome segments. Finally, there are two regions of crest known as the vagal and sacral, which contribute cells to the network of enteric ganglia.

Formation and migration of neural crest

If individual cells within the neural folds are labeled, it is found that their clonal descendants can include epidermal, neural, or neural crest cells, or any combination of the three. This suggests that there is no specific neural crest rudiment at this stage, but that there is an environment in this region that promotes the formation of neural crest cells. When the neural folds are removed and the neuroepithelium joined directly to the epidermis, neural crest can be generated from cells on both sides of the junction, indicating that an interaction between epidermis and neuroepithelium can lead to its formation. BMP4 from the epidermis is an important ingredient in this signal, but may not be sufficient on its own to induce neural crest formation. Experiments in *Xenopus* and chick suggest that FGFs, Wnts, Notch, and retinoic acid are all likely required as well.

It seems that the molecular pathway for neural crest determination has some properties of a linear sequence, but also with some feedback effects. The initial response to BMP signaling is the expression of the Msx1 and 2 homeodomain transcriptional repressors. Following this are two zinc-finger transcriptional repressors called Snail and Slug, and further transcription factors including Sox9 and 10. Evidence for the pathway is derived from overexpression, inhibition, and **epistasis** experiments. The *Xenopus* results are described here but evidence from mouse knockouts and chick retroviral infection experiments are broadly consistent. In *Xenopus* it is found that overexpression of the Msx factors will induce expression of *Snail*,

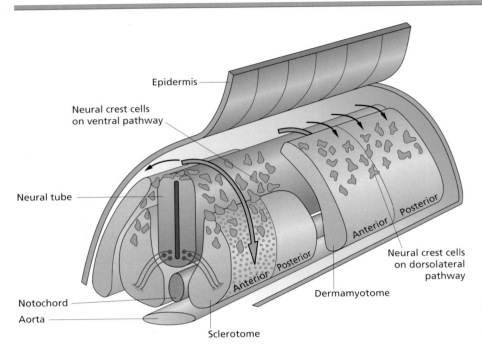

Epidermis

Neural crest cells
on ventral pathway

Neural tube

Neural crest cells
on dorsolateral
pathway

Notochord

Aorta

Dermamyotome

Sclerotome

Anterior Posterior

Anterior Posterior

Fig. 14.18 Dorsolateral (black arrows)
and ventral (red arrows) pathways of trunk
neural crest migration.

and overexpression of Snail will induce expression of *Sox10* and overexpression of Sox10 will induce expression of *Slug*. All these factors can drive neural crest formation and injection of the appropriate antisense **morpholinos** or dominant negative **domain-swap** constructs of any of them will inhibit neural crest formation. The sequence based on this overexpression data is therefore Msx→Snail→Sox10→Slug. However, there are also many results inconsistent with a simple linear pathway and it is likely that there is considerable parallel function and feedback function of these transcription factors.

A glucocorticoid-inducible version of the dominant negative Slug will block crest formation if induced early, and if induced later, after the crest has formed, it will block the cell migration. This shows that Slug function is required for an extended time to control the properties of the crest cells. Sox10 is initially expressed in the whole crest, but is soon replaced by Sox9 in the cranial crest (Fig. 14.17). Sox9 is known to be important for cartilage differentiation from the mesoderm and the importance of Sox9 for cartilage formation from the neural crest is shown by a tissue-specific knockout in the mouse. In embryos of the constitution *Wnt1-cre x floxed Sox9* the **Cre** recombinase is expressed in the neural crest but not in the mesoderm. The resulting ablation of *Sox9* in the neural crest inhibits formation of the cranial skeleton without affecting the formation of the trunk skeleton. Snail inhibits the production of the adhesion molecule E-cadherin, and of components of tight junctions, hence facilitating the epithelial–mesenchymal transition prior to migration.

The migration of crest cells depends on a wide variety of extracellular matrix components and the production of pro-

teases including matrix metalloproteases and ADAMs enables penetration by the cells. There is also a positive role played by extracellular matrix components such as fibronectin or laminin as shown by the fact that migration of crest cells can be blocked by neutralizing antibodies to these proteins. *In vivo* the migration of the truck crest of the ventral pathway is specifically inhibited by the posterior sclerotome, leading to the preferred migration route through anterior sclerotome. The reason is that the crest cells express EphB3 and the posterior sclerotome expresses ephrin B1 and -2. These make up a repulsive combination such that when the crest cell processes touch the ephrin B-containing surface they collapse, due to actin depolymerization mediated by the RhoGTPase system (see Appendix). Addition of an ephrin inhibitor to chick embryos enables the crest cells to migrate through the posterior as well as anterior sclerotome. In rhombomeres 3 and 5 of the hindbrain the outward migration of the cranial crest is also inhibited by the Eph system.

Commitment and differentiation of neural crest

The overall anteroposterior commitment of the neural crest has been investigated by interchanging segments of neural tube and examining the fate of the crest cells that emerge from the graft. At least in higher vertebrates, this type of experiment has shown that the distinction between cranial and trunk crest arises early, as only cranial crest can generate skeletal tissues. Trunk crest will not generate cartilage even when grafted to the neural tube of the head, whereas cranial crest will contribute to skeletal structures of the body after grafting to the trunk, in addition to forming the various neuronal cell types of the trunk. However, trunk crest cells can form cartilage if they are actually grafted into the facial primordia themselves, rather than into the hindbrain, so the distinction may be more to do with migration capacity than the actual competence to form cartilage. In general, any region of the crest will form parasympathetic, sympathetic, or sensory ganglia if grafted to the appropriate position along the axis. For example the vagal crest normally produces cholinergic (parasympathetic) neurons and the thoracic crest adrenergic (sympathetic/ adrenal medulla). But if these regions are interchanged the cell types formed are appropriate to the new position of the graft.

Labeling of individual crest cells *in vivo*, by microinjection or by retroviral infection, shows that those leaving the neural tube early may be **multipotent**, as some labeled clones include several cell types such as sensory neurons, pigment, adrenal medulla, and glial cells. This shows that at least some of the cells in the crest are multipotent. Those leaving the neural tube late are more likely to populate just a single terminal cell type, for example the cells following the dorsolateral pathway that become melanocytes. However, the population of just one cell type in normal development does not prove that the injected cell was not capable of forming other cell types (see clonal analysis in Chapter 4). To prove this it is also necessary to show that the clonal descendents are exposed at an early enough stage to a wide range of different environments.

The final differentiation of the cells certainly does in some instances depend on the environment through which they migrate (Fig. 14.19). For example the formation of dorsal root ganglia depends on exposure to brain-derived neurotrophic factor (BDNF) from the neural tube. If a barrier is inserted between the neural tube and the crest cells, then they fail to form the dorsal root ganglia; however, they will do so if supplied with BDNF protein. Mice lacking the *Bdnf* gene do not form dorsal root ganglia. The actual segmental arrangement of dorsal route

ganglia is not just a passive consequence of exclusion from the posterior sclerotomes as ganglia can still form in mutants where the repulsive influence is removed. However, it does depend on the Semaphorin guidance system, as removal of the genes for both Neuropilin 1 and -2 will prevent segmentation.

The decision whether to form autonomic neurons, glia, or smooth muscle also depends on environmental signals. Cultured trunk crest cells will preferentially form **Schwann cells** when exposed to Neuregulin, smooth muscle when exposed to TGFβ, and autonomic neurons when exposed to BMP2 or -4. These effects are primarily **instructive** although the factors may also contribute to differential survival of their own cell type. These factors are expressed in the appropriate regions of the embryo through which the cells migrate, neuregulin in the nerve sheaths, TGF-β in the major outflow vessels of the heart, and BMP2/4 in the dorsal aorta and gut. Mouse knockouts have the predicted effects, for example a knockout of the neuregulin gene (*Nrg*) reduces the number of Schwann cells, and knockouts of the *Tgfβ* genes interfere with heart development. The autonomic neurons express the transcription factor gene *Mash1*, encoding a bHLH transcription factor which is a homolog of the *Drosophila* Achaete–Scute family and the knockout of *Mash1* lacks most autonomic neurons. Commitment to form the pigment cells and enteric ganglia depends on the inducing factor Endothelin 3. Mice lacking this factor or its receptor are deficient in both cell types. The requirement for activity is in the time interval 9.5–12.5 days of gestation, shown by making transgenic mice for an inducible form of the receptor and inducing its activity at different times.

In addition to these data showing control of differentiation of neural crest cells by environmental signals, there are also various experiments in which crest cells do show differentiation into just a single cell type even when they are exposed to a wide range of environmental conditions, suggesting that there is also an element of autonomous commitment to particular fates that occurs progressively with time.

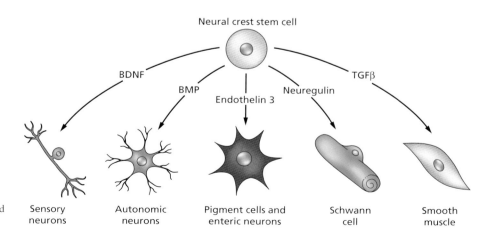

Fig. 14.19 Preferred differentiation pathways of trunk neural crest cells exposed to different inducing factors.

Development of neuronal connectivity

The growth cone

The axons of neurons grow out and innervate their targets. Motor neurons in the ventral spinal cord innervate muscles or glands, and sensory neurons in the dorsal root ganglia innervate a variety of peripheral receptors and sense organs. The final pattern of connections is one of exquisite complexity, so how do they all know where to go? There is, in fact, a whole hierarchy of different mechanisms controlling the specificity of connections. Each of them is of quite low specificity but together they yield a very precise final result.

Each developing axon has at its tip a **growth cone** (Fig. 14.20). This is a structure somewhat resembling a migrating fibroblast. It is constantly extending and retracting **filopodia** and it crawls actively across the substrate. The cytoskeleton of the growth cone is in a dynamic steady state. Its **lamellipodia** and filopodia are built around microfilaments that are constantly elongating at the plus (exterior) ends. At the same time the filaments are being drawn back into a central depolarized domain by myosin-type **motor proteins**. In this region terminate the microtubule bundles that maintain the structure of the axon. Stimuli that cause growth cone advance will increase the rate of actin polymerization, decrease the activity of the myosin motors, and increase the elongation of the microtubules. As the growth cone advances new material becomes laid down to elongate the axon. The cell body and proximal axon do not move, but they are necessary to the process as they carry out much of the synthesis of the new materials required for elongation.

Because of the relatively large size of the growth cone, it is possible for different parts of it to be exposed to different signals, and a unilateral stimulus will bring about a change in the direction of growth. The effect is that local depolymerization of microfilaments causes the growth cone to turn away while local stabilization causes it to turn toward the stimulus. The coupling between receptors for signals and the cytoskeleton depends on the small GTP exchange proteins: RhoA causes growth cone collapse in its active form, but active Rac and Cdc42 promote elongation.

The extracellular factors known to be most potent at promoting axonal extension are the neurotrophins. These comprise nerve growth factor (NGF), brain-derived neurotrophic factor (BDNF), neurotrophin 3 (NT3), and neurotrophin 4/5 (NT4/5). The receptors are called Trk (pronounced "track") proteins, which are receptor tyrosine kinases. NGF binds TrkA, BDNF binds TrkB, and NT3 and NT4/5 bind predominantly TrkC, providing some specificity to the system. The neurotrophins also all bind to a lower-affinity receptor called p75. This has a tonic function of activating RhoA, and in the presence of the neurotrophin the activation is stopped, thus enabling growth cone elongation. In addition, extracellular matrix proteins such as fibronectin and laminin often provide a permissive environment for elongation. The neurotrophins can maintain survival if applied only to the cell body, but for elongation they must be applied to the growth cone. It is thought that endocytotic vesicles containing activated receptors are transmitted to the cell body by retrograde transport in order to bring about the changes in gene expression.

Guidance molecules

The guidance of growth cones is facilitated by some factors and antagonized by others. Four general categories of effect can be recognized: contact attraction, contact repulsion, long-range attraction, and long-range repulsion. These were formerly each associated with a particular class of guidance molecules but it is now recognized that the same molecule may be attractive or repulsive depending on the context. The identification of guidance molecules was achieved mainly by the use of *in vitro* assays

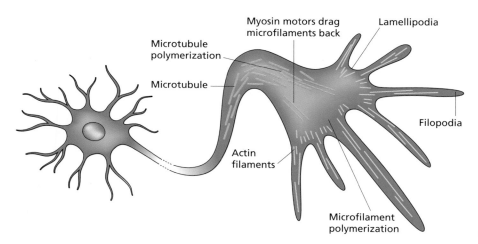

Myosin motors drag microfilaments back

Lamellipodia

Microtubule polymerization

Microtubule

Actin filaments

Filopodia

Microfilament polymerization

Fig. 14.20 A growth cone at the tip of a developing axon.

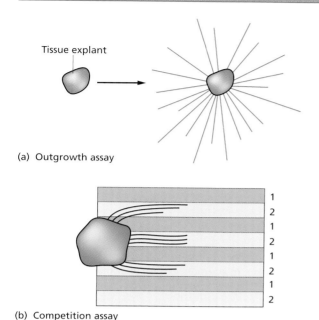

(a) Outgrowth assay

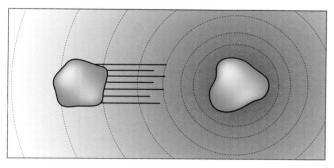

(b) Competition assay

(c) Chemoattraction assay

Fig. 14.21 Assays for attraction and repulsion. (a) Simple assay for outgrowth of fibers by plating a tissue explant on a coated surface, or by treating it with factors. (b) Competition assay. Two substrates, 1 and 2, are coated onto the thin alternating strips so that the axons growing out of the explant can choose between them. (c) Chemoattraction assay. The fibers from one explant grow towards the second explant, which is emitting a chemoattractant.

(Fig. 14.21). The contact-active factors can simply be coated onto a culture dish to see whether they will support outgrowth of axons from an explant of neural tube. However, a more discriminating assay is provided by offering a choice between alternative substrates (Fig. 14.21b), since growth cones in vivo will be constantly choosing between alternative environments. For example ephrins with growth cone-repelling activity were isolated from the optic tectum using this sort of assay. For long-range factors it is necessary to show that they are diffusible and that growth cones can sense a concentration gradient. This is done by setting up two explants in a collagen gel (Fig. 14.21c). If axons grow directly from one to the other this suggests that there must be a long-range chemoattractant secreted by the

target explant. To obtain a conclusive proof of chemoattraction also requires that the source be moved and produce a change of direction of the growth of the axons.

The contact attractive factors include extracellular matrix components such as laminin and fibronectin, and also adhesion molecules on other neurons such as N-CAM, NgCAM, or N-cadherin. The contact repulsive factors are ephrins, which bind receptors of the Eph group, and some extracellular matrix components such as tenascin. Long-range guidance molecules are netrins, which are diffusible extracellular factors related to laminin, semaphorins, and Slits. In general, netrins are attractive while semaphorins and Slits are repellent to axon growth, but this does depend on the competence of the individual cell population. Receptors for the netrins are called DCC (deleted in colon carcinoma), receptors for semaphorins are called neuropilins and plexins, and receptors for Slits are called Robos (encoded by homologs of the *Drosophila roundabout* gene). Although some semaphorins are secreted and act as long-range factors, others are transmembrane molecules and act at short range through cell contacts.

In addition to the specialist neural guidance molecules, effects on growth cone growth are also exerted by inducing factors familiar from many other developmental processes, such as FGFs, BMPs, Hhs, and Wnts.

Neuronal pathways

Studies of the initial pathways have been made mainly in the limb regions where the spinal nerves fuse into the **brachial** or **lumbosacral plexus** and then branch again into a few principal nerves supplying the limb. The spinal nerves innervating the leg first grow through the dorsoanterior sclerotome surrounding the neural tube. As is the case for neural crest cells, the ventral and the posterior sclerotome carry a repulsive factor and prevents growth. The mesenchyme of the plexus region forms a permissive environment for growth, but the nerves are obliged to pass through two holes in the forming pelvic girdle. If more holes are created, then the nerves will grow through. If a segment of neural tube is rotated or displaced by a small amount, the nerves will still find their correct normal destinations. But if it is displaced by a large amount then the nerves from a region that does not normally innervate the limb can grow into the limb and form an approximately normal-looking pattern. These experiments show that the crudest level of control must be the presence of pathways through the tissues that can be followed by any axon. These pathways may simply be spaces, or regions rich in suitable substrates such as laminin, or regions free from repulsive factors.

The most discriminating studies on this scale of connectivity have been made in insects, some in *Drosophila* but also particularly in the grasshopper. However, it is essential to remember an important difference between insect and vertebrate neural development. In vertebrates most neurons arise in the CNS and

Classic experiments

Specificity of nerve connections

Kennedy, T.E., Serafini, T., Delatorre, J.R., & Tessier-Lavigne, M. (1994) Netrins are diffusible chemotropic factors for commissural axons in the embryonic spinal-cord *Cell* **78**, 425–435.
This paper marked the first discovery of a guidance molecule that controlled the direction of growth zone growth.
Drescher, U., Kremoser, C., Handwerker, C., Loschinger, J., Noda, M. & Bonhoeffer, F. (1995) In-vitro guidance of retinal ganglion-cell axons by RAGS, a 25 kda tectal protein related to ligands for Eph receptor tyrosine kinases. *Cell* **82**, 359–370.
Purification of a molecule controlling retino-tectal specificity, later known as ephrin 5.
Cheng, H.J., Nakamoto, M., Bergemann, A.D. & Flanagan, J.G. (1995) Complementary gradients in expression and binding of ELF-1 and MEK4 in development of the topographic retinotectal projection map. *Cell* **82**, 371–381.
Visualization of ephrin and Eph gradients in the retinotectal system.

send axons out to peripheral organs, while in insects many neurons arise in peripheral organs and send axons into the CNS. Studies on the development of insect pathways have shown that the complete pathway is divided into short segments of about 100 μm. Over this length, growth cones will grow along the pathway. When they get to a choice point, which may be a pre-existing neuron, or a change in the character of the substrate, they will stop, and extend processes in all directions to explore the surroundings and locate the most favorable pathway for the next segment of growth. These initial connections are made by "pioneer neurons" when the whole embryo is quite small, so the overall length of a pathway is also small. Later pathways mainly depend on growth along pre-existing axonal tracts, often called **fascicles**. These provide a "highway" along which subsequent axons will grow. Whether axons remain bundled into a fascicle or not depends on a local balance of attraction and repulsion. If axons attract each other more than the substrate then the fascicle will persist; if they attract the substrate more than each other then the fascicle will break up. For example defasciculation in vertebrates can arise because of addition of polysialic acid to N-CAM on the axonal surface. This introduces a high surface negative charge and so makes the axons repel each other.

In the CNS there is a population of **commissural neurons** in the dorsal part of the spinal cord. Their axons grow down to the ventral midline, then most of them cross to the other side, and either contact a target neuron immediately, or turn and grow longitudinally along the midline to make eventual synapse at another anteroposterior level (Fig. 14.22). The ventral direction of growth depends on the secretion of netrins and Shh from the **floor plate**. Evidence for this is that a floor plate explant will attract the growth of commissural axons *in vitro*, as will cells transfected with *Netrin1*. Furthermore, the mouse knockout of *Netrin1*, or its receptor *DCC*, shows a deranged pattern of commissural axon growth. When the axons have reached the midline, those that cross do so because they are expressing an NgCAM capable of binding a similar molecule on the floor plate cells. Addition of a neutralizing antibody to NgCAM will cause most of the axons to grow longitudinally without crossing. In addition, there is an active repulsive mechanism within the floor plate region consisting of the presence of Slit, secreted by midline glia. The receptor for Slit, Robo, undergoes a splice

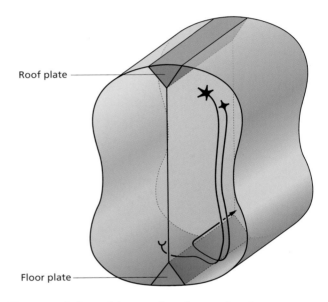

Roof plate

Floor plate

Fig. 14.22 Pathway of the axons from the commissural neurons.

variant change to a repulsive form once the growth cones have entered the midline region. So Slit causes repulsion of the growth cones from the floor plate and it also inactivates the guidance (but not the growth promoting) function of DCC. This ensures that once the axons have exited the floor plate the netrin signal continues to promote axonal elongation but does not attract the axons back towards the midline.

Neurotrophins and cell survival

During the postnatal growth of the CNS there is a substantial level of neuronal death by apoptosis. Although substantial neuron loss occurs in both central and peripheral nervous systems, the mechanisms are better understood in the peripheral system, perhaps because the level of redundancy of gene action is lower.

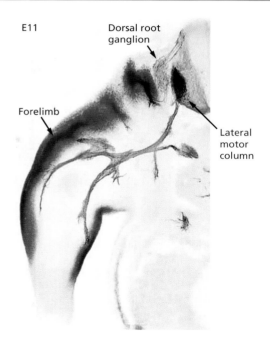

E11

Dorsal root
ganglion

Forelimb

Lateral
motor
column

Fig. 14.23 Expression of neurotrophin 3 (NT3). This shows a reporter mouse with one allele of *NT3* replaced by *E. coli lacZ*. Transverse section through thoracic region of E11 embryo showing areas of NT3 expression in turquoise and spinal nerves in brown. Reproduced from Fariñas *et al. Neuron* 1996; 17: 1065–1078, with permission from Elsevier.

Several growth factors can function as survival factors for neurons in culture. They include the NGF group of neurotrophins (NGF, BDNF, NT3, NT4/5), some FGFs, hepatocyte growth factor, and some other factors, including glial-derived neurotrophic factor (GDNF) and ciliary neurotrophic factor (CNTF). The survival of neurons depends on absorbing enough of the relevant neurotrophin from the target organ (Fig. 14.23). If the target organ is removed, then most of the neurons that would normally form connections to it die off. For example initially all of the dorsal root ganglia are the same size. But there is more cell death in those that do not innervate the limbs, and so they become smaller. If a limb bud is removed, then its ganglia will suffer more cell death and will shrink. If an extra limb bud is grafted in, then more cells will survive in the ganglia that supply it, and they will remain larger (Fig. 14.24).

NGF and the other neurotrophins have two distinct actions on sensory and sympathetic neurons in culture. They will promote the outgrowth of axons and also support the survival of the cell bodies. If an experiment is set up so that only the growth cones and the apical part of the axons are exposed to the factor, the cell bodies still survive, showing that the factor can be absorbed at the growth cone and its effects are transmitted down to the cell body by retrograde transport. The protective effect on dorsal root ganglia of grafting an extra limb bud into their vicinity can be mimicked by an injection of NGF into the tissue of the flank.

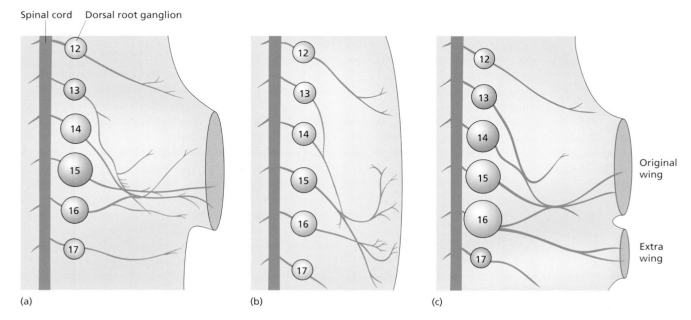

Spinal cord   Dorsal root ganglion

Original
wing

Extra
wing

(a)   (b)   (c)

Fig. 14.24 Effects of neurotrophins on the dorsal root ganglia. (a) Normal brachial plexus in a chick embryo; (b) reduced ganglion size after removal of the wing bud; (c) increased ganglion size after grafting of extra wing bud. Reproduced from Alberts *et al.* (1989), *Molecular Biology of the Cell*, 2nd edn, with permission from Garland Science/Taylor & Francis Group.

The mouse knockouts of genes encoding the neurotrophins, or their receptors, show an excessive loss of sensory and sympathetic neurons. For example knockout of TrkA, which preferentially binds NGF, leads to substantial shrinkage of the dorsal root and trigeminal ganglia. Knockout of TrkB, which preferentially binds BDNF, leads to shrinkage of nodose, vestibular, cochlear, trigeminal, and dorsal root ganglia.

Axonal competition

Once the initial synaptic connections have been made and the central neurons have been thinned out by cell death, there is then a reshuffling of connections. For innervation of muscle this means that an initial situation of several neurons projecting to one multinucleate myofiber becomes reorganized so that each myofiber is innervated by only one neuron, but with a greater number of synapses (the specialized synapses on muscle are called neuromuscular junctions). Unlike the initial, rather crude and diffuse, pattern of connections, this later phase occurs postnatally and depends on electrical activity. It is governed by a rule which states that each neuromuscular junction becomes strengthened by muscle excitation if it has recently been active itself, and is weakened by muscle excitation if it has not been recently active. Therefore an impulse from one motor neuron will provide mutual reinforcement for all of its own neuromuscular junctions and will weaken all those from other neurons. This process will automatically proceed until each myofiber has eliminated the neuromuscular junctions from all but one neuron.

A similar principle operates within the CNS, where the initial arbors of connections are very wide and become narrowed and defined according to activity.

Neuronal connections in the visual system

The visual system represents one of the most complex examples of neuronal specificity and the elucidation of the mechanism in recent years is a very elegant chapter of developmental neuroscience. As described above, the retina is formed from the inner surface of the optic cup. Because of the camera-like optics of the vertebrate eye, each point on the retina receives light from a particular point in the visual field. The retinal neurons send axons down the optic tube, which becomes the optic nerve (the second cranial nerve). The fibers grow back into the midbrain when they form synapses on the **optic tecta** (or, in mammals, the lateral geniculate nuclei). Each point on the retina sends fibers to a particular point on the tectum, so that the surface of the tectum has a one-to-one, or **topographic**, relationship to the retina and therefore also to the external visual field. This projection can be visualized by illuminating a point in the visual field and recording electrical activity from the corresponding point on the tectum. In lower vertebrates and birds there is a complete crossing of optic fibers at the optic chiasma such that the right retina projects to the left tectum and vice versa, while in mammals there is a projection from both retinas to both sides of the brain. The nature of the projection is such that the anterior retina projects to the posterior tectum and the dorsal retina projects to the ventral tectum (Fig. 14.25). Note that in the neuroscience literature, the anteroposterior axis of the retina is called the nasal – temporal axis, but the usage here is harmonized with the rest of developmental biology.

It is clear that the specificity of connections requires some system of cell labeling in both retina and tectum, and some means for matching the two sets of labels. This principle is known as the "chemoaffinity hypothesis" of Sperry. The pattern in the eye is set up at an early embryonic stage. If a *Xenopus* eye is rotated before the tailbud stage, then the projection to the tectum will still grow normally. If it is rotated later, the projection will form but remains inverted. In lower vertebrates the optic nerve will regenerate if cut and it is also possible to examine the topography of the regenerated map. In normal regeneration the connections re-establish the correct projection. If half of the eye is removed at the time that the nerve is cut, then the other half will project across the whole tectum, with normal orientation. Likewise, if half the tectum is removed when the nerve is cut, the whole eye will project to the remaining half with normal orientation. Numerous experiments of this sort showed that the labeling systems are quantitative but with

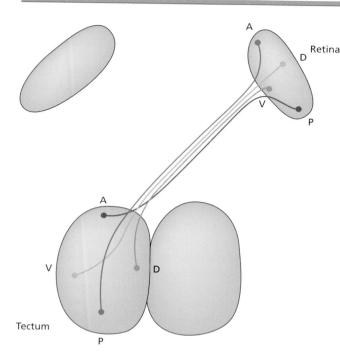

Fig. 14.25 Normal retinotectal projection in a lower vertebrate. Anterior (nasal) retina projects to posterior (caudal) tectum; dorsal retina projects to ventral (lateral) tectum.

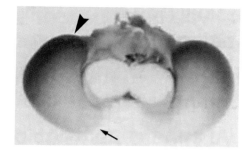

Fig. 14.26 Gradient of ephrin A visualized on the optic tectum of a 13-day chick embryo by the binding of an alkaline phosphatase conjugate of Eph A3. The anterior end of the tectum is indicated by an arrow and the posterior end by an arrowhead. Reproduced from Chen *et al.* (1995), with permission from Elsevier.

a preferred hierarchy of connectivity. They also showed that the projection must be dynamic and be continually rearranging itself. This is because the retina grows radially, adding new cells from a germinative zone at the margin. By contrast the tectum grows at its posterior–medial surface, adding cells at one end. These different modes of growth mean that the connections must be continually readjusting themselves in order to maintain the overall topographic projection.

The understanding of how the visual connections are set up has depended on work on a variety of vertebrate species. In fish and frogs, in which regeneration is possible, much work on tracing connections was done using anterograde transport of substances such as ³H-proline or HRP, and by electrophysiological mapping. More recently DiI has been preferred for tracing pathways because of superior visualization in wholemounts. In the chick embryo retroviral overexpression has been used for gain-of-function experiments; and in the mouse, knockouts have been the basis for loss-of-function experiments. There are slight differences between the different models but the principles are the same.

First, the retinal fibers have to get to the tectum. Initially, they are attracted into the optic nerve by the presence of Slit. At the optic chiasma, expression of the transcription factor Zic1 maintains an ipsilateral path and its suppression by Isl1 causes a contralateral path to be adopted.

The basis for the topographic projection is now known to be encoded by gradients of cell-surface molecules across both the

retina and the tectum (Fig. 14.26). Disaggregated cells from a small region of the retina will adhere to tectal explants with the same specificity as the normal *in vivo* neuronal projection (i.e. anterior retina to posterior tectum and dorsal retina to ventral tectum). Using the stripe assay (see Fig. 14.21b), it was shown that an extract of posterior tectum could inhibit the growth of posterior retinal axons. Anterior tectum had no repulsive activity and anterior retina was not responsive to the tectal factor. The factors were purified and found to be members of the ephrin A family of adhesion molecules, with a gradient of expression from high posterior to low anterior. There is a corresponding gradient of Eph A-type receptor on the retinal cells, again with high expression posterior and low anterior. So the molecular basis of the system is a short-range repulsion due to the Eph–ephrin system which causes posterior fibers preferentially to adhere to anterior tectum. Although the anterior fibers could also adhere to anterior tectum, the competition between fibers displaces them to the posterior tectum, resulting in the observed specificity. Evidence for this mechanism depends on both gain and loss-of-function experiments. For example in the chick if ephrin A2 is misexpressed in the tectum and patches in the contralateral retinal are labeled with diI, it may be seen that the posterior retinal fibers now avoid the ectopic patches of ephrin A2. In the mouse, where the principal ephrins expressed on the tectum are ephrin 2 and 5, a double knockout of the corresponding two genes provokes considerable disruption to the map.

The expression of the ephrins is controlled by the transcription factor Engrailed, which itself forms a posterior–anterior gradient across the tectum, initially established by the action of the isthmic organizer on the midbrain (see above). If *En* is misexpressed in chick embryos using a retrovirus, then the regions of ectopic expression on the tecta show an increased expression of ephrins and an exclusion of posterior retinal fibers.

In the dorsoventral axis a similar system operates but based on attraction rather than repulsion. EphB molecules are expressed on the retina in a ventral high to dorsal low gradient

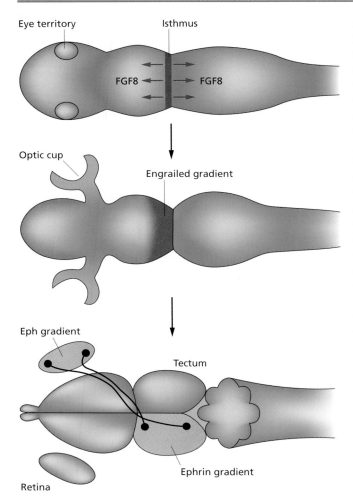

Fig. 14.27 Stages in the acquisition of retinotectal specificity. The axonal projection pattern is controlled by ephrin–Eph interactions, and the initial polarity of the tectal ephrin gradient depends on the gradient of Engrailed set up earlier in response to the signal from the isthmic organizer.

and ephrin B in the tectum is dorsal high to ventral low (usually called medial to lateral in neuroscience literature). Since the system is attractive this means that ventral retinal axons go to the dorsal tectum.

Later refinement

The initial retinotectal projection can be formed without any neuronal activity. This may be shown by treating embryos with tetrodotoxin, which blocks voltage-gated Na channels, to suppress neuronal activity. However, the initial projection is quite crude and it becomes refined later on by processes that do require neuronal activity.

A similar principle underlies the projection from the primary visual centers to the cerebral cortex. In mammals where each retina projects to both sides of the brain, different layers of cells in the lateral geniculate nuclei (the equivalent of the optic tecta)

receive inputs from the two retinas, but both project to the same layer of cells in the visual cortex. At birth the receptive fields for the two eyes in the visual cortex are mixed. But with exposure to visual stimuli, they sort out into stripes of about 0.5 mm wide, called ocular dominance columns. These columns can be visualized by injecting a tracer, such as radioactive proline, into one eye. It is taken up by the retinal cells and transported all the way to the cortex by anterograde transport, first by the retinal cells and then by midbrain cells. In the cortex the stripes can be visualized by autoradiography. Neighboring points on adjacent stripes correspond to the same point in visual space, perceived through the two eyes. These columns do not form if neuronal activity is blocked with tetrodotoxin. Nor do they form if the young animal is exposed to strobe light, which gives a uniform, simultaneous stimulus through both eyes. If one eye is blocked, the input from the neighboring eye will expand to fill the cortical space. The explanation seems to be the same type of principle that refines the neuromuscular connections. If two midbrain cells are connected to the same cortical cell, then the connection will be strengthened when the inputs are simultaneous, but weakened when they are not simultaneous. If the midbrain cells respond to adjacent points on the retina of one eye they are likely to fire simultaneously most of the time. But if they are responding to the same visual point viewed through different eyes the input will never be quite simultaneous, and the competitive principle will force the sorting into adjacent files of cortical cells. Although this is the conventional explanation, recent work has also shown that columns can form in the simple absence of input from one eye so there may be an element of programmed development as well.

From simplicity to complexity

The development of the visual system is now quite well understood from the initial neural induction to the final physiological function. It demonstrates at some stage all the principles of neural development and neuronal specificity. The initial territories of neuroepithelium that become the retina and the tectum are specified by gradients of inducing factors upregulating specific combinations of transcription factors. The pathway for the optic nerve depends on anatomical spaces with various chemotactic and matrix-based stimuli to guide the retinal fibers. The ephrin gradients that determine the topographic mapping onto the tectum depend on a gradient of Engrailed, which in turn depends on the FGF8 emitted from the isthmic organizer (Fig. 14.27). The early stages of the map are constructed without the need for neuronal activity or visual experience, while the later refinement does require an input of visual experience. The visual system nicely illustrates the overall principle that there are numerous processes and interactions involved in neural development, each one of which is intrinsically quite simple. But in combination and in sequence they can combine to build up a structure of quite extraordinary complexity and sophistication.

Key points to remember

- Initial neural induction depends on inhibition of BMP activity by factors secreted from the organizer, and sometimes on additional inducing factors. Subsequent regional specification depends on a posterior to anterior gradient of FGF and Wnt, which upregulates the Hox genes at different anteroposterior levels. Dorsoventral patterning depends on a gradient of Sonic hedgehog from the floor plate and of BMPs from the roof plate.
- Local regional specification depends on signals from centers such as the anterior neural rim, the cortical hem, and the isthmic organizer. The signals include FGFs, Wnts, and BMPs. The segmental identity of rhombomeres in the hindbrain depends on a gradient of retinoic acid from the posterior mesoderm.
- Primary neurogenesis in both *Drosophila* and vertebrates depends on a symmetry breaking process of lateral inhibition whereby incipient neurons suppress neuronal differentiation of their neighbors by stimulation of the Notch pathway. Most neurogenesis in mammals arises from radial glia active in late embryogenesis. The mammalian CNS contains two regions of neural stem cells that persist throughout adult life. Neural stem cells can differentiate to form both neurons and glia, and can be grown in culture as neurospheres.

- A variety of tissues arise from the migratory cells of the neural crest. The cephalic crest produces skeletal tissues while the trunk crest does not. Commitment to particular cell types depends partly on the environment through which the cells migrate, but there also seems to be a degree of commitment of cells before migration.
- Growth of axons depends on the elongation of the growth cone. This occurs in response to neurotrophins and other extracellular factors. The direction of growth depends on permissive components of the environment and also on specific attractants and repellents. A succession of simple decisions can generate a very complex pattern of neuronal connectivity. Some systems of connection, such as that from the retina to the brain, maintain a topographic mapping determined by matching of adhesive gradients on the axons and the target brain regions.
- Initial nerve connections are quite crude. They are refined by differential neuronal survival, depending on the supply of neurotrophins from the target tissues; and by axonal competition for innervation of the same target cell. At late stages this may include the effects of sensory experience.

Further reading

General

Brown, M., Keynes, R. & Lumsden, A. (2001) *The Developing Brain.* New York: Oxford University Press.

Sanes, D.H., Reh, T.A. & Harris, W.A. (2005) *Development of the Nervous System,* 2nd edn. New York: Academic Press.

Guillemot, F. (2007) Spatial and temporal specification of neural fates by transcription factor codes. *Development* **134**, 3771–3780.

Hochman, S. (2007) Spinal cord. *Current Biology* **17**, R950–R955.

Kiecker, C. & Lumsden, A. (2009) Recent advances in neural development. *F1000 Biology Reports* **1**.

Bellen, H.J., Tong, C. & Tsuda, H. (2010) 100 years of Drosophila research and its impact on vertebrate neuroscience: a history lesson for the future. *Nature Reviews Neuroscience* **11**, 514–522.

Neural induction and anteroposterior patterning

Lumsden, A. & Krumlauf, R. (1996) Patterning the vertebrate neuraxis. *Science* **274**, 1109–1115.

Marshall, H., Morrison, A., Studer, M., Pöpperl, H. & Krumlauf, R. (1996) Retinoids and Hox genes. *FASEB Journal* **10**, 969–978.

Blumberg, B. (1997) An essential role for retinoid signalling in anteroposterior neural specification and in neuronal differentiation. *Seminars in Cell and Developmental Biology* **8**, 417–428.

Simeone, A. (2000) Positioning the isthmic organizer. Where otx2 and Gbx2 meet. *Trends in Genetics* **16**, 237–240.

Wilson, S.I. & Edlund, T. (2001) Neural induction: toward a unifying mechanism. *Nature Neuroscience* **4**, 1161–1168.

Dasen, J.S., Liu, J.-P. & Jessell, T.M. (2003) Motor neuron columnar fate imposed by sequential phases of Hox-c activity. *Nature* **425**, 926–933.

Briscoe, J. & Wilkinson, D.G. (2004) Establishing neuronal circuitry: Hox genes make the conection. *Genes and Development* **18**, 1643–1648.

Lumsden, A. (2004) Segmentation and compartition in the early avian hindbrain. *Mechanisms of Development* **121**, 1081–1088.

O'Leary, D.D.M., Chou, S.-J. & Sahara, S. (2007) Area patterning of the mammalian cortex. *Neuron* **56**, 252–269.

Dorsoventral patterning

Placzek, M. & Furley, A. (1996) Patterning cascades in the neural tube. *Current Biology* **6**, 526–529.

Jessell, T.M. (2000) Neuronal specification in the spinal cord: inductive signals and transcriptional codes. *Nature Reviews Genetics* **1**, 20–29.

Marquandt, T. & Pfaff, S.L. (2001) Cracking the transcriptional code for cell specification in the neural tube. *Cell* **106**, 651–654.

Strähle, U., Lam, C.S., Ertzere, R. & Rastegar, S. (2004) Vertebrate floor plate specification: variations on common themes. *Trends in Genetics* **20**, 155–162.

Dessaud, E., McMahon, A.P. & Briscoe, J. (2008) Pattern formation in the vertebrate neural tube: a sonic hedgehog morphogen-regulated transcriptional network. *Development* **135**, 2489–2503.

Roessler, E. & Muenke, M. (2010) The molecular genetics of holoprosencephaly. *American Journal of Medical Genetics Part C: Seminars in Medical Genetics* **154C**, 52–61.

Neurogenesis and gliogenesis

Sanes, J.R. (1989) Analysing cell lineage with a recombinant retrovirus. *Trends in Neuroscience* **12**, 21–28.

McKay, R.D.G. (1997) Stem cells in the central nervous system. *Science* **276**, 66–71.

Gross, C.G. (2000) Neurogenesis in the adult brain: death of a dogma. *Nature Reviews Neuroscience* **1**, 67–73.

Cayouette, M. & Raff, M. (2002) Asymmetric segregation of Numb: a mechanism for neural specification from Drosophila to mammals. *Nature Neuroscience* **5**, 1265–1269.

Henrique, D. & Schweisguth, F. (2003) Cell polarity: the ups and downs of the Par6/aPKC complex. *Current Opinion in Genetics and Development* **13**, 341–350.

Götz, M. & Huttner, W.B. (2005) The cell biology of neurogenesis. *Nature Reviews Molecular Cell Biology* **6**, 777–788.

Malatesta, P., Appolloni, I. & Calzolari, F. (2007) Radial glia and neural stem cells. *Cell and Tissue Research* **331**, 165–178.

Miller, F.D. & Gauthier-Fisher, A. (2009) Home at last: neural stem cell niches defined. *Cell Stem Cell* **4**, 507–510.

Nishiyama, A., Komitova, M., Suzuki, R. & Zhu, X. (2009) Polydendrocytes (NG2 cells): multifunctional cells with lineage plasticity. *Nature Reviews Neuroscience* **10**, 9–22.

Suh, H., Deng, W. & Gage, F.H. (2009) Signaling in adult neurogenesis. *Annual Review of Cell and Developmental Biology* **25**, 253–275.

Rowitch, D.H. & Kriegstein, A.R. (2010) Developmental genetics of vertebrate glial-cell specification. *Nature* **468**, 214–222.

Neural crest

Le Douarin, N.M. & Smith, J. (1988) Development of the peripheral nervous system from the neural crest. *Annual Reviews of Cell Biology* **4**, 375–404.

Anderson, D.J. (1997) Cellular and molecular biology of neural crest cell lineage determination. *Trends in Genetics* **13**, 276–280.

LaBonne, C. & Bronner-Fraser, M. (1999) Molecular mechanisms of neural crest formation. *Annual Reviews of Cell and Developmental Biology* **15**, 81–112.

Le Douarin, N.M. & Kalcheim, C. (1999) *The Neural Crest*, 2nd edn. Cambridge: Cambridge University Press.

Graham, A. & Smith, A. (2001) Patterning the pharyngeal arches. *Bioessays* **23**, 54–61.

Krull, C.E. (2001) Segmental organization of neural crest migration. *Mechanisms of Development* **105**, 37–45.

Le Douarin, N.M. (2004) The avian embryo as a model to study the development of the neural crest: a long and still ongoing story. *Mechanisms of Development* **121**, 1089–1102.

Sauka-Spengler, T. & Bronner-Fraser, M. (2008) A gene regulatory network orchestrates neural crest formation. *Nature Reviews Molecular Cell Biology* **9**, 557–568.

Gammill, L.S. & Roffers-Agarwal, J. (2010) Division of labor during trunk neural crest development. *Developmental Biology* **344**, 555–565.

Neuronal growth and connectivity

Colman, H. & Lichtman, J.W. (1993) Interactions between nerve and muscle: synapse elimination at the developing neuromuscular junction. *Developmental Biology* **156**, 1–10.

Tessier-Lavigne, M. & Goodman, C.S. (1996) The molecular biology of axon guidance. *Science* **274**, 1123–1133.

Bibel, M. & Barde, Y.A. (2000) Neurotrophins: key regulators of cell fate and cell shape in the vertebrate nervous system. *Genes and Development* **14**, 2919–2937.

Lisman, J.E. & McIntyre, C.C. (2001) Synaptic plasticity: a molecular memory switch. *Current Biology* **11**, R788–R791.

Wilkinson, D.G. (2001) Multiple roles of Eph receptors and ephrins in neural development. *Nature Reviews Neuroscience* **2**, 155–164.

Goldberg, J.L. (2003) How does an axon grow? *Genes and Development* **17**, 941–958.

Schneider, V.A. & Granato, M. (2003) Motor axon migration: a long way to go. *Developmental Biology* **263**, 1–11.

Lemke, G. & Reber, M. (2005) Retinotectal mapping: new insights from molecular genetics. *Annual Review of Cell and Developmental Biology* **21**, 551–580.

Rüdiger, S. (2006) The dual nature of neurotrophins. *BioEssays* **28**, 583–594.

Butler, S.J. & Tear, G. (2007) Getting axons onto the right path: the role of transcription factors in axon guidance. *Development* **134**, 439–448.

Tran, T.S., Kolodkin, A.L. & Bharadwaj, R. (2007) Semaphorin regulation of cellular morphology. *Annual Review of Cell and Developmental Biology* **23**, 263–292.

Ypsilanti, A.R., Zagar, Y. & Chédotal, A. (2010) Moving away from the midline: new developments for Slit and Robo. *Development* **137**, 1939–1952.

Gibson, D.A. & Ma, L. (2011) Developmental regulation of axon branching in the vertebrate nervous system. *Development* **138**, 183–195.

Sun, K.L.W., Correia, J.P. & Kennedy, T.E. (2011) Netrins: versatile extracellular cues with diverse functions. *Development* **138**, 2153–2169.

This chapter contains the following animations:

Animation 6   Lateral inhibition.
Animation 19   Neuroblast formation.

 For additional resources for this book visit www.essentialdevelopmentalbiology.com

Chapter 15

Development of mesodermal organs

In higher vertebrates the mesoderm becomes partitioned at an early stage into four zones from medial to lateral. The **noto-chord** occupies the midline; then the **paraxial mesoderm**, which will become the somites; then the **intermediate meso-derm**, which will form the gonads, kidneys, and adrenals; then the **lateral plate** mesoderm, which forms body wall, limb buds, blood vessels, and heart. In lower vertebrates the arrangement is the same, but runs from dorsal to ventral since in these organisms the mesoderm surrounds the yolk mass.

The lateral plate becomes subdivided by the coelom into an outer **somatic** mesoderm, later forming the limb buds, and the inner **splanchnic mesoderm**, forming the mesenteries and the heart. The skeleton originates from three regions: most of the skull is formed from the **neural crest**; the vertebrae are formed from the **somites**; and the bones and girdles of the limbs arise from the limb buds and associated lateral plate.

Invertebrate mesoderm does not show the same regional subdivision as vertebrate mesoderm, although some similar tissue types are formed including muscle, gonad, connective tissues, and blood cells.

In this chapter the mouse gene and protein nomenclature convention is followed in those sections dealing mainly with mice (kidney, gonad, germ cells), and the general convention applied elsewhere.

Somitogenesis and myogenesis

Normal development of the somites

Along with the rhombomeres of the hindbrain, the pattern of somites represents the clearest indication of a segmental arrangement of the body pattern in vertebrates. Somites arise in anteroposterior sequence from the mesoderm flanking the notochord, which is called the **paraxial mesoderm** or **segmental plate** (Fig. 15.1). This is distinguished from the intermediate mesoderm by expression of the forkhead transcription factors Foxc1 and c2. The double knockout of these genes in the mouse lacks somites, and overexpression in the chick by retroviral infection increases the width of the somite files. As somite formation is progressive and clearly visible, the number of somites formed is often used as an accurate indication of the stage of a vertebrate embryo. The total number formed over the course of development is a property of the species and can vary widely between animal groups, from some fish with about 30 up to snakes with several hundred.

Although all vertebrates form somites, most experimental work has utilized the chick embryo, with some additional information coming from mouse, *Xenopus*, and zebrafish. In the chick, somites first become visible as loose cell associations that condense into epithelial somites, vesicles of cells enclosing a cavity. Expression of fibronectin and N-cadherin increases with the formation of the epithelial somite and is responsible for the changes in cell adhesion driving the process. The epithelial somite is a transient structure as it soon undergoes an epithelial-to-mesenchymal transformation on the medial side to form the **sclerotome**, which is a mesenchymal condensation that subsequently forms the vertebrae and ribs (Fig. 15.2). The dorsal part of the sclerotome also forms tendons and has been called the "syndetome." Like other parts of the skeleton outside the skull, the vertebrae are formed initially of cartilage, which is later replaced by bone. Each vertebra is formed by the sclerotome of two somites, the posterior half of one somite combining with the anterior half of the next, so the vertebrae end up half a

Essential Developmental Biology, Third Edition. Jonathan M.W. Slack.
© 2013 John Wiley & Sons, Ltd. Published 2013 by John Wiley & Sons, Ltd.

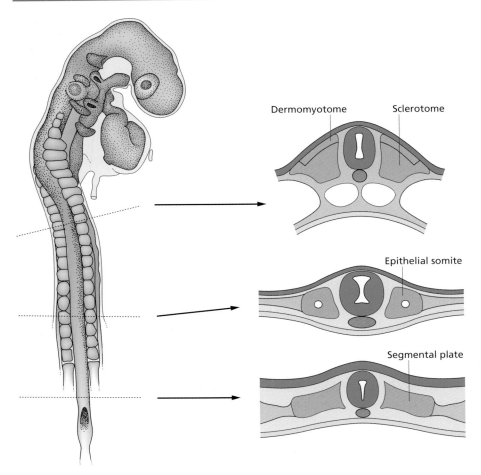

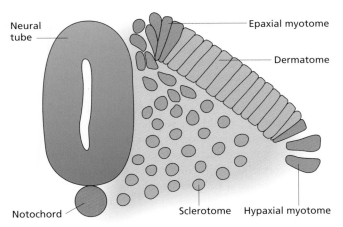

Fig. 15.1 Somitogenesis in the chick embryo. The process starts in the anterior and new segments are cut off at the posterior end of the file for a considerable time. The morphology is initially epithelial and then changes to an outer dense dermomyotome and an inner loose sclerotome.

Fig. 15.2 Differentiation of the amniote somite.

segment out of phase with the original somite repeating pattern. This is known as "resegmentation" and is somewhat reminiscent of the relationship between parasegments and definitive segments in *Drosophila*.

The lateral part of the epithelial somite forms a plate called the **dermomyotome**. This in turn becomes subdivided into two parts, the **dermatome**, which forms the dermis of the dorsal midline, and the **myotome**, which forms skeletal muscle. The epaxial myotome, or region near the midline, forms the segmental muscles of the main body axis, and the hypaxial myotome, which is more lateral, forms the muscles of the ventral body wall, limbs, and diaphragm.

Segmentation mechanism

The segmental repeating pattern of the somites arises from the operation of a molecular oscillator, or clock, operating in conjunction with a spatial gradient (see Animation 20: Operation of somite oscillator). One cycle of the clock causes the formation of one somite, and the gradient determines that the somites are formed in anterior to posterior sequence. The period of the clock depends on the species, about 30 minutes in the zebrafish, 90 minutes in the chick, and 120 minutes in the mouse. In the chick the clock starts operating in the early (H&H 3) embryo, and the third cycle corresponds to formation of the first somite. The clock can be visualized as periodic expression of components of the Notch pathway, including its targets the Hes genes and *Lunatic fringe*. The Hes genes encode

Classic experiments

Somitogenesis and myogenesis

A clock-based model for somitogenesis was first proposed by Cooke and Zeeman in the first paper, but it was largely ignored and it was not until many years later that the somite oscillator was discovered (second paper).

Meanwhile, the understanding of somite cell differentiation got a boost from the discovery of MyoD, the first "master controller" among transcription factors, that was able to reprogram other cell types to become muscle (third paper).

Cooke, J. & Zeeman, E.C. (1976) A clock and wave front model for control of the number of repeated structures during animal morphogenesis. *Journal of Theoretical Biology* **58**, 455–476.

Palmeirim, I., Henrique, D., Ish-Horowicz, D. & Pourquie, O. (1997) Avian hairy gene expression identifies a molecular clock linked to vertebrate segmentation and somitogenesis. *Cell* **91**, 639–648.

Davis, R.L., Weintraub, H. & Lassar, A.B. (1987) Expression of a single transfected cDNA converts fibroblasts to myoblasts. *Cell* **51**, 987–1000.

bHLH transcription factors and are homologs of *Drosophila hairy*, a pair-rule gene also involved in segmentation. *Lunatic fringe* is one of a group of genes homologous to *Drosophila fringe* and codes for a glycosyl transferase. *Drosophila fringe* is involved in the formation of compartment boundaries in imaginal discs.

The mRNA levels of these genes oscillate with a periodicity corresponding to the frequency of formation of new somites. In most of the presomite plate this oscillation is synchronous, indicating that cell communication is occurring to maintain the same phase of the cycle in adjacent cells. The periodic nature of this gene expression was discovered because the static *in situ* hybridization patterns appeared very variable between specimens until it was eventually realized that they were rapidly

changing in time. If embryos were cut longitudinally and fixed at different times, then the change in phase could be observed by comparing *in situs* for left and right sides. In the region adjacent to the last-formed somite, the oscillation slows such that it lags relative to the rest of the presomite plate. Eventually, it stops altogether in the region that will form the next somite. Because the slowdown of the clock in this region is progressive rather than instantaneous, the anterior part of the newly formed somite is left with a low level of mRNAs and the posterior part with a high level (Fig. 15.3).

The oscillator mechanism depends on autoregulation of transcription by the Hes factors. These proteins will repress their own transcription and are very unstable, so they decay rapidly in the absence of new synthesis. Transcription is allowed

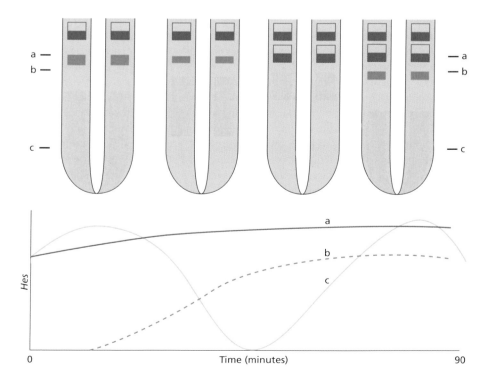

Fig. 15.3 Operation of the somite oscillator over one cycle of somite formation in the chick. The diagrams show the expression pattern of *Hes* at four times in the cycle, and the graphs show how the level of transcript varies at the points a, b, and c.

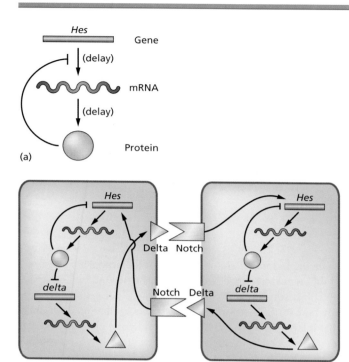

(a)

(b)

Fig. 15.4 (a) The simple feedback and delay model for the oscillator. (b) Coupling between cells by the Delta–Notch system.

when the level of the product is low; this causes the protein to build up so transcription becomes repressed, but because the protein is unstable this means its level then falls, and so the transcription increases again (Fig. 15.4a). The main evidence for this negative-feedback-type mechanism is that the inhibition of Hes synthesis by mutation (in the mouse) or **morpholino** treatment (in the zebrafish) will stop the oscillator. In zebrafish the oscillations probably arise simply from this mechanism, with the Delta–Notch system maintaining synchrony between adjacent cells. In amniotes the oscillation has a longer period and involves the Delta–Notch system as an integral component of the oscillator. In this more complex cycle Hes proteins repress *Delta* transcription, which reduces Notch signaling in adjacent cells, but Notch signaling increases *Hes* transcription, so Hes proteins go down in adjacent cells. This leads to more *Delta* transcription so more Notch signaling in the original cell, so *Hes* transcription goes up again (Fig. 15.4b). Loss of *Delta* in mice suppresses the oscillations, whereas loss of *Delta* in zebrafish permits the oscillations to continue, but abolishes the synchrony in adjacent cells. This indicates that in the mouse the Delta–Notch system is part of the oscillator itself whereas in zebrafish it serves simply to maintain the same phase in the population of presomite plate cells.

The Notch pathway drives expression both of the Hes genes and of *Lunatic fringe*. Lunatic fringe can modify the components of the Delta–Notch signaling system such that Notch activation occurs only along a boundary between a *Lunatic fringe*-expressing and nonexpressing region. The mouse knockouts of these components all have similar phenotypes with respect to somitogenesis, such that the segmentation pattern is disrupted, although the subsequent cell differentiation of the somite derivatives is fairly normal.

In order to generate a series of segments in space the operation of the oscillator needs to interact with a gradient. The gradient part of the system is the posterior-to-anterior gradient of fibroblast growth factors (principally FGF4 and FGF8) that is responsible for other aspects of posterior body pattern. The gradient gradually decays, such that the same concentration of FGF is experienced at progressively more posterior positions with time. The clock and gradient cooperate to generate segments because both are involved in regulating genes that control somite segmentation, such as the transcription factor gene *Mesp2*. When the FGF level has fallen below a certain threshold, the clock components upregulate *Mesp2*. This stabilizes a steep gradient of Fringe expression across the posterior part of each prospective somite, which then causes activation of Notch in the posterior part. Notch signaling in turn drives expression of other components, such as ephrins, which contribute to the cell adhesion changes required to form the epithelial somite. Mesp2 also represses expression of *Snail*, whose product normally represses cell adhesion molecules such as integrins and cadherins. The result is upregulation of these cell adhesion molecules and the mesenchymal-to-epithelial transition, leading to the formation of an epithelial somite. In zebrafish, chick, and mouse a gradient of *Fgf8* mRNA can be visualized by *in situ* hybridization, and a corresponding gradient of phosphorylated ERK detected by immunostaining. Segmentation is disrupted by addition of the inhibitor of FGF signaling, SU5402. This generates large somites because a larger than normal region in each cycle experiences FGF signaling below threshold and is then incorporated into the somite. Conversely, the administration of FGF protein on a slow-release bead, or by electroporation, gives rise to smaller than normal somites.

In descriptions of somitogenesis reference is often made to a "wavefront," a term which baffles almost everyone. The wavefront is simply the moving boundary of mesenchymal-to-epithelial transition, and it reflects the threshold concentration of the gradient of FGF, the position of which moves posteriorly as the gradient decays.

Since all vertebrates have a similar clock and gradient process for somitogenesis, it is presumed that humans share this mechanism. Interestingly, some types of congenital scoliosis (curvature of the spine, sometimes with other segmental abnormalities) are due to mutations in genes known to have key roles in the segmentation process, including *DELTA-LIKE 3, LUNATIC FRINGE*, and *MESP2*.

Subdivision of the somite

The internal anterior–posterior subdivision of the somite set up at the time of segmentation remains important in later

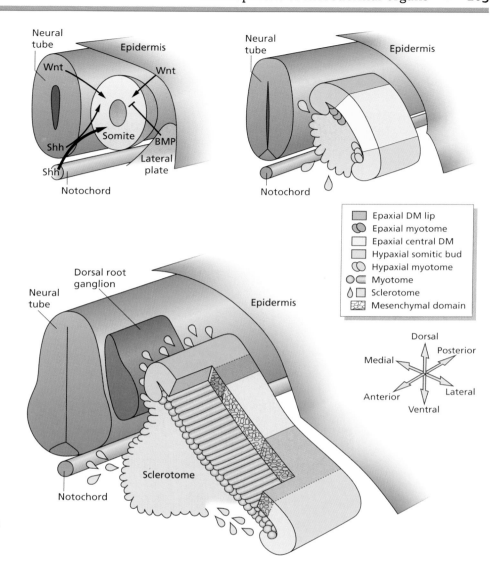

Fig. 15.5 Subdivision of somite by signals from surrounding tissues. DM=dermomyotome. Reproduced from Ordahl (2000), with permission from Academic Press.

development. As we have seen in Chapter 14, the neural crest cells and the spinal nerves are both repelled from the posterior sclerotome because of its high ephrin level, and therefore preferentially grow through the anterior sclerotome of each segment.

The subdivision of the somite in the transverse axes generates the sclerotome, dermatome, and myotome and depends on interactions with the surrounding tissues (Fig. 15.5). This may be shown by the fact that the subdivision will occur normally with respect to the whole body axes, even after the newly formed epithelial somites have been microsurgically rotated or inverted. Each of the specific inductions was identified by showing that it failed to occur when the presomite tissue was isolated, and that it occurred at the site of contact when the signaling tissue was cultured together with presomite tissue. The **sclerotome** is induced by the notochord and the ventral part of the neural tube. The epaxial part of the **myotome** is induced by the noto-

chord and dorsal neural tube, and the spread of myotomal induction is limited by the lateral plate mesoderm. The **dermatome** is induced by the neural tube. The signals responsible for each of these interactions have been deduced from the usual criteria for proof described in Chapter 4: namely, expression, activity, and inhibition.

The principal signal for sclerotome induction is Sonic hedgehog (Shh). This is expressed in the notochord and the floor plate of the neural tube and is also required for ventrodorsal patterning of the neural tube, as explained in Chapter 14. Cells expressing Shh will induce cartilage from any part of the epithelial somite. The early events involve upregulation of the transcription factor genes *Nkx3.2*, *Pax1* and *9*, *Scleraxis*, and also of genes for myogenic inhibitors of the Id class (see below). The mouse knockout for *Shh* lacks most derivatives of the sclerotome, showing that it really is needed for sclerotome formation *in vivo*. In the course of normal development the notochord itself is not lost, but ends up within the intervertebral discs as the nucleus pulposus.

As mentioned above, each vertebra is later formed from the posterior sclerotome of one somite combined with the anterior

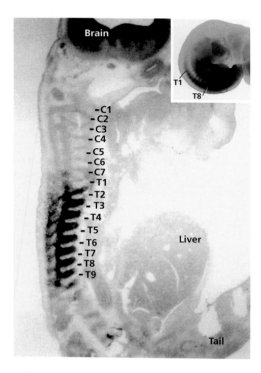

Fig. 15.6 Expression of Hoxc6 in the future thoracic vertebrae of a mouse embryo. C1-7 cervical vertebrae, T1-9 thoracic vertebrae. Reproduced from Robertis. Cell 2008; 132: 185–195, with permission from Elsevier.

sclerotome of the next most posterior somite, a process called "resegmentation." The regional character of vertebrae (cervical, thoracic, lumbar, sacral, and caudal) is controlled by the combination of Hox genes expressed at that level (Fig. 15.6). The mouse and the chick have different numbers of vertebrae in each of these body regions, but the Hox codes for the regions are the same (Fig. 15.7). For example the anterior boundary of *Hoxc6* expression marks the transition from cervical to thoracic vertebrae, the anterior boundary of *Hoxd10* the transition from lumbar to sacral, and the anterior boundary of *Hoxd12* the transition from sacral to caudal. Transgenic overexpression and knockout of Hox genes can bring about predictable changes in vertebral character, as mentioned in Chapter 10.

The formation of the myotome is more complex. Some pioneer muscle cells arise from the dorsal medial region before the somite has segregated into dermomyotome and sclerotome. Subsequently, **myoblasts** continue to arise from cells at the medial (epaxial) and lateral (hypaxial) edges of the dermomyotome, which migrate around the edges of the dermomyotomal plate to join the growing myotome underneath. Once these cells have moved into the myotome they become postmitotic and become oriented in anteroposterior direction. Although the signaling environment is somewhat different in epaxial and hypaxial regions, they both generate myoblasts. This is one of many examples in organogenesis where the same cell type arises from different rudiments by molecular pathways that are initially somewhat different but eventually converge on the same genes, in this case the myogenic transcription factors of the bHLH class described below.

In the epaxial region myogenesis requires an early exposure to Shh at the presegmental stage, followed by a Wnt signal from the dorsal neural tube. This results in the induction of myogenic

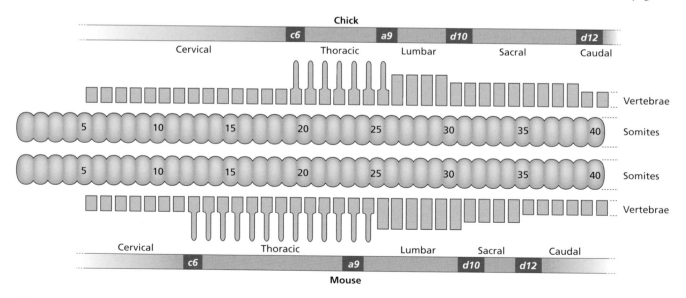

Fig. 15.7 Formation of vertebrae from the somites of the chick (above) and the mouse (below). Each vertebra is formed from two half somites, and the regional character of vertebrae correlates with the expression domains of Hox genes. *c6, a9, d10, d12* indicate the anterior boundaries of these Hox genes in the two species.

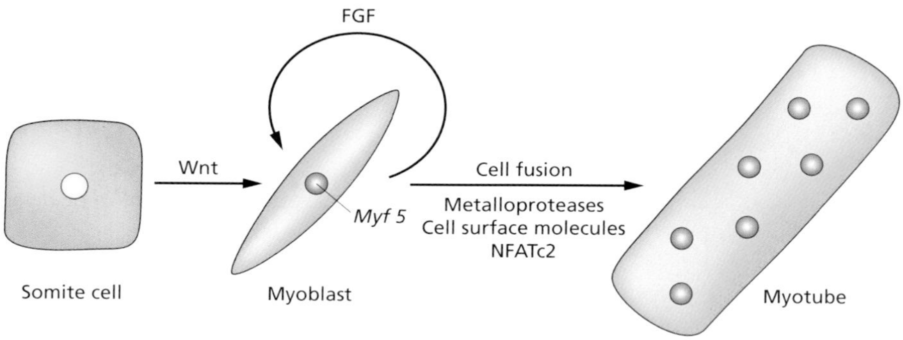

Fig. 15.8 Factors controlling myogenesis.

genes such as *Myf5* and the repression of *Pax3*, which is initially expressed in the whole epithelial somite. A similar effect is shown by removal of the lateral plate mesoderm. The lateral plate expresses BMP4, and this will repress myogenesis if supplied by one of the standard overexpression methods (cell pellet, slow-release bead, or retroviral overexpression). The spatial extent of the epaxial myotome seems to be determined by a balance between the effects of the Wnt signal from the dorsal neural tube, and the BMP signal from the lateral plate. One function of the Wnt signal is the induction of *Noggin* expression in the medial region, which inhibits the lateral plate-derived BMP4.

The induction of the hypaxial myotome proceeds through a different route, as the primordium is much further from the neural tube and out of range of the Shh and Wnt signals. Wnt7A from the dorsal epidermis plus a permissive level of BMP from the epidermis is believed to be the main inductive signal. The hypaxial myotomes expresses the transcription factor Lbx1 (LIM class) and the receptor c-Met, later required for migration of myoblasts to the limb buds.

The **dermatome** arises in response to a Neurotrophin 3 signal from the dorsal neural tube. A neutralizing antibody to this factor will prevent the epithelial-to-mesenchymal transition, which is involved in the formation of the dermis. The dermatome generates the dorsal dermis, but the lateral and ventral regions of the dermis come from the lateral plate mesoderm. Its properties depend in part on a bHLH protein, Dermo1, which represses expression of the myogenic transcription factors.

Myogenesis

Vertebrate muscle is classified into skeletal muscle, smooth muscle, and cardiac muscle. All skeletal muscle is derived either from the myotomes of the somites, which form axial muscles, limb muscles, and body wall muscles, or from the corresponding regions of the unsegmented head mesoderm, which form the muscles of the eyes and jaw. Smooth muscle is formed from both lateral plate mesoderm and from the hypaxial part of the

somites, and cardiac muscle arises from the myocardium of the early heart, derived from anterior splanchnic mesoderm.

Skeletal muscle is the familiar voluntary muscle, making up a large proportion of the mass of a vertebrate animal. It consists of multinucleate **myofibers**, which contain myofibrils composed of the contractile proteins actin and myosin. Differentiation of skeletal muscle involves first the commitment of cells to become **myoblasts**, then a period of multiplication and/or migration, then a cessation of division and fusion of the myoblasts to form the multinucleate myofibers (Fig. 15.8). In normal development the cessation of proliferation occurs after myoblasts have migrated from dermomyotome to the underlying myotomes. Once postmitotic, they should strictly be called "myocytes," but the name "myoblast" still persists.

Myoblast fusion

The fact that several myoblasts fuse to form one myofiber was proved to occur *in vivo* by the use of **chimeric** mice. If a chimera is made from embryos from strains expressing different electrophoretic variants of the enzyme isocitrate dehydrogenase, then the muscle is found to contain not only the two separate forms but also an intermediate form. This is because the enzyme is dimeric. Single cells can produce only AA or BB dimers, depending on the embryo from which they are descended, but fused cells will also produce the heterodimer AB because both forms are present in the same cytoplasm.

The mechanism of myoblast fusion has been investigated both through *Drosophila* genetics and through *in vitro* work on mammalian models of myogenesis.

In *Drosophila* embryos 30 multinucleate fibers arise, each comprising a different muscle. Each is initiated by a founder cell, which fuses with several fusion-competent myoblasts. The founder myoblasts express a cell-surface molecule called Kin-of-irreC (Kirre), which binds a receptor on the competent cells encoded by the gene *sticks-and-stones*. Loss of function of either component will prevent fusion. The significance of this is that the zebrafish expresses homologs (called Kirre-like) on fusing myoblasts, and morpholino blockage of these will also prevent fusion, indicating a likely commonality of mechanism.

Mammalian muscle fusion has been studied mostly with mouse C2C12 cells, which are a myoblast cell line easy to study in tissue culture. These cells keep growing while they are supplied with FGF, but when this is removed they stop growing and start fusing into **myotubes**, which are similar to the myofibers found *in vivo*. FGF maintains expression of the transcriptional repressor Msx1, which inhibits myogenesis. Fusion is thought to depend on various molecules, as shown by acceleration of fusion by transfection with the appropriate genes, or inhibition of fusion by treatment with specific antibodies. One group are metalloproteases of the ADAM class (also called meltrins), which are also involved in mammalian fertilization (see Chapter 10). In addition, the protease calpain and the cell-surface molecule caveolin have been implicated. A rise of intracellular Ca^{2+} activates the transcription factor NFATC2, and knockout of the gene for this factor causes mice to have smaller fibers with fewer nuclei.

Myogenic transcription factors

Central to skeletal muscle differentiation is the family of **myogenic** proteins. These are all transcription factors of the bHLH class, and are called MyoD, Myf5, Myogenin, and MRF4 (Muscle regulatory factor 4, also known as Myf6). They were identified because, when the genes were transfected into various types of tissue-culture fibroblast, they could cause the cells to become myogenic. The transfected cells would start to express genes for muscle proteins and fuse into multinucleate myotubes. MyoD has been shown directly to activate genes for some muscle proteins such as creatine phosphokinase or the acetylcholine receptor. It also activates its own transcription, so, once turned on, it stays on (see bistable switches, Chapter 4). bHLH factors act as dimers, usually as heterodimers with another ubiquitous bHLH protein called an E protein. There also exist inhibitory HLH proteins containing the dimerization but not the transcriptional activation domain, and these work by sequestering the bHLH proteins in unproductive dimers. For example the Id protein, upregulated in the sclerotome as a response to Shh, is an HLH-type inhibitor that forms unproductive dimers with myogenic bHLH proteins and ensures that this part of the somite does not form muscle.

In normal development *Myf5* is initially upregulated in the epaxial myotome and *MyoD* in the hypaxial myotomes, although the other factor is also soon upregulated in both zones. Knockouts of either gene have only slight effects on muscle development because if one is removed the other will upregulate and take its place. However, knockout of both together leads to a severely affected mouse lacking both myoblasts and skeletal muscle. This double knockout also loses activity of *Mrf4*, which lies close to *Myf5* in the genome. *Myogenin* is normally expressed all over the myotome later than *Myf5* and *MyoD*. When it is knocked out there is a serious defect of skeletal muscle formation, showing that the effects of *MyoD* and *Myf5* must to a large extent operate through Myogenin.

Once they have differentiated, mammalian myofibers may survive for the entire life of the animal, and muscles can usually increase in size only by enlargement of pre-existing fibers. Growth is controlled by a feedback system involving a circulating inhibitor called myostatin, which is described in Chapter 19. Although the myofibers are postmitotic there are associated with them some persistent progenitor cells called **muscle satellite cells**. These are small, mononuclear cells located under the basement membrane of individual myofibers. They originate from the central part of the dermomyotome and are characterized by expression of the transcription factor Pax7. Muscle satellite cells are responsible for growth in fiber number while the animal is growing, and also for repair following damage in the adult. They are described further in Chapter 18.

The kidney

The kidney is familiar as the excretory organ of the vertebrate body and because of the shortage of human kidneys for transplantation it is one of the principal targets for replacement through **tissue engineering** or some other form of applied developmental biology. The embryonic development of the kidney is representative of a large class of organs, as it involves reciprocal inductive interactions between an **epithelium** and a **mesenchyme**. Congenital abnormalities of the kidney are common and often arise from mutations in genes involved in its development. For example complete agenesis (absence) can result from loss of *Fgfr1*; hypoplasia (fewer nephrons than normal) from *Pax2* mutations; and WAGR syndrome (Wilm's tumor, aniridia, genitourinary abnormalities, and mental retardation) from a contiguous defect affecting the *Wt1* and *Pax6* genes.

The mature kidney has a reiterated, small-scale structure consisting of many nephrons. A nephron comprises a renal corpuscle with associated tubule and collecting duct. The renal corpuscle serves to filter fluid from the blood into the tubule. The fluid is processed into urine as it passes down the tubule and ions are added or removed, with corresponding movements of water. Eventually, it passes into the collecting duct system, and thence to the ureter, and, in mammals, fish, and amphibians, a urinary bladder. The renal corpuscle itself consists of a convoluted ball of capillaries (the glomerulus) invested with kidney epithelial cells (podocytes), within a Bowman's capsule (Fig. 15.9). Its development can readily be studied in **organ culture**, so it has been possible to do most of the experimental work on mammals. An explanted rudiment will grow and differentiate for several days in culture and this enables various types of procedure to be carried out that cannot be done on a mammalian embryo *in utero*. In particular, tissues can be separated and recombined, components can be added to or removed from the culture medium, growth factors can be locally applied on slow-release beads, and the expression of individual genes can be inhibited by the addition of specific antisense oligonu-

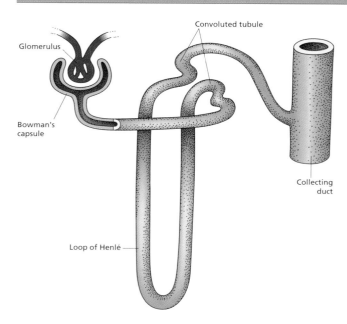

Glomerulus

Convoluted tubule

Bowman's
capsule

Loop of Henlé

Collecting
duct

Fig. 15.9 Structure of a nephron.

cleotides or RNAi. More recently, the use of mouse genetic labeling and conditional knockout technologies have added much information about mechanisms *in vivo*.

Normal development of the kidney

Lateral to the paraxial mesoderm that forms the somites lies the **intermediate mesoderm**, which gives rise to the kidney and the gonads. Amniotes (reptiles, birds, and mammals) have three kidneys, from anterior to posterior there is the **pronephros**, the **mesonephros**, and the **metanephros**. The pronephros is vestigial and nonfunctional, the mesonephros may function transiently in embryonic life, and the metanephros is the definitive kidney. In lower vertebrates (fish and amphibians) there is no metanephros. The pronephros may function in larval life and the mesonephros forms the definitive kidney.

The kidney develops from two components, both arising within the intermediate mesoderm, in addition to the recruitment of some stromal cells from the adjacent paraxial mesoderm. The **nephric duct (= Wolffian duct)** originates in the anterior, at the level of somite 10 in the chick, and grows posteriorly to the cloaca. It is characterized by expression of *Pax2* and -8, and the double knockout of these genes prevents nephric duct formation. Labeling of the duct with **DiI**, or by local introduction of a *lacZ* gene, shows that it elongates by intrinsic growth rather than by recruitment of surrounding cells. The duct is initially solid and requires the presence of the neighboring epidermis in order to form a tube, a requirement that can be substituted by a BMP slow-release bead. A part of the surrounding intermediate mesoderm is already committed to become kidney, and is known as the **nephrogenic** mesoderm. As the duct grows

posteriorly, tubules differentiate from the nephrogenic mesoderm and join up with it. The mesonephros, in particular, shows a segmented arrangement with nephrons arranged in anteroposterior sequence. The metanephros differs from the other parts of the kidney in that its collecting system is not formed from the main nephric duct, but from an outgrowth called the **ureteric bud**. This originates from a pseudostratified posterior section of the nephric duct, which grows into the nephrogenic mesenchyme and starts to branch extensively. In the mouse there are about 10 generations of branching. Ureteric bud elongation depends on oriented mitoses and the loss of planar polarity genes such as *Fat4* and *Wnt9b* will cause cyst formation. In their later differentiation the collecting ducts consist mainly of "principal" cells, responsible for water reabsorption, with some "intercalated" cells, which regulate pH by secretion of H^+ or HCO_3^-. Loss of Notch pathway components, such as the E3 ubiquitin ligase encoded by *mindbomb1(Mib1)*, causes an increase of intercalated cells at the expense of the principal cells. This suggests that the different cell types originate through the operation of a Notch-based lateral inhibition system similar to that responsible for primary neurogenesis (see Chapter 14).

The nephrogenic mesenchyme differentiates into nephrons, which join up with the branches of the ureteric bud to form a compact, nonsegmental organ. This process is an example of a mesenchymal-to-epithelial transition. First the mesenchyme aggregates and forms dense caps around the ureteric bud tips. These generate ovoid renal vesicles, which fuse with the bud tip as comma-shaped bodies, then they acquire an S-shape, and then elongate into tubules (Fig. 15.10). The condensed cap region expresses the homeobox gene *Six2*, and labeling of mice with *Six2-CreER x R26R* indicates that this domain forms the whole nephron. Initially, the mesenchyme secretes a typical matrix-containing fibronectin and collagen I and III. As the transition occurs, these products are replaced by laminin and collagen IV, typical components of an epithelial basement membrane. At the same time, N-CAM on the cell surfaces becomes replaced by E-cadherin. Loss of *Six2* leads to premature formation and differentiation of the renal vesicles. In the mouse nephron formation continues until shortly after birth, after which the *Six2*-positive cells are lost. These cells have been called **stem cells** as they divide about four times during the second half of gestation while retaining the capacity to form renal vesicles. However, this behavior does not satisfy the normal definition of a tissue-specific stem cell, which should have a much longer lifetime (see Chapter 18) and the *Six2*-positive cells are better thought of as **progenitor cells**. In lower vertebrates such as the zebrafish, the kidney continue to make new nephrons throughout life.

Tissue interactions in kidney development

If the tip of the nephric duct is destroyed, its further growth is inhibited. Mesonephric tubules develop normally at the level

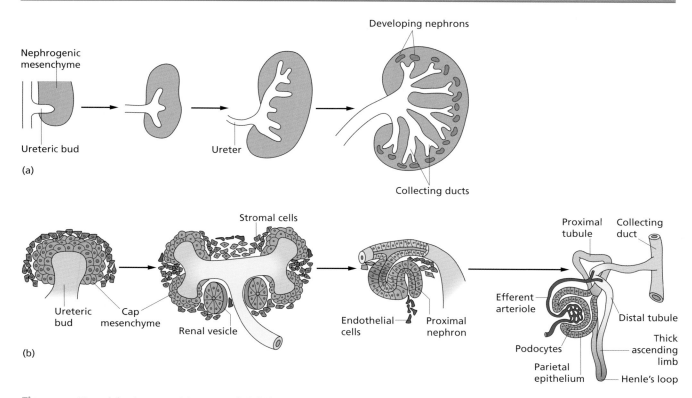

Fig. 15.10 Normal development of the metanephric kidney. (a) Growth and branching of ureteric bud into metanephrogenic mesenchyme; (b) induction of a nephron. Adapted from Dressler. *Development* 2009; 136: 3863–3874, with permission from Company of Biologists Ltd.

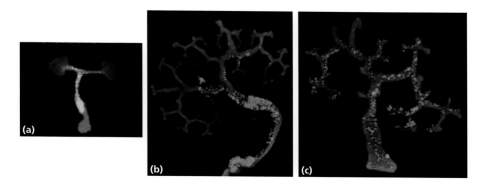

Fig. 15.11 Development of the collecting system of the kidney in chimeric embryos in which ES cells lacking the *Ret* gene were injected into normal blastocysts. The *Ret*⁻ cells are also GFP labeled. (a) A ureteric bud at E11.5; the *Ret*⁻ cells fail to colonize the tips. (b) The branching pattern after 3 more days of culture *in vitro*. The *Ret*⁻ cells are excluded from the terminal branches. (c) Control with Ret⁺ donor cells. Scale bars 250 μm. Reproduced from Costantini and Shakya. *BioEssays* 2006; 28: 117, with permission from John Wiley & Sons Ltd, and Shakya *et al. Dev Cell* 2005; 8: 65–74, with permission from Elsevier.

where the duct is present, but they do not develop more posteriorly. This shows that the duct is needed for formation of the nephrons. The same is true for the metanephros. It is possible to culture the metanephrogenic mesenchyme *in vitro*, but without the ureteric bud there is no tubule formation, showing that the bud emits an inductive signal necessary for tubule formation. This is an example of a **permissive** induction, because the mesenchyme cannot form anything other than kidney tubules. In fact, the induction is a reciprocal process because there is also an effect of the mesenchyme on the ureteric

bud. Without the presence of the mesenchyme, the bud will not arise from the nephric duct and it will not continue to grow and branch. So there are at least two inducing factors at work, one from the mesenchyme making the bud grow and branch, and one from the bud making the mesenchyme differentiate into tubules.

The signal from the mesenchyme is glial-derived neurotrophic factor (GDNF). This is expressed in the mesenchyme and its receptor RET is expressed in the ureteric bud (Fig. 15.11). RET is a tyrosine kinase working through both ERK and

PI3K pathways. Mice with either the *Gdnf* or *Ret* gene knocked out form no kidney. If a GDNF slow-release bead is placed on a culture of nephrogenic mesenchyme from these embryos, then branching of the duct is restored in the *Gdnf* knockout, which lacks the factor, but not in the *Ret* knockout, which lacks the capacity to respond to it. Conversely, a *Ret* gain-of-function mutant with constitutive receptor signaling activity shows an unregulated growth of the ureteric bud. Hence, this system passes all three tests of appropriate expression, activity, and inhibition. Various tissues other than the ureteric bud will induce tubules when cultured in contact with metanephrogenic mesenchyme, and much of the earlier experimental work on this problem used spinal cord as the inducer. This is very active, and may work through secretion of FGFs, which are also active *in vitro*.

There has been much controversy over the identity of the inducing factor from the ureteric bud leading to the formation of tubules. The current front runner is WNT9B; this is expressed in the ureteric bud, it will induce condensations in the organ culture system, and the knockout prevents the mesenchymal-to-epithelial transition *in vivo*. An early consequence of the induction is the upregulation of *Wnt4* in the aggregating mesenchyme, also needed for kidney development. In this context WNT4 seems to operate through the Wnt-Ca pathway, and calcium ionophore will also provoke tubule formation.

The competence of the mesenchyme to become induced depends on expression of a zinc-finger transcription factor WT1 (Wilm's tumor 1). The mouse knockout for *Wt1* shows a complete failure of metanephric development. WT1 also has later functions. Its level increases in the mesenchyme following induction and it is required to turn off *Pax2*, which is upregulated at the condensation stage but whose removal is necessary to enable differentiation of the nephron. Wilm's tumor itself is a human pediatric tumor in which renal progenitor cells continue to proliferate, and some but not all cases are caused by somatic loss of the *Wt1* gene. This probably reflects loss of the late function, repression of *Pax2*, rather than the early function, competence of the mesenchyme.

In the absence of the inducing signals there is massive cell death in the mesenchyme. FGF2 is produced by the ureteric bud and it will both maintain survival of the mesenchyme and also induce aggregation. But it will not support further development. BMP7 is also expressed in the bud and will maintain survival of isolated mesenchyme and induce tubules. The knockout of *Bmp7* has severely dysplastic kidneys containing many nephrons blocked at the comma or S stage of development, suggesting that the main requirement occurs shortly after the initial induction of tubules. WNT4, which is expressed in the developing tubules themselves, will also induce tubulogenesis *in vitro* and the knockout is blocked at the aggregation stage.

To summarize these data: the nephrons are initially induced from the nephrogenic mesenchyme by WNT9B from the ureteric bud, and further development depends on WNT4 expressed in the mesenchyme itself, with FGFs and BMP7 probably being required as survival factors.

Germ cell and gonadal development

The **gonads** are both the repositories of the gametes and important endocrine organs. They have a dual embryonic origin. The somatic tissues of the gonads arise from the genital ridges, which are formed from the **intermediate mesoderm** in the vicinity of the mesonephros. The germ cells that produce the gametes are derived from **primordial germ cells (PGCs)**. In most embryo types these form at a distant site in early development and undergo a migration to the genital ridges. Both the mature germ cells and the gonad have a different morphology depending on whether the individual is male or female.

Germ cell development

In the mouse embryo the PGCs originate from the **proximal** part of the **egg cylinder**. This was shown by fate mapping using **horseradish peroxidase (HRP)** injection (see Chapter 10). The PGCs are induced by the effect of the extraembryonic ectoderm since they do not develop in the absence of this tissue, and in recombination experiments extraembryonic ectoderm is able to induce PGCs from both proximal and distal epiblast. The signal consists of BMP2,-4, and -8b. These BMPs are biologically active, they are expressed in the extraembryonic ectoderm, and the knockouts show various degrees of failure to form PGCs. The initial product of induction is a zone of tissue around the posterior epiblast expressing the gene *fragilis*, which encodes a transmembrane protein. This region forms both PGCs and the allantois. A few cells within the region upregulate *blimp1*, encoding a zinc-finger transcription factor. BLIMP1 represses various early developmental genes such as brachyury (*T*) and *Snail*, and allows persistent expression of a set of pluripotency associated genes including *Oct4*, *Nanog*, and *Sox2*, which are characteristic both of primitive ectoderm and PGCs. The pluripotency genes *Oct4* and *Nanog* normally become repressed in all somatic tissue, while *Sox2* remains active in the neuroepithelium. About 40 PGCs are formed by E7.5, partly by induction and partly by multiplication of those already formed. They can readily be identified by their expression of high levels of the enzyme alkaline phosphatase. There is a temporary cessation of all transcription around E8, but this is not prolonged as it is in *Drosophila* or *C. elegans*.

From the base of the allantois, the PGCs enter the hindgut at about E9–10 and migrate up the mesentery, reaching the genital ridges, which form the gonads, by E11.5 (Fig. 15.12). In addition to Oct4, certain gene products discovered in *Drosophila* as being necessary for germ cell development are also required in mice. These include VASA (an RNA helicase) and NANOS (an RNA-binding protein), both mentioned in Chapter 11. The migration of the PGCs is controlled by a **chemokine**, SDF1 (stromal cell derived factor 1, = CXCL12), which is expressed on the lateral plate mesoderm. Its receptor, CXCR4, which is a G-protein coupled receptor, is expressed on the PGCs. In the knockout for

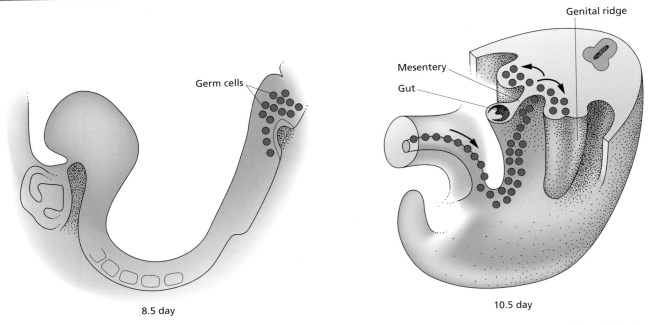

Fig. 15.12 Origin and migration of germ cells. Reproduced from Hogan *et al.* (1994) *Manipulating the Mouse Embryo*, 2nd edn, with permission from Cold Spring Harbor Laboratory Press.

Cxcr4 the PGCs fail to colonize the genital ridges. During migration various factors are required for survival, including SCF (stem cell factor), encoded by the *steel* gene, which binds to the receptor called KIT expressed on the germ cells.

The parental DNA imprints (see Chapter 10) are erased shortly after arrival at the gonads, as indicated by nuclear transplantation cloning from these germ cells, and observing the allele specificity of expression in the resulting embryos. The new sex-specific imprints are established subsequently by *de novo* DNA methylation.

Normal development of the gonads

Male and female **gonads** are different from each other and so the development of the gonads is intimately associated with sex determination. Unlike most other aspects of development, the mechanism of sex determination is not at all well conserved between animal groups, so the account that follows applies only to mammals. In the mouse the gonads can develop normally in various mutants defective in germ-cell migration, so we know that gonadal development does not necessarily require the presence of the germ cells themselves.

The gonads develop from the intermediate mesoderm in close proximity to the kidneys. The lateral part of the intermediate mesoderm forms the mesonephros and the medial part forms the gonad. In the early phase before sexual dimorphism appears there is a requirement for several gene products, which include WT1, the zinc-finger transcription factor also required

for kidney development, and SF1 (steroidogenic factor 1), a member of the nuclear hormone receptor family. Knockouts of genes for either of these will prevent gonadal development. The **genital ridge** appears at E9.5 on each side of the trunk, projecting slightly into the coelom (Fig. 15.13 and see also Fig. 9.22 for the layout in the avian embryo). The ridges express the Lim-homeodomain transcription factor LHX9, and the knockout mouse for the *Lhx9* gene fails to form any gonads. Slightly after appearance of the genital ridge, cords of cells begin to form from the coelomic lining epithelium and to grow into the underlying mesenchyme.

Two ducts also arise in this period which are important elements in the formation of the sex organs. The **Wolffian duct** is another name for the **nephric duct**. Once the development of the metanephros is underway it has no further excretory function but will eventually become the vas deferens in the male. The **Müllerian (= paramesonephric) duct** appears alongside the Wolffian duct, and will eventually become the oviduct, uterus, and proximal vagina of the female. Up to this point (E12.5) there is no difference in the visible differentiation of the two sexes.

In the male, the cords of coelomic lining cells growing into the genital ridge form a complex system of seminiferous tubules composed of **Sertoli** cells, into which the germ cells are integrated. The Sertoli cells express *Sry*, and later *Sox9*, key elements of the sex determination program (see below). Under their influence, cells responsible for secretion of testosterone, called **Leydig** cells, differentiate from the mesenchyme between the tubules. The tubules become connected to the Wolffian duct

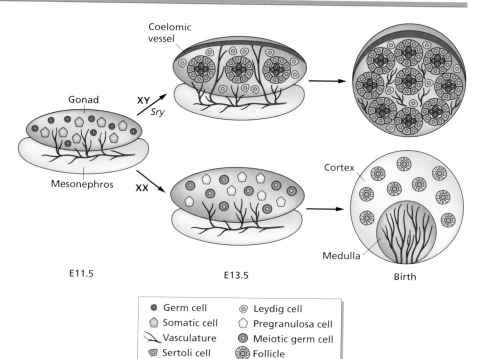

Coelomic vessel

Gonad

XY *Sry*

Mesonephros

XX

Cortex

Medulla

E11.5

E13.5

Birth

Germ cell   Leydig cell
Somatic cell   Pregranulosa cell
Vasculature   Meiotic germ cell
Sertoli cell   Follicle

Fig. 15.13 Formation of gonads from the genital ridge. Reproduced from Dressler. *Development* 2009; 136: 3863–3874, with permission from Company of Biologists Ltd.

and the developing gonad shortens and becomes encapsulated to form the testis. The germ cells are mitotically quiescent from E13.5 to birth. Meanwhile the Müllerian duct regresses. The tubules of the testis remain solid until after birth they start to hollow out and the **spermatogonia**, derived from the germ cells, appear (Fig. 15.14). Spermatogenesis in the mature male involves the continuous production of sperm from a stem cell population and this is further considered in Chapter 18.

In the female, the cords of cells from the coelomic lining do not grow so far into the mesenchyme but stay near the surface as granulosa (**= follicle**) cells, in proximity to the germ cells. Following operation of the sex determination system, the absence of SRY allows upregulation of *Wnt4* and *Foxl2* in the follicle cells. These both repress *Sox9* expression and lead, indirectly, to the formation of **thecal** cells from the mesenchyme, which are responsible for estrogen production. Following this the gonad becomes encapsulated as an ovary. The two Müllerian ducts fuse at their posterior ends and form the proximal (i.e. deeper) part of the vagina and the uterus, which is Y-shaped in the mouse. At the anterior end they form the oviducts, whose collecting funnels are closely apposed to the ovaries. Meanwhile the Wolffian duct degenerates.

Within the ovaries the **oogonia**, derived from the germ cells, become primary **oocytes** following their final mitotic division. Meiotic prophase commences about E13.5 and it is generally believed that there is no subsequent formation of new oocytes in mammals after this stage. Each primary oocyte becomes invested by granulosa cells to form a primordial follicle. Once sexual maturity has been attained, a number of follicles become activated during each reproductive cycle, then the granulosa cells multiply and the follicle with its oocyte expands (Fig. 15.15). In the mouse 8–12 oocytes are likely to be ovulated at one time. Each ruptured follicle becomes a **corpus luteum**, which secretes progesterone, helping to prepare the uterus for implantation.

Sex determination

Whether the gonad forms an ovary or a testis ultimately depends on the chromosome constitution: in mammals this is XY for males and XX for females. However, there is an asymmetry to the situation because the Y-chromosome contains a region not homologous to the X-chromosome. Within this lies the gene *Sry* (sex-determining region of Y), which, in mammals, is the critical switch controlling the pathway of sexual development. It codes for a transcription factor of the HMG (High Mobility Group) class and is the prototype of the Sox family of transcription factors. The evidence for its role is as follows. If *Sry* is deleted from the Y chromosome, then an XY mouse will develop as a female, even producing viable oocytes. If *Sry* is introduced as a transgene, then an XX mouse will develop as a male, although it does not produce sperm. Similar although not identical effects are found in naturally occurring chromosomal

Type A spermatogonia   Type B spermatogonia   Mature sperm   Blood vessel

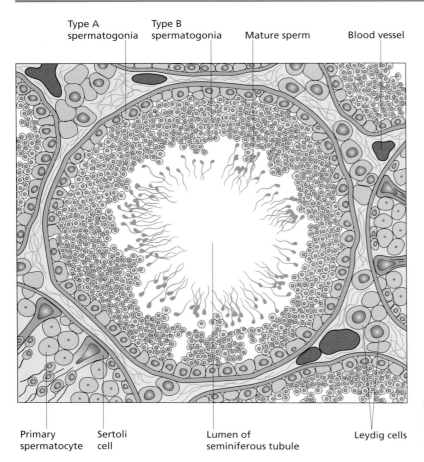

Fig. 15.14 Organization of a seminiferous tubule. Reproduced from Hildebrand (1995) *Analysis of Vertebrate Structure*, 4th edn, with permission from John Wiley & Sons Ltd.

Primary spermatocyte   Sertoli cell   Lumen of seminiferous tubule   Leydig cells

abnormalities of humans. XXY individuals are male (Klinefelter's syndrome) while XO individuals are female (Turner's syndrome), so it follows that possession of a Y-chromosome must confer maleness. Occasional XX individuals who are phenotypically male are found to have an *Sry* gene on one X-chromosome due to an aberrant recombination at meiosis. Conversely, occasional XY individuals who are female are found to have a mutated *Sry* gene.

In males, *Sry* is expressed in the gonad for a short period just before the sexual dimorphism appears (Fig. 15.16). Its only function is probably to activate expression of *Sox9*, another member of the HMG box family. Unlike *Sry*, *Sox9* expression persists through embryogenesis and it is associated with maleness in all vertebrates. It upregulates a set of genes whose products are needed for male functions, including *Fgf9* and *Amh* (encoding anti-Müllerian hormone, AMH), and represses a set required for female development including *Wnt4* and *Dax1*. AMH, also known as Müllerian inhibitory substance (MIS), is secreted by the Sertoli cells. It is a member of the TGFβ superfamily and it has a very specific role in causing regression of the Müllerian duct in the male, as witnessed by the fact that the knockout mouse for *Amh* retains its Müllerian duct. Another

important male function is exerted by the *Sf1* gene, which is expressed in early gonadal development, is involved in *Sry* upregulation, and is then expressed further, sustained by SOX9, during testis differentiation. SF1 participates in the upregulation of *Amh* and also of the genes for the testosterone synthesis pathway in the Leydig cells. Circulating testosterone is the key to the remainder of male development. This, metabolized to the active form dihydrotestosterone, brings about the differentiation of the male external genitalia, and is also responsible for the various secondary sexual characteristics, such as the facial hair and deeper voices of human males.

An early step in the female pathway is the upregulation of *Wnt4*, also required for development of the kidney. *Wnt4* is normally expressed in the early genital ridges, then is repressed in males but persists in females. If *Sry* upregulation is delayed for some reason then the level of WNT4 signaling is sufficient to suppress *Sox9* expression and the female pathway prevails. The *Wnt4* knockout has an ovary containing few oocytes, with cells secreting male-type steroids, and also lacks the Müllerian ducts.

Although the germ cells belong to a different cell lineage from the gonads, their pathway of development does depend on the

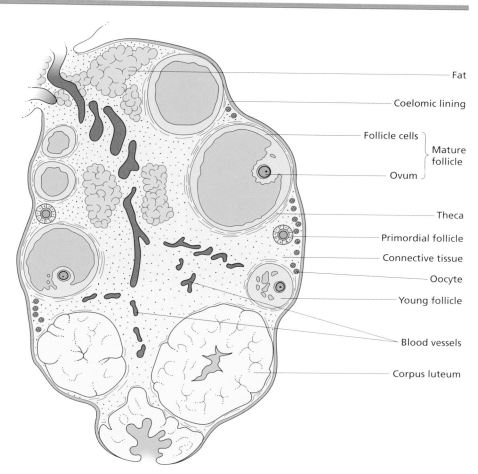

Fat

Coelomic lining

Follicle cells ⎤
          ⎬ Mature follicle
Ovum ⎦

Theca

Primordial follicle

Connective tissue

Oocyte

Young follicle

Blood vessels

Corpus luteum

Fig. 15.15 A mammalian ovary. Reproduced from Hildebrand (1995) *Analysis of Vertebrate Structure*, 4th edn, with permission from John Wiley & Sons Ltd.

Classic experiments

Mammalian sex determination

Since the discovery of sex chromosomes it was known that possession of a Y chromosome was necessary to be male, as human individuals of XO constitution were female while those with one Y and multiple X chromosomes were male. But there was a very small minority of apparently XX individuals who were male. They had experienced a translocation of the sex-determining gene from the Y to the X chromosome. Conversely, there was a very small minority of apparent XY females who had lost the sex-determining gene from the Y chromosome. The application of positional cloning techniques to the DNA of these unusual cases resulted in the discovery of the *Sry* gene. Studies on the mouse showed that the gene was expressed in the correct location, and that the human gene could masculinize genetically female mice.

Gubbay, J., Collignon, J., Koopman, P., Capel, B., Economou, A., Munsterberg, A., Vivian, N., Goodfellow, P. & Lovell Badge, R. (1990) A gene-mapping to the sex-determining region of the mouse Y-chromosome is a member of a novel family of embryonically expressed genes. *Nature* **346**, 245–250.

Koopman, P., Munsterberg, A., Capel, B., Vivian, N. & Lovell Badge, R. (1990) Expression of a candidate sex-determining gene during mouse testis differentiation. *Nature* **348**, 450–452.

Sinclair, A.H., Berta, P., Palmer, M.S., Hawkins, J.R., Griffiths, B.L., Smith, M.J., Foster, J. W., Frischauf, A.M., Lovell Badge, R. & Goodfellow, P.N. (1990) A gene from the human sex-determining region encodes a protein with homology to a conserved DNA-binding motif. *Nature* **346**, 240–244.

Koopman, P., Gubbay, J., Vivian, N., Goodfellow, P. & Lovell Badge, R. (1991) Male development of chromosomally female mice transgenic for sry. *Nature* **351**, 117–121.

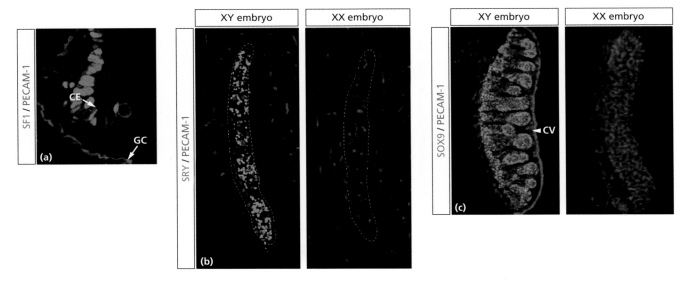

Fig. 15.16 Mouse gonads during the period of sex determination. (a) E10.5 the gonad of male and female are the same. Coelomic lining cells are immunostained green with antibody to SF1. Germ cells and endothelial cells of blood vessels are stained red. (b) At E11.5, Sry (green) is present in cells of the male but not female gonad. (c) At E12.5, Sox9 (green) is present in the male but not the female gonad and testis cords are forming. CE, coelomic epithelium; GC, germ cells; CV, coelomic vessel. Reproduced from Polanco and Koopman. *Dev Biol* 2006; 302: 13, with permission from Elsevier.

route followed by the gonad. In XY animals any germ cells that fail to find the gonad during their migration will persist for a while in their ectopic location and will develop as oocytes, not spermatogonia. The converse situation can be seen in a chimeric mouse formed from both XX and XY cells. So long as they possess more than about 35% of XY cells, such mice are male, because the *Sry* gene becomes activated in the XY Sertoli cells and sets off the male program, leading to release of AMH and testosterone. These hormones exert the same effects on cells whatever their chromosome constitution, so the chimeric mouse develops with a male morphology but with both testes and germ cells of mixed XY and XX chromosome constitution. The XX germ cells that enter the testis do become spermatogonia, but are incapable of developing all the way to mature sperm.

Limb development

Vertebrate limbs are of special interest to the developmental biologist because their visible nature makes congenital defects affecting the limbs of humans particularly obvious and distressing. Accordingly, limb development is one of the classic topics of organogenesis research. The developmental mechanisms are well conserved between vertebrate species, although there are some differences of detail. Amphibians, reptiles, birds, and mammals are known as **tetrapods** because they have four limbs; a pair of forelimbs, which may be legs, arms, or wings, and a pair of hindlimbs. The few tetrapods that have no limbs, such as snakes, have lost them secondarily in the course of evolution.

The bulk of the microsurgical experimental work has been done on the chick embryo, and this has in recent years been combined with extensive genetic experimentation using the mouse. In the chick embryo it is possible to operate on the limb bud *in ovo* through a window cut in the egg shell. The shell can then be sealed and the egg returned to the incubator to allow further development. Genes may be ectopically expressed using a replication-competent retrovirus, or by electroporation, or substances can be administered from slow-release beads or from pellets of transfected cells. However, it is not routinely possible to knock out genes in the chick and so the loss-of-function information has mostly come from the mouse.

Normal development of the limb

Following the phylotypic stage of the embryo, limb buds arise in the flank, each consisting of a core of mesenchyme overlaid by epidermis (Fig. 15.17). The region of mesoderm from which the limbs arise lies lateral to the intermediate mesoderm and is known as the **somatic mesoderm**. The somatic mesoderm together with the overlying epidermis is called the **somatopleure**. The early limb bud consists of an undifferentiated mass of loose mesenchyme surrounded by epidermis. At the distal edge the epidermis is thickened to form the **apical ectodermal ridge (AER)**, which is an essential structure controlling outgrowth (see below). The limb-bud mesenchyme forms all of the skeletal structures and connective tissues of the limb: the cartilages, tendons, ligaments, dermis, and the sheaths surrounding the muscles. However, the **myofibers** of the muscles are derived

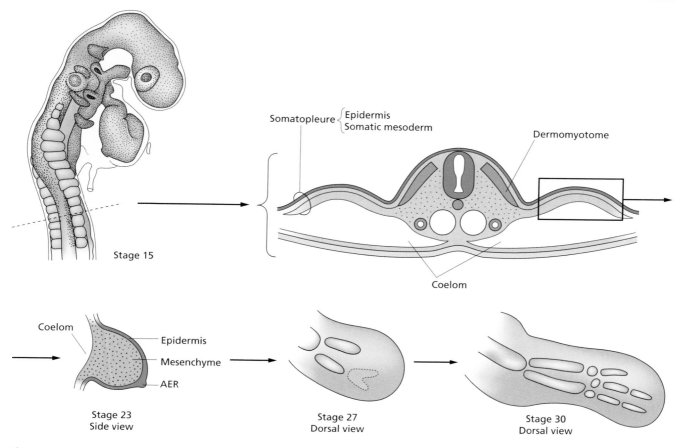

Fig. 15.17 Origin and growth of limb buds.

from the somites. This may be shown by making grafts of somites from quail to chick at a stage before the limb bud starts to grow out. When the limbs have developed it can be seen that all of the myofiber nuclei are quail type, while all other cell types in the limb are of chick type (Fig. 15.18).

During their outgrowth the limb buds elongate, flatten, and the various muscles and cartilage elements differentiate in a proximal to distal sequence, that is from the body outwards. Fate maps of the limb bud have been constructed by applying small marks of **DiI** and **DiO** and these show that the proximo-distal order of cells is maintained during outgrowth, but that the distal parts undergo much more elongation (Fig. 15.19).

The skeleton of the differentiated forelimb consists of a **humerus** in the upper limb, a **radius** and **ulna** in the lower limb, **carpals** in the wrist, and **metacarpals** and **phalanges** in the digits (Fig. 15.20). In the hindlimb the skeleton comprises the **femur** in the upper limb, **tibia** and **fibula** in the lower limb, **tarsals** in the ankle, **metatarsals** and **phalanges** in the toes. In terms of anatomical description, the **proximal–distal** axis runs from humerus/ femur to phalanges. In accounts of proximodistal patterning the terms **stylopod**, zeugopod, and **autopod** are often used, referring respectively to the upper limb, lower limb,

and hand or foot. The **anteroposterior** axis runs from digit 1 to the highest numbered digit. This is the same as the thumb-to-little finger axis of a human hand and corresponds to the anteroposterior axis of the whole body when the arm is extended with palm facing ventrally. The vertebrate limb is described as **pentadactyl** because the number of digits rarely exceeds five. However, the number is often reduced, so for example the chick wing only has three digits, which are confusingly numbered 2, 3, and 4, and the chick leg has four digits. The **dorsal–ventral** axis runs from the upper to the lower surface of the limb (from back of the hand to the palm). The results of experiments are often scored by wholemount staining to reveal the pattern of cartilage elements. This is a useful guide to the proximal–distal and the anterior–posterior pattern, but does not reveal the dorsal–ventral pattern because all the skeletal elements are in one plane. There are, however, various other features that differ along the dorsal–ventral axis and can be used to analyze unusual patterns, in particular the muscle pattern and the types of epidermal specializations such as feathers, scales, claws, and nails.

A differentiated limb contains many other cell types in addition to the skeleton and the muscles. All the skin specializations, such as scales and claws, are formed from the epidermis. Blood

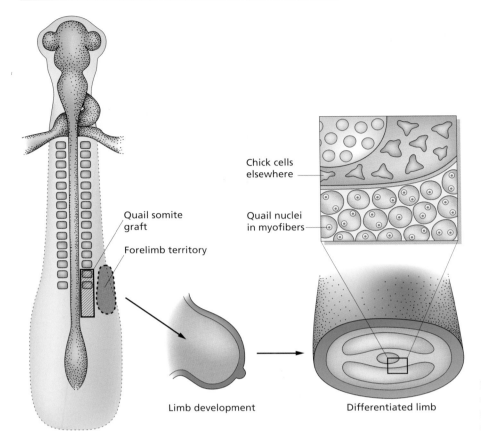

Fig. 15.18 Evidence that the limb muscles derive from the somites was obtained by grafting somite mesoderm from quail to chick and analyzing the resulting limbs.

vessels grow from the capillary network already present in the mesoderm at the time of limb bud outgrowth. Nerves grow in from the spinal cord, arriving as the limb bud is elongating but still undifferentiated. The relevant spinal nerves join together to form the **brachial plexus** to innervate the forelimb and the **lumbosacral plexus** to innervate the hindlimb. As elsewhere in the body, the pigment cells are derived from the neural crest. The skeleton initially differentiates as cartilage and this is later replaced by bone. Unlike the initial phase of limb development, ossification of the cartilage elements does not proceed from proximal to distal, but it does follow a predetermined sequence, generally starting earlier in the larger skeletal elements and later in the small ones.

Experimental analysis of limb development

Limb determination

The limb rudiment is initially specified as a territory in the mesoderm. This may be shown by transplantation of the prospective limb tissues to a site on the flank in between the limb buds. In this position the mesoderm will provoke formation of a limb bud whereas the prospective limb epidermis will not. However, there is a stage before this when even the prospective limb mesoderm will not produce a limb unless it is accompanied by somite tissue from the same anteroposterior level, sug-

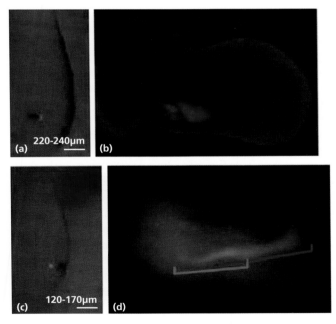

Fig. 15.19 Fate mapping of the chick limb bud. Small marks of DiI (red) and DiO (green) are applied at early bud stage (a,c) and the labeled cells are located 2 days later (b,d). The results show considerable elongation of the more distal parts of the early bud, and a lack of any proximodistal boundaries at early stages. (The dyes appear different colors in (c) because of the lighting conditions.) Reproduced from Sato *et al.* (2007), with permission from Company of Biologists Ltd.

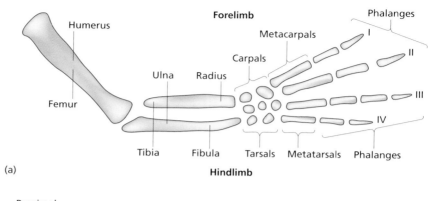

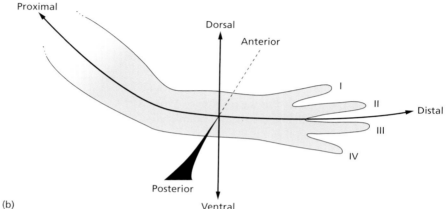

Fig. 15.20 Anatomy of the tetrapod limb. (a) Nomenclature of the skeletal elements in the forelimb and the hindlimb; (b) anatomical axes used for description of the limb.

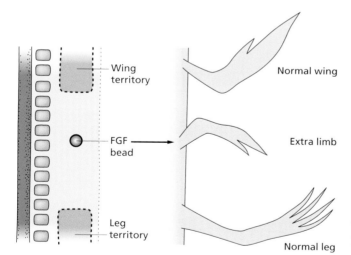

Fig. 15.21 Induction of a limb bud from the flank by an FGF bead. The induced limb has an inverted anteroposterior polarity compared to the normal ones.

gesting that an early induction from the somites is necessary for the acquisition of limb commitment.

It is possible to induce a new limb from the region of the flank between the normal limb buds by implantation of a slow-release bead of FGF (Fig. 15.21). Forelimbs arise from the ante-

rior part of the flank and hindlimbs from the posterior flank in response to such a bead implantation. *Fgf10* is expressed in the early presomite plate and subsequently in the prospective limb mesoderm itself. Furthermore, we know that FGF10 is necessary for limb development because the *Fgf10* knockout mouse forms no limb buds. FGF10 therefore satisfies the criteria of expression, activity, and inhibition for being responsible for the initial induction of the limb-forming territories in the somatic mesoderm. The *Fgf10* gene is upregulated by Wnt signaling. The forelimb-bud region expresses *Wnt2b* and the hindlimb region *Wnt8b*, and these factors can both induce expression of *Fgf10* and outgrowth of limbs. Furthermore, the Wnt inhibitor Axin can suppress limb development, and mouse knockouts of *Lef1* or *Tcf1*, both components of the Wnt signal transduction pathway, form no limb buds.

The fore- and hindlimb are distinguished by expression of T-box genes, with *Tbx5* expressed in the early forelimb and *Tbx4* in the early hindlimb, and these genes are regulated, directly or indirectly, by the combination of Hox genes active at different body levels. Overexpression of *Tbx4* can bring about a partial conversion of the chick wing bud to a leg bud, suggesting that the T-box genes do contribute to determining the difference between forelimb and hindlimb.

The migration of myogenic cells from the somites occurs some hours before the commencement of outgrowth of the bud,

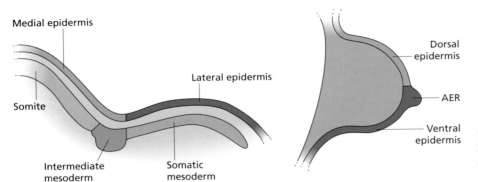

Fig. 15.22 Origin of dorsal and ventral limb-bud epidermis. The left section shows a pre-limb bud stage, the right one shows a stage 23 limb bud.

and somites from any level can contribute cells if they are grafted to the limb levels, not just those which normally supply the limbs. A major component of the chemotactic signal for migration is hepatocyte growth factor (HGF, = scatter factor). This is expressed in the prelimb-bud mesenchyme, and its receptor, called Met, is expressed in the myotomes. When *Met* or *Hgf* are knocked out in the mouse there is no migration of myoblasts into the limbs and they develop without muscles.

Proximal–distal outgrowth and patterning
The earliest visible event in limb-bud development is the formation of the **apical ectodermal ridge** (AER) in the epidermis. The AER is induced by the mesoderm, as may be shown by grafting an extra prelimb mesoderm under the flank epidermis. But it also requires for its formation an interaction between two epidermal territories: a medial territory overlying the somites and intermediate mesoderm, and a lateral territory overlying the somatic mesoderm. Fate mapping of the prospective limb epidermis using small marks of DiI shows that the medial epidermis becomes the dorsal epidermis of the limb bud, and the lateral epidermis becomes the ventral epidermis of the limb bud, and also forms the AER itself (Fig. 15.22). AER formation requires BMP signaling and can be prevented in transgenic mice by expressing a dominant negative BMP receptor under the control of an epidermis-specific promoter. The lateral epidermis expresses the transcription factor gene *Engrailed-1*, which probably functions to repress AER formation, as the mouse knockout of *Engrailed-1* has an AER extended onto the ventral side.

The AER persists during the outgrowth of the limb bud. If it is surgically removed at any stage this leads to a distal truncation of the final pattern (Fig. 15.23). The earlier it is removed the more structures are lost and it is presumed that this represents a sequence of determination of successive proximal-to-distal levels of the pattern in the course of limb-bud outgrowth. The AER itself needs factors from the mesenchyme for its survival as if it is grafted to a nonlimb-bud site it will quickly regress. Thus, the relationship between AER and mesenchyme is a reciprocal one, an inducing factor from each being required for the function of the other. The tip region about 250 μm from the

Stage of AER removal

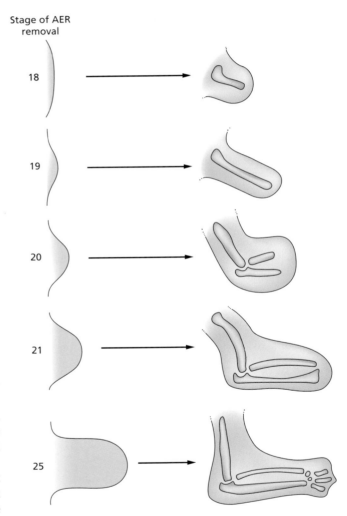

18

19

20

21

25

Fig. 15.23 Increasing extent of distal truncation the earlier the AER is removed.

AER remains undifferentiated during outgrowth. This region expresses several specific transcription factor genes including *Msx1*, *Ap2*, and *Lhx2*. Overexpression of an inhibitory **domain swap**, *Lhx-engrailed*, will inhibit limb outgrowth, suggesting that at least this gene is necessary for the continued formation

Classic experiments

The zone of polarizing activity in the limb

The ZPA is one of the classic inductive signaling centers of developmental biology. Anteroposterior respecification was first described by J.W. Saunders in the course of tip rotation experiments on the chick limb bud. But the interpretation of this effect as a graded morphogen was made by Tickle et al. in the first of these papers. In the second they came up with retinoic acid as a candidate morphogen,

but in the third the true morphogen was shown to be Sonic hedgehog.

Tickle, C., Summerbell, D. & Wolpert, L. (1975) Positional signalling and specification of digits in chick limb morphogenesis. *Nature* **254**, 199–202.

Tickle, C., Alberts, B., Wolpert, L. & Lee, J. (1982) Local application of retinoic acid to the limb bond mimics the action of the polarizing region. *Nature* **296**, 564–566.

Riddle, R.D., Johnson, R.L., Laufer, E. & Tabin, C. (1993) Sonic-hedgehog mediates the polarizing activity of the ZPA. *Cell* **75**, 1401–1416.

of structures. Interestingly, *Lhx2* is a homolog of the gene *apterous*, required for wing outgrowth in *Drosophila* (see Chapter 17).

The AER itself expresses various fibroblast growth factor genes (*Fgf 4*, *8*, *9*, *17*). If the AER is removed, a slow-release FGF bead can substitute for it and support continued outgrowth and development of distal parts. An AER-specific knockout of *Fgf8* using the Cre-lox system has distal limb defects, and the AER-specific knockout of *Ffg4* and *-8* together has no limb development. So the FGFs satisfy the usual three criteria of expression, activity, and inhibition as the active factors released by the AER.

The AER maintenance activity from the mesenchyme is partly accounted for by Sonic hedgehog (Shh) from the **zone of polarizing activity** (**ZPA**, see below), and partly by a secreted BMP inhibitor called Gremlin. Several BMPs are expressed both in the limb mesenchyme and the AER. At this stage, if the BMP activity is locally inhibited the AER will expand, indicating that the BMPs have a suppressive effect on the AER during outgrowth. Later in limb development, BMPs are also required for the cell death that occurs in between the forming digits. If this is prevented in transgenic mice by using an epidermis-specific promoter to drive expression of *Noggin*, then webbed feet will result, resembling those of a duck.

The differentiation of proximodistal structures is foreshadowed by expression of various genes at different levels. The most proximal territory, the stylopod, is marked by expression of the homeobox genes *Meis1* and *-2*, which are induced by retinoic acid coming from the main body. Overexpression of these genes can cause proximalization of the more distal parts indicating that they are causally involved in specifying proximal character. The middle section, or zeugopod, expresses *Shox* (short stature homeobox gene), and humans with mutations in this gene have a shortened ulna and radius. There is also a nested expression in the early limb bud of the *Hoxa10*, *-11*, and *-13* genes, with all three expressed in the future autopodium (Fig. 15.24a). This is often thought to control proximal–distal differentiation, but the expression patterns change considerably during development and the exact role of Hox genes in limb patterning

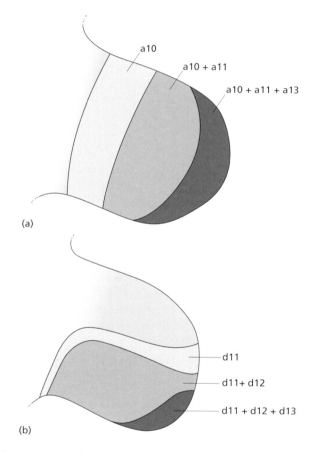

Fig. 15.24 Nested expression of Hox genes in the limb bud. (a) *Hoxa* genes; (b) *Hoxd* genes.

remains uncertain. Knockouts of individual Hox genes have limited effect on the limb phenotype. Knockouts of whole paralog groups can cause spectacular, segment-specific deletions, although not homeotic effects (Fig. 15.25). So Hox genes are certainly necessary for limb development but their main role may be concerned with growth control rather than regional specification. In terms of signals, a combination of the gradient

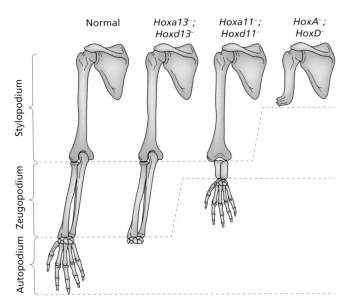

Normal   Hoxa13⁻;   Hoxa11⁻;   HoxA⁻;
         Hoxd13⁻   Hoxd11⁻   HoxD⁻

Stylopodium · Zeugopodium · Autopodium

Fig. 15.25 Limb phenotypes resulting from loss of Hox genes of the a and d clusters. In the case shown on the right the entire Hoxa and d clusters have been deleted. Reproduced from Zakany and Duboule. *Curr Opin Genet Dev* 2007; 17: 359–366, with permission from Elsevier.

of FGFs from the AER and of retinoic acid from the body axis may be sufficient to specify the different zones of gene upregulation.

The joints that form between the different cartilage elements are prespecified from an early stage. The initial cartilage condensation of the upper limb encompasses the future femur, ulna, and radius. Microsurgical ablation of the prospective elbow joint from this condensation prevents its later formation, so joints do not form simply because two cartilage condensations become apposed. The prospective joint regions express various genes including *Gdf5, -6, and -7* (BMP family) and *Wnt14* (now called *Wnt9a*). Loss of function of *Gdf5* and *-6* cause joint defects. Ectopic expression of *Wnt9a* will cause the formation of extra joints. *Wnt9a* works by upregulating its own expression at short range but repressing it at long range, hence causing formation of a periodic pattern of expression prefiguring the arrangement of joints.

Anteroposterior patterning

The **anteroposterior** pattern is controlled by the **zone of polarizing activity (ZPA)** located at the posterior margin of the limb bud. This behaves like the source of a diffusible morphogen, which is held at a constant concentration in the ZPA itself, can diffuse into the surrounding tissues, and is there destroyed or removed. Such a mechanism will establish an exponential concentration gradient from posterior to anterior (see Chapter 4). If a second ZPA is grafted to the anterior margin of the bud, it induces from the anterior tissue the formation of a second set

of structures in a **mirror symmetrical** arrangement to the original set (a double posterior duplication; Figs 15.26, 15.27). The structure of the duplication can vary from the six-digit 4–3–2–2–3–4 to the three-digit 4–3–4. The number of structures included in the duplication increases with distance between the two ZPAs. This is because if they are far apart the central minimum of the U is lower and so the thresholds for the formation of more anterior-type structures will be triggered.

The ZPA signal is the same in different vertebrate species, so grafts of lizard or mouse ZPA into the anterior side of the chick limb bud will also provoke the formation of a duplicated limb. All parts of the duplicate limb have a chick-type of morphology, rather than that of the donor species, because they are made from chick tissue. Limbs induced in the flank by FGF beads have an anteroposterior polarity that is inverted with regard to the normal limbs (Fig. 15.21). This is because the ZPA of the induced limb bud develops from the same region that forms the ZPA of the normal forelimb, so it is on the anterior rather than the posterior side of the bud.

Implantation of a slow-release bead containing retinoic acid (RA) has the same effect as a ZPA graft, inducing an extra set of posterior parts. Retinoic acid and retinoic acid receptors are present in the limb bud, and retinoid-deficient quail embryos, made by depleting the mothers' diets of vitamin A, show limb-bud abnormalities. However, retinoic acid is not the endogenous ZPA signal because transgenic mice containing a retinoic acid **reporter** (consisting of a retinoic acid response element linked to *lacZ*) do not show a gradient of retinoic acid, or indeed any retinoic acid activity at all, in the limb bud itself. There is retinoic acid signaling activity in the adjacent trunk of the body and its *in vivo* role is now thought to be the induction of the *Meis* genes, defining the most proximal part of the limb bud.

The morphogen from the ZPA is actually the product of the *Sonic hedgehog* gene (*Shh*). *Shh* is expressed in the region known to have ZPA activity on the basis of grafting experiments. Measurement of the amount of active Shh in explants of limb tissue using a tissue culture-based bioassay shows about five times higher concentration on the posterior than the anterior side. Slow-release beads or cells producing Shh can induce duplications just like the ZPA. In the chick Shh seems to behave just like the simple exponential gradient model would predict. But the mouse has a more complex mechanism, as progressive reduction of Shh dose leads to loss of digits in a different order from the expected posterior to anterior.

Shh can also support the continued existence of the AER and its *Fgf* expression, the activity formerly known as "AER maintenance factor." *Shh* expression can be upregulated by retinoic acid, but maintenance of its expression by the ZPA requires continued contact with the AER or with an FGF bead. This shows that the genes for Shh and FGF make up a positive-feedback loop. FGF from the apical ridge maintains the ZPA, and Shh from the ZPA maintains the apical ridge. The *Shh-*

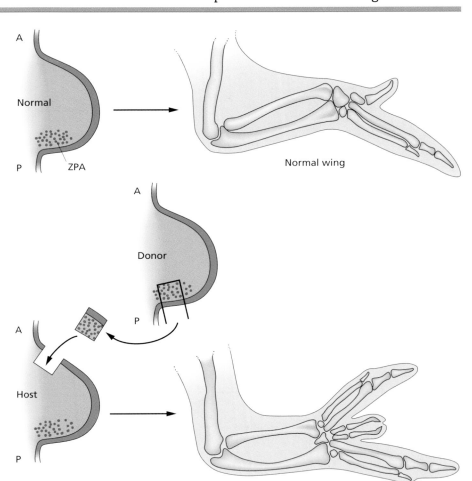

Fig. 15.26 The ZPA graft leads to the formation of a double posterior duplicated limb. Only the distal parts (digits) are usually affected because the more proximal parts are already determined when the graft is carried out.

Normal wing

Double posterior duplication

knockout mouse has abnormal limbs with severe distal defects and lacking anteroposterior polarity. As might be expected, the knockout limb buds contain no ZPA activity as assessed by grafting to the anterior side of normal limb buds. The knockout phenotype is due to the combination of loss of anteroposterior patterning and loss of support for the AER. The initial upregulation of *Shh* expression can be brought about by retinoic acid, but the mechanism that establishes it in normal development seems to be a mutual repression between the transcription factors dHAND and Gli3. dHAND is a bHLH factor, which is initially expressed in the whole limb mesenchyme, while Gli3 is in the Hedgehog signal transduction pathway. In normal development the two factors achieve an equilibrium with dHAND being expressed in the posterior of the future limb bud and activating expression of *Shh*, and Gli3 being expressed elsewhere. Evidence for the mutual repression is that if either factor is knocked out in the mouse, the other becomes expressed in a wider domain.

Gli3 is one of three Gli-type transcription factors lying in the Shh signaling pathway (see Appendix). In principle, Shh signaling stabilizes the Gli factors against degradation and enables them to enter the nucleus and regulate target genes. But in fact only Gli2 behaves this way. Gli1 is not very active and Gli3 is predominantly repressive, antagonizing the expression of the target genes. This explains why the mouse knockout of *Gli3* has too many digits (**polydactyly**). This is partly because the *dHAND*, and therefore also *Shh* domain, is bigger than usual, and also because the Shh signal is more effective than usual and induces digits from more of the limb bud. Loss or reduction of Gli3 is also a cause of polydactyly in humans, as seen for example in Greig syndrome.

The genes *Hoxd9, -10, -11, -12,* and *-13* have expression patterns showing nested territories running across the limb bud from posterior to anterior (Fig. 15.24b). They can be upregulated ectopically in this nested pattern by a ZPA graft, or by application of Shh. It has been widely thought that these genes encode

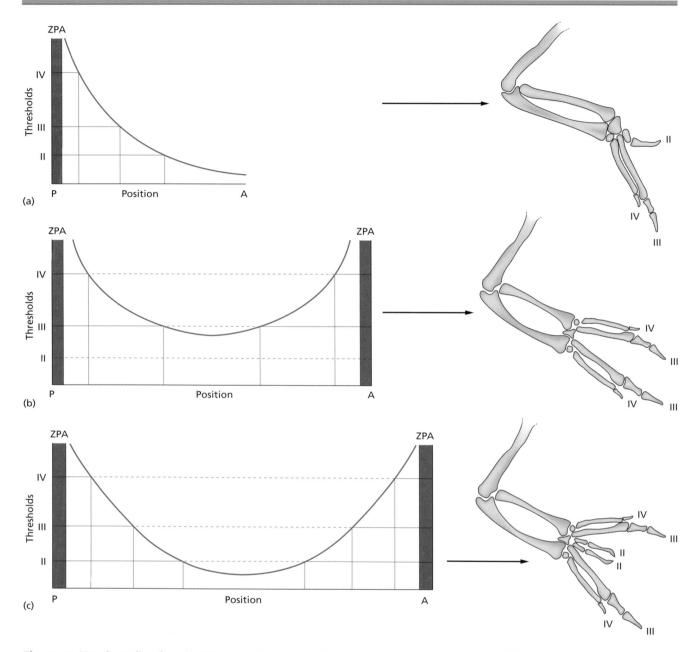

Fig. 15.27 How the gradient from the ZPA controls the pattern of digits: (a) normal development; (b) and (c) ZPA grafts. Because the separation of graft and host ZPAs is greater in (c), the minimum of the U-shaped gradient is lower and the duplication contains more elements than in (b).

states of determination for different parts of the limb. However, the results of mouse knockouts are not consistent with this idea as loss of subsets of the *Hoxd* cluster tend to give distal deletions rather than homeotic transformations. The *Hox* genes have dynamic expression patterns that change at different stages, and the nested patterns of Fig. 15.24 are found only at the stage indicated. As for the Hoxa genes, it remains unclear at which stage in limb development the Hoxd genes really are required,

and whether or not they really code for states of anteroposterior determination.

Dorsoventral patterning

The dorsoventral pattern is apparent mainly in the arrangement of muscles or of epidermal specializations such as feather germs and claws. The dorsoventral pattern of the internal limb tissues can be inverted by inverting the epidermis at an early stage. This

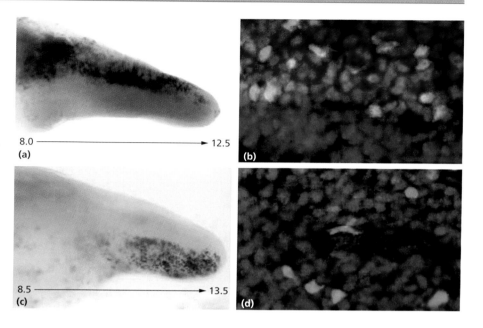

Fig. 15.28 Identification of a compartment boundary between dorsal and ventral mesenchyme in the mouse limb bud. Clones are induced early (E8.0 or 8.5) by a low dose of 4-hydroxytamoxifen to *Pol2-CreER x R26R* mice. In parts (a) and (c) clones are visualized in wholemounts by XGal staining, at E12.5. In parts (b) and (d), in section at higher power, clones are visualized by GFP fluorescence. Cells expressing the transcription factor LMX1B are immunostained red, nuclei are stained blue. Reproduced from Arques *et al.* (2007), with permission from Company of Biologists Ltd.

is achieved by removing the limb bud from the embryo, separating the epidermis from the mesenchyme by treatment with trypsin, recombining them with the epidermis inverted, then reattaching to the correct position on the embryo with small platinum pins.

The initial polarity of the epidermis comes from the fact that the dorsal epidermis was originally above the somites while the ventral epidermis overlay the lateral plate (see Fig. 15.22). *Engrailed-1* is normally expressed in the ventral epidermis. The *Engrailed-1* knockout mouse has a partial dorsalization of the ventral paw, suggesting that *Engrailed-1* encodes the ventral state, and, as we have seen, the knockout also has a ventral extension of the AER, suggesting that Engrailed-1 activity limits the extent of the AER. If the prospective ventral epidermis is separated from the underlying mesoderm by a barrier, then *Engrailed-1* does not become upregulated, and the resulting limbs are double-dorsal in character. This shows that an inducing factor from the lateral mesoderm is ultimately responsible for the dorsoventral polarity of the limb bud. The signal may be BMP4 which is produced by the lateral mesoderm and is also partly responsible for regionalizing the somites (see somitogenesis section above).

One of the functions of Engrailed-1 is to repress expression of *Wnt7a*, which is normally expressed in the dorsal but not the ventral epidermis. Wnt7a contributes to the maintenance of the *Shh* expression by the ZPA, and this indirectly helps maintain the AER. But it also has a key role in dorsoventral specification. Overexpression of *Wnt7a* produces double-dorsal limb and the *Wnt7a* knockout mouse has double-ventral paws. Hence, the evidence is good that Wnt7a is the signal from the epidermis conferring dorsal identity on the underlying limb-bud mesen-

chyme. Wnt7a turns on a LIM-homeobox gene, *Lmx1*, in dorsal mesenchyme. This has consequences for cell adhesion as the Lmx1 zone is the only known example of a vertebrate compartment boundary that does not correspond to a visible anatomical boundary. A compartment boundary is a line that is not crossed by clones of cells during normal development (see Chapters 4 and 17). This particular boundary has been visualized with a clonal labeling techniques based on using a low dose of tamoxifen to activate Cre recombinase in a very small proportion of cells. The *CreER* is driven by a ubiquitous promoter and when the Cre enzyme is active it causes recombination at the *R26R* reporter locus, generating a small number of well-separated clones (see Chapter 13). Using this method it is possible to see that, despite considerable cell dispersion during limb-bud outgrowth, the dorsal clones do not cross out of the Lmx1 zone and ventral clones do not cross into it (Fig. 15.28).

The limb differs from most other organs in that it is asymmetrical in three dimensions and because it is formed from a solid mass of tissue rather than by morphogenetic movements of cell sheets. The evidence is good that the pattern in each of the three axes is controlled by a separate process. The proximal–distal patterning is still not fully understood but relies on a continuous supply of FGF from the apical ridge and a retinoic acid signal from the proximal body region. The anteroposterior pattern is controlled by a gradient of Shh from the ZPA. The dorsoventral pattern is controlled by Wnt7a from the dorsal epidermis. Development in the three axes is tied together by the fact that the apical ridge requires both Shh and Wnt7a for its own continued survival and function (Fig. 15.29), and this means that the three processes work together to make an integrated organ.

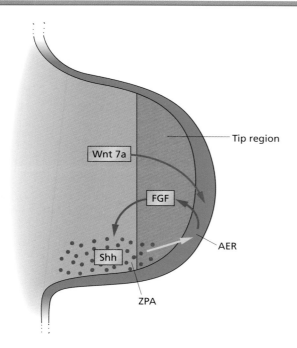

Fig. 15.29 The three signaling systems driving limb elongation and patterning.

Blood and blood vessels

Blood

The formation of blood is known as **hematopoiesis**. We shall deal here with developmental hematopoiesis and defer consideration of adult hematopoiesis until Chapter 18. The first development of blood cells in the mouse embryo is apparent by E7.5 as islands of cells in the yolk sac. Subsequently, hematopoiesis takes place in the **endothelium** of the major arteries (dorsal aorta, vitelline, and umbilical arteries), then in the liver, and finally in the bone marrow (Fig. 15.30). The yolk sac islands represent primitive hematopoiesis, while the later phases represent "definitive" hematopoiesis, which generates all cell types of the blood and immune system. Primitive hematopoiesis generates mostly primitive erythrocytes. These are much larger than the normal, definitive erythrocytes, and are mostly nucleated, although the nuclei do become lost over a few days. They contain an embryonic type of hemoglobin that has a higher oxygen affinity than the later fetal or adult forms. The primitive blood islands also give rise to some megakaryocytes and some primitive macrophages, from which are derived the **microglia** of the brain.

New directions for research

Much organogenesis research is motivated by practical medical objectives. There is an obvious interest in heart development in terms of isolating cardiogenic progenitor cells that could perhaps be expanded and used for cell therapy of damaged hearts. The same applies to the kidney, where there is a serious shortage of organs for transplantation and an ability to regenerate damaged organs would be very valuable.

The knowledge of the molecular differences between arteries and veins should be of importance for tissue engineering. It will not be possible to create real artificial organs without a vascular supply and the principles of molecular recognition at work in the formation of capillary beds will be critical for achieving this.

Better understanding of germ cell development should lead eventually to successful preservation and maturation of immature germ cells from children or young adults. This is important if the individuals are subject to cancer treatment, as preserved cells could be grafted back after treatment to restore fertility.

The genes encoding the three transcription factors GATA1 (Zn finger), SCL (bHLH type), and LMO2 (LIM type) are all required for primitive as well as definitive hematopoiesis. If mRNAs for these three factors are injected into *Xenopus* embryo blastomeres they can cause ectopic blood to form from most parts of the mesoderm. Although the early blood islands are enveloped in endothelial cells, at this stage these two cell types are though to arise independently. If "four-color" mice are made by injection of three differently labeled types of ES cell into the blastocyst, followed by reimplantation, then it may be seen that most blood islands are polyclonal, and that the blood progenitors and surrounding endothelium are usually differently labeled (Fig. 15.31), indicating origin from different cells.

The blood circulation commences in the mouse embryo about E8.5. From about E10.5 to 12.5 hematopoiesis is detected in the dorsal aorta and other arteries. This site has been called the **AGM (aorta–gonad–mesonephros) region**, but this designation was based on imprecise microsurgery of embryos, and it is actually the main central arteries that are relevant. In the dorsal aorta the blood cells can be observed delaminating from the ventral region of the vessel lining, or endothelium. It is possible to label the endothelium with DiI-Ac-LDL (acetylated low-density lipoprotein, preferentially taken up by endothelial cells) injected into the circulation. This was initially done with avian embryos and later with mice and in both cases shows that label passes from endothelial cells expressing endothelial markers

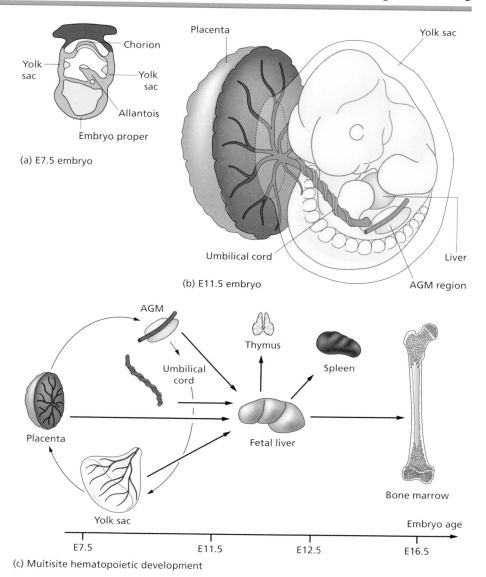

Fig. 15.30 Developmental hematopoiesis. (a) Primitive hematopoiesis takes place in the yolk sac. (b) Definitive hematopoiesis commences in the dorsal aorta and other central arteries. (c) Definitive progenitor cells migrate to the fetal liver, then the bone marrow and lymphoid organs. Reproduced from Medvinsky *et al. Development* 2011; 138: 1017–1031, with permission from Company of Biologists Ltd.

(such as Flk1, a VEF receptor), to cells of adherent clumps expressing both endothelial and hematopoietic markers, and thence to cells of the circulating blood. In addition, an adhesion molecule called VE-cadherin is expressed on endothelia. In a *VE-Cadherin-Cre x R26R* mouse, the label is found in a high proportion of bone marrow cells of the later adult. Furthermore, there is a test for the presence of definitive hematopoietic stem cells (see Chapter 18), which consists of injection into a lethally irradiated mouse to see if the graft can repopulate the blood of the host. Early embryo cells will not do this. From E10.5 to 11, a few cells in the region of the dorsal aorta will repopulate. In the *VE-Cadherin-Cre x R26R* mouse the repopulating cells are among those labeled, indicating their endothelial origin. The origin of the definitive blood has been controversial, but the evidence for an origin from the endothelium of central

arteries is now conclusive. A cell with the potential to produce either endothelium or blood is known as a **hemangioblast**, and an existing endothelium producing blood progenitors is called a **hemogenic endothelium**.

The transcription factor Runx1 is required for definitive, but not for primitive, hematopoiesis. *Runx1* is expressed in all the definitive hematopoietic cell populations. A *Runx1-CreER x R26R* mouse, labeled by injection of 4-hydroxytamoxifen at different times, indicates that the establishment of the definitive hematopoietic stem cells occurs in the period E9.5–10.5, just before the visible emergence of the hematopoietic clusters in the dorsal aorta, and just before the ability to recover repopulating cells from this region.

The Cre labeling experiments also confirm that the definitive hematopoietic cell populations found in dorsal aorta, fetal liver,

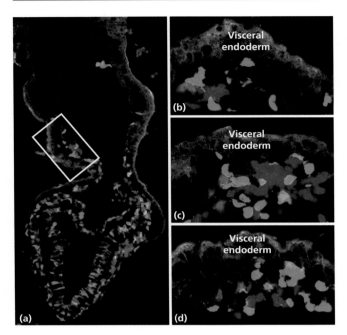

Fig. 15.31 Clonal analysis of a blood island in the yolk sac of a mouse embryo. The embryo was made by injecting ES cells labeled with red, green, and blue proteins into an unlabeled blastocyst. (a) The conceptus at neurula stage. The three-color labeling is apparent and a blood island is outlined with a box. (b–d) Three sections through this blood island indicating a polyclonal composition. Reproduced from Ueno and Weissman. *Int J Dev Biol* 2011; 54: 1019–1031, with permission from Elsevier.

and bone marrow are actually the same cells, which migrate from one site to another at different developmental stages. There has been considerable controversy about whether cells from the primitive blood islands contribute to the definitive populations. In *Xenopus* embryos it is possible to label three separate populations by injecting lineage labels into different early blastomeres (see Chapter 7). A D4 injection labels the posterior ventral blood islands, an erythropoietic population like that of the amniote yolk sac cells. A C1D1 injection labels anterior ventral blood islands, which form primitive myeloid cells. A C3 injection labels the AGM region and the definitive blood-forming tissue. This sharp segregation argues against a lineage relationship between primitive and definitive cells at least in *Xenopus*. In mouse some yolk sac cells can be traced to later hematopoietic populations, but these may already be definitive cells that migrated to the yolk sac just after the establishment of the circulation.

Blood vessels

De novo formation of blood vessels in the embryo is called **vasculogenesis** while the formation of new capillaries from existing ones by cell division and cell movement is known as **angiogen-**esis. Vasculogenesis occurs from lateral plate and extraembryonic mesoderm over a wide area to form primary capillary networks. Capillaries are composed of a single layer of endothelial cells with a basement membrane on the exterior surface. Larger blood vessels have a layer of smooth muscle and a layer of connective tissue outside the endothelium. Endothelial cells can divide throughout life and there is usually a low level of growth associated with tissue remodeling. The formation of new capillaries by cell division and cell movement may occur quite actively in wound healing or during the growth of tumors.

One of the most important findings about the developmental biology of the blood vessels is that the endothelia of arteries and veins are intrinsically different. When the main blood vessel rudiments for dorsal aorta and cardinal veins arise, the arterial cells produce ephrin B2 and the venous cells produce EphB4. The capillary beds of the later embryo arise by fusion of capillaries growing from the arterial and the venous sides, and the molecular complementarity between these molecules enables the two components to fuse together. The initial difference is though to depend on Sonic hedgehog from the notochord, which upregulates expression of the gene for VEGF (vascular endothelial growth factor, see below) in the region of the forming dorsal aorta, which in turn upregulates Notch pathway components. In the zebrafish the mutant *mindbomb*, which inhibits Notch signaling, shows a substitution of veins for arteries. In the mouse, knockout of the gene encoding the transcription factor COUP TFII will substitute arteries for veins.

A number of growth factors are active in promoting vasculogenesis and angiogenesis, particularly VEGF, FGFs, and angiopoietins. VEGF stimulates a set of receptors including Flk1 (= VEGFR2, Kdr), which are tyrosine kinases and are connected to several signaling pathways including ERK, p38, PI3K, and PLCγ (see Appendix). The extensive branching, which is characteristic of the vascular system, is promoted by VEGF. Tip cells are induced, which are highly polarized with many filopodia and each of these cells initiates a new branch. Their formation depends on the Notch lateral inhibition system (Fig. 15.32; and see Chapter 4). Following VEGF exposure the cell with the highest level of Delta4 in the vicinity becomes a tip cell and stimulates Notch on the surrounding cells, thus suppressing formation of additional tip cells.

Anatomists have known for centuries that major nerve tracts tend to be associated with major blood vessels. Now it is known that they are bound together by common signaling systems. This is because VEGF binds not only to Flk1 but also to neuropilins, which are receptors for semaphorins. Neuropilin 1 is expressed on arteries and neuropilin 2 on veins. The double knockout of both neuropilins not only has axonal growth defects but also has no blood vessels. VEGF acts as a positive signal and semaphorin 3A as a negative one for the growth of both nerves and vessels, ensuring that they will tend to run together. Moreover, nerves often secrete VEGF and so can be expected to attract growth of blood vessels directly.

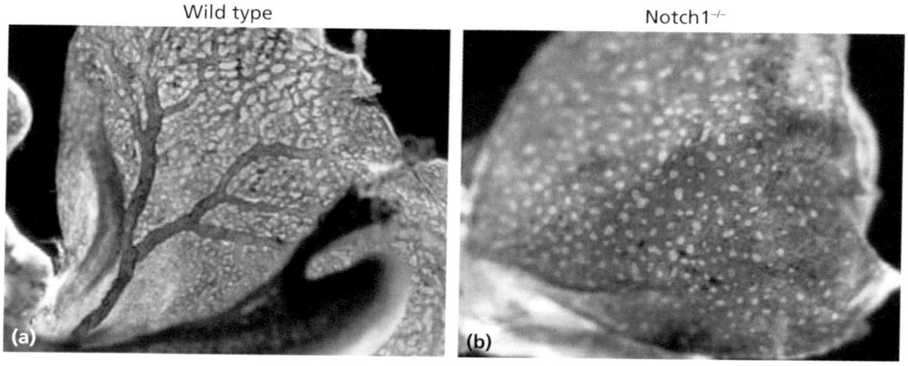

Fig. 15.32 Vascular system in the yolk sac of a mouse embryo. In (a) normal remodeling has occurred; in (b) due to loss of function of the Notch system, the primitive capillary plexus persists. Reproduced from Gridley. *Development* 2007; 134: 2709–2718.

The heart

The heart is very important in human embryology because as many as 1% of live births have a congenital defect of heart development, so heart development has become a very active area of research. The hearts of the different vertebrate classes all originate in a similar way although the end product is rather different. Fish have a heart consisting of a single muscular tube, divided into sinus venosus, atrium, ventricle, and outflow tract. This drives a single circulation through the gills to the body organs and back to the heart. The same basic regions exist in the hearts of higher vertebrates although the sinus venosus is reduced to the sinoatrial node in mammals. Amphibians have separate pulmonary and systemic circulations. The atria are separated but there is a single ventricle which somehow seems to be able to keep the two blood flows fairly separate. Birds and mammals have a complete double circulation with the right atrium receiving blood from the organs, the right ventricle sending it to the lungs, the left atrium receiving back from the lungs and the left ventricle sending out to the organs.

Since the double circulation cannot work until the lungs are functional, the developmental system has evolved to enable a rapid transition at the time of birth or hatching. In humans the main organ of fetal respiratory exchange is the placenta, whose blood returns through the umbilical vein to the right atrium. It then proceeds through the **foramen ovale**, a gap in the interatrial septum, and thence to the left ventricle and the systemic circulation. The output of the right ventricle is also diverted into the systemic circulation through a connection between the pulmonary artery and the aorta called the **ductus arteriosus**. At birth both the foramen ovale and the ductus arteriosus are closed, preventing fluid flow from right to left atrium and making the right ventricular output go through the lungs.

Heart tube formation and regionalization

The earliest stages of heart development are similar in the chick and mouse (Figs 15.33–15.35). Fate mapping by DiI labeling shows that the cardiogenic mesoderm originates from the epiblast lateral to the node. Cells from here pass through the anterior third of the primitive streak, forming lateral territories which then move anteriorly to form two elongated strips on either side of the embryonic axis. The more posterior part forms the epicardium and coronary arteries. As the head fold forms, the prospective cardiac mesoderm resolves into a crescent around its anteroventral edge, known as the cardiac crescent. Various transcription factor genes become expressed in this region. These include *Nkx2.5* (homeodomain), *Gata4-6* (zinc finger), *Mef2c* (MADS box), and *Tbx5* (T box). This is also the stage from which microsurgical interchange of regions or removal of explants causes later heart defects, so is taken to correspond to the time of determination. There is no real master regulator gene for formation of the heart, although the homolog of *Nkx2.5*, called *tinman*, does play this role in *Drosophila*, and *Nkx2.5* is very important in vertebrates. The expression pattern of *Nkx2.5* is shown for the chick embryo in Fig. 15.36. Cre-lox labeling experiments in mouse embryos shows that all the layers of the heart tube are formed from *Nkx2.5*-positive cells. In *Xenopus* it has been shown that overexpression of *Nkx2.5* will enlarge the heart, and a dominant negative version will suppress heart development completely. The mouse knockout for *Nkx2.5* does have a heart but its development is arrested at the looping stage with defects in the inflow and outflow tracts.

The cardiac mesoderm is induced by signals from the adjacent endoderm. Removal of the anterior endoderm leads to failure of heart development, and endoderm can induce cardiac differentiation when combined with mesoderm from areas outside the normal fate map region of the heart. An essential ingredient of the endoderm-derived signal is BMP. BMPs are expressed in the anterior endoderm, they will induce heart from at least some ectopic positions, and application of the BMP inhibitor Noggin will prevent heart development. Another ingredient is FGF8, also expressed in the anterior endoderm. This too will induce heart formation in mesoderm so long as BMP signaling is also present.

As the head lifts off the blastoderm surface and the foregut begins to form, the heart rudiments move underneath it towards

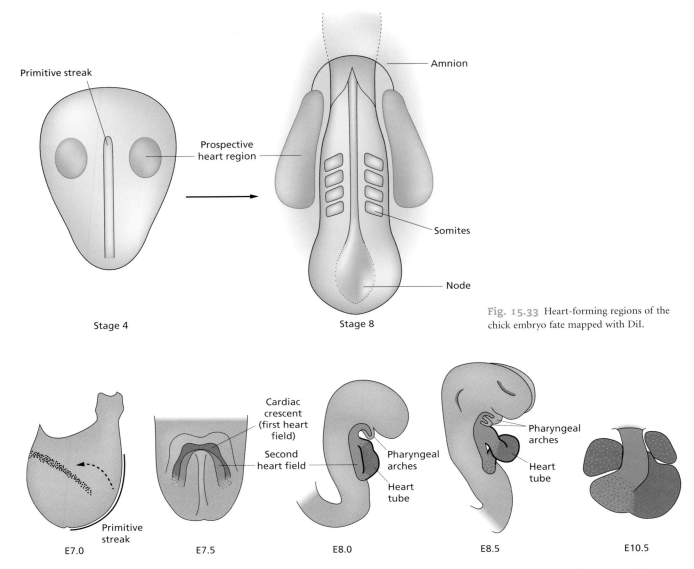

Fig. 15.33 Heart-forming regions of the chick embryo fate mapped with DiI.

Fig. 15.34 In the mouse embryo, cells from the lateral territories migrate through the primitive streak (ps) to become the cardiac crescent. The two sides fuse midventrally to form the cardiac tube. Cells are subsequently added to this from the second heart field. Reproduced from Rosenthal and Harvey (eds) (2010), *Heart Development and Regeneration*, with permission from Academic Press.

the midline (Fig. 15.35). This migration depends on fibronectin secreted by the anterior endoderm. A failure of migration leads to the condition of **cardia bifida** where two separate hearts form side by side. Normally, the two rudiments fuse to form a single tube, which has four layers: the **endocardium** within, a layer of extracellular matrix called the cardiac jelly, the **myocardium**, which forms the actual cardiac muscle, and the pericardium, which becomes the thin outer connective tissue sheath. The cardiogenic cells begin to segregate into endocardial and myocardial populations during the migration phase.

Shortly after fusion of the two rudiments, the heart tube begins to undergo slow pulsations. This occurs from stage 11 in the chick and E8.5 in the mouse. These contractions are an intrinsic property of the differentiating cardiac muscle and will occur in isolation in tissue or organ culture. *In vivo* the heartbeat will later become controlled by the pacemaker in the sinoatrial node. Also concurrently with fusion, the heart tube starts to become asymmetrical by looping to the right. This is seen in all vertebrates and depends on the left–right asymmetry system described in Chapter 9, which has the left-sided expression of *Nodal* as the principal common factor. The initial asymmetry leads to the asymmetric expression of various transcription factors affecting the heart.

The early formed heart tube is not the whole heart, but forms primarily the atria and the left ventricle. Further tissue continues to be recruited into its anterior end as regionalization and

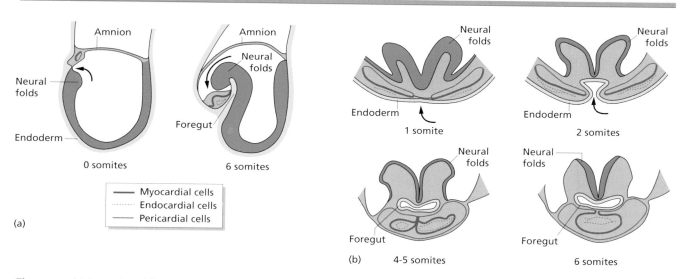

Fig. 15.35 (a) Formation of the anterior body fold in the mouse (arrows) brings the heart rudiment to an anteroventral position. (b) Transverse section through the early mouse heart showing fusion of the lateral mesodermal rudiments. The arrows indicate the formation of the foregut pocket. Reproduced from Rosenthal and Harvey (eds) (2010), *Heart Development and Regeneration*, with permission from Academic Press.

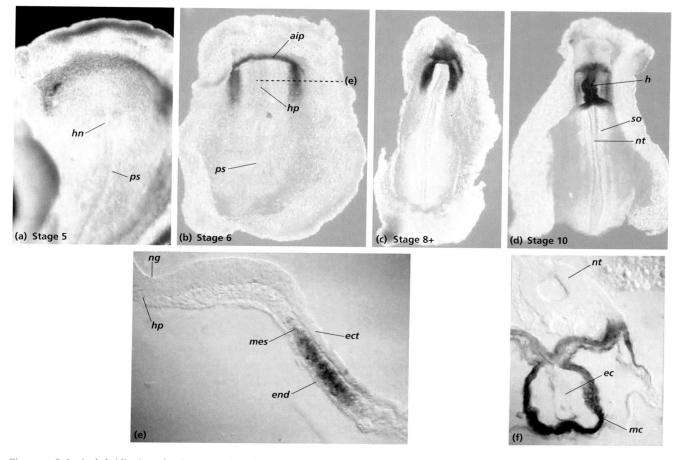

Fig. 15.36 *In situ* hybridizations showing expression of Nkx2.5 in the chick embryo: (a–d) wholemounts; (e) transverse section at stage 6, as shown on (b); (f) transverse section of heart at stage 10. hn, Hensen's node; ps, primitive streak; aip, anterior intestinal portal; hp, head process; h, heart; so, somites; nt, neural tube; ng, neural groove; ec, endocardium; mc, myocardium. Reproduced from Schultheiss *et al. Development* 1995; 121: 4203–4214, with permission from Company of Biologists Ltd.

looping proceeds, and forms most of the right ventricle and outflow tract. This comes from what is variously called the anterior, or second, or secondary, heart field. Definitions for these regions differ slightly but they represent tissue that is initially medial to the first heart field prior to gastrulation, then anterior to the region initially expressing *Nkx2.5*. They approximate to the region expressing the LIM-homeodomain transcription factor gene *Islet1*. Although *Nkx2.5* has been associated with the primary heart field and *Islet1* with the secondary heart field, it now seems from Cre-lox labeling studies that the whole heart rudiment expresses both *Islet1* and *Nkx2.5* at some level, and also that all three main cell types, namely cardiomyocytes, smooth muscle cells, and endothelial cells, are formed from progenitors expressing these genes.

Although most of the heart originates from anterior mesoderm, there is also an important contribution of **neural crest** cells. These mainly enter the developing outflow tract and the migration is controlled by signals from surrounding pharyngeal endoderm and secondary heart field, partly dependent on the T box transcription factor Tbx1. The neural crest-derived cells are responsible for forming the septum that divides the pulmonary and aortic circulation, and also part of the interventricular septum (see below).

Many transcription factor genes have been discovered to be critically important in heart development. *Nkx2.5* and *Islet1* have been mentioned already. Other key transcription factor genes include: *Mef2c*, whose knockout is an early lethal with severe heart defect; *COUP TFII*, required for formation of the atria; *dHAND*, required for formation of the right ventricle; *eHAND*, whose knockout is an early lethal because of a requirement in the placenta, but which is expressed in the forming left ventricle; and *Tbx5*, which is normally graded in expression from posterior to anterior and whose knockout has vestigial atria.

Development of cardiac septa

An aspect of heart development that has been very thoroughly studied in the human embryo is the process that converts the simple early heart tube into the four-chambered heart with separate right- and left-sided circulation (Fig. 15.37). This involves looping, asymmetric growth, movement of blood vessel insertion sites, and, most important, the formation of internal septa to divide up the lumen. More recently, mouse genetic and chick organ culture experiments have added to the picture.

First, the looping of the heart tube brings the atria to the anterior and the ventricles to the posterior (Fig. 15.37a). Then the insertion of the main veins move such that they run into the right side of the atrium, and a new pulmonary vein sprouts from the left side of the atrium. Then two atrial septa grow down towards the junction with the ventricles. These contain openings that will allow blood through until the time of birth. At the same time, four **endocardial cushions** appear at the atrioventricular junction. These arise from the endocardium by epithelial-to-mesenchymal transition, with the cells invading the hyaluronan-rich cardiac jelly, the transition being induced by TGFβ from the myocardium. Evidence for this is that addition of TGFβ will cause the transition, and addition of TGFβ inhibitors will prevent it. Two of the endocardial cushions meet to form a **septum intermedium**, dividing the ventricle into right and left sides. The other cushions later contribute to the atrioventricular valves: the biscupid or mitral on the left side, and the triscupid on the right side.

Initially, the primitive atrium is connected only to the prospective left ventricle and the prospective right ventricle is joined to the whole outflow tract. During the period of formation of the septum intermedium, a remodeling occurs that aligns what are now right and left atria with the corresponding side of the ventricle (Fig. 15.37b). At the same time the outflow tract becomes symmetrically related to the prospective ventricles. Then a muscular **ventricular septum** grows from the ventricular wall to separate the right and left sides. This is controlled by Tbx transcription factors, as demonstrated by some experiments done at the corresponding stage in the chick. Normally, *Tbx5* is expressed in the left ventricle and *Tbx20* in the right. If *Tbx5* is electroporated into a region of the right side, it will suppress *Tbx20* expression and a ventricular septum starts to form at the junction. Simultaneous with these events, longitudinal **truncoconal swellings** develop in the outflow tract and grow towards each other, eventually separating the pulmonary artery from the aorta (Fig. 15.37c).

All the various ingrowths, the atrial septa, the ventricular septum, and the septa of the outflow tract, eventually meet at the septum intermedium such that the heart has become fully divided into two. This process is generally similar in the chick as compared to mammals, except that there is just a single atrial septum and this has many small perforations instead of one large one.

The postnatal heart

In the mouse the cardiomyocytes of the heart have mostly ceased dividing by the time of birth. In the first postnatal week there is another round of DNA synthesis followed by nuclear, but not cell, division to make most of the cardiomyocytes binucleate. Immediately after birth damage to the mouse heart, either by trauma or following coronary artery ligation, can be regenerated by division of cardiomyocytes. But after 1 week or more, damage to the heart leads to permanent scarring rather than regeneration of cardiac muscle. In the human heart the same is true although there is evidence for a very low level of cell renewal (see Chapter 18).

In lower vertebrates (fish and amphibians) it is possible in the adult to remove the apex of the heart surgically and to obtain complete regeneration. Studies in zebrafish have shown that release of retinoic acid from the epicardium is a signal for

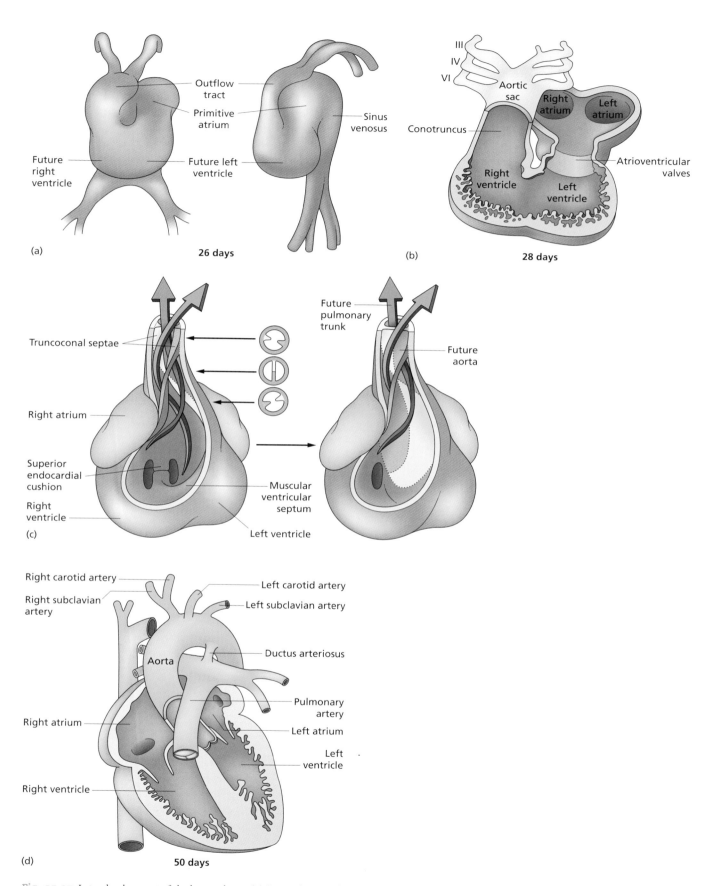

Fig. 15.37 Later development of the human heart. (a) External views of the heart after looping, 26 days. (b) Arrangement of future chambers at 28 days. (c) Formation of the longitudinal septum in the outflow tract. (d) Human embryo heart at 50 days. The four chambers and major blood vessels are all formed. (a) and (c) reproduced from Larsen 1993, *Human Embryology*. Churchill Livingstone, with permission from Elsevier. (b) and (d) reproduced from Srivastava. *Cell* 2006; 126: 1037–1048, with permission from Elsevier.

regeneration and the muscle is replaced from a population of cardiomyocytes lying near the epicardium.

Congenital heart defects

As heart development has become better understood, it is possible to explain at least some of the congenital human heart defects in terms of mutations found in the genes encoding one or other of the cardiac transcription factors. It should be noted that mouse knockouts of these genes are mostly lethal at embryonic stages, but the heterozygotes are usually apparently normal. By contrast, in human genetics many congenital heart problems are associated with heterozygotes of the same genes, the mutations being dominant and haploinsufficient. In other words, a viable but defective heart arises from a halving of the normal gene dosage. For this reason, although mouse mutants have been critical for understanding heart development, they do not necessarily model the human conditions very accurately.

A few examples are as follows. Mutations in the human *NKX2.5* gene are associated with several different heart defects, including defects of looping, of the conduction system, of atrial septa, of the triscupid valve, and the tetralogy of Fallot. As its name suggests, this last syndrome involves four defects: a narrowing of the pulmonary artery, a displacement of the aorta towards the right ventricle, a hole in the interventricular septum, and an enlargement of the right ventricle (which is secondary to the other problems). Mutations of *TBX5* cause Holt–Oram syndrome, which is notable as it involves defects both in the heart and the upper limbs, where *TBX5* is also normally required. Mutations of the gene encoding the transcription factor AP2 are associated with a patent ductus arteriosus between the aorta and the pulmonary artery. The involvement of the Notch system in the processes of septation and of remodeling to form the major blood vessels is apparent from Alagille syndrome, which involves the loss of one copy of the gene for the Notch ligand JAGGED1, and is associated with the tetralogy of Fallot, pulmonary artery stenosis, and also biliary atresia.

DiGeorge syndrome is a complex condition arising from a deletion on chromosome 22, which comprises maybe 45 genes. One of these is *TBX1*, required indirectly for the correct migration of neural crest cells into the outflow tract of the heart, and this lack is thought to account for the cardiac aspects of the syndrome.

Key points to remember

- The somites of the vertebrate body arise in anterior to posterior sequence, controlled by the posterior to anterior gradient of FGF. The number of cells in each somite is controlled by an oscillator based on the feedback inhibition of transcription of bHLH transcription factors of the Hes/Her family.
- The somite becomes divided into regions forming the muscle, the vertebrae, and the dermis in response to inducing signals from the surrounding tissues.
- Muscle differentiation is driven by bHLH transcription factors of the MyoD group. Individual myoblasts fuse to form multinucleate myofibers. Although the myofibers are postmitotic, satellite cells associated with them can divide and contribute to muscle growth.
- The functional kidney in higher vertebrates is the metanephros. This is formed by a reciprocal interaction between the ureteric bud from the nephric duct, and the metanephrogenic mesenchyme. The mesenchyme causes branching of the ureteric bud by a GDNF signal, and the ureteric bud induces tubules from the mesenchyme by signals including Wnt9b.
- Germ cells in mammals are induced from the egg cylinder epiblast by BMP from the extraembryonic ectoderm. They then migrate to the base of the allantois, round the hindgut, and up the mesentery to the genital ridges.
- The gonads develop from the genital ridges. In the male the SRY protein upregulates *Sox9*, which controls a battery of genes required for male development, including those required to suppress the Müllerian duct and to activate testosterone production from the Leydig cells. In the female there is no *Sry* gene, so no male pathway development. Instead, the Müllerian duct persists and becomes the female reproductive tract.

- The limb buds originate in the somatic mesoderm of the flank. They require a Wnt signal, an FGF signal, and expression of T-box transcription factors (Tbx5 or 4) to become established.
- The limb buds are asymmetrical in three dimensions and need three signals to control the formation of the pattern. One is associated with FGFs from the apical ectodermal ridge (AER). The second is Sonic hedgehog (Shh) from the zone of polarizing activity (ZPA). The third is Wnt7a from the dorsal epidermis. Shh and Wnt7a both help maintain FGF expression in the apical ridge, establishing a reciprocal feedback system.
- Primitive hematopoiesis, leading to formation of embryonic erythrocytes, occurs in the yolk sac of mammals or the homologous ventral mesoderm of lower vertebrates. Definitive hematopoiesis, leading to formation of the adult blood and immune system, commences from the endothelium of the dorsal aorta and other central arteries. The hematopoietic progenitors then migrate to the fetal liver and eventually to the bone marrow.
- Initial formation of blood vessels occurs from lateral plate mesoderm (vasculogenesis), with later formation by growth from pre-existing vessels (angiogenesis). Vascular endothelial growth factor (VEGF) is a key signal for angiogenesis. Arteries and veins differ in the expression of various receptors and adhesion molecules and this complementarity is essential to development of the capillary beds.
- The heart arises as bilateral mesodermal primordia that fuse in the midline to form a simple tube. There is a contribution of neural crest cells to the outflow tract. The four-chambered heart of higher vertebrates arises from the simple tube through the formation of various septa combined with remodeling of the proportions of the heart tube and of the insertion sites of major vessels.

Further reading

Somitogenesis

Burke, A.C., Nelson, C.E., Morgan, B.A. & Tabin, C.J. (1995) Hox genes and the evolution of vertebrate axial morphology. *Development* **121**, 333–346.

Christ, B., Huang, R. & Wilting, J. (2000) The development of the avian vertebral column. *Anatomy and Embryology* **202**, 179–194.

Ordahl, C.P., ed. (2000) *Somitogenesis.* Current Topics in Developmental Biology, Vol. 48, Parts 1 and 2. San Diego: Academic Press.

Brent, A.E. & Tabin, C.J. (2002) Developmental regulation of somite derivatives: muscle, cartilage and tendon. *Current Opinion in Genetics and Development* **12**, 548–557.

Giudicelli, F. & Lewis, J. (2004) The vertebrate segmentation clock. *Current Opinion in Genetics and Development* **14**, 407–414.

Dequéant, M.-L. & Pourquié, O. (2008) Segmental patterning of the vertebrate embryonic axis. *Nature Reviews Genetics* **9**, 370–382.

Pourquié, O. (2011) Vertebrate segmentation: from cyclic gene networks to scoliosis. *Cell* **145**, 650–663.

Myogenesis

Rudnicki, M.A. & Jaenisch, R. (1995) The MyoD family of transcription factors and skeletal myogenesis. *Bioessays* **17**, 203–209.

Pownall, M.E., Gustafsson, M.K. & Emerson, C.P. (2003) Myogenic regulatory factors and the specification of muscle progenitors in vertebrate embryos. *Annual Reviews of Cell and Developmental Biology* **18**, 747–783.

Horsley, V. & Pavlath, G. K. (2004) Forming a multinucleated cell: Molecules that regulate myoblast fusion. *Cells Tissues Organs* **176**, 67–78.

Buckingham, M. & Relaix, F. (2007) The role of Pax genes in the development of tissues and organs: Pax3 and Pax7 regulate muscle progenitor cell functions. *Annual Review of Cell and Developmental Biology* **23**, 645–673.

Bryson-Richardson, R.J. & Currie, P.D. (2008) The genetics of vertebrate myogenesis. *Nature Reviews Genetics* **9**, 632–646.

Buckingham, M. & Vincent, S.D. (2009) Distinct and dynamic myogenic populations in the vertebrate embryo. *Current Opinion in Genetics and Development* **19**, 444–453.

Kidney

The kidney development database: http://www.ana.ed.ac.uk/anatomy/database/kidbase

Sariola, H. & Sainio, K. (1997) The tip top branching ureter. *Current Biology* **9**, 877–884.

Kuure, S., Vuolteenaho, R. & Vainio, S. (2000) Kidney morphogenesis: cellular and molecular regulation. *Mechanisms of Development* **92**, 31–45.

Shah, M.M., Sampogna, R.V., Sakurai, H., Bush, K.T. & Nigam, S.K. (2004) Branching morphogenesis and kidney disease. *Development* **131**, 1449–1462.

Dressler, G.R. (2006) The cellular basis of kidney development. *Annual Review of Cell and Developmental Biology* **22**, 509–529.

Costantini, F. & Kopan, R. (2010) Patterning a complex organ: branching morphogenesis and nephron segmentation in kidney development. *Developmental Cell* **18**, 698–712.

Germ cells, gonads, and sex determination

Goodfellow, P.N. & Lovell-Badge, R. (1993) SRY and sex determination in mammals. *Annual Review of Genetics* **27**, 71–92.

Wylie, C. (1999) Germ cells. *Cell* **96**, 165–174.

Capel, B. (2000) The battle of the sexes. *Mechanisms of Development* **92**, 89–103.

Matova, N. & Cooley, L. (2001) Comparative aspects of animal oogenesis. *Developmental Biology* **231**, 291–320.

McLaren, A. (2003) Primordial germ cells in the mouse. *Developmental Biology* **262**, 1–15.

Raz, E. (2003) Primordial germ cell development: the zebrafish perspective. *Nature Reviews Genetics* **4**, 690–700.

Saga, Y. (2008) Mouse germ cell development during embryogenesis. *Current Opinion in Genetics and Development* **18**, 337–341.

Sasaki, H. & Matsui, Y. (2008) Epigenetic events in mammalian germ-cell development: reprogramming and beyond. *Nature Reviews Genetics* **9**, 129–140.

DeFalco, T. & Capel, B. (2009) Gonad morphogenesis in vertebrates: divergent means to a convergent end. *Annual Review of Cell and Developmental Biology* **25**, 457–482.

Sekido, R. & Lovell-Badge, R. (2009) Sex determination and SRY: down to a wink and a nudge? *Trends in Genetics* **25**, 19–29.

Kashimada, K. & Koopman, P. (2010) Sry: the master switch in mammalian sex determination. *Development* **137**, 3921–3930.

Limb development

Johnson, R.L. & Tabin, C.J. (1997) Molecular models for vertebrate limb development. *Cell* **90**, 979–990.

Martin, G.R. (1998) The roles of FGFs in the early development of vertebrate limbs. *Genes and Development* **12**, 1571–1586.

Capdevila, J. & Izpisúa-Belmonte, J.C. (2001) Patterning mechanisms controlling vertebrate limb development. *Annual Reviews of Cell and Developmental Biology* **17**, 87–132.

Christ, B. & Brand-Saberi, B. (2002) Limb muscle development. *International Journal of Developmental Biology* **46**, 905–914.

Niswander, L. (2003) Pattern formation: old models go out on a limb. *Nature Reviews Genetics* **4**, 133–143.

Arques, C.G., Doohan, R., Sharpe, J., & Torres, M. (2007) Cell tracing reveals a dorsoventral lineage restriction plane in the mouse limb bud mesenchyme. *Development* **134**, 3713–3722.

Sato, K., Koizumi, Y., Takahashi, M., Kuroiwa, A. & Tamura, K. (2007) Specification of cell fate along the proximal-distal axis in the developing chick limb bud. *Development* **134**, 1397–1406.

Tabin, C. & Wolpert, L. (2007) Rethinking the proximodistal axis of the vertebrate limb in the molecular era. *Genes and Development* **21**, 1433–1442.

Towers, M. & Tickle, C. (2009) Growing models of vertebrate limb development. *Development* **136**, 179–190.

Blood and blood vessels

Risau, W. & Flamme, I. (1995) Vasculogenesis. *Annual Reviews of Cell and Developmental Biology* **11**, 73–91.

Walmsley, M., Ciau-Uitz, A. & Patient, R. (2002) Adult and embryonic blood and endothelium derive from distinct precursor populations which are differentially programmed by BMP in *Xenopus*. *Development* **129**, 5683–5695.

Adams, R.H. (2003) Molecular control of arterial-venous blood vessel identity. *Journal of Anatomy* **202**, 105–112.

Ruhrberg, C. (2003) Growing and shaping the vascular tree: a multiple role for VEGF. *Bioessays* **25**, 1052–1060.

Orkin, S.H. & Zon, L.I. (2008) Hematopoiesis: an evolving paradigm for stem cell biology. *Cell* **132**, 631–644.

Ciau-Uitz, A., Liu, F. & Patient, R. (2010) Genetic control of hematopoietic development in *Xenopus* and zebrafish. *International Journal of Developmental Biology* **54**, 1139–1149.

Palis, J., Malik, J., McGrath, K.E. & Kingsley, P.D. (2010) Primitive erythropoiesis in the mammalian embryo. *International Journal of Developmental Biology* **54**, 1011–1018.

Medvinsky, A., Rybtsov, S. & Taoudi, S. (2011) Embryonic origin of the adult hematopoietic system: advances and questions. *Development* **138**, 1017–1031.

Adamo, L. & García-Cardeña, G. (2012)The vascular origin of hematopoietic cells. *Developmental Biology* **362**, 1–10.

The heart

Fishman, M.C. & Chien, K.R. (1997) Fashioning the vertebrate heart: earliest embryonic decisions. *Development* **124**, 2099–2117.

Harvey, R.P. (2002) Patterning the vertebrate heart. *Nature Reviews Genetics* **3**, 544–556.

Brand, T. (2003) Heart development: molecular insights into cardiac specification and early morphogenesis. *Developmental Biology* **258**, 1–19.

Garry, D.J. & Olson, E.N. (2006) A common progenitor at the heart of development. *Cell* **127**, 1101–1104.

Srivastava, D. (2006) Making or breaking the heart: from lineage determination to morphogenesis. *Cell* **126**, 1037–1048.

Abu-Issa, R. & Kirby, M.L. (2007) Heart field: from mesoderm to heart tube. *Annual Review of Cell and Developmental Biology* **23**, 45–68.

Dyer, L.A. & Kirby, M.L. (2009) The role of secondary heart field in cardiac development. *Developmental Biology* **336**, 137–144.

Musunuru, K., Domian, I.J. & Chien, K.R. (2010) Stem cell models of cardiac development and disease. *Annual Review of Cell and Developmental Biology* **26**, 667–687.

Rosenthal, N. & Harvey, R.P., eds (2010) *Heart Development and Regeneration*, 2 vols. London: Academic Press.

This chapter contains the following animation:

Animation 20 Operation of somite oscillator.

 For additional resources for this book visit www.essentialdevelopmentalbiology.com

Chapter 16

Development of endodermal organs

The endoderm is the innermost of the three germ layers formed during gastrulation. It forms the epithelial lining of the gut together with its outgrowths, which include the liver, pancreas, and respiratory system. The outer coats of these organs, comprising smooth muscle, connective tissue, and blood vessels, is formed from the **splanchnic mesoderm**, which is the inner subdivision of the lateral plate following the opening of the **coelomic** cavity. Just as the combination of somatic mesoderm and epidermis is called the **somatopleure**, the combination of splanchnic mesoderm and endoderm is known as the **splanchnopleure**.

Normal development

Formation of the gut tube in amniotes

In the chick the endoderm is formed during gastrulation as a lower layer of **definitive endoderm,** which emerges from the primitive streak and displaces the **hypoblast** cells to the periphery of the blastoderm. The gut lumen originates from the space beneath the blastoderm. The gut tube itself is formed by the folding of the body away from the blastoderm, which can be thought of as a sort of evagination of all three germ layers (Fig. 16.1). The formation of the head fold resembles the situation that would arise if a hypothetical miniature finger were pushed up from below to deform the blastoderm, then pushed sideways to elongate the projection. Thus, the anterior part of the body becomes lifted off the blastoderm, and the endoderm in this region becomes a tube of **foregut**. The junction of foregut and midgut, or **anterior intestinal portal**, is the level where the tube opens out into the flat sheet of endoderm. Somewhat later a similar process happens at the rear end and the **hindgut** is formed, with the junction between midgut and hindgut being the **posterior intestinal portal**. The mouth forms at the start of the third day by fusion of the ventral foregut with the overlying epidermis, and a **cloaca** forms at the posterior end, although does not become patent until hatching. The whole process is depicted in Animation 21: Early stage amniote gut morphogenesis.

Growth and morphogenetic movements increase the size of the extremities of the embryo such that the residual area of open endoderm rapidly shrinks. By 4 days this has become reduced to a **vitellointestinal duct** connecting the midgut to the yolk mass. As the gut closes, the splanchnic mesoderm from the two sides fuses to form dorsal and ventral mesenteries (Fig. 16.2a,b). The dorsal mesentery persists and holds the mature gut while the ventral mesentery is lost except in the vicinity of the heart, liver, and cloaca. The process continues by the epidermis and mesoderm closing around the gut to form the ventral body wall (Fig. 16.2c). This eventually becomes complete except for the umbilical tube, which is the connection to the extraembryonic tissues enclosing the vitellointestinal duct, the vitelline blood vessels, and the **allantois**. The allantois is a ventral outgrowth from the hindgut (Fig. 16.1b). From the second day it grows out of the embryo into the amniotic cavity and fuses with the chorion to form the **chorioallantoic membrane** (see Fig. 9.7). This is the richly vascularized membrane that is visible immediately on opening the egg of a late embryo and it serves as an important respiratory organ before hatching. The stalk and

Essential Developmental Biology, Third Edition. Jonathan M.W. Slack.
© 2013 John Wiley & Sons, Ltd. Published 2013 by John Wiley & Sons, Ltd.

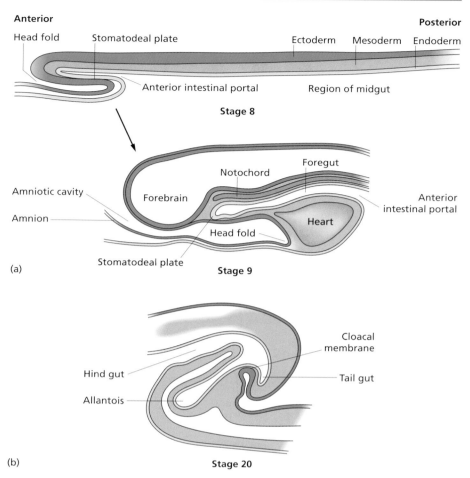

Fig. 16.1 Formation of the fore- and hindgut in the chick embryo. (a) Formation of headfold and foregut. (b) Formation of hindgut (later stage). Reproduced from Bellairs and Osmond (1998) *The Atlas of Chick Development*, with permission from Academic Press.

blood vessels of the allantois become incorporated into the umbilical tube. By about 4 days of incubation the chick gut is fully formed as a tube and has commenced asymmetric coiling, controlled by the left–right asymmetry system described in Chapter 9. It is also starting to show some regional pattern in terms of the shape and size of the various organ rudiments (Fig. 16.3).

The course of events in other higher vertebrates is fairly similar. In the mouse the definitive endoderm starts out as a ventral strip of tissue on the outside of the egg cylinder, bordered by visceral endoderm. It becomes incorporated into the fore- and hindgut pockets in a similar way to the chick, assisted by the "**turning**" process described in Chapter 10. The overall structure of the gut is also similar. The allantois of the mouse embryo also contributes to the placenta, although it is composed only of mesoderm.

Organization of the gut tube

In the region of the pharynx are formed the **pharyngeal pouches**, which are aligned with the branchial arches formed from alternating rhombomeres (Chapter 14). There are four major pouches, which are associated with a segmental arrangement of endodermal outgrowths, and in some species one or two more rudimentary ones (Fig. 16.4). Buds from the first pouch form the cavities of the middle ear and the Eustachian tubes. The ventral midline region opposite the second arch forms the thyroid gland. The second pair of pouches (in mammals) form the tonsils, and both the third and fourth pairs the thymus and the parathyroids. In the ventral midline, opposite the fourth pouches, arises the laryngotracheal groove, which becomes the trachea and produces paired buds generating the bronchi and tissue of the lungs. Much of the floor of the pharynx becomes the tongue, which is mostly composed of muscle derived from the head mesoderm.

Regional differentiation of the gut tube is indicated in Fig. 16.5. In anteroposterior sequence, the pharyngeal region comes first, then the esophagus, which is initially lined with columnar and later with stratified squamous epithelium. Then comes the stomach, where the epithelium becomes glandular and specialized to secrete hydrochloric acid and pepsin. At the exit of the stomach is the pyloric sphincter, leading into the small intestine where the embryonic epithelium is columnar but on

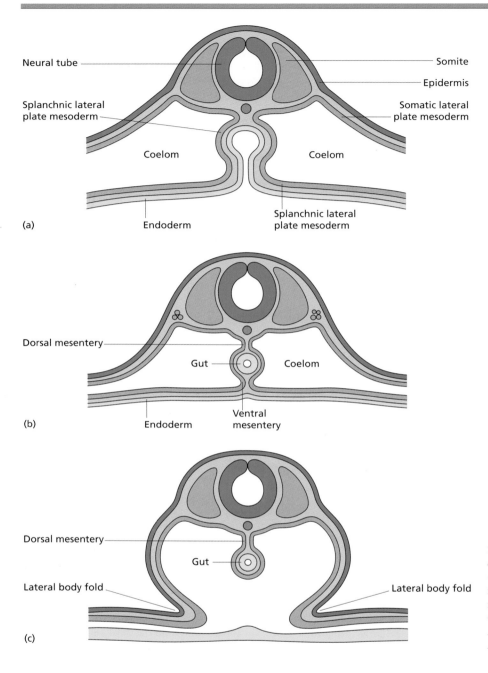

(a)

(b)

(c)

Fig. 16.2 Enclosure of the gut and ventral closure of the body in the chick embryo. (a–c) represent transverse sections between midgut and hindgut levels at 3 days of incubation. Reproduced from Bellairs and Osmond (1998) *The Atlas of Chick Development*, with permission from Academic Press.

maturation differentiates into the crypt and villus arrangement described in Chapter 18 (Fig. 16.6). In the mature animal the junctions between the esophagus and stomach, and the stomach and intestine, are two of a very small number of places in the body where there is a sharp discontinuity of epithelial type which has arisen *in situ* from a common cell sheet.

The first section of small intestine is called the duodenum and this forms a loop to which are attached the liver and the pancreas. The liver is a ventral outgrowth of the endodermal epithelium that expands into the adjacent ventral mesentery shortly after formation of this part of the foregut. This region of the mesoderm is called the **septum transversum** in mammals. The pancreas arises from a large dorsal bud and a small ventral one (paired ventral buds in the chick). These move together and fuse in later development to form a single organ. The ventral pancreas is closely associated with the liver and the ventral pancreatic duct fuses with that of the liver to form the common bile duct. The more posterior parts of the small intestine are called the jejunum and the ileum. The junction between small and large intestine (or colon) is often marked by the pres-

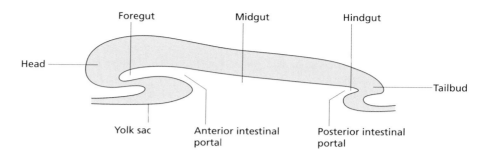

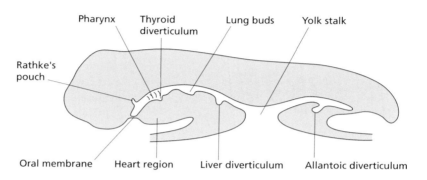

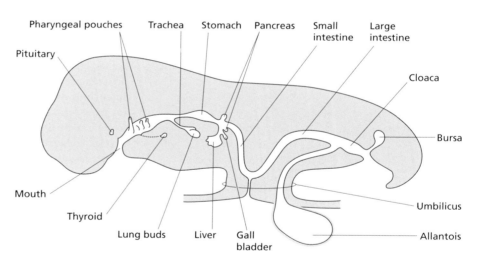

Fig. 16.3 Formation of the regions of the gut in an amniote embryo. Reproduced from Hildebrand (1995) *Analysis of Vertebrate Structure*, 4th edn, with permission from John Wiley & Sons Ltd.

ence of outgrowths called ceca, which are blind-ended sacs lined with intestinal epithelium. The mature large intestinal epithelium resembles that of the small intestine but there are no villi and no Paneth cells in the crypts. At the posterior end the large intestine becomes a rectum and in mammals joins to the exterior through the anal sphincter.

In the mature gut the terms fore-, mid-, and hindgut are still used, but they no longer relate to the gut segments enclosed by the anterior and posterior intestinal portals because these are moving boundaries during the formation of the gut tube. Conventionally, the **foregut** at late stages is taken to extend as far as the pancreas and liver, and the **midgut** as far as the junction between small and large intestine.

This account is generic for most amniote vertebrates, but experimental work is largely performed on the chick and the mouse, and they are not quite the same. The following check list is provided to keep track of the differences:

• The mouse has tonsils in the second pharyngeal pouch, the chick does not.

• The mouse incorporates the ultimobranchial body into the thyroid as C (calcitonin) cells, the chick keeps them separate.

• The mouse lungs have alveoli, the chick lungs have anastomosing channels running into large distal air sacs.

• The chick has a specialization of the esophagus forming the crop, the mouse does not.

- The chick has a division of the stomach into a glandular **proventriculus** and a muscular **gizzard**, the mouse has no gizzard
- The mouse embryo liver is a major hematopoietic organ while the chick liver is not.
- The chick has paired ventral pancreatic buds, the mouse has one.

- The mouse has one large cecum between small and large intestines, the chick has two small ones.
- The chick has a bursa of Fabricius, which produces B lymphocytes, near the cloaca, and the mouse does not.
- The chick has a **cloaca** while the mouse has separate urinary/reproductive and alimentary orifices.

Fate map of the endoderm

In the chick the endoderm has been fate mapped by several methods, including **orthotopic grafting** of quail tissue and small extracellular injections of **DiI**, followed by *in vitro* culture of the whole embryo until the gut tube has closed. Comparable studies have also been made in mouse and *Xenopus*. As might be expected, there is a broad conservation of anterior–posterior relationships, so the anterior endoderm becomes the foregut, the middle section becomes the midgut, and the posterior endoderm becomes the hindgut (Fig. 16.7a). Because the early endoderm is an open sheet, its midline becomes the dorsal midline of the gut tube and therefore structures like the dorsal pancreatic bud are represented on the midline of the fate map. The ventral midline structures, like the liver, arise from the edges of the definitive endoderm, which come together as the anterior and posterior portals move towards one another. This is why there are often paired buds for ventral structures.

One very important aspect of the fate map is the relationship between prospective regions in the endoderm and those in the **splanchnic mesoderm** (Fig. 16.7b). The prospective regions in the mesoderm are quite different from those in the endoderm, tending to be arranged longitudinally. This shows that there is a considerable relative movement of the two germ layers during the process of closure of the gut and closure of the ventral body wall. An important theme in gut development is the inductive interactions between mesoderm and endoderm and the fate map shows clearly that a particular region of endoderm experiences contact with different regions of the mesoderm as the

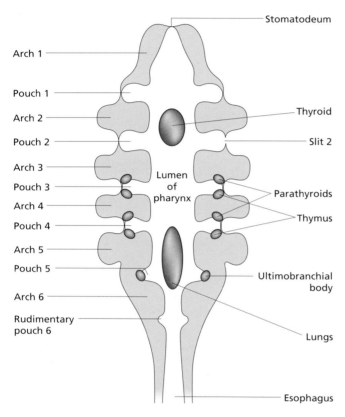

Fig. 16.4 Location of outgrowths of the endoderm in the pharyngeal region of the chick embryo.

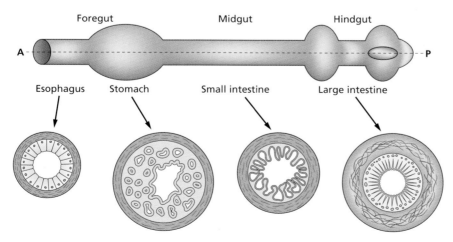

Fig. 16.5 Cytodifferentiation of regions of the chick gut tube.

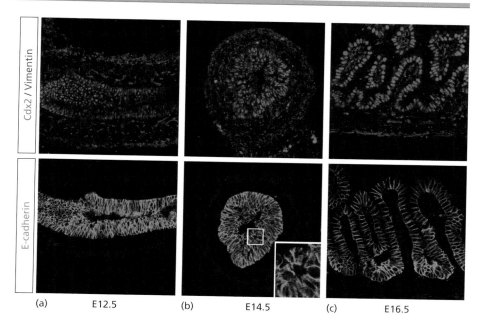

Fig. 16.6 Development of the mouse intestine. Red is immunostain for the transcription factor Cdx2, which encodes intestinal character, blue is vimentin, indicative of mesenchyme, green is E-cadherin, a cell adhesion molecule characteristic of epithelia. (a) Initially, the intestine is a simple columnar epithelial tube surrounded by mesenchyme. (b) Multilayered epithelium. (c) Beginning of resolution of the epithelium into a pattern of crypts and villi. Reproduced from Spence *et al. Dev Dynamics* 2011; 240: 501–520, with permission from John Wiley & Sons Ltd

process of gut tube formation takes place. This means that the state of specification of the endoderm may change as it contacts different mesenchymes.

Experimental analysis of endoderm development

Determination of the endoderm

The initial formation of the endoderm has been investigated mainly in *Xenopus* and zebrafish. In *Xenopus* the endoderm forms from the tissue roughly corresponding to the eight most vegetal blastomeres at the 32-cell stage. The endoderm is formed because of the presence of mRNA for the transcription factor VegT. This activates the expression of various further transcription factor genes including *Sox17α* and *β*, *Mix1*, *Mixer*, and *GATA1-4* (Fig. 16.8). Some of these are upregulated directly but others require inductive signals. In particular, the expression of *Mixer* and *GATA4* require Nodal signaling. In zebrafish there is more dependence on Nodal signaling as the endoderm is not formed at all in the *cyclops/squint* double mutant that lacks both of the *Nodal* genes. A Sox family gene called *casanova* and a POU domain homolog of *Oct4* called *spiel-ohne-grenzen* are both controlled by Nodal. Likewise in mice, mutants that are defective in Nodal signaling do not form the definitive endoderm. Nodal is initially expressed throughout the epiblast and then becomes confined to the primitive streak. The spread of activity is restricted because one of the targets of Nodal is *Lefty*, which encodes an inhibitor that is more diffusible than Nodal, so Nodal prevails in the region near the source but is inhibited further away. Work with embryonic stem cells has revealed the

transient existence of cells expressing both mesodermal and endodermal markers. This has been called **mesendoderm**. Its transitory nature is due to the strong mutual inhibition exerted by several transcription factors; for example Brachyury, characteristic of mesoderm, represses expression of *Mixer*, characteristic of endoderm, and vice versa.

The various endodermal transcription factors show a considerable mutual interdependence. They will mostly upregulate the other factors if overexpressed in *Xenopus* animal caps, and the introduction of dominant negative constructs made with the engrailed repression domain will generally suppress endoderm development. A similar set of transcription factors are expressed in the early endoderm of zebrafish and mouse, and mutants show endodermal defects of various degrees of severity, for example the mouse knockouts of *Mixl1* lacks definitive endoderm, of *Foxa2* lacks the foregut, and of the *Sox17* genes lack the mid- and hind gut.

Regional specification

The transcription factors mentioned above specify endoderm as a germ layer rather than any specific type of epithelium or organ. There are many other transcription factors that are expressed at specific levels within the endoderm (Fig. 16.9). Some of these are known to be important for regional specification because of their mutant phenotype. *Pax9* (paired domain) is expressed in all the pharyngeal pouches and the associated knockout mouse has no thymus, parathyroid, or ultimobranchial body. *Nkx2.1* (homeodomain, = *TTF1*) is expressed in the thyroid bud and in the tracheoesophageal septum. The knockout mouse has no thyroid or lungs. *Sox2* is expressed in

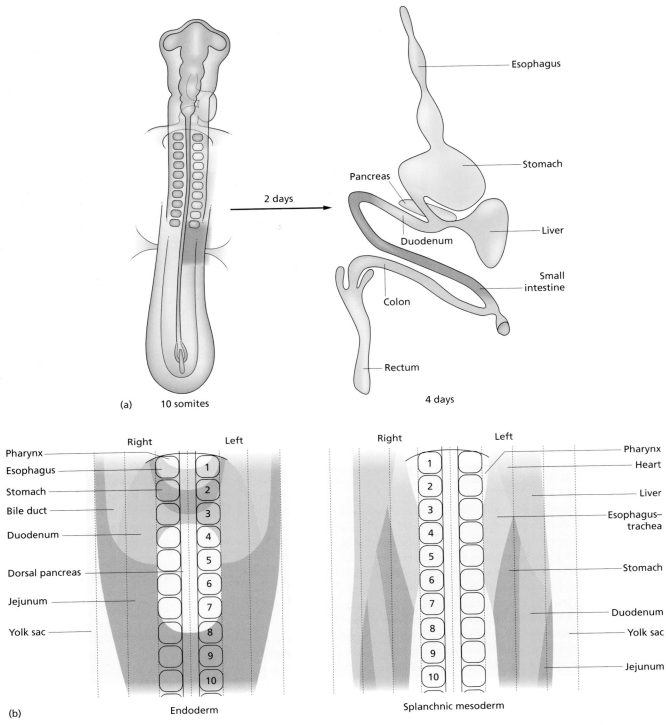

Fig. 16.7 (a) Fate map of the chick endoderm. (Modified from Kumar et al. *Developmental Biology* 2003; **259**, 109–122.) (b) Mismatch between the fates of endoderm and splanchnic mesoderm at 1.5 days. (Modified from Matsushita. *Roux's Archives of Devel Biol* 1996; 205; 225–231.)

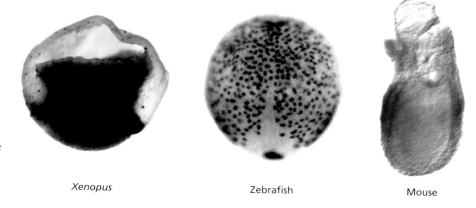

Fig. 16.8 Expression of *Sox 17*, visualized by *in situ* hybridization, marks the early endoderm of Xenopus, zebrafish and mouse embryos. Reproduced from Zorn and Wells. *Ann Rev Cell Dev Biol* 2009; 25: 221–251, with permission from Annual Reviews.

Xenopus   Zebrafish   Mouse

Classic experiments

Master genes of the pancreas

Pdx1 (then called *IPF1*) was discovered as a transcription factor controlling *insulin* expression. But when its gene was knocked out it was found to be necessary for development of the whole pancreas (first paper). How a fraction of the endodermal cells were diverted to become endocrine cells was mysterious until the discovery of a lateral inhibition system similar to that controlling neurogenesis. The second paper shows how *Neurogenin3* is

essential for endocrine cell formation and how Notch signaling represses endocrine development among the neighbors of *Neurogenin3*-expressing cells.

Jonsson, J., Carlsson, L., Edlund, T. & Edlund, H. (1994) Insulin-promoter-factor-1 is required for pancreas development in mice. *Nature* **371**, 606–609.

Apelqvist, A., Li, H., Sommer, L., Beatus, P., Anderson, D.J., Honjo, T., de Angelis, M.H., Lendahl, U. & Edlund, H. (1999) Notch signalling controls pancreatic cell differentiation. *Nature* **400**, 877–881.

part of the foregut and is needed for formation of the esophagus. *Hex* (= *HHex*, homeodomain) is expressed in the foregut region and is required for formation of the liver. *Pdx1* (homeodomain, = *XlHbox8* in *Xenopus*) is expressed in a region of the proximal intestine including the future pancreatic buds and the knockout mouse has no pancreas (see below). *Foxa1 and -2* (forkhead domain) are expressed in the whole endodermal epithelium, and *Foxa3* has an anterior boundary at the level of the future liver bud. The knockout of *Foxa2* shows a loss of fore- and midgut. *HNF4α* (nuclear hormone receptor) is expressed posterior to the future esophageal-gastric junction. *Cdx2* (homeodomain, = *CdxC* in chick) is expressed in the posterior part of the gut corresponding to most of the intestine. The knockout of *Cdx2* is an early lethal because it is also needed in the trophectoderm, but gut-specific ablation using *Foxa3-Cre x floxed Cd2* generates a mouse with esophagus-like epithelium in the intestine. This is notable as it is a rare example in vertebrates of a clear **homeotic** transformation, of posterior to anterior gut epithelium, depending on loss of function of a single gene.

Several of the endodermal transcription factor genes are expressed again at a later stage and have functions during the terminal differentiation of the organs concerned. For example *Pdx1* not only controls formation of the whole pancreas, but

later on it is specifically expressed in β-cells and helps to control *insulin* expression. *Nkx2.1* not only controls formation of the whole thyroid gland, but later controls expression of the thyroid-specific genes *thyroglobulin* and *thyroperoxidase*.

A number of transcription factors are regionally expressed in the splanchnic mesoderm rather than the endodermal epithelium. Some affect differentiation of the mesoderm directly. For example loss of the whole HoxD cluster (except *d1* and *d3*) will provoke loss of the ileocecal sphincter, a muscular structure formed from the mesoderm. Others are probably important in controlling regional inductive signals from the mesoderm that control regional pattern in the endoderm (see below). For example misexpression using retrovirus of *Hoxd13* in the midgut mesenchyme causes the midgut epithelium to develop as large intestine rather than small intestine.

Epithelial–mesenchymal interactions

By the stage of the primitive gut tube, the splanchnic mesoderm is referred to as a **mesenchyme**, since it has become a layer of loosely packed, undifferentiated cells surrounding the epithelial tube. Classical recombination experiments in which

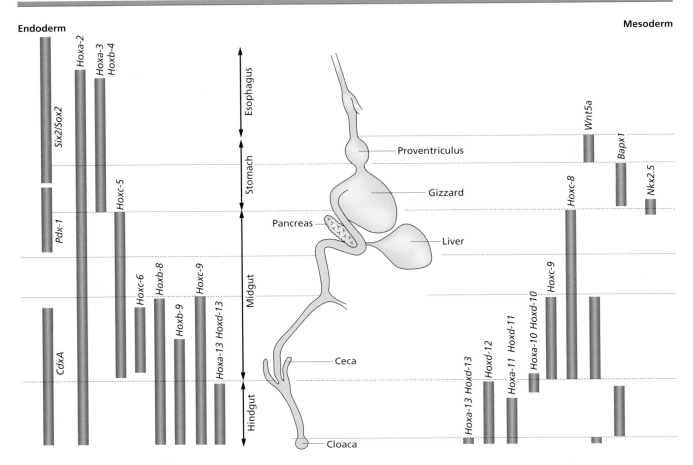

Fig. 16.9 Expression domains of transcription factors in the endodermal and mesodermal layers of the early chick gut. (Modified from Roberts. *Developmental Dynamics* 2000; 219: 109–120.)

mesenchyme and epithelium from different parts of the developing gut were recombined indicated that the mesenchyme imparted regional specificity to the endoderm (Fig. 16.10). However, the situation is more complex than this; partly because there is a continuous relative displacement of mesenchyme and epithelium, shown by the fate mapping experiments mentioned above, and partly because there is also considerable signaling from endoderm to mesenchyme, so each gut region becomes determined to form its specific tissue type as a result of a complex set of local and reciprocal interactions.

The initial interactions are long range and are mediated by the same posteriorizing signals that are responsible for the whole body pattern, mainly FGFs and Wnts. These induce and maintain expression of the *Cdx* genes, and thereby of posterior-type Hox genes in the future mid- and hindgut. Recombination experiments with the chick embryo gut show a "posterior dominance," meaning that anterior endoderm can be posteriorized by contact with posterior mesoderm, but posterior endoderm cannot be reprogrammed by anterior mesoderm (Fig. 16.11). In particular, the future intestinal region, expressing *CdxA*, appears fixed in its commitment from before 1.5 days of incubation,

while the prospective esophagus and stomach regions remain competent to form intestinal epithelium up to about 4.5 days of incubation. This type of behavior is to be expected in response to graded signals that activate genes in a bistable manner (see Chapter 4). Because of the multiplicity of functions of these factors and the redundancy of different gene products, the knockout phenotypes are often not informative, but often they do show the expected posterior defects. Retinoic acid, formerly considered a posteriorizing factor, exerts most of its effects in the foregut region. The mouse knockout of *Raldh2*, encoding the enzyme for retinoic acid production, lacks lungs, stomach, and dorsal pancreas, and has a defective liver.

The chick stomach, unlike that of the mammal, is divided into two quite different parts. The anterior **proventriculus** has a glandular epithelial lining similar to the mammalian stomach, while the posterior **gizzard** has a **squamous** epithelial lining and a thick muscular wall, designed for churning and macerating the food. These two areas of epithelium remain interconvertable until about 9 days of incubation if they are cultured with mesenchyme from the other. The mesenchyme of the proventriculus expresses several BMPs, while that of the esophagus

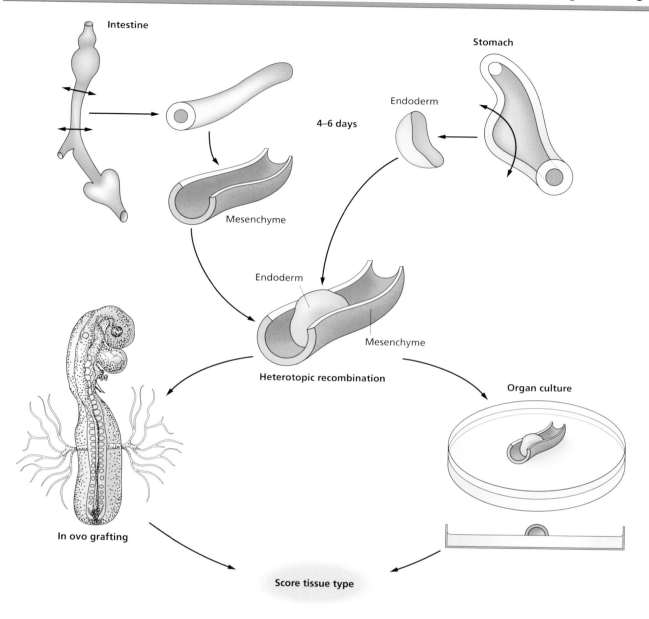

Intestine

Stomach

Endoderm

4–6 days

Mesenchyme

Endoderm

Mesenchyme

Heterotopic recombination

Organ culture

In ovo grafting

Score tissue type

Fig. 16.10 Protocols for recombination experiments between endoderm and mesenchyme of the chick embryo gut.

and gizzard does not, and there is good evidence of the importance of BMP as the inductive signal. Treatment of either esophagus or gizzard epithelium with BMP will cause them to differentiate proventriculus-like glands, and this effect is inhibited by Noggin. The transcription factor Nkx2.3 (homeodomain, = Bapx1) is normally expressed in the gizzard mesenchyme and prevents Shh from the endoderm from activating BMP synthesis in the mesenchyme. Introduction of this factor into the proventriculus mesenchyme will cause the adjacent epithelium to develop as gizzard rather than proventriculus. At the distal end of the gizzard the BMP signaling causes a ring of mesenchyme to express Nkx2.5 (the same transcription factor

that controls heart development), and this controls formation of the pyloric sphincter.

Another example of mesenchymal–epithelial signaling is the formation of the liver and its distinction from the ventral pancreas. The liver arises from a ventral diverticulum of the foregut that proliferates into the ventral mesenchyme, this mesenchyme being called the **septum transversum** in mammalian embryos (Fig. 16.12). But the liver diverticulum will not form unless it receives a signal from the adjacent cardiogenic mesoderm. There is good evidence that this signal is composed of FGF, since the cardiogenic mesoderm makes FGFs, FGF inhibitors will block the signal, and FGFs will mimic the effect on isolated

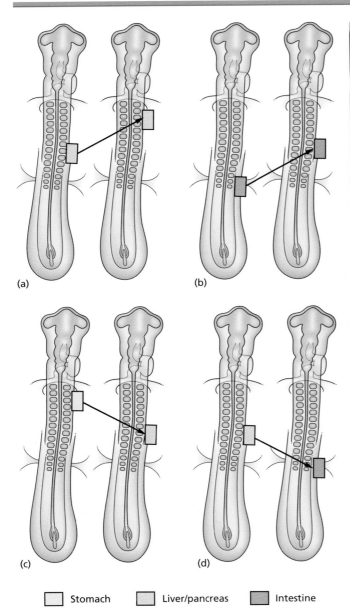

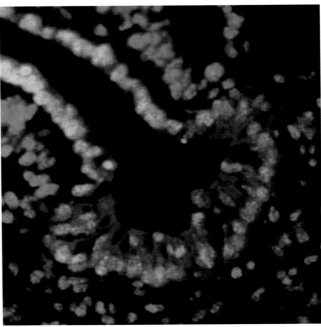

Fig. 16.12 Early liver bud in a mouse embryo. Hepatoblasts are blue due to a reporter gene (*Hex-lacZ*), and the gut endoderm is orange because of immunostaining for Foxa2. Cell nuclei are green. Reproduced from Zaret and Grompe. *Science* 2008; 322: 1490–1494, with permission from the American Association for the Advancement of Science.

Stomach ☐ **Liver/pancreas** ☐ **Intestine** ▉

Fig. 16.11 *In ovo* grafting experiments showing posterior dominance. (a) Anterior granft of prospective liver/pancreas. (b) Anterior graft of prospective intestine. (c) Posterior graft of prospective stomach. (d) Posterior graft of prospective liver/pancreas. The colors indicate the structures as shown in Fig. 16.7a.

hepatic endoderm (Fig. 16.13). Although FGFs are also involved in the early regionalization of the endoderm as posteriorizing factors, posteriorization occurs at the primitive streak stage, which is very much earlier than the local interaction involved in the formation of the liver, so it is possible to use the same signal again for a different purpose. In addition, the liver-inducing signal probably includes BMP and also a component emitted by the blood vessels. The evidence for this is that in recombination experiments normal hepatic endoderm will not develop well if the mesenchyme comes from an embryo lacking

Bmp4, or lacking blood vessels as a result of knockout of the gene for the VEGF receptor Flk1. Once the liver bud is formed its later growth requires Wnt activity, as shown by the fact that retroviral overexpression of β-catenin in the chick liver will promote growth, and overexpression of Wnt inhibitors will reduce growth. The early epithelial cells are called hepatoblasts and have the ability to form both hepatocytes, the main cell type of the mature liver, and also the cells of the bile duct epithelium (cholangiocytes).

In addition to inductive signals leading to regional specificity, it is generally found that mesenchyme provides **permissive** signals that is required for growth, morphogenesis, and maturation of endodermal explants. An important component of this type of signal is FGF10. For example this is known to be required for the outgrowth of the lung buds and the cecum of the intestine. The *Fgf10* and *Fgfr2b* knockouts both lack lungs and cecum as well as lacking limb buds (see Chapter 15). The **branching morphogenesis** of the developing lungs involves a lateral inhibition-type system whereby new tips produce FGF10 and suppress the formation of other tips in their immediate neighborhood.

As the gut tube matures the mesenchyme layer resolves into several layers of tissue: lamina propria, muscularis mucosa, submucosa, and smooth muscle. This radial pattern is controlled by Sonic hedgehog (Shh) emitted from the whole of the endodermal epithelium. Shh forms a radial gradient across the mesenchyme and controls its differentiation into the various layers.

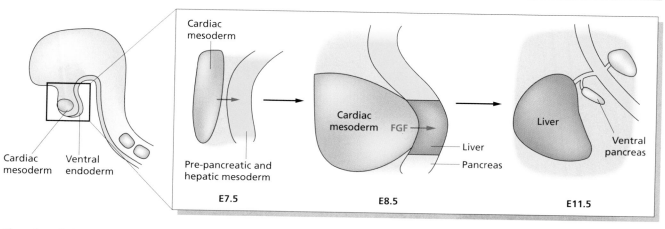

Fig. 16.13 Induction of the liver in the mouse embryo by FGF from the cardiac mesoderm.

This may be demonstrated by the fact that application of Shh will induces this pattern locally, while the inhibitor of Shh, cyclopamine, will block its development.

Maturation also involves the formation of glands in the stomach epithelium and crypts and villi from the intestinal epithelium (Fig. 16.6 above). Each part of the gut has a specific set of differentiated cell types in the epithelium, which are continuously renewed from a population of stem cells. This process is described in Chapter 18.

The pancreas

The development of the pancreas has attracted substantial research interest in recent years because of the considerable importance of two major diseases: diabetes and pancreatic cancer. The mature pancreas (Fig. 16.14) is an organ with two quite distinct physiological functions. One is the synthesis of digestive enzymes. These are made in the **exocrine** cells, which make up most of the mass of the pancreas. They are bunched into **acini** (singular: **acinus**) and secrete their products into a duct system that runs into the duodenum. The second is the synthesis of hormones that are released into the blood stream. These are made by **endocrine** cells, which are mostly grouped into the islets of Langerhans, although some may be found isolated or in small clusters. The majority of the islet cells are **β-cells**, which secrete insulin. In addition, are smaller numbers of α-cells that secrete glucagon, δ-cells that secrete somatostatin, PP cells that secrete pancreatic polypeptide, and ε-cells that secrete ghrelin. In the embryo the proportion of endocrine cells is about 10% but it falls to 1–2% in the adult pancreas.

All the epithelial cell types of the pancreas arise from the endoderm (see below) while the blood vessels and the rather small amount of pancreatic connective tissue comes from the mesenchyme. The pancreatic mesenchyme also forms the spleen. Development of the pancreas is fairly similar in all vertebrates (Figs 16.15, 16.16). It originates from the duodenum as

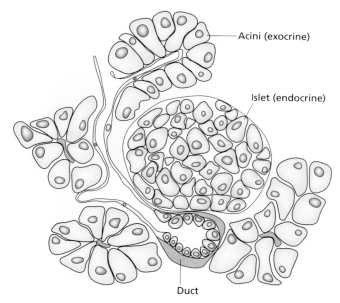

Fig. 16.14 Cell types in the mature pancreas.

a large dorsal bud and a smaller ventral bud (paired ventral buds in the chick). These buds expand and move towards each other until they fuse together. Each of the original buds has a duct and in humans the duct of the ventral bud becomes the principal duct and that of the dorsal bud an accessory duct. In the zebrafish the dorsal bud forms mostly endocrine cells and the ventral bud mostly exocrine cells, with later generation of endocrine cells.

Most experimental work has been done in the mouse, with some in the chick and a little in *Xenopus* and zebrafish. In the mouse the dorsal bud appears at E9.5 and the ventral bud about 2 days later. Cells containing endocrine hormones are visible in the early buds and often make more than one hormone, unlike the mature cells. But it is now known that they are not precursors of the mature endocrine cells (see below). Growth of the

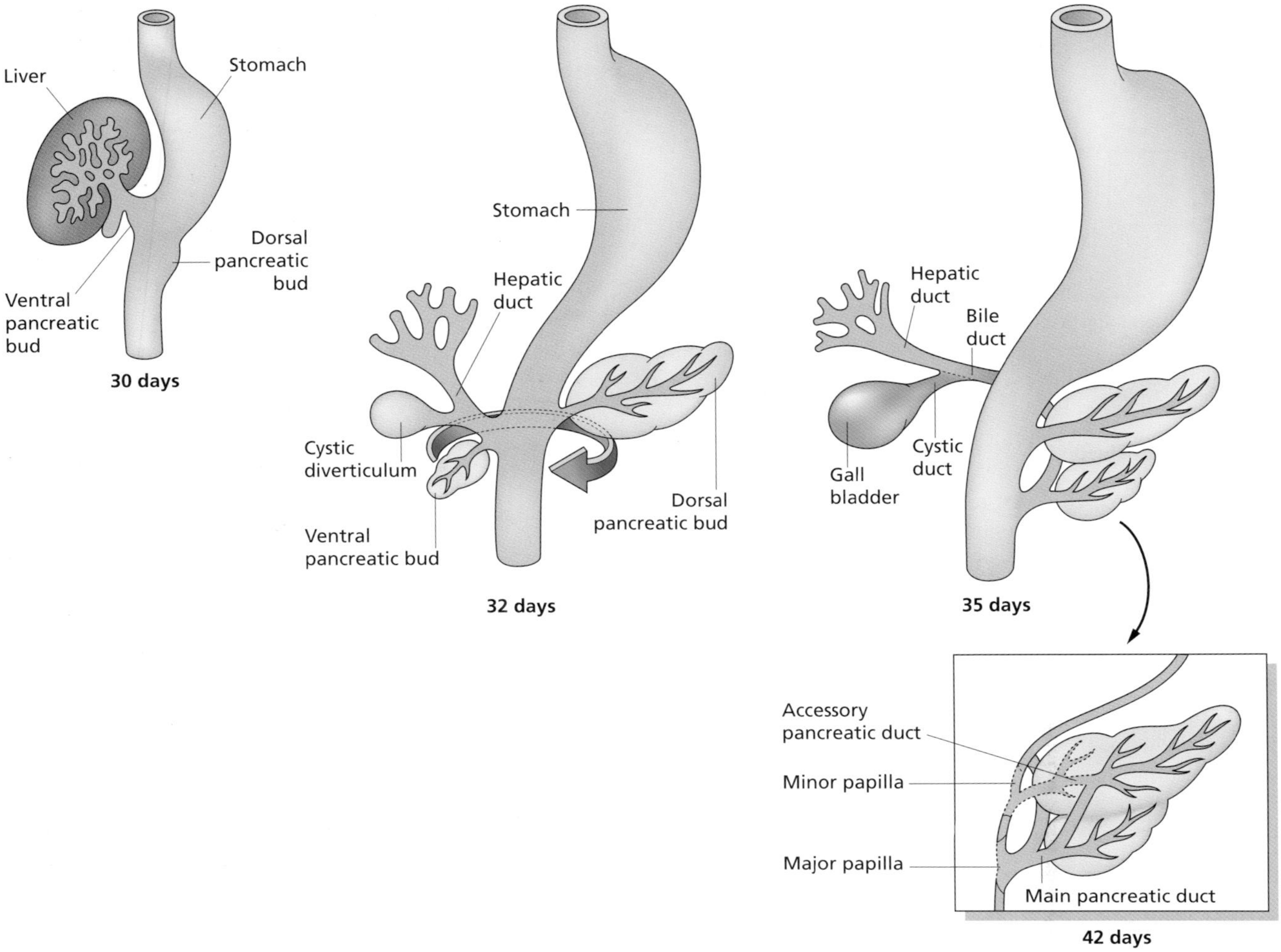

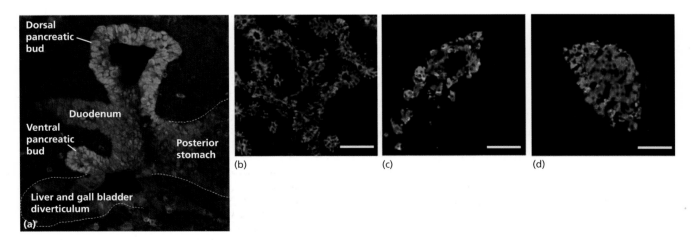

Fig. 16.15 Pancreatic bud development in the human embryo. Reproduced from Larsen (1993) *Human Embryology*. Churchill Livingstone, with permission from Elsevier.

Fig. 16.16 Development of the mouse pancreas visualized by immunostaining. (a) At E10 the pancreatic bud and its surroundings express Pdx1 (green). Primitive endocrine cells express glucagon (red). (b) At E15.5 the definitive endocrine precursors express Ngn3 (red). The pancreatic epithelium is visualized with E-cadherin (green). (c) At E17.5 the islets are beginning to form: insulin (red) glucagon (green). (d) In the mature islet the β-cells are the majority and lie in a central position: insulin, red; glucagon, green. Scale bars 50 μm. Reproduced from Jensen. *Developmental Dynamics* 2004; 229: 176–200, with permission from John Wiley & Sons Ltd, and Murtaugh. *Development* 2007; 134: 427–438, with permission from Company of Biologists Ltd.

buds is rapid in the period 14.5–18.5 days, during which there is a rapid increase of endocrine cell numbers and differentiation of much of the rest of the pancreas into exocrine cells. The exocrine acini form somewhat before birth and the islets around the time of birth.

Induction of pancreatic buds

Although both the dorsal and the ventral buds give rise to the same range of pancreatic cell types, they arise in different ways. The dorsal bud appears in a region where the notochord contacts the gut roof. Here the expression of Shh, which occurs all over the rest of the endoderm, becomes suppressed. The effect of the notochord can be mimicked by administration of activin or FGF so these may be the inducing factors responsible. *Activin receptor* knockout mice have small pancreases and fewer endocrine cells, although they have many other defects to the gut as well. Application of the Shh inhibitor **cyclopamine** to the endoderm will provoke the formation of ectopic pancreatic buds from the zone expressing Pdx1. However, neither contact with the notochord nor application of cyclopamine can induce ectopic pancreas from more posterior endoderm, and so the role of the notochord is considered **permissive** rather than **instructive**.

The ventral pancreas is formed from the adjacent region of endoderm to the liver and in the absence of the FGF signal that normally induces the liver bud, the prospective liver also becomes ventral pancreas. The FGF signal maintains *Shh* expression in the endoderm that would otherwise be turned off and allow formation of the ventral pancreatic bud. Normally, the region that forms the ventral pancreas is too far from the cardiac mesoderm to receive the FGF signal. However, in mouse knockouts of *Hex* (encoding a homeodomain transcription factor) the ventral pancreas is missing. This is not because the Hex factor codes for pancreas. It is because Hex is needed for the cell growth and movement that normally extends the ventral foregut such that the region forming the ventral pancreas is not exposed to FGF. The upshot is that the dorsal pancreatic bud is induced by FGF whereas the ventral bud develops because of an absence of FGF, although the common factor for the formation of both buds is suppression of Shh expression in the endoderm.

Experiments in both zebrafish and *Xenopus* indicate that retinoic acid is also needed at an early stage for formation of the pancreatic buds. The inhibitor BMS453 will prevent the buds forming and exogenous retinoic acid will promote endocrine rather than exocrine development. This is consistent with the requirement for retinoic acid for development of other foregut organs seen in the mouse.

Once the buds are formed, their continued outgrowth and the differentiation of exocrine cells depends on close proximity of the pancreatic mesenchyme. This is a permissive effect and mesenchyme from other parts of the gut can be successfully substituted. The LIM domain transcription factor Islet-1 is expressed in the pancreatic mesenchyme and in the knockout mouse for *Islet-1* the dorsal pancreas does not grow. But if the endoderm from the knockout is cultured with mesenchyme from a normal mouse then the bud will grow normally. As in other examples of organogenesis, FGF10 is an important component of the permissive signal. FGF10 will stimulate proliferation of isolated pancreatic endoderm, and the knockout of *Fgf10*, or the gene for its receptor *Fgfr2b*, show reduced pancreatic growth. Although the mesenchymal factors are permissive they can also affect the proportions of the cell types that differentiate from the endoderm. In organ culture experiments treatment with TGFβ tends to produce more endocrine cells, and treatment with follistatin, an inhibitor of activin, tends to suppress endocrine differentiation.

Endocrine development is also favored by a signal from blood vessels. The two pancreatic buds have in common the proximity of a major blood vessel: the dorsal aorta near the dorsal bud and the vitelline veins near the ventral bud. Combinations of pancreatic buds with blood vessels can increase the proportion of endocrine cells and removal of the vessels reduces the proportion of endocrine cells. However, in the zebrafish, removal of all blood vessels in the *cloche* (*Vegfr2*) mutant does not seem to affect pancreatic endocrine cell development.

New directions for research

Research on pancreatic development is dominated by the search for a method of creating and growing human β-cells. This is because diabetes, in which β-cells are either destroyed or insufficiently active, is such a massive international health problem. Insulin therapy is very effective but many diabetics still suffer serious and life-threatening complications from their disease. Islet cell transplantation has become a successful method for treating some types of diabetes but at present the only source of human islets is cadaveric organ donors, and the supply of these will always be hopelessly inadequate. So it is hoped to use an understanding of β-cell development to design new methods to make β-cells for transplantation.

Somewhat similar arguments apply to the liver, as the supply of human livers for transplantation is seriously short of demand. Although the liver will regenerate following damage, it is not possible to cultivate human hepatocytes *in vitro* as they rapidly lose their differentiated properties.

Pancreatic transcription factors

In recent years, a large number of transcription factor genes have been identified as having important roles in pancreatic development. Some of the most important are discussed here.

Pdx1 (LIM-homeodomain) is expressed in a ring around the duodenum, including the prospective territories of the pancreatic buds and it is essential for formation of the pancreas. The knockout mouse forms rudimentary buds, which do not grow or differentiate. Pdx1 is expressed in all cells of the early buds and is then downregulated to a low level, with persistent high expression in β-cells where it serves as a transcription factor for the *insulin* genes (rodents, unlike humans, have two *insulin* genes). A postnatal knockout of *Pdx1*, using a doxycycline-regulated gene in place of the normal one, shows a loss of insulin from the β-cells. Despite its apparent status as a "master control gene" for the pancreas, *Pdx1* does not cause the formation of pancreas from other parts of the gut when misexpressed, although an activated **domain swap** version, *Pdx1-VP16* is capable of reprogramming the developing liver to pancreas in *Xenopus*.

Hb9 (homeodomain, encoded by *Hlxb9*) is expressed in a wider region than Pdx1 and at a slightly earlier stage. Its expression is promoted by retinoic acid. The knockout mouse has no dorsal pancreas and a ventral bud that is normal except that β-cells do not mature properly, since Hlxb9 is also needed in differentiated β-cells.

PTF1 is a three-subunit transcription factor required to upregulate various exocrine genes. It is also expressed in the pancreatic buds. When the p48 subunit is knocked out this prevents formation of the ventral bud. The dorsal bud forms but develops poorly, lacking exocrine tissue. In *Xenopus* embryos, the overexpression of both p48 and Pdx1 from injected RNA can induce ectopic pancreas from endoderm more posterior than the normal domain.

Sox9 (SRY domain) is required for proliferation of the pancreatic buds. It maintains expression of the FGF receptors needed for the response to FGF10 from the mesenchyme.

Pax4 and Pax6 (paired box) are both needed for endocrine cell development. The *Pax4* knockout mouse lacks β and δ-cells, while the *Pax6* knockout lacks α cells. The homeodomain factors Nkx6.1 and Nkx2.2 are also required for β–cell development but at later stages. MafB (Leu zipper) is required for development of α-cells and MafA for maturation of β-cells. Arx (homeodomain) is required for formation of α-cells, and the knockout has an excess of β and δ-cells.

Hox11 (= Tlx1; homeodomain, but not part of the Hox clusters despite its name) is expressed in pancreatic mesenchyme and the knockout mouse has a normal pancreas but completely lacks the spleen.

Just as lateral inhibition mechanisms controls cell differentiation in neurogenesis, so they do also in the pancreas. Early stage endocrine cells make Delta, and repress endocrine differentiation by their neighbors by stimulating the Notch pathway (see Animation 6: Lateral inhibition). The key transcription factor promoting endocrine development is Neurogenin 3 (bHLH family). In mouse embryos this is expressed in scattered cells of the pancreatic buds that are not currently making the hormones, with a maximal number at 13.5–15.5 days. The knockout of *Ngn3* loses all endocrine cells. Overexpression of *Ngn3* in transgenic mice, using the *Pdx1* promoter, will drive a high proportion of pancreatic cells to become endocrine, mostly α-cells. Knockouts of components of the Notch pathway, including *Delta like 1, RBPJκ,* and *Hes1,* all have a similar effect to the overexpression of *Ngn3* and increase the proportion of endocrine cells.

The function of Neurogenin 3 is to control the formation of endocrine progenitor cells. The decision as to which type of endocrine cell to become (α, β, δ, PP, ε) depends on other factors and on interactions that are still not well understood. One example of the process is provided by the reciprocal relationship between Arx and Pax4. As mentioned above, the *Arx* knockout lacks α-cells, while the *Pax4* knockout lacks β and δ-cells. If *Arx* is overexpressed in β-cells, using *Pdx1-Cre, Pax6-Cre* or *Insulin-Cre*, as driver, then β-cell formation is greatly reduced. If *Pax4* is overexpressed in the same way, then α-cell formation is reduced. Interestingly, these experiments work even in postnatal islets and can bring about the transdifferentiation of already differentiated α or β-cells.

Pancreatic cell lineage

The pancreatic endocrine cells are very neuron-like in morphology, behavior, and gene expression patterns. At one time they were thought to be derived from the **neural crest**, but quail to chick neural crest grafts showed that this was not the case. That they are indeed formed from the endoderm may be shown by culturing explants of isolated endoderm *in vitro*, which results in the formation of a few endocrine cells, or by recombining labeled endoderm with mesenchyme in order to get good development in organ culture, and then showing that all the resulting cell types are endoderm-derived. It has also been shown by retroviral labeling at clonal multiplicity that a single endoderm cell can produce both exocrine and endocrine progeny.

To understand the lineage at higher resolution many Cre-lox lineage experiments have been carried out. When the *Pdx1* or the *p48* promoter is used to drive the *Cre*, and are combined with the *R26R* reporter, all pancreatic cell types become labeled. Use of the tamoxifen-activated CreER enables labeling to be carried out at a particular developmental stage. In *Pdx1-CreER x R26R* mice, exocrine and endocrine cells are labeled if the tamoxifen is given from 8.5 days of development onwards. But ducts are only labeled in the interval 9.5–11.5 days, suggesting some special mode of formation of this lineage at this stage. Use of the *Sox9* promoter to drive *CreER* shows that, at embryonic stages, the proliferating *Sox9*-positive cells generate both exo-

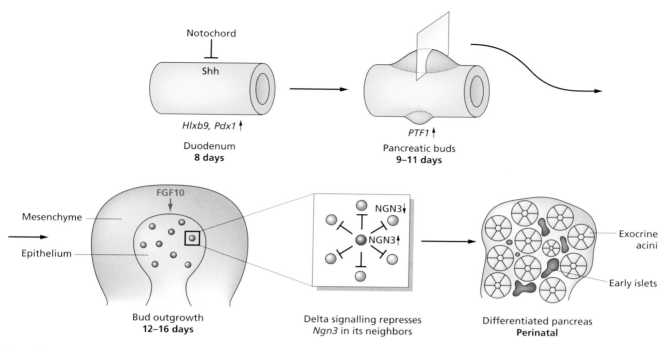

Fig. 16.17 Overall mechanism of pancreatic development.

crine and endocrine progeny. But in adult life they generate only ductal cells. When the *Neurogenin 3* promoter is used to drive the *Cre*, then all endocrine cells are labeled although the expression of the *Ngn3* itself is only transient.

Because of the existence in the early buds of multihormone cells producing two or three of the hormones simultaneously, it was considered that these might be precursors for all the later endocrine cells. But in an *insulin-Cre × R26R* mouse the later-formed α-cells are not labeled, showing that they had not expressed *insulin* in the past. Likewise, when the *glucagon* promoter was used to drive the *Cre*, the β-cells were not labeled, showing that they had not expressed glucagon in the past. However, β-cells are labeled if the *pancreatic polypeptide* promoter is used to drive the *Cre*, so cells expressing this hormone may indeed be precursors to other endocrine cell types.

Another issue that has been extensively addressed by Cre labeling is whether β-cells are formed from undifferentiated stem or progenitor cells during postnatal life. When the *insulin* promoter is used to drive *CreER*, and activity is induced after

birth, then it is found that it is possible to label virtually all the β-cells. This label is then retained indefinitely, indicating that there is no recruitment of new β-cells from ducts or from other cells in the islets that were not themselves making insulin at the time of tamoxifen administration. Other labeling experiments in which ductal cell promoters are used sometimes show small numbers of β-cells arising from ducts, but only in conditions of major tissue regeneration, following partial surgical ablation of the pancreas or pancreatic duct ligation. So it appears that most β-cells normally arise from pre-existing β-cells and that recruitment of β-cells from other cell types in the postnatal pancreas is an unusual event.

The considerable efforts devoted to the study of pancreatic development in recent years have made it one of the best understood examples of organogenesis. The overall course of events, deduced from a combination of embryological experiments and mouse genetics, is summarized in Fig. 16.17 and a consensus pancreatic cell lineage based on the Cre-lox fate mapping experiments is shown in Fig. 16.18.

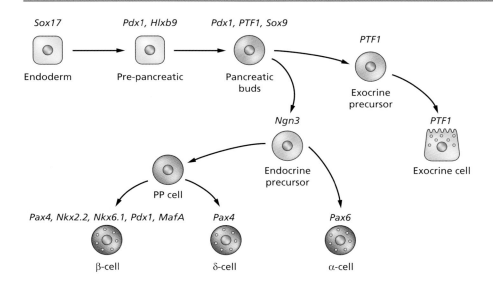

Fig. 16.18 Pancreatic cell lineage.

Key points to remember

• The endoderm of amniotes starts as a flat sheet of cells below the ectoderm and mesoderm layers. The gut tube is formed by the folding of the blastoderm to create a headfold and tailfold, which respectively contain pockets of fore- and hindgut. Subsequently, the anterior and posterior intestinal portals move towards each other until they join. The gut tube and ventral body wall are closed by lateral body folds which meet each other on the ventral side of the embryo.

• The basic pattern of the gut is the same in all vertebrates with a pharynx, esophagus, stomach, liver, pancreas, and small and large intestine. Each is lined with a characteristic epithelium derived from the endoderm.

• In the fate map, the endoderm and mesenchyme of each part of the gut come from different positions in the embryo. They move together in the course of gut closure.

• Nodal signaling is necessary to activate expression of the early endodermal transcription factors.

• Regional specification depends on inductive signals. The initial patterning occurs in response to the whole body posterior to anterior gradients of FGF and Wnt. Subsequent regional specification occurs in response to reciprocal local signaling events between endoderm and mesoderm.

• The pancreas consists of exocrine cells that secrete digestive enzymes into the intestine, and endocrine cells that secrete hormone, including insulin, into the bloodstream. Both cell types arise from the endoderm.

• The pancreas originates as dorsal and ventral buds from the duodenum. Suppression of Shh expression in the endoderm is necessary for bud formation and FGF is necessary for continued growth.

• Pdx1 and PTF1 are essential for the development of all pancreatic cell types. Neurogenin 3 is essential for the development of endocrine cells. The ratio of various other transcription factors control which type of endocrine cell is formed.

Further reading

General endoderm

Grapin-Botton, A. & Melton, D.A. (2000) Endoderm development from patterning to organogenesis. *Trends in Genetics* **16**, 124–130.

Roberts, D.J. (2002) Molecular mechanisms of development of the gastrointestinal tract. *Developmental Dynamics* **219**, 109–120.

Shivasani, R.A. (2002) Molecular regulation of vertebrate early endoderm development. *Developmental Biology* **249**, 191–203.

Yasugi, S. & Mizuno, T. (2008) Molecular analysis of endoderm regionalization. *Development, Growth and Differentiation* **50**, S79–S96.

Burn, S.F. & Hill, R.E. (2009) Left-right asymmetry in gut development: what happens next? *BioEssays* **31**, 1026–1037.

Zorn, A.M. & Wells, J.M. (2009) Vertebrate endoderm development and organ formation. *Annual Review of Cell and Developmental Biology* **25**, 221–251.

Spence, J.R., Lauf, R. & Shroyer, N.F. (2011) Vertebrate intestinal endoderm development. *Developmental Dynamics* **240**, 501–520.

Liver and lung

Warburton, D., Schwartz, M., Tefft, D., Flores-Delgado, G., Anderson, K.D. & Cardoso, W.V. (2000) The molecular basis of lung morphogenesis. *Mechanisms of Development* **92**, 55–81.

Lemaigre, F. & Zaret, K.S. (2004) Liver development update: new embryo models, cell lineage control, and morphogenesis. *Current Opinion in Genetics and Development* **14**, 582–590.

Cardoso, W.V. (2006) Regulation of early lung morphogenesis: questions, facts and controversies. *Development* **133**, 1611–1624.

Rawlins, E. & Hogan, B.L.M. (2006) Epithelial stem cells of the lung: privileged few or opportunities for many. *Development* **133**, 2455–2465.

Si-Tayeb, K., Lemaigre, F.P. & Duncan, S.A. (2010) Organogenesis and development of the liver. *Developmental Cell* **18**, 175–189.

Pancreas

Slack, J.M.W. (1995) Developmental biology of the pancreas. *Development* **121**, 1569–1580.

Herrera, P.L. (2002) Defining the cell lineages of the islets of Langerhans using transgenic mice. *International Journal of Developmental Biology* **46**, 97–103.

Gu, G., Brown, J.R. & Melton, D.A. (2003) Direct lineage tracing reveals the ontogeny of pancreatic cell fates during mouse embryogenesis. *Mechanisms of Development* **120**, 35–43.

Lammert, E., Cleaver, O. & Melton D.A. (2003) Role of endothelial cells in early pancreas and liver development. *Mechanisms of Development* **120**, 59–64.

Murtaugh, L.C. (2007) Pancreas and beta-cell development: from the actual to the possible. *Development* **134**, 427–438.

Gittes, G.K. (2009) Developmental biology of the pancreas: A comprehensive review. *Developmental Biology* **326**, 4–35.

Wandzioch, E. & Zaret, K.S. (2009) Dynamic signaling network for the specification of embryonic pancreas and liver progenitors. *Science* **324**, 1707–1710.

This chapter contains the following animations:

Animation 6   Lateral inhibition.

Animation 21   Early stage amniote gut morphogenesis.

 For additional resources for this book visit
www.essentialdevelopmentalbiology.com

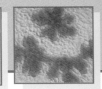

Chapter 17

Drosophila *imaginal discs*

Metamorphosis

Drosophila belongs to one of the families of insects (Diptera) that undergo a complete **metamorphosis** (Fig. 17.1). After hatching, the larva passes through three **instars**, separated by **molts**. The first two instars last about 1 day and the third about 2 days. During this period the larva feeds voraciously and grows substantially in size, but without much change in body form. The third-instar larva becomes a **pupa** and during pupation most of the larval body is resorbed and replaced by adult structures derived from the **imaginal discs**. The contributions are shown in Table 17.1 and the positions of disc rudiments in the embryo in Fig. 17.2.

Imaginal discs are infoldings of the epidermis. They are composed of columnar epithelium one cell thick, which is covered with a squamous peripodial membrane. Associated with each disc are some mesoderm-derived adepithelial cells which later differentiate into muscle cells. During larval life the discs grow extensively, for example the wing disc starts with about 40 cells and expands to about 50,000 cells by the time of differentiation. In the pupa, discs evaginate to produce the appropriate appendage (Fig. 17.3), and differentiate to give the cuticle with its characteristic appendage-specific pattern of hairs and bristles. Unlike the situation in vertebrates, imaginal discs produce their own sensory neurons whose axons grow inwards to connect with the central nervous system .

Molting in insects is controlled by a surge of the steroid hormone ecdysone (of which the active form is 20-hydroxyecdysone). Neurosecretory cells in the brain release a peptide hormone called prothoracicotropic hormone (PTTH), which stimulates the prothoracic gland to secrete ecdysone. A pair of glands called the corpora allata, attached to the brain, secrete the terpenoid juvenile hormone (JH). When JH is high,

a rise in ecdysone causes a larval molt; when JH is low, ecdysone initiates metamorphosis. The ecdysone receptor is a typical nuclear hormone receptor (see Appendix) and among its targets is its own gene, so following stimulation the responding tissues become still more responsive to the hormone.

The growth of discs can be prolonged beyond the normal duration of larval life by implanting fragments into the abdomen of an adult fly. If they are transplanted repeatedly they will grow indefinitely without differentiating. On the other hand, discs or disc fragments can be caused to differentiate by implantation into a larva about to metamorphose (Fig. 17.4), or by culture *in vitro* in the presence of ecdysone.

Genetic study of larval development

Many of the genes affecting the development of the imaginal discs are also important in embryonic development. Therefore null mutants often die as embryos and never display a disc-specific phenotype. Mutations allowing survival to adulthood are usually **hypomorphic** rather than **null**. They are often **temperature sensitive**, meaning that the protein product becomes inactivated when the organism is moved to the nonpermissive temperature. This makes it possible to find the time of action of the gene during larval life or metamorphosis. Groups of mutant larvae are subjected to pulses of the nonpermissive temperature at different times. The temperature pulse will inactivate the gene product and if the gene is required at that time then the mutant phenotype will develop.

Overexpression experiments in *Drosophila* are often done by introducing a transgene driven by a heat-shock promoter. Heat-shock proteins exist in all cells and function to assist in correct folding of other proteins. Their expression is greatly increased

Essential Developmental Biology, Third Edition. Jonathan M.W. Slack.
© 2013 John Wiley & Sons, Ltd. Published 2013 by John Wiley & Sons, Ltd.

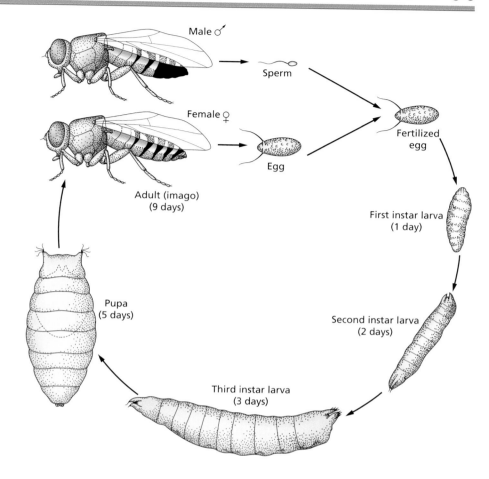

Fig. 17.1 Life cycle of *Drosophila*.

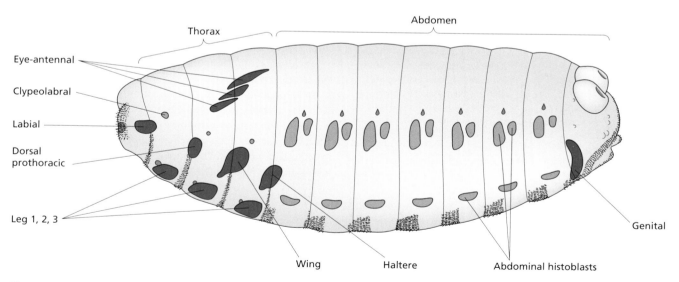

Fig. 17.2 Position of imaginal disc rudiments in the late embryo. Reproduced from Hartenstein (1993) *Atlas of Drosophila Development*, with permission from Cold Spring Harbor Laboratory Press.

at elevated temperature, and this is a function of the promoter. When a heat-shock promoter is joined to another gene then this gene can be activated by raising the temperature. This technique should not, of course, be confused with the use of temperature-sensitive mutants.

Developmental genes often have several **enhancers**, each activated in a different expression domain. It is possible to take just one enhancer, link it to a **reporter gene**, such as *Gfp* or *lacZ*, and make a transgenic fly line with this construct. Fluorescence microscopy (for GFP) or XGal staining (for the β-galactosidase product of *lacZ*) of embryos, larvae, or discs, will then reveal the domain of activity of this particular enhancer, usually representing one of several expression domains of the normal gene. The use of such reporters is both more specific and more sensitive than looking at the messenger RNA of the endogenous gene by *in situ* hybridization.

Enhancer trap lines (see also Chapters 3 and 10) are transgenic flies in which a reporter gene is expressed with a particular spatial pattern that reflects the activity of an endogenous enhancer, but in this case one identified by random insertion. They are made by injecting a construct in which the reporter is under the control of a weak promoter. The promoter is too weak to drive expression unless the construct happens to integrate in the genome near to an enhancer sequence, in which case it is expressed with a pattern characteristic of the enhancer. Using this method, numerous lines have been created having particular patterns of *lacZ* expression, many resembling expression patterns of known genes but also many with novel patterns. They can be useful experimentally because particular cell populations can be highlighted and easy to observe simply using XGal staining.

A variation of the enhancer trap method is used to drive **ectopic** expression of any gene of interest. This is based on the GAL4 regulatory system from yeast. GAL4 is a transcriptional activator of the zinc-finger class and can drive expression of any gene sequence cloned downstream of the "upstream activating sequence" (*UAS*) to which it binds. Enhancer trap lines are made in which the gene introduced is not *lacZ*, but *Gal4*. Another line of transgenic flies is made in which the gene of interest is linked to *UAS*. When individuals of the two lines are crossed together, the GAL4 is expressed under the control of its enhancer and it will activate the target gene via the *UAS*. The spatial pattern of expression of the target gene will depend on the enhancer used to drive the *Gal4* (Fig. 17.5). Along with "FLP-out" (see below), this is the usual way to bring about regionally specific overexpression in imaginal discs.

Table 17.1 Origin of adult body parts from imaginal discs and abdominal histoblasts.

Disc/ histoblast	Body part
Clypeolabral disc	Some mouthparts
Eye–antenna disc	Eyes, antennae, rest of head
Labial disc	Proboscis
Dorsal prothoracic disc	Dorsal prothorax
Prothoracic leg disc	First leg and ventral prothorax
Mesothoracic leg disc	Second leg and ventral mesothorax
Metathoracic leg disc	Third leg and ventral metathorax
Wing disc	Wing and dorsal mesothorax
Haltere disc	Haltere and dorsal metathorax
Dorsal abdominal histoblasts	Tergites (dorsal abdominal cuticle)
Ventral abdominal histoblasts	Sternites (ventral abdominal cuticle)
Genital disc*	Genital apparatus

*There is a single genital disc, but the others are paired on right and left sides.

Mitotic recombination

Although recombination normally occurs at meiosis, it also possible to force recombination between homologous chromosomes at mitosis. When recombination is induced in a cell heterozygous for a mutation (+/−) then it can generate a pair of cells of which one is +/+ and the other is −/− (Fig. 17.6). These grow to form two adjacent **clones**, called a **twin spot**. The +/+ clone is not usually visible but the ability to generate the mutant (−/−) clones is important for at least four purposes, all of which will be discussed below.

1 The **clonal analysis** of determination in imaginal discs. No clonal restriction means no determination.

2 Examination of the **autonomy** of mutations. If the mutation affects the surroundings as well as the clone itself then it is said to be nonautonomous. This implies that it affects some intercellular signaling mechanism, either directly or indirectly.

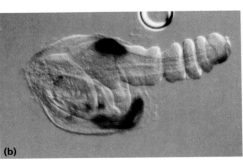

(a) (b)

Fig. 17.3 Eversion of a leg disc during pupation: (a) two third-instar leg discs and (b) one everting leg disc from a pupa. Reproduced from Bate and Martinez-Arias (eds) (1993), *The Development of Drosophila Melanogaster,* with permission from Cold Spring Harbor Laboratory Press.

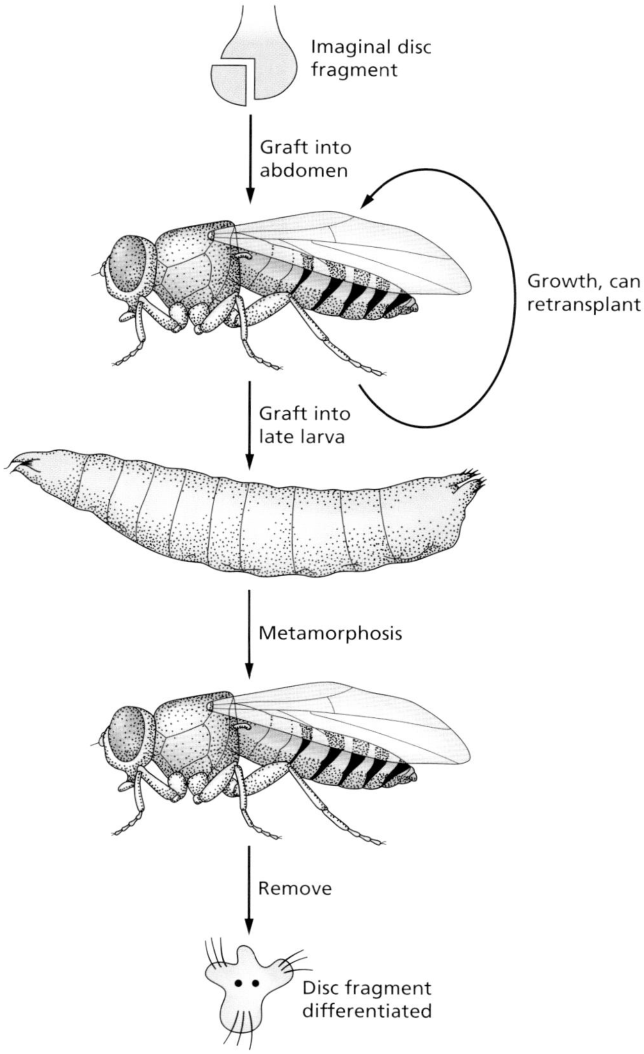

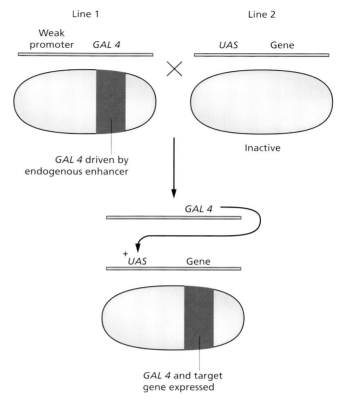

Fig. 17.4 Experimental methods for propagation of discs in an adult host and differentiation in late larva.

Fig. 17.5 GAL4 method for ectopic overexpression of a transgene. When the two transgenic lines are mated, the offspring will show tissue-specific expression of the target gene.

3 The study of late phenotypes of mutations that are lethal during embryonic development. It is often the case that a mutation that arrests development of the whole embryo is not lethal at the cell level. Therefore a mutant clone in an otherwise normal larva or adult will probably be viable. Such a clone may display morphological defects indicating a late function which could not have been observed otherwise.

4 Localized overexpression of individual genes. This is complementary to the GAL4 method described above. GAL4 generates one specific pattern of overexpression depending on the promoter that is used to drive it, while mitotic recombination can generate a random set of overexpressing clones. The two methods are both heavily used in research on disc development.

If the object of induced mitotic recombination is just to visualize clones in normal development, then the mutation should not perturb development but simply change the appearance of the cells. Mutations originally used for this purpose were cuticular markers such as *yellow* and *multiple wing hairs*, although these have the limitation of only being visible in the adult and not during embryonic or larval stages. More recently, a number of transgenic labels, encoding GFP and other fluorescent proteins, have been introduced and are of general utility.

In early experiments mitotic recombination was induced using X-irradiation, which causes chromosome breaks. But because of its low efficiency this has now been superseded by the **FLP** (pronounced "Flip") system. The basic elements are the FLP recombinase, which is a yeast enzyme, and the FRT sites in the target DNA. The FLP recombinase recognizes two FRT sites and catalyzes a recombination event between them. A line of flies carrying the FRT sites is crossed to one carrying the FLP recombinase, driven by a heat-shock promoter. Recombination is initiated by a brief temperature shock, which causes transient transcription of the recombinase, and hence brings about recombination in just a few cells. If there is a marker transgene such as *Gfp* or *lacZ* further from the **centromere** than the FRT site, then this will be lost in the recombinants, generating clones

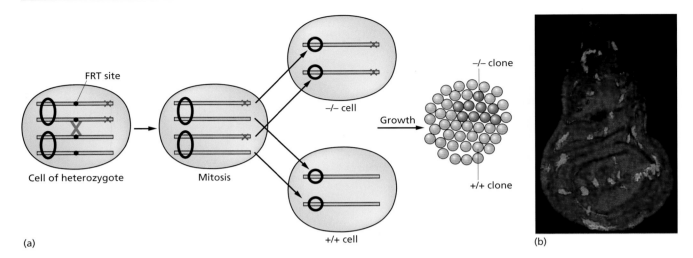

(a)

(b)

Fig. 17.6 (a) Mitotic recombination generates a twin spot consisting of a −/− and a +/+ clone. The position of the mutation is indicated by the red cross, and of the centromere by the black circle. Recombination takes place at the FRT site, catalyzed by FLP recombinase. (b) A wing disc containing several clones. This picture was obtained with the MARCM technique such that the green label is in the induced clones and not the background. Reproduced from Lee and Luo. *Neuron* 1999; 22: 451–461, with permission from Elsevier.

that are identifiable by the absence of the marker but are otherwise developmentally normal.

In this way identifiable clones can be produced, but the more usual application is to make clones that are not just visible but in which a particular functional gene is ablated or ectopically expressed. If the intention is to ablate the gene in the clones, then the FRT sites need to lie on the centromeric side of the functional mutation so that recombination generates a −/− clone, as shown in Fig. 17.6. The method as described suffers from the fact that the marker transgene is normally *lost* in the mutant clone, so may be hard to see against a background of ubiquitous staining. A way around this is the MARCM procedure (mosaic analysis with repressible cell marker). Here the expression of the marker is continuously repressed by a locus on the same chromosome arm as the FRT site and the mutant gene. When the recombination is induced the repressor segregates into one clone and the marker and mutant locus into the other. This means that the marker becomes derepressed and easily visible against a uniform or unlabeled background (Fig. 17.6b). The normal repressor used is GAL80, a yeast protein that represses GAL4. The marker itself is GAL4 driven by a ubiquitous promoter, driving a UAS-reporter gene.

A transgene intended for ectopic expression will contain a constitutive promoter to drive the gene in question and a transcriptional or translational stop sequence causing the product to be truncated and functionless. This stop sequence is flanked by two FRT sites. When induced, the recombinase can excise the stop sequence by recombination between these adjacent FRT sites, leading to the production of clones in which the full-length gene is expressed (Fig. 17.7). This variant of the method is known as "FLP-out." It is not always necessary to have a separate marker gene for identification of the clone because, if the gene under study is a transgene, an identification sequence can

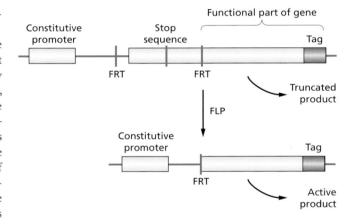

Fig. 17.7 Upregulation of a transgene in a clone produced by the FLP-out method.

be incorporated into its coding region. For example the "Myc tag" is a short peptide from the Myc oncoprotein that can be recognized by immunostaining with a specific antibody and enables the mutant clone to be visualized.

Disc development

Origin of discs

The first sign of the thoracic imaginal discs is at about 5 hours of embryonic development, when the germ band is maximally extended. They are not visible by ordinary microscopy but can be visualized in embryos transgenic for *lacZ* driven by an early enhancer of the gene *Distal-less*. *Distal-less* codes for a homeodomain transcription factor and is expressed in all the ventral appendages. Early disc development, and activation of the *Distal-less* enhancer, is dependent on the action of the *wingless* gene. As discussed in Chapter 11, *wingless* is required for embryonic segmentation, and null mutants are lethal. However, there are also viable temperature-sensitive mutations of *wingless*, and use of these shows several sensitive periods for wing development, the first of which is just before 5 hours of development, when *wingless* is required to turn on the *Distal-less* enhancer.

In the embryo, *wingless* is initially expressed in circumferential stripes at the posterior margin of each parasegment. Another gene required for disc formation is *decapentaplegic* (*dpp*). Early on, *dpp* is expressed in the dorsal 40% of the embryo, following operation of the Dorsal protein gradient (see Chapter 11) but later the expression pattern resolves into several longitudinal bands. The three thoracic discs arise at the intersections of the circumferential bands of *wingless* with a longitudinal band of *dpp* expression (Fig. 17.8). During germ-band retraction a group of cells separates from two of the disc rudiments and moves dorsally. These dorsal cell clusters are the wing and haltere discs while the three ventral clusters are the leg discs. By 10 hours the dorsal clumps express *vestigial*, which encodes a transcriptional cofactor needed for wing and haltere development. At about this stage all the imaginal discs and abdominal histoblasts begin to express the *escargot* gene, coding for a zinc-finger transcription factor. At least in the histoblasts, its function is to prevent **polyteny** of the DNA, and hence maintain the cells in a state capable of division.

Compartments and selector genes

The progressive formation and subdivision of the discs can be followed by **clonal analysis**. If mitotic recombination is induced at the blastoderm stage, individual clones do not cross segment boundaries nor an invisible anteroposterior boundary running through each of the thoracic imaginal discs. This line of clonal restriction is called the anteroposterior **compartment** boundary

and is now known to be the **parasegment** boundary of the early embryo, defined by the *cubitus interruptus* and *engrailed* systems (Fig. 17.9a and see Chapter 11). This boundary is inherited through cell divisions and maintained in each of the thoracic imaginal discs. Although confined to a compartment,

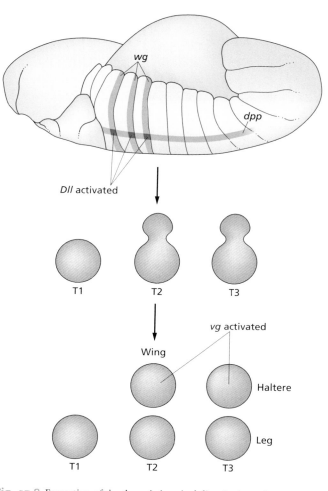

Fig. 17.8 Formation of the thoracic imaginal discs in the embryo.

Classic experiments

Discovery of developmental compartments

The idea of the developmental compartment has been a basic concept in developmental biology, by no means confined to *Drosophila*. Compartments were discovered by clonal analysis of imaginal discs and were defined as units of body structure that were produced by the progeny of a small number of cells. Through work on the properties of *engrailed* mutants they also became identified as domains of activity of homeotic genes.

The first paper describes the behavior of labeled clones and the second introduces the *Minute* technique allowing production of larger and more informative clones. The third paper is actually a

review but is unusual because of the messianic certainty of the authors that the work of the Spanish school was so important that it just had to receive wider publicity.

Garcia-Bellido, A., Ripoll, P. & Morata, G. (1973) Developmental compartmentalisation of the wing disk of *Drosophila*. *Nature New Biology* **245**, 251–253.

Morata, G. & Ripoll, P. (1975) Minutes: mutants of *Drosophila* autonomously affecting cell division rate. *Developmental Biology* **42**, 211–221.

Crick, F.H.C. & Lawrence P.A. (1975) Compartments and polyclones in insect development. *Science* **189**, 340–347.

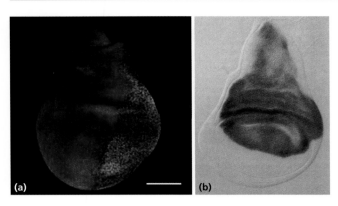

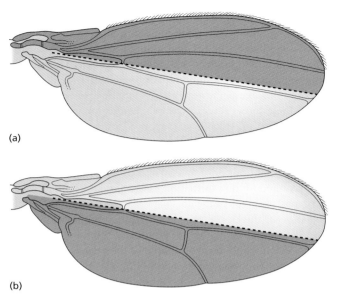

Fig. 17.9 Anteroposterior and dorsoventral compartments in the third instar wing disc. (a) Immunostaining of Engrailed (En, green) in the posterior, and Cubitus interruptus (Ci, red) in the anterior compartment. Scale bar 100 μm. (b) A *lacZ* reporter shows *apterous* expression in the dorsal compartment. Reproduced from Smith-Bolton *et al. Dev Cell* 2009; 16: 797–809, with permission from Elsevier (part a); and Bate and Martinez-Arias (1993) (eds), *The Development of Drosophila Melanogaster*, with permission from Cold Spring Harbor Laboratory Press (part b).

Fig. 17.10 Visualization of the anterior and posterior compartments using the *Minute* technique. A normal clone is induced on a *Minute* background, and fills up the whole compartment. (a) Anterior compartment; (b) posterior compartment. Reproduced from Lawrence (1992) *The Making of a Fly: The Genetics of Animal Design*, with permission from John Wiley & Sons Ltd.

blastoderm-stage clones in the thorax can still cross between wing and leg disc, whereas clones induced in larval life are confined to either wing or leg. This is because each thoracic disc rudiment originates as a single cell clump that divides into two.

The anteroposterior compartment boundary can be visualized more easily if the labeled clone is very big so that it abuts the boundary over much of its length. This can be achieved by the use of *Minute* mutants, which are a class of dominant mutants in genes for ribosomal proteins. Flies carrying one copy of a *Minute* mutant grow slowly during larval life because of a defect in protein synthesis but they are viable and do form adults. By inducing +/+ clones on a Minute/+ background, it is possible to give the wild-type clones a growth advantage over the surrounding tissue, with the result that a single clone can fill a large fraction of the final organ (Fig. 17.10). Remarkably, this differential growth of clones does not disturb the final size or proportions of the organ. In the case of the wing disc (but not the leg disc), there is also a dorsal–ventral compartment boundary, which becomes established in the second larval instar. In other words, clones induced in the first instar will cross between dorsal and ventral parts of the disc, whereas those induced in late second instar are confined to either dorsal or ventral parts.

Each of the compartmental subdivisions depends on the expression of a particular homeotic gene, sometimes called a **selector gene** in this context. The posterior compartment is defined by expression of the *engrailed* gene. *engrailed* activity defines the anterior stripe of cells in each embryonic parasegment, but since discs arise at the parasegment boundary, this becomes the posterior part of the disc. Null mutations of *engrailed* are embryo lethal, but there are also hypomorphic mutations that are viable and give a partial transformation of posterior to anterior wing. Clones experimentally induced by

mitotic recombination may fall, at random, either in the anterior or the posterior compartment. If clones lacking *engrailed* function are induced in the anterior compartment, they remain confined to the anterior and produce a normal anatomy. If such clones are produced in the posterior, they produce anatomical defects, and the affected cells may cross over into the anterior compartment (Fig. 17.11). These experiments show that *engrailed* function defines the character of the posterior compartment, and lack of *engrailed*, by default, defines the anterior. The actual barrier to cell movement is based on a difference of cell adhesion between the compartments due to integins and other components whose expression depends on *engrailed*. Although the details of the cell adhesion molecules involved remain uncertain, anterior and posterior cells will sort out from each other in the manner described in Chapter 2 (Fig. 2.18).

The dorsal compartment of the wing disc is defined by expression of *apterous*, a gene coding for a transcription factor of the Lim class (Fig. 17.9b). *apterous* expression starts in the second instar, coincident with the time of dorsoventral clonal restriction, and is found in the dorsal but not the ventral compartment. It is activated by a neuregulin-like signal encoded by the *vein* gene. Neuregulins act through the EGF receptor and loss of function of the *Egfr* gene will prevent the activation of *apterous*. If clones are induced lacking *apterous* function, they develop normally on the ventral side, but on the dorsal side an ectopic wing margin is formed around the boundary of the clone. The cells lacking *apterous* lie within the clone producing ventral bristles, while the cells around the clone that express *apterous* form the dorsal bristles. Thus, the ectopic

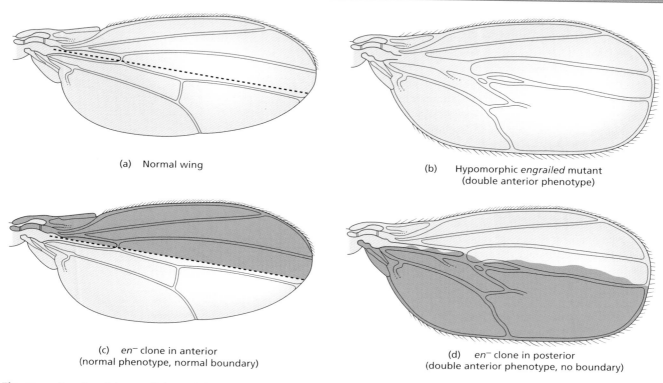

Fig. 17.11 Function of the *engrailed* gene. (a,b) A viable but hypomorphic mutant of *engrailed* gives a wing with structure tending toward double anterior. (c,d) Clones lacking *engrailed* respect the compartment boundary on the anterior but not on the posterior side. Reproduced from Lawrence (1992) *The Making of a Fly: The Genetics of Animal Design*, with permission from John Wiley & Sons Ltd.

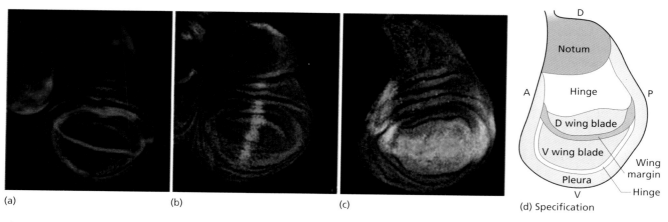

Fig. 17.12 Gene expression domains in the third instar wing disc. (a) *wingless* expressed along the DV boundary, and elsewhere. (b) *dpp* expressed along the AP boundary. (c) *vestigial* expressed in the prospective wing blade. (d) Specification map showing what structures develop from each part. Reproduced from Neto-Silva *et al. Ann Rev Cell Devel Biol* 2009; 25: 197–220, with permission from Annual Reviews.

dorsal–ventral boundary, like the normal one, arises at the boundary of *apterous* activity

Although they remain visibly undifferentiated, by the third instar the discs display a complex pattern of gene expression domains (Fig. 17.12). *In vitro* culture of discs and disc fragments has not proved very successful, although using a special viscous medium it is possible to keep them alive long enough to observe

the events of eversion. But disc fragments will survive long term when grafted to other larvae or to adults. When disc fragments are implanted into metamorphosing larvae, they will differentiate and form all of the structures formed in normal development from that part of the disc. This has been used to construct **fate maps**, to show which part of the adult body derives from which part of each disc, although strictly speaking these should be called **specification** maps because they employ the method of culture in isolation (see Chapter 4). The epidermis of the adult *Drosophila* is very rich in various types of specialization such as hairs and bristles that can be identified by the skilled observer, and therefore the maps can have quite high resolution. In the wing disc the dorsal part forms the notum (dorsal cuticle), the ventral part the pleura (cuticle between wing and leg), and the central part the wing hinge and wing blade (Fig. 17.12d).

Regional patterning of the wing disc

Anteroposterior pattern

The patterning mechanism for several of the imaginal discs in normal development is now well understood. The wing disc is described here and although the principles are similar for all the discs, the actual developmental events do differ considerably. The basic model for the wing disc can be summarized as follows: the transcription factor Engrailed, expressed in the posterior compartment, activates the *hedgehog* gene. The extracellular Hedgehog protein stimulates the anterior cells along the antero-posterior (AP) compartment boundary and causes them to upregulate the *decapentaplegic* (*dpp*) gene. The Dpp protein forms an extracellular gradient with the high concentration along the anterior–posterior boundary and the low concentration at the disc margins. This gradient brings about patterning of both anterior and posterior compartments by activation of transcription factors at different concentration thresholds.

The evidence for this model comes from a variety of experiments mainly involving the ablation of gene activity in clones, or the ectopic expression of the gene, either in FLP-out clones or by the GAL4 method. The anterior–posterior compartment boundary arises in early embryogenesis because *engrailed* is on in the anterior of each parasegment and off in the remainder. The discs form around the parasegment boundaries such that *engrailed* activity defines the posterior compartment. Ectopic expression of *engrailed* in the posterior gives normal clones, but ectopic expression in the anterior gives abnormal clones. These produce pattern duplications in the surrounding tissue (Fig. 17.13a). Since engrailed is a transcription factor, the nonautonomy of these clones shows that a signal must be emitted from *engrailed*-overexpressing cells. If *engrailed* is removed by mitotic recombination, the anterior clones are normal but posterior clones show excess proliferation and pattern duplication (Fig. 17.13a).

As in embryogenesis, the signal emitted by the *engrailed*-expressing cells is Hedgehog, a signaling molecule with a short diffusion range. Cells of the posterior compartment are not competent to respond to Hedgehog but those of the anterior compartment are. Therefore the effect of the Hedgehog signal is to stimulate a band of cells running along the anterior–posterior compartment boundary. As described in Chapter 11, the effect is to activate expression of a Gli-type transcription factor, Cubitus interruptus (Ci). The target genes of Ci include

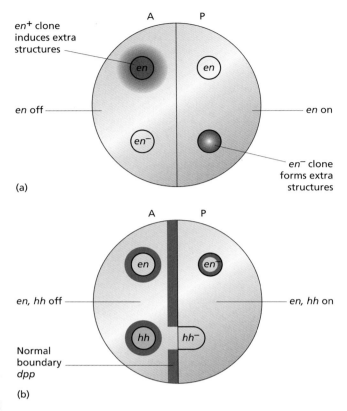

Fig. 17.13 (a) Behavior of clones with altered *engrailed* expression. Clones overexpressing *engrailed* in the anterior compartment induce pattern duplications in their surroundings; clones lacking *engrailed* activity in the posterior compartment themselves form ectopic structures. (b) Effects of clones with altered *engrailed* or *hedgehog* expression on activity of the *dpp* enhancer. Posterior clones lacking *engrailed* activate the *dpp* enhancer but only where they abut surrounding posterior tissue. Anterior clones expressing *engrailed* activate the *dpp* enhancer in the surroundings. Posterior clones lacking *hedgehog* prevent *dpp* activation across the normal boundary. Anterior clones expressing *hedgehog* activate *dpp* both internally and in the surroundings.

dpp and also *patched*, which encodes the Hedgehog receptor. Thus, the response to the Hedgehog signal is for the responding cells to secrete Dpp and also to increase their own sensitivity to Hedgehog. Ci activity is antagonized by protein kinase A, the cAMP-dependent protein kinase (see Appendix). Therefore, ablating PKA function has a similar effect to activation of Hedgehog signaling, and a number of experiments in this area are carried out using clones in which PKA activity has been removed.

Evidence for the process as described comes from the study of ectopic *hedgehog*-expressing clones and of clones lacking *hedgehog* activity. Clones producing hedgehog are normal if they lie in the posterior. If they lie in the anterior they cause pattern duplications in tissues surrounding the clone, just like *engrailed*-expressing clones. If *hedgehog*-expressing clones are induced in hosts carrying a *lacZ* reporter coupled to the *dpp* promoter, it can be seen that clones in the anterior activate the

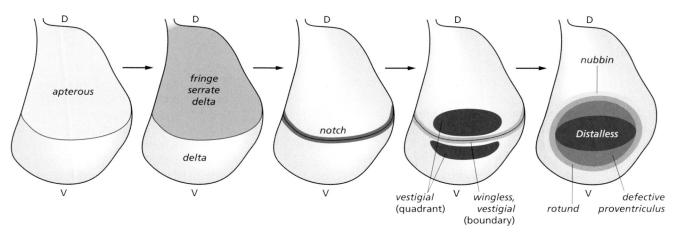

Fig. 17.14 Dorsoventral patterning in the wing disc. *apterous* in the dorsal compartment activates *fringe*. Notch becomes activated at the *fringe* boundary and turns on *wingless* and *vestigial*. *vestigial* is also turned on by Dpp through the quadrant enhancer. Finally, expression of a concentric set of transcription factors become activated that control the differentiation of the proximodistal pattern of the wing.

dpp promoter, while those in the posterior do not (Fig. 17.13b). By comparison, *engrailed*-expressing clones in the anterior compartment do not themselves activate *dpp*, but do cause an activation of *dpp* in the surrounding cells (Fig. 17.13b). Clones lacking *hedgehog* function only have a phenotype if they abut the anterior–posterior border. In this case there is no activation of *dpp* in neighboring anterior cells and there are local defects in the resulting pattern (Fig. 17.13b).

The Dpp protein forms a gradient from the anterior–posterior border, which patterns both anterior and posterior compartments. Evidence that Dpp is the principal effector of patterning comes from the study of ectopic expression of *dpp*. In the anterior compartment this has a similar effect to *hedgehog*, while in the posterior compartment it induces pattern duplications and overgrowth.

Dorsoventral pattern

In the wing disc there is an independent signaling system responsible for establishing the regional pattern in the dorsoventral axis. It depends on the stimulation of Notch along the dorsal–ventral compartment border, which leads to activation of a transcription factor gene, *vestigial*, and an inducing factor gene, the familiar *wingless* (Fig. 17.12).

The genes responsible for this process, *wingless*, *Notch*, *Delta*, *Serrate*, and *fringe*, all have broadly similar effects to each other when ectopically expressed, or when ablated from clones. Ectopic expression tends to give ectopic patches of wing margin, while ablation tends to lead to defective wing margin or even complete absence of the wing blade. The wing margin of the adult is of developmental significance because it derives from the junction between the dorsal and ventral compartments in the larval wing disc.

Notch has two ligands, Serrate, which is expressed just in the dorsal compartment, and Delta, which is present all over the disc. However, Notch is mostly stimulated along the dorsal–ventral border, rather than in the remainder of the disc. This is achieved by two properties of the system. Firstly, there is a positive feedback built in since Notch signaling activates the expression of genes encoding components of the Notch pathway, in particular *Delta*, *Serrate*, and *Notch* itself. Secondly, there is the action of *fringe*, which is expressed only in the dorsal compartment. *fringe* encodes a glycosyl transferase that modifies Notch by addition of sugar residues and makes it more sensitive to Delta and less sensitive to Serrate. Because of the action of Fringe, Notch on the dorsal side is more sensitive to Delta so there is modest activation of Notch all over the dorsal compartment, resulting in production of slightly more Delta and Serrate. But the Serrate has no effect on the dorsal compartment because of the presence of Fringe. It has most effect on the first file of cells in the ventral compartment, which lack Fringe. So this file of cells produces more Serrate and Delta. This then becomes autocatalytic because the localized increase of Serrate and Delta cause increased expression of the genes encoding components of the Notch pathway. The effect is that both gene expression and signaling activity of Notch pathway components are upregulated just along the dorsoventral boundary. The evidence for this mechanism was based on the use of the GAL4 system to misexpress *Delta* and *Serrate*. It was shown that Delta could drive an increase in *Serrate* transcription in the dorsal compartment while Serrate could drive an increase in *Delta* transcription in the ventral compartment. Furthermore, the key function of the dorsal compartment selector gene *apterous* was shown to be to upregulate *fringe*. If a transgene that drives *fringe* off the *apterous* promoter is introduced into *apterous* mutants then this will rescue wing formation.

vestigial codes for a transcriptional cofactor required for proliferation of wing cells and is normally transcribed in the whole prospective wing blade. *vestigial* has the status of a "master control gene" because if it is activated in other imaginal discs then it can force the formation of wing blade from these discs.

vestigial has two enhancers, each of which is responsive to one of the signaling systems active in the wing disc. The boundary enhancer is turned on by Notch signaling and is normally activated along the dorsal–ventral border. The quadrant enhancer is activated by Dpp signaling and this activates *vestigial* in the remainder of the prospective wing blade.

The end result of the intercompartmental interactions is the activation of *wingless* along the prospective wing margin and of *vestigial* over the whole prospective wing blade (Fig. 17.14). The proximodistal pattern of the wing is represented by a nested set of expression domains of transcription factor genes arranged concentrically from distal (central) to proximal (circumferential): *Distal-less* (homeodomain), *defective proventriculus* (homeodomain), *rotund* (zinc finger), *nubbin* (homeo-POU domain). These are activated at least partly by the combination of Wingless and Notch, and lead to the final proximodistal pattern of visible structures consisting of hinge, proximal wing, and wing blade.

Regeneration and transdetermination

If a disc fragment is cultured not in another larva, but in the abdomen of an adult female fly, it will continue to survive and, under certain conditions, to regenerate. Localized damage can also be effected on discs *in situ* by using the GAL4 system to upregulate a cell-lethal product in a specific domain. Over about a week, a disc fragment will either regenerate the missing part or it will regenerate a copy of what is already present, a type of regeneration called "duplication." The consequences of regeneration can be observed directly using *in situ* hybridization or expression of reporter genes. But the traditional method was to infer what had happened by grafting the regenerated fragment to a metamorphosing larva and observing the structures produced on differentiation.

Regeneration has been studied most intensively in the leg discs, whose development differs somewhat from the wing (Fig. 17.15). There is still an expression of *engrailed* and *hedgehog* in

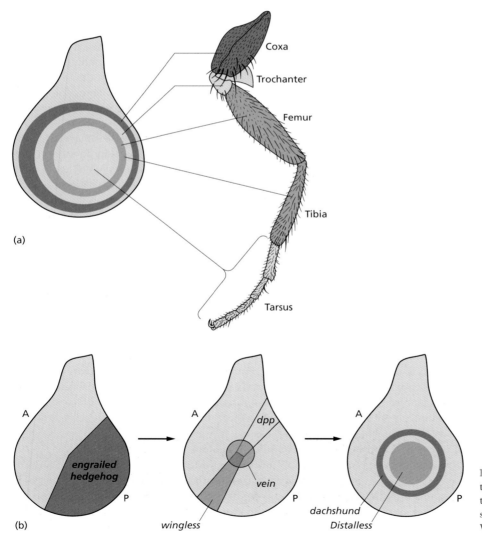

Fig. 17.15 The leg disc. (a) Fate map of the leg disc. (b) Upregulation of transcription factors in proximodistal sequence by the combined action of Wingless and Dpp.

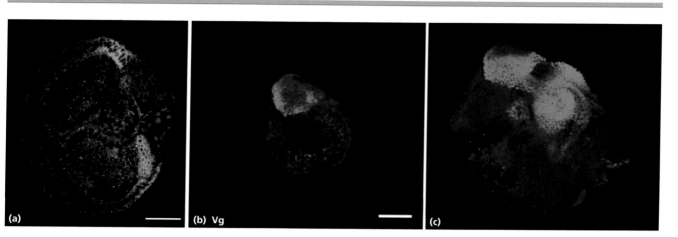

Fig. 17.16 Regeneration of the first (prothoracic) leg disc. Proteins are visualized by immunostaining. (a) Regeneration of posterior cells from an anterior fragment after 2 days of *in vivo* culture. Engrailed (posterior compartment) is magenta and Cubitus interruptus (anterior compartment) is green. Scale bar 25 μm. (b,c) Transdetermination at the "weak point," visualized by expression of the wing marker Vestigial: (b) 4 days, scale bar 100 μm; (c) 7 days of culture. Reproduced from Schubiger *et al. Developmental Biology* 2010; 347: 315–324, with permission from Elsevier.

the posterior compartment, but in the leg disc there is no dorsoventral boundary. Hedgehog induces expression of both *wingless* (in the ventroanterior region) and *dpp* (in the dorsoanterior region). The central point, where the joint effect of Wingless and Hedgehog signaling together is maximal, becomes the distal tip of the leg and activates expression of *vein*, encoding a neuregulin-like factor. The resulting distal to proximal gradient of EGF receptor activation controls the pattern of the tarsus, the most distal part of the leg, and perhaps more proximal parts.

When a disc is cut, it initially heals such that the columnar epithelium joins to the peripodial membrane, but then readjusts such that the cut edges of the columnar epithelium fuse with each other. This leads to a local activation of the JNK branch of the MAP kinase signaling pathways, with consequent stimulation of cell division in a region about 15 cells wide (Fig. 17.16). Excess cell division persists until the normal steepness of Wingless or Dpp gradients has been restored. Duplications will arise where one gradient is re-established in a new region of the disc, because structures are formed in accordance with the local concentration, which will now be in two places instead of one. However, regeneration of distal parts only occurs where *hedgehog*, *wingless*, and *dpp* zones abut. So, for example, induction of *wingless*-expressing clones only produces distal parts if the clones abut the *dpp*-expressing zone. In the prothoracic leg discs, (but not those of the second and third thoracic segments), the peripodial membrane itself expresses *hedgehog*. This means that additional *wingless* or *dpp* zones may be induced by Hedgehog following the initial healing phase in which the

peripodial membrane joins to the columnar epithelium and this provides a mechanism for regeneration from relatively small fragments.

In the early days of experimentation with disc fragments cultured in adult hosts, it was sometimes observed that the disc type would switch to a different disc type, for example from leg to wing. This was called **transdetermination** (Fig. 17.16b). It is now known that transdetermination can be produced reliably if discs are cut through a "weak point." In the case of the leg disc the weak point is in the center of the *dpp* domain and is a region that is sensitive to autocatalytic upregulation of *wingless*. There is a low level of wingless expression all over the disc, and in this region a wound will increase sensitivity such that the wingless expression activates itself and builds up to a high level. Since Wingless and Dpp are now expressed at high levels together, the result is upregulation of *vestigial* and a homeotic transformation of part of the leg disc to a wing character.

It has been confirmed that ectopic expression of *wingless* can provoke transdetermination even in intact leg discs, and at the same position. Ectopic expression of *vestigial* has also been shown to cause wing structures to develop from the leg or haltere discs. This explains the homeotic effects of *Ultrabithorax* (*Ubx*) mutations. Ubx is a member of the Hox cluster and loss-of-function mutations transform haltere to wing, while ectopic expression transforms wing to haltere. The reason for this is that Ubx suppresses the upregulation of *vestigial* by both the Wingless and the Notch pathways, and hence the absence of Ubx allows expression of *vestigial* and development of a wing.

New directions for research

Like the early *Drosophila* embryo, the imaginal discs are potentially useful for research into a variety of cell biology topics because of the powerful genetic methods that are available. Areas of current interest include the study of intracellular trafficking, of planar polarity, and of growth control.

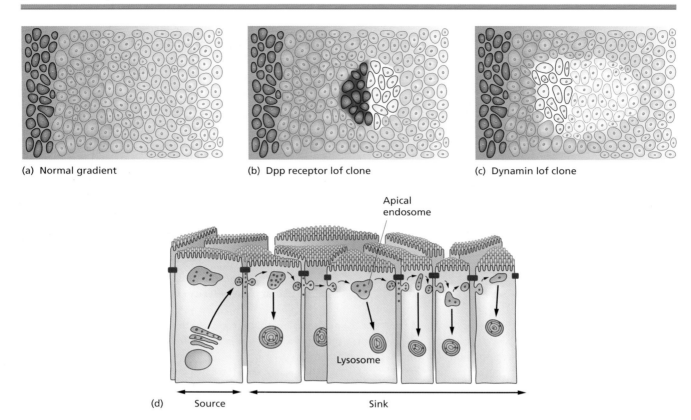

(a) Normal gradient

(b) Dpp receptor lof clone

(c) Dynamin lof clone

Apical endosome

Lysosome

(d) Source Sink

Fig. 17.17 Dpp gradient in the wing disc: (a) normal; (b) effect of a clone lacking the Dpp receptor; (c) effect of a clone lackingdynamin; (d) model of planar transcytosis. Reproduced from González-Gaitán. *Mech Dev* 2003; 120: 1265–1282, with permission from Elsevier.

Morphogen gradients and polarity

The *Drosophila* imaginal discs provide some of the best evidence for the existence of extracellular **gradients** of **morphogens**. In particular, the gradients of Dpp and Wingless have been observed directly by antibody staining or by the visualization of GFP fusion proteins in transgenic flies. But studies on the Dpp gradient in the wing disc have shown that transport is not by simple diffusion of protein in the extracellular space, as formerly supposed, but actually involves receptors and the intracellular trafficking systems (Fig. 17.17). A gradient of the fusion protein Dpp-GFP can be initiated by the use of a heat shock-regulated transgene. The gradient builds up over about 7 hours and has an exponential form across about 25 cell diameters, as would be predicted by the local source–dispersed sink model (see Chapter 2). The phosphorylation of MAD (the *Drosophila* Smad) occurs in concert, indicating biological activity. However, no gradient is formed if the transgene-encoded Dpp is a construct unable to bind to its receptor, and also it can be seen that most of the protein constituting the gradient is intracellular rather than extracellular. If a clone lacking the Dpp receptor is induced then the Dpp-GFP does not diffuse across it but accumulates on the source side of the clone (Fig. 17.17b). Internali-

zation of ligand–receptor complexes requires the activity of the motor protein dynamin and the small GTPase Rab5 (see Appendix). The gradient is also unable to cross clones lacking dynamin (Fig. 17.17c), and mutants of *Rab5* will alter the range of the gradient: loss of function reducing the range, and gain of function increasing it. So it seems that the gradient is set up and maintained by an active process of receptor binding, internalization, and re-release. It is thought that vesicles carry the receptor–Dpp complexes across the cell and then release the Dpp on the other side in a process called transcytosis (Fig. 17.17d).

The Wingless gradient is also exponential, but much steeper, with an effective range more like 10 cells. The mechanism of transport is different from Dpp as it is unaffected by removal of dynamin. These careful studies on two morphogens in the disc system indicate that different morphogens may be transported by quite different processes.

There is another aspect to the pattern of any body part, which is well displayed in *Drosophila* appendages such as the wing: the **polarity** of individual cells (Fig. 17.18). Wing cells each have one actin prehair, which are oriented such that they point distally. This cell polarity uses a number of components of the Wingless signaling pathway, which behave in a symmetry-

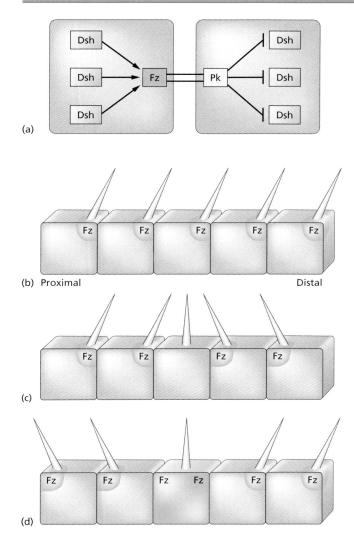

(a)

(b) Proximal   Distal

(c)

(d)

Fig. 17.18 Planar cell polarity system. (a) the formation of complexes involving Frizzled and Prickle-Spiney leg leads to a propagating polarization of cells. (b) Normal orientation of wing hairs. (c) Orientation of hairs around a cell lacking Frizzled. (d) Orientation of hairs around a cell overproducing Frizzled.

breaking manner. Frizzled, the wingless receptor, recruits Dishevelled. This both stabilizes the accumulation of Frizzled and also attracts Prickle-Spiney leg (a LIM domain protein) on the adjacent cell, which repels Dishevelled to the other side of its cell. The accumulation of Dishevelled on the other side of the second cell encourages the formation of another complex with Frizzled at this location and the attraction of Prickle on the same side of the next neighbor. The net result is that, starting from a homogeneous situation, the polarity can spontaneously develop and propagate from cell to cell. The main evidence for this mechanism comes from studying the effects of mutants of *frizzled*. If clones lacking *frizzled* are induced then the hairs on the surrounding normal cells tend to point towards the clone. On the other hand if clones overexpressing *frizzled* are produced, the hairs on the surrounding cells tend to point away. The accumulations of the various proteins can also be seen within the cells by immunostaining using confocal microscopy. Similar **planar cell polarity** systems also operate in other animals, for example in the control of the orientation of cells during *Xenopus* gastrulation, although in vertebrates the overall polarity is controlled by members of the Wnt family such as Wnt5 and -11.

In *Drosophila* there is an entirely separate cell polarity system based on two atypical cadherins: Fat and Dachsous, and on a kinase located in the Golgi body, Fourjointed, which can activate Fat by phosphorylation. Both Dachsous and Fourjointed are graded in concentration, in response to earlier morphogen gradients. Fat can bind Dachsous carried on an adjacent cell. If a lot of Fat or Dachsous is sequestered at one side of the cell by such binding then the other side becomes depleted and thus a cell polarity can propagate. This system is also connected to the Hippo pathway controlling growth (see Chapter 19) and is thought to provide information about pattern discontinuities, which result in excess local cell division.

Key points to remember

- The *Drosophila* larva undergoes two molts and then forms a pupa. In the pupa the majority of the larval body becomes replaced by differentiation of a number of imaginal rudiments. The cuticle of the head, thorax, and genitalia is formed from imaginal discs that arise in the embryo and grow in size during larval life, but remain visibly undifferentiated until metamorphosis.

- Mitotic recombination has been extensively used to analyze disc development. It can be used for clonal analysis or to ablate or overexpress particular genes of interest in random clones.

- The GAL4 method is used to overexpress genes of interest in specific domains.

- The thoracic imaginal discs become subdivided by an invisible anteroposterior border, which can be detected by clonal analysis, and corresponds to the parasegment border of the embryo. The posterior compartment in each disc is defined by activity of the *engrailed* gene.

- The wing disc also becomes subdivided into dorsal and ventral compartments in the second instar larva. The dorsal compartment is defined by activity of the *apterous* gene.

- The anteroposterior pattern of the wing disc arises by Engrailed upregulating *hedgehog*, and

Hedgehog upregulating *decapentaplegic* (*dpp*) along the anteroposterior boundary. This generates a gradient of Dpp protein from the boundary towards both margins.

- The dorsoventral pattern of the wing disc is controlled by upregulation of *Notch* expression and Notch signaling along the dorsoventral border. This upregulates *wingless*. The combination of Dpp, Notch, and Wingless signals generates the pattern of cell differentiation over the whole disc.

- When damaged, imaginal discs can regenerate. The cut faces heal together, and a region of cell proliferation forms in the vicinity. If the junction establishes a new distal tip, the result will be regeneration of missing parts. If the junction extends a morphogen gradient into new territory, the result will be a duplicated structure.

- Transdetermination is the switch of imaginal disc type, consequent on regeneration. Some transdifferentiation events can be provoked by overexpression of *wingless* or *vestigial* genes.

- Study of the Dpp gradient in imaginal discs have shown that it depend on receptor-mediated endocytosis for transport, whereas the Wingless gradient does not.

Further reading

Website for data on *Drosophila* genes: http://www.sdbonline.org/fly/aimain/1aahome.htm

General

Lawrence, P.A. (1992) *The Making of a Fly*. Oxford: Backwell Science.

Cohen, S.M. (1993) Imaginal disc development. In: *The Development of Drosophila melanogaster*. Bate, M. & Martinez-Arias, A., eds. Cold Spring Harbor, NY: Cold Spring Harbor Laboratory Press, pp. 747–841. Reprinted 2009.

Brook, W.J., Diaz-Benjumea, F.J. & Cohen, S.M. (1996) Organizing spatial pattern in limb development. *Annual Reviews of Cell and Developmental Biology* **12**, 161–180.

Methods

Brand, A.H. & Perrimon, N. (1993) Targeted gene expression as a means of altering cell fates and generating dominant phenotypes. *Development* **118**, 401–415.

Harrison, D.A. & Perrimon, N. (1993) Simple and efficient generation of marked clones in *Drosophila*. *Current Biology* **3**, 424–433.

Lee, T. & Luo, L. (1999) Mosaic analysis with a repressible cell marker for studies of gene function in neuronal morphogenesis. *Neuron* **22**, 451–461.

Blair, S.S. (2003) Genetic mosaic techniques for studying *Drosophila* development. *Development* **130**, 5065–5072.

Disc regeneration and transdetermination

Bryant, P.J. (1978) Pattern formation in imaginal discs. In: *Biology of Drosophila*. Ashburner, M. & Wright, T.R.F., eds, Vol. 2c. New York: Academic Press, pp. 229–335.

Campbell, G. & Tomlinson, A. (1995) Initiation of proximodistal axis in insect legs. *Development* **121**, 619–628.

Gibson, M.C. & Schubiger, G. (2001) *Drosophila* peripodial cells, more than meets the eye? *Bioessays* **23**, 691–697.

Maves, L. & Schubiger, G. (2003) Transdetermination in *Drosophila* imaginal discs: a model for understanding pluripotency and selector gene maintenance. *Current Opinion in Genetics and Development* **13**, 472–479.

Bergantiños, C., Vilana, X., Corominas, M. & Serras, F. (2010) Imaginal discs: renaissance of a model for regenerative biology. *BioEssays* **32**, 207–217.

Compartments

Lawrence, P.A. & Struhl, G. (1996) Morphogens, compartments and pattern: Lessons from *Drosophila*? *Cell* **85**, 951–961.

Serrano, N. & O'Farrell, P.H. (1997) Limb morphogenesis: connection between patterning and growth. *Current Biology* **7**, R186–R195.

Blair, S.S. (2003) Lineage compartments in *Drosophila*. *Current Biology* **13**, R548–R551.

Wing

Zecca, M., Basler, K. & Struhl, G. (1995) Sequential organizing activities of engrailed, hedgehog and decapentaplegic in the *Drosophila* wing. *Development* **121**, 2265–2278.

Cohen, S.M. (1996) Controlling growth of the wing: Vestigial integrates signals from the compartment boundaries. *Bioessays* **18**, 855–858.

Eaton, S. (2003) Cell biology of planar polarity in the *Drosophila* wing. *Mechanisms of Development* **120**, 1257–1264.

González-Gaitán, M. (2003) Endocytic trafficking during *Drosophila* development. *Mechanisms of Development* **120**, 1265–1282.

Kicheva, A., Pantazis, P., Bollenbach, T., Kalaidzidis, Y., Bittig, T., Julicher, F. & Gonzalez-Gaitan, M. (2007) Kinetics of morphogen gradient formation. *Science* **315**, 521–525.

Growth, regeneration, evolution

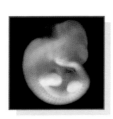

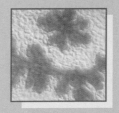

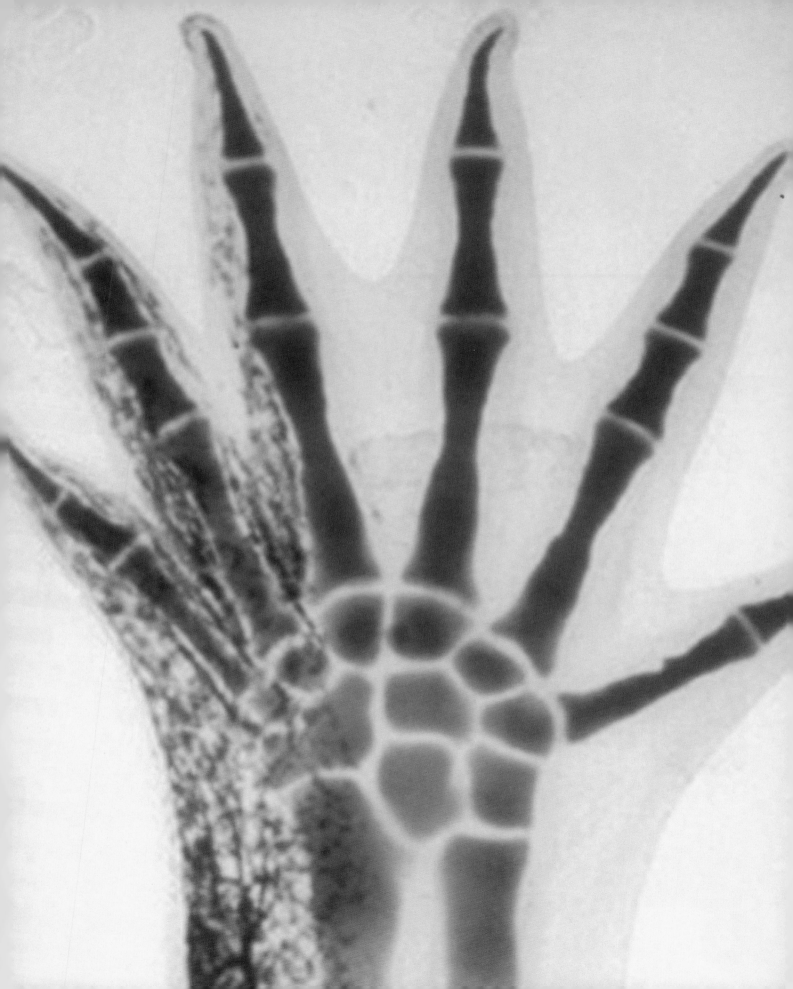

Chapter 18

Tissue organization and stem cells

The process of organogenesis generates a number of organ rudiments composed usually of more than one cell layer, and containing cell populations committed to form particular tissue types. In some tissue types development terminates as those cell populations differentiate. In others, development continues throughout life with the continuous production of new differentiated cells from populations of stem cells. This chapter deals with the chief tissue types found in the vertebrate body with special attention to their cellular renewal.

On the basis of light microscopy there are about 200 types of differentiated cell, although *in situ* hybridization and immunostaining reveal many more, especially in the central nervous system and the immune system. They are arranged in tissues, each of which normally contains several different cell types. In terms of developmental biology, a tissue is the set of cell types formed from a particular type of progenitor cell or stem cell. An organ or body part normally contains several tissue types arranged to fulfill a common function, and these are usually derived from more than one embryonic origin. Traditional histology distinguishes various types of tissue: epithelia, connective tissues, muscle, neural tissues, blood/blood vessels, and germ cells, some of which have already been introduced in Section 3.

Types of tissue

Epithelia

An **epithelium** (plural **epithelia**) is a sheet of cells, arranged on a **basement membrane** with each cell joined to its neighbors by specialized junctions. About 60% of the 200 named cell types are constituents of epithelia. The cells show a distinct apical–basal **polarity**, where the basal surface is that next to the basement membrane and the apical surface is on the opposite side, often facing a fluid-filled lumen. The basement membrane consists of a basal lamina secreted by the epithelium itself, together with some additional extracellular material from the underlying connective tissue. It is composed of laminin, type IV collagen, entactin, and heparan sulfate proteoglycan. The junctional complexes consist of three components: **tight junctions**, adherens junctions, and desmosomes (Fig. 18.1). The tight junction belt prevents liquid leaking between the cells and also isolates the components of the apical and basolateral membranes; adherens junctions are attachment points joining the microfilament networks of the cells; and desmosomes are point contacts joining bundles of cytokeratin filaments. Cell–cell contacts through adherens junctions and desmosomes are made by cadherins, with their homophilic calcium-dependent binding. The cells are anchored to the basement membrane by cell–matrix adherens junctions and by hemidesmosomes. These are similar to the cell–cell junctions but utilize integrins for attaching the cell to the matrix components. Apical surfaces often bear cilia, and may bear **microvilli** if the epithelium has an absorptive function.

Epithelia may be simple, with one layer of cells, stratified, with many layers of cells, or pseudostratified, meaning that they look stratified but in fact all cells contact the apical and basal surfaces. They may be **squamous**, with flattened cells, cuboidal, or columnar. Many epithelia are glandular and secrete materials into their surroundings. Glands may be simple or branched, and tubular or **acinar** (Fig. 18.2). The duct of a gland represents its original site of invagination from the cell layer of origin during development. **Exocrine** glands secrete into the duct, **endocrine** glands have lost their ducts and secrete into the bloodstream. **Myoepithelial** cells are often found surrounding the acini, and

Essential Developmental Biology, Third Edition. Jonathan M.W. Slack.
© 2013 John Wiley & Sons, Ltd. Published 2013 by John Wiley & Sons, Ltd.

their contraction helps drive the secretion down the duct. The terms mucous membrane or **mucosa** are often used to refer to a moist internal epithelium together with the immediately underlying connective tissue layer.

Although commonly thought to be ectodermal in origin, epithelia are, in fact, derived from all three of the germ layers of the embryo. The organization and cell renewal in epidermis (ectodermal) and intestinal epithelium (endodermal) are described below.

Connective tissues

The term **connective tissue** usually refers to those tissues dominated by **fibroblasts**, such as the dermis of the skin and the fibrous capsules surrounding most organs. In some histology or

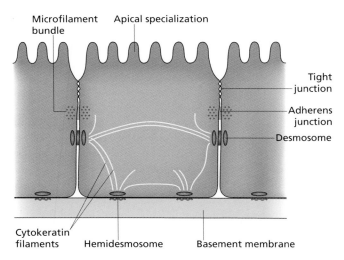

Fig. 18.1 Diagram of an epithelial cell showing the types of cell junction.

biology textbooks it may be used in a wider sense to include the skeletal tissues, muscle, and even the cells of the blood.

Much of the connective tissue is derived from the mesoderm of the embryo, although some is also formed by the neural crest. Mature connective tissue consists of fibroblasts embedded in an extracellular matrix. Fibroblasts are cells specialized to secrete the matrix components, which include hyaluronan, proteoglycans, fibronectin, type I collagen, type III collagen (reticulin), and elastin. Also found in connective tissue are histiocytes, which are macrophages resident in the tissues, and mast cells, which are histamine-secreting cells similar to the basophils of the blood but also resident in the tissues. Both these types originate from the bone marrow.

The skeletal tissues are composed of cartilage and bone and arise both from embryonic mesoderm and the neural crest. Much of the skeleton is formed initially as cartilage, which is then gradually replaced by bone. Skeletal parts formed in cartilage are known as cartilage **models**. Some parts, particularly in the skull, differentiate directly from mesenchyme into bone, and these are known as **membrane bones**. Skeletal tissues are discussed further in Chapter 19 in connection with growth.

Adipose tissue is also a type of connective tissue, and arises from embryonic mesoderm. Cells known as **mesenchymal stem cells** (**MSC**) can be isolated from various connective tissues and expanded in tissue culture. They are distinct from fibroblasts and display the ability to become fibroblasts or skeletal tissues or adipocytes or smooth muscle, under suitable culture conditions. They are more abundant in younger animals, suggesting persistence of a type of progenitor cell present in the late embryo. Whether they actually function as stem cells *in vivo* is unclear.

A term causing much confusion is **mesenchyme**. This is not a synonym for connective tissue nor for mesoderm. It is a descriptive term for scattered stellate cells embedded in a loose **extracellular matrix** (see also Chapter 2). Mesenchyme, derived either from mesoderm or from neural crest, fills up much of the

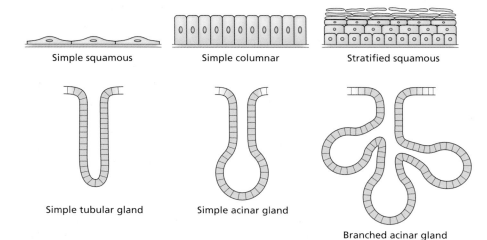

Simple squamous   Simple columnar   Stratified squamous

Simple tubular gland   Simple acinar gland

Branched acinar gland

Fig. 18.2 Types of epithelium.

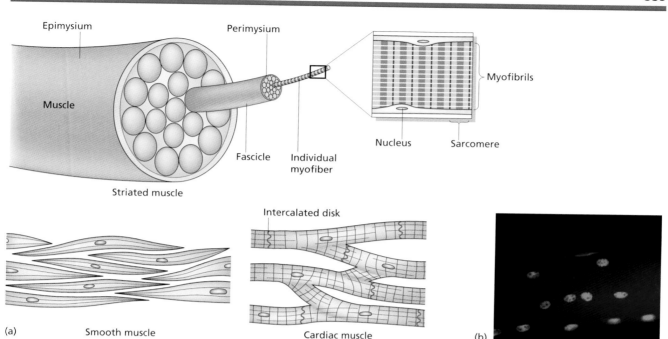

Fig. 18.3 (a) Types of muscle. (b) A myofiber with an associated muscle satellite cell with nucleus immunostained red for Pax7. The green nuclei belong to the myofiber.

embryo and forms fibroblasts, adipose tissue, smooth muscle, and skeletal tissues; however, these tissues should not be referred to as "mesenchymal" once they are differentiated.

Other tissue types

Muscle

There are three main types of muscle (Fig. 18.3). The development of skeletal muscle from the myotomes has been considered in Chapter 15. It is composed of elongated multinucleate cells called **myofibers**. Bundles of myofibers are gathered together in fascicles surrounded by a fibrous sheath, the perimysium, and the whole muscle is surrounded by another sheath called the epimysium. Myofibers are postmitotic, but can increase in size as a result of exercise. There are also some small cells lying beneath the basement membrane called **muscle satellite cells** (Fig. 18.3). These arise from the myotomes and multiply without differentiation as the animal grows. In adult muscle satellite cells remain quiescent unless there is some form of injury. Then they can become activated, upregulate myogenic transcription factors, become released from the myofibers, and become myoblasts. These can then either fuse with each other to generate new fibers, or fuse with existing fibers to augment their size. Evidence that muscle satellite cells are essential for regeneration comes from ablation experiments using *Pax7-CreER* to drive expression of diphtheria toxin. Following tamoxifen treatment the diphtheria toxin is produced and the

satellite cells killed. This renders the mice unable to repair muscle damage. When satellite cells proliferate, they also generate additional, undifferentiated satellite cells and for this reason they are considered to be a type of stem cell.

Smooth (= visceral) muscle exists as bundles of individual spindle-shaped, mononuclear cells. These contain a similar contractile apparatus to skeletal muscle but it is not arranged as visible sarcomeres. Smooth muscle is derived from the lateral plate of the embryo and is found mainly around the gut, blood vessels, and the ducts of glands, where inherent rhythmic contraction is required. Smooth muscle is usually mitotically quiescent but can be stimulated to grow following tissue damage.

Cardiac muscle, composed of **cardiomyocytes**, occurs only in the heart. As described in Chapter 15, it derives from the anteroventral margin of the lateral plate mesoderm of the embryo. Like skeletal muscle it has visible myofibrils, but like smooth muscle it remains as individual cells. The cells are joined end to end by intercalated discs, which contain structural junctions (adherens and desmosomes), together with gap junctions that allow rapid spread of electrical signals through the myocardium. Cardiomyocytes, like skeletal muscle fibers, are almost entirely postmitotic, although growth occurs by cell enlargement when the heart is placed under prolonged stress.

Neural tissues

Development of the neural tissues is described in Chapter 14. They comprise those cell types formed from the **neural tube** and some of those formed from the **neural crest**. The neural tube is composed of a specialized epithelium, the **neuroepithelium**, and produces both central neurons and glial cells (oligodendrocytes and astrocytes), while the neural crest produces autonomic neurons of the peripheral nervous system, together with **Schwann cells** and pigment cells.

Blood and blood vessels

The embryonic development of blood and blood vessels is considered in Chapter 15. Adult blood contains a variety of cell types. In addition to the red cells, there are granulocytes, monocytes, platelets, and lymphocytes. The platelets, essential for clotting, are produced from large cells called megakaryocytes. The nonlymphocytes are collectively called **myeloid cells**, meaning related to the bone marrow. Both myeloid cells and lymphocytes, as well as other cells such as histiocytes, osteoclasts, and Langerhans cells of the skin, arise from the **hematopoietic** stem cells in the bone marrow. The hematopoietic system is a state of continuous cell production and renewal throughout life, and is further described below.

The circulatory system consists of arteries taking blood from the heart to the tissues, capillaries supplying the tissues, and veins returning blood to the heart (Fig. 18.4). Macroscopic blood vessels have three layers. The inner layer is composed of a single layer of **endothelial cells**, which may have a little delicate underlying connective tissue. The middle layer is composed of smooth muscle, which may be very thick in arteries and thinner in veins, and the outer layer is composed of fibrous connective tissue. The capillaries consist of a single layer of endothelial cells with a basal lamina on the exterior surface. There is no smooth muscle but there may be some associated contractile cells called pericytes, sometimes considered to be the cell of origin of the **mesenchymal stem cells** that may be grown in tissue culture. Usually, the capillary wall is continuous but sometimes, as in the sinusoids of the liver, it contains gaps. Endothelial cells can divide throughout life and there is usually a low level of growth associated with tissue remodeling. The formation of new capillaries by endothelial cell division and cell movement is known as **angiogenesis**. A number of growth factors are active in promoting angiogenesis, particularly vascular endothelial cell growth factor (VEGF) and the fibroblast growth factors (FGFs).

Tissue renewal

With the exception of genomic DNA, most parts of an animal's body turn over in the sense that the constituent molecules are continuously synthesized and degraded. However, this does not mean that the cells of a tissue are necessarily replaced. It turns out that there are many types and degrees of cell replacement, the study of which makes up much of postnatal developmental biology. Some tissue types become largely postmitotic

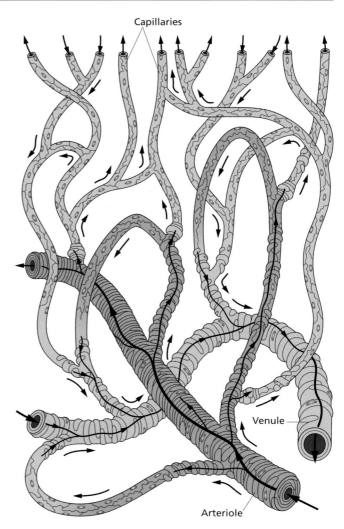

Fig. 18.4 A microcirculatory unit, showing joining of terminal arteriole and venule by capillaries. Reproduced from Warwick and Williams (eds) (1973) *Gray's Anatomy*, 35th edn, with permission from Elsevier.

after birth, or once the growth of the animal is complete. Others do not normally undergo cell renewal, but have some regeneration capacity following damage, while others undergo continuous, lifelong renewal. Much attention today is focused on **stem cells**. These are defined as cells that can both renew themselves and generate differentiated progeny. Much of this chapter will deal with the properties of the **tissue-specific stem cells**, about which there are two common misconceptions. Firstly, it is often thought that stem cells can turn into any cell type. This is not correct. Like any cell after the earliest stages of embryonic development, stem cells within the body have a commitment to form a specific subset of differentiated cell types appropriate to their own tissue, although this may be somewhat altered by *in vitro* culture. Secondly, every tissue is often thought to have its own stem cell. This is also not correct. The tissues that undergo continuous renewal have stem cells but those that undergo little or no renewal do not. Furthermore, not every cell that divides is a stem cell. Most dividing cells in the body are actually committed **progenitors**, or **transit amplifying cells**, not stem cells. The study of stem cells should be carefully rooted in an understanding of normal cell turnover in the body.

Measurement of cell turnover

Although tissue culture cells in optimal medium may grow exponentially, this is rarely true of cells within the body, where exponential growth is confined to some tissues during fetal life. Once the final size of the animal has been reached, cell division needs to be balanced by cell removal. In the renewal tissues, where cell production is significant, this is associated with compartmentalization of the tissue into separate proliferative and differentiating zones.

The multiplication of cell populations can be estimated by counting the proportion of visible mitoses to obtain a **mitotic index**. However, the M phase is usually a short period within the cell cycle, and a population has to be growing fast to show a significant mitotic index. More sensitive are methods that identify cells in **S phase**, which represents a longer fraction of the cell cycle and hence enables more cycling cells to be observed. A simple method is immunostaining of tissue sections for proteins associated with the cell cycle. One of these is a nuclear protein recognized by the antibody Ki67, which is involved with ribosomal RNA transcription. This is present in nuclei of cells throughout the cell cycle, so indicates the fraction of cells of a tissue that are in cycle. Another is proliferating cell nuclear antigen (PCNA). This is a cofactor of DNA polymerase and is found in cell nuclei during S phase, when DNA is being replicated. Its presence will give an estimate of the proportion of cells in S phase at the time of fixation.

Alternatively, cells, tissues, or whole animals can be labeled by administration of a DNA precursor, usually **bromodeoxyuridine (BrdU)**, a thymidine analogue that is incorporated into DNA and can be detected by immunostaining with a specific antibody (Fig. 18.5). This will reveal which cells were in S phase during the period that the label was administered. A short label with BrdU labels cells currently in S phase, and so gives similar information to PCNA staining. If BrdU is administered regularly for a prolonged period of time, then eventually all the cells in cycle will become labeled (similar information to Ki67 staining). It is also possible to give a short (pulse) label at one time and then at a later time to identify the position and differentiation class of the cells that were labeled. If a short label is given followed by a delay, then the labeling pattern will depend on what the cells are doing. If they stop dividing soon after the label was given then their DNA will remain labeled thereafter. The final S phase during development, normally followed by terminal differentiation, is called the cell's **birthday** and many studies have been made of cell birthdays, particularly of neurons in the

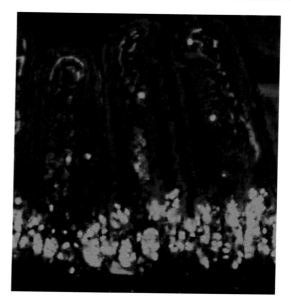

Fig. 18.5 BrdU staining of the intestine. A short BrdU administration labels all cells in S phase. In the intestinal epithelium dividing cells are found only in the crypts. One or two labeled cells are also visible in the connective tissue within the villi. Source: http://www.abcam.com.

nervous system. If a short label is followed by continued division then the BrdU content of the nuclear DNA will be diluted out by incorporation of thymidine during the subsequent rounds of replication. After five cell divisions the BrdU will have been diluted by a factor of 32 and will probably no longer be visible by immunostaining. In a complex tissue *in vivo*, the persistence of label in some dividing cells and not in others usually indicates that these are dividing more slowly.

Before BrdU became available, many similar studies were conducted using ^{3}**H-thymidine** (^{3}HTdR), which is also incorporated into DNA in S phase and can subsequently be localized by **autoradiography** (see Chapter 5). A recent addition to this set of reagents is **ethynyl deoxyuridine (EdU)**, which has similar properties to BrdU but involves a simple histochemical detection method instead of immunostaining. Also choro- and iododeoxyuridine can be useful where more than one DNA label needs to be applied, as they are recognized by different antibodies.

Much of the developmental biology of postnatal life concerns the behavior of the **renewal tissues**, which are described below. Renewal involves cell death as well as cell birth. The index of apoptotic cell death can be measured by several methods. The most popular are immunostaining for the presence of one of the apoptosis-associated proteins, such as the caspase enzymes, and the detection of DNA breaks by a method called TUNEL (TdT-mediated dUTP nick end labeling). Here the enzyme terminal nucleotidyl transferase is used to add a modified nucleotide, usually biotin-labeled, to the fragmented DNA of the dying cell. This is then detected with a fluorescent or enzyme-linked streptavidin.

Although most measurements of cell turnover look at the proportion of cells in cycle, or the proportion of cells in apoptosis, what is really required is a measure of cell production rate and cell removal rate. To obtain a cell production rate it is necessary, as well as the S-phase labeling index, to know the duration of the cell cycle and the proportion of the cycle spent in S phase. For example if cells divide on average once per 24 hours and the S phase lasts 6 hours, then a short pulse of BrdU would enable the observation of about 6/24, or one-quarter, of the cells in cycle. So in this case the cell production rate per day is about four times the cell labeling index. Even for a simple tissue this would a very rough approximation because in a real animal there is a broad distribution of cycle times across the dividing cells, as well as circadian or other physiological effects on cell division.

The cell removal rate is rarely calculated, and often underestimated. Because the duration of apoptosis is quite short (1–4 hours) the flux to cell death per 24 hours is a high multiple of the apoptotic labeling index. For example if the apoptotic index is 1% and the dying cells are observable for only 2 hours, then the flux to cell death is actually about $1 \times (24/2)$ or 12% per day.

Measuring the cell cycle time *in vivo* can be done by various methods. A recently introduced procedure utilizes the ability to label cells with two distinguishable DNA precursors, for example BrdU and EdU. It needs to be assumed that the cell population is completely asynchronous, so that the proportions of cells in particular cell cycle states is proportional to the duration of those states. For example suppose that the organism is injected with BrdU at $T = 0$, and then with EdU at $T = 1.5$ hours, then is killed at $T = 2$ hours and the tissue of interest is sectioned and stained for BrdU and EdU. The 30-minute period is sufficient for the EdU to be incorporated at detectable level into all cells currently in S phase. Let the proportion of EdU-labeled cells be designated S. These cells will all also be labeled with BrdU, which was still available during the last 30 minutes. But some cells will be labeled only with BrdU. These are the cells that have exited the S phase during the 1.5 hours before the EdU was administered. Let the proportion of cells labeled with BrdU but not with EdU be called L. Because the proportions of cells represent the proportions of time spent in each phase of the cell cycle, we know that:

$$\frac{1.5}{T_s} = \frac{L}{S}$$

where T_s is the duration of S phase. Therefore:

$$T_s = 1.5S/L$$

If, for example, L is 0.05 and S is 0.167 then:

$$T_s = (1.5 \times 0.167)/0.05 = 5 \text{ hours}$$

To calculate T_c, the duration of the whole cell cycle, we need another piece of information, which is the proportion of the whole cell population that is proliferating. This can be determined by administering a prolonged label of BrdU until saturation is reached, or by staining for Ki67 antigen. Suppose that this proportion is P. By the same logic we know that:

$$\frac{T_s}{T_c} = \frac{S}{P}$$

Therefore:

$$T_c = T_s P / S$$

If, for example, P is 0.5, then:

$$T_c = (5 \times 0.5)/0.167 = 15 \text{ hours}$$

There are other method for estimating the duration of the cell cycle and the rate of cell production and loss *in vivo*. However, all the methods involve difficulties and attention needs to be paid to controlling for a lot of artifacts and confounding variables.

Proliferative behaviors of tissues

Tissue types in the body can be classified on the basis of their proliferative behavior, as visualized with DNA labels. It is important to note that a cell type within a tissue may be postmitotic, but still capable of replacement from stem or progenitor cells in that tissue. This classification refers only to mammals, as lower vertebrates may show very different patterns of cell renewal.

1 Tissues where the principal functional cell types are postmitotic and where they are mostly not replaced when lost. If they are replaced, this is from stem or progenitor cells, not by division of differentiated cells. For example neurons, skeletal muscle fibers, and cardiomyocytes are all postmitotic. Once formed they do not divide again. There is a limited new formation of neurons of the olfactory bulbs and the hippocampus from neuronal stem cells, but in other parts of the central nervous system there is no renewal of neurons. In skeletal muscle there is some new formation of myofibers from satellite cells following injury. In the heart the levels of cardiomyocyte renewal are very low and perhaps even zero.

2 Tissues consisting of differentiated cells that grow by cell division while the animal is growing and stop when the adult size is attained. In the adult such tissues are mostly quiescent, although there may be a slow turnover of cells. In addition, the differentiated cells remain capable of division to a greater or lesser degree if stimulated by wounding. In this category are some connective tissues, smooth muscle, and liver.

3 Renewal tissues. Here the tissue is in a constant state of cell turnover. There is a permanent, active proliferative zone containing stem cells and this generates a population of differentiated cells, which itself has a finite lifetime and is constantly

dying and being repopulated. The chief examples are the hematopoietic system, the epidermis of the skin, the epithelium of the gut, and the spermatogonia of the testis.

Measuring slow cell turnover

Even BrdU/EdU labeling is not sensitive enough to detect very slow rates of cell turnover, and there have been many disputes about whether tissues like the heart turn over at all. But there is now a method capable of measuring slow cell turnover, which is admirably suited to use in humans. It relies on the fact that in the 1950s and early 1960s there were many atmospheric tests of nuclear weapons, which caused release into the atmosphere of large amounts of radioactive isotopes. After the Test Ban Treaty of 1963, all nuclear tests were conducted underground and the releases stopped. The isotope that is used for cell turnover analysis is carbon-14 (^{14}C). This is familiar for its use in archaeology, where dating studies make use of its radioactive decay. However, the use of ^{14}C for cell turnover studies does not involve radioactive decay but on the much faster loss of CO_2 from the atmosphere due to chemical and physical processes. Plants absorb CO_2 from the atmosphere for photosynthesis, and people eat these plants, or the animals that were fed on them, so the intake of carbon in a given year has an isotopic composition very similar to that of the atmosphere at the same time. The ratio of $^{14}C/^{12}C$ in each year is known precisely by isotopic analysis of tree rings. This shows a rapid increase from 1955 to 1963 due to the nuclear tests, followed by a slower exponential decline, due to the gradual removal of ^{14}C from the air (Fig. 18.6). Radioactive decay of ^{14}C is negligible over this period, and does not affect the results, since the radioactive half life is 5730 years.

Cell turnover is established by analysis of the carbon isotope composition of nuclear DNA in the cell types of interest. This is done in human postmortem tissue samples. The DNA of the cell nucleus is unique in that, once it is formed, it can be stable for life, assuming that there is no further DNA replication and that the level of repair synthesis is negligible. So the nuclear DNA of cells that do not divide after birth should have the isotope composition corresponding to the birth year. Conversely, an actively dividing cell population, as found in a renewal tissue, will have the isotopic composition of the year of death. Because all tissues contain mixtures of cell types it is necessary to purify nuclei of the specific cell type of interest, then extract the DNA, then measure the isotope ratio. The nuclei are isolated by staining with a specific antibody followed by fluorescence-activated cell sorting. The DNA is then extracted, extensively purified, pyrolyzed to CO_2, and the isotope ratio is measured using an accelerator mass spectrometer.

Application of this method has confirmed that in the cerebral cortex, which is not an area fed by neural stem cells, there is no detectable renewal of neurons at all. All the cerebral cortex neurons are formed during fetal life and they are not added to after this time, or not enough to alter the isotope ratio. In the heart the results suggest that there may be a little cell turnover, of the order of 0.5% per annum. This is a very low level, meaning for example that the cardiomyocytes of the heart would only be partly replaced over an 80-year lifespan. In adipose tissue, the rate of turnover is faster, about 10% per annum.

Stem cells

Tissue-specific stem cells are defined as cells that continue to divide for the lifetime of the organism, are visibly

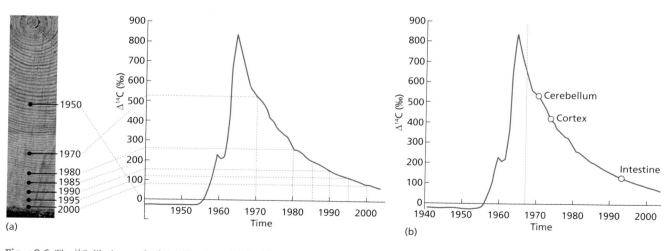

(a) (b)

Fig. 18.6 The ^{14}C dilution method. (a) The rise and fall of ^{14}C in the atmosphere during the era of nuclear bomb tests. The values can be determined for each year from tree ring samples. (b) The ^{14}C abundance of DNA in three tissues from an individual who was born in 1967 and died in 2003. These results are averages for whole tissues comprising many cell types. Reproduced from Spalding *et al. Cell* 2005; 122: 133–143, with permission from Elsevier.

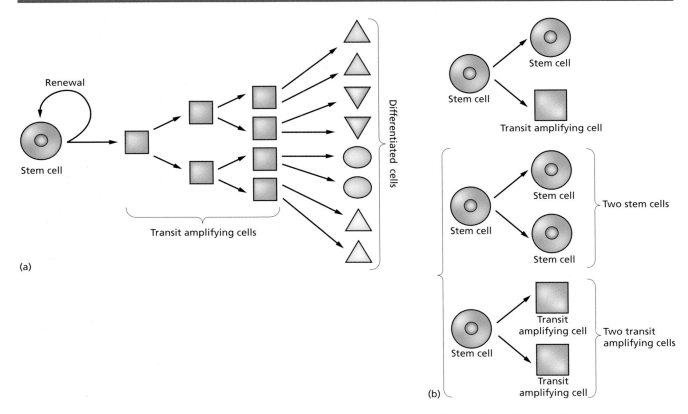

Fig. 18.7 (a) Cell lineage in a renewal tissue, showing stem cells, transit amplifying cells, and differentiated cells. (b) Stem cells can maintain themselves either by repeated asymmetrical division, or by generating stem cell and transit daughters with equal frequency but at separate divisions.

undifferentiated, and whose progeny include both further stem cells and cells destined to differentiate (Fig. 18.7a). They share the name "stem cell" with **embryonic stem cells** (**ES cells**, see Chapters 10 and 19) but they are somewhat different. Tissue-specific stem cells are defined by their role in the intact organism, while the stem cell character of ES cells is defined by long-term renewal in tissue culture. Tissue-specific stem cells are committed to form the cell types of a specific tissue (= uni- or **multipotent**), while ES cells can form all cell types (= **pluripotent**). The genes responsible for stem cell behavior, as well as other markers of the stem cell state, are generally different in ES cells and tissue-specific stem cells.

By definition, stem cells are capable of unlimited division. However, they are by no means the only dividing cells in the tissue because their direct progeny are **transit amplifying cells**, capable of dividing only a few times but whose division can be regulated to correspond to the demand for new differentiated cells. Transit amplifying cells are one class of **progenitor cell**, which also includes all the cells found in early embryos. The distinction is that stem cells self-renew indefinitely while progenitor cells have a finite division lifetime and will develop into a terminally differentiated cell type in due course, precisely the behavior of early embryo cells. Transit amplifying cells are normally the majority of dividing cells in a renewal tissue. The stem

cells are a minority and generally grow more slowly, hence sometimes being identified by label retention following a pulse of BrdU. Although they must repopulate the stem cell compartment as well as feed cells to the transit amplifying and differentiated compartments, this does not necessarily mean that every individual stem cell division need be an asymmetrical one. It is simply required that, on average, the progeny of the stem cell consists of 50% stem cells and 50% cells destined to differentiate (Fig. 18.7b). Stem cells are often considered to be radiosensitive, and whole-body radiation does selectively damage renewal tissues, leading to bone marrow failure, intestinal breakdown, sterility, and skin damage. A possible evolutionary rationale for this is that a modest degree of cell death arising from naturally occurring toxins and radiation is better than survival with mutations, which in a stem cell may well initiate a cancer and kill the organism.

Tissue-specific stem cells are committed to form just one particular tissue type. An intestinal stem cell can form only intestinal types and an epidermal stem cell can form only epidermis. This corresponds to the idea that embryonic development is hierarchical, with the cells of the early blastula or epiblast being able to form anything, and then being progressively restricted in their potency by a succession of inductive signals and responses to them. For example in the course of development the precursors of an intestinal stem cell would have been committed to endoderm and then to intestine, while the embryonic precursors of a hematopoietic stem cell would have been committed to mesoderm and then to the blood-forming tissue. Most tissue-specific stem cells do produce more than one cell type. Of the examples considered here, only the spermatogonia of the testis produce a single cell type (unipotent), while the hematopoietic stem cell produces a wide range of cell types in the blood and immune system, the intestinal stem cell produces all cell types of the intestinal epithelium, and even the epidermal stem cell produces a small minority of Merkel (sensory) cells as well as the predominant keratinocytes of the epidermis.

The behavior of cells in tissue culture may differ significantly from behavior *in vivo*. So the ability to culture stem-like cells from a particular tissue does not mean that the tissue normally contains stem cells. For example neural stem cells are found only in two regions of the brain, but neurospheres, containing stem cell-like cells, may be cultured from many other regions. The ability to form large clones *in vitro* is often taken as a property of stem cells, but again the potential change of properties in the tissue culture environment means this is not an infallible guide.

Many tissues have a histological substructure such that they consist of many small repeating modules or units, for example glandular acini or intestinal crypts. These units are not only the functional units of the tissue, but are also often the units within which cell proliferation and turnover are organized. The places where stem cells are found contain a specific microenvironment known as the **stem cell niche** suitable for their persistence and growth. If the stem cells are displaced from this microenvironment they will grow no more. Conversely, if they are reintroduced to the niche then they will grow again. The best kind of characterization for tissue-specific stem cells is to label a single cell and identify its progeny at later stages. According to the principles of clonal analysis, the stem cell and all its descendants

should remain labeled indefinitely. The most usual method to do this is the transgenic *CreER x R26R* system, described in Chapter 13. A promoter is used to drive the *CreER* that is active only in the stem cells of a particular tissue, so only the stem cells contain the CreER protein. A pulse of tamoxifen (or 4-hydroxytamoxifen) will activate the Cre, and recombination will ensure that these cells subsequently express *lacZ*. At later time points the expansion of the clone can be visualized by its *lacZ* expression. Eventually a steady state will be reached, where a band of cells are labeled, leading from the stem cell, through the transit amplifying and differentiated compartments. In humans it is not possible to make transgenics, but a similar result can be obtained by examination of clones lacking the COX (cytochrome c oxidase) enzyme. This is a mitochondrial enzyme encoded by a gene on the mitochondrial DNA. Mitochondrial genes are often mutated because of the relatively poor DNA repair mechanisms of mitochondrial as compared to nuclear DNA. However, there are many copies of the genome per mitochondrion and many mitochodria per cell, so it takes years for genetic drift to create a cell completely lacking the enzyme. But once this has occurred it acts as a genetic marker just like *lacZ*, and the architecture of structural proliferative units in human tissues can be mapped out using histochemistry for COX.

Intestinal epithelium

The gastrointestinal tract of vertebrates consists of a muscular tube running from the mouth to the anus. It is lined by a number of different epithelia: pharyngeal, esophageal, gastric, small intestinal, and colonic, separated by abrupt discontinuities of cell type. This epithelium is derived from the **endoderm** of the early embryo, while the other cell layers of the gut are derived from the **splanchnic mesoderm**. The epithelium, together with underlying connective tissue called the lamina

Classic experiments

Cell turnover in tissues

The first paper is a study of mitoses in the epithelium of the small intestine and arrives at the conclusion that cells must be being continuously produced in the crypts and shed from the villi. The second paper uses the incorporation of radioactive debris from ³H-thymidine-labeled cells that have died as a cell marker for neighboring cells. From studying the subsequent distribution of these radioactive debris it was postulated that a single type of stem cell produces all four cell types of the small intestinal epithelium.

The third paper provides a completely novel method for estimating cell turnover rate over long time periods.

Leblond, C.P & Stevens, C.E. (1948) The constant renewal of the intestinal epithelium in the albino rat. *Anatomical Record* **100**, 357–377.

Cheng, H. & Leblond, C.P. (1974) Origin, differentiation and renewal of the four main epithelial cell types in the mouse small intestine. V. Unitarian theory of the origin of the four epithelial cell types. *American Journal of Anatomy* **141**, 537–562.

Spalding, K.L., Bhardwaj, R.D., Buchholz, B.A., Druid, H. & Frisen, J. (2005) Retrospective birth dating of cells in humans. *Cell* **122**, 133–143.

propria and a thin muscle layer called the muscularis mucosa, is often known as the **mucosa**. Outside the mucosa lie further thick layers of connective tissue and smooth muscle.

The program of cell renewal is best understood for the small intestine. This contains regions called the duodenum, jejunum, and ileum, although the difference of cell type between them is not great. On a microscopic scale the intestinal epithelium is arranged on finger-like villi projecting into the lumen, and between the villi lie crypts of Lieberkuhn sunk below the surface (Figs 18.8, 18.9). Cell proliferation takes place only in the crypts,

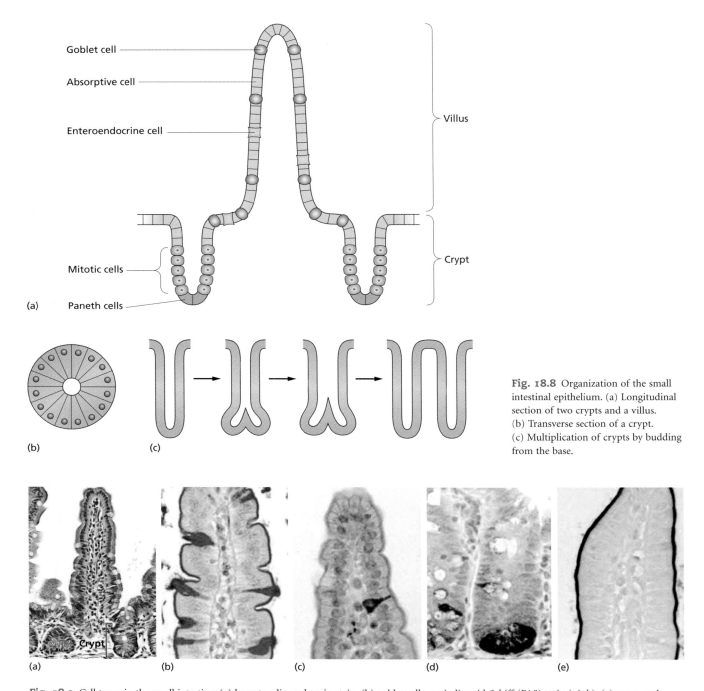

Fig. 18.8 Organization of the small intestinal epithelium. (a) Longitudinal section of two crypts and a villus. (b) Transverse section of a crypt. (c) Multiplication of crypts by budding from the base.

Fig. 18.9 Cell types in the small intestine: (a) hematoxylin and eosin stain; (b) goblet cells, periodic acid-Schiff (PAS) stain (pink); (c) enteroendocrine cell, synaptophysin immunostaining (brown); (d) Paneth cells, lysozyme immunostaining (brown); (e) absorptive cells (enterocytes), alkaline phosphatase histochemistry (dark blue). Reproduced from van der Flier and Clevers. *Ann Rev Phys* 2009; 71: 241–260, with permission from Annual Reviews.

and differentiated cells are continuously moving out from the crypts, moving up the villi, then dropping off into the gut lumen. The main cell types are **enterocytes** (absorptive cells) and **goblet cells**. The absorptive cells are characterized by a brush border at their apical surface, consisting of numerous, close-packed microvilli. Goblet cells contain a large vesicle filled with mucins. In addition, there are **Paneth cells** located at the base of the crypts, which secrete antibacterial substances including lysozyme and cryptdins, and several types of **enteroendocrine cells**, each secreting a particular peptide hormone. Collectively, the goblet, Paneth, and enteroendocrine cells are known as secretory cells. Some experimental studies have also been performed on the colon because of its importance in terms of colonic cancer. Its structure is similar to that of the small intestine, but without villi and without Paneth cells.

The experimental study of intestinal organization has depended heavily on certain types of transgenic mice. The *Villin* promoter can be used to drive genes just in the intestinal epithelium and so is used in transgenics for tissue-specific overexpression. The *Cyp1a* promoter (*Cyp1a* encodes one of the cytochrome p450 enzymes) is inducible and may be upregulated by treatment of the animal with β-naphthoflavone, via the aryl hydrocarbon receptor, so can be used to induce transgene activity in the epithelium at a specific time. These promoters may also be used to drive *Cre* or *CreER*, which can bring about overexpression or ablation of selected genes that are appropriately floxed.

In mouse the villi start to form from about E15, and the crypts themselves are set up shortly after birth. This occurs by a lateral inhibition type of process involving BMPs and Hedgehogs. Initially Sonic and Indian hedgehog are expressed in the whole epithelium, and are needed as trophic signals for formation of the mesenchymal coat of the gut. Then the expression of Hedgehogs becomes concentrated in certain places (the future crypts). At this stage, transgenic inhibition of Hedgehog signaling will inhibit crypt formation. The growth of villi is promoted by BMP2 and 4, which is produced in the mesenchyme and acts on the epithelium. BMP expression is stimulated by the Hedgehogs but because its range of diffusion is greater its effects are felt more widely. So the zones where Hedgehog prevails over BMP become the crypts and the surrounding areas where BMP prevails over Hedgehog become the villi. The actual morphogenesis of villi and crypts occurs by folding of the endodermal epithelium, which is initially a simple columnar epithelium. It depends on the expression of ephrin B1 in the villi and EphB2 and B3 in the crypts. Production of these proteins is, respectively, repressed and upregulated by β-catenin. If *EphB2* and -*3* are removed then the intestine develops with proliferative cells and differentiated cells all mixed together rather than segregated as normal. New crypts and villi continue to form during the postnatal growth of the animal. The crypts divide by budding, starting at the base (Fig. 18.8). The signal for crypt division is not known but probably requires an increase in the number of stem cells.

Cell division in the crypts is rapid. The stem cells are located near the crypt base, with several layers of transit amplifying cells lying above them. A mouse crypt contains about 250 cells of which roughly 160 are dividing, with a cycle time of about 13 hours. The progeny move up and out of the crypts and the tissue is arranged such that each crypt feeds more than one villus, and each villus draws cells from several crypts. Wnt-β–catenin signaling is critical to the maintenance of cell division in the crypts. Wnt ligands and receptors are expressed in the crypt epithelium. Knockouts that reduce Wnt-β–catenin activity, especially that of the transcription factor gene *Tcf4*, cause cessation of all division, of transit amplifying cells as well as stem cells. The *c-Myc* gene is an important target of Wnt, and also essential for continued proliferation. If *c-Myc* is ablated using an intestine-specific Cre excision method, growth stops in the affected crypts, which are rapidly overgrown by normal, nonexcised ones by the process of crypt fission.

APC, the product of the *adenomatous polyposis coli* gene, is a cytoplasmic protein required to enable the phosphorylation of β-catenin by GSK3. In *APC* loss-of-function mutants β-catenin is not inactivated, and is therefore constitutively active. This leads to the inability to shut off production of Eph B2 and B3, and to the formation of **polyps**, which are projections into the lumen of differentiated but abnormally organized intestinal tissue. Human patients suffering from adenomatous polyposis coli have many such polyps and a high risk of a polyp developing to cancer (see below). The disease is hereditary and due to loss of one copy of the APC gene. When the other, good, copy is lost from an individual cell due to occasional somatic mutations, then that cell will have constitutively active β-catenin and develop into a polyp.

Intestinal stem cells

The intestinal stem cells can now be identified by the presence of a cell-surface marker: Lgr5, which is an orphan G protein-coupled receptor. About six cells per crypt express this molecule and are located at the crypt base in between the Paneth cells. If they are labeled by tamoxifen treatment of a *Lgr5-CreER x R26R* mouse, initially the cells themselves are labeled, then ribbons of descendants become visible leading up from the stem cells out of the crypts and onto the villi, and after a few days as far as the villus tips where the cells die and drop off (Fig. 18.10). All four cell types, absorptive, goblet, enteroendocrine, and Paneth, are present in monoclonal ribbons, which arise following low-dose tamoxifen treatment which only causes recombination in rare isolated cells. This cell-labeling behavior is exactly what is expected of a stem cell. The same patterns were formerly observed by mutagenesis studies on the intestine in which rare stem cells became labeled as a result of a mutation of a cell surface carbohydrate binding *Dolichos* lectin (Fig. 18.11).

Although *in vitro* culture is not an infallible guide to cell properties *in vivo*, it is notable that isolated Lgr5-positive cells, isolated by FACS, can establish organoid cultures in Matrigel,

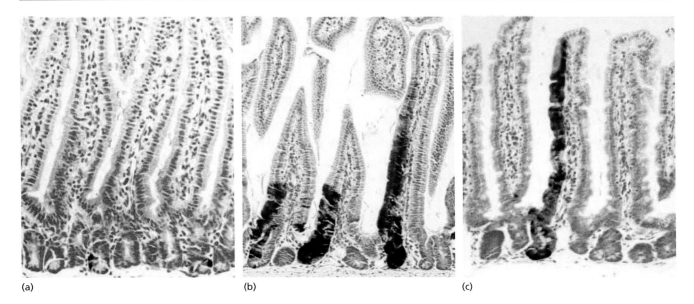

(a)   (b)   (c)

Fig. 18.10 Lineage of Lgr5⁺ cells revealed in *Lgr5-CreER x R26R* mouse. The mice are sacrificed and the intestine stained with XGal 1, 5, and 60 days after tamoxifen induction of the label. Initial label is in Lgr5⁺ cells themselves; subsequently, ribbons of descendant cells up the crypts and villi become labeled. Reproduced from Barker *et al. Nature* 2007; 449: 1003–1007, with permission from Nature Publishing Group.

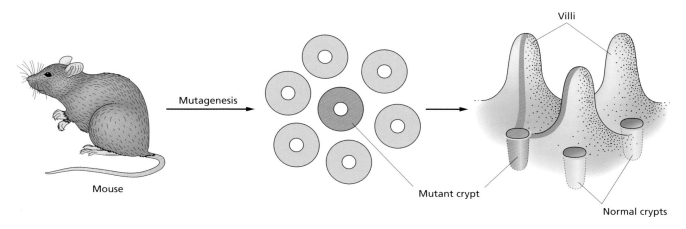

Fig. 18.11 Mutagenesis produces occasional cells that can be visualized by binding of *Dolichos* lectin. One mutant stem cell may populate an entire crypt, and its progeny form streams up to the tips of the adjacent villi.

in which the crypt–villus pattern is reconstituted in a closed vesicle. Even single cells can do this to some degree (Fig. 18.12). The niche for the small intestinal stem cells *in vivo* appears to be provided by the Paneth cells. Normally the Lgr5 cells lie in between Paneth cells and in close contact with them. The Paneth cells express two of the *Wnt* genes (*Wnt3* and *9b*) providing the required mitogenic signal for stem and transit amplifying cells. If Lgr5 cells are cultured after mixing with isolated Paneth cells, also isolated by FACS using the CD24 cell-surface marker, then the yield of organoids is much higher than for stem cells alone. Paneth cells can be removed from the animal by targeted dele-

tion of *Sox9* from the intestine using *Cyp1a-Cre*, which leads to a fall in Paneth cell number from 4–8 weeks after induction. Over this period there is a concomitant drop in stem cell numbers. Eventually, the number of viable crypts declines and the intestine regenerates by rapid budding of the minority of normal, unrecombined, crypts.

Interestingly, there is at least one other class of cell in the intestinal epithelium that exhibits stem cell behavior, which are those expressing *Bmi1*, encoding a protein which is part of the PRC1 polycomb complex. When labeled with *Bmi1-CreER x R26R*, these give rise to long-lived ribbon patterns in the same

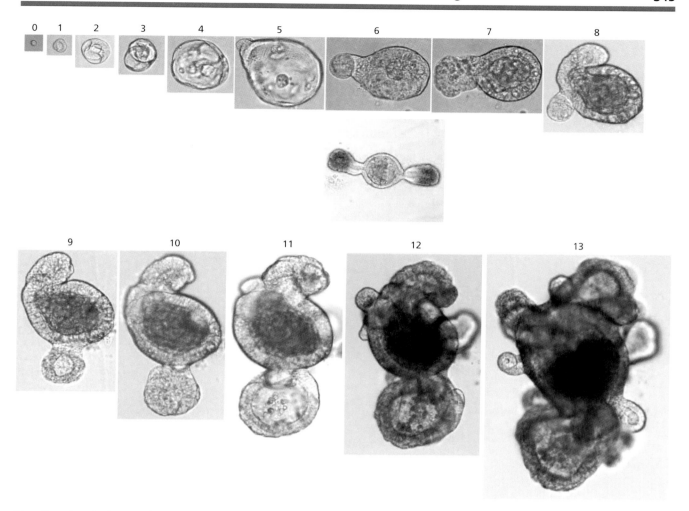

Fig. 18.12 Intestinal organoid growing in Matrigel. This colony was founded by a single Lgr5$^+$ stem cell, and growth is recorded each day (number of days of development are shown). Inset: an early stage organoid from a Wnt reporter mouse, the region of Wnt signaling is revealed by XGal staining; as *in vivo*, it is seen in the crypts. Reproduced from Sato *et al. Nature* 2009; 459: 262–265, with permission from Nature Publishing Group.

way as the Lgr5 cells. However, the Bmi1 cells are located above the Paneth cell zone. They are more abundant in the proximal small intestine (jejunum) while the Lgr5 cells are more abundant in the distal small intestine and colon, so they may reflect the presence of two type of stem cell, which may be partly responsible for differences between these intestinal regions. Lgr5 cells have been ablated by knocking a gene for the human diphtheria toxin receptor into the *Lgr5* locus. Mice are normally resistant to diphtheria toxin so an administered dose will just eliminate the *Lgr5* cells expressing the receptor. In this case the intestinal cell renewal continues as normal, from the *Bmi1* cells.

There has been some debate about whether the stem cell divisions in the intestine are or are not asymmetric. Mathematical analysis of the process of drift from a crypt containing a single labeled stem cell, to a uniformly labeled crypt, suggests that stem cell divisions are random and may generate either two stem cells, or one, or none. This process of drift is apparent in two situations: in the 2 weeks following birth, and in crypts in which a single stem cell has been labeled at random by mutagenesis. The postnatal period has been analyzed by making **aggregation chimeras** between mouse strains that differ in the expression of a marker, *Dolichos* lectin receptor (Fig. 18.13). This carbohydrate is expressed by intestinal cells in some mouse strains and is absent in others. In an aggregation chimera the cells of the two donor embryos become intimately mixed, so at the time of crypt formation most crypts will receive cells of both donor types. However, over the first 1–2 weeks of postnatal life the crypts lose one of the two types, such that all cells in each crypt are either one type or the other. In other words, the crypts become **monoclonal**. Initially, it was thought that this meant

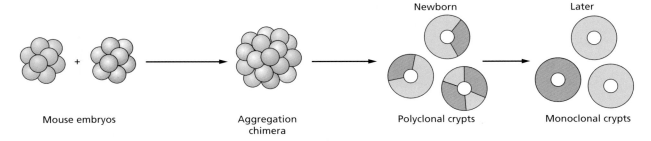

Fig. 18.13 Aggregation chimeras. Early crypts are polyclonal but later become monoclonal.

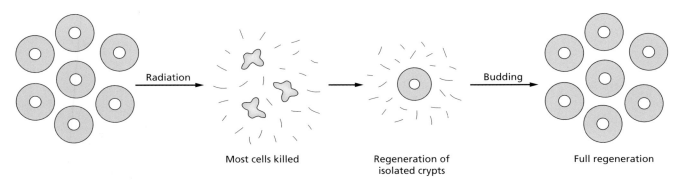

Fig. 18.14 A dose of X-radiation that destroys most cells leads to regeneration of whole crypts from individual clonogenic survivors.

there was only one stem cell per crypt. However, it is now clear that there are several stem cells per crypt and that the monoclonality arises by a random cell selection process. This arises because the stem cells, on division, are able to produce either two stem cells or one stem and one transit amplifying, or two transit amplifying cells. Over several cell cycles the genetic diversity of the stem cells within a crypt will become progressively reduced because any stem cell that produces two transit amplifying cells will be lost to the stem cell pool. Eventually, this random loss of stem cell diversity will result in monoclonality. An additional reason for the acquisition of monoclonality is the fact that the crypts themselves multiply by budding, sharing their stem cells between the two daughter crypts. By this process it is possible to lose all the stem cells of one genotype in one or both of the daughter crypts.

The number of stem cells per crypt is fairly constant in the steady state. But in a situation of tissue damage and regeneration it appears that some transit amplifying cells can be promoted to a stem cell state. This is deduced from the effects of radiation (Fig. 18.14). A given dose of X- or gamma rays will sterilize a proportion of cells in the epithelium. Cells that are capable of growing and repopulating the tissue are known as **clonogenic** cells. It is assumed that a crypt can only regenerate if it includes at least one clonogenic cell that survived the radiation. Measurements of crypt survival following various dose regimens suggest that there are about 80 clonogenic cells per

crypt. This is substantially greater than the number of stem cells per crypt observed from *Lgr5* or *Bmi1* expression, or calculated from mutagenesis, and the difference in the two estimates suggests that some transit amplifying cells must be able to become stem cells. This may be associated with migration into the stem cell niche.

In the intestine, as is generally the case for renewal tissues, the stem cells are responsible for producing several types of differentiated cell. It now seems that this is achieved using the Delta-Notch **lateral inhibition** mechanism (see also Chapters 4, 14, and 16). There is a "master switch" at the level of the decision whether to become an ordinary absorptive cell or one of the secretory types (goblet, enteroendocrine, or Paneth), and this is controlled by a bHLH-type transcription factor called Math1, which also promotes the formation of Delta (Fig. 18.15). All the cells express Notch and initially have a similar level of Math1. Cells that by chance have a slightly higher level of Math1 produce a little more Delta and signal to surrounding cells. Notch is stimulated in these surrounding cells, leading to inhibition of expression of *Math1*, and hence reduction of the level of Delta. The process will run on until there are a few high Math1–high Delta cells surrounded by a larger number of low Math1–low Delta cells. The cells with low Math1 become absorptive cells, while those with high Math1 become secretory. The main evidence for this process is that the knockout of *Math1* has an intestinal epithelium that is normal in overall structure, and contains normal absorptive cells, but totally lacks all the types of specialized cell. Another knockout, of the transcription factor gene *Hes1*, shows an opposite phenotype with an elevation of *Math1*-expressing cells and of the proportion of secretory cell types in the epithelium. Hes1 is on the Notch signaling pathway and its loss will reduce the inhibition of Math1 expression by Notch signaling.

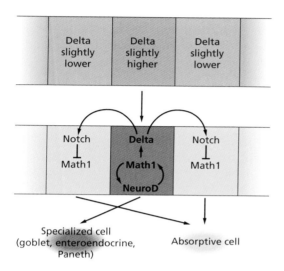

Fig. 18.15 Control of cell differentiation in the intestinal epithelium by lateral inhibition.

Epidermis

Skin consists of a squamous stratified epithelium, the **epidermis**, on top of a connective tissue, the **dermis** (Fig. 18.16). The main cell type in the epidermis is the **keratinocyte** and cell division is confined to the basal layer. In humans the epidermis is renewed from the basal layer about every 2 weeks. The entire basal layer of the epidermis, and of other squamous epithelia such those of the esophagus or vagina, depends on the activity of a transcription factor called p63. This is expressed throughout the basal layer and the gene is switched off when cells migrate upwards. The knockout mouse lacking p63 is unable to form any squamous epithelia and dies soon after birth (Fig. 18.17).

The dermis is a dense fibroelastic connective tissue derived from the dermatome and neural crest of the embryo. At deeper levels it is largely adipose tissue. The junction between dermis and epidermis is marked by a basement membrane and, in humans, this is wrinkled with epidermal ridges and corresponding dermal papillae (note that the specialized dermal core of the hair bulb is also called a dermal papilla). The dermis contains the usual nerves and blood vessels, and pressure receptors called Pacinian corpuscles, as well as the epidermis-derived hair follicles, sebaceous glands, and sweat glands.

The factors required for proliferation of the epidermal basal layer include both factors produced by the epidermis itself, such as TGFα, and factors secreted by the underlying dermis,

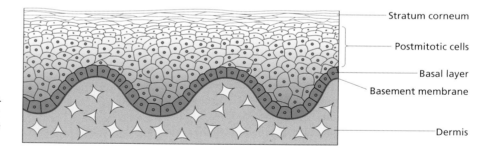

Fig. 18.16 Organization of the epidermis. All keratinocytes are born in the basal layer and differentiate progressively as they move up to the surface.

Fig. 18.17 Skin of (a) normal and (b) *p63* knockout newborn mice. In the absence of *p63*, stratification does not develop. Reproduced from Koster and Roop. *Ann Rev Cell Dev Biol* 2007; 23: 93–113.

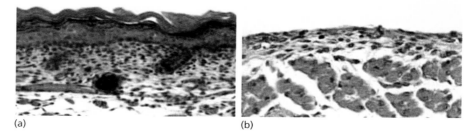

including FGF7. Keratinocyte cultures can be grown *in vitro* on feeder layers of fibroblasts and form the same stratified arrangement as the natural skin.

Labeling studies show that about 60% of basal layer cells are in cycle, but because there are no well-established molecular markers for the stem cells it is not clear what fraction of these are stem cells. Some consider that there is no distinction between stem cells and transit amplifying cells in the epidermis, so all the dividing cells are effectively stem cells. However, there is considerable heterogeneity of the cells when placed in tissue culture. Some form large self-sustaining colonies, which grow for many passages, while others form small colonies of a few cells, which soon stop growing. By this criterion, about 10% of basal layer cells are thought to be stem cells. A similar proportion, presumed to be the same stem cells, are capable of forming large colonies that can repopulate the epidermis following severe radiation damage.

Stem cells defined by this criterion are characterized by a higher level of β-1 integrin than the transit amplifying cells. This is a cell adhesion molecule involved in the recognition of collagen, laminin, and fibronectin. *In vivo*, in human foreskin, the high-integrin cell clusters are found at the tips of the dermal papillae, suggesting that this may be the stem cell niche for the epidermis. However, ablation of the *β-1 integrin* gene does not cause loss of stem cells.

Once cells leave the basal layer they stop dividing and enter a program of further differentiation. The progress of maturation is reflected by the names given to successive layers of the epidermis: stratum germinativum (the basal layer), stratum spinosum (the "prickle" layer—the apparent prickles are abundant desmosomes), stratum granulosum (with granules composed of profilaggrin), and stratum corneum (cells have lost nuclei and have become flat sacs of keratin crosslinked by transglutaminase). Keratin is a generic name for the large family of fibrous proteins forming the cytokeratin **intermediate filament** family found in all epithelial cells. There are many different keratins encoded by different genes, and those that are expressed change as cells move up from the basal layer. For example when cells leave the basal layer and enter the prickle layer they downregulate K5 and K14 and upregulate K1 and K10. In the granular and cornified layers the cells also contain a tough internal sheath of an insoluble protein called involucrin.

When the epidermis is formed in the early embryo it is a single layer of cells, with a temporary outer squamous layer called periderm, which is later shed. At about E14 of mouse development the pattern of cell divisions shifts from symmetrical to asymmetrical, and this shift generates the stratified pattern of the mature epidermis. As in other cases of cell polarity, the cells have the Par3/Par6/aPKC complex at the apical end. This interacts with other components such as LGN (Leu-Gly-Asn repeat-enriched protein) and NuMA (nuclear mitotic apparatus protein), which bind microtubules and reorient mitotic spindle orientation from horizontal to vertical. About half of the cell divisions in the embryonic epidermis are vertical. If lentivirus carrying shRNA against LGN or NuMA is injected into the amniotic fluid at E9.5, it can effectively infect the whole epidermis and the result is a great reduction of vertical cleavages and a consequent failure of good stratification and development of effective permeability properties. Some vertical divisions are also found postnatally, although by this stage most are in the plane of the basal layer.

Many experiments on the epidermis and hair follicles involve the use of the *Keratin14* (*K14*) promoter to drive transgenes specifically in the basal layer. This may also be used to ablate genes in the form of *K14-Cre*, combined with a floxed version of the target gene.

Hair follicles

The structure of the hair follicle is shown in Fig. 18.18a. The hair shaft is composed of dead keratinocytes, which are produced in the epidermal **matrix** region at the base. This lies in close proximity to the **dermal papilla**, a projecting bud of fibroblastic cells, also containing **melanocytes**, which transfer pigment granules to the keratinocytes of the hair. Surrounding the whole is a layer of cells continuous with the surface epidermis called the outer root sheath. High up, near the junction with the surface epidermis, lies a sebaceous gland. The entire region at the base of the follicle, comprising the dermal papilla, the proliferative epidermal zone, and the outer layers, is known as the hair bulb.

Hair does not grow continuously but in a cycle (Fig. 18.18b). The active growth phase is known as **anagen**, which lasts about 3 weeks in mice but 3–6 years in humans. The period of regression of the follicle is called **catagen** and the period of quiescence is called **telogen**. In mice the hair cycle is synchronized, with waves of each new phase commencing at the head. The hair cycle is associated with a cyclic expression of BMP2 and 4 in the dermis, which is high during the late anagen–early telogen period, and low during late telogen–anagen, indicting a repressive effect on hair growth. In humans the different follicles are not synchronized and behave independently. Mice also have four different kinds of coat hair (guard hairs, awls, auchenes, zigzags), which show slight differences in their development.

Hair follicle development

Hair follicles start life as epidermal placodes, which form in the mouse from E14.5. Then a dermal condensation appears below the placode. Then the placode grows downwards as a peg-shaped body, and envelopes the condensation, which becomes the **dermal papilla** of the hair follicle (Fig. 18.19).

Hair follicle formation requires both Wnt signaling and also inhibition of BMP signaling. BMPs are produced by the epidermis, and the BMP inhibitor Noggin by the dermis. A combination of Wnt3A and Noggin will induce new follicle buds. The BMP inhibition upregulates transcription of *Lef1* while the Wnt

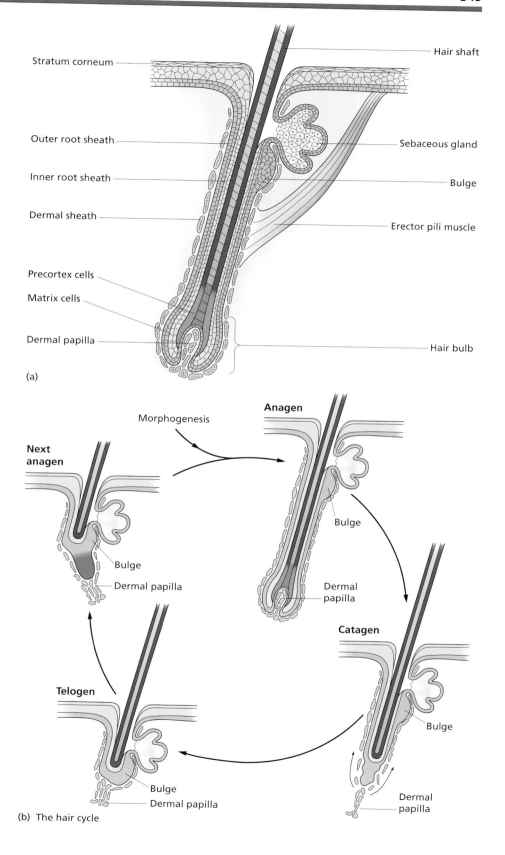

Stratum corneum

Hair shaft

Outer root sheath

Sebaceous gland

Inner root sheath

Bulge

Dermal sheath

Erector pili muscle

Precortex cells

Matrix cells

Dermal papilla

Hair bulb

(a)

Morphogenesis

Anagen

Next anagen

Bulge

Bulge

Dermal papilla

Bulge

Dermal papilla

Catagen

Telogen

Bulge

Bulge

Dermal papilla

Dermal papilla

(b) The hair cycle

Fig. 18.18 (a) Structure of the hair follicle. (b) The hair growth cycle.

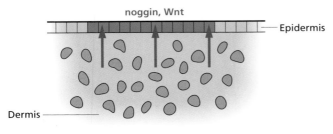

First mesenchymal signal

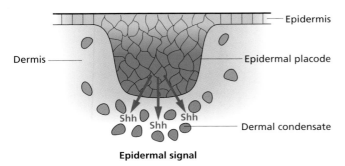

Epidermal signal

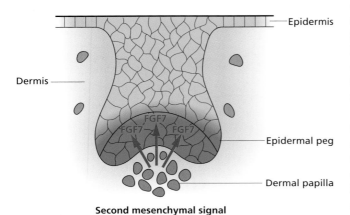

Second mesenchymal signal

Fig. 18.19 Initial formation of hair follicles. There are at least three stages of signaling between epidermis and dermis.

signal stabilizes β-catenin and the combination carries the Lef1 into the nucleus to control target genes. One target is *E-cadherin*, whose expression becomes repressed, thus reducing the mutual adhesion of the cells and leading to invagination to form the bud. Evidence for this mechanism is based on various lines of evidence. Wnt reporters are active with both the early placode and the dermal condensate showing Wnt signaling activity. Mouse knockouts of components of the Wnt pathway, or of *Noggin*, have few or no hair follicles. Invaginating buds are produced by adding Wnt3A + Noggin to keratinocyte cultures, or by overexpressing a constitutive form of *β-catenin*. Bud formation is suppressed in mice transgenic for the Wnt-inhibitor *Dickkopf*. Overexpression experiments of this type make use of

the *K14* promoter to confine effects to the epidermis. Varying the dose of Dickkopf in overexpression causes changes in the spacing of the forming placodes, indicating a probable type of lateral inhibition process at work.

Although the initial signal for bud formation comes from the dermis, the formation of the **dermal papilla** depends on a second signal from the invaginating bud to the dermis (Fig. 18.19). Shh is needed during this stage as the hair follicles in the *Shh* knockout arrest at the peg stage, with the dermal papilla small or absent. The dermal papilla secretes growth factors, especially FGF7, needed by the proliferative zone of the epidermal matrix region. Isolated papillae will induce new epidermal proliferative zones from the upper halves of follicles, and in some situations can induce complete new follicles from epidermis.

Not only is it possible to induce extra follicles in embryonic life, but also in adult life, by overexpression of Wnt pathway components, especially stabilized β-catenin. The effects are dose dependent, high doses promoting new follicle formation, intermediate doses new sebaceous gland formation, and low doses suppressing follicle formation. New hair follicles are often induced as supernumerary growths from existing follicles but can also be formed *de novo* in interfollicular epidermis. New follicles can also arise (in mice) in wounds, being formed from interfollicular epidermis and being unpigmented because of a lack of melanocytes in the new follicles. This process is also Wnt dependent, as upregulation of Wnt pathway components will increase the number of new follicles and downregulation will reduce the number.

The understanding of hair follicle initiation makes it possible to contemplate a possible "cure" for human baldness by the overexpression of Wnt pathway components. The main potential problem is that the Wnt pathway is often upregulated in cancers and methods of stimulating it to provoke new hair follicle formation would need to avoid provoking the formation of tumors.

The bulge

The dividing epidermal cells of the of the hair follicle bulb are not stem cells because they are only active as long as the follicle is in the anagen phase. The real stem cell population lies in a lateral bulge half way up the outer root sheath. This is the most distal part of the follicle to survive permanently through the hair cycle. It contains slow-dividing (label retaining) cells, which migrate to repopulate the hair bulb at the beginning of each new anagen phase. In the large vibrissal follicles of rodents the bulge is large enough to be dissected out manually. If a bulge region from a *lacZ*-positive (**Rosa26**) mouse is grafted to a unlabelled vibrissal follicle and then cultured under the kidney capsule of a **nude mouse**, the β-galactoside-expressing cells are seen to repopulate the entire follicle. Moreover, individual *lacZ*-positive bulge regions grafted to the back of unlabelled late embryos will generate labeled surface epidermis, hair follicles, and sebaceous glands (Fig. 18.20). The bulge cells secrete the

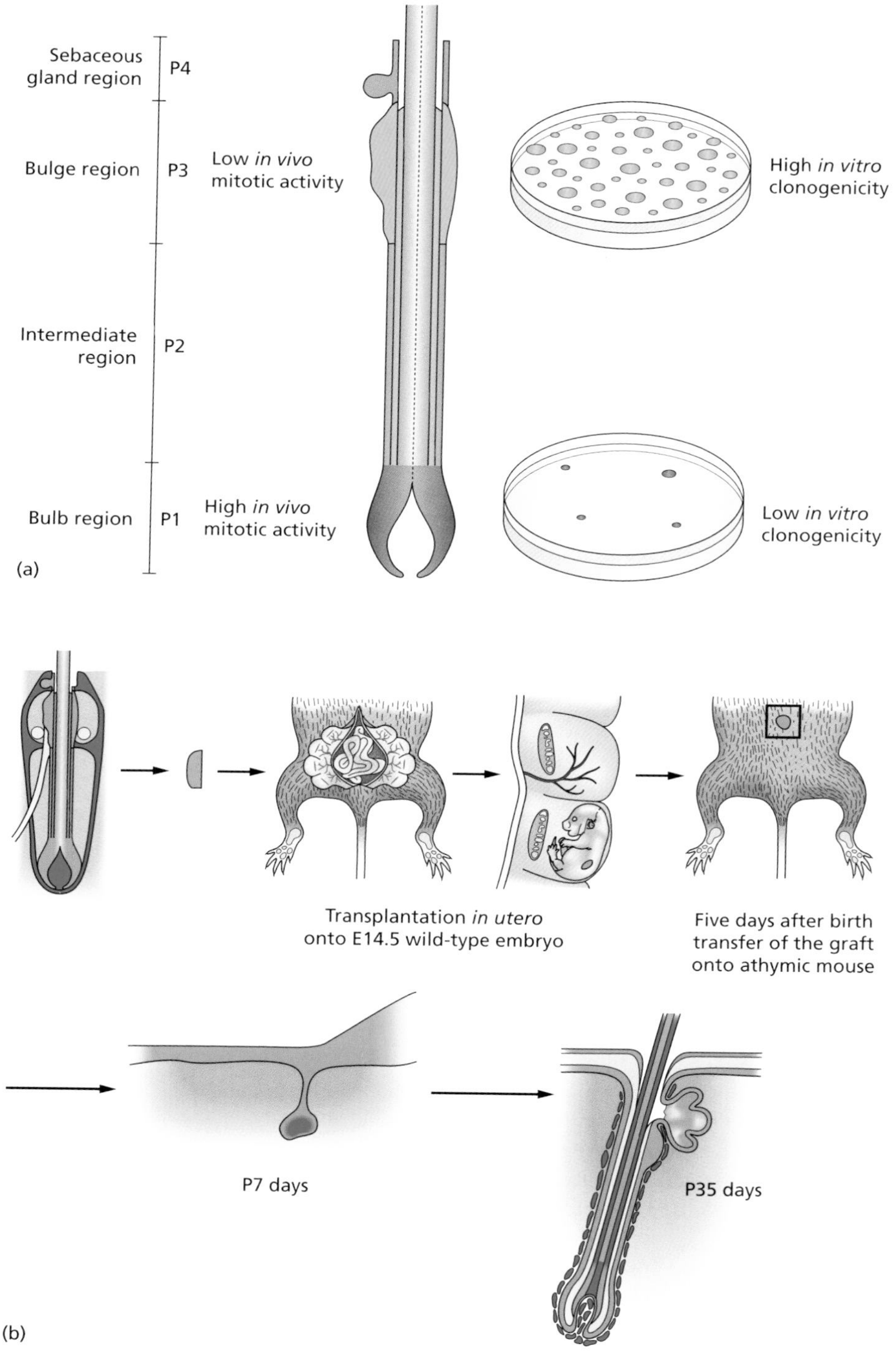

Fig. 18.20 Evidence that the bulge region of the hair follicle contains the stem cells. (a) The bulb region contains more cells in cycle but the bulge region contains more clonogenic cells. (b) A β-galactosidase-labeled graft from the bulge region can repopulate the surface epidermis and other epidermal structures as well as the entire hair follicle. These experiments were done with large vibrissal follicles. Scale bars 10 μm.

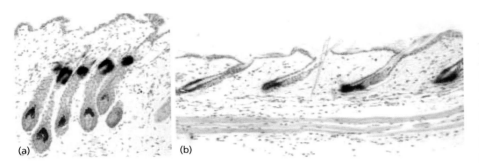

Fig. 18.21 *K15-lacZ* reporter mouse. Activity of the promoter in the bulge is shown by XGal staining. In (a) the hairs are in anagen and the bulge region is half way up the follicle. In (b) the hairs are in telogen and the bulge is at the tip of the follicle. Reproduced from Liu *et al. J Invest Dermatol* 2003; 121: 963–968, with permission from Nature Publishing Group.

extracellular protein nephronectin into their basement membrane and this serves to induce differentiation and anchorage of the erector pili muscle. In the human there is no visible bulge, but the stem cells also reside in the lowest permanent part of the outer root sheath.

Various markers have been associated with the stem cells of the bulge. One is the promoter for Keratin 15 (*K15*). This is active in the whole epidermal basal layer in the embryo and early postnatal period, and subsequently confined to the bulb (Fig. 18.21). It has been used to ablate the bulge region by driving the *thymidine kinase* "suicide" gene, which metabolizes the drug ganciclovir to a toxic product. In fact, the *K15-TK* mouse is killed by ganciclovir because of toxicity to the gut, but the skin can be studied in isolation after grafting to an immuno-deficient host mouse. In this situation ganciclovir causes loss of hair follicles over a few days, but does not affect the interfollicular epidermis, indicating that the two stem cell populations are different. Another marker, probably more intimately associated with stem cell function, is the transcription factor SOX9. This is an interesting example of where the actual origin of a tissue-specific stem cell can be observed in the embryo (Fig. 18.22). SOX9-positive cells appear associated with the epidermal placode, and as the follicle develops they enter the bulge. *Sox9-Cre x R26R* mice show that descendants of the SOX9-positive cells populate the entire hair follicle. In *K14-Cre x Sox9 (floxed)* mice, *Sox9* is ablated in the epidermis. These mice do form hair follicles, but they do not last long and they do not undergo any subsequent phases of the hair cycle. They also lack sebaceous glands.

Interestingly, the Lgr5 molecule, characteristic of intestinal stem cells, is also expressed in the hair follicle. Labeling experiments suggests that the cells are transit amplifying cells of high potency. A related molecule, Lgr6, is expressed above the bulge. Its expression commences in the epidermal placode, although in adult life labeling experiments indicate contributions to sebaceous gland and surface epidermis rather than the active part of the hair follicle.

Although the surface epidermis and the hair follicles have their own stem cells and do not populate each other in the steady state, this is not the case following wounding. It has been indicated above that some new hair follicles do arise from inter-follicular mouse epidermis in wounds, and that they can also be induced by Wnt stimulation. The various methods for labeling the bulge (microdissected and grafted Rosa26 tissue, or *K15-Cre* or *Sox9-Cre* label), all indicate that following a wound, some bulge-derived cells do contribute to the surface epidermis. This is one example of the fact that the behavior of stem cells in regeneration situations may be different from that in normal homeostasis.

Hematopoietic system

In the adult mammal, the **hematopoietic** system resides in the bone marrow within the larger bones of the skeleton. In the embryo it is found at various other sites. As described in Chapter 15, initially it is extraembryonic in the yolk sac, then in the region of the mesoderm around the dorsal aorta ("AGM"), then in the liver, then in the spleen and lymph nodes, and finally in the bone marrow. At least from the dorsal aorta onwards, the same cell population migrates from one site to the next.

In the postnatal mammal the hematopoietic system is in a state of continuous cell production and renewal throughout life. There exists a **hematopoietic stem cell** (**HSC**) that can both renew itself and also differentiate into a wide variety of cell types. These include all cells of the blood and immune system together with histiocytes, osteoclasts, and Langerhans cells of the skin.

The cellular components of the blood are as follows.
1 **erythrocytes** (red cells);
2 **granulocytes**, comprising neutrophils (phagocytes), eosinophils, and basophils (similar to mast cells);
3 **monocytes** (similar to macrophages/ histiocytes);
4 **megakaryocytes**, giant cells that break up to become platelets.
The above four cell types are collectively known as **myeloid cells**.
5 **lymphocytes** (T and B cells).

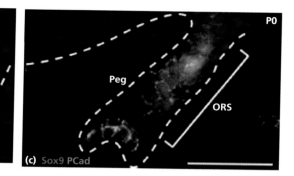

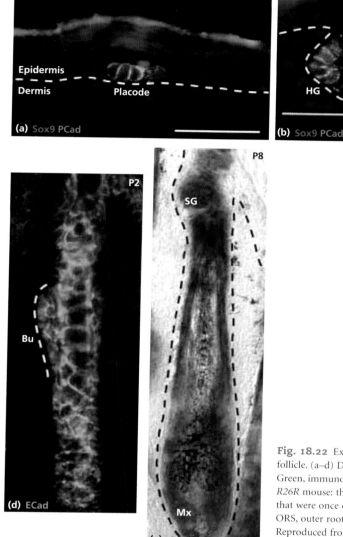

Fig. 18.22 Expression of SOX9 (red immunostain) during development of the hair follicle. (a–d) Development from epidermal placode stage to young follicle with bulge. Green, immunostain for P or E cadherin; blue, DAPI stain for DNA. (e) *Sox9-Cre x R26R* mouse: the whole hair follicle, visualized by X-Gal staining, is derived from cells that were once expressing *Sox9*. Epi, epidermis; Der, dermis; Pc, placode; HG, hair germ; ORS, outer root sheath; Bu, bulge; SG, sebaceous gland; Mx, matrix; P, days after birth. Reproduced from Nowak *et al. Cell Stem Cell* 2008; 3: 33–43, with permission from Elsevier.

Much of the evidence for the existence of HSCs comes from reconstitution experiments. The bone marrow is the most sensitive tissue in the body to irradiation, and so there is a dose range that will kill by bone marrow failure while other tissues are still potentially able to recover. If a mouse, lethally irradiated with such a dose, is given a graft of marrow cells by injection into the bloodstream, then the graft will colonize the marrow of the host, proliferate extensively, and enable survival of the host. After a few weeks the counts of the various blood cell types have returned to normal and the observation of genetic markers shows that they are all derived from the graft. The ability per-

manently to rescue lethally irradiated mice is often taken to be the defining feature of the HSC.

It is possible to isolate HSCs directly from bone marrow using the technique of fluorescence-activated cell sorting (**FACS**, see Chapter 13). Mouse HSCs are characterized by high levels of Sca1 and c-Kit, low but finite levels of Thy1, and the absence of all other differentiation markers (i.e. Sca1$^+$, c-Kit$^+$, Thy1^{lo}, Lin$^-$). Sca1 (stem cell antigen 1) and Thy1 (thymus 1) are both cell-surface glycoproteins, Thy1 being abundant on mature T cells. c-Kit is a tyrosine kinase type receptor for Stem Cell Factor, otherwise known as Steel Factor or Mast Cell Growth Factor.

Classic experiments

The hematopoietic stem cell and its cell lineage

In the 1940s, it was known that irradiated mice could be rescued by a bone marrow graft. But it was thought that some substance or hormone present in the marrow was responsible. The paper of Ford et al. showed, by identification of a chromosomal marker, that the rescue activity of the marrow graft was actually due to colonization by blood-forming cells.

The second paper describes the ability of hematopoietic cells to form monoclonal colonies in the spleen of irradiated animals. This provided a method for quantifying numbers of particular types of progenitor and showed the existence of multipotent cells forming clones of mixed cell type.

The third paper established an *in vitro* assay for colony formation, which resulted in the characterization of further multipotent cell types and was used to purify the hematopoietic growth factors.

Finally, the paper of Spangrude et al. describes the isolation of pure HSCs from mouse bone marrow by fluorescence activated cell sorting (FACS).

Ford, C.E., Hamerton, J.L., Barnes, D.W.H. & Loutit, J.F. (1956) Cytological identification of radiation chimaeras. *Nature* **177**, 452–454.
Till, J.E. & McCulloch, E.A. (1961) A direct measurement of the radiation sensitivity of normal mouse bone marrow cells. *Radiation Research* **14**, 213–222.
Bradley, T.R. & Metcalf, D. (1966) The growth of mouse bone marrow cells in vitro. *Australian Journal of Experimental Biology and Medical Science* **44**, 287–300.
Spangrude, G.J., Heimfeld, S., & Weissman, I.L. (1988) Purification and characterization of mouse hematopoietic stem-cells. *Science* **241**, 58–62.

Mouse HSCs can also be isolated by the presence of the glycoprotein CD150, also found as a coreceptor on lymphocytes, and also because they preferentially exclude certain fluorescent dyes such as Hoechst 33324, so after exposure give a lower fluorescence signal than all other cells in the marrow. The best methods of isolation, which may combine cell sorting for the presence and absence of several different surface molecules, yield preparations that are about 50% pure HSCs. It has been possible to show that a single HSC is able to reconstitute the entire blood and immune system of an irradiated mouse if grafted together with an excess of other, inactive, cells. Inspection of the sorted single cell plus the use of genetic markers to distinguish the HSC from the accompanying inactive cells and the host cells, provides good evidence that this really works.

HSCs make up about $1/10^5$ of bone marrow cells. BrdU labeling followed by isolation of HSCs indicates that most of them are slow dividing, perhaps once per month, although there is also a subpopulation that divides more rarely. It is possible to separate these cells using the H2B-GFP cell labeling method, which imparts a green fluorescence to labeled cells. The dormant fraction are much better at reconstituting irradiated hosts, and will also support serial transplantations to further irradiated hosts, something the more actively dividing HSC fraction will not do. Serial transplants of unfractionated bone marrow can be carried out up to about five times, after which they are no longer effective. The reasons for this are not clear but the situation does involve hugely more divisions than would occur in a normal lifespan. Knockout of the cell cycle inhibitors p16, p19, and p53, which often accumulate in the course of cellular senescence (see Chapter 19), improves the reconstitution activity significantly.

The human hematopoietic system is similar to that of the mouse. HSCs are the active cells in bone marrow transplants, which are used especially for treatment of leukemias and lymphomas. Human HSCs can be enriched because of possession of a cell surface sialomucin CD34, which is not present on mouse HSCs. Experimental work on human HSCs makes use of reconstitution assays with immunodeficient mice, such as the NOD-SCID line, as hosts. These mice will tolerate grafts of tissue from other species. If irradiated and transplanted with human HSC, the HSC will colonize the bone marrow and produce a full range of human blood and immune cells in the mouse host.

The self-renewing properties of the HSC depend on the proto-oncogene *Bmi1*, which encodes a *Polycomb*-type transcriptional repressor. Knockouts for *Bmi1* develop HSCs but the numbers are greatly reduced postnatally and they have little reconstitution activity. HSCs are difficult to grow in culture but the numbers, measured by the mouse reconstitution assay, can be increased by introduction of certain genes using retroviruses. These include stabilized (constitutively active) β-catenin, suggesting that Wnt signaling may be necessary for HSC self-renewal, as it is for the stem cells of the skin and the intestine. A compound called Stemregin1, which inhibits the aryl hydrocarbon nuclear hormone receptor, has been reported to enable significant expansion of HSCs *in vitro*.

The HSC niche

The reason that the reconstitution assay for HSCs works is that there are a number of stem cell niches for the HSC within the marrow, and these become vacant following heavy irradiation.

New directions for research

It is important to establish exactly how tissue-specific stem cells arise during development from progenitor cells that are not themselves stem cells.

The nature of stem cell niches needs to be clarified and the signals from the niches that maintain self-renewal behavior need to be better understood.

We also need to understand whether the Notch lateral inhibition system controls the differentiation of multiple cell types in all tissues, or whether there are other molecular systems for achieving the same objective.

The same cells may behave differently in their normal tissue environment, during tissue regeneration, and when placed in tissue culture. The reasons for these differences need to be better understood.

HSCs have the remarkable property that they can become mobilized from the marrow into the circulation and then re-enter the marrow and occupy a vacant niche. But the identification of niches has given rise to much controversy.

There is some evidence for the importance of **osteoblasts** lining the trabecular bone surfaces in the marrow cavity. Various treatments that increase the number of trabecular osteoblasts, including downregulation of BMP receptor 1A or injection of parathyroid hormone, also increase the number of HSCs. There is also evidence for the importance of vascular endothelial cells, which can be seen adjacent to HSCs immunostained with CD150. Mice lacking the *Bis* gene (BCL2 interacting cell death suppressor) have defective endothelia in the bone marrow. They can provide HSC grafts that will thrive in normal mice, but grafts of normal HSC into *Bis⁻* mice do not take. It is likely that the niche provides the chemokine SDF1, as the HSCs carry its receptor, CXCR4, and this system is found repeatedly to be involved in chemotaxis and cell trafficking in many other developmental situations.

Figure 18.23 shows the presence of grafted HSCs in the bone marrow of a live host by imaging through the thin skull bone with two-photon microscopy. The cells are located close to both osteoblasts and vasculature. **Mesenchymal stem cells** (see below) have also been considered a key part of the niche. These cells produce SDF1, angiopoietin, and other trophic and attractive substances for HSCs.

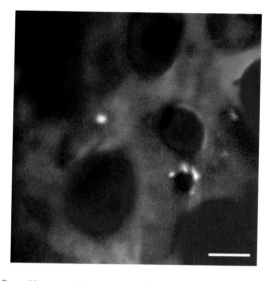

Fig. 18.23 Hematopoietic stem cell (HSC) in the bone marrow. The image shows grafted cells from an HSC preparation lodged in the marrow of a skull bone in a live mouse. The HSCs are labeled white with DiD dye, the vasculature is red, the bone matrix blue, and the osteoblasts green. One isolated HSC and a small group of four, perhaps a clone formed by two divisions, are visible. Scale bar 50 µm. Reproduced from Mendez-Ferrer *et al. Nature* 2010; 466: 829–834, with permission from Nature Publishing Group.

Committed progenitors

In addition to the HSC, the marrow contains other cells that can be isolated by different combinations of cell-surface markers. In the reconstitution assay they give rise to only a subset of the complete HSC repertoire. This criterion shown the existence of various multipotent progenitors, including the common lymphoid progenitor, the common myeloid progenitor, the granulocyte–macrophage progenitor, and the megakaryocyte–erythrocyte progenitor. There are also pluripotent stem cells that have only a temporary repopulating ability, which are believed to represent the next step of maturation after the permanent HSC. A current consensus model for the cell lineage of the hematopoietic system is shown in Fig. 18.24.

A second line of evidence for this model comes from *in vitro* colony assays. It is possible to obtain clones of hematopoietic cells *in vitro* by plating marrow cells in soft agar or methyl cellulose in the presence of the appropriate growth factors. Since most single colonies will be clones derived from a single cell, the production of multiple cell types by a single colony indicates the existence of a multipotent progenitor, and the same potency classes are recovered in these assays as in the whole mouse reconstitution assay.

Although the model shown in Fig. 18.24 is generally accepted, it has not yet been confirmed by prospective labeling, which would require the insertion of a permanent genetic marker into HSCs *in vivo*, followed by identification of each type of progeny cell as it is formed. Moreover, the assays on which it is based are

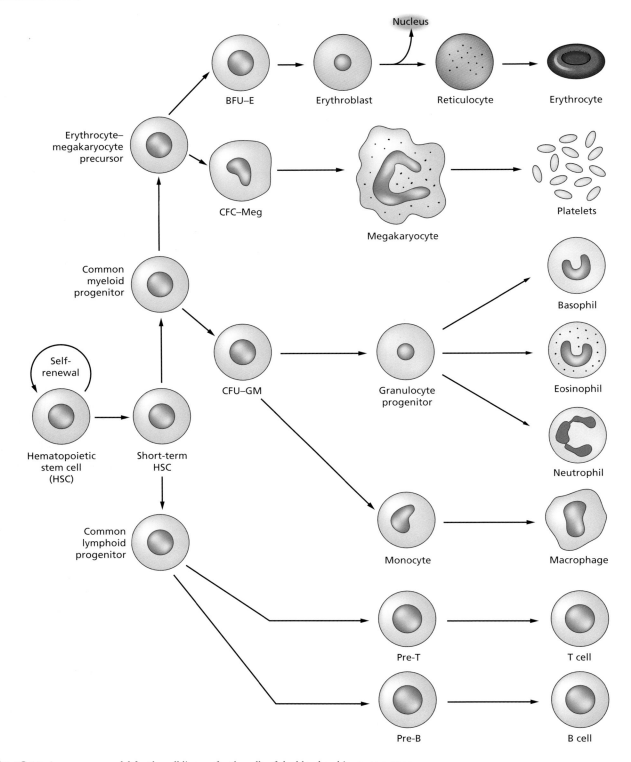

Fig. 18.24 A consensus model for the cell lineage for the cells of the blood and immune system.

ones involving very abnormal conditions, either large amounts of regeneration in the reconstitution of an irradiated mouse, or *in vitro* culture in the soft agar assay. Since cells often behave differently in such situations compared to the condition of normal homeostasis within the animal, the model should be regarded as provisional. In particular it is possible that the committed progenitors self-renew sufficiently *in vivo* to be considered as the stem cells, and the dormant HSCs are only called upon to replenish them in severe conditions such as heavy dose of radiation.

Hematopoietic growth factors

The *in vitro* colony formation assay has been used to isolate a number of colony-stimulating factors (CSFs), otherwise known as hematopoietic growth factors. Interleukin 3, and stem cell factor can stimulate proliferation of the HSCs, while granulocyte–macrophage CSF (GM-CSF), granulocyte CSF (G-CSF), macrophage CSF (M-CSF), and erythropoietin work in combinations to stimulate the division of the various transit amplifying lineages. Most of the feedback control over the production of the various cell types is exerted by varying the growth rate at the transit amplifying cell level. This is because there are large numbers of such cells and a rapid response can be obtained to changing demand. By contrast, it would take several weeks to alter the rate of production by regulation at the level of the HSC. Several of the hematopoietic growth factors have been prepared in therapeutic quantities by recombinant DNA methods and are now very useful in clinical practice, particularly for treatment of various types of anemia and for enabling people to rebuild their marrow after cancer therapy.

Mesenchymal stem cells and "transdifferentiation"

In addition to the HSCs, the bone marrow contains another type of stem cell. These are called mesenchymal stem cells, or marrow stromal cells, both conveniently abbreviating to **MSC**. They adhere to plastic and long-term cultures can be grown in which the other cell types of the marrow are selected out and disappear. *In vitro*, MSCs will form adipocytes, chondrocytes, or osteocytes when cultured in appropriate media. The normal function of MSCs *in vivo* is probably as the stem cells that feed the gradual turnover of bone.

A number of studies around the year 2000 suggested that bone marrow cells are capable of colonizing a wide variety of other tissue types when transplanted into irradiated hosts. Some of these were performed with unfractionated marrow, some with enriched or purified HSCs or MSCs. The tissues colonized comprised virtually everything including numerous epithelia, muscle, and neurons. This work generated considerable controversy because it suggested a very different model of development from the conventional one. Instead of cell populations undergoing a series of decisions during embryonic development, in each of which their competence is restricted, the idea

is that the whole body is continuously being renewed by highly pluripotent cells from the bone marrow. The phenomenon is known as "transdifferentiation" although this is unfortunate as the term was previously used to refer to the rare but well-established cases of direct transformation between differentiated cell types. It now appears that some of the results were due to lodgement of cells in these tissue but without actual differentiation, and others were due to cell fusion, whereby the genetic markers from donor cells became incorporated into host cells. There may be some genuine reprogramming of marrow-derived cells to various other tissue types, but this certainly only occurs at very low frequency. Because the hosts are nearly always irradiated, and therefore have considerable tissue damage and a situation of widespread tissue regeneration all over the body, it is thought that this situation allows favorable circumstances for the occasional reprogramming event. It is not, however, at all likely that reprogramming occurs on a large scale, or that the bone marrow is a repository for cells that can regenerate the rest of the body.

Spermatogonia

Spermatogenesis has already been described in Chapter 2 from the point of view of gametogenesis. Mitotic **spermatogonia** give rise to spermacytes, which each undergo meiosis to form four spermatids, which then mature into sperm. But it has another aspect because the male mammalian reproductive cells are continuously created throughout life from a population of stem cells. In this respect they differ radically from the female mammal, in which all of the postmitotic oocytes arise during embryonic life.

Testicular development was described in Chapter 15. The testis has its origin from the intermediate mesoderm, into which primordial germ cells migrate, and gives rise to a system of seminiferous tubules. The tubules contain Sertoli cells and Leydig cells, which are of somatic origin, as well as the reproductive cells, which are of germ-line origin. The process of spermatogenesis becomes established in the mouse shortly after birth. In the newborn mouse the germ cells are called gonocytes and are directly descended from the primordial germ cells. These become spermatogonia in the first postnatal week. Like the other stem cell systems, spermatogenesis is structured, with the mitotic spermatogonia occupying the basal layer of the tubule, the meiotic spermacytes move toward the lumen as they mature, in contact with the Sertoli cells, and the postmeiotic spermatids occupying an apical, or lumenal, position (Fig. 18.25a). In the mouse the complete process from spermatogonium to mature sperm takes about 35 days.

Morphological studies show that there are several kinds of spermatogonia in the basal layer: isolated ones, and cysts consisting of 2–16 cells joined by cytoplasmic bridges. It was thought that the single cells, called As, were the stem cells and

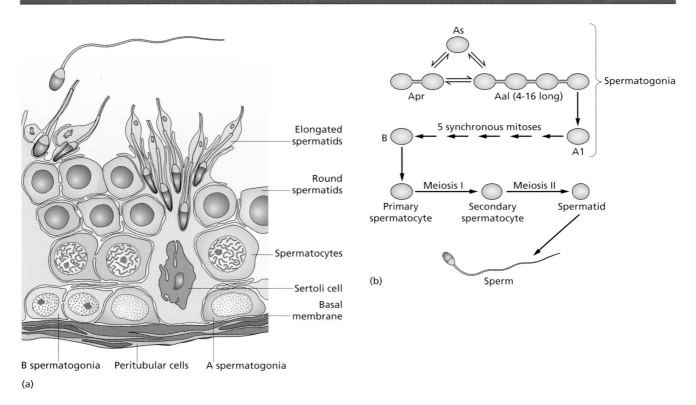

(a)

(b)

Fig. 18.25 Spermatogenesis. (a) The spermatogonia are in the basal layer of the testicular tubule in contact with Sertoli cells. They give rise to meiotic spermacytes, which migrate up the Sertoli cells. The elongated spermatids are near the tubule lumen and are about to finish differentiating into mature sperm. (b) Diagram of spermatogenesis showing the possible relationship between the various types of spermatogonia. Reproduced from de Rooij and Mizrak. *Development* 2008; 135: 2207–2213, with permission from Company of Biologists Ltd.

the cysts (called Apaired or Aaligned) were the transit amplifying descendants. However, a fraction of the cysts express *Ngn3*, and if these cells are labeled with the *CreER* system, a few of the labeled descendants form persistent colonies in the tubules, indicting that they are stem cells. Moreover, the breakdown of cysts to single cells has been observed by time-lapse filming. So it seems that the true stem cells are a subset of the As + Apr + Aal populations, perhaps those expressing GFRa1, which is a receptor for glial-derived neurotrophic factor (GDNF; see below). Once the 16-cell cysts have broken down they undergo an irreversible set of further divisions, to form A1, A2, A3, A4, intermediate, and B spermatogonia. These then enter the meiotic pathway as spermacytes (Fig. 18.25b). The niche for the stem cells involves the Sertoli cells, and perhaps also the Leydig and other somatic cells. Among other products, Sertoli cells produce GDNF. Knockouts of *GDNF*, or the gene for its receptor *c-Ret*, disrupt spermatogenesis and cause sterility. GDNF is also necessary for the *in vitro* culture of spermatogonial stem cells.

A useful assay for the stem cells consists in grafting them into the tubules of a host mouse. This can either be done between mice of the same inbred strain, in which case they are immu-

nocompatible, or across strains or even species, in which case immunodeficient hosts are needed. The hosts are normally sterilized with busulfan, an alkyl sulfonate drug formerly used for cancer treatment, which destroys spermatogonia and liberates niches for the transplant cells to occupy. The donors are generally labeled with a genetic marker such as *GFP* or *lacZ*. Over a few weeks, each stem cell can set up a colony in the host tubule, generating a population of cells at all stages of spermatogenesis (Fig. 18.26). This assay enables experiments to be done in which the number of stem cells needs to be counted or compared. The analysis of the distribution of clone sizes indicates that there is no obligatory pattern of asymmetric division among stem cells, but rather a random choice of renewal versus commencement of differentiation, and, as in the intestine, some drift such that one stem cell clone can take over the territories of its neighbors, or be taken over by them.

As with all such assays, it does involve tissue damage and regeneration so it may well be that the number of colony-forming cells is greater than the number of stem cells in the undisturbed steady state situation. The host mice can be mated and will produce offspring carrying the genetic marker intro-

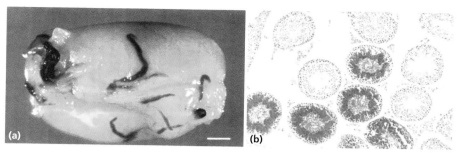

Fig. 18.26 Spermatogonial stem cell transplantation. (a) A host testis into which stem cells were injected from a mouse labeled with *lacZ* reporter. The patches of tubule colonized by donor-derived cells are revealed by XGal staining. (b) Section through a similar testis. In the colonized patches, all of the germ line, but not the somatic cells, of the tubule are graft-derived. Reproduced from Brinster and Avarbock. *Proc Natl Acad Sci USA* 1994; 91: 11303–11307, with permission from National Academy of Sciences.

duced in the grafted cells, proving that they do develop all the way to viable sperm. This is not, however, the case for the interspecies grafts. Here the cells commence differentiation but do not form viable sperm.

The availability of the transplantation assay has enabled studies to be carried out on the *in vitro* culture of spermatogonial stem cells. This can be achieved long term using feeder cells and a medium containing GDNF. The stem cells share some properties with embryonic stem cells, such as expression of Oct4, Sox2, and alkaline phosphatase, but they do not express some other typical ES genes, such as Nanog, and they do not form teratomas when grafted into adult hosts.

Key points to remember

- The organs of the body are mostly composed of more than one tissue layer. Each tissue layer contains multiple cell types.
- Cell types within tissues may be classified as postmitotic, quiescent but capable of growth, and continuously renewing. In renewal tissues there is a balance of cell production and cell death.
- Cell turnover rates may be assessed by immunostaining for proteins present during specific parts of the cell cycle, by labeling with DNA precursors such as BrdU, and, in humans, by measuring ^{14}C loss in DNA since the year of birth.
- Tissue-specific stem cells are cells that can both renew themselves and produce offspring destined to differentiate. They persist for the lifetime of the animal and are found in niches defined by signals from surrounding cells.
- The small intestine has its proliferative zone in the lower part of the crypts. The stem cells are found

adjacent to the Paneth cells at the crypt base. Four types of differentiated cells are produced, which move continuously to the upper crypt and villi.
- The epidermis is a stratified squamous epithelium. It has proliferative cells only in the basal layer, and the successive higher layers represent degrees of postmitotic cell maturation. The stem cells are located in the basal layer.
- Hair follicles contain stem cells in the bulge of the outer root sheath that reconstitute the hair bulb in each active phase of the hair cycle.
- The bone marrow contains hematopoietic stem cells (HSCs), which continuously renew the cells of the blood and immune system. HSCs, and various committed progenitors descended from them, can be isolated by FACS using specific cell-surface markers.
- Spermatogonial stem cells adjacent to Sertoli cells in the tubules of the testes produce sperm continuously in male mammals.

Further reading

Histology and tissue organization

Le Gros Clark, W.E. (1971) *The Tissues of the Body*. Oxford: Clarendon Press.

Young, B., Lowe, J.S., Stevens, A. & Heath, J.W. (2006) *Wheater's Functional Histology: A Text and Colour Atlas*, 5th edn. Philadelphia: Churchill Livingstone, Elsevier.

Alberts, B., Johnson, A., Lewis, J., Raff, M., Roberts, K. & Walter, P. (2008) Specialized tissues, stem cells and tissue renewal, Chapter 23. In: *The Molecular Biology of the Cell*, 5th edn. New York: Garland.

Eroschenko, V.P. (2008) *diFiore's Atlas of Histology with Functional Correlations*. Baltimore and Philadelphia: Lippincott, Williams and Wilkins.

Mescher, A.L. (2009) *Junqueira's Basic Histology: Text and Atlas*, 12th edn. Lange, McGraw Hill.

Ross, M.H. & Pawlina, W. (2010) *Histology: A Text and Atlas with Correlated Cell and Molecular Biology*, 6th edn. Baltimore and Philadelphia: Lippincott Williams and Wilkins.

Stem cells and their niches

Potten, C.S. & Loeffler, M. (1990) Stem cells: attributes, cycles, spirals, pitfalls and uncertainties. Lessons for and from the crypt. *Development* **110**, 1001–1020.

Raff, M. (2003) Adult stem cell plasticity: fact or artifact. *Annual Reviews of Cell and Developmental Biology* **19**, 1–22.

Robert, J.S. (2004) Model systems in stem cell biology. *BioEssays* **26**, 1005–1012.

Wagers, A.J. & Weissman, I.L. (2004) Plasticity of adult stem cells. *Cell* **116**, 639–648.

Li, L. & Xie, T. (2005) Stem cell niche: structure and function. *Annual Review of Cell and Developmental Biology* **21**, 605–631.

Moon, K.A. & Lemischka, I.R. (2006) Stem cells and their niches. *Science* **311**, 1880–1885.

Martinez-Agosto, J.A., Mikkola, H.K.A., Hartenstein, V. & Banerjee, U. (2007) The hematopoietic stem cell and its niche: a comparative view. *Genes and Development* **21**, 3044–3060.

Slack, J.M.W. (2008) Origin of stem cells in organogenesis. *Science* **322**, 1498–1501.

Jones, D.L. & Wagers, A.J. (2008) No place like home: anatomy and function of the stem cell niche. *Nature Reviews Molecular Cell Biology* **9**, 11–21.

Fellous, T.G., McDonald, S.A.C., Burkert, J., Humphries, A., Islam, S., De-Alwis, N.M.W., Gutierrez-Gonzalez, L., Tadrous, P.J., Elia, G., Kocher, H.M., Bhattacharya, S., Mears, L., El-Bahrawy, M., Turnbull, D.M., Taylor, R.W., Greaves, L.C., Chinnery, P.F., Day, C.P., Wright, N.A. & Alison, M.R. (2009) A methodological approach to tracing cell lineage in human epithelial tissues. *Stem Cells* **27**, 1410–1420.

Fuchs, E. (2009) The tortoise and the hair: slow-cycling cells in the stem cell race. *Cell* **137**, 811–819.

Barker, N., Bartfeld, S. & Clevers, H. (2010) Tissue-resident adult stem cell populations of rapidly self-renewing organs. *Cell Stem Cell* **7**, 656–670.

Losick, V.P., Morris, L.X., Fox, D.T. & Spradling, A. (2011) Drosophila stem cell niches: a decade of discovery suggests a unified view of stem cell regulation. *Developmental Cell* **21**, 159–171.

Morin, X. & Bellaïche, Y. (2011) Mitotic spindle orientation in asymmetric and symmetric cell divisions during animal development. *Developmental Cell* **21**, 102–119.

Simons, B.D. & Clevers, H. (2011) Strategies for homeostatic stem cell self-renewal in adult tissues. *Cell* **145**, 851–862.

Epidermis

Millar, S.E. (2002) Molecular mechanisms regulating hair follicle development. *Journal of Investigative Dermatology* **118**, 216–225.

Schmidt-Ullrich, R. & Paus, R. (2005) Molecular principles of hair follicle induction and morphogenesis. *BioEssays* **27**, 247–261.

Watt, F.M., Celso, C.L. & Silva-Vargas, V. (2006) Epidermal stem cells: an update. *Current Opinion in Genetics and Development* **16**, 518–524.

Fuchs, E. (2007) Scratching the surface of skin development. *Nature* **445**, 834–842.

Koster, M.I. & Roop, D.R. (2007) Mechanisms regulating epithelial stratification. *Annual Review of Cell and Developmental Biology* **23**, 93–113.

Blanpain, C. & Fuchs, E. (2009) Epidermal homeostasis: a balancing act of stem cells in the skin. *Nature Reviews Molecular Cell Biology* **10**, 207–217.

Driskell, R.R., Clavel, C., Rendl, M. & Watt, F.M. (2011) Hair follicle dermal papilla cells at a glance. *Journal of Cell Science* **124**, 1179–1182.

Intestinal epithelium

Winton, D.J. & Ponder, B.A.J. (1990) Stem cell organization in mouse small intestine. *Proceedings of the Royal Society London [Biology]* **241**, 13–18.

Gordon, J.I., Schmidt, G.H. & Roth, K.A. (1992) Studies of intestinal stem cells using normal, chimeric and transgenic mice. *FASEB Journal* **6**, 3039–3050.

Crosnier, C., Stamataki, D. & Lewis, J. (2006) Organizing cell renewal in the intestine: stem cells, signals and combinatorial control. *Nature Reviews Genetics* **7**, 349–359.

Barker, N., van Es, J.H., Kuipers, J., Kujala, P., van den Born, M., Cozijnsen, M., Haegebarth, A., Korving, J., Begthel, H., Peters, P.J. & Clevers, H. (2007) Identification of stem cells in small intestine and colon by marker gene Lgr5. *Nature* **449**, 1003–1007.

Barker, N., van de Wetering, M. & Clevers, H. (2008) The intestinal stem cell. *Genes and Development* **22**, 1856–1864.

van der Flier, L.G. & Clevers, H. (2009) Stem cells, self-renewal, and differentiation in the intestinal epithelium. *Annual Review of Physiology* **71**, 241–260.

Hematopoietic

Kondo, M., Wagers, A.J., Manz, M.G., Prohaska, S.S., Scherer, D.C., Beilhack, G.F., Shizuru, J.A. & Weissman, I.L. (2003) Biology of hematopoietic stem cells and progenitors: implications for clinical application. *Annual Review of Immunology* **21**, 759–806.

Metcalf, D. (2007) Concise review: hematopoietic stem cells and tissue stem cells: current concepts and unanswered questions. *Stem Cells* **25**, 2390–2395.

Purton, L.E. & Scadden, D.T. (2007) Limiting factors in murine hematopoietic stem cell assays. *Cell Stem Cell* **1**, 263–270.

Orkin, S.H. & Zon, L.I. (2008) Hematopoiesis: an evolving paradigm for stem cell biology. *Cell* **132**, 631–644.

Wilson, A., Laurenti, E. & Trumpp, A. (2009) Balancing dormant and self-renewing hematopoietic stem cells. *Current Opinion in Genetics and Development* **19**, 461–468.

Oh, I.-H. & Kwon, K.-R. (2010) Concise review: multiple niches for hematopoietic stem cell regulations. *Stem Cells* **28**, 1243–1249.

Mesenchymal stem cells

Prockop, D.J. (1997) Marrow stromal cells as stem cells for nonhematopoietic tissues. *Science* **276**, 71–74.

da Silva Meirelles, L., Caplan, A.I. & Nardi, N.B. (2008) In search of the in vivo identity of mesenchymal stem cells. *Stem Cells* **26**, 2287–2299.

Nombela-Arrieta, C., Ritz, J. & Silberstein, L.E. (2011) The elusive nature and function of mesenchymal stem cells. *Nature Reviews Molecular Cell Biology* **12**, 126–131.

Spermatogonia

de Rooij, D. (2001) Proliferation and differentiation of spermatogonial stem cells. *Reproduction* **121**, 347–354.

Brinster, R.L. (2007) Male germline stem cells: from mice to men. *Science* **316**, 404–405.

Oatley, J.M. & Brinster, R.L. (2008) Regulation of spermatogonial stem cell self-renewal in mammals. *Annual Review of Cell and Developmental Biology* **24**, 263–286.

Chapter 19

Growth, aging, and cancer

Growth: control of size and proportion

Although superficially familiar to all, growth is one of the less well understood aspects of development. By the time the general body plan is formed the whole embryo will be about 1 mm long and each organ is present as a small rudiment. In a large animal, such as the human, there will be an approximately 10^9-fold increase in volume from this stage to that of the full grown adult. This enormous change is driven largely by cell division, but also by secretion of extracellular matrix and by some increase in cell size. The intracellular events controlling cell division are now reasonably well understood (see Chapter 2), as is the biochemical mode of action of the growth factors that act on cells from their environment (see Appendix), but the overall control of growth is still somewhat mysterious.

One problem relates to final size, or why the animal stops growing at a particular point. All animals do grow to reach a finite size, although many fish take such a long time to reach it that they may not do so in a normal lifetime and so their growth appears to be indeterminate. The growth of cells in tissue culture is typically exponential in the presence of excess nutrients. However, exponential growth is rarely found in animals. The growth curve for an animal, or a part of an animal, will normally resemble Fig. 19.1, with a rate that is rapid at the beginning and tails off to zero as the final size is reached. Various equations have been proposed that give a mathematical description of overall growth but none is entirely satisfactory because of the large number of individual processes that all contribute in some way to increase of size.

The other main issue is the coordination of growth between parts of the body. Small differences in growth rate between body parts over a 10^9-fold expansion would lead to very large changes in relative proportions. This suggests that there must exist mechanisms that can sense the overall size of the organism and regulate the growth of individual parts accordingly. In fact, relative proportions do usually change systematically during growth, for example the human head becomes smaller relative to the body as a child matures. The relationship between the growth of two parts of an organism, or the relationship of one part to the whole, can often be represented by the equation $y = ax^b$, or $\log y = \log a + b \log x$, where y is the size of one part, x the size of the other, and a and b are constants. This is called "**allometry.**" However, there is no real theoretical basis to the equation and it is often not followed, so should be regarded as a convenience for representing data rather than an explanation of the growth process. An example of an allometric relationship is shown in Fig. 19.2 where the palp of *Drosophila* is shown to be proportional to overall size, while the genitalia are proportionately smaller with increasing body size.

Growth nearly always involves increase of cell number. Usually, cell division occurs at a threshold cell size and so average cell size is conserved, such that the volume of a cell population is proportional to cell number. In the rare cases where cell size is not conserved, as for example in embryonic cleavage, cell division on its own does not lead to growth because the same volume of tissue becomes divided into smaller and smaller cells (Fig. 19.3). Cell size may also increase without cell division, especially when the chromosome number, or ploidy, increases, or where cells become binucleate. This often occurs in postmitotic or slow dividing cell types in mammals, such as hepatocytes and cardiomyocytes.

Biochemical pathways of growth control

Two signaling pathways are critically important for growth control, although they have other developmental functions as

Essential Developmental Biology, Third Edition. Jonathan M.W. Slack.
© 2013 John Wiley & Sons, Ltd. Published 2013 by John Wiley & Sons, Ltd.

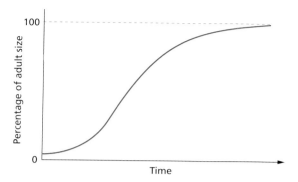

Fig. 19.1 Typical growth curve of an animal.

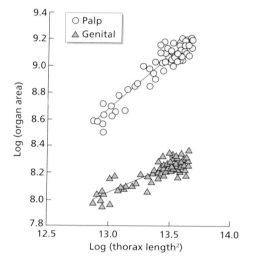

Fig. 19.2 An example of an allometric relationship between two body parts. In individuals of adult *Drosophila* the size of palps is proportionate to total body size, while the size of genitalia is proportionately smaller in large flies. The data fit the allometry equation $y = ax^b$. Reproduced from Alexander *et al. BioEssays* 2007; 29: 536–548, with permission from John Wiley & Sons Ltd.

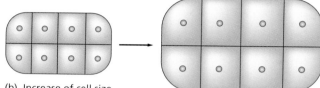

(a) Increase of cell number

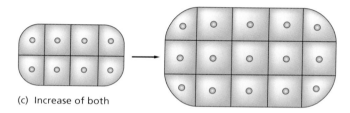

(b) Increase of cell size

(c) Increase of both

Fig. 19.3 Effect of increasing the cell division rate and the cell number.

well: the insulin–TOR system and the Hippo pathway. In general, the insulin system seems to control overall size whereas the Hippo pathway controls growth in response to local contact-based signals from neighboring cells.

The insulin pathway

The insulin/IGF signaling system, and the associated intracellular TOR (target of rapamycin) system (Fig. 19.4) are essentially similar in all animals although the number of parallel components can vary, for example *C. elegans* has no fewer than 37 *insulin*-like genes compared with three or four in mammals. TOR gets its name because it is the site of action of the drug rapamycin, used as an immunosuppressant to support human patients who have received organ grafts. The operation of the system is best understood by considering the sequence of events following binding of insulin or IGF to its receptor. The receptors are heterotetramers of two α and two β subunits. The β subunits

are transmembrane proteins with internal tyrosine kinase domains. When activated by ligand binding they transphosphorylate and activate their downstream targets via adapter proteins. There are two key signal transduction pathways activated. The familiar ERK pathway is coupled to the receptor via the adapter proteins Shc, Grb2, and SOS. These lead to activation of the GTP exchange protein Ras, this activates the kinase Raf by recruitment to the cell membrane, and this phosphorylates MEK, which phosphorylates ERK. Active ERK enters the nucleus and turns on target genes in cooperation with other transcription factors. The ERK pathway is activated by many other growth factors besides the insulin group, for example EGF, FGF, PDGF, and has effects on a large number of processes, often including stimulation of cell division.

The other main pathway activated by the insulin-like receptors is the PI3 kinase (PI3K) pathway. This is more concerned with metabolism, and particularly control of the rate of protein synthesis, which is the main determinant of cell size. PI3K is activated via an adapter protein called the insulin receptor substrate (IRS). When activated, it phosphorylates the membrane lipid phosphatidyl inositol-4,5-diphosphate (PIP_2) to phosphatidyl inositol-3,4,5-trisphosphate (PIP_3). This then activates a serine–threonine kinase called protein kinase B (PKB or Akt). PIP_3 does not accumulate indefinitely because it becomes dephosphorylated by a phosphatase called PTEN. The PI3K pathway unleashes a number of well-known metabolic consequences, in particular the increased uptake of glucose, of fatty acids, and increased synthesis of glycogen. In terms of growth regulation, the active PKB phosphorylates, and thereby inactivates, a complex of two proteins encoded by the *Tuberous*

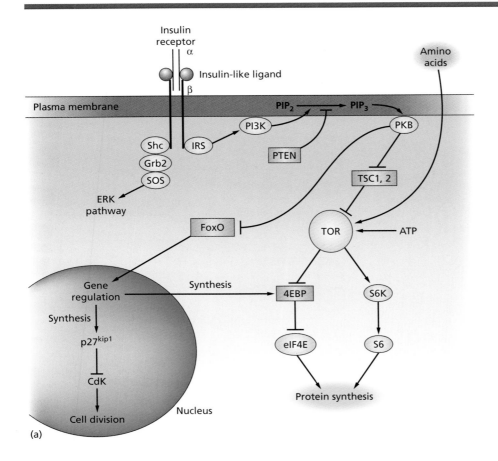

Fig. 19.4 The insulin/TOR pathway. (a) The components promoting growth are shown as ovals and those inhibiting growth as rectangles. (b) State of each component in the presence or absence of insulin-like ligand.

	Receptor	PI3K	PKB	TSC	TOR	FoxO	Protein synthesis
+ Ligand	Active	Active	Active	Inactive	Active	Inactive	Up
– Ligand	Inactive	Inactive	Inactive	Active	Inactive	Active	Down

(b)

sclerosis genes 1 and 2, which in turn normally inactivate TOR. Because two inhibitions equals one activation this means that insulin signaling activates TOR. TOR is an intracellular serine–threonine kinase, which phosphorylates another kinase, S6 kinase, which in turn phosphorylates and activates the ribosomal protein S6. This enables ribosomes to translate a class of mRNAs encoding ribosomal proteins, thus increasing the overall protein synthesis capacity of the cell. It also stimulates protein synthesis by another route: inhibiting the protein 4EBP which is an inhibitor of the protein synthesis initiation factor eF1a. The pathway has an effect on gene expression via the transcription factor FoxO (forkhead class), which is inactivated by PKB. In the absence of insulin signaling FoxO becomes dephosphorylated, goes to the nucleus, and activates transcription of certain components (including 4EBP) that reduce protein synthesis and others (including the Cdk inhibitor p27^{kip1}) that reduce cell division.

This biochemical system ensures that the amount of growth will be determined by the appropriate growth-promoting signals, the available nutrients, and the energy supply. The point at which all these factors are brought together is TOR, because it is activated by ATP and by amino acids as well as by insulin signaling. It thus integrates the sensing of growth factors, energy supply, and nutrient availability.

The Hippo pathway

This was discovered in *Drosophila* but is very widespread in the animal kingdom and is upregulated in many human cancers. In Fig. 19.5 the pathway is shown in its *Drosophila* version, the vertebrate one is given in the Appendix. Hippo itself is a kinase and loss-of-function mutations of its gene lead to overgrowth, as shown in Fig. 19.6a,b. Hippo activates another kinase, called Warts, and this phosphorylates Yorkie, a nuclear protein that is retained in the cytoplasm when phosphorylated. When unphosphorylated, Yorkie migrates to the nucleus, serves as a cofactor for the transcription factor Scalloped, and upregulates various growth-related genes including those for cyclin E and Myc. Overexpression of *yorkie* causes overgrowth, rather like loss of function of *hippo* (Fig. 19.6c). This effect is due both to increased cell division and restriction of apoptosis. The pathway is not known to be regulated by an extracellular growth factor but there are controls from cell surface proteins that suggest that the pathway can respond to cues from neighboring cells. In particular, the protein Fat is an atypical cadherin involved in the control

of planar polarity (see Chapter 17). When bound to another atypical cadherin, Dachsous, on neighboring cells, it can be phosphorylated and will then sequester a myosin like protein, Dachs, and prevent it from inhibiting Warts. Fat can also activate Hippo via a complex composed of Expanded, Merlin, and Kibra, adapter proteins present in a subapical position in epithelia. In either event, activated Fat reduces activation of Yorkie so is antagonistic to growth. Fat is also regulated by Fourjointed, a kinase found in the Golgi body, which phosphorylates it on the extracellular domain. The effects of Dachsous and Fourjointed on Fat depend not on their absolute level but on the difference of content in the neighboring cells. The more difference there is, the more inhibition of the Hippo kinase pathway and the more stimulation of growth via Yorkie. This is of great potential significance in the context of regeneration where surgically created discontinuity of pattern usually results in a local stimulation of growth (see Chapter 20).

The mammalian Hippo pathway is the same in its core components (Hippo homologs are called Mst1 and 2; Warts homologs are called Lats 1 and 2; Yorkie homologs are called Yap and Taz). Homologs to Fat, Dachsous, and Fourjointed exist in mammals but at present similar connections from them to the Hippo pathway are not clearly established. There is, however, a connection between the cell surface and YAP in the form of α-catenin, which normally links the cadherins to the cytoskeleton. This can also bind phosphorylated YAP, via a protein called 14-3-3, and prevent it from being dephosphorylated.

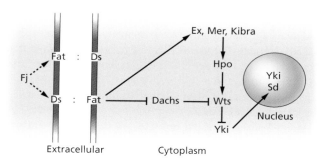

Fat	Dachs	Hpo	Wts	Yki
+	−	+	+	−
−	+	−	−	+

Fig. 19.5 The Hippo pathway in *Drosophila*. The effects of Fj and Ds on Fat depend not on the absolute levels but on the difference between neighboring cells. Because of the inhibitory steps in this pathway, when Fat is activated, Yorkie is inactive.

Control of final size

Insects
Control of final size is understood best in holometabolous insects. Much work has been done with the tobacco hornworm, *Manduca sexta*, although the mechanisms in *Drosophila* appear to be similar. The adults of these insects are almost entirely postmitotic, so the final size depends on the amount of growth of the imaginal discs during the larval and pupal stages. There

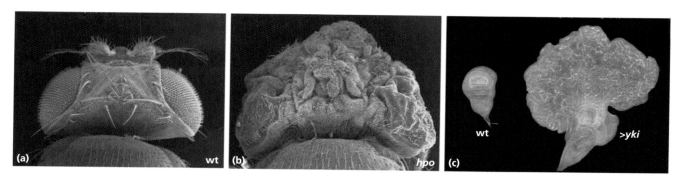

Fig. 19.6 (a,b) Loss of Hippo (*hpo*) from the *Drosophila* head leads to disproportionate overgrowth. (c) Overexpression of Yorkie (*yki*) in the wing disc leads to profound overgrowth. wt, wild type. Reproduced from *Pan. Genes Dev* 2007; 21: 886–897, with permission from Elsevier.

is a critical size during larval growth which, when reached, allows the insect to go through metamorphosis even if starved thereafter. This indicates a close relationship between the hormonal control of metamorphosis and the control of growth. Growth of the whole body is promoted by insulin-like peptides released from a few cells in the brain. In *Drosophila* imaginal discs, loss or reduction of function of any of the positive components of the insulin–TOR system (the insulin receptor, PI3K, PKB, TOR) will reduce both the size and number of cells, hence reducing the size of the structures they make up. Conversely, loss or reduction of function of the inhibitory components (PTEN, TSC1 and 2, FoxO) will increase the size and number of cells and increase the size of resulting structures.

The sensing of overall body size depends specifically on the prothoracic gland, which makes ecdysone. Ecdysone suppresses juvenile hormone production by the corpora allata and, eventually, initiates metamorphosis. If the prothoracic gland is genetically manipulated to be extra large by overexpression of PI3 kinase, then the critical size of the larva is reduced, metamorphosis occurs early, and the adults are small. Conversely, if the prothoracic gland is made too small by overexpression of PTEN, then metamorphosis occurs later and adults are large. These effects only work if the method of adjusting the size of the prothoracic gland involves insulin signaling. If size is manipulated via some other route then the systemic effects are not seen.

Drosophila wing discs have been much used for the investigation of growth control because of the ease of overexpressing genes in clones by the FLP-out method, or removing genes from clones by mitotic recombination (see Chapter 17). It is possible to bypass the insulin system by interfering directly with cell cycle controls. For example a local increase in cell division rate can be caused by overexpressing the transcription factor E2F, which activates expression of a number of genes required for S phase of the cell cycle. Conversely, the cell division rate can be slowed by overexpressing the Retinoblastoma (RB) protein, which inhibits the endogenous E2F. But neither of these manipulations have much effect on overall size of the wing disc because the rate of cell size increase does not change. In other words, the faster cell cycle generates a clone containing more, smaller cells and the slower cell cycle generates a clone containing fewer, larger cells. In order to change the rate of growth it is necessary also to change the rate of cell size increase, as is done in response to insulin signaling.

Mammals

The adult mammal is very different from the adult insect. Mammals are composed of various tissues some of which show continuous cell turnover while others are mostly quiescent but can display regenerative growth. Some cell types, such as neurons and skeletal muscle fibers, are postmitotic but few if any whole tissues are. There is no counterpart to the juvenile hormone–ecdysone system for regulating molting and metamorphosis in arthropods. However, in terms of final size control there is a key similarity in the central role of the insulin pathway.

It is not so much insulin itself but the two insulin-like growth factors, IGF1 and IGF2, that are principally responsible for growth control in mammals. They both operate through a receptor called the "IGF1 receptor," which is similar to the generic insulin receptor described above. IGF2 is more important for supporting growth during the fetal period, largely through its effects on the placenta, while IGF1 is more important in supporting postnatal growth. IGF1 is made in a wide variety of tissues and is under the control of growth hormone (GH or somatotrophin) secreted by the somatotrophs of the anterior pituitary gland. Animals or humans lacking sufficient growth hormone grow and mature slowly and their final size is reduced. Similarly, a lack of the GH receptor leads to a type of dwarfism (Laron's syndrome). Conversely, individuals suffering a pituitary tumor that secretes excessive growth hormone will grow faster during childhood and adolescence and reach a larger than normal final size, or gigantism. Pituitary-deficient animals can be restored to near normal growth rates by administration of either growth hormone or IGF1. There are some normal hereditary differences in these components, for example there is a considerable difference of IGF1 levels in the serum of large and small breeds of dog. This is due to genetic variation at the *IGF1* locus itself and is responsible for the enormous size differences between dog breeds, about 80-fold in weight between the largest and smallest.

Mouse knockouts of *Igf1* or *Igf2* genes are about 60% normal size at birth, and the double knockout is only 30% normal size and dies at birth. Likewise, the knockout of the IGF1 receptor, or of two of the three *PKB* genes, are about 50% normal size and die at birth. The knockout of *Igf2* alone grows normally after birth, which is why it is thought that its functions are mostly exerted in fetal life. Conversely, transgenic overexpression of either IGF1 or of growth hormone, which stimulates the synthesis of IGF1, can increase size up to about twofold.

Control of relative proportion

This is a somewhat neglected but very important problem in growth control. Even if the overall size of the organism is correct, how can it be guaranteed that different body parts expand in proportion to one another? Or, in cases where there are allometric deviations from proportional growth, how is this controlled?

A popular model for proportion control involves a feedback inhibition of each body part on itself. For example if each body part produces an inhibitor that is diluted into the total blood volume of the organism then its concentration will measure the size of the part relative to the size of the whole. If the part gets a bit too big, the inhibitor will slow it down, and if the part is too small the inhibitor level will be low, so its growth will accelerate (Fig. 19.7; and see Animation 22: Chalone action). Such a hypothetical negative-feedback signal is called a **chalone** (pronounced "kaylone"). The obvious method of investigation is to

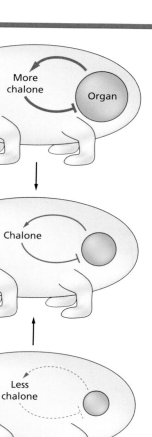

Fig. 19.7 Operation of a hypothetical chalone system.

slow down protein synthesis such that the individuals develop slowly and are small in adult size. Recently, it has been established that there is an active process of cell competition going on within imaginal discs such that the slower growing "loser" clones preferentially undergo apoptosis compared to the more rapidly growing "winner" clones. This process requires a cell membrane protein called Flower, of which certain isoforms become upregulated in loser cells and lead to apoptosis.

In contrast to these systems, the overexpression of Yorkie, or removal of its inhibitors, from clones of cells does cause disproportionate growth. The reason for the absence of cell competition in this case is not known. Yorkie is also required for imaginal disc regeneration following the loss of parts and subsequent healing of the cut surfaces. This probably arises from a discontinuity in the gradients of Dachsous and Fourjointed, which senses the absence of part of the structure, and leads to inhibition of Fat with consequent inhibition of the Hippo kinase cascade and movement of Yorkie to the nucleus.

The original control of regional specification in the discs during development lies with morphogen gradients, especially that of Decapentaplegic (Chapter 17), which also has a strong growth-promoting function. However, normal discs grow uniformly, not in a manner corresponding to the Decapentaplegic gradient. This is probably because the gradient activates different downstream systems at different concentrations and these have different local effects on the growth rate such that the overall growth is fairly uniform.

graft organs from young to adult animals and vice versa. While this sounds simple it is actually difficult because of the complex effects of the immune system and the difficulties of establishing equivalent vascular supplies for grafted and host organs. In general, the chalone model is probably not correct since there is a tendency for grafts between animals to behave in an autonomous way, neither adapting to the size of the host, nor affecting the relative growth of corresponding parts within the host. But there are some examples of chalone-like systems, for example in the control of size of the muscles and the liver (see below).

Insects

In *Drosophila* it has been know for some time that an individual clone can be given a growth advantage over its neighbors and this does not necessarily cause a disproportion. Instead, it fills up most of the compartment at the expense of other clones. This is the basis of the clonal analysis of imaginal discs described in Chapter 17. This principle operates for wild-type clones in a background of *Minute* mutant tissue and also for clones overexpressing the *d-myc* oncogene on a wild-type background. *Minute* mutations are in genes for ribosomal proteins and they

Mammals

The best-characterized example of chalone action in vertebrates is the control of muscle growth exerted by the TGFβ superfamily member myostatin (= Growth and Differentiation Factor 8; GDF8). This is secreted by developing muscle cells and it reduces both the cell division of myoblasts and the enlargement of myofibers. The *Myostatin* knockout mouse has a two to threefold increase in muscle mass, arising from the presence of both more and larger myofibers. Some breeds of domestic cattle, such as the Belgian Blue, have been shown to carry hypomorphic alleles for *Myostatin*, indicating that traditional animal breeding has already anticipated rational genetic modification in this area (Fig. 19.8). In the embryo *Myostatin* is expressed in the central part of the dermomyotome, in between the epaxial and hypaxial myoblast-generating areas, and is later expressed in the limb buds, both by the myoblasts and some other mesenchymal tissue. The early expression in tissues neighboring the developing muscle indicates that it may have a role in defining the boundaries of the myogenic areas as well as the later negative-feedback regulation.

Another well-documented chalone-like molecule is GDF11 (Growth and Differentiation Factor 11) in the olfactory system. This is expressed in olfactory neurons and inhibits further neurogenesis by upregulating production of the Cdk inhibitor p27[kip1]. The knockout mouse for *Gdf11* has an increased number of progenitor cells and neurons in the olfactory epithelium.

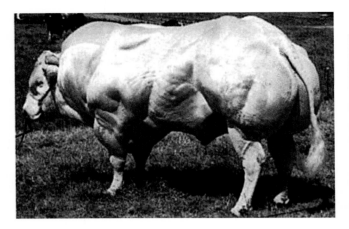

Fig. 19.8 A Belgian Blue bull. Note the considerable muscle mass.

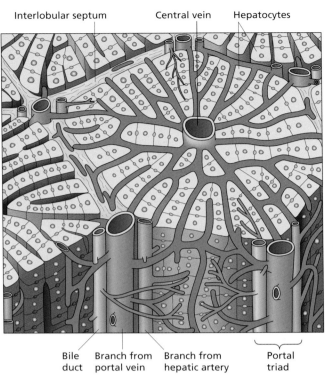

Fig. 19.9 Histological structure of the liver.

Interestingly, both myostatin and GDF11 belong to the same subgroup of TGFβ-like factors, and both are inhibited by follistatin. This suggests that there could be global controls that modulate the behavior of a number of tissue-specific negative-feedback mechanisms.

In addition to pituitary dwarfism and gigantism, components of the GH-IGF pathway appear in some types of disproportionate growth found in humans. Pituitary tumors in adult life do not lead to gigantism because the growth zones of the long bones are already shut down (see below). But they do lead to a disproportionate enlargement of the jaws, hands, and feet, which is known as acromegaly. The gene for the phosphatase *PTEN* is often mutated to inactivity in human cancer. It is a **tumor suppressor gene** (see below) so the absence of one copy of the gene in the germline leads to a high risk of cancer because cells may easily lose the other copy through somatic mutation during postnatal life. Complete loss of function of *PTEN* leads to "Proteus syndrome." This is exceedingly rare but can occur as a very early somatic mutation in embryonic development, creating a mosaic in which part of the body lacks the gene entirely. The result is numerous, chaotic outgrowths of bone and connective tissue, and this condition may have been experienced by the famous "Elephant Man," Joseph Merrick. The *Tuberous sclerosis* genes are also tumor suppressor genes and get their name from a dominant hereditary disease in which loss of one copy of the gene predisposes to the formation of numerous benign tumors, again because of occasional somatic mutations in the good copy.

Liver regeneration

The liver is unusual among mammalian organs in being able to regenerate to full size within a few days following surgical removal of a part. It is considered here rather than in the regeneration chapter because it really illustrates a phenomenon of growth control. If one liver lobe is removed, as is usual in liver regeneration experiments, the others increase in size to restore the original organ volume, but they do not regenerate the missing lobe, or the shape of the original liver. This is also true in embryonic development. If the number of progenitor cells in mouse liver or pancreas is reduced by ablation with diphtheria toxin driven by respective tissue-specific promoters, then the developing liver adjusts to normal volume within a few days while the pancreas remains permanently small into adult life.

The histological structure of the liver consists of a reiterated patters of plates of hepatocytes, extending from a periportal location where a small bile duct runs alongside a branch of the hepatic artery and the hepatic portal vein (the portal triad), to the perivenous location, adjacent to a branch of the hepatic vein (Fig. 19.9). The region lying around one central vein is known as a lobule. Following the removal of tissue the hepatocytes all over the remaining liver start to proliferate within 24 hours. Other cell types in the liver (biliary epithelium, Kupffer, and endothelial cells) commence proliferation somewhat later, such that all of the liver cell populations eventually increase in number in a coordinated manner. It has long been known that there are circulating factors controlling the size of the liver, as removal of part of the liver from one of a pair of animals with surgically combined circulations (**parabiosis**) will lead to additional growth in the intact liver of the other animal. The circulating factors are probably a complex mixture, but are likely to include bile acids that are released into the blood from the damaged liver. In general, the candidate factors are considered to be activators rather than inhibitors, so do not entirely fit the chalone model, but do constitute circulating factors titrating the size of the liver against that of the whole animal. This regulatory feedback system is also operative in the situation of liver transplantation. When a small liver is grafted into a large host the graft grows rapidly until it achieves the size of the liver that was removed.

Although overall hepatic size is unchanged following regeneration, it appears that the lobules of the regenerated liver are larger than those of the original; in other words, the average distance from the portal triad to the central vein increases. This difference may, however, resolve as the tissue is remodeled in the course of the normal slow turnover.

The size of the liver can be manipulated via the Hippo pathway. In Fig. 19.10 is shown an experiment in which a gain-of-function mutation of *Yap1* (a mammalian *yorkie* homolog) is expressed in the liver using the Tet-off system. In 35 days of transgene induction the liver increased in size by a factor of 4, but when the transgene was turned off, the size gradually reverted to normal. The effects are due both to changes in frequency of cell division and of cell death. This level of plasticity is not seen in other vertebrate organs and indicates that the liver is specialized to adjust cell numbers in a very specific way. This capability has probably arisen in vertebrate evolution because the liver is the primary destination for foodstuffs absorbed by the intestine, so any environmentally derived toxins will hit the liver first and are often likely to cause some cell damage.

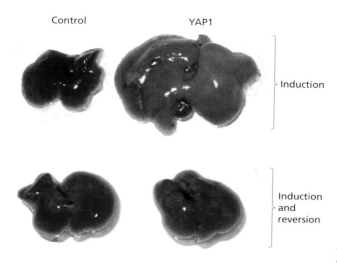

Fig. 19.10 Enlargement of mouse liver provoked by overexpression of YAP1 for 5 weeks. After a further 5 weeks without induction the liver returns to its normal size. Reproduced from Camargo *et al. Curr Biol* 2007; 17: 2054–2060.

Growth in stature

The height of a human being, or overall stature of a vertebrate animal, depends to a large extent on the length of the long bones in the limbs. These long bones, like most of the embryonic skeleton, are initially formed as a cartilage **model**, which become replaced by bone in later development (Fig. 19.11). Cartilage consists of **chondrocytes** embedded in holes (lacunae) in a clear extracellular matrix without blood vessels, nerves, or lymphatics. The matrix is characterized by the presence of type II collagen and aggrecan, a specific cartilage-type proteoglycan. Sox9 serves as a key transcription factor in cartilage development, whether derived from mesoderm or neural crest, and activates transcription of the specific cartilage-type collagens. The human genetic disease campomelic dysplasia, which causes various skeletal abnormalities because of cartilage defects, is a dominant haploinsufficient condition arising from the loss of one copy of *Sox9*. The periphery of the cartilage, or **perichondrium**, is composed of fibroblast-like cells that divide to generate **chondroblasts**. These divide a few more times and then differentiate to become chondrocytes embedded in the matrix.

The bones of the face and cranium are known as **membrane bones** as they are not formed from a cartilage model but differentiate directly from the dermis, which in the head is of neural crest origin. Bone formation by either method requires the action of the bone morphogenetic proteins (BMPs), which are normally expressed in differentiating bone and in healing fractures. Because the BMPs are also required for various earlier developmental events, *Bmp* knockout mice may not develop far enough to form a skeleton. However, the knockout of *Bmp7* is viable and does have skeletal defects, as do naturally occurring null mutants of *Bmp5* and *Growth and differentiation factor 5 (Gdf5)*, also a member of the BMP family.

The shaft of a long bone is called the **diaphysis** and the knobs at either end are called the **epiphyses**. Primary ossification starts in a collar around the center of the diaphysis where the perichondrium becomes converted to a **periosteum** containing **osteoblasts**. Combinations of perichondrium from **Rosa26** mice with unlabelled cartilage, grown under the kidney capsule of **nude mice**, shows that the osteoblasts are derived from the perichondrium. The key transcription factor driving osteogenesis is called Cfba1 (homeodomain, = Runx2). The mouse knockout of *Cbfa1* has a fully cartilaginous skeleton with a complete absence of bone. The heterozygote is known in the human syndrome of cleidocranial dysplasia, and has defects of membrane bone ossification. During ossification, the osteoblasts invade the cartilage in association with blood vessels and secrete a matrix (osteoid) mainly composed of type I collagen. As this thickens, some cells become trapped in it and they, now called **osteocytes**, secrete calcium phosphate, which precipitates in the matrix as hydroxyapatite. Unlike chondrocytes, the osteocytes remain in contact with one another by means of thin cellular processes running through channels called canaliculi. The bone continues to grow at the surface and becomes eroded

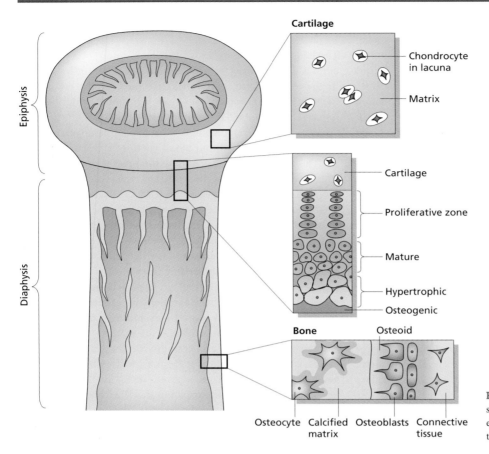

Cartilage

Chondrocyte in lacuna

Matrix

Cartilage

Proliferative zone

Mature

Hypertrophic

Osteogenic

Bone   Osteoid

Osteocyte Calcified Osteoblasts Connective
           matrix   tissue

Fig. 19.11 A developing long bone, showing the cartilaginous epiphysis, the epiphyseal growth plate, and the ossification of the diaphysis.

from the inside by **osteoclasts**, which are giant multinucleate cells formed in the bone marrow from the hematopoietic system. Osteoclasts also make channels in the bone matrix for blood vessels, nerves, and lymphatics. Additional ossification centers appear in the epiphyses, but growth continues for some time at the ends of long bones in a zone between the remaining cartilage of each epiphysis and the bone of the diaphysis.

This region is called the epiphyseal **growth plate**. Within the growth plate the most distal part consists of actively dividing chondrocytes, and more proximally the division ceases and the cartilage cells become **hypertrophic** (enlarged) and the matrix becomes calcified. The hypertrophic chondrocytes eventually become replaced by osteocytes and the matrix becomes converted to bone. When full adult stature has been attained, the ossified epiphysis fuses with the diaphysis and the growth plate disappears. The ultimate size of an individual depends both on the activity of the growth plates and on the length of time that they remain active. Because of the timing of the shutdown of the growth plates, excess growth hormone given during childhood will increase stature, but after epiphyseal fusion it leads to acromegaly.

The most common form of human dwarfism, achondroplasia, is rather different from pituitary dwarfism as it leads to a disproportion in which the long bones are abnormally short. It is caused by a dominant gain-of-function mutation of Fgf receptor 3 (*FGFR3*) that reduces proliferation of the growth plate. Interestingly, the mouse knockout of *Fgfr3* has the opposite phenotype, namely abnormally long bones caused by a prolongation of activity of the growth plate. These results are explained because in the chondrocytes of the growth plate FGF acts as a growth inhibitor rather than a growth promoter. This inhibitory activity is exerted via the STAT signal transduction pathway, rather than the ERK pathway normally activated by the FGFs. Stronger gain-of-function mutants of *Fgfr3* lead to the more severe and fatal thanatophoric dysplasia, in which there is a truncation of growth of much of the skeleton other than the skull. Gain-of-function mutations of other FGF receptor genes cause defects in membrane bone formation, particularly resulting in craniosynostosis, in which there is premature fusion of the bones of the skull.

Within the growth plate, proliferation of chondrocytes is maintained by a feedback loop involving a member of the hedgehog family of signaling molecules, Indian hedgehog (Ihh), and parathyroid hormone-related protein (PTHrP), a protein similar to parathyroid hormone (Fig. 19.12). Ihh is produced by the cartilage cells that have finished dividing and become com-

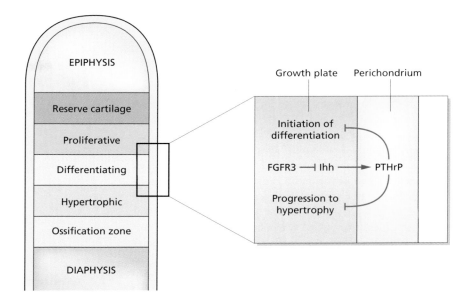

Fig. 19.12 Feedback loop for control of chondrocyte hypertrophic differentiation. Ihh from the committed, but nonhypertrophic, cartilage represses differentiation via PTHrP.

mitted to hypertrophic differentiation. It acts on the perichondrium to upregulate *Pthrp*. PTHrP then acts on the dividing chondrocytes to maintain division and hence inhibit their progression to the hypertrophic compartment. Knockouts of genes for either factor, or of the *Pthrp receptor*, lead to reduced proliferation of the growth plates, premature ossification, and hence another type of short-limbed dwarfism. Overexpression of *Ihh* prolongs the phase of chondrocyte division and hence suppresses the formation of hypertrophic cartilage. This also leads to a type of dwarfism because the hypertrophic differentiation itself normally drives elongation of the limbs to some extent. However, the phenotype is actually the opposite of that found in the *Pthrp* knockout, because the formation of hypertrophic cartilage is delayed instead of advanced. The overexpression of Ihh does not rescue the phenotype of the *Pthrp* knockouts, but a constitutive mutation of *Pthrp receptor* will do so, confirming that the PTHrP acts downstream of the Ihh. Therefore, in essence, this is another chalone-like system in which the rate of production of a cell type (the hypertrophic cartilage) is regulated by the size of the cell population committed to this pathway.

Aging

There has long been debate about whether aging is really a developmental process or not. The argument in favor is that senescence and death arise from pre-programmed events, which are under developmental control. The argument against is that such events are simply due to progressive mutational and other damage, which have an inherently random character. In fact,

both positions have proved correct to some extent. Aging is developmental insofar as it responds to genes that can be manipulated, and remarkably it turns out that one of the important systems controlling it is the same as that regulating growth, namely the insulin signaling pathway. On the other hand, the genes regulated by this pathway mostly encode components that minimize the consequences of oxidative stress and other types of macromolecular damage, indicating that the multiple pathologies that lead to the eventual destruction of the organism do depend largely on random and nondevelopmental damage.

Cell autonomous processes

Studies on mammalian cells grown in primary culture show that they tend to slow down and eventually stop growing after a certain number of passages. This is known as the **Hayflick limit**. It is still debated whether this type of process contributes to aging of the whole organism but it is generally thought that it does, at least in terms of stem cell depletion, that is a reduction in the number of stem cells in renewal tissues and sometimes in their proliferative capability. However, stem cells do not become exhausted in a single lifetime. Hematopoietic stem cells, epidermis, and spermatogonia have all been grafted to host mice and continued to function well in excess of one mouse lifespan.

There are several explanations for the Hayflick limit. In addition to random processes leading to DNA damage, there are two systematic processes. One is the gradual accumulation of proteins encoded by the *INK4a/ARF* locus, the other is the gradual shortening of telomeres. Permanent cell lines that grow indefinitely in culture have undergone mutational changes similar to those undergone by cancers, which enable them to avoid both types of change.

INK4a/ARF locus

This encodes two proteins, which are transcribed in different reading frames: p16^{INK4a} and ARF. P16^{INK4a} is an inhibitor of the cyclin-dependent kinases Cdk4 and 6 and brings about cell cycle arrest. ARF inhibits a ubiquitin ligase that modifies p53, and thereby renders it liable to proteolysis. ARF therefore increases the activity of p53, which has a range of functions generally concerned with cell cycle arrest or the initiation of apoptosis in response to stress or DNA damage. p53 is a very important protein in cancer (see below) where loss of function or dominant negative mutations are commonly found. A progressive increase of p16^{INK4a} and ARF levels is found in mammalian cells in primary culture and are among the factors that slow down their growth rate. The two proteins are also found to be elevated in old compared to young animals. This occurs in all tissues whether or not they are rapidly turning over, so seems related to elapsed time rather than to cell division number. Normal repression of the locus is a function of BMI1, a member of the Polycomb group of chromatin repressors. BMI1 is needed for the function of hematopoietic and other stem cells, and its upregulation can restore cell division to cultured cells near the Hayflick limit. Mice lacking the *INK4a/ARF* locus are relatively normal when young, but have a substantially elevated risk of developing cancer as they mature, suggesting that the evolutionary reason for the existence of this system is protection against cancer.

Telomere shortening

It is often believed that cells can count their divisions over many cycles. The only known mechanism that approximates to this ability is the shortening of **telomeres**, which are the regions at the ends of chromosomes. Usually, at each DNA replication cycle the telomeres become slightly shorter and eventually the deletions encroach so far that they behave like a double-stranded DNA breaks activating p53, with consequent apoptosis of the cell. The telomeres consist of thousands of repeats of a six-base pair sequence, TTAGGG, bound by a multiprotein complex called Shelterin. In normal cell division 50–100 base pairs are lost in each cycle. However, the ends can be repaired by an enzyme–RNA complex called **telomerase**, normally present in germ cells and often upregulated in cancer cells and permanent tissue culture cell lines. This includes a piece of RNA (Terc) encoding the telomere sequence, and a reverse transcriptase (TERT) to make the DNA.

Does telomerase shortening contribute to aging of whole animals? Old mice and humans do have shorter telomeres. Tissue-specific stem cells have some telomerase activity but not enough to maintain telomere length completely, so there is some shortening even in stem cells. The knockout of the RNA component of telomerase causes mice to have a reduced lifespan, which interestingly gets shorter each generation showing that the germ cells are not maintaining their telomeres as usual. It also reduces the capacity of hematopoietic stem cells to support serial grafts. Normal HSCs will sustain about four serial transplants to irradiated host mice but *Terc$^-$* mice will only do two. This indicates that stem cell depletion is at least partly due to telomere shortening. Overexpression of telomerase in mice causes cancers, and so generally shortens life. But if mice are engineered to resist cancer by insertion of extra copies of *INK4a/ARF* and *p53*, the surviving animals do live longer than controls. Cloned animals made by somatic cell nuclear transfer tend to have normal telomere lengths, indicating that the telomerase becomes reactivated in the early embryo following reprogramming of the somatic nucleus.

The insulin pathway and aging

The role of the insulin pathway in aging was discovered in *C. elegans*. If conditions are bad in terms of limited nutrients or overcrowding during the first larval stage then it can form a **dauer larva** instead of the third larval stage. The dauer larva is a nonfeeding, impermeable, dispersal phase. It can survive for months but can still develop to an adult if conditions improve. When mutants were screened to find what caused this developmental switch, it was found that loss-of-function mutants of components of the insulin pathway will cause dauer formation. Furthermore, hypomorphic alleles in the same genes do not spontaneously form dauers but do live longer than the normal 2–3 weeks. The *C. elegans insulin receptor* gene is called *daf2*, the *PI3kinase* gene is called *age1/daf23*, and the homolog of *FoxO* is called *daf16*. It was found that the insulin pathway would not exert its effect unless *FoxO* was present suggesting that, unlike growth control which operates largely through the protein synthesis level, the aging pathway mostly operates through transcriptional control. FoxO promotes transcription of genes for a variety of products that reduce oxidative stress, such as the enzymes catalase and superoxide dismutase. Many other hypomorphic mutants of *C. elegans* that extend lifespan are in components of the mitochondria, suggesting that aerobic metabolism, with its concomitant production of free radicals, causes the damage that leads eventually to death. So FoxO prolongs life, at least partly, by causing production of products that reduce oxidative stress. Because the insulin pathway represses FoxO it follows that conditions that normally stimulate insulin signaling, namely an abundance of food, will tend to reduce lifespan. Treatment of *C. elegans* with RNAi against the insulin receptor makes it possible to determine when the protective effect is exerted, and it is found that the receptor needs to be downregulated soon after the final molt in order to get the maximum degree of prolongation of life (Fig. 19.13).

The general, significance of the results with *C. elegans* is suggested by comparisons with other animals. *Drosophila* adults

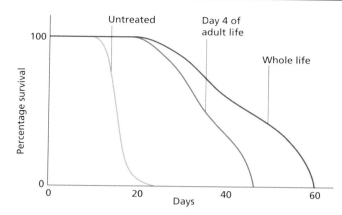

Fig. 19.13 Effects of downregulating the insulin pathway on lifespan of *C. elegans*. Feeding with RNAi to insulin receptor will extend life substantially if started before the last molt, but the effect falls off in adult life.

have a life span of about 3–4 months and this can be extended by about 40% in heterozygotes for *chico*, which encodes the insulin receptor substrate (IRS). Mice live 2–3 years but the Snell and the Ames dwarf mice can live 40% longer. The Snell dwarf mouse lacks the transcription factor Pit1 (POU domain) and the Ames dwarf mouse the transcription factor Prop1 (paired homeodomain). Both of these strains fail to form three cell types in the anterior pituitary: the somatotrophs, lactotrophs, and thyrotrophs. As described above, the reduction of growth hormone arising from the absence of somatotrophs leads to reduction of IGF1 production and so reduction of insulin pathway signaling. The Laron dwarf mouse, a knockout of the GH receptor, lives even longer, about 55% longer than usual.

Caloric restriction

It has also been known for many years that the lifespan of animals can be extended significantly by "caloric restriction," in other words supplying about two-thirds of food compared with the amount normally consumed when it is supplied in excess. Such individuals mature more slowly and live a long time. The effect was initially discovered for rats but it is now known to occur in primates as well (Fig. 19.14). Lower animals, including *Drosophila* and *C. elegans*, also show the effect. Even the yeast, *Saccharomyces cereviseae*, which is not an animal at all, shows a comparable phenomenon. This is an increase in the number of daughter cells that can be produced by one mother cell when cultured in a reduced concentration of glucose.

At least in part, caloric restriction is another way of reducing insulin signaling. The *Drosophila chico* heterozygote achieves about the same level of lifespan increase as can be achieved with caloric restriction, and caloric restriction does not extend it further. However, there are probably also some parallel or synergistic effects as the pituitary dwarf mice and some *C. elegans* mutants in the insulin system can have their lives extended even longer by imposing caloric restriction. Caloric restriction has innumerable effects and generally reduces all of the events and processes associated with aging, including diabetes, cancer, cardiovascular disease, and brain atrophy. The reduction of cancer risk is particularly significant since knockout of *INK4a/ARF* or overexpression of *telomerase* both increase cancer risk.

The yeast experiments showed that the lifespan extension depended on a gene *sir2*, encoding a NAD-dependent histone deacetylase. Histone deacetylases are generally associated with inhibition of gene activity, and the dependence on NAD (nicotinamide adenine dinucleotide, of central importance in intermediary metabolism) indicates a means of linking the cell's metabolic state to a global control of gene activity. While yeast may seem a long way from animals in terms of developmental biology, remarkably, the overexpression of *sir2* homologs in *C. elegans* and *Drosophila* has been claimed to extend lifespan. However, the effect may be due to linked mutations and has not yet been shown for mammals, which have several *sir2* homologs, the most similar being called *Sirt1*. A compound called resveratrol has been considered to activate SIRT1 activity and to counteract the adverse long-term effects of a high fat diet in mice. But this also does not extend mammalian lifespan.

Classic experiments

Aging

These are three classic experiments that discovered aspects of the mechanisms underlying aging. The first is the original study of caloric restriction in rats. The second is the first report about the enzyme telomerase. *Tetrahymena* is a ciliate protozoan used for a variety of cell biology studies. The third paper is the first report of the *daf2* and *daf16* mutations of *C. elegans*. These lie in genes later identified as encoding the insulin receptor and FOXO homologs, respectively.

McCay, C.M., Crowell, M.F. & Maynard, L.A. (1935) The effect of retarded growth upon the length of life span and upon the ultimate body size. *Journal of Nutrition* **10**, 63–79.

Greider, C.W. & Blackburn, E.H. (1985) Identification of a specific telomere terminal transferase-activity in *Tetrahymena* extracts. *Cell* **43**, 405–413.

Kenyon, C., Chang, J., Gensch, E., Rudner, A. & Tabtiang, R. (1993) A *C. elegans* mutant that lives twice as long as wild type. *Nature* **366**, 461–464.

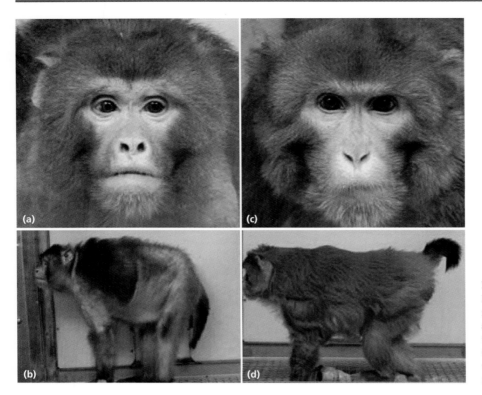

Fig. 19.14 Effects of caloric restriction on Rhesus macaques. The restricted diet was maintained for 20 years. (a,b) Control at age 27; (c,d) age-matched animal subjected to caloric restriction. Reproduced from Colman *et al. Science* 2009; 325: 201–204, with permission from American Association for the Advancement of Science.

Cancer

Cancer is of huge interest to human beings since about 40% of us will get some form of cancer and about half this number will die from it. Accordingly, cancer biology is a huge subject and this section will just touch on the aspects that relate to developmental biology. Cancer is fundamentally a growth of the body's own tissue to an inappropriate extent, or in an inappropriate place. But it is not the only type of inappropriate postnatal development. There are many other disorders of growth and differentiation that may be less life-threatening but are of great biological interest.

Postnatal defects of growth and differentiation

Hyperplasias involve an excessive cell production but with fairly normal cell differentiation and turnover. Some hyperplasias represent normal physiological events, for example the growth of the mammary epithelium at puberty or in pregnancy. Others are pathological, for example a goiter is a hyperplasia of the thyroid due to a lack of iodine or other thyroid hormone insufficiency. An inadequate level of thyroid hormones causes production of more thyroid-stimulating hormone (TSH) by the pituitary and this causes increased growth of the thyroid.

Although not usually described as such, this is another **chalone**-like system. Where hyperplasia affects a tissue type that stops growing when the animal reaches final size, it can lead to a disproportionate increase in final size of that tissue. By contrast, in a renewal tissue hyperplasia may not increase size, but instead causes a faster flux of cells through the system, as for example in psoriasis, a skin disease in which there is a massive, continuous overproduction of keratinocytes associated with an elevated level of TGFα.

Metaplasia means the conversion of one tissue type into another. It occurs quite commonly, usually in association with tissue regeneration provoked by trauma or infection, and represents a switch in developmental commitment of a stem or progenitor cells from one tissue type to another. Many glandular epithelia can develop **squamous** metaplasia, which means that they acquire patches of tissue organized as a stratified squamous epithelium. For example squamous metaplasia of the bronchus is common in smokers. Patches of intestinal metaplasia can arise in the stomach, often in association with ulcers (Fig. 19.15). The cells of these patches express *Cdx2*, important in normal intestinal development, and overexpression of *Cdx2* in the stomach of developing mice can provoke the formation of patches of heterotopic intestine. Metaplasias can also arise in connective tissues. Patches of cartilage or bone are often formed in ectopic locations, for example in surgical scars or in traumatized muscle.

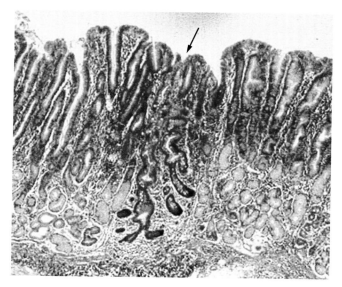

Fig. 19.15 Intestinal metaplasia in the stomach. The patch of metaplasia (arrowed) has arisen in a stomach of a patient suffering from an inflammatory condition called atrophic gastritis. Reproduced from Owen. *Mod Pathol* 2003; 16: 325–341, with permission from Nature Publishing Group.

They are thought to derive from fibroblasts that have been caused to enter the skeletal development pathway. **Transdifferentiation** is a term originally coined to describe direct conversions of one type of differentiated cell into another, thus being a subdivision of metaplasias. It has more recently been applied to transformations from one type of tissue-specific stem cell such that they form a different tissue type, for example the rare occurrence of donor-derived cells in nonhematopoietic tissues following a graft of hematopoietic stem cells to an irradiated host. Because of the controversy about whether such transformations really happen at all (see Chapter 18), the well-established examples of transdifferentiation in other areas of biology have tended to be unnecessarily discredited. For example Wolffian regeneration of the lens in some amphibians is a well-documented phenomenon that involves transformation of the pigmented cells of the iris to replace a lens that has been removed.

The terms **neoplasm**, meaning "new growth," and tumor, originally meaning "swelling," are more or less synonymous, as a tumor is any type of neoplastic growth. Tumors may be benign or malignant (Fig. 19.16). A benign tumor will be well differentiated, localized to the site of origin, and often surrounded by a fibrous capsule. A malignant tumor sends cells invading into the

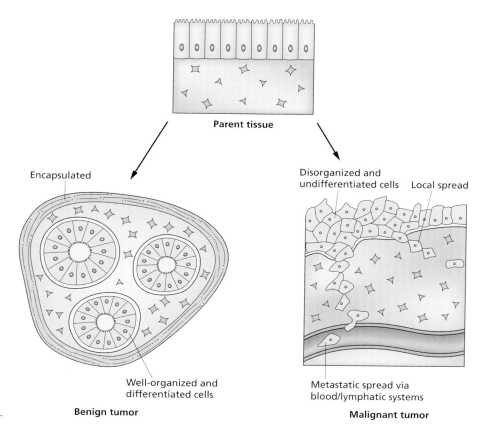

Fig. 19.16 Tumors: benign and malignant.

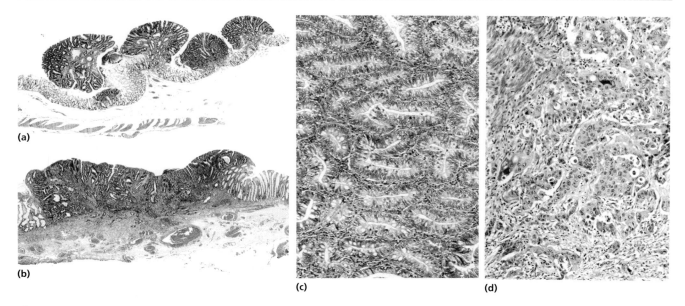

Fig. 19.17 Adenoma and carcinoma of the colon. (a) Adenomas in a patient with adenomatous polyposis coli. (b) A carcinoma, note the invasion of the tumor into underlying connective tissue. (c) A high-power view of a well-differentiated carcinoma of the colon. (d) A poorly differentiated carcinoma of the colon. Reproduced from *Histological Typing of Intestinal Tumours*, No.15 of International Histological Classification of Tumours. WHO 1976, with permission from WHO Press.

locality and, via the bloodstream or lymphatic system, to establish secondary tumors, or **metastases**, at distant sites in the body. It is also likely to be less well differentiated and faster growing. The spread to other parts of the body makes malignant tumors particularly difficult to treat by surgery. The term **cancer** tends to be reserved for malignant tumors, although benign tumors can also, on occasion, cause death.

The defects that cause all of these abnormalities of postnatal development arise from several causes: through somatic mutation, through inherited germ-line mutation, or through changes in gene expression by nonmutational means. For generation of cancer, somatic mutation is the most important, and the mutation frequency often accelerates as the cancer develops because the cells acquire mutations at an early stage that increase the likelihood of further mutations. For example these can include defects in the systems responsible for DNA repair, and defects promoting chromosomal breaks and rearrangements. Although most human cancer is due to random somatic mutation acquired throughout life, predispositions to various types of cancer exist where there are already **germ-line mutations** of the same target genes. For example if six mutations are normally needed for formation of a tumor, then an individual with one pre-existing germ-line mutation will only need to collect five more and this will represent a hereditary predisposition, increasing the lifetime risk of cancer.

Histological types of cancer

Most cancers in the human body arise ultimately from a single clone of cells that has acquired several somatic mutations in its growth and differentiation machinery. As it takes a long time to collect several relevant mutations in one clone, most cancers increase in frequency with age and are commonest in older people. In many cases there exist visible precursor lesions that have some altered growth control but are not in themselves dangerous. For example Fig. 19.17a shows adenomas in a patient suffering from adenomatous polyposis coli. This has a fairly normal tissue architecture, but the growth is abnormal enough to generate excess crypts and to produce an outgrowth. Other precursor lesions include some of the metaplasias, for example most lung cancer arises within areas of squamous metaplasia, and many gastric cancers arise within areas of intestinal metaplasia. A general term for regions of altered growth and differentiation is **dysplasia**. Some dysplasias are precursor lesions to cancer but others may regress spontaneously. An example is the dysplasia of the uterine cervix detected by cervical smear cytology tests. This is now usually called cervical intraepithelial neoplasia (CIN) and may, but does not necessarily, progress to cervical cancer.

Even malignant tumors usually retain a good deal of the differentiated histological characteristics of the parent cell, and the histological classification of tumors is based on this fact. This means that there should be at least as many kinds of cancer as there are cell types in the body. In fact, at a molecular level there are many more, because there are many ways in which each cell type can lose its normal growth control capability. Figure 19.17b and c show well-differentiated cancers of the colon. These retain the crypt structure characteristic of the intestine, but by contrast with the adenoma, the cells are abnormal, are not being removed at the normal rate, and the tumor is beginning to invade the underlying connective tissue. In Fig. 19.17d is shown another carcinoma of the colon, but this one has lost most visible differentiation and would be difficult to recognize as intestinal epithelium if it were not for its position.

Cancers of epithelial tissues are known as **carcinomas**. If they are glandular in histology they are called adenocarcinomas, and if they are squamous in histology they are called squamous cell carcinomas. Tumors of the connective tissues (in the broad sense) are called **sarcomas**. Thus, an osteosarcoma is a malignant tumor of bone, and a liposarcoma of adipose tissue. The corresponding benign tumors carry the suffix -oma, for instance an adenoma is a benign neoplasm of glandular epithelium, a papilloma of squamous epithelium, an osteoma of bone, or a lipoma of adipose tissue. If a tumor is so undifferentiated that its parent tissue cannot be recognized, it is described as anaplastic.

Leukemias are cancers of the hematopoietic tissue, which result in an overproduction of one or more types of blood cell. Acute myeloid leukemia is thought to arise from hematopoietic stem cells, and other types arise from one of the **transit amplifying cell** types that has acquired indefinite multiplication ability. The type of leukemia depends on exactly where in the pathway of hematopoietic differentiation the parent cell is located, so granulocytic leukemias are diseases of a **granulocyte** progenitor cell and lymphocytic leukemias of a **lymphocyte** progenitor cell. Lymphomas are solid overgrowths of lymphocytes, although this may also be accompanied by circulating malignant cells.

There are, in addition, various miscellaneous types of cancer. These include some that probably only need one or two mutations for their causation as they are most commonly found in children, for example neuroblastoma is a tumor of immature neuroblasts and Wilm's tumor an overgrowth of progenitor cells in the kidney. A tumor of particular interest to developmental biologists is the teratoma or **teratocarcinoma**, which arises from germ cells and closely resembles the embryonic stem cells of mice (see Chapter 10). Like ES cells, teratocarcinomas can form a wide variety of differentiated tissues and may contain a jumbled mixture of structures in addition to the stem cells themselves.

Cancer stem cells

Pathologists have long believed that some cancers have a stem cell type of organization, where the bulk of the tumor is fed from a small population of stem cells. More recently, this concept has become popular among experimental biologists. Cancer stem cells are usually defined as a subset of cells that can reconstitute the whole tumor on transplantation to an immunodeficient mouse. For example the subset of cells from human acute myelogenous leukemia (AML) that is CD34$^+$ and CD38$^-$ will reconstitute the tumor in an immunodeficient mouse while other cell fractions will not. CD34 is a sialomucin found on normal hematopoietic stem cells (HSCs), and CD38 is a glyco-

protein found on the surfaces of differentiated lymphocytes. Because these, and other characteristics of the cancer stem cells, match those of normal HSC, the HSC is thought to be the cell of origin for AML. Similarly, in certain human gliomas the subset of cells positive for CD133, a pentaspan cell surface glycoprotein also known as Prominin1, will initiate a tumor in immunodeficient mice while the CD133$^-$ cells will not. CD133/Prominin1 is a marker for neural stem cells, raising the possibility that they are the parent cells for the gliomas. This cell fraction will also initiate neurospheres in culture (see Chapter 14). However, it is known that neurospheres, and also gliomas, can arise from regions of the CNS not containing neural stem cells, so it is also possible that the tumor stem cells arise from more mature cells that have acquired their stem cell characteristics as a result of somatic mutations.

Rather few studies have attempted to trace the lineage of the stem cells through to the mature tumor, in the manner of developmental biology studies of tissue-specific stem cells *in situ*. One example of this is the ABCB5-positive subset of a human melanoma. ABCB5 is an ABC type transporter glycoprotein. As with the above examples, this cell fraction will initiate a new tumor in an immunodeficient mouse. Moreover, if the ABCB5$^+$ cells are genetically labeled by stable transfection of a marker gene, then they can be seen to overgrow the unlabeled cells and repopulate the tumor.

The significance of cancer stem cells for the design of therapy is that this is the cell population that needs to be eradicated. If the differentiated progeny are eradicated, the tumor will soon recur from surviving stem cells. Although some cancers do certainly have a stem-cell-like organization this is not true of all tumors. Sometimes there is no subfraction responsible for repopulation behavior. Also, it is not necessary that cancer stem cells must arise from tissue-specific stem cells. Cancers have always experienced many somatic mutations and it is perfectly possible for cancers to arise from transit amplifying cells or other cells capable of cell division. It is, however, very rare for cancers to arise from completely postmitotic cell types such as mature neurons or skeletal muscle fibers.

Cancer progression

Cancers are not just undifferentiated cell populations growing like micro-organisms. Although one tissue type, usually an epithelium, gives the tumor its name, tumors normally also contain a stroma, consisting of connective tissue, blood vessels, and immune cells. Interactions between epithelium and stroma are very important for the progression of the tumor. The fibroblasts of the stroma, which may also sometimes carry relevant somatic mutations, can secrete factors promoting growth, angiogenesis, inflammation, and invasive behavior. Vascularization is particularly important for tumor growth as this cannot occur without a sufficient supply of nutrients. Oxygen is also required, although

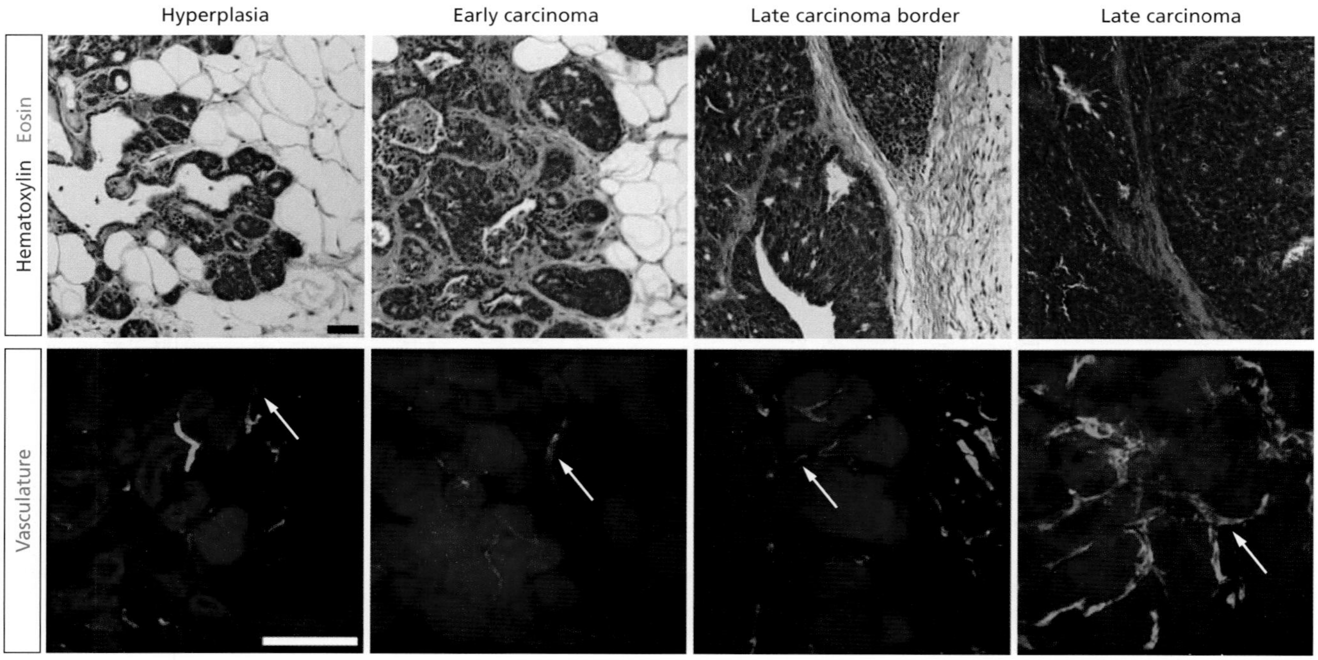

| Hyperplasia | Early carcinoma | Late carcinoma border | Late carcinoma |

Fig. 19.18 The top row shows the progression of a mouse mammary tumor with accompanying changes in the stroma. The bottom row shows development of the vasculature (arrows), which is labeled by intravenous injection of a fluorescent conjugated lectin. Scale bar 100 μm. Reproduced from Egeblad *et al. Dev Cell* 2010; 18: 884–901.

many tumors are somewhat hypoxic with a largely anaerobic metabolism.

The progression that tumors normally undergo is more like biological evolution than like normal development. This is because tumor progression arises from continual genetic change in the cells with selection of the fastest growing variants. Figure 19.18 shows the progression of an experimental mouse tumor: a mammary carcinoma induced by murine mammary tumor virus, which is a DNA virus that integrates at various oncogenic sites such as the *Wnt1* or *Fgf3* genes. Initially, the tumor preserves good histological structure, but this breaks down, and at the same time the stroma loses adipose tissue and acquires more fibroblasts and immune cells. This is accompanied by considerable proliferation of the vasculature.

Ultimately, malignant tumors progress to the point of invading other tissues in their vicinity, then dispatching cells to become secondary, or metastatic, tumors in the local lymph nodes and eventually all over the body. This involves a considerable number of processes, which include some familiar to developmental biologists. The early stages of invasion often bear hallmarks of the epithelial to mesenchymal transitions seen in development of the neural crest or sclerotome. As in development, this process is driven by transcription factors such as Snail, Slug, and Twist. These bring about a suppression of E-cadherin production, responsible for calcium-dependent adhesion and very important for the maintenance of epithelial structure. There is a corresponding increase of vimentin, the intermediate filament protein characteristic of mesenchymal cells. Metastatic tumors often express CXCR4, the receptor for the chemokine SDF1, and this is abundant in lung and bone marrow, which are common sites for metastases to lodge. The same SDF1-CXCR4 system is responsible for guiding cell migration in development, for example of germ cells to the gonads and of HSCs to the bone marrow.

In addition to having local destructive effects on their surroundings, tumors also secrete factors that have systemic effects. These include TNFα (tumor necrosis factor) and IL6 (an interleukin), which are among the causes of cachexia, a pathological loss of fat and skeletal muscle caused by the tumor and a common accompaniment of advanced cancer.

Molecular biology of cancer

The molecular biology of cancer has become quite well understood in recent years. However, it is not possible to understand the molecular pathology simply by identifying all the somatic mutations acquired by a tumor. This can be done with modern sequencing technology, indeed it is feasible to do a complete RNA-seq profile and ChIP-seq profiles as well. But even normal somatic cells contain 10^4–10^5 mutations of various kinds compared to the germ line. So in a tumor only a tiny handful of all the mutations found are really essential to the cancer phenotype. Epidemiological studies of the increase of cancer incidence with age suggest that something like five to seven rate-limiting steps need to be passed to transform a normal cell into a cancer.

Classic experiments

Molecular basis of cancer

There have been many discoveries building our picture of the molecular biology of cancer but these papers focus on two key genes with developmental relevance. In the early 1980s, it was found that DNA from human tumors could cause oncogenic transformation of tissue culture cells. When isolated, the genes responsible were found to be similar to the mysterious "oncogenes" already discovered in oncogenic retroviruses. The first three papers show that a single base change was all that distinguished an oncogene from the normal homolog. This oncogene was later found to encode the Ras protein in the key ERK signaling pathway.

Tabin, C.J., Bradley, S.M., Bargmann, C.I., Weinberg, R.A., Papageorge, A.G., Scolnick, E. M., Dhar, R., Lowy, D.R. & Chang, E.H. (1982) Mechanism of activation of a human oncogene. *Nature* **300**, 143–149.

Reddy, E.P., Reynolds, R.K., Santos, E. & Barbacid, M. (1982) A point mutation is responsible for the acquisition of transforming properties by the T24 human bladder-carcinoma oncogene. *Nature* **300**, 149–152.

Taparowsky, E., Suard, Y., Fasano, O., Shimizu, K., Goldfarb, M. & Wigler, M. (1982) Activation of the T24 bladder-carcinoma transforming gene is linked to a single amino-acid change. *Nature* **300**, 762–765.

The other two papers deal with a tumor suppressor gene, one that is normally required to suppress growth and so the loss-of-function mutant is oncogenic. The gene, *APC*, is defective in the human hereditary disease of adenosis polyposis coli. Individuals with this condition have one good copy of the gene, and whenever this is lost in one cell by somatic mutation an adenoma is initiated. These papers identified the gene in mice and showed that loss of the good copy was an early step in development of cancer. APC is now known to be a component of the Wnt pathway, needed for targeting β-catenin for degradation.

Su, L.K., Kinzler, K.W., Vogelstein, B., Preisinger, A.C., Moser, A.R., Luongo, C., Gould, K. A. & Dove, W.F. (1992) Multiple intestinal neoplasia caused by a mutation in the murine homolog of the *APC* gene. *Science* **256**, 668–670.

Powell, S.M., Zilz, N., Beazerbarclay, Y., Bryan, T.M., Hamilton, S.R., Thibodeau, S.N., Vogelstein, B. & Kinzler, K.W. (1992) *APC* mutations occur early during colorectal tumorigenesis. *Nature* **359**, 235–237.

This corresponds quite well with the number of cell biology systems that are usually deranged in cancer, although the formation of a precursor lesion, like a metaplasia, will require fewer changes than this. Most of the mutations arise from random processes such as DNA base mispairing arising from tautomeric shifts, or DNA damage from background radiation or from oxidative damage arising from the products of metabolism. The incidence of cancer can be accelerated by exposure to chemical carcinogens, such as those found in cigarette smoke or in food contaminated by fungi. Ultraviolet light from the sun is also a significant cause of human skin cancers. A minority of cancers are caused by viruses, either because they bring an oncogene (see below) into the cells, or because they activate an endogenous oncogene, or simply because they increase tissue damage and the consequent regeneration, and hence chances for establishment of mutations.

The first type of defect needed to establish a cancer is independence from growth factors for continued division. The mutations bringing this about are those that generate **oncogenes**, which are genes that can cause changes in tissue culture cells that resemble cancerous behavior, such as loss of contact inhibition. Oncogene products are often elements of the signaling systems that control growth and other aspects of development, either the growth factors themselves, or growth factor receptors, or components of the signal transduction pathways, or transcription factors activated by these pathways. Gain-of-function mutations may create constitutively active forms of the proteins, for example human tumors often contain constitutive mutations of the GTP exchange protein Ras, a component of the ERK pathway activated by receptor tyrosine kinases such as EGFR and FGFR. Oncogenic mutations may also act by causing an inappropriate expression of otherwise normal gene products. For example many tumors secrete growth factors, and these may be factors normally supplied by neighboring tissues in order to create a stem cell niche. When the tumor itself makes the factors, or activates the pathway independent of the presence of the factors, then its growth will no longer be controlled by the neighboring tissue and will become autonomous. Another very important oncogene is *Myc*, a transcription factor which upregulates a great many genes by recruitment of histone acetyl transferases. Among its numerous functions is a stimulation of cell division, and it is often found as a gain-of-function mutation in human cancers.

The next change is the insensitivity to growth inhibitors. The genes involved here are known as **tumor suppressor genes**, and normally code for inhibitors of growth. When they are inactivated by mutation the normal inhibition is lifted and this can contribute to the development of a cancer. An example is the product of the *Retinoblastoma (Rb)* gene. This is an inhibitor of the cell cycle, normally upregulated by factors of the TGFβ family. If both copies of the *Rb* gene are lost from a retinal cell then it will develop into a retinoblastoma. Under normal circumstances this is an exceedingly rare event since it requires an independent loss-of-function mutation of both copies of the gene in the same cell. However, there is a hereditary form of retinoblastoma in which affected individuals are highly susceptible to development of the disease and often suffer from multiple foci forming in both eyes. This is because they are heterozygotes who have already inherited one defective copy of the gene through the germ line. In such individuals the disease will be initiated in any retinal cell that suffers a single somatic mutation inactivating the remaining good copy, and since there are tens of thousands of retinal cells it highly probable that a few cells will experience this mutation. Retinoblastoma is unusual because the loss of function of a single gene is enough to initiate formation of the tumor, but it exemplifies very clearly the nature of tumor suppressor genes. There are many other tumor suppressor genes. These include the *p16INK4a/ARF* locus, which is described above.

The third change is the ability to avoid apoptosis. This often depends on a loss of function of the gene coding for p53, a protein that detects DNA damage and other stress conditions normally leading to cell death. p53 has numerous functions but essentially it will promote DNA repair, arrest the cell cycle in order to facilitate DNA repair, or, if damage is severe, initiate apoptosis. If control mediated by p53 is removed then many cells that have acquired defects of various kinds will fail to die and will instead survive and proliferate. The mutations in *p53* are often dominant negative, so just a single dominant mutation is enough to reduce cell death substantially.

The fourth change is the ability to proliferate without limit. This is due to an upregulation of **telomerase** to evade the consequences of chromosome shortening. Telomerase activity in normal cells is at a very low level and telomere shortening is one of the factors that leads to stem cell depletion. But telomerase expression is upregulated in a high proportion of human cancers.

The next element that is essential for the growth of a tumor is the ability to promote **angiogenesis**. Unless the tumor can attract the inward growth of capillaries from the surroundings it cannot obtain nutrients and cannot grow. Many tumors secrete angiogenesis factors such as VEGF or FGFs and are thereby able to attract their own abundant blood supply.

Finally, there are the changes required for invasion and metastasis. These include the downregulation of various cell adhesion molecules like E-cadherin and the secretion of proteases that degrade the surrounding extracellular matrix and enable cells to move more freely.

Colon cancer

An example where the mutational sequence is reasonably well known is the development of carcinoma of the colon (Fig. 19.19). The histological structure of the colon is similar to the small intestine described in Chapter 13, but without the villi or the Paneth cells. The sequence described is a common pathway, but is not the only possible one.

The first step in development of cancer is the loss of the tumor suppressor gene *APC* (adenomatous polyposis coli), which leads to the formation of a small adenoma. *APC* encodes a cytoplasmic protein that assembles the complex responsible for the phosphorylation of β-catenin by GSK3. Since phosphorylation leads to degradation of β-catenin, APC is an inhibitor of Wnt signaling. As discussed in Chapter 18, the Wnt pathway

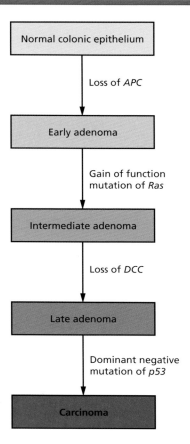

Fig. 19.19 One sequence of mutational changes involved in the formation of a carcinoma of the colon.

is normally involved in controlling the size of the proliferative zone in the intestinal crypts. Individuals suffering from the hereditary disease familial adenomatous polyposis coli are heterozygous for loss-of-function mutations of *APC*. They suffer multiple colonic adenomas due to repeated somatic loss of the good allele, and have a very high lifetime risk of cancer.

The next step is the acquisition of a constitutive mutation of *RAS*, involved in the ERK signal transduction pathway. This increases the size of the adenoma. Then there is the loss of the netrin receptor called *DCC* (deleted in colon carcinoma), which leads to a large adenoma. Finally, there is a dominant negative mutation of *p53*, which increases net growth by the suppression of apoptosis in the tumor. Having collected all these changes, the tumor finally becomes invasive and hence malignant.

This sequence of events as described is not complete, and not all colon carcinomas show all four mutations, or all in the same order. However, it does represent a common mutational pathway for generating this type of cancer, and this example illustrates that multiple developmental events involving growth, differentiation, and cell adhesion need to be perturbed in order for a cancer to develop.

Cancer therapy

Cancer therapy has long depended on surgery to remove the offending tumor. This can be successful so long as no vital structures are destroyed and the tumor has not yet metastasized. Benign tumors, that are still confined within a capsule, are often curable by surgery alone. But most cancer patients are not operated on until local spread or metastasis has already occurred. This is why most also receive radiotherapy or chemotherapy intended to destroy rapidly dividing cells. Because tumors can regrow from small numbers of cells, these methods need to destroy all of the cells, or at least all of the stem cells, in the tumor. Because normal dividing cells are also attacked, these treatment methods have severe side effects and lead to much morbidity. Most sensitive are the normal renewal tissues, so bone marrow failure, hair loss, damage to the gut, and sterility are common complications of radio- or chemotherapy.

Immunotherapy has become important in the treatment of leukemia and lymphoma in the form of bone marrow transplants, which contain and generate T lymphocytes that mount an immune attack on the cancer cells. This is a form of stem cell therapy and will be described in Chapter 21.

The vast amount of research on the molecular biology of cancer has produced some new drugs that attempt to target the various processes necessary for cancer development. Among these are kinase inhibitors that reduce the activity of pathways such as the ERK pathway and antiangiogensis agents that inhibit VEGF signaling. The clinical impact of these is so far only modest but it does indicate that rational therapies can now be introduced for a group of diseases whose biology was completely mysterious only a few decades ago. Since many of the molecular processes underlying cancer are also developmental processes, developmental biology can claim some share of the credit for this.

New directions for research

This chapter covers a vast area of modern biological research and touches on many priorities outside developmental biology. But some specifically developmental problems are as follows:

1 Where no known systemic factor exists, how do cells know how fast the rest of their organ, or the rest of the body, is growing?

2 To what extent do telomerase shortening and CDK inhibitor accumulation contribute to organismal aging, given that stem cells can survive more than one lifetime after transplantation

3 To what extent are the properties of precancerous lesions (hyperplasias, dysplasias, metaplasias) explicable on the basis of change of activity of small numbers of developmental control genes?

Key points to remember

- The control of growth requires both a control of the final size and a control of the relative proportions of different body parts.
- Systemic growth is controlled to a large extent by the insulin/IGF signaling pathway. Loss-of-function mutations will reduce growth and gain-of-function procedures will increase growth.
- TOR (target of rapamycin) integrates insulin signaling with the availability of nutrients and energy.
- The Hippo pathway exerts a negative control on growth and can integrate growth rate with information from neighboring cells.
- In postnatal life of mammals growth hormone from the pituitary controls synthesis of IGF1, and loss of any component of this system can lead to dwarfism.
- There is some evidence for chalone-like feedback systems controlling the growth of individual tissues, especially muscle and liver.
- The growth of stature in vertebrates depends mainly on the elongation of long bones. This depends on the activity of the terminal cartilage growth plates. Their growth is under negative control from FGFs and their differentiation from Indian hedgehog.

- Senescence of cells in culture depends on erosion of telomeres, and on the accumulation of p16^{INK4a} and ARF. These processes also occur in whole animals.
- The insulin pathway affects lifespan, by negative regulation of genes for proteins that reduce oxidative stress, and partly explains the life-extending properties of caloric restriction.
- There is a variety of pathological abnormalities of normal growth and differentiation, classified as hyperplasias, dysplasias, metaplasias, and neoplasias.
- The distinction between benign and malignant tumors mostly depends on their propensity to invade surrounding tissues, and establish metastases elsewhere in the body.
- Cancers retain many of the properties of their parent tissue, but change properties as the disease progresses.
- Development of a cancer normally requires several mutational changes to occur in a clone derived from a single cell. Key steps are gain-of-function mutations of oncogenes, loss of tumor suppressor genes, evasion of apoptosis, upregulation of telomerase, promotion of angiogenesis, and changes leading to metastatic spread.

Further reading

Growth

Robertson, E.J. (1995) Insulin-like growth factors, imprinting and epigenetic growth control. *Seminars in Developmental Biology* **6**, 293–299.

Lee, S.J. & McPherron, A.C. (1999) Myostatin and the control of skeletal muscle mass. *Current Opinion in Genetics and Development* **9**, 604–607.

Day, S.J. & Lawrence, P.A. (2000) Measuring dimensions: the regulation of size and shape. *Development* **127**, 2977–2987.

Lupu, F., Terwilliger, J.D., Lee, K., Segre, G.V. & Efstratiadis, A. (2001) Roles of growth hormone and insulin-like growth factor 1 in mouse postnatal growth. *Developmental Biology* **229**, 141–162.

Kozma, S.C. & Thomas, G. (2003) Regulation of cell size in growth, development and human disease: PI3K, PKB and S6K. *Bioessays* **24**, 65–71.

Nijhout, H.F. (2003) The control of body size in insects. *Developmental Biology* **261**, 1–9.

Hay, N. & Sonenberg, N. (2004) Upstream and downstream of mTOR. *Genes and Development* **18**, 1926–1945.

Lecuit, T. & Le Goff, L. (2007) Orchestrating size and shape during morphogenesis. *Nature* **450**, 189–192.

Mirth, C.K. & Riddiford, L.M. (2007) Size assessment and growth control: how adult size is determined in insects. *BioEssays* **29**, 344–355.

Shingleton, A.W., Frankino, W.A., Flatt, T., Nijhout, H.F. & Emlen, D.J. (2007) Size and shape: the developmental regulation of static allometry in insects. *BioEssays* **29**, 536–548.

Reddy, B.V.V.G. & Irvine, K.D. (2008) The Fat and Warts signaling pathways: new insights into their regulation, mechanism and conservation. *Development* **135**, 2827–2838.

Halder, G. & Johnson, R.L. (2011) Hippo signaling: growth control and beyond. *Development* **138**, 9–22.

Skeletal development

Kronenberg, H.M. (2003) Developmental regulation of the growth plate. *Nature* **423**, 332–336.

Zelzer, E. & Olsen, B.R. (2003) The genetic basis for skeletal diseases. *Nature* **423**, 343–348.

Hartmann, C. (2009) Transcriptional networks controlling skeletal development. *Differentiation and Gene Regulation* **19**, 437–443.

Karsenty, G., Kronenberg, H.M. & Settembre, C. (2009) Genetic control of bone formation. *Annual Review of Cell and Developmental Biology* **25**, 629–648.

Long, F. (2012) Building strong bones: molecular regulation of the osteoblast lineage. *Nature Reviews Molecular Cell Biology* **13**, 27–38.

Ageing

Kenyon, C. (2001) A conserved regulatory system for aging. *Cell* **105**, 165–168.

Carter, C.S., Ramsey, M.M. & Sonntag, W.E. (2002) A critical analysis of the role of growth hormone and IGF-1 in aging and lifespan. *Trends in Genetics* **18**, 295–301.

Helfand, S.L. & Rogina, B. (2003) Genetics of aging in the fruit fly *Drosophila melanogaster. Annual Reviews of Genetics* **37**, 329–348.

Blasco, M.A. (2005) Telomeres and human disease: ageing, cancer and beyond. *Nature Reviews Genetics* **6**, 611–622.

Bordone, L. & Guarente, L. (2005) Calorie restriction, Sirt1 and metabolism: understanding longevity. *Nature Reviews Molecular Cell Biology* **6**, 298–305.

Kim, W.Y. & Sharpless, N.E. (2006) The regulation of INK4/ARF in cancer and aging. *Cell* **127**, 265–275.

Collado, M., Blasco, M.A. & Serrano, M. (2007) Cellular senescence in cancer and aging. *Cell* **130**, 223–233.

Kenyon, C.J. (2010) The genetics of ageing. *Nature* **464**, 504–512.

Houtkooper, R.H., Pirinen, E. & Auwerx, J. (2012) Sirtuins as regulators of metabolism and healthspan. *Nature Reviews Molecular Cell Biology* **13**, 225–238.

Cancer

Vogelstein, B. & Kinzler, K.W. (1993) The multistep nature of cancer. *Trends in Genetics* **9**, 138–142.

Kinzler, K.W. & Vogelstein, B. (1996) Lessons from hereditary colorectal cancer. *Cell* **87**, 159–170.

Garcia, S.B., Novelli, M. & Wright, N.A. (2000) The clonal origin and clonal evolution of epithelial tumours. *International Journal of Experimental Pathology* **81**, 89–116.

Hanahan, D. & Weinberg, R.A. (2000) The hallmarks of cancer. *Cell* **100**, 57–70.

Frank, S.A. & Nowak, M.A. (2004) Problems of somatic mutation and cancer. *Bioessays* **26**, 291–299.

Gupta, G.P. & Massague, J. (2006) Cancer metastasis: building a framework. *Cell* **127**, 679–695.

Slack, J.M.W. (2007) Metaplasia and transdifferentiation: from pure biology to the clinic. *Nature Reviews Molecular Cell Biology* **8**, 369–378.

Schatton, T., Frank, N.Y. & Frank, M.H. (2009) Identification and targeting of cancer stem cells. *BioEssays* **31**, 1038–1049.

Egeblad, M., Nakasone, E.S. & Werb, Z. (2010) Tumors as organs: complex tissues that interface with the entire organism. *Developmental Cell* **18**, 884–901.

Hanahan, D. & Weinberg, R.A. (2011) Hallmarks of cancer: the next generation. *Cell* **144**, 646–674.

Rubin, H. (2011) Fields and field cancerization: The preneoplastic origins of cancer. *BioEssays* **33**, 224–231.

Greaves, M. & Maley, C.C. (2012) Clonal evolution in cancer. *Nature* **481**, 306–313.

This chapter contains the following animation:

Animation 22   Chalone action.

For additional resources for this book visit www.essentialdevelopmentalbiology.com

Chapter 20

Regeneration of missing parts

Distribution of regenerative capacity

The ability to regrow missing parts is something that has fascinated humanity for millennia. But only when the molecular basis is understood will it be possible to know whether we humans could ever regenerate missing limbs, or defective kidneys, or new heart muscle. Like many of the other topics in this section, regeneration is at the frontier of developmental biology and still involves some fundamental unsolved problems.

The most dramatic type of regeneration is the ability to regrow the main body **axis** following transection of the whole body. **Hydroids**, including the familiar freshwater *Hydra*, can often regenerate new hydranths from a distal facing surface and new basal parts from a proximal facing surface. Nemertean worms (ribbon worms) can also regenerate whole bodies from small tissue fragments. Although often supposed to have this ability, annelids (earthworms and bristle worms) show it to a much lesser degree. In annelids, tail regeneration is common, but head regeneration is less common, and a high level of regeneration of both heads and tails is confined to a few polychaete species. The "highest" animal in evolutionary terms known to be able to regenerate the head is the hemichordate *Ptychodera flava*, a marine acorn worm. This is of interest insofar as the hemichordates are a **deuterostome** phylum, much more closely related to the vertebrates than are the other types of head-regenerating worm. In terms of experimental study, whole-body regeneration has been examined in most detail using the **planarian** worms, which can often regrow a new head from an anterior-facing cut surface and a new tail from a posterior-facing cut surface (see below).

Whole-body regeneration is usually associated with the type of asexual reproduction in which the body can spontaneously fragment into two parts, which each develop to reconstitute a new individual. The requirements for asexual reproduction by fission and for whole-body regeneration are obviously very similar. Whole-body regeneration is also associated with a continuous turnover of cells in which there is a flux from undifferentiated stem cells to the principal body parts. This is found in both planarians and hydroids. An essential feature of whole-body regeneration is that it is bidirectional. The same cut surface can regenerate either a head or a tail, and normally what it forms depends on its spatial relationship to the underlying tissues. This means that there is an essential polarity control that makes a decision about whether the regenerate is to be a head or a tail, and does so before the process of new part formation itself begins.

In terms of organisms familiar to developmental biologists, *Caenorhabditis elegans* cannot regenerate at all. Neither vertebrates nor insects, which are the animals most studied by developmental biologists, show any bidirectional regeneration at all, and regeneration of the whole body axis is confined to the tail of some vertebrates. They do, however, show some monodirectional (i.e. distal) regeneration of appendages, an ability that is not linked to asexual reproduction in the same way as whole-body regeneration. Regeneration of external appendages in insects is confined to the **Hemimetabola**, those families which show a gradual progression to adulthood through a series of larval forms. It is not found in the **Holometabola**, which are those, like *Drosophila*, showing an abrupt metamorphosis in a pupal phase. Accordingly, *Drosophila* does not show any regeneration of adult structures, although the imaginal discs can regenerate if they are damaged prior to metamorphosis (see Chapter 17). Regeneration in vertebrates is displayed mainly by the amphibians. Urodeles (newts and salamanders) are usually able to regenerate limbs, tails, and jaws, both before and after

Essential Developmental Biology, Third Edition. Jonathan M.W. Slack.
© 2013 John Wiley & Sons, Ltd. Published 2013 by John Wiley & Sons, Ltd.

metamorphosis. Anurans (frogs and toads) can often do so in the tadpole stage but lose the ability at metamorphosis. Lizards are well known for their ability to regenerate tails but the new tail does not include all of the tissues and structures of the original one and is a somewhat defective copy.

Among mammals, the only regeneration of significant external appendages is the regrowth of antlers in deer. However, there are some interesting phenomena that are usually called regeneration but are more properly referred to as hyperplastic growth. One of these is liver regeneration, mentioned in Chapter 19. A somewhat similar hyperplasia is shown by the remaining kidney following removal of the other kidney. This involves a rapid increase in the cell number but, except in very young animals, no increase in the number of nephrons.

This quick survey of regeneration reveals that one of the problems for research is that extensive regenerative abilities are found mainly in animals that are not the familiar laboratory model species for developmental biology. This means that there are not pre-existing collections of cloned genes, or methods for introducing or removing genes, or methods for cell labeling or transplantation. New experimental approaches have needed to be introduced for individual experiments and this has been a factor reducing the rate of progress. However, important results have been obtained in regeneration research in recent years and the intrinsic interest of the phenomena ensures that they will stay high on the agenda.

Regeneration: a pristine or adaptive property?

Another important aspect of the distribution of regenerative capacity is that it often varies considerably between otherwise quite similar species. For example one species of planarian will regenerate heads, while another will not; or one species of salamander will show good limb regeneration for the whole life cycle, while another will only regenerate prior to metamorpho-

sis. Such variations have led regeneration specialists to conclude that the ability to regenerate is probably an intrinsic, or pristine, biological property that is easily lost in evolution, presumably because it is associated with some significant cost in terms of overall reproductive fitness. The alternative view, that each example of regenerative behavior is a specific, newly acquired, adaptation to a specific life history, is not favored, although is hard to disprove altogether. Regenerative mechanisms often recapitulate aspects of the normal development, consistent with the notion of a pristine potential that needs the right conditions to be expressed. But if some regenerative processes are specific adaptations then their mechanism need not necessarily correspond to normal development at all.

Regeneration necessarily involves regional specification of pattern and as the relevant interactions are local ones this means that cryptic information about pattern must be present throughout the mature structure. This is sometimes called **positional information**. The term was once widely used to indicate developmental pattern information but has fallen out of use as understanding of its molecular basis has increased. However, it is still frequently encountered in the regeneration literature.

Planarian regeneration

Planarians are free-living freshwater worms belonging to the phylum Platyhelminthes. They are the simplest animals showing bilateral symmetry. They have three germ layers: ectoderm, mesoderm, and endoderm, but no coelom and no segmentation. They have a head with eyes, but they do not have any circulatory or respiratory system. The gut opens at a muscular pharynx in the central part of the body and is a bilobed, blind-ended sac, with no anus.

Although simple, planarians do display the basic features of animals, with Hox genes expressed in a nested way from posterior to anterior in regenerating animals, an *Otx* homolog

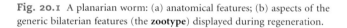

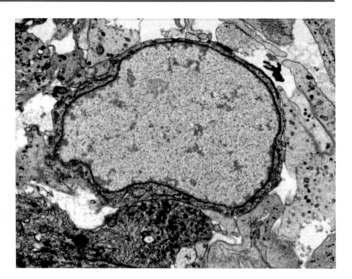

Fig. 20.2 A planarian neoblast viewed with the transmission electron microscope. The nucleus and cytoplasm are pseudocolored in light red and blue, respectively. Reproduced from Reddien and Alvarado. *Ann Rev Cell Dev Biol* 2004; 20: 725–757 with permission from Annual Reviews.

Fig. 20.1 A planarian worm: (a) anatomical features; (b) aspects of the generic bilaterian features (the **zootype**) displayed during regeneration.

expressed in the regenerating head, and a *Pax6* homolog expressed in the eyes (Fig. 20.1). Like other **protostome** invertebrates, planarians show expression of *Bmp* homologs on the dorsal side and the principal nerve cord is located ventrally. A number of different species of planarian have been used for regeneration work, although *Schmidtea mediterranea* has gradually become favored and now has a fully sequenced genome. Planaria are not suitable for genetic experiments involving sexual crosses but the **RNAi** method for ablation of specific messenger RNA (see Chapter 3) works well. The required dsRNA is encoded in a plasmid with two T7 promoters facing each other. Bacteria containing this plasmid, and also the phage T7 polymerase, will transcribe the sequence in both directions and the two complementary RNA strands hybridize to form double-stranded RNA. The bacteria can be mixed with the food fed to the planarians and when the worms eat it some of the dsRNA will be absorbed into their cells. Extensive screens have been carried out with this method and have led to the identification of many genes needed for regeneration (see Chapter 3, Fig. 3.12c).

Cell turnover and neoblasts

Planarians are in a state of continuous cell turnover and can grow or shrink on a daily basis depending on the food supply. There is a population of cells called **neoblasts** that are small with large nuclei (Fig. 20.2) and characterized by expression of an ATP-dependent RNA helicase similar to the *Drosophila* Vasa

protein. In other animals Vasa homologs are generally associated with germ cell development and therefore with developmental **pluripotency**. Neoblasts can be separated from the other cell types by **FACS** on the basis of size and granularity. They are the only cells that normally divide, and they are considered to comprise, or to include, the **stem cells** that produce the 12–15 histologically distinguishable cell types making up the tissues of the worm. In the steady state, a worm contains about 20% neoblasts and 80% differentiated cells.

Because of the continued cell turnover, the regional pattern of the entire animal can sometimes be modified by RNAi treatment. For example RNAi treatment directed against elements of the BMP signaling pathway will induce the expression of ventral markers on the dorsal side of intact worms, including occasional duplications of the nerve cord, which is normally a ventral structure. Likewise, RNAi directed against elements of the Wnt/β-catenin pathway, normally associated with the posterior, can lead to formation of anterior structures in the tail of intact worms.

Regeneration

Planarians have long been known for their high regenerative capacity (see Animation 23: Planarian regeneration). If cut in half, the anterior half will regenerate a new tail region from its posterior cut surface, while the posterior half will regenerate a new head region from its anterior cut surface (Fig. 20.3). Following transection, there is a muscular contraction limiting the area of the cut, and a rapid formation of a thin wound epithelium across the cut surface. There is an accumulation of undifferentiated cells under the wound epithelium, which makes up a regeneration blastema. **Blastema** is a general term for a regeneration bud containing undifferentiated proliferating cells, although in planarians the term confusingly refers to an unpigmented distal region in which there is no cell division. The cells populating the planarian blastema come from the dividing cells in the proximal region. The blastema enlarges by cell recruitment and redifferentiates over a few days to form the missing structures.

The evidence that neoblasts give rise to the regenerate comes from two sources. Firstly, BrdU labeling of neoblasts before amputation leads to a blastema containing many BrdU-labeled cells. Secondly, the ability of worms to regenerate is destroyed by X-irradiation, which is followed by a rapid disappearance of neoblasts. There is no further production of new cells and the worms will die a few weeks later. If the radiation is not quite lethal, then a few neoblasts survive. They can found colonies, similar in principle to the colonies founded by stem cells of the mammalian intestine or testis following high doses of radiation. These colonies can repopulate all tissues of the worm, indicating that at least some planarian neoblasts are fully pluripotent, in the manner of mammalian embryonic stem cells. It has also proved possible to isolate single neoblasts, by micromanipulation after FACS purification, then inject them into irradiated host worms of a different genetic strain. Although few such grafts survive, occasionally the host becomes repopulated by the descendants of the grafted cell and can establish a new asexual colony of worms. This is a formal proof of the existence of some genuinely pluripotent stem cells among the neoblast population.

Head–tail polarity

The first decision that the cells of the blastema have to make is that of **polarity**: whether to be a head or a tail blastema. The critical controlling element for this is the Wnt-β–catenin pathway. Following amputation, Wnt-type ligands are preferentially expressed at the posterior surface. RNAi directed against components of this pathway will cause a posterior-facing amputation surface to generate a new head rather than a tail (Fig. 20.4). The normal polarity control may depend on an asymmetry of healing at anterior and posterior surfaces. The wound epithelium is derived from dorsal epidermis in an anterior blastema and from ventral epidermis in a posterior blastema. Certainly, a surgically created dorsal–ventral confrontation is sufficient to initiate the formation of a new blastema. All of the Hox genes become upregulated in the first hour of regeneration in blastemas of both types, so they are unlikely to determine the polarity decision.

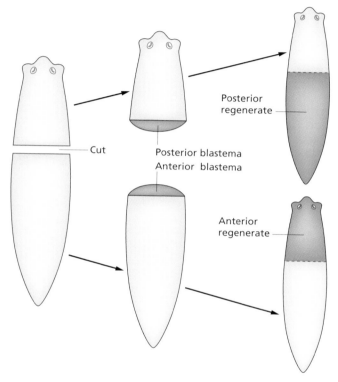

Fig. 20.3 Planarian anterior and posterior regeneration.

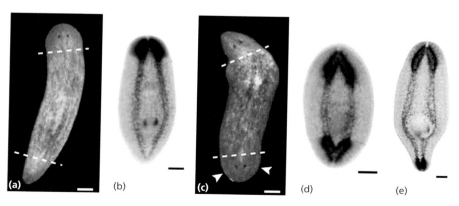

Fig. 20.4 Polarity control by β-catenin. (a) Normal regeneration of head and tail. (b) Normal central nervous system, visualized by *in situ* hybridization for the gene encoding prohormone convertase 2. (c,d) Regeneration of double head after treatment with dsRNA against β-catenin. Arrow heads indicate regenerated photoreceptors. (e) Reprogramming of the tail towards head in a nonregenerating worm by dsRNA against β-catenin. Dashed lines indicate amputation levels. Scale bars 200 μm. Reproduced from Gurley *et al. Science* 2008; 319: 323–327, with permission from American Association for the Advancement of Science.

Errors can arise in polarity control. In several species of planarian it is quite common for short segments from the central part of the worm to form anterior blastemas at both ends and regenerate into a Janus-headed **bipolar form** (Fig. 20.5), and this correlates with activation of β-catenin at both surfaces.

Although events in different regions of a worm can occur with some autonomy, there is also evidence for some signaling processes associated with regeneration (Fig. 20.6). If a second head is grafted close to the first head, and then the first head is removed, its regeneration is now suppressed because of the presence of the second. If a head is grafted into the posterior, postpharyngeal region, then a pair of new pharynxes can regenerate from the junction. The nature of the signals involved in these interactions is presently unknown.

Insect limb regeneration

After metamorphosis, *Drosophila* and other holometabolous insects are mostly postmitotic, so there is no regeneration. But appendages do regenerate in hemimetabolous insects such as locusts or cockroaches. Members of this group hatch from the egg as nymphs and then develop gradually to the adult via a succession of molts. If a leg or antenna is removed, it will grow again and will become visible at the next molt when the old cuticle is shed. Often several cycles of growth and molting are required for complete regeneration.

Leg regeneration has recently been studied in some detail in the cricket, *Gryllus bimaculatus*. The legs consist of a single layer epidermis with muscles and tracheae (air tubes) internally. The overall shape of the leg is asymmetrical in three dimensions, as is the distribution on its surface of various appendages such as spikes and hairs (Fig. 20.7a). There are some visible differences between the pairs of legs on the first, second, and third thoracic segments. Following amputation in the mid region of the tibial segment at the early third instar, complete regeneration takes place over 35–40 days, or four to six molts (Fig. 20.7b). Immediately following amputation, the hemocytes of the blood form a scab. Epidermis then grows over the wound, underneath the scab, by about 2 days. The wound epidermis then forms a blastema and grows to form distal epidermal parts. Although there has been no study of cell lineage it is probable that the muscle and tracheae regenerate from their own respective tissues remaining in the stump.

Experiments over many years on the cricket and other hemimetabolous insects have shown certain interesting properties of leg regeneration in terms of regional specification. First, if there is a gap in the pattern, as for example produced by grafting a distal part to a more proximal stump, this becomes filled in, by a process known as **intercalary regeneration**. In the cricket, the new tissue comes from the side of the junction that was more distal in character. This can be determined because grafts can be made between legs from different thoracic segments, which

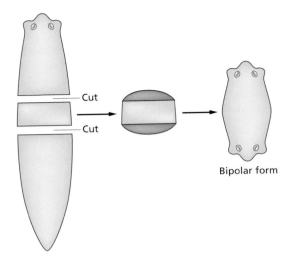

Fig. 20.5 Regeneration of a double anterior bipolar form from a short body section.

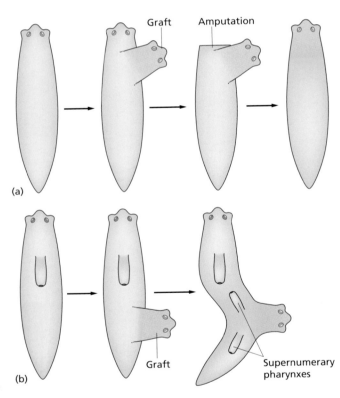

Fig. 20.6 Evidence for long-range interactions in planarian regeneration. (a) Suppression of head regeneration by grafted head; (b) induction of extra pharynxes by a grafted head.

carry characteristic specializations such as spikes and hairs, and then the nature of such features in the regenerate indicates to a first approximation from which side the regenerated tissue originated. Intercalary regeneration also occurs if different regions are apposed from around the circumference of a limb.

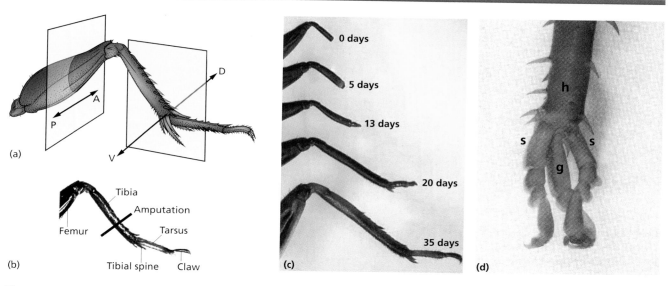

Fig. 20.7 Regeneration of the insect leg. (a) Structure of the metathoracic leg of the cricket, showing dorsal–ventral (D–V) and anterior–posterior (A–P) anatomical axes. (b,c) Regeneration of the cricket leg following a mid-tibial amputation. (d) Paired supernumerary limbs resulting from an axial inversion experiment on the cockroach leg. h, host; g, graft; s, supernumerary. Reproduced from Mito *et al. Mech Dev* 2002; 114: 27–35, with permission from Elsevier.

This is most dramatically illustrated by a famous type of experiment in which a left leg is grafted to a right stump (or vice versa) with one transverse axis aligned. For example if the dorsoventral axis is aligned, it follows that the anteroposterior axis of graft and host will be opposed in a graft from left to right limbs. This leads to the formation of complete supernumerary limbs growing from the points of maximum pattern discontinuity. After such grafts the epidermis of graft and host join up after 3 days. After the first molt (9 days), two blastemas appear, and after the second molt (14 days), the two supernumerary limbs are clearly visible (Fig. 20.7d).

Molecular basis of supernumerary limb generation

The molecular understanding of these remarkable phenomena arises from two sources: the understanding of normal *Drosophila* leg disc development, and the use of RNAi to ablate gene activity in the cricket. In *Drosophila*, the posterior compartment of each leg disc is defined by the presence of the transcription factor Engrailed. The posterior compartment secretes Hedgehog and this induces expression of the gene encoding Decapentaplegic (*Dpp*) in the dorsal part of the anterior compartment, and of Wingless (*Wg*) in the ventral part of the anterior compartment (see Chapter 17, Fig. 17.15). The region where maximal levels of Dpp and Wg proteins combine shows upregulation of *vein*, which encodes a ligand for the EGF receptor (EGFR) and promotes distal outgrowth.

The cricket does not have imaginal discs. Instead there are limb buds formed in the embryo. But the gene expression patterns in the buds indicate that the mechanism for regional specification is identical to the *Drosophila* discs. In developing larval limbs these domains continue to exist although at a lower level of gene expression. Following amputation, all of the key genes are upregulated in the blastema in essentially the same domains. So it can be predicted that normal regeneration occurs because the Hedgehog from the posterior compartment re-induces expression of Dpp and Wg from the adjacent anterior tissue, and that the point of maximum Dpp + Wg induces expression of an EGFR ligand and grows out as the distal tip. In the anteroposterior inversion experiment, it may be predicted that these events will happen twice, as each side of the junction apposes an Engrailed-positive posterior compartment to anterior tissue. Hence two supernumeraries are initiated, each with opposite polarity to the host limb (Fig. 20.8).

These predictions have been tested by knocking down expression of individual genes using RNAi. In the cricket, dsRNA is injected into the abdomen of the nymph. It may be shown by *in situ* hybridization that the target RNA level is decreased, and that the effect is gone after about three molts. Administration of dsRNA against *hedgehog*, *β-catenin*, or *Egfr* will inhibit regeneration. They do not inhibit normal nymphal limb development, perhaps because only low levels of gene expression are required for this.

What enables the blastema to detect proximodistal regional discontinuities and to initiate intercalary regeneration? Probably it is the Fat–Dachsous system, which connects events at the cell surface to the Hippo pathway and to the growth-promoting effects of Yorkie (see Chapter 19). In normal limb buds Fat is present proximally, Dachsous distally, and Dachs all over. When segmentation of the developing limb becomes visible, there is a

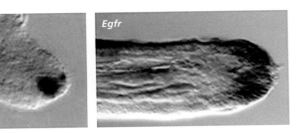

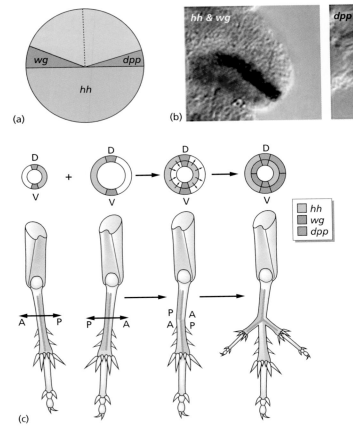

Fig. 20.8 Genes controlling cricket leg regeneration. (a) A leg bud shown schematically, viewed from the tip. (b) *In situ* hybridization of *hh* (red) and *wg* (purple) expression, *dpp* expression, and *EgfR* expression in a regeneration blastema. (c) Explanation for the formation of supernumerary limbs following an axial inversion. Reproduced from Campbell and Tomlinson. *Development* 1995; 121: 619–628, with permission from Springer.

reiterated pattern such that *fat* and *four-jointed* mRNAs are seen to be proximal in each segment, while *dachsous* and *dachs* are seen to be distal.

Knock down of *fat* or *dachsous* with RNAi increases cell division in the blastema and gives expanded, short, thick regenerates. Knockdown of *dachs* or *yorkie* suppresses the RNAi effects on *fat* or *dachsous*, consistent with the downstream position of these components in the pathway. Similar experiments show that *fat* and *dachsous* are also needed for intercalary regeneration, but not for formation of the supernumerary limbs arising from axial inversion experiments. This confirms that the Fat system is concerned with proximodistal rather than with anteroposterior patterning.

Vertebrate limb regeneration

Among the vertebrates only certain amphibian species can regenerate limbs after surgical removal. **Anuran** tadpoles can often regenerate limbs during the period of outgrowth and differentiation but the ability is lost during **metamorphosis**. By contrast, many **urodele** species can regenerate limbs throughout larval and adult life. Newts or axolotls are normally used as experimental models for regeneration, the axolotl being a type of salamander that never metamorphoses and retains features of the larval anatomy for life. Although a complete genome sequence is not yet available for the axolotl there is a good inventory of coding sequences. Transgenics can be made by injection of DNA into fertilized eggs, and genes can also be introduced into the limb by electroporation.

The process of limb regeneration

After amputation of a limb, the stump rapidly forms a wound epithelium by migration of epidermal cells over the cut surface. Then the internal tissues de-differentiate to a depth of about 1 mm from the cut. The cells from this region form a **blastema**, consisting of loose-packed mesenchymal cells surrounded by a thick epidermal jacket. Unlike the planarian blastema, the amphibian limb blastema is composed of proliferating cells. The blastema proliferates for a while and then the structures of the limb re-differentiate in a proximal to distal sequence. The regenerate is completed in miniature and then undergoes a long period of growth to get back to its original size. The stages of regeneration are referred to as de-differentiation, cone, palette, early digit, and late digit (Fig. 20.9). The time required to regenerate the pattern in miniature varies with the age and size of the animal but is typically 2–3 weeks for small larvae and a few months for large adults.

Because of their importance in limb development the expression of Hox genes of the a- and d-paralog groups have both been studied in amphibian limb development and regeneration. During larval limb bud development the expression of the genes is similar to higher vertebrates. Expression of the a genes is nested from distal to proximal, and of the d genes is nested from posterior to anterior. In regeneration they are expressed very early following amputation. Initially, genes *Hoxa9* and *Hoxa13* come on all over the blastema, then as the blastema grows their patterns resolve into a nested arrangement with *Hoxa13* distal. Some of the d genes are activated as part of a wound response. This can occur in any part of the body and is not specific to regeneration. Fibroblast growth factors (FGFs) are found to be expressed in the epidermal cap of the regeneration blastema and are mitogenic for isolated blastemas. Sonic hedgehog is expressed at the posterior margin of the blastema, as in the larval limb bud.

The source of cells for regeneration

When considering the source of cells for regeneration, there are three distinct issues that are often confused with one another:
• Are progenitor cells local or distant in origin?
• Are they formed by de-differentiation of functional cells or from undifferentiated "reserve cells"?
• Are they uni-, multi-, or pluripotent?

In relation to the first question it is known that the origin of cells is local to the amputation surface. This may be shown by X-irradiation, which can completely suppress regeneration following a dose that inhibits most, but not all, cell division. The radiation beam can be shielded so that only a small part of a limb is irradiated. If the level of the future amputation site is irradiated, there is no regeneration, while if the region of the future amputation is protected, but neighboring tissues are irradiated, regeneration proceeds normally. This indicates that the blastema must arise from cells located no further than a few millimeters from the amputation surface, rather than from stem cells in the bone marrow or other distant sites.

The second question relates to the possible existence of "reserve cells," which are supposed to be pluripotent cells left over from early embryonic development and scattered throughout the tissues. Such hypothetical cells are often called **neoblasts**, following the planarian usage. However, neoblasts are clearly visible in planarians but are not visible at all in the amphibian limb. While it is never easy to prove the nonexistence

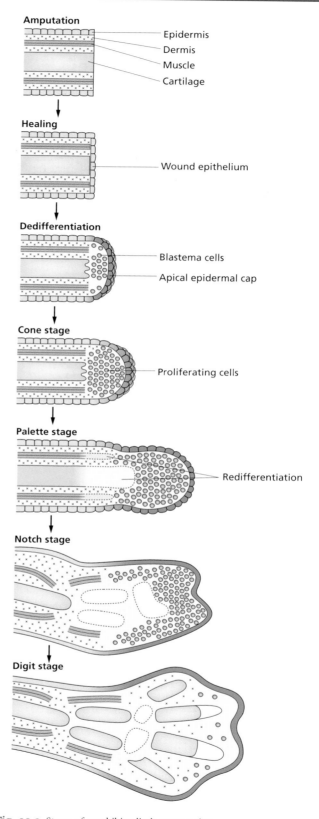

Fig. 20.9 Stages of amphibian limb regeneration.

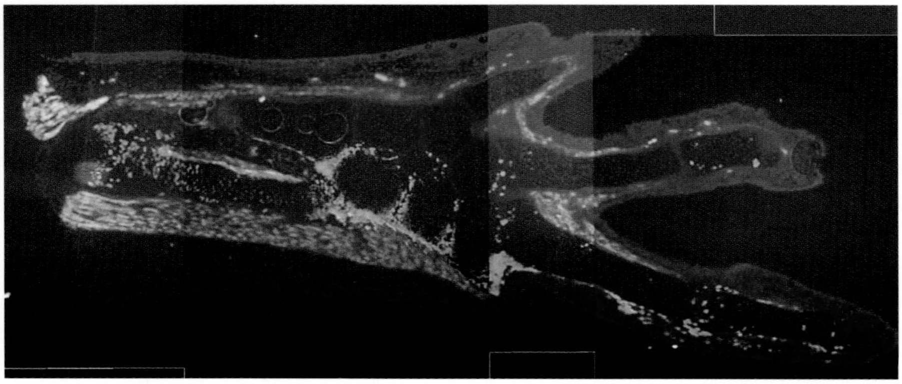

Fig. 20.10 Cell lineage tracing in regeneration. An explant of cartilage was grafted from a GFP transgenic to a normal animal, and the limb was amputated through the graft region. The section shows that the green graft-derived cells populate only the skeletal and connective tissues, but not the muscle fibers or epidermis. Reproduced from Kragl *et al. Nature* 2009; 460: 60–65, with permission from Nature Publishing Group.

of an entity like the reserve cell, there is no positive evidence that they do exist.

The final question relates to the occurrence or otherwise of **metaplasia** during regeneration. Metaplasia was mentioned in Chapter 19 and means a conversion in the mature organism of one tissue type into another. In regeneration its occurrence has been investigated by grafting tissues from transgenic axolotls labeled with GFP to unlabeled hosts. In order to ensure that only one tissue type is labeled, these experiments are best performed at embryonic stages such that cell migration does the job of purification. For example a limb containing labeled myofibers can be made by grafting GFP-labeled mesoderm to the somite region of host embryos. When the limbs develop, their musculature is derived from the somites so the muscles will have GFP + myofibers surrounded by unlabeled connective tissue components.

Experiments in which a limb is made containing one labeled tissue type, then amputated and allowed to regenerate, give the following results. Epidermis becomes only new epidermis. Muscle becomes only new muscle. Schwann cells of the nerve sheaths become only new Schwann cells. However, the connective tissue family (dermis, cartilage, tendons and ligaments, fibrous capsules) show extensive metaplasia one to the other (Fig. 20.10). The cell lineage restrictions in the regenerating limb are therefore similar to those of the developing limb where the connective tissue types derived from the limb-bud mesenchyme, the muscle cells from the somites, the Schwann cells from the neural crest, and the epidermal cell types from the limb-bud epidermis.

The "neurotrophic" factor

Limb regeneration is absolutely dependent on a nerve supply to the limb. In animals in which the nerve supply has been transected, the early events occur as normal resulting in the formation of a blastema, but this blastema fails to grow and regeneration is aborted. The function of the nerves is to release mitogenic factors, often referred to as **neurotrophic factor(s)**. Note that the term "neurotrophic factor" normally means a growth factor required for the growth or survival of neurons, whereas in the regeneration context it means a mitogenic activity derived from neurons, required for the growth of the blastema. The neurotrophic factors probably include neuregulins, which are abundant in nerve axons and are mitogenic for isolated blastemas. Neuregulins are similar to EGF and are ligands for the receptor tyrosine kinase ErbB2. In addition, there is a substance known as anterior gradient (AG) homolog, a glycoprotein originally discovered in the *Xenopus* embryo cement gland, and often overexpressed in human tumors. AG is synthesized by nerve sheaths and by gland cells of the wound epidermis following amputation and is also mitogenic for blastema cells. Electroporation of AG into the blastema of a denervated limb will rescue regenerative ability to some extent.

A remarkable fact about the neurotrophic effect is that it is possible to make limbs that lack almost all innervation and these "aneurogenic" limbs do not require a neurotrophic factor for regeneration. One way of making them is to remove the neural tube from the limb region and to join the operated embryo to another in **parabiosis** (joint blood circulation) to ensure survival (Fig. 20.11). A limb will then grow out in the operated region in which the innervation is limited or nonexistent. These limbs are of normal appearance although there is a degeneration of muscle at a later stage. Despite the lack of nerves, they can regenerate normally. If an aneurogenic limb is grafted back to a normal host in the larval stage when innervation is still occurring, then it can be innervated by the host, and will then become nerve dependent. The explanation is that the AG substance is made constitutively by the epidermis of aneurogenic limbs. But once exposed to nerves it is down regulated and becomes dependent on nerve damage.

Another bizarre aspect of the role of nerves is the phenomenon of "paradoxical regeneration." If a limb is given a sufficient dose of X-radiation this will suppress regeneration. If the brachial plexus is crushed to induce denervation, this will also suppress regeneration. However, if both treatments are given,

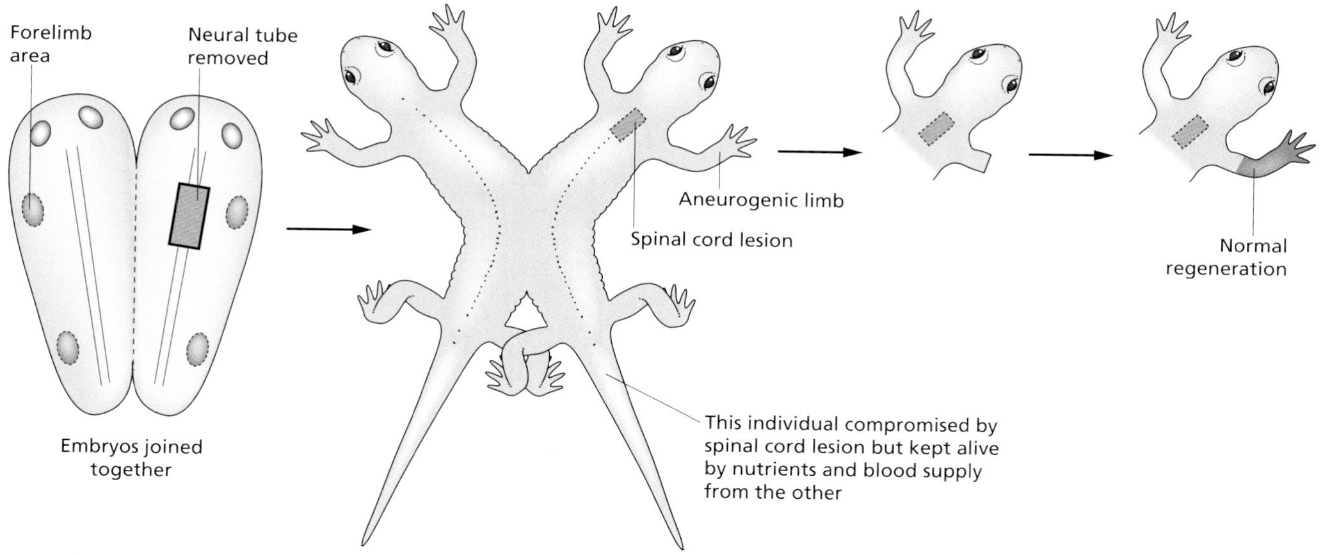

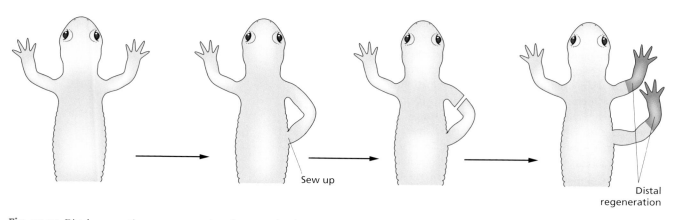

Fig. 20.11 Aneurogenic limb regeneration. The limb lacks innervation because of the removal of part of the neural tube from the embryo. Despite this, it regenerates normally.

Fig. 20.12 Distal regeneration occurs even when the cut surface is proximal facing, as shown in this classic experiment.

then after a certain time the limb regains the ability to regenerate. The explanation for this is the regeneration of the nerves down the old tracts into the limb. The nerve axons grow down these tracts and are therefore able to supply the necessary mitogenic factors. In addition, fibroblasts grow down the regenerating tracts and introduce a population of nonirradiated cells which, because of their connective tissue character, can form the skeleton of the regenerate.

Regeneration of regional pattern

Proximodistal pattern

The pattern of the regenerate conforms with that of the stump such that all parts **distal** to the cut are replaced. Proximal regen-

eration does not occur from an amputation surface even when the cut surface is proximal facing, as achieved by the procedure shown in Fig. 20.12, or Experiment I in Animation 24: Amphibian limb regeneration. This fact is generalized as the "law of distal transformation," which holds for all vertebrate and arthropod appendages capable of regeneration. In the amphibian limb, but not the insect limb, it also holds for **intercalary regeneration**, which means regeneration occurring at a tissue junction between parts that are not normally neighbors. When blastemas are transplanted distal to proximal, the potential gap in the pattern of the regenerate is filled in by intercalary regeneration from the stump side. However, if a proximal blastema is grafted to a distal stump, there is no reverse intercalation to fill in the gap, instead each component regenerates what it normally would, leaving a discontinuity in the pattern (Fig.

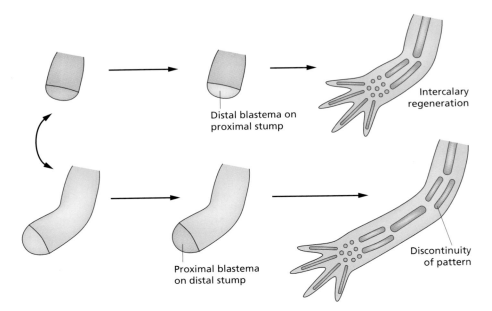

Fig. 20.13 Intercalary regeneration in the proximodistal axis. The stump can generate more distal tissue, but the blastema cannot generate more proximal tissue.

20.13). In such experiments the tissue contributions can be determined by using grafts between animals genetically labeled by skin pigmentation, triploidy, or transgenic markers.

The results of many experiments on regeneration suggest that the tissues of the limb carry positional information, or a code for regional identity, which specify what structures shall be formed on differentiation. This may be illustrated by the following type of experiment: a limb is irradiated to suppress regeneration; a cuff of skin is grafted from the upper arm of an unirradiated animal to the lower arm of the irradiated one; then the composite limb is amputated through the graft. A regenerate will form, derived from the graft, and the regenerated pattern will start at the upper arm rather than at the lower arm, since this represents the proximodistal identity of the graft. The regional identity is associated with differences in cell adhesion. This may be shown by grafting a blastema from one limb to the junction of the blastema and stump on another. Following the outgrowth and differentiation of this graft–host combination, the donor blastema will form an ectopic limb branching off the host limb. Remarkably, the proximodistal level at which the ectopic limb branches off corresponds to its own proximodistal character. In other words, if it is a proximal blastema then it will end up proximally, while if it is a distal blastema it will end up distally (Fig. 20.14; and also see Experiment II of the Animation 24: Amphibian limb regeneration).

One probable molecular component of positional information is a substance called Prod1. This protein may be unique to urodele amphibians although it is biochemically related to mammalian CD59, an inhibitor of complement activation. Prod1 is more abundant in the proximal than the distal blastema and its expression is upregulated by retinoic acid (see below). It is a cell-surface protein with a GPI anchor. Normally, a proximal blastema will engulf a distal one if they are cultured

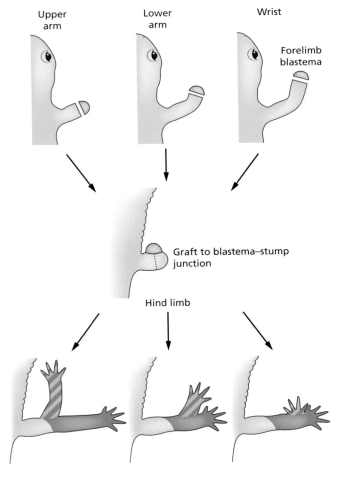

Fig. 20.14 Migration of a blastema, transplanted to the blastema–stump junction of the hindlimb. Proximal blastemas end up in proximal positions, distal blastemas end up in distal positions.

in contact, indicating that distal cells adhere more strongly to each other than do proximal cells (see Chapter 2). But a neutralizing antibody to Prod1 will prevent this, suggesting that the reduced adhesivity of the proximal cells is due to the higher level of Prod1. If *Prod1* is introduced into a blastema by electroporation, the cells that receive it will end up in a proximal position, in the same way as a grafted proximal blastema, while control electroporated cells end up in a distal position (Fig. 20.15).

Transcription factors associated with positional information are the Meis proteins, which are expressed in the proximal (stylopodial) segment of the vertebrate limb during development, and also in the proximal part of the *Drosophila* leg disc. If *Meis* genes are introduced by electroporation into cells of an axolotl blastema, the cells containing them will move to more proximal positions than control cells, in the same way as *Prod1*-expressing cells.

Regeneration of pattern in the transverse axes

Because the limb is asymmetrical in three dimensions it should, in principle, require three sets of regional codes to specify the pattern of differentiation in the three anatomical axes (proximodistal, anteroposterior, and dorsoventral) and in some respects it does behave in this way. In the transverse axes the codes can only be detected when de-differentiation and blastema formation are provoked by amputation. A common experimental protocol involves rearranging a limb surgically and sewing it together so that it heals up. This in itself does not provoke regeneration. But if the modified limb is then amputated, a blastema will form at the cut surface and the pattern of structures regenerated depends both on the representation of codes at the cut surface and on the nature of interactions between them.

There are two reasonably well-established principles that summarize the operation of these interactions in the transverse axes. First is the principle of **intercalation**. This states that where a blastema is generated from two tissue regions with a discontinuity between their codes, the regenerate will fill in the gap with structures that would normally form in between the two regions. This may be shown by grafting skin from one part of the limb to another and then amputating through the graft. If neither component was irradiated, the intervening structures are derived from both graft and host (Fig. 20.16). Such experiments can also reveal which tissues in particular carry the regional codes. The answer is that they are carried by the connective tissues, but not by the muscle or by the epidermis. Hence, although skin grafts are often used in such experiments, it is only the **dermis** that is active. This is shown by the fact that grafts of pure dermis have a similar effect whereas grafts of pure epidermis are without effect.

The other principle is that some degree of discontinuity of codes at the cut surface is necessary to initiate distal regeneration. For example if a small piece of skin is grafted to an irradiated limb followed by amputation through the graft, then only a very limited regenerate will be formed. But if two similar-sized pieces from opposite sides of the limb are grafted, then the regenerate will be much more substantial and contain many pattern elements (Fig. 20.17). The same applies to the formation of extra regenerates following nerve deviation. If the brachial nerve is surgically deviated through the skin of the upper arm

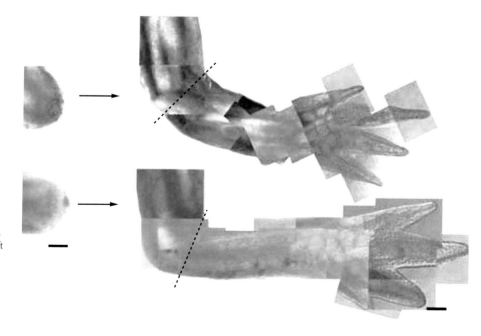

Fig. 20.15 Evidence that Prod1 controls proximal–distal patterning. *Prod1* was electroporated into a blastema along with *Gfp*. After regeneration the green cells end up in a proximal position. In control blastemas electroporated with *DsRed* alone, the red cells end up in a distal position. Left scale bar 200 μm, right scale bar 1 mm. Dashed lines indicate plane of amputation. Reproduced from Echeverri and Tanaka. *Dev Biol* 2005; 279: 391, with permission from Elsevier.

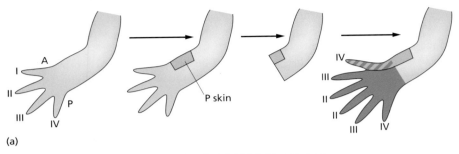

(a)

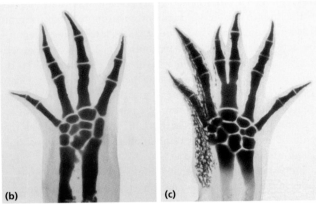

(b)   (c)

Fig. 20.16 Intercalary regeneration in the anteroposterior axis. (a) A posterior skin graft provokes the formation of the intervening digits from host tissue. (b,c) A duplicated axolotl limb arising from a skin graft and regeneration. (b) A control regenerate, note the different appearance of the four digits. (c) A regenerate following a graft of posterior skin from a pigmented animal to the anterior side of an unpigmented host limb. Judging from the pigmentation, the extra digit 4 and 3 arise from the graft, and the duplicated digit 2 from the host.

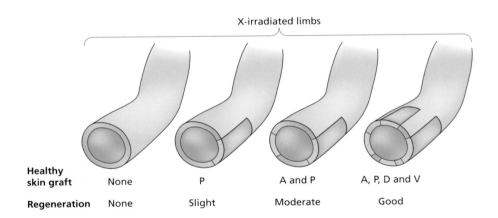

Fig. 20.17 Need for pattern discontinuity at the cut surface for distal regeneration.

then this will provoke the formation of a blastema, but this then regresses. If the nerve deviation is accompanied by a posterior to anterior skin graft then a complete limb will form at the junction. It is not, however, necessary that *all* codes need be present at the cut surface, as partial limbs or abnormal limbs lacking some structures can both regenerate well. For example double-posterior limbs made by ZPA (**zone of polarizing activity**) grafting in the embryo (see Chapter 15) will regenerate double-posterior regenerates.

Both principles are seen at work in the axial inversion experiment, very similar to the same procedure described above in the insect limb. This is the graft of a blastema to a stump on the contralateral side of the animal. Because limbs are asymmetrical in three dimensions, it is not possible to superimpose a right limb on a left limb. If a blastema is grafted from right to left, or left to right, side, then it will inevitably have one axis inverted in relation to the host stump (Fig. 20.18). This graft reliably gives rise to a pair of supernumerary regenerates at the points of maximum disparity. Therefore in total there are three limbs regenerated (see Experiment III of the Animation 24: Amphibian limb regeneration). One arises from the blastema, with its original polarity, and two arise from the discordant blastema–stump junctions, with polarity inverted in relation to the central limb. A simple rotation of the blastema on its own stump can also give rise to supernumeraries, but the number, position, and anatomical completeness is much

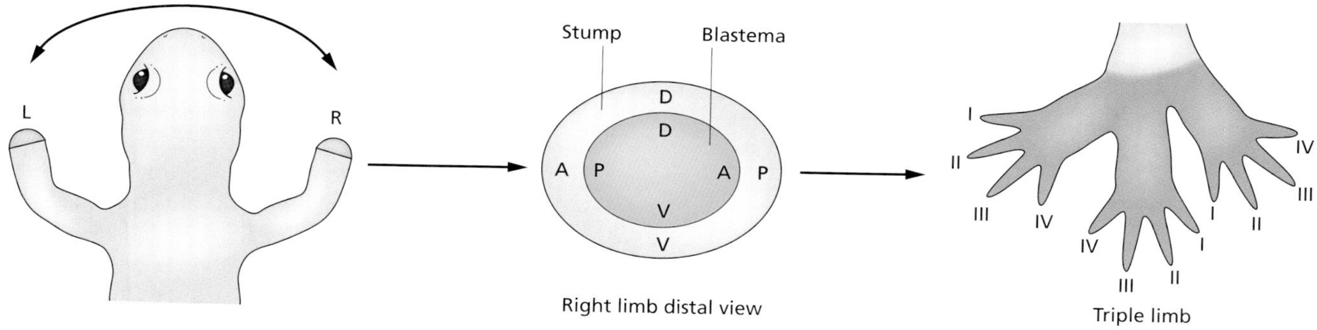

Fig. 20.18 Axial inversion of a blastema onto a stump produces a triple limb with one central member and two supernumeraries.

Fig. 20.19 Proximalization of blastema by retinoic acid treatment.

more variable because the disparity of codes is similar all round the circumference.

There is limited molecular information about the molecular identity of the positional information around the circumference of the limb. Sonic hedgehog (Shh) is expressed in the posterior part of the blastema and introduction of *Shh* into the anterior side of a blastema using a viral vector will produce a double-posterior duplication. This shows at least some conservation of mechanism from limb development.

Retinoic acid effects

Several different types of re-specification of code can be achieved by treating cone-stage blastemas with retinoic acid. These include effects on each of the three pattern axes, leading to proximalization, posteriorization, and ventralization. Since in these experiments the whole animal is immersed in a solution of retinoic acid, the blastema is probably exposed to a uniform dose of the reagent during treatment.

When an amputated, but otherwise anatomically normal limb is treated, the result is proximalization, such that the further development of the blastema leads to serial duplication of structures. So, for example, a wrist-level blastema treated with retinoic acid can regenerate a complete arm (Fig. 20.19). This of course violates the normal "law" of distal transformation. There is some dose–response effect, with higher

doses giving more proximalization, but above a certain dose a growth inhibition is found, causing a total suppression of regeneration.

The posteriorization effect is shown when blastemas from half limbs are treated. Normally, a posterior half limb will show more regenerative capacity than an anterior half limb, the former often regenerating a complete distal limb, and the latter little or nothing. But this is altered by retinoic acid, which causes normal limb regeneration from anterior halves and suppresses regeneration from posterior halves. If limbs are surgically created with double-anterior or double-posterior morphology, then normally the double posteriors produce some distal regenerate while the double anteriors produce nothing. Again, in retinoic acid this is reversed and the double-anterior limbs produce two complete regenerates while the double-posterior limbs produce nothing (Fig. 20.20). These results can be explained on the assumption that retinoic acid posteriorizes the blastema. The uniform treatment with retinoic acid makes the blastema posterior in character all over. Interactions between the blastema and newly dedifferentiated tissue from the stump then give rise to a limb wherever there is a discontinuity between anterior and posterior codes.

Comparable results are obtained when retinoic acid is used to treat dorsal and ventral half limbs. After retinoic acid treatment the dorsal halves regenerate and double-dorsal halves form double limbs. This suggests that the retinoic acid acts to ventralize the blastema, and the regenerates arise after interaction between blastema and newly dedifferentiated stump tissue. Evidence that this type of interpretation is correct comes from the fact that late treatment with retinoic acid, after dedifferentiation has finished, leads to pattern truncation. This is because by this stage the stump can no longer provide further dedifferentiated tissue to enable the interactions to occur that are needed to initiate distal regeneration.

Thus, the overall effects of retinoic acid are rather complex, involving a simultaneous proximalization, posteriorization, and ventralization of the blastema. The molecular basis of this remains unclear except for the proximalization process, where it is known that both *Prod1* and *Meis* genes are upregulated by retinoic acid. The initial step in its action requires binding of the retinoic acid to its receptors, which belong to the nuclear receptor class. The urodele limb contains several types of retinoic acid receptor, including some forms specific to urodeles called δ-receptors. To identify which receptor was responsible for the recoding effects, **domain swaps** were made with thyroid receptor, such that the chimeric molecules had the thyroid hormone-binding domain but the retinoic acid receptor domains for DNA binding and gene activation (Fig. 20.21). This means that the genes normally regulated by retinoic acid can now be regulated by thyroid hormone, but only if the particular receptor introduced into the limb is the one that can transduce the signal. Genes for the receptors, together with a gene for cell identification, *alkaline phosphatase*, were introduced into retinoic acid-treated distal blastemas by biolistic particle bombardment. The blastemas were then grafted to proximal stumps. Proximalization is revealed by the presence of alkaline phosphatase-positive cells in the intercalary regenerate, which is normally formed from the stump tissue alone. These experiments showed that proximalization was only mediated by one out of the five possible receptors, one of the urodele-specific types called retinoic acid receptor δ2.

Although the spectacular effects of experimentally applied retinoic acid are well described, this does not necessarily mean that retinoic acid also has an endogenous role in the control of regeneration. Levels of endogenous retinoic acid have been measured in one of two ways, either by chemical extraction and separation, or by observing the response to the amphibian tissues of mammalian reporter cells containing a retinoic acid response element (RARE) coupled to a reporter gene. These methods have shown that retinoic acid is present in the wound epidermis of regeneration blastemas. It is actually the 9-*cis* isomer of retinoic acid, which is more potent than the all-*trans* form. It is found to be enriched in proximal over distal blastemas, and in posterior over anterior region within a blastema. So retinoic acid is present in the wound epidermis and it has biological activity, but the effects of retinoic acid inhibitors on regeneration are quite modest and its exact *in vivo* role remains uncertain.

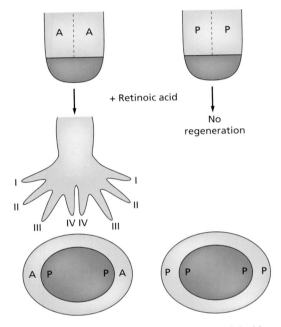

Fig. 20.20 Effect of retinoic acid on surgically produced double-anterior and double-posterior limbs. The double-anterior limb is caused to form a double regenerate because of interaction between the posteriorized blastema and the unmodified stump.

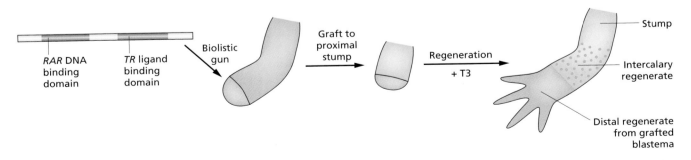

Fig. 20.21 Experiment to find which retinoic acid receptor causes proximalization. A domain-swapped receptor is transfected into a distal blastema. This is grafted to a proximal stump and the animal treated with thyroid hormone. The presence of donor cells in the intercalary regenerate is an indication of proximalization induced by the thyroid hormone.

New directions for research

The sky is the limit in regeneration research. Some key questions are:

1 Why can some structures regenerate and other, apparently similar ones not do so (e.g. urodele and anuran limbs)?

2 In planarians, what is the hierarchy of developmental decisions leading from the pluripotent neoblasts to specific cell types?

3 What is the molecular basis of de-differentiation?

4 How can cells regenerate a particular pattern when this may be quite different from what is present at the cut surface (e.g. the pattern of a hand versus the upper forelimb)?

General properties of regeneration

Certain general features are apparent in the regeneration of missing parts. The most important is that the part in question should still contain dividing cells and that the genes that control growth and regional specification should either still be active or at least be capable of upregulation following wounding. Regeneration does tend to reuse mechanisms used in embryonic development, although there are always some differences dependent on the different starting point and scale of events. Regeneration of individual body parts, like limbs, does not require that cells should lose all developmental commitment and become like embryonic stem cells, and in fact the cells seem largely to conserve their tissue types. However, in the case of planarians, where the whole body can be regenerated, at least some neoblasts do have pluripotent properties comparable to embryonic stem cells. Genes that need to be re-expressed for regeneration are typically hard to upregulate in nonregenerating situations. For example in urodeles an enhancer of the *Shh* gene, responsible for anteroposterior patterning, is unmethylated at the DNA level, and presumably accessible to the relevant regulatory proteins, while in anurans, which cannot regenerate limbs post metamorphosis, the same enhancer is methylated and silent.

Key points to remember

• Some invertebrates can regenerate whole bodies from small tissue fragments. In such cases a transected body can form a new head from the anterior-facing cut surface and a new tail from the posterior-facing cut surface. This is known as bidirectional regeneration. Insects and vertebrates can only regenerate appendages, and this is monodirectional, always proceeding distally from the cut surface.

• Planarians are animals capable of bidirectional regeneration. Regenerative capacity is associated with the existence of neoblasts, which are small, undifferentiated, mitotic cells making up about 20% of the worm's body. Cells within the neoblast population act as stem cells both for normal growth and for regeneration.

• Most urodele amphibians can regenerate limbs throughout life. The regenerate is formed from the cells near the cut surface. There is some dedifferentiation, in particular of multinucleate myofibers which can re-enter the cell cycle and break up into mononucleate cells.

• Following amputation of a urodele limb, a regeneration blastema of visibly undifferentiated cells is formed. It grows and differentiates in proximal to distal sequence to reform the missing parts. The miniature organ then grows to the normal size.

• There is some metaplasia in limb regeneration, in particular the dermis and cartilage interconvert extensively. However, myofibers, Schwann cells, and epidermis do not convert to other cell types.

• Growth factors secreted by nerves are needed for the growth of the blastema.

• The limb behaves in response to grafting experiments as though it contains a system of positional information, labeling each part, which control the pattern of structures formed during regeneration. Some discontinuity of positional information is necessary for distal regeneration.

• Retinoic acid can proximalize, posteriorize, and ventralize the blastema.

Further reading

General

Goss, R.J. (1969) *Principles of Regeneration.* New York: Academic Press.

Scadding, S.R. (1977) Phylogenetic distribution of limb regeneration capacity in adult amphibia. *Journal of Experimental Zoology* **202**, 57–68.

Brockes, J.P., Kumar, A. & Velloso, C.P. (2001) Regeneration as an evolutionary variable. *Journal of Anatomy* **199**, 3–11.

Agata, K., Saito, Y. & Nakajima, E. (2007) Unifying principles of regeneration I: Epimorphosis versus morphallaxis. *Development Growth and Differentiation* **49**, 73–78.

Tanaka, E.M. & Reddien, P.W. (2011) The cellular basis for animal regeneration. *Developmental Cell* **21**, 172–185.

Planarian regeneration

Agata, K. (2003) Regeneration and gene regulation in planarians. *Current Opinion in Genetics and Development* **13**, 492–496.

Reddien, P.W. & Alvarado, A.S. (2004) Fundamentals of planarian regeneration. *Annual Review of Cell and Developmental Biology* **20**, 725–757.

Sánchez Alvarado, A. (2006) Planarian regeneration: its ends is its beginning. *Cell* **124**, 241–245.

Forsthoefel, D.J. & Newmark, P.A. (2009) Emerging patterns in planarian regeneration. *Current Opinion in Genetics and Development* **19**, 412–420.

Wagner, D.E., Wang, I.E. & Reddien, P.W. (2011) Clonogenic neoblasts are pluripotent adult stem cells that underlie planarian regeneration. *Science* **332**, 811–816.

Insect limb regeneration

Nakamura, T., Mito, T., Bando, T., Ohuchi, H. & Noji, S. (2008) Dissecting insect leg regeneration through RNA interference. *Cellular and Molecular Life Sciences* **65**, 64–72.

Amphibian limb regeneration

General

Iten, L.E. & Bryant, S.V. (1973) Forelimb regeneration from different levels of amputation in the newt *Notophthalamus viridescens*: length, rate and stages. *Wilhelm Roux's Archives of Developmental Biology* **173**, 263–282.

Mescher, A.L. (1996) The cellular basis for limb regeneration in urodeles. *International Journal of Developmental Biology* **40**, 785–795.

Brockes, J.P. (1997) Amphibian limb regeneration: rebuilding a complex structure. *Science* **276**, 81–87.

Nye, H.L.D., Cameron, J.A., Chernoff, E.A.G. & Stocum, D.L. (2003) Regeneration of the urodele limb: a review. *Developmental Dynamics* **226**, 280–294.

Nacu, E. & Tanaka, E.M. (2011) Limb regeneration: a new development? *Annual Review of Cell and Developmental Biology* **27**, 409–440.

Dedifferentiation and metaplasia

Namenwirth, M. (1974) The inheritance of cell differentiation during limb regeneration in the axolotl. *Developmental Biology* **41**, 42–56.

Brockes, J.P. & Kumar, A. (2002) Plasticity and reprogramming of differentiated cells in amphibian regeneration. *Nature Reviews in Molecular and Cellular Biology* **3**, 566–574.

Tanaka, E.M. (2003) Cell differentiation and cell fate during urodele tail and limb regeneration. *Current Opinion in Genetics and Development* **13**, 497–501.

Kragl, M., Knapp, D., Nacu, E., Khattak, S., Maden, M., Epperlein, H.H. & Tanaka, E.M. (2009) Cells keep a memory of their tissue origin during axolotl limb regeneration. *Nature* **460**, 60–65.

Positional information and retinoic acid

Tank, P.W. & Holder, N. (1981) Pattern regulation in the regenerating limbs of urodele amphibians. *Quarterly Reviews of Biology* **56**, 113–142.

Bryant, S.V. & Gardiner, D.M. (1992) Retinoic acid, local cell–cell interactions, and pattern formation in vertebrate limbs. *Developmental Biology* **152**, 1–25.

Stocum, D.L. (1996) A conceptual framework for analysing axial patterning in regenerating urodele limbs. *International Journal of Developmental Biology* **40**, 773–783.

Maden, M. (1997) Retinoic acid and its receptors in limb regeneration. *Seminars in Cell and Developmental Biology* **8**, 445–453.

Chernoff, E.A.G., Stocum, D.L., Nye, H.L.D. & Cameron, J.A. (2003) Urodele spinal cord regeneration and related processes. *Developmental Dynamics* **226**, 295–307.

This chapter contains the following animations:

Animation 23 Planarian regeneration.

Animation 24 Amphibian limb regeneration.

 For additional resources for this book visit www.essentialdevelopmentalbiology.com

Chapter 21

Applications of pluripotent stem cells

Although developmental biology has already contributed to human welfare in various ways, in the long term the area that will probably represent its major contribution is the technology of making differentiated cells for transplantation therapy. This means making new cells to replace ones that have died, and introducing them into the appropriate part of the body in a functional state; for example heart muscle cells into the myocardium of a failing heart, or neurons into a part of the brain affected by a stroke. This is part of a wider suite of technologies called regenerative medicine, which consists of all the treatments devoted to the repair and replacement of cells, tissues, and organs. Along with cell transplantation therapy this includes gene therapy and tissue engineering. Gene therapy is the technology of introducing good copies of genes that are missing or defective into appropriate cells of the patient. Tissue engineering is the technology of growing three-dimensional assemblages of cells on scaffolds, sometimes multiple cell types in cooperative assemblies.

Because of their central role as potential sources of cells for transplantation, this chapter will focus on the properties and applications of **pluripotent stem cells**, which comprise **embryonic stem cells** (ES cells) and **induced pluripotent stem cells** (iPS cells).

Cell transplantation therapy

Most of the useful cell types required for cell transplantation therapy do not divide in tissue culture, or if they do divide they rapidly lose the characteristics and properties that make them useful. This is true, for example, of hepatocytes from the liver and of β-cells from the pancreas. Because of this problem, very little of the successful cell therapy carried out today involves expansion of cell populations in tissue culture prior to grafting but instead mostly consists of grafting of cells directly from one person to another, or from one part of the body to another. For example in **bone marrow transplantation** it is possible to harvest bone marrow (and the **hematopoietic stem cells** it contains) in sufficient quantity from living donors. Other types of cell therapy, such as grafts of pancreatic islets or hepatocytes, are currently done using human cadaver donors, which greatly limits the cell supply and the utility of the methods. Because they can be expanded without limit *in vitro*, the use of pluripotent stem cells can potentially make available large quantities of any required cell type.

A major problem confronting cell transplantation therapy is that of host immunity. The immune system, evolved to combat infection, also causes the rejection of cell, tissue, or organ grafts from one individual to another. The cells of the graft are recognized as "not self" by T **lymphocytes**. These destroy target cells by exposing them to **cytokines**, among which are the interferons, also important in the response to virus infection, and tumor necrosis factors, which are often secreted by tumors and account for many of their systemic toxic effects. The T lymphocytes can recognize almost any novel molecule that they do not normally encounter in the body, but the bulk of the immune response is directed against a family of cell-surface molecules called **HLA (human leucocyte antigen)** factors. These are cell-surface glycoproteins that show enormous variability between individuals. They are encoded by two clusters of genes, called ABC and DR, each of which may be occupied by any of a large number of possible alleles. The total number of combinations of these alleles is enormous, which is why grafting of tissue from one individual to another (an **allogeneic graft** or **allograft**) usually provokes a rejection reaction by the T lymphocytes of the host. In clinical practice, these reactions are controlled by immunosuppressive drugs, which reduce the response of

Essential Developmental Biology, Third Edition. Jonathan M.W. Slack.
© 2013 John Wiley & Sons, Ltd. Published 2013 by John Wiley & Sons, Ltd.

activated T cells. For example cyclosporine or tacrolimus do so by inhibiting calcineurin, which mediates the effects of elevated calcium following the activation of T-cell receptors; rapamycin (sirolimus) does so by inhibition of TOR (see Chapter 19). But the use of immunosuppressive drugs to control the rejection of allografts comes at considerable cost. The drugs often have side effects causing various types of organ damage, and because they reduce the ability of the individual to mount an immune reaction against invading micro-organisms they also make immunosuppressed patients liable to contract infections. Successful immunosuppression is easier to attain if there is not too much mismatch to begin with between the HLA alleles of graft and host. Determining the degree of HLA match between donor and host is the basis of tissue typing, which is very important both for organ transplantation and for the currently practiced types of cell therapy.

It has been known since the 1950s that tissue grafts can be tolerated if they are carried out between identical twins. This is because identical twins have exactly the same set of alleles for all of the HLA genes, and for any other genes involved with graft rejection. The same is also true for any graft taken from the individual him- or herself and reimplanted back into another position in the same individual (an **autologous graft**, or **autograft**).

The above considerations explain the potential importance of pluripotent stem cells in relation to the problem of immune rejection. One solution is to create banks of pluripotent stem cells such that there is always a reasonably good HLA match available for any individual. The number of cell lines required is calculated to be of the order of a few hundred. The other possibility is to use iPS cells grown from the individual patient (see below).

Hematopoietic stem cell transplantation

Various sorts of cell transplantation therapy have been practiced in the past but overwhelmingly the most important in terms of numbers is **hematopoietic stem cell transplantation**. This is more familiarly known as bone marrow transplantation, but the source of hematopoietic stem cells may be peripheral blood, or umbilical cord blood, as well as bone marrow itself. About 50,000 hematopoietic cell transplants are performed worldwide each year. They are mostly for treatment of leukemia or lymphoma, with a small proportion for treatment of other cancers or for genetic diseases of the blood. As described in Chapter 18, hematopoietic stem cells continuously give rise to all cell types of the blood and immune system. They exist in niches in the bone marrow, formed by osteocytes and blood vessels, and they persist for the lifespan of the individual. Following severe irradiation or chemotherapy, the stem cells are killed and their niches become vacant. Under these circumstances a graft containing stem cells can become established by occupying the vacant niches and can then repopulate the blood and immune

system for life with donor-derived cells. Because of the radiation or chemotherapy that is given, the host usually has little ability to mount an immune response against the graft and to reject it. However, in hematopoietic cell grafts, the graft itself contains many T lymphocytes, which can generate an immune reaction against the host tissues. This causes graft-versus-host disease, which is the main cause of morbidity and mortality in allogeneic transplants. For the treatment of cancer the original rationale was to enable larger doses to be given of radio- or chemotherapy in the hope of completely eradicating the cancer cells. Complete eradication is necessary because the cancer will soon grow back if any of its cells survive. The normal limitation to the intensity of treatment is the sensitivity of normal stem cell populations, particularly those in the bone marrow. But if the bone marrow of the patient could be sacrificed then a larger dose of treatment could be given. Following this, the patient is saved from inevitable marrow failure and death by an infusion of healthy bone marrow. Today the rationale for the procedure has changed somewhat in that the main intention is to utilize the antitumor effect of the graft itself as a means to remove the last few cancer cells. This effect is due to an immune reaction by lymphocytes in the graft against determinants on the tumor. It is related to, but not entirely identical to, the normal graft-versus-host effect.

Mostly because of the problems of graft-versus-host disease, and the accompanying immunosuppression, allogeneic hematopoietic stem cell transplantation remains a dangerous procedure with a significant treatment-associated mortality. It is therefore used only for treating lethal diseases such as leukemias, lymphomas, and also some genetic defects of the hematopoietic system itself. It is not used for the numerous immune disorders that could theoretically be remedied by replacing the cells of the immune system, but which are not so lethal in their effects.

Considerable use is also made of autologous hematopoietic grafts. In this case the bone marrow sample is taken from the patient, then the he or she is treated with radio- or chemotherapy, then the bone marrow is reinfused. This has the advantage that there is no graft-versus-host disease, but the disadvantage that there is also no graft-versus-tumor effect. Another problem with autologous hematopoietic transplants is that it is quite likely that some cancer cells will be present in the bone marrow and these can be reintroduced with the graft and act as seeds for regrowth of the cancer.

Although the success of hematopoietic stem cell transplantation is a pointer to what might be achieved with other cell types, there are a number of differences from what is currently being planned with pluripotent stem cells. Firstly, the graft is a stem cell graft, which aims to repopulate stem cell niches and produce life-long re-population of the target tissue, in this case the blood and immune system. Cell transplantation therapy based on pluripotent stem cells will not be of the stem cells themselves, as these would form dangerous teratomas when implanted into patients, but of differentiated cells derived from them. Because

Classic experiments

Human embryonic stem cells

This paper describes the growth of cells from inner cell masses of human blastocysts, which resemble the embryonic stem cells previously made from mice and from nonhuman primates.

Thomson, J.A., Itskovitz-Eldor, J., Shapiro, S.S., Waknitz, M.A., Swiergiel, J.J., Marshall, V. S. & Jones, J.M. (1998) Embryonic stem cell lines derived from human blastocysts. *Science* **282**, 1145–1147.

At the same time, similar cells were grown from primordial germ cells from human fetuses.

Shamblott, M.J., Axelman, J., Wang, S., Bugg, E.M., Littlefield, J.W., Donovan, P.J., Blumenthal, P D., Huggins, G.R. & Gearhart, J.D. (1998) Derivation of pluripotent stem cells from cultured human primordial germ cells. *Proceedings of the National Academy of Sciences USA* **95**, 13726–13731.

the grafts will not be stem cells there will be no need to clear the host niches before grafting. Another difference is that the hematopoietic grafts are not cultured *in vitro*, they are taken from one person and grafted directly into another (or in the cause of autografts, to the same patient). This avoids problems of mutation of the cells in tissue culture or of the inadvertent introduction of pathogenic viruses. Lastly, the main complication of hematopoietic cell grafts is graft-versus-host disease, but this will not occur for other types of graft that do not contain lymphocytes. For these reasons transplants of differentiated cells derived from pluripotent stem cells are likely to be safer than hematopoietic grafts and, once the numerous technical and regulatory problems have been solved, they may become even more widely used.

Embryonic stem cells

Human embryonic stem cells

Mouse **ES cells** were introduced in Chapter 10 in the context of their use as a vehicle for creating genetically modified mice. Here we shall consider the role of human ES cells and their potential role in human medicine. Like mouse ES cells, human ES cells are cultivated from the inner cell masses of preimplantation blastocysts and grow as refractile colonies composed of small cells (Fig. 21.1). **Feeder** cells are normally used for cultivation, although some media for feeder-free cultivation are now available. This is particularly important for cells destined for transplantation into humans because the use of any animal cells or products in their cultivation raises a small potential risk of introducing viruses that could harm the patient.

Human ES cells resemble mouse ES cells in a number of respects. Both cell types express a number of pluripotency genes, such as those encoding the transcription factors OCT4, SOX2, and NANOG. These proteins form a network upregulating the genes for each other, and also repressing the expression of genes required for the early steps of differentiation, resulting

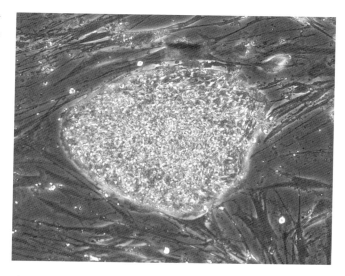

Fig. 21.1 A colony of human ES cells in culture (phase contrast). The colony contains a few hundred small cells and is surrounded by a feeder layer of large, elongated fibroblasts. Courtesy of Lucas Greder, University of Minnesota Stem Cell Institute.

in a reasonably stable pluripotent state. Both cell types can be grown without limit in tissue culture; and both are pluripotent as indicated by the formation of embryoid bodies *in vitro* and teratomas *in vivo* (Fig. 21.2).

However, there are a number of significant differences between mouse and human ES cells. The morphology of the colonies is different, with human ES cells forming rounder, flatter colonies having a defined edge. Mouse ES cells can be cloned from single cells, while human ES cells cannot, and are usually subcultured as clumps of cells. Mouse ES cells require LIF (leukemia inhibitory factor) for stabilization of the pluripotent state, while human ES cells instead require FGF2 and activin. In terms of X-inactivation, which in mouse occurs at the egg cylinder stage, mouse ES cells from female embryos have both X chromosomes active, while human ES cells often have

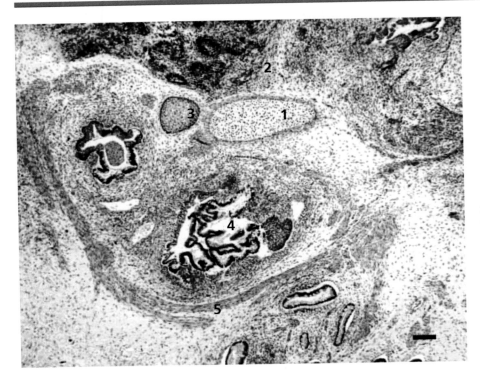

Fig. 21.2 Micrograph of a typical teratoma following the grafting of human ES cells into the testis capsule of an immunodeficient mouse, showing the presence of cartilage (1), primitive neural cells (2), stratified squamous epithelium (3), glandular epithelium (4), muscle (5), and other cell types. Hematoxylin–Eosin stain. Scale bar 800 μm. Reproduced from Pera and Trounson. *Development* 2004; 131: 5515–5525, with permission from Company of Biologists Ltd.

one inactive X-chromosome. Mouse ES cells do not easily form primordial germ cells *in vitro*, whereas human ES cells do so when treated with BMP4. Given that they seem to be more mature in their other properties, a curious feature of human ES cells is that differentiation of trophectoderm can occur, whereas this does not happen from mouse ES cells.

It is possible to culture a type of ES-like cell from epiblasts of postimplantation mouse embryos. These are called EpiSC, and they have properties very similar to human ES cells. In addition, they are poor at forming chimeras when reintroduced into the blastocyst, while ES cells do this readily. It is not known whether human ES cells can repopulate early human embryos as such an experiment would be unethical. The idea that human ES cells and mouse EpiSC are similar to each other is also supported by gene expression patterns from microarrays and RNAseq studies.

The actual cell of origin of mouse ES cells is thought to be the early preimplantation epiblast, although this is not known for sure, while mouse EpiSC and human ES cells are probably derived from the postimplantation epiblast. Whichever cell type is the true *in vivo* counterpart of the ES cells, it is not a stem cell. All the early cell populations in the embryo are short lived and soon develop into other cell types committed to form specific body parts or tissue types. Because of this, ES cells have often been described as an "*in vitro* artifact." However, this fact does not detract from their great importance for the various applications that we shall consider.

The teratoma assay (Fig. 21.2) is particularly important for testing new lines of human ES cells to establish their pluripo-

tency, since it would not be ethical to implant them into human embryos for this purpose. It is carried out by injecting cells subcutaneously or intramuscularly into a mouse. The cells are judged to be pluripotent if their teratomas form tissues characteristically derived from all three of the embryonic **germ layers: ectoderm, mesoderm,** and **endoderm**. The mouse hosts for teratomas need to be a severely immunocompromised strain, such as the NOD-SCID (nonobese diabetic, severe combined immunodeficiency) otherwise the graft of human tissue would be rejected by the immune system of the host. The NOD (nonobese diabetes) defect lies in the innate immune system and on its own will predispose to the development of an autoimmune diabetes resembling human type 1 diabetes. The SCID (severe combined immunodeficiency) defect is of a DNA repair enzyme that is required for gene rearrangements during maturation of T and B lymphocytes. Despite their name, NOD-SCID mice are not actually diabetic because the immune deficiency caused by the SCID defect prevents development of the autoimmune diabetes normally suffered by NOD mice. The combination of both immune defects makes them able to tolerate human or other species grafts, not just for teratoma testing but for many other purposes as well. However, because of their immunodeficiency they are also very susceptible to infection and need careful rearing.

The word **pluripotent** has now become the standard term for describing the ability of ES cells to form a broad range of cell types. ES cells used to be described as "**totipotent**," but this usage was discontinued because mouse ES cells do not normally form any trophectoderm when differentiated *in vitro* or in chimeras. Ironically, human ES cells do form trophectoderm *in vitro*, despite the fact that they seem to represent a more mature developmental state than mouse ES cells. But the term "totipotent" is not now used for human ES cells either, and has become reserved just for the fertilized egg itself.

The embryos for the production of human ES cells come from *in vitro* fertilization (IVF) clinics. Normally, the mother receives a hormone treatment that causes several of her **oocytes** to mature simultaneously, instead of the normal one per month. So as many as 10–12 oocytes may be harvested from one cycle, fertilized with sperm, and allowed to develop for a few days as preimplantation embryos. It is now considered unethical to implant more than two embryos at a time because of the risk of multiple pregnancy, so the remainder will be cryopreserved for future use. If the first implantation is not successful, two further embryos can be thawed and reimplanted without the need for another cycle of hormone stimulation and oocyte harvest. But very often not all the frozen embryos are used. The parents may succeed with their pregnancy or pregnancies and not want any more, or they may decide for some other reason not to have further rounds of implantation. So all IVF clinics have large numbers of embryos stored in liquid nitrogen. At some stage the surplus embryos have to be disposed of. One option is simply to discard them. Another is to donate them for research purposes, including the establishment of new ES cell lines, and many parents are happy to do this. There is a set of ethical guidelines for this type of tissue donation used by the US National Institutes of Health and similar bodies in other countries. They include the fact that the embryos should not have been created specifically for research purposes, there should be no payment for donation, and the parents understand the use to which the cells will be put.

Notwithstanding these careful regulatory measures, some people consider the use of human preimplantation embryos for any type of research to be unethical. This issue has led to a fierce debate, and to restrictive legislation in various countries. At the present time, human ES cells may be created and used under license in some countries, such as the UK, Sweden, Australia, and Singapore, while in others there are bans on certain types of procedure, most notably the creation of new human ES cell lines. In the particular case of the USA, there are no federal statutory prohibitions on any research activity, but there is a prohibition of the use of federal research funds for any procedure involving human embryos.

Applications of human ES cells

Attention is usually focused on the potential for cell transplantation therapy using differentiated cells derived form human ES cells. Most of the common diseases that afflict the western world, such as heart failure, cancer, diabetes, arthritis, or neurodegenerative conditions, involve the loss of or damage to certain specific cell populations and human ES cells offer a source of healthy cells that could potentially be transplanted to repair damaged tissues or organs. This is indeed an important prospect and the first clinical trials are now underway. But there are several other important applications as well.

First is the investigation of normal development. Human development is still very poorly studied, both because of the difficulty of obtaining early postimplantation embryos, and because of the ethical issues over embryo procurement for research. Although it is certain that the overall course of development is the same as that of other vertebrates, the details do differ and can only be established by direct studies of human development. The availability of human cells that will carry out developmental processes *in vitro* offers a method of investigating at least some aspects of normal human development that would not otherwise be accessible.

Second is the possibility of obtaining normal human cells for drug screening. Some cell types, such as cardiomyocytes, are normally very difficult to obtain, and even human hepatocytes are in very short supply. Cardiomyocytes are very important because many drugs have side effects on the heart, and it is important to screen for such effects at an early stage of drug development. Hepatocytes are important because drug action depends a lot on metabolism in the liver, which is the first organ encountered by drugs given by mouth and absorbed through the intestine. This metabolic transformation may inactivate the drug, or convert it to an active form, or otherwise alter it in a significant way.

Third, there is also the possibility of studying cellular pathology for those human genetic diseases where the pathology is cell autonomous and the relevant cell types can be obtained *in vitro* from ES cells. However, this is more likely to be an application of iPS cells, which can be obtained from individual patients of known genotype. For such studies ES cells will be needed to set the baseline for the specific cell type.

The strategy for causing human ES cells to become a particular cell type depends on a good understanding of normal development, and on the assumption that human development is similar to that of other vertebrates. The formation of any particular cell type involves a hierarchy of developmental decisions, for example if the aim is to make pancreatic β-cells then the ES cells need to be converted to endoderm, then to anterior endoderm, then to pancreatic bud tissue, then to endocrine progenitors, and finally to β-cells. To achieve this, the cells are exposed to a series of different inducing factors at the same concentrations and times as they would normally experience during embryonic development. Typically, this involves four to six steps of treatment and each response from the ES cells makes them competent to respond to the next treatment. The final result is a population of the desired type of differentiated cell. In reality, the normal events are not understood with complete precision, the environment of the culture dish is somewhat different from that of the intact embryo, and different lines of ES cell behave slightly differently. For all these reasons there is some need for empirical testing as well as rational design, and different

Stages of induced development:

Endoderm →	Gut tube →	Posterior foregut →	Pancreatic bud →	Endocrine cells

Substances used in culture:

Activin Wnt	FGF10 Cyclopamine	FGF10 Cyclopamine Retinoic acid	DAPT Exendin 4	IGF1 Exendin 4 HGF

Media used for culture:

RPMI RPMI+0.2% serum	RPMI 2% serum	DMEM 1% B27	DMEM 1% B27	CMRL 1% B27

Maximum duration of stage:

4 days	4 days	4 days	3 days	3+ days

Fig. 21.3. An example of a procedure used to control differentiation of human ES cells, in this case towards pancreatic β-cells. There are five stages to the procedure, each involving a different culture medium and the addition of different factors and inhibitors to control the next developmental decision. IGF, insulin like growth factor 1; HGF, hepatocyte growth factor; DAPT is an inhibitor of Notch signaling; RPMI, DMEM, and CMRL are tissue culture media; B27 is a serum substitute. Reproduced from Slack, *Stem Cells. A Very Short Introduction*, Oxford University Press 2012, with permission from Oxford University Press.

Fig. 21.4 A somatic cell nuclear transfer experiment performed on *Xenopus*. The albino frogs are all descended from blastula nuclei from one male albino embryo, and are clones of this embryo. The host eggs were pigmented, from the female shown. Reproduced from Gurdon and Colman. *Nature* 1999; 402: 743–746, with permission from Nature Publishing Group.

laboratories may produce slightly different protocols. However, the essential strategy is common to all labs and may be illustrated by the example shown in Fig. 21.3, which shows a protocol for the production of insulin-producing cells from human ES cells.

Personalized pluripotent stem cells

Somatic cell nuclear transfer

The first nuclear transplantation experiments were carried out with frog embryos, the original motivation being to prove that all genes were retained in the nucleus of all cell types. It was shown that nuclei from blastula cells could be injected into secondary oocytes whose own nucleus had been removed, and would become reprogrammed and support normal embryonic development. A clone of genetically identical frogs can be created by injecting the nuclei taken from a single blastula into several different enucleated oocytes (Fig. 21.4). Blastula nuclei give a high proportion of viable embryos but when nuclei are taken from later stages they work much less well, and nuclei from adult differentiated cells hardly work at all. For example nuclei from differentiated skin keratinocytes of *Xenopus* can, with very low efficiency, support the development of tadpoles, although not of mature frogs.

Classic experiments

Somatic cell nuclear transfer

The paper of Briggs and King is the first example of successful nuclear transplantation into enucleated secondary oocytes, resulting in the formation of viable embryos. Gurdon extended this work and examined in detail the behavior of nuclei from differentiated cells. The paper of Wilmut et al. describes extension of the methodology to mammals with the creation of Dolly the Sheep.

Briggs, R. & King, T.J. (1952) Transplantation of living nuclei from blastula cells into enucleated frogs' eggs. *Proceedings of the National Academy of Science USA* **38**, 455–463.
Gurdon, J.B., Laskey, R.A. & Reeves, D.R. (1975) The developmental capacity of nuclei transplanted from keratinized cells of adult frogs. *Journal of Embryology and Experimental Morphogy* **34**, 93–112.
Wilmut, I., Schnieke, A.E., McWhir, J., Kind, A.J. & Campbell, K.H.S. (1997) Viable offspring derived from fetal and adult mammalian cells. *Nature* **385**, 810–813.

Following the well-known creation of Dolly the sheep, nuclear transplantation into enucleated secondary oocytes has been carried out with a variety of mammalian species, including mice, rats, and cattle (Fig. 21.5). As with *Xenopus*, if the donor nuclei are taken from after the earliest developmental stages they function with very low efficiency, but given enough oocytes it is possible to produce some viable embryos. The method is now known as "somatic cell nuclear transfer" or SCNT for short, a **somatic cell** being any cell that is not a germ cell. SCNT has not been successfully carried out in humans, except for the formation of triploid cells in which the haploid oocyte nucleus is retained. Even if the method were to work it would be considered highly unethical to attempt to use it to clone human beings, for example as a type of fertility treatment. One reason is the high probability of generating developmental abnormalities. The studies of animal embryos that have been made by SCNT have indicated many abnormalities, especially in the state of epigenetic marks across the genome, including DNA methylations and histone modifications, and also in the expression of imprinted genes (see Chapter 10).

Although very few people support the idea of using SCNT to clone human beings, there is widespread support in the scientific community for using the early embryos resulting from SCNT to establish ES cell lines. These would be genetically identical to the donor of the original nucleus, and so the cell lines could serve as a source of immunologically compatible grafts for their own donors. Because this procedure involves no formation of an actual cloned human being, it has become known as **therapeutic cloning**. Although the procedure has generated much ethical debate, it has also not actually been successfully achieved for humans, and a more practical as well as ethically noncontentious route to personalized pluripotent cells has now been developed in the form of **induced pluripotent stem cells**.

Induced pluripotent stem cells

iPS cells, which are extremely similar to ES cells, are made by introducing certain genes into normal cells, and then applying selective conditions such that those few cells that become reprogrammed to an ES-like state will grow into colonies (Figs 21.6, 21.7). The original gene set used was *Oct4, Sox2, Klf4,* and *Myc,* called "Yamanaka factors" after the originator. OCT4 and SOX2 are key transcription factors regulating the pluripotency of ES cells. The function of *Klf4* and *Myc* are less well defined but they are both expressed in ES cells and known as **oncogenes** that can cause cancer if activated inappropriately. Their effect may be exerted by promoting cell division and/or by making sites in the chromatin more accessible to the OCT4 and SOX2. Delivery of the Yamanaka factors to mouse fibroblasts using retroviral vectors generates about one ES cell colony per 10,000 input cells. This frequency is typically lower for human cells, but it can be increased by including extra genes or by treating with substances, such as the histone deacetylase inhibitor valproic acid, that open chromatin and increase accessibility of transcription factors to the genome. Of the four genes in the Yamanaka set, only *Oct4* is really obligatory, the others can be substituted by other genes, usually members of the ES cell pluripotency factor group, such as *Nanog*, or genes that stimulate cell division. The critical requirement is that overexpression of the genes can upregulate the autocatalytic network of pluripotency factors characteristic of ES cells. The efficiency of the process can also be increased by antagonizing other components with miRNA. For example downregulation of the *Ink/Arf* locus, which encodes antagonists of cyclin-dependent kinases, and of p53 (see Chapter 20), both improve efficiency.

iPS cells can be made from a variety of mammalian species. Mouse iPS cells will integrate into the mouse embryo following injection into blastocysts and are capable of contributing to all tissues including the germ line. This is the "gold standard" test for pluripotency. It has also been found possible to make iPS cells from cell types other than fibroblasts. In particular, lymphocytes isolated from blood and stimulated to divide by treatment with appropriate growth factors or lectins can serve as the starting material. This is potentially important because blood samples are so easily obtained from individual people. iPS cells have been made from lymphocytes in which gene rearrangement of antibody or T-cell receptor genes has already occurred, which shows that it is possible to reprogram even

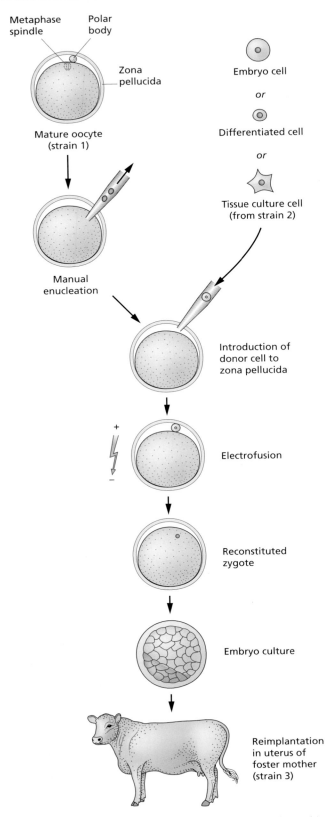

Metaphase spindle
Polar body
Zona pellucida

Mature oocyte (strain 1)

Manual enucleation

Embryo cell

or

Differentiated cell

or

Tissue culture cell (from strain 2)

Introduction of donor cell to zona pellucida

Electrofusion

Reconstituted zygote

Embryo culture

Reimplantation in uterus of foster mother (strain 3)

Fig. 21.5 Procedures used for cloning of whole mammals. The resulting animal will have the genetic constitution of the cell nucleus that was introduced into the oocyte.

well-differentiated cells by the introduction of suitable transcription factors.

The genes are delivered using retroviral or lentiviral vectors that insert their DNA into the chromosomes of the cells and enable high-level expression of the products. Eventually, the viral-encoded genes usually become "silenced," that is they cease to be active, probably because the chromatin around them becomes impermeable to the transcription machinery. Continued activity of the introduced genes is detrimental, because, being pluripotency factors, they will inhibit differentiation of the cells.

There is a set of standard tests that is used to characterize iPS cells and to establish whether a particular new line has a genuinely ES cell-like phenotype or not. This includes: an appropriate gene expression pattern, presence of the correct cell-surface markers and alkaline phosphatase activity, colony morphology, the removal of DNA methylation from promoters of endogenous pluripotency gene, and formation of embryoid bodies capable of differentiating into derivatives of all three germ layers. Also the cells should have a normal karyotype and viral genes should be fully silenced. One key property of mouse ES cells is the ability to form **chimeras** when injected into early mouse embryos. Ideally, these should be germ-line chimeras, that is some of the donor cells become germ cells (sperm or eggs according to the sex of the animal) and the resulting mice should be capable of reproducing and generating offspring of the ES cell genotype. Even more demanding is the tetraploid rescue test. Here, the host embryo is caused to become **tetraploid** (i.e. the normal chromosome number is doubled) by electric pulse-induced fusion of the first two cells into a single cell. For reasons that are not well understood, tetraploid cells cannot contribute to the fetus although they can still form placental structures. When good-quality ES cells are injected into a tetraploid host embryo, they can form the entire fetus with no significant contribution from host cells. This has also been achieved with iPS cells, showing that it is possible to generate a complete animal from a pure population of such cells, given the environment of the early mouse embryo.

For human iPS cells it is not possible for ethical reasons to inject the cells into embryos, so the standard approach is the teratoma assay where the cells are injected into an immunodeficient host animal. Good-quality iPS cells should grow to form a teratoma, and this should contain tissues characteristically derived from all three embryonic **germ layers**: ectoderm, mesoderm, and endoderm.

Mice have been produced from iPS cells in which the Yamanaka gene set is driven by the tetracycline response element. With the additional presence of rtTA in the mice, the Yamanaka genes can be expressed following treatment with doxycycline. For primary fibroblast cultures this gives a much higher yield of about 5% of "secondary" iPS cells and enables some mechanistic investigation of the process to be carried out. Such experiments show that the viral vector-encoded gene needs to be active for about 12 days to generate iPS cells,

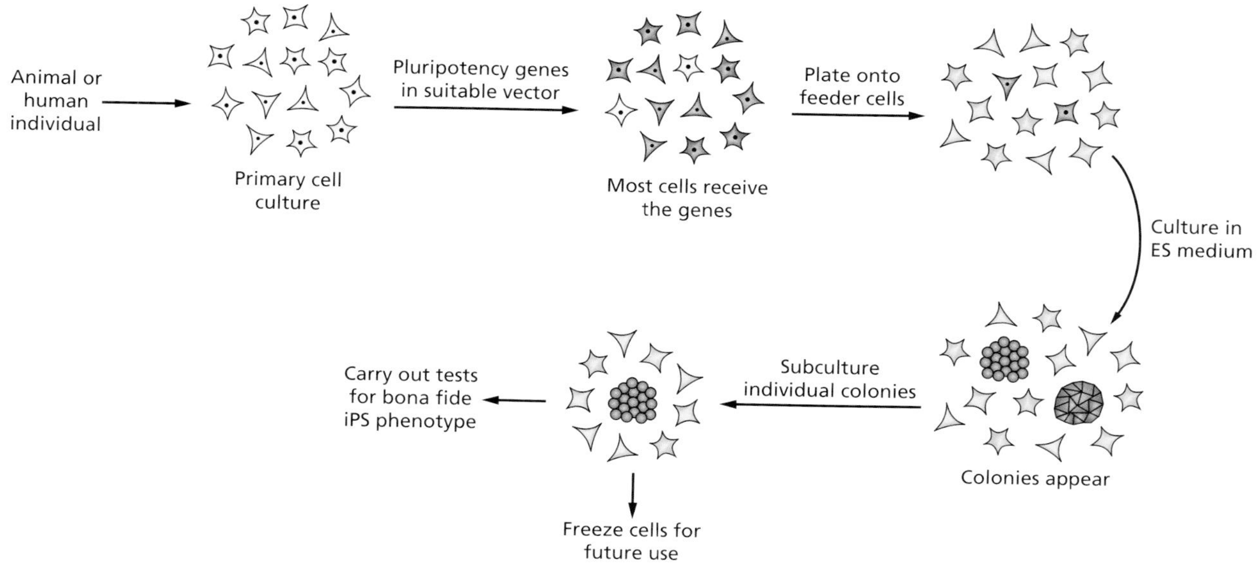

Animal or human individual → Primary cell culture

Pluripotency genes in suitable vector →

Most cells receive the genes

Plate onto feeder cells →

Colonies appear

Culture in ES medium

Subculture individual colonies →

Carry out tests for bona fide iPS phenotype ←

Freeze cells for future use

Fig. 21.6 Procedure for making induced pluripotent stem cells (iPS cells). Reproduced from *Coronary Heart Disease: Clinical, Pathological, Imaging and Molecular Profiles*, Vlodaver (ed.), New York 2011: Springer, with permission.

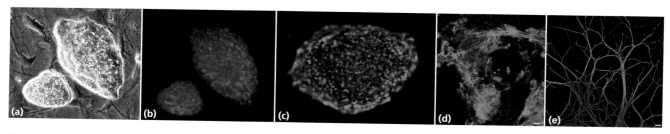

Fig. 21.7 Induced pluripotent stem (iPS) cells. (a,b). Mouse iPS cells viewed with phase contrast, and showing expression of NANOG (immunofluorescence, red). (c) A colony of human iPS cells. TRA1-81 immunostaining (green). (d,e) Differentiation of mouse iPS cells into (d) cardiomyocytes, visualized by immunostaining for cardiac troponin (green), and (e) neurons, visualized by immunostaining for neurofilament protein (green). Courtesy of Lucas Greder (part c), and James Dutton (parts a, b, d and e), University of Minnesota Stem Cell Institute.

Classic experiments

Induced pluripotent stem cells

The first paper describes the creation of cells resembling embryonic stem cells by introducing a cocktail of candidate genes into mouse fibroblasts and applying a selective medium. The next two papers describe creation of human iPS cells using similar methods. The last was an early *tour de force* indicating the type of cell transplantation therapy that iPS cells could make possible.

Takahashi, K. & Yamanaka, S. (2006) Induction of pluripotent stem cells from mouse embryonic and adult fibroblast cultures by defined factors. *Cell* **126**, 663–676.

Yu, J., Vodyanik, M.A., Smuga-Otto, K., Antosiewicz-Bourget, J., Frane, J.L., Tian, S., Nie, J., Jonsdottir, G.A., Ruotti, V., Stewart, R., Slukvin, I.I. & Thomson, J.A. (2007) Induced pluripotent stem cell lines derived from human somatic cells. *Science* **318**, 1917–1920.

Takahashi, K., Tanabe, K., Ohnuki, M., Narita, M., Ichisaka, T., Tomoda, K. & Yamanaka, S. (2007) Induction of pluripotent stem cells from adult human fibroblasts by defined factors. *Cell* **131**, 861–872.

Hanna, J., Wernig, M., Markoulaki, S., Sun, C.-W., Meissner, A., Cassady, J.P., Beard, C., Brambrink, T., Wu, L.-C., Townes, T.M. & Jaenisch, R. (2007) Treatment of sickle cell anemia mouse model with iPS cells generated from autologous skin. *Science* **318**, 1920–1923.

although the visible appearance of iPS cell colonies starts from about 7 days. Experiments with this system, as well as the ability to make iPS cells from a variety of differentiated types, lead most workers to believe that any cell type can be reprogrammed, and that the low efficiency arises from stochastic factors such as the accessibility in the chromatin of regulatory sites for the pluripotency network genes.

Although good-quality iPS cells closely resemble ES cells, the random elements in their formation mean that it is easy to isolate imperfect cell lines. It is commonly found that iPS cell lines carry some "memory," probably encoded in DNA methylation, of the cell type that they used to be. This also leads to a subsequent bias in the ease of differentiation in favor of this cell type.

Patient-specific iPS cell lines

Although there has been discussion of the use of the SCNT technique to make ES cell lines from individual patients ("therapeutic cloning"), it is not a practical possibility because of the low efficiency of the process and the difficulty of obtaining human oocytes. In contrast, patient-specific iPS lines are easy to make since the starting material required is a small skin biopsy or blood sample. Fibroblasts can be cultured from the dermal layer of the skin biopsy, or white cells can be isolated from the blood and activated to divide by treatment with growth factors or lectins. Many iPS lines have been established from patients carrying specific mutations or genetic diseases. A further attraction of the patient-specific cell line is that any differentiated cells made from it will be a perfect immunological match to the donor and can therefore potentially be grafted back to that individual without the use of immunosuppressive drugs.

The problem of gene insertion

The original methods for making iPS cells used replication-deficient retroviruses, or the related lentiviruses, as vectors to deliver the genes (see Chapter 13, Fig. 13.4). These types of virus have an RNA genome that is reverse transcribed to a DNA copy postinfection. The DNA can then integrate into the host genome and so the iPS cells continue to carry copies of the viral inserts, which usually become silenced after a period.

Gene insertions can create mutations during the integration process because they may interrupt host genes, or cause host genes to be inappropriately activated. Moreover, the silenced genes encoding the pluripotency factors may become reactivated at low frequency and may subsequently cause the formation of tumors. This has been noticed in mice grown from iPS cell chimeras, especially when the *Myc* gene was one of those used to make the iPS cells. For these reasons, in order to make clinical delivery of patient-specific cells a reality it is necessary to use methods that do not involve gene integration. At present the most promising methods are those that still retain replication of the vectors postdelivery. One is the use of episomes:

circles of DNA with their own origin of replication. These rarely integrate into the genome and eventually become lost from the cells as they divide. Another is the Sendai vector, a replication-defective vector derived from the animal pathogen Sendai virus. This has a single-stranded RNA genome which replicates via a double-stranded RNA in the cytoplasm. Again, the virus is eventually lost from the cell line in the course of cell division. A third is the use of selected microRNAs that remove repressors to upregulation of the pluripotency genes.

Direct reprogramming

A spin-off from the iPS cell technology has been the identification of groups of transcription factors capable of reprogramming cells directly from one differentiated type, usually dermal fibroblasts, to another. In each case the successful gene combination was identified by testing many genes encoding transcription factors involved in the normal development of the cell type in question, and refining the selection to a small group that gave the optimal result. The screens usually use a reporter of the desired cell type, for example *Tau-GFP* uses the promoter of the gene encoding the neuronal Tau protein to label cells with a neuron-like phenotype. The genes are introduced using replication-defective viruses and successful transformations have been brought about from fibroblasts to neurons, cardiomyocytes, or hepatocytes, using appropriate sets of factors. One problem with these cell types is that they are not proliferative. So, unlike the situation with iPS cells, it is not possible to expand the new cell type *in vitro* in order to scale it up. However, reprogrammed hepatocytes transplanted to regenerating liver can proliferate, and with suitable fluorescent reporters or cell-surface markers, nonproliferating cells can be purified in small quantities by fluorescence-activated cell sorting. Use of doxycycline-inducible lentivirus to express the factors enables them to be up- and downregulated at will. Using such vectors, it is found that after a few days the transformed cells have activated an endogenous program of gene expression for the new cell type and no longer need the vector-encoded products.

Direct reprogramming does not recapitulate normal development. Instead it causes "jumps" in an unnatural manner from one cell type to another. In terms of Waddington's "epigenetic landscape" (Chapter 4, Fig. 4.13), direct reprogramming would represent a movement of a cell from one terminal valley to another, while the formation of iPS cells might be regarded as moving backwards up the network of valleys to the origin. Since direct reprogramming is brought about by the deliberate alteration of regulatory gene expression, there is no particular reason why unnatural jumps should not be achieved. On the other hand, it is known that such jumps are very rare when cells are simply exposed to different environmental conditions, as was established in the investigation of transdifferentiation events following a hematopoietic stem cell graft (see Chapter 18).

Proposed cell transplantation therapies using pluripotent stem cells

Future cell transplantation therapies based on pluripotent stem cells will involve making the required differentiated cells *in vitro* and then implanting them into the appropriate site in the patient. They will not involve implanting the pluripotent stem cells themselves because one of the properties of both ES and iPS cells is to form teratomas when implanted into a host. Like most tumors, teratomas are very dangerous and a primary safety requirement for this type of therapy is to ensure that there are no pluripotent cells left in the cell population to be implanted, so that the danger of forming a teratoma is negligible.

The most sophisticated type of procedure envisaged, combining iPS cells, gene therapy, and hematopoietic transplantation, is illustrated by the animal experiment depicted in Fig. 21.8. Here a strain of mouse was used that had its own hemoglobin β genes replaced by the human hemoglobin βS allele causative of sickle cell anemia. iPS cells were prepared from tail-tip fibroblasts. Then the gene defect was repaired by gene targeting, using methods described in Chapter 10, such that one sickle cell allele was replaced by a normal human hemoglobin β gene.

Then hematopoietic stem cells were generated from the iPS cells. This requires the formation of embryoid bodies and the overexpression of *Hoxb4*. Finally, the sickle cell mice were irradiated to destroy their own hematopoietic stem cells and grafted intravenously with the differentiated iPS cells. This resulted in repopulation of the bone marrow by hematopoietic stem cells from the graft and hence repopulation of the defective red blood cells by normal ones.

β-cells

Diabetes is a common disease characterized by high and uncontrolled levels of blood glucose. Under normal circumstances, a rise in blood glucose following a meal will stimulate the β-cells of the pancreas to secrete insulin and this will lead to uptake of the glucose by adipose tissue and muscle, and also by the liver where it is converted into glycogen and fats. In the absence of insulin, the concentration of glucose in the blood will rise because it cannot be absorbed into the tissues.

There are two main types of the disease. Type 1 diabetes usually strikes during childhood or young adulthood and is caused by a loss of β-cells due to an autoimmune attack. The

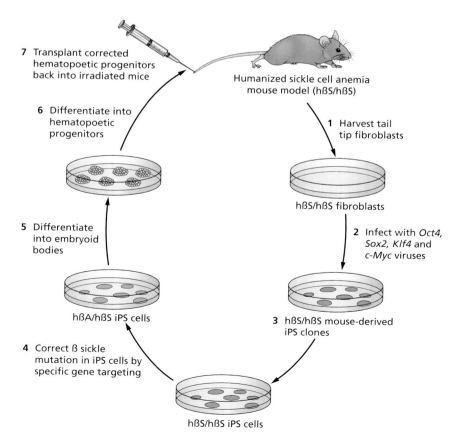

Fig. 21.8 A protocol used in a model experiment to cure mice of sickle cell disease. The mice have humanized hemoglobin β genes carrying the sickle cell allele. iPS cells are prepared, and the gene defect repaired by homologous recombination with a good copy. Then the cells are caused to differentiate to hematopoietic stem cells by overexpression of *Hoxb4*. Then the sickle cell mice are irradiated to remove endogenous hematopoietic stem cells and grafted with the iPS cells. The result is colonization of the blood with the genetically repaired cells. hβA, normal allele; hβS, sickle cell allele. Reproduced from Hanna *et al. Science* 2007; 318: 1920–1923, with permission from American Association for the Advancement of Science.

7 Transplant corrected hematopoetic progenitors back into irradiated mice

Humanized sickle cell anemia mouse model (hβS/hβS)

6 Differentiate into hematopoetic progenitors

1 Harvest tail tip fibroblasts

hβS/hβS fibroblasts

5 Differentiate into embryoid bodies

2 Infect with *Oct4, Sox2, Klf4* and *c-Myc* viruses

hβA/hβS iPS cells

3 hβS/hβS mouse-derived iPS clones

4 Correct β sickle mutation in iPS cells by specific gene targeting

hβS/hβS iPS cells

initial causes of the autoimmunity remain unclear but may be triggered by a viral infection or by genetic predisposition. Although β-cells can regenerate to a limited extent, patients with type 1 diabetes will eventually lose all their β-cells and will then die rapidly without treatment. They are absolutely dependent on injected insulin and need to monitor their diet carefully and inject the correct dose of insulin regularly to achieve reasonable control of the blood glucose level.

Type 2 diabetes is more common and tends to develop at older age. It is a complex and multifactorial disease, but usually seems to involve some pathology in the β-cells such that they cannot adapt to an increased demand for insulin, for example following the development of obesity. There is also often an insulin resistance of the peripheral tissues such that the available insulin does not have a sufficient effect and cannot drive uptake of the excess glucose. Therapy for type 2 diabetes usually involves treatment with various drugs to slow the release of glucose from the gut, to enhance the secretion of insulin from β-cells, and to enhance the activity of the available insulin. Often, as the disease progresses, type 2 patients also need to take insulin by injection.

With good care, diabetes can now be managed so that the patients expect a lifespan approaching normal, but the inability to control blood glucose concentration with precision causes damage to blood vessels. In the long term this damage leads to complications, which can be severe and distressing. They include heart disease, stroke, blindness, and peripheral vascular disease leading to difficulty healing minor wounds and sometimes even to amputations. Diabetes is one of the top targets for cell transplantation therapy based on pluripotent stem cells because it can build on an existing form of therapy: islet transplantation.

Islet transplantation for type 1 diabetes

A partly effective form of cell transplantation therapy for diabetes has been developed in which islets taken from the pancreases of deceased human organ donors are grafted into patients. This therapy is mostly used for patients with unawareness of low blood glucose, which can be very dangerous because it can quickly lead to unconsciousness or death. The technique involves taking islets from the pancreas of a deceased organ donor, and infusing them into the liver through the hepatic portal vein, which normally connects the intestine and liver. The islets lodge in the liver and will secrete appropriate amounts of insulin to manage the glucose levels that are encountered.

The therapy has proved quite successful. All treated patients do recover from their unawareness of hypoglycemia and some are able to stop taking insulin altogether. The effectiveness can be monitored by measuring C-peptide in the serum. This molecule is excised from the proinsulin polypeptide during maturation of the two chain insulin protein, and its level serves as a measure of β-cell activity.

But islet transplantation suffers from two major problems. Firstly, as for all types of graft that are dependent on organ donors, the supply of donor cells is hopelessly inadequate to meet the demand. Secondly, these grafts are **allografts** and subject to rejection by the immune system of the host, meaning that the recipients have to receive immunosuppressive drugs for the rest of their lives. The drugs can cause unpleasant side effects and they also damage the β-cells of the graft and reduce their lifespan and effectiveness.

The potential of stem cell research in this field is twofold. If β-cells could be made *in vitro* from pluripotent stem cells then the supply will become potentially infinite. Furthermore, if the stem cell source is a line of iPS cells made from the individual patient, then the cells will be a perfect genetic match and no immunosuppression should be required to suppress graft rejection. Because type 1 diabetes is an autoimmune disease, the patient is still likely to mount an autoimmune attack on the graft even if it is a perfect genetic match. It will still, therefore, require some immunosuppression to control it, although probably a less intense regime will be sufficient than is required to protect an allograft from rejection. Another method for avoiding both graft rejection and autoimmunity is the encapsulation of the cells in a partially permeable material such as alginate, that will allow passage of insulin and nutrients but not of cytotoxic lymphocytes from the host.

Making β-cells from pluripotent stem cells

The methods for making β-cells depend on a good understanding of normal β-cell and pancreatic development (see Chapter 16). Starting from the primitive streak stage of the embryo, first the endoderm germ layer is formed, this then becomes subdivided and a foregut territory is established, the foregut produces two pancreatic buds, which grow out from the primitive gut tube and subsequently fuse together to become a single organ. Within the pancreatic buds, endocrine progenitor cells develop. Some of these will become β-cells, and the rest will become the other endocrine cell types found in the islets of Langerhans. The protocols designed to direct differentiation of β-cells from human pluripotent stem cells therefore involve five steps: to endoderm, to foregut, to pancreatic bud, to endocrine precursor, and finally to β-cell. Each of these steps is achieved by treatment with specific inducing factors or hormones or inhibitors of signaling pathways. The success of each step is monitored by detecting the activation of the key genes that are known to be required: for example *SOX17* (endoderm), *HNF4A* (foregut endoderm), *PDX1* (pancreatic bud), *NGN3* (endocrine precursor), *INSULIN* (β-cell). Different labs have devised slightly different protocols but they are all similar in principle and an example was shown in Fig. 21.3 above. At present, the differentiation protocols are not perfect, as the proportion of insulin-producing cells falls well short of 100% of the cells in the dish. Moreover the cells are not fully mature, resembling rather an immature type of β-cell found in the fetal pancreas, which is not responsive to glucose.

It is possible to test the effectiveness of such cells by implanting them into immunodeficient mice whose own β-cells have been ablated by treatment with the drug streptozotocin. Immunodeficient mice must be used to avoid rejection of human cells that are implanted. The strain usually used is called NOD-SCID, although even more immunodeficient models are available. The NOD-SCID mouse is not diabetic because the SCID defect prevents development of autoimmunity, hence streptozotocin treatment is needed to obtain a diabetic mouse able to tolerate a graft of human cells. Grafts are usually inserted into the kidney, which allows for good vascularization and cell survival.

It has been shown that relief of diabetes can be obtained in such mice following implantation of β-cell precursors derived from human ES cells. If the grafted kidney is later removed, the blood glucose goes up again. The animal can survive with one kidney, and the removal of the grafted kidney is an important control indicating that the effects were due to the graft and not to something else. Based on the animal experiments and the existing clinical procedure of islet transplantation, it is likely that pluripotent stem cell-produced β-cells will be used for islet transplantation in the near future.

Dopaminergic neurons

Parkinson's disease is a progressive disorder affecting movement. It is the second most common neurodegenerative condition after Alzheimer's disease. The average age of onset is 60 and the disease is characterized by rigidity, tremor, slow movements, and, in extreme cases, inability to move. Although Parkinson's disease primarily affects movement it can also affect language and cognitive ability. The movement problems arise from a decreased stimulation of the motor region of the cerebral cortex, which means that the motor cortex cannot properly control movement and coordination. This is caused by a loss of dopaminergic neurons from the substantia nigra region in the brainstem. In Parkinson's disease, the degenerating neurons die and in the course of their degeneration they display structures called Lewy bodies. Parkinson's disease patients have lost 80% or more of their dopamine-producing cells by the time symptoms appear.

Dopamine is normally made from L-DOPA (L-3,4-dihydroxyphenylalanine). From the 1960s, L-DOPA, along with other drugs, has been administered to treat Parkinson's disease. It does have beneficial effects but these are accompanied by some problems with uncontrolled movements, and side effects such as mood disorders and sleep disturbances. As the disease progresses, the freezing to immobility and the loss of cognitive functions may cease to respond to L-DOPA.

Fetal midbrain grafts
There have been a number of small trials involving the implantation of grafts of tissue from the midbrains of 6 to 9-week-old human fetuses into the caudate nucleus and anterior putamen, which are the brain regions to which the dopaminergic neurons normally project. These grafts generate abundant dopaminergic neurons, which can survive long term and connect to other neurons. However, the effectiveness of the treatment is disputed. Also, rather surprisingly, some post mortem studies after many years have indicated that cells of the graft can develop the Lewy bodies, indicating that the disease can spread from the host to the graft cells. The mechanism for this spread is unknown.

One problem with this method is that the supply of early human fetuses as graft donors is limited. Also, they are obtained as a result of elective abortion and so there are inevitably some ethical issues. In order to increase cell supply and to avoid the use of aborted fetuses, various methods have been developed for the differentiation of human pluripotent stem cells into dopaminergic neurons. They generally involve a stage of embryoid body formation, followed by a neural induction process and the cultivation of cells resembling neural stem cells as aggregates in suspension. These are then treated with inducing factors such as FGF8, known to induce formation of the midbrain in normal embryonic development, and differentiated as a monolayer culture.

The standard animal model for cell therapy experiments is a rat that has been injected on one side of the brain with 6-hydroxydopamine, which causes destruction of the dopaminergic neurons. This one-sided lesion leads to various asymmetrical behaviors. The putative therapeutic cells are injected into the affected side of the brain, usually into the striatum which is the region to which dopaminergic neurons normally connect, and the efficacy is assessed by a battery of behavioral tests. After a suitable period of time the rats are sacrificed and their brains are analyzed for the persistence and differentiation state of the transplanted cells. Experiments of this sort do indicate effectiveness of the treatment in terms of improving behaviors. They also indicate long-term persistence of the grafted cells, including many dopaminergic neurons, and there are usually no teratomas formed.

Cardiomyocytes

The most common form of sudden death in modern developed societies is the heart attack (myocardial infarction), in which one or more coronary arteries becomes blocked and the sector of heart muscle that it normally supplies becomes deprived of oxygen. Unless the blockage resolves spontaneously, or as a result of emergency treatment, the affected region of heart muscle will die in about 1 hour. If the damage is extensive enough to abolish most cardiac function, the individual will also die. If the affected region is more limited, the patient will survive but with a permanently damaged heart. There is some dispute about whether cardiomyocytes are normally replaced during life, but if they are replaced it is at a very slow rate, and there does not appear to be any significant regeneration in areas of damage. Instead, the damaged areas fill up with cardiac fibroblasts, which secrete extracellular matrix material, and so the region of dead muscle becomes replaced by a scar. This has

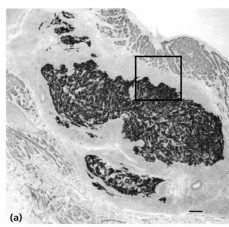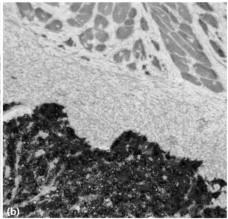

Fig. 21.9 Cell therapy of the heart. Immunodeficient rats were subject to coronary artery ligation to cause an infarct. Four days later the damaged regions were injected with cardiomyocytes prepared from human ES cells, and analyzed after 4 weeks. The red is immunostain for myosin heavy chain, the brown (visible at high power(b)) is *in situ* hybridization for a human pan-centromeric sequence. All the cardiac muscle visible is graft derived. Scale bar 100 μm. Reproduced from Laflamme *et al. Nature Biotech* 2007; 25: 1015–1024, with permission from Nature Publishing Group.

mechanical integrity but is inactive when it comes to contraction, and so the heart function is correspondingly diminished. If the heart is subjected to a lot of excess stress then surviving cardiomyocytes will become enlarged and less mechanically competent, leading eventually to heart failure. The loss of function due to heart attacks is the most common cause for this, although there are other causes such as high blood pressure or valve disease. By definition, heart failure means an inability to supply the rest of the body with enough blood. It leads to numerous problems and is likely to prove fatal in a shorter or longer period, depending on severity.

There is currently no clinical cell transplantation therapy of the heart based on the introduction of new cardiomyocytes. However, in the belief that grafts of healthy heart muscle cells could help reverse the effects of heart attacks, many groups are developing methods to make cardiomyocytes from human pluripotent stem cells, and testing protocols for cell replacement in animal experiments. As in the other cases, the differentiation protocols depend on guiding the pluripotent cell through a series of developmental steps resulting from exposure to a sequence of inducing factors. First mesoderm needs to be formed, then anterior type mesoderm, then cardiac progenitor cells, and finally the cardiomyocytes themselves. The production of human cardiomyocytes is in itself a very valuable objective. These cells are important for testing new drugs, not just those intended to act on the heart, but also all other drugs in line for human clinical trials because adverse side effects on the heart are so common. Animal experiments have some value in predicting these effects, but more reliable results may be obtained with human cells, and for obvious reasons live human cardiomyocytes are generally not available.

The animal models for cell transplantation therapy of the heart usually involve making a coronary artery ligation to cause damage in a specific sector of the heart muscle. Then the cells are injected into the affected part, and the heart function is studied with a set of physiological tests. Eventually, the animal is sacrificed and the presence, differentiation state, and spatial distribution of the graft-derived cells can be established. Some success has been achieved, although there are experimental difficulties associated with the fact that rat or mouse heart beats at several hundred beats per minute, rather than the 60–80 of the human heart. Figure 21.9 shows the results of implanting cardiomyocytes made from human ES cells into the heart of an athymic (immunodeficient) rat. In general, published results of animal experiments show persistence of the grafted cells and some improvement in cardiac function. The interpretation is complicated by the fact that the grafts generally contain progenitor cells for blood vessels as well as cardiomyocytes themselves. Indeed, it is widely believed that the normal cardiac progenitor cells in embryonic development are multipotent and can form either the endothelial lining of blood vessels or smooth muscle, which is found in the outer layer of arteries, or the cardiomyocytes themselves. More blood vessels may improve blood supply to the damaged region and improve the function of residual heart muscle.

Retinal pigment epithelium

In the eye, in the center of the retina, lies a small (5 mm diameter) pigmented area called the macula. It contains the highest density of cone-type photoreceptors in the retina and is responsible for the high level of detail and discrimination in the central part of the visual field. It is quite common for this part of the retina to deteriorate with age, with loss of photoreceptors, and about 10% of individuals over 65 have some degree of age-related macular degeneration (ARMD). In severe cases this can lead to loss of central vision, which prevents reading, recognition of faces, and other tasks requiring high visual acuity. Although some peripheral vision remains, many patients with severe ARMD are legally classified as blind.

There are two main forms of ARMD. The "dry" form is associated with the appearance of debris in the region and is thought to be due to a defect in the retinal pigment epithelium (RPE),

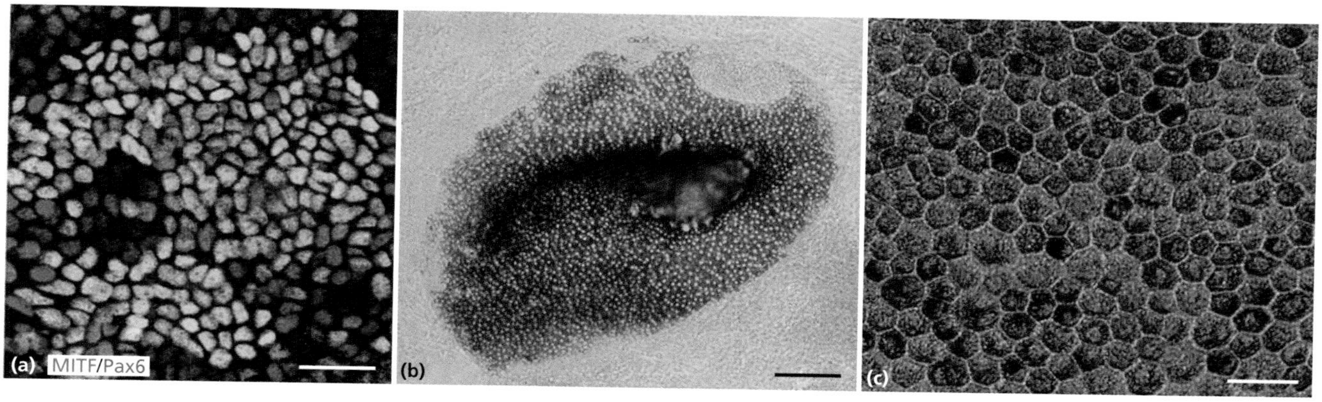

Fig. 21.10 Differentiation of retinal pigment epithelium from cynomolgus monkey ES cells. The cells were allowed to differentiate as small clumps in serum-free medium containing Lefty and Dkk, then transferred to a polylysine–laminin–fibronectin substrate. (a) Immunostaining of the transcription factors MITF and PAX6, characteristic of pigment cell precursors. (b) Appearance of pigment cells in a colony. (c) More mature pigment cells forming an orthogonal array. Scale bars (a) 30 μm; (b) 100 μm; (c) 20 μm. Reproduced from Osakada *et al. Nature Biotech* 2008; 26: 215–224, with permission from Nature Publishing Group.

which is a layer of pigmented cells lying beneath the photoreceptors, originally formed from the outer layer of the optic cup. The "wet" form is a type of overgrowth of blood vessels from the eye capsule (the choroid) into the subretinal space. There is no treatment for the dry form. The wet form can be treated by laser ablation of the new blood vessels and/or by injection of specific antibodies that antagonize blood vessel growth.

Experiments have been done using an animal model called the Royal College of Surgeons (RCS) rat. This carries a spontaneous mutation which prevents the RPE from removing debris normally arising from the photoreceptor layer. The resulting accumulation of debris leads to death of photoreceptors and to vascular changes. A graft of RPE into the subretinal space of the RCS rat can preserve the neighboring photoreceptor layer and can maintain visual ability, as assessed both by electrophysiological recording and by behavioral tests. A human RPE cell line also works in this animal assay. Although the eye is often considered a site relatively inaccessible to cells of the immune system, this ceases to be the case as soon as there is any damage present, so immunosuppression is still required.

These experiments have led to the idea that RPE cells could be made from human pluripotent stem cells and used as subretinal grafts to treat ARMD. It turns out that RPE is a relatively easy cell type to obtain from pluripotent stem cells, and is very obvious because of its pigmented character (Fig. 21.10). When grafted to the subretinal space of the RCS rat, such RPE cell preparations do improve visual function and they do not form teratomas. The first clinical trial of this technique was approved by the US Food and Drug Administration in 2011. RPE will be prepared from human ES cells and used to treat dry ARMD and also another type of macular degeneration called Stargardt's disease. This is a favorable opportunity for cell transplantation therapy for several reasons. One is the lack of any alternative treatment for the dry form of macular degeneration. Another is the ability to monitor events after grafting by observation

through the pupil of the eye. A third is the concept that if a teratoma were to arise, the patient could still be saved from its effects by the removal of the affected eye.

Spinal repair

The first pluripotent stem cell-derived clinical trial started in 2009 for the use of remyelinating cells to treat spinal trauma. Following spinal trauma the paralysis and loss of sensation below the level of the injury is typically caused by damage to descending (motor) or ascending (sensory) nerve fiber tracts. The cell bodies from which the fibers of the tracts originate are likely to be at remote locations and still alive. Spinal trauma also causes local cell death of both neurons and glial cells, and the formation of a glial scar which has the effect of inhibiting any regrowth of nerve fibers through the region. In addition some fibers that are not otherwise damaged may be subject to loss of their myelin sheaths, and this loss prevents efficient electrical signal transmission.

The trial is examining grafts of **oligodendrocytes**, made from human ES cells. These are the class of glial cell normally responsible for myelination of axons, and the rationale is that some of the surviving nerve fibers will be remyelinated and will then recover functional activity.

Introduction of new therapies

The opportunities from the applications of pluripotent stem cells are considerable. But progress is likely to be slow. There are several reasons for this. All new therapies are subject to intense regulatory scrutiny from the Food and Drug Administration in the USA, and its counterparts in other countries, and obtaining approval to commence a clinical trial is a long and expensive process.

New directions for research

To advance the aims presented in this chapter some key scientific and technological advances will be needed:

1 A reliable method for making iPS cells without the use of integrating vectors.

2 More robust and reliable methods for bringing about directed differentiation of ES and iPS cells.

3 A better understanding of the extent to which "direct reprogramming" of cell type is possible.

4 More sophisticated methods of cell delivery that maintain viability and function of the graft after transplantation.

One issue is safety. Cell transplantation therapy is relatively new and involves new potential hazards. One obvious risk is the formation of teratomas from persistent pluripotent cells present in the graft. It is important to show that that the methods employed remove all pluripotent cells with a high degree of confidence. Another type of risk arises from the use of animal products during the preparation of the graft, for example feeder cells or animal serum. These may contain viruses that could cause disease in humans. It means that all cells used for therapy have to have a history of cultivation in highly defined media free from all animal products. To address this and other issues there are exacting standards known as "Good Manufacturing Practice" (GMP), which must be adhered to during preparation of the cells. It is not necessary that a new therapy should be 100% guaranteed safe, rather the level of risk that may be tolerated depends on the severity of the disease. For example when bone marrow transplantation was first introduced, the patients suffered from leukemias that were known to be rapidly fatal, and so testing a high-risk new treatment was acceptable. For diabetes, where the existing treatment is good, the level of tolerable risk is very much lower.

Secondly, there is the issue of effectiveness. It is not uncommon for procedures to work well in experimental animals and then not to work in humans, so treatments are often abandoned after the first clinical trials. Furthermore, a new treatment has to be not only effective, but also more effective than existing treatments. There are existing treatments for most of the conditions described above, for example sophisticated insulin pumps for type 1 diabetes, which are very effective, or drug treatments and deep electrical stimulation for Parkinson's disease, which are moderately effective. Given the likely costs and the possible risks, stem cell therapy has to do better that the best current treatment in order to be adopted.

Finally, there is the issue of cost. Large-scale cell culture under GMP conditions is very expensive. Patient-specific cell culture will be even more expensive because it will require establishment of a new iPS cell line from the patient, its full characterization, its growth to large scale, and its directed differentiation to the cell type required. Analysts currently consider that this is too costly to be feasible. However, previous technologies have become much cheaper as they are developed and pluripotent stem cells may well follow this trend.

Key points to remember

- Most current stem cell therapy is hematopoietic stem cell transplantation.
- Allogeneic grafts require immunosuppression while autografts (or grafts from an identical twin) do not.
- Pluripotent stem cells (ES and iPS cells) can be grown *in vitro* without limit, and can differentiate into all cell types of the body.
- Human ES cells are grown from blastocysts which have been stored frozen in fertility clinics, and are donated for the purpose by the parents.
- Human ES cells share some properties with mouse ES cells, but certain differences indicate that they correspond to a more mature cell type.
- Somatic cell nuclear transplantation into enucleated secondary oocytes can, with poor efficiency, lead to viable embryos.
- Induced pluripotent stem cells (iPS cells) are made by introducing pluripotency genes into differentiated cell types and applying selection. iPS cells can be made from samples taken from individual human patients.
- Direct reprogramming of other cell types to neurons, cardiomyocytes, or hepatocytes can sometimes be achieved by overexpression of selected groups of transcription factors.
- In addition to clinical transplantation, human pluripotent stem cells can be used to study human development, to investigate cellular pathology of genetic diseases, and to test drugs on human differentiated cells.
- Directed differentiation of pluripotent stem cells can be achieved by exposing the cells to a sequence of factors to mimic the normal hierarchy of developmental decisions involved in formation of that cell type.
- Clinical trials are underway of embryonic stem cell-derived cells for treatment of spinal injury (oligodendrocytes) and retinal degeneration (retinal pigment epithelium). Active research is directed towards similar programs using pancreatic β-cells, cardiomyocytes, dopaminergic neurons, hepatocytes, and other cell types.

Further reading

General

Cohen, C.B. (2007) *Renewing the Stuff of Life Stem Cells, Ethics, and Public Policy*. New York: Oxford University Press.

Goldstein, L.S.B. & Schneider, M. (2010) *Stem Cells for Dummies*. Indianapolis, Indiana: Wiley Publishing Inc.

Mummery, C., Wilmut, I., van de Stolpe, A. & Roelen, B. (2011) *Stem Cells: Scientific Facts and Fiction*. London: Academic Press.

Slack, J.M.W. (2012) *Stem Cells: A Very Short Introduction*. Oxford: Oxford University Press.

Mouse ES and iPS cells

Niwa, H. (2007) How is pluripotency determined and maintained? *Development* **134**, 635–646.

Nichols, J. & Smith, A. (2009) Naive and primed pluripotent states. *Cell Stem Cell* **4**, 487–492.

Hanna, J.H., Saha, K. & Jaenisch, R. (2010) Pluripotency and cellular reprogramming: facts, hypotheses, unresolved issues. *Cell* **143**, 508–525.

Marión, R.M. & Blasco, M.A. (2010) Telomere rejuvenation during nuclear reprogramming. *Current Opinion in Genetics and Development* **20**, 190–196.

Stadtfeld, M. & Hochedlinger, K. (2010) Induced pluripotency: history, mechanisms, and applications. *Genes and Development* **24**, 2239–2263.

González, F., Boué, S. & Belmonte, J.C.I. (2011) Methods for making induced pluripotent stem cells: reprogramming à la carte. *Nature Reviews Genetics* **12**, 231–242.

Nichols, J. & Smith, A. (2011) The origin and identity of embryonic stem cells. *Development* **138**, 3–8.

Plath, K. & Lowry, W.E. (2011) Progress in understanding reprogramming to the induced pluripotent state. *Nature Reviews Genetics* **12**, 253–265.

Human ES cells

Thomson, J.A., Itskovitz-Eldor, J., Shapiro, S.S., Waknitz, M.A., Swiergiel, J.J., Marshall, V.S. & Jones, J.M. (1998) Embryonic stem cell lines derived from human blastocysts. *Science* **282**, 1145–1147.

Pera, M.F. & Trounson, A.O. (2004) Human embryonic stem cells – prospects for development. *Development* **131**, 5515–5525.

Adewumi, O., Aflatoonian, B., Ahrlund-Richter, L., Amit, M., Andrews, P.W., Beighton, G., Bello, P.A., Benvenisty, N., Berry, L.S., et al. (2007) Characterization of human embryonic stem cell lines by the International Stem Cell Initiative. *Nature Biotechnology* **25**, 803–816.

Lensch, M.W., Schlaeger, T.M., Zon, L.I. & Daley, G.Q. (2007) Teratoma formation assays with human embryonic stem cells: a rationale for one type of human–animal chimera. *Cell Stem Cell* **1**, 253–258.

Yu, J. & Thomson, J.A. (2008) Pluripotent stem cell lines. *Genes and Development* **22**, 1987–1997.

Somatic cell nuclear transfer

Armstrong, L., Lako, M., Dean, W. & Stojkovic, M. (2006) Epigenetic modification is central to genome reprogramming in somatic cell nuclear transfer. *Stem Cells* **24**, 805–814.

Gurdon, J.B. (2006) From nuclear transfer to nuclear reprogramming: the reversal of cell differentiation. *Annual Review of Cell and Developmental Biology* **22**, 1–22.

Byrne, J.A., Pedersen, D.A., Clepper, L.L., Nelson, M., Sanger, W.G., Gokhale, S., Wolf, D.P. & Mitalipov, S.M. (2007) Producing primate embryonic stem cells by somatic cell nuclear transfer. *Nature* **450**, 497–502.

Yang, X., Smith, S.L., Tian, X.C., Lewin, H.A., Renard, J.-P. & Wakayama, T. (2007) Nuclear reprogramming of cloned embryos and its implications for therapeutic cloning. *Nature Genetics* **39**, 295–302.

Gurdon, J.B. & Melton, D.A. (2008) Nuclear reprogramming in cells. *Science* **322**, 1811–1815.

Direct reprogramming

Zhou, Q. & Melton, D.A. (2008) Extreme makeover: converting one cell into another. *Cell Stem Cell* **3**, 382–388.

Ieda, M., Fu, J.-D., Delgado-Olguin, P., Vedantham, V., Hayashi, Y., Bruneau, B.G. & Srivastava, D. (2010) Direct reprogramming of fibroblasts into functional cardiomyocytes by defined factors. *Cell* **142**, 375–386.

Vierbuchen, T., Ostermeier, A., Pang, Z.P., Kokubu, Y., Südhof, T.C. & Wernig, M. (2010) Direct conversion of fibroblasts to functional neurons by defined factors. *Nature* **463**, 1035–1041.

Cohen, D.E. & Melton, D. (2011) Turning straw into gold: directing cell fate for regenerative medicine. *Nature Reviews Genetics* **12**, 243–252.

Huang, P., He, Z., Ji, S., Sun, H., Xiang, D., Liu, C., Hu, Y., Wang, X. & Hui, L. (2011) Induction of functional hepatocyte-like cells from mouse fibroblasts by defined factors. *Nature* **475**, 386–389.

Jopling, C., Boue, S. & Belmonte, J.C.I. (2011) Dedifferentiation, transdifferentiation and reprogramming: three routes to regeneration. *Nature Reviews Molecular Cell Biology* **12**, 79–89.

hES cell differentiation and transplantation

Nistor, G.I., Totoiu, M.O., Haque, N., Carpenter, M.K. & Keirstead, H.S. (2005) Human embryonic stem cells differentiate into oligodendrocytes in high purity and myelinate after spinal cord transplantation. *Glia* **49**, 385–396.

Daley, G.Q. & Scadden, D.T. (2008) Prospects for stem cell-based therapy. *Cell* **132**, 544–548.

Murry, C.E. & Keller, G. (2008) Differentiation of embryonic stem cells to clinically relevant populations: lessons from embryonic development. *Cell* **132**, 661–680.

Mooney, D.J. & Vandenburgh, H. (2008) Cell delivery mechanisms for tissue repair. *Cell Stem Cell* **2**, 205–213.

Passier, R., van Laake, L.W. & Mummery, C.L. (2008) Stem-cell-based therapy and lessons from the heart. *Nature* **453**, 322–329.

Harlan, D.M., Kenyon, N.S., Korsgren, O. & Roep, B.O. (2009) current advances and travails in islet transplantation. *Diabetes* **58**, 2175–2184.

McKnight, K.D., Pei Wang & Kim, S.K. (2010) Deconstructing pancreas development to reconstruct human islets from pluripotent stem cells. *Cell Stem Cell* **6**, 300–308.

Ronaghi, M., Erceg, S., Moreno-Manzano, V. & Stojkovic, M. (2010) Challenges of stem cell therapy for spinal cord injury: human embryonic stem cells, endogenous neural stem cells, or induced pluripotent stem cells? *Stem Cells* **28**, 93–99.

Smith, K. (2010) Outlook: Parkinson's disease. Treatment frontiers. *Nature* **466**, S15–S18.

Regulatory and financial issues

Ährlund-Richter, L., De Luca, M., Marshak, D.R., Munsie, M., Veiga, A. & Rao, M. (2009) Isolation and production of cells suitable for human therapy: challenges ahead. *Cell Stem Cell* **4**, 20–26.

Goldring, C.E.P., Duffy, P.A., Benvenisty, N., Andrews, P.W., Ben-David, U., Eakins, R., French, N., Hanley, N.A., Kelly, L., Kitteringham, N.R., Kurth, J., Ladenheim, D., Laverty, H., McBlane, J., Narayanan, G., Patel, S., Reinhardt, J., Rossi, A., Sharpe, M. & Park, B.K. (2011) Assessing the safety of stem cell therapeutics. *Cell Stem Cell* **8**, 618–628.

Maziarz, R.T. & Driscoll, D. (2011) Hematopoietic stem cell transplantation and implications for cell therapy reimbursement. *Cell Stem Cell* **8**, 609–612.

Other applications

Nishikawa, S.-I., Goldstein, R.A. & Nierras, C.R. (2008) The promise of human induced pluripotent stem cells for research and therapy. *Nature Reviews Molecular Cell Biology* **9**, 725–729.

Saha, K. & Jaenisch, R. (2009) Technical challenges in using human induced pluripotent stem cells to model disease. *Cell Stem Cell* **5**, 584–595.

Rubin, L.L. & Haston, K.M. (2011) Stem cell biology and drug discovery. *BMC Biology* **9**, 42.

Zhu, H., Lensch, M.W., Cahan, P. & Daley, G.Q. (2011) Investigating monogenic and complex diseases with pluripotent stem cells. *Nature Reviews Genetics* **12**, 266–275.

Robinton, D.A. & Daley, G.Q. (2012) The promise of induced pluripotent stem cells in research and therapy. *Nature* **481**, 295–305.

Chapter 22

Evolution and development

The interface between developmental biology and evolutionary biology has a long history. At the beginning of the nineteenth century, the German embryologist Karl Von Baer noticed that the early stages of different types of vertebrate embryo were very similar to each other and proposed that the more general features of animals develop before the more specific. In the mid-nineteenth century, Ernst Haeckel formulated the theory that "ontogeny recapitulates phylogeny," in other words the developmental stages of an individual organism resemble the sequence of its evolutionary ancestors. This view made sense in the context of the then current Lamarkian theory of evolution in which heritable changes could arise from life experience and would therefore be "added on" to the end of a developmental sequence, but was more difficult to account for in terms of natural selection.

In the early twentieth century, neo-Darwinism became generally accepted. This is the synthesis of Darwin's theory of natural selection, Mendel's genetics, and the quantitative mathematical theory of genetics produced by Fisher, Wright, and Haldane. The mechanism of evolutionary change is considered to arise from mutations each of which confers a reproductive advantage on the individuals carrying them such that the mutation spreads through the population to become the wild-type allele. According to neo-Darwinism, morphological change will occur gradually and result from the action of many mutations each having a small effect. Change arising from natural selection is called **adaptive evolution** or adaptation. In the second half of the twentieth century, studies in molecular biology made it clear that a great deal of change in the primary sequence of DNA was not adaptive but **neutral evolution**, consisting of an accumulation of mutations of no selective consequence, which spread through the population by the effects of random sampling of alleles from one generation to the next (genetic drift).

The molecular understanding of development has had an important impact on evolutionary biology because it enables an approach to certain questions that could not be answered before. The first is the issue of long-range homologies. **Homology** between the parts of different animals was traditionally determined by methods of comparative anatomy, including a study of developmental stages. But our present knowledge of the molecular basis of developmental mechanism often makes it possible to determine homology even when there is no morphological similarity remaining. The study of early developmental mechanisms, especially around the **phylotypic stage**, has enabled us to be sure that all animals are derived from a single common ancestor. It has also enabled a credible reconstruction of the main gene expression domains of the primordial animal.

The second issue is the investigation of **developmental constraints** on evolutionary change. Given a particular early developmental mechanism, we can predict that mutations leading to loss or gain of function of particular key components will always be deleterious because the genes concerned have many different functions (**pleiotropic**). A potentially advantageous change to one function is highly likely to be accompanied by deleterious changes to others. This means that the available mutational variation in a breeding population is always limited. Selection cannot operate without variation and so this leads to the idea that there are developmental constraints preventing certain pathways of evolution.

The third issue is to understand what actually happened in evolution. In particular, we can now investigate the developmental basis of morphological changes of evolutionary significance. This has two aspects. Firstly, there is the identification of the change in the relevant developmental system that has caused the observed change of morphology. Secondly, there is

Essential Developmental Biology, Third Edition. Jonathan M.W. Slack.
© 2013 John Wiley & Sons, Ltd. Published 2013 by John Wiley & Sons, Ltd.

the discovery of the actual mutation(s) that became established in the population and brought about the change. These are not necessarily the same thing. For example it may be that a change of segment identity is caused by a change of Hox expression, but that the mutations bringing this about are not themselves in the Hox genes. This approach offers the possibility of solving the long standing evolutionary puzzle of whether **macromutations** (mutations of large effect), generally not favored in the neo-Darwinian tradition, can contribute to evolutionary change.

The study of evolutionary developmental biology necessarily involves studying organisms other than the "big six" laboratory model species. Examples met in this chapter are: amphioxus, as a "sister group" to the vertebrates; cnidarians, as a "sister group" to bilaterian animals; crustacea, as an example of extreme segmental diversity originating from a common ancestor; and snakes as an example of the loss of limbs. But evolutionary developmental biology may end up studying just about any organism. Fortunately, the technology for doing this has advanced considerably. Firstly, the relative ease of whole-genome sequencing now makes it possible to get a complete inventory of genes in any organism, as well as information about the chromosomal position of genes relative to the well-studied laboratory models. Secondly, *in situ* hybridization enables a study of gene expression patterns. Thirdly, electroporation may often be used to introduce genes where no more sophisticated method of transgenesis is available, and, finally, RNAi may often be used to knock down gene activity. Even so, it is usually important to have embryonic stages of the organism available and this means that those animals that can be domesticated such that they will undergo the whole life cycle in the lab. are favorable material for study.

Macroevolution

Living organisms are classified in a hierarchical way, and the subdivisions of life are called **taxa** (singular **taxon**). Animal species are grouped into genera, families, orders, classes, and the highest-grade taxa are the **phyla** (singular **phylum**) of which there are about 35, the exact number depending on the author. They vary in size from large phyla such as the Mollusca, containing tens of thousands of species, to tiny phyla such as the Loricifera, with only a handful of species. The vertebrates and insects, which are the most important groups from a developmental biology point of view, are not actually phyla but classes. The vertebrates make up by far the largest class of the phylum Chordata, and the insects are the largest class of the phylum Arthropoda.

A classification system (=**taxonomy**) can be completely arbitrary and still be useful in the sense that it allows for the unambiguous identification of specimens. However, there has long been an attempt in biology to make the taxonomy congruent with the actual evolutionary history of the organisms con-

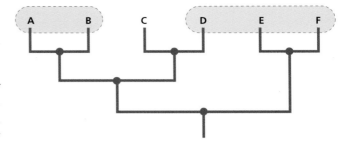

Fig. 22.1 Phylogenetic (=cladistic) taxonomy. A–F represent taxa arising from the phylogeny shown. A + B represents an acceptable higher taxon because it is a complete set of organisms derived from a common ancestor. D + E + F is not an acceptable higher taxon because some descendants of the common ancestor are omitted.

cerned. According to this tradition, each taxon should be a **clade** consisting of the complete set of descendants from a common ancestor (Fig. 22.1). Certain well-known groups are not clades but are tolerated for reasons of familiarity. For example the reptiles are not a clade because they do not contain the birds, which are descendants from within the reptiles and not a separate group derived from the vertebrate stem line. Although any number of arbitrary classification schemes are possible, there is only one "true tree," which follows exactly the bifurcating branches that actually occurred in evolution. This is called a **phylogenetic tree** whether it deals with the phyla themselves or with lower level taxa. It follows that each node of a phylogenetic tree should correspond to a population of real ancestral organisms that split to give rise to the two branches. Of course, this will only be literally true at the species level of detail, because what actually evolves are species in the sense of interbreeding populations of organisms. Evolution above the species level is called **macroevolution**. It must be remembered that there is no motor of evolution for higher taxa because this would involve parallel changes being selected simultaneously across many different species, which is exceedingly unlikely. Macroevolution is simply the consequence of a lot of species-level evolution, including extinction events, which over long time periods shifts the overall composition of life on Earth. So a node on a phylogenetic tree depicting higher taxa still corresponds to one specific ancestor, namely the individual species that split to give two lines of descent, but there will be many other organisms of related taxa around at the same time and represented on the same tree.

The construction of phylogenetic trees is a complex business, outside the scope of this book. But there are two approaches in principle. One involves scoring large numbers of characters, including continuous quantitative characters, and constructing the tree on the basis that more similar specimens should be closer together in descent. This is called numerical, or phenetic, taxonomy. The other involves applying certain assumptions about the likelihood of different types of change to try to deduce an actual phylogenetic tree. This is called cladistic, or phylogenetic, taxonomy. There is no guarantee that application of either method will produce the "true tree," but the more data that are used, the more accurate the trees are likely to be.

Zoologists have long drawn a distinction between two types of similarity between different organisms. They are called **homology** and **analogy**. If two structures are homologous, it means not only that they look similar, but that this similarity is due to descent from a common ancestor, which possessed the ancestral version of the part in question. An example is the **tetrapod** limb. Many types of vertebrate, from humans to crocodiles, have limbs that are obviously similar in the number and arrangements of their bones and muscles, and in their position on the body. The reason they are similar is because there was once an ancestral tetrapod, from which all existing tetrapods are descended, and its limbs were like that. Because they share common descent, homologous structures should be formed by similar developmental mechanisms. By contrast, if two structures are analogous it means that there is no common ancestry and the parts look similar because the pressure of natural selection has forced a convergence of structure to meet the need for a similar function. An example is the insect wing versus the wing of a bird. Obviously, these cannot have a winged common ancestor, because we know from the fossil record that there was a long history both of vertebrates and of insects before any winged members appear in either group. Furthermore, the common ancestor of insects and vertebrates must have been a marine organism as it would have existed before colonization of the land by any animal group. Note that although the wings themselves cannot be homologous, various individual genes or genetic pathways involved in making the wings may still be homologous, an issue which is discussed below. In general, characters that are thought to resemble those of ancestors are called primitive or **basal** while those thought to result from adaptive evolution are called **derived**.

Molecular taxonomy

A great deal of modern taxonomy uses the primary sequences of genes as the raw data rather than morphological characters. Because only a fraction of the amino acids in a protein are necessary for its biological activity, the sequences of genes and proteins change gradually over evolutionary time. Most of the changes are **neutral**, or nearly neutral, that is they do not affect the survival or reproduction of the individual organism and they are neither favored nor discriminated against by natural selection. For any individual mutation there is a very small chance that it will eventually spread through the whole species and become the normal version of the gene or protein. Because there are many such mutations, and evolutionary time is very long, there are many substitutions of amino acids in those positions in the protein where they do not compromise the biochemical function. Population genetic theory predicts the existence of a **molecular clock**, indicating an approximately

linear relationship between the time since the divergence of two lineages and the number of differences between the sequences. Phylogenetic trees can be constructed from primary sequence data using various different types of algorithm, including ones that simply compare the number of differences (phenetic) and those that try to minimize the number of substitution events (cladistic). The ideal genes to use for these studies are those that can clearly be identified as homologous and that show some variation over the range of organisms being studied, but not so much that substitutions are piled on top of each other. For long-range phylogeny much use has been made of the large and small ribosomal RNA genes as these have retained the same function since before the divergence of prokaryotes and eukaryotes. Among protein -coding genes some favorites have been RNA polymerases, glycolytic enzymes, and cytochromes, as they all have essentially the same function in any organism. Molecular phylogenetic methods have the attraction that they are completely independent of morphology and can in principle be applied to a set of sequences from present-day organisms without knowing anything else about them. However, they do not necessarily predict a unique phylogenetic tree and the closest approach to the true tree is likely to be obtained by combining information from comparative morphology, molecular sequences, and fossil evidence.

The rise of molecular taxonomy has extended and refined our concepts of homology and analogy. Although many amino acids at inessential positions in a protein sequence will change because of neutral mutation, the amino acids at many other inessential positions remain the same. It is the presence of these inessential identities that show that two genes from different organisms are really homologous. But it is not entirely straightforward identifying homologous genes in different organisms because of the widespread occurrence of **gene duplication** in evolutionary time. If a gene becomes duplicated then only one copy is required to perform the original function, and the other copy is free to assume a different, although often related, function. In particular, among the vertebrates a high proportion of genes belong to multigene families. Two genes in the same organism that result from a gene duplication are called **paralogs**. Two genes in different organisms are called **orthologs** if they are directly related by lineage and preserve a common function. To make proper comparisons between species it is essential to identify the true ortholog, otherwise the comparison that is made will relate to the time of the gene duplication event, which may be much earlier than the time of the splitting of the ancestral species.

If two genes converge to the same function starting from totally different sequences they may have no sequence identity at all, or any they do have is confined to a small region of the molecule actually responsible for its catalytic or other biochemical activity. A molecular example of analogy is provided by the crystallins, which are the proteins making up the lens of the eye. In different types of vertebrate quite different sorts of protein have been brought into service to make a transparent lens and

Classic experiments

Long-range developmental homologies

The first two papers describe the breakthrough in which it was found that the sequence of anterior expression limits of the mouse Hox genes was the same as that for *Drosophila*. This was the key discovery implying the existence of a common developmental plan for all animals. However, readers should note that the nomenclature for Hox genes used in these papers is not the same as that used today.

The third paper showed that the *Small eye* gene in mice was the homolog of *eyeless* in *Drosophila*, and encoded Pax6. Overexpression in other imaginal discs of *Drosophila* of either

Drosophila or mouse *Pax6* could induce formation of ectopic eye tissue.

Duboule, D. & Dolle, P. (1989) The structural and functional organization of the murine Hox gene family resembles that of *Drosophila* homeotic genes. *EMBO Journal* **8**, 1497–1505.

Graham, A., Papalopulu, N. & Krumlauf, R. (1989) The murine and *Drosophila* homeobox gene complexes have common features of organization and expression. *Cell* **57**, 367–378.

Halder, G., Callaerts, P. & Gehring, W.J. (1995) Induction of ectopic eyes by targeted expression of the *eyeless* gene in *Drosophila*. *Science* **267**, 1788–1792.

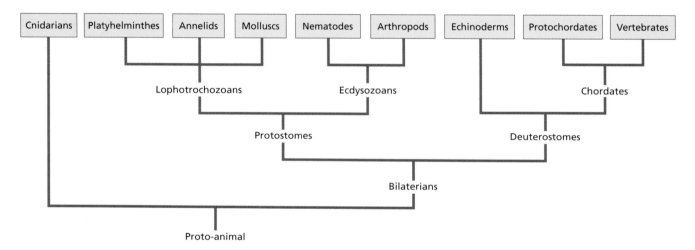

Fig. 22.2 A consensus phylogenetic tree for animals showing some key animal phyla.

although they serve the same physiological function, they have no primary sequence in common.

Phylogeny of animals

Figure 22.2 shows a modern consensus phylogenetic tree for a few of the most important animal groups. Because of the lack of morphological characters in common between the adult animals of different phyla, information from embryos and larvae has traditionally been used to attempt to draw up the phylogenetic tree.

The first of these characters is the number of germ layers. As we have seen, most animals have formed three tissue layers by the end of gastrulation, the ectoderm, mesoderm, and endoderm. They are called **triploblasts**. Only three phyla are not

triploblastic, the Porifera (sponges), which do not have well-defined tissue layers at all, the Cnidaria (*Hydra*, jellyfish, sea anemones), and the Ctenophora (comb jellies), which have two layers. The Cnidaria and Ctenophora are called **diploblasts**. Animals of these three phyla also, in the main, show radial symmetry whereas most of the triploblastic animals show bilateral symmetry and are therefore also known as the **Bilateria**.

The next important feature is the presence or otherwise of a **coelom**. This is the cavity formed within the mesoderm, lined with a mesoderm-derived epithelium, the peritoneum, and often making up the principal body cavity. Platyhelminthes (flatworms) and Nemertea (ribbon worms) do not have a coelom and are described as acoelomate. Many other invertebrate phyla, including the Nematoda, have a body cavity that is only partly surrounded by mesoderm, and are described as pseudocoelomate. Annelida (segmented worms), Mollusca,

Arthropoda, Echinodermata (sea urchins, starfish, etc.), and Chordata all have a coelom.

The coelomate phyla were traditionally divided into two "super phyla" called the **Protostomia** and the **Deuterostomia**. The latter group contains the Echinodermata and Chordata. It was defined by **radial cleavage**, formation of the anus from the blastopore, and **enterocoely**, which is formation of the coelom by budding of the mesodermal rudiment from the gut. In fact, true enterocoely is not found in vertebrates but it is found in some protochordates and echinoderms and thus serves to link these two phyla together. The defining features of protostomes were **spiral cleavage**, early-acting **cytoplasmic determinants**, and schizocoely (formation of the coelom by splitting of the mesodermal layer). Older textbooks may also include formation of the mouth from the blastopore, but this is incorrect. In protostomes the blastopore typically narrows to a ventral slit from which both mouth and anus derive.

Recently, the use of molecular taxonomy has changed this picture. The deuterostomes remain as a taxon, but the significance of the coelom has disappeared and there are now two new superphyla making up the protostomes. The annelids and molluscs, which share a type of larva called the **trochophore**, are grouped with some other phyla in the **Lophotrochozoa**, while the arthropods and nematodes, together with all other molting animals, comprise the **Ecdysozoa**.

The advantage of molecular taxonomy is that it can look back further in time than morphology and can, in principle, tell us when the last common ancestor of all animals was flourishing. In practice, this has proved rather difficult as the estimates depend on the calibration dates derived from the fossil record. Although the fossil dates themselves are reasonably reliable, when applied to molecular trees it turns out that the vertebrates have been accumulating neutral mutations more slowly than invertebrates. The reason for this is not known, but because of the calibration uncertainty, estimates of the last common ancestor for all animals span quite a long period from about 1000 million years ago to about 600 million years ago. The latter date is obtained using invertebrate divergence times and does correspond quite well to the earliest metazoan fossils dating from the Ediacaran period.

The fossil record

The fossil record provides a real view of the past and is the only source of time calibration for phylogenetic trees. A fossil is any residue of an organism preserved in the rocks. Virtually all fossils are found in sedimentary rocks, those formed by accretion of sediments in the sea or fresh water. Most fossils consist of only the hard parts of organisms, for example the shells of marine invertebrates or the bones of vertebrates, and these are often mineralized such that the original organic material has been replaced by rock. Fossils may also be "trace fossils," for example burrows or tracks, which show evidence of activity but

without the actual remains. A very small number of geological sites contain organisms in an exceptional state of preservation in which soft parts are also visible, for example the well-known Burgess shale and Chengjiang deposits from the Cambrian period.

Figure 22.3 shows the geological periods and the first occurrence of fossils of various taxonomic groups. It may be noticed that most of these first occurrences are of vertebrates. Most, and possibly all, of the invertebrate phyla were already in existence during the Cambrian period, which is the earliest geological period for which the rocks contain significant numbers of fossils. The "Cambrian explosion," during which the invertebrate phyla appeared, occupies a geologically rather short period of 5–20 million years. This rapid diversification poses problems for attempts to find the true phylogenetic tree of the animal kingdom. Trees are constructed from characters of modern

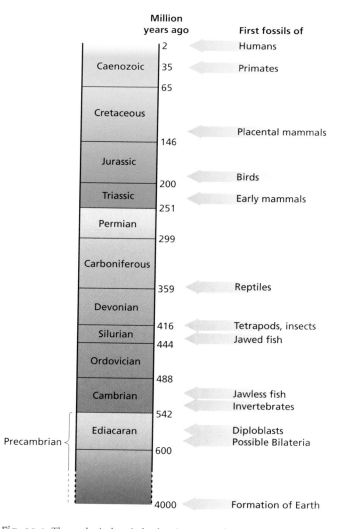

Fig. 22.3 The geological periods, showing currently accepted dates of their commencement and the first occurrence of fossils for major animal groups.

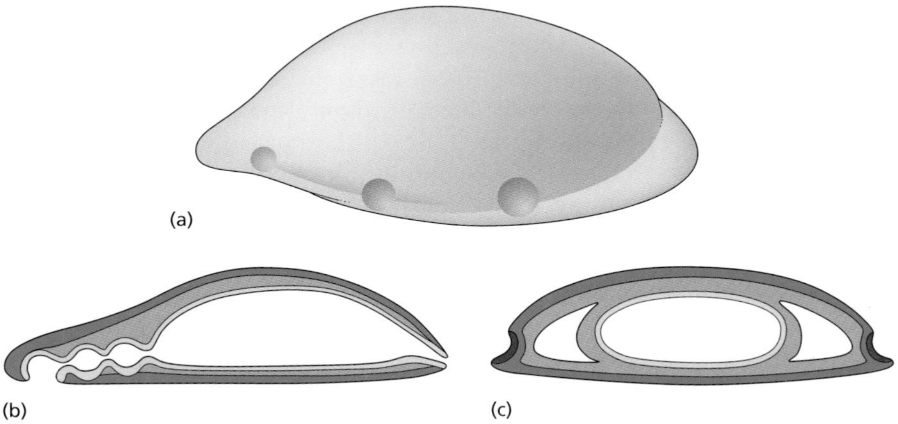

Fig. 22.4 A possible bilaterian fossil from the Doushantuo formation of SW China: (a) reconstruction; (b) sagittal section; (c) transverse section. The creatures appear to have three germ layers, an anteroposterior pattern, and a through gut.

organisms, whether morphological or gene sequences, and almost all the changes that have happened were in the lines of descent established in the Cambrian explosion. For neutral characters that should evolve in a clock-like manner only about 2% are likely to relate to the time of lineage diversification during the explosion and the remaining 98% have accumulated since. This means that it is hard to find an informative signal among all the noise of later evolution.

The fossil record cannot prove a phylogeny because it is impossible to know whether a particular fossil is really an ancestor of a fossil found at a higher stratigraphic level. However, some possibilities can be excluded because a late-arising group obviously cannot be ancestral to an early-arising group. The fossil record also provides valuable time calibration for phylogenetic trees in the form of divergence dates, which are the points at which a new group appears for the first time. For example the divergence of birds and mammals is about 300 million years ago and of *Drosophila* from mosquitoes about 235 million years ago. These times cannot actually be deduced from the fossils themselves. What the fossils do is enable the correlation of different geological sites into a coherent sequence. Absolute times can only be obtained by radioactive dating of the rocks, and this itself is a complex business since sedimentary rocks rarely contain radioactive elements with suitable half lives for this sort of measurement. However, correlation of results from many sites around the world has led to the time scale shown on Fig. 22.3

Preceding the Cambrian is a period called the Ediacaran, or Vendian, in which are found fossils that resemble cnidarians. There is some uncertainly about this since jellyfish are notoriously difficult to preserve as fossils and the Ediacaran fauna may possibly represent some other type of radially symmetrical life, which had hard parts and which died out without leaving descendants. Evidence of recognizable metazoan animals has been hard to find in the Precambrian rocks although tiny 100–200 μm long worm-like creatures have been found in the

Doushantuo formation of SW China, which is of Ediacaran age. Sections of these indicate that they may be coelomate animals with a through gut and some anterior sense organ (see below and Fig. 22.4).

The primordial animal

The most important novel result of evolutionary developmental biology is the ability to uncover previously invisible homologies through the study of expression and function of the genes that make the body plan. In order to explain this it is necessary to consider first the concepts of the **body plan** itself and of the **phylotypic stage**. The body plan (or **Bauplan** in the original German) refers to the idea that it is possible to abstract the essential features of anatomical organization from a wide range of animals. This is done in zoology textbooks whenever they show a diagram of a generalized vertebrate or mollusc. Evidence that the body plan is real rather than an arbitrary abstraction has come from developmental biology because of the fact that the key transcription factors and signaling molecules are found to have very similar expression domains within a taxon. For example the expression of the *Brachyury* gene in vertebrates occurs in rather differently shaped regions in zebrafish, *Xenopus*, chick, and mouse, but they all correspond to what was previously thought to be the mesoderm, and indeed they reinforce the concept of mesoderm as a real cell state definable by the activity of a combination of transcription factors, and not just an arbitrary label for an embryo region.

The phylotypic stage

The **phylotypic stage** is the stage of development at which all members of a taxon show the maximum morphological similarity. This concept has been most widely applied to the insects

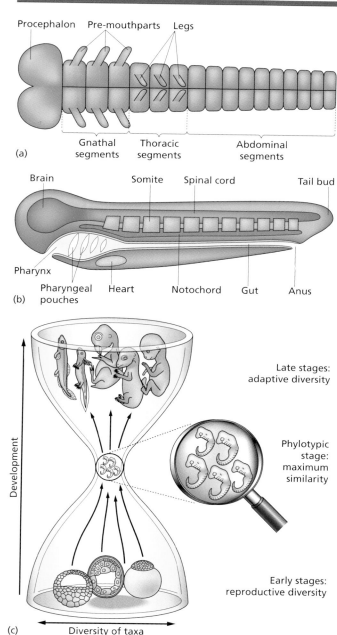

(a)

Procephalon Pre-mouthparts Legs

Gnathal segments Thoracic segments Abdominal segments

(b)

Brain Somite Spinal cord Tail bud

Pharynx

Pharyngeal pouches Heart Notochord Gut Anus

(c)

Late stages: adaptive diversity

Phylotypic stage: maximum similarity

Early stages: reproductive diversity

Development

Diversity of taxa

Fig. 22.5 The phylotypic stage. (a) Phylotypic (extended germband) stage of insects. (b) Phylotypic (tailbud) stage of vertebrates. (c) "Phylotypic hourglass" for vertebrates. Reproduced from Duboule. *Dev Suppl* 1994: 135–142.

most details of the future epidermal structures are yet to appear, and the endoderm, in terms of the anterior and posterior midgut invaginations, has yet to form.

Among vertebrates the phylotypic stage is the **tailbud** stage (Fig. 22.5b), when all vertebrates have a dorsal nerve cord with anterior specialization, segmented somites, a ventral heart and a set of pharyngeal arches.

Phylotypic stages for groups other than the insects and vertebrates might include the **veliger** larva of molluscs. This has a head, a posterior, a through gut, a dorsal shell gland, and a ventral muscular foot. A phylotypic annelid might be considered to be a stage at which a certain number of segments have arisen and the organism has a head, a ventral nerve cord, a through gut, and segments involving all the germ layers. The echinoderms are a remarkable phylum in which metamorphosis from the larva to the adult is particularly dramatic. Here it would seem that the **echinus rudiment** of sea urchins, or its equivalent in the other classes, should be considered phylotypic, since the earlier-stage larvae are typically bilaterally rather than radially symmetrical and do not have echinoderm features such as a water vascular system or tube feet. Finally, the **nauplius** larva, which has three pairs of head appendages, has been considered a phylotypic stage for crustaceans. The concept of the phylotypic stage has been disputed on various grounds, but it has been shown that, on average, the genes expressed at the phylotypic stage did originate further back in evolutionary time than genes expressed at earlier or later developmental stages.

It is helpful to have some idea of why conserved body plan features should be displayed at one stage. Although it is often stated that the earliest stages of development cannot change in evolution because this would compromise later events, it is nonetheless a familiar fact that early embryonic morphology is very diverse. The main reason for this is the diversity of reproductive lifestyle and strategy. The eggs of even quite closely related animals can differ markedly in yolk content. This is the result of evolutionary trade-offs, such that organisms come to occupy different niches in which they may produce a lot of small eggs with a poor chance of survival or a few large ones with a good chance of survival. The presence of more yolk drives various changes in early development including the disposition of cleavages (**meroblastic** rather than **holoblastic**) and the nature of gastrulation movements (**epiboly** rather than **invagination**). **Viviparity** imposes even more drastic changes on early embryonic life and is necessarily accompanied by the early formation of a variety of extraembryonic membranes and supporting structures arising from the zygote as well as the mother. So early development is diverse because reproductive behavior is diverse. Late development is also diverse, but for quite different reasons. By late development the embryo will be becoming quite similar to the postembryonic organism, whether larval or adult. Free-living organisms are subject to selection and must acquire distinct niches for their survival to the reproductive stage, so the late embryos have to be diverse because the organisms to which they give rise are diverse. It follows inexorably

and vertebrates although phylotypic stages can also be defined for other groups. Insects all look rather similar at the **extended germ band** stage. At this stage the segmental arrangement of the body is very obvious, with three gnathal, three thoracic, and a variable number of abdominal segments (Fig. 22.5a). The extended germ band is reached before dorsal closure of the embryo and consists of a plate composed of ectodermal and mesodermal components. There is a ventral nerve cord but

that the stage of maximum similarity within a large taxon will be the early–middle period of embryonic development, after the constraints imposed by reproductive strategy and before the constraints imposed by the adaptation of the free-living organism. The phylotypic stages are early–middle stages that exist within egg cases, or jelly layers, or a uterus, and so will not be interacting with the environment to more than a minimal degree. According to this way of thinking there is no specific cause for the phylotypic stage, it is just the stage in the middle at which the selective pressures for change are minimized, hence is most likely to retain the features of the common ancestor. This idea is encapsulated in the "phylotypic egg timer" diagram formulated by Duboule (Fig. 22.5c).

The zootype

Animal phyla have been defined in such a way that each of them corresponds to a different body plan, which means that it is inevitably difficult to find any morphological homologies between them apart from the developmental features mentioned above. However, the discoveries of developmental biology make it possible to compare the expression patterns of key developmental genes between phyla. Some of these seem to be conserved across most of the animal kingdom. Figure 22.6 is a diagram of an animal embryo showing the expression domains of various key genes involved in regional specification. They are active around the phylotypic stage for all the main bilaterian groups that have been examined. The totality of common expression domains has been called the **zootype**, to signify that this cryptic anatomy of developmental gene expression patterns defines what an animal actually is. The vertebrate nomenclature is used, but homologs of all these genes exist in *Drosophila* and

several other invertebrates and perform comparable tasks. It is intriguing to speculate whether this constellation of gene expression domains, deduced from modern organisms, is actually what was active 600 million years ago in the tiny creatures of the Doushantuo formation (Fig. 22.4). The term "zootype" in now known not to be quite appropriate, as this set of gene expression domains is characteristic of bilaterian animals but not of sponges or cnidarians. It might more accurately be called a "triplotype," characteristic of the triploblastic, or bilaterian, animals.

Hox genes

The Hox genes have already been described in Section 2 in the context of each of the model organisms. Current data suggest that all bilaterian animals use the same set of Hox genes to control anteroposterior pattern. They fall into homologous groups, and often, but not always, comprise a cluster on one chromosome, sometimes interrupted by other genes. In terms of expression pattern there is a general trend for each gene to be expressed in a different anteroposterior domain and for the sequence of anterior expression limits to be similar to the sequence of genes on the chromosome. In those systems where experimentation is possible it has been shown that loss of function of a Hox gene generally causes a homeotic transformation such that the affected body level is anteriorized, while gain of function generally leads to a posteriorization of the affected part.

From an evolutionary point of view the critical thing about Hox genes, and indeed almost all the components of the developmental toolbox, is that they function simply as switches, upregulating or repressing other genes. Any transcription factor

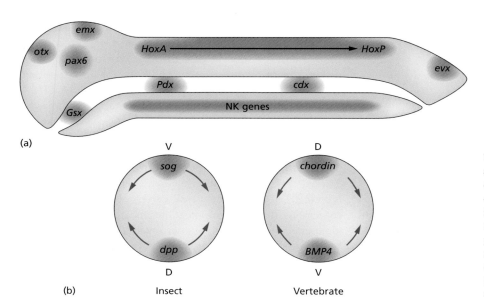

Fig. 22.6 The "zootype," now more appropriately called the "triplotype," a common constellation of expression domains for developmental genes found in the body plan of bilaterian animals. (a) Transcription factors active at the phylotypic stage. *HoxA→HoxP* indicates a set of Hox genes with nested expression domains having different anterior limits. (b) Homologous dorsoventral signaling systems in insects and vertebrates.

could do this job given appropriate connections from the signaling pathways. If animals had evolved from a variety of different ancestral lower eukaryotes then it is virtually certain that different sets of transcription factors would have been co-opted to deal with the task of assembling an anteroposterior pattern. The fact that the same set of genes does this job in all bilaterians proves that they must have arisen from a single common ancestor.

Study of the vertebrate Hox genes illustrates a remarkable event that is thought to have occurred around the time of the vertebrate origin, which is a quadruplication of the entire genome. There is a creature called amphioxus which is a **proto-** chordate and has long been regarded as the modern animal most similar to a putative vertebrate common ancestor. It has a dorsal nerve cord, notochord, somites, pharyngeal gill slits, and a through gut (Fig. 22.7). Like other invertebrates, it has a single Hox cluster, but unlike them it has increased the number of genes to 15 (Fig. 22.8). Vertebrates have four Hox clusters on different chromosomes (Fig. 22.8). Each cluster contains a subset of orthologs of the amphioxus cluster, making it look as though the vertebrate common ancestor acquired four copies of the cluster by two rounds of chromosome duplication, and then individual genes were lost from each cluster independently within the various vertebrate lineages. Although the similarity

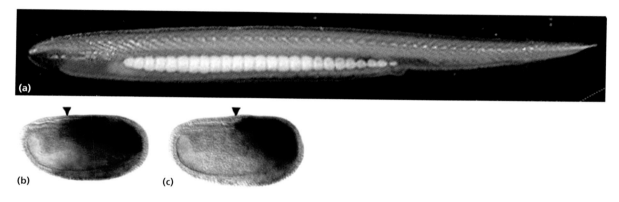

Fig. 22.7 The primitive chordate amphioxus. (a) Adult amphioxus of the species *Branchiostoma lanceolatum*. (b) *Hox1* and (c) *Hox3* expression postgastrulation demonstrated by *in situ* hybridization (dark blue). Arrowheads indicate the anterior limits of gene expression. Photograph by Christian König © www.konig-photo.com (part a); reproduced from Koop *et al. Dev Biol* 2010;338:98–106, with permission from Elsevier (parts b and c).

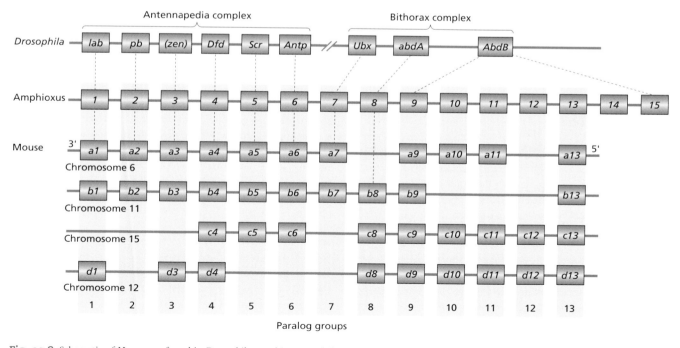

Fig. 22.8 Schematic of Hox genes found in *Drosophila*, amphioxus, and the mouse.

between Hox clusters in different animal groups is striking, it should be noted that, particularly among invertebrates, the spacing and orientation of Hox genes can vary widely, and in some cases, including *Drosophila*, the cluster has broken up to a greater or lesser degree.

The vertebrate clusters are called a, b, c, and d, and individual genes are numbered 1–13, starting from the 3′ end of the cluster. Because the expression domains in the embryo lie in the same order as the position on the chromosome, the numbers also reflect the sequence of anterior expression boundaries. The individual genes with the same number, for example *a4*, *b4*, *c4*, *d4*, are **paralogs** (as explained above), and the set of paralogs is called a **paralog group**. The reason why it is thought that the whole genome, and not just the Hox cluster, underwent two chromosome duplications, is that many other gene families in vertebrates correspond to single genes in invertebrates. The easiest way to double the genome is to become **tetraploid**, that is fail to segregate the products of mitosis such that each nucleus now contains twice as many chromosomes. Tetraploidization is not uncommon in evolution, and one of the model organisms, *Xenopus laevis*, is a tetraploid of about 30 million years standing, and the ancestor of teleost fish underwent tetraploidization about 420 million years ago. Based on genomic sequencing of amphioxus and of a number of vertebrates, it is thought that the origin of vertebrates was marked by two tetraploidization events, which increased the total gene number from about 15,000 to about 60,000, and was followed by gene loss back to about 25,000. This huge increase in potentially useful genetic material may have helped provide the opportunity for the spectacular adaptive radiation of the vertebrates.

In vertebrates the Hox expression domains do not extend anterior to the hindbrain. The regional pattern of the fore- and midbrain is controlled by other homoebox genes of the *Otx* and *Emx* groups. These also have orthologs in invertebrates that are expressed at the extreme anterior and so are considered to be part of the zootype/ triplotype.

Other gene clusters

Although Hox genes are expressed in all germ layers in vertebrates, they are confined to the ectoderm in most invertebrates and the anteroposterior pattern in the other germ layers is controlled by other genes. The endoderm is regionalized by a sister cluster of genes called the Parahox cluster. This may represent a very early duplication of the Hox cluster and contains three genes *Gsx*, *Pdx* (*Xlox*), and *Cdx*. The latter two are well known as being very important in vertebrate gut development: *Pdx1* is essential for pancreatic development and *Cdx2 is* essential for development of the intestine (see Chapter 16). The *Gsx* group is expressed in the brain of vertebrates, but often in the anterior endoderm of invertebrates.

The regional pattern of the mesoderm is controlled to at least some degree by the NK gene cluster. The group of homeobox genes known as Nkx in vertebrates form a single cluster in

Drosophila. This cluster is broken into three pairs in amphioxus and is even more dispersed in vertebrates. An example of their role in regional patterning of the mesoderm is that one of these genes, called *Nkx2.5* in vertebrates and *tinman* in *Drosophila*, is especially associated with the development of the heart (see Chapter 15). The primordial bilaterian would probably not have had a heart because if it was as small as the organisms of the Doushantuo formation its tissues would have been oxygenated by diffusion. Only in the Cambrian, when animals became large, would the need for a circulatory system have arisen. So the primordial role of *Nkx2.5* was probably to specify an anteroventral mesodermal territory, rather than to specify a heart as such.

Dorsal–ventral pattern

Work on *Xenopus* showed that the main activity of the organizer was the secretion of BMP inhibitors such as Chordin (see Chapter 7). This establishes a ventral-to-dorsal gradient of BMP signaling activity that controls the dorsoventral pattern in both the ectoderm and mesoderm. Work on *Drosophila* showed that there was a dorsal to ventral gradient of activity of the BMP homolog decapentaplegic (Dpp) (see Chapter 11). This is antagonized by Short-gastrulation which is a Chordin homolog secreted from ventrolateral ectoderm. Likewise, in regenerating planarians BMP homologs are expressed dorsally and RNAi knock down of components of BMP pathways causes ventralization. So a similar system controls formation of the dorsoventral pattern, but dorsal in vertebrates corresponds to ventral in invertebrates.

This information can be correlated with the long-known fact that vertebrates have their principal nerve cord on the dorsal side, while at least some invertebrates, such as arthropods and annelids, have it on the ventral side. In those invertebrate phyla that have hearts, this tends to be dorsal while it is ventral in vertebrates. This all suggests that there is a primordial system of dorsoventral patterning that operates in all animals but it became inverted at some point such that the vertebrate polarity is now opposite to that of the invertebrate groups (Fig. 22.6b). This might literally have occurred by the vertebrate common ancestor inverting its habit. Alternatively, it might be that the key change was a rotation of the trunk relative to the head. This would also explain the crossover of sensory tracts such as the optic nerves, typical of vertebrate neuroanatomy.

Sense organs and cell types

One of the defining characteristics of bilaterian animals is that they have some sort of concentration of sense organs at one end, representing a "head," and referred to as the anterior end. The *Pax6* gene is intimately associated with the formation of eyes. It is defective in the *small eye* mutant of the mouse; and in *Drosophila Pax6* is called *eyeless* and its loss of function prevents eye development completely. *Pax6* is also expressed in the eyes of cephalopod molluscs and of planarians. Cephalopods are significant because they have image-forming eyes structurally quite similar to those of vertebrates except for the orientation of the retina. This had always been a textbook case of **analogy** because it was felt impossible for vertebrates and cephalopods to have a common ancestor with an image-forming eye. However, the involvement of the same transcription factor in making the two eyes suggests that they are homologous at least to the extent that the ancestor had some sort of photoreceptor made with the help of *Pax6*. In planarians RNAi ablation during regeneration shows that *Pax6* itself is not required for eye regeneration but another homeobox gene, *Sine oculis*, is required. *Sine oculis* is immediately downstream of *Pax6* in the *Drosophila* eye development pathway; and its vertebrate homolog *Six3* is needed for eye development in mice. These results show a remarkable degree of conservation of function of a small group of transcription factors involved with the creation of light-sensitive organs across the whole animal kingdom.

The above genes are all concerned with regional specification during development. But there is also a long-range homology of individual differentiated cell types and the genes responsible for their formation. Again, this is most apparent for the Bilateria. The sponges and diploblasts tend to have a smaller number of cell types than the bilaterian animals, and they often have multiple functions, such as light-sensitive steering rudder cells in sponges, or myoepithelial cells in cnidarians.

An example of genes important for making a cell type is give by the myogenic gene family, of which the prototype member is *MyoD*. These exist in all bilaterians and are needed for the differentiation of muscle cells. Another is the two families of *Opsin* genes, which are essential for the formation of photoreceptors. Also the Delta–Notch signaling system is responsible for controlling the differentiation of neurons from neurogenic epithelia, and, as we have seen in Chapter 16, also of specialized cell types within endodermal epithelia.

Basal animals

Now that many entire genomes have been sequenced it is possible to get some idea of what genes are needed to make an animal and when they arose in evolution. Some of the families of developmental control genes, encoding transcription factors and signaling systems, are found in choanoflagellates, which are unicellular eukaryotes having some characteristics of animal cells. But most of the systems, including the Hox genes and some other homeobox genes, Wnts, TGFβs, and nuclear hormone receptors, appear first among the animals.

The most basal animals in modern phylogenies are the Porifera or sponges. These are traditionally thought to have few cell types and no real tissue grade of organization. They contain an NK-like cluster of homeobox genes, but no true Hox genes. Neither does the placozoan *Trichoplax*, which is another simple creature with no tissue grade of organization.

The phylum Cnidaria comprises sea anemones and jellyfish, and also the familiar freshwater polyp *Hydra*, used in regeneration research. Cnidarians have traditionally been considered as basal animals that differ from most others in that they possess only two germ layers (ectoderm and endoderm) instead of three, and that they are radially rather than bilaterally symmetrical. Cnidarians do have Hox genes although they do not form a linked cluster, being scattered singly or in pairs. The sea anemone *Nematostella* (Fig. 22.9) has recently been domesticated for experimental work. It can be reared in the lab. and undergoes a life cycle in which a **planula** larva hatches from the egg, and after a phase of mobility become sessile and forms a **polyp**. Expression data for Hox genes indicates rather little staggering of expression limits, making it difficult to establish that the long axis of a polyp corresponds to the anterior–posterior axis of a bilaterian (Fig. 22.9f).

Analysis of gene expression patterns has also shown that cnidarians are not entirely radially symmetrical. This is apparent from the study of BMP and Chordin homologs in *Nematostella*. Unlike in Bilateria, these BMP and Chordin are both expressed on the same side of the polyp (Fig. 22.9d,e), making it unclear whether the second ("directive") axis is homologous to the bilaterian dorsoventral axis or not. Furthermore, cnidarians may not really be diploblastic. Jellyfish can possess striated muscle cells and their formation has been analyzed in the hydrozoan *Podocoryne carnea* (Fig. 22.10). This species shows an alternation of generations between polyp and **medusa** (plural **medusae**), and will undergo its whole life cycle in laboratory conditions. The polyps generate asexual buds, which produce the medusae. The medusae produce gametes and undergo sexual reproduction to form motile planula larvae, which eventually settle down to become new polyps. The muscle cells of the medusa develop from a cell layer called the entocodon, which arises in the asexual buds and expresses mesoderm-type genes including homologs of *twist*, *snail*, and *Mef2* (22.10a). In other types of cnidarians without a medusa phase the same genes may be expressed in regions of the endoderm (Fig. 22.10b), which has given rise to the suggestion that the mesoderm originated as a subdivision of the endoderm.

The study of basal animals has shown that most of the genes needed to make an animal were present when the cnidarians originated. On their own these genes do not lead to a bilaterian body plan. But once such a body plan arose, with various developmental control genes co-opted into setting up its key features, then it has proved very stable and led to the long-range homologies that can be detected today by genomic sequencing and expression analysis.

What really happened in evolution?

The other main area where developmental biology can assist in the solution of evolutionary problems is to find out what actually happened in evolution. Some evidence about this can be obtained by comparing the expression patterns of key

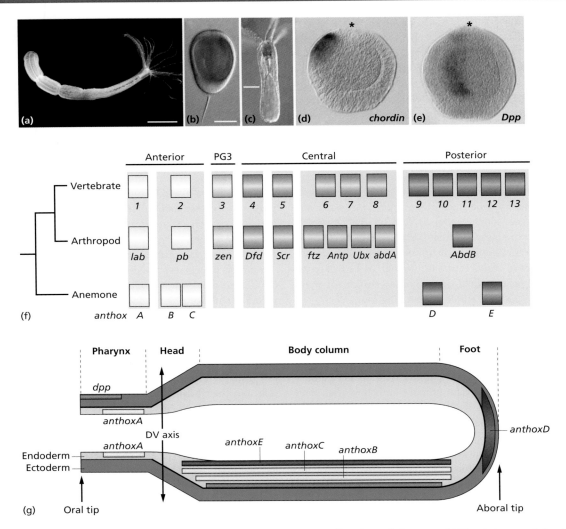

Fig. 22.9 Developmental gene expression in the anemone *Nematostella*. (a) *Nematostella* polyp, scale bar 1 cm. (b) Planula larva, scale bar 70 mm. (c) Primary polyp, scale bar 80 mm. (d,e) In the planula, *chordin* and decapentaplegic (*dpp*) are expressed on the same side of the body. Asterisks mark the blastopore. (f) Comparison of *Nematostella* Hox genes with those of vertebrate and *Drosophila*. (g) Expression of Hox genes in primary polyp. Reproduced from Technau and Steele. *Development* 2011; 138: 1447–1458, with permission from Company of Biologists Ltd (parts a–e); and from Finnerty *et al. Science* 2004;304:1335, with permission from American Association for the Advancement of Science (part f and g).

developmental genes that give rise to two different morphologies, one ancestral and one derived. However, this is not sufficient. As when elucidating developmental mechanisms it is also necessary to establish the actual biological activity of the genes in question by performing overexpression and loss-of-function experiments. This may sometimes be difficult as the organism under study will not be the standard laboratory models and so it may even be difficult to obtain embryos let alone perform successful experiments on them.

When trying to decide what happened, it is important to avoid the "fallacy of design," in other words once a developmental system is understood there is a tendency to say "extra legs can be produced very easily by allowing gene A to turn on gene B, so this must be what happened in evolution." Firstly, it may

not have happened. Maybe the extra legs were formed without A turning on B. Secondly, even if A does turn on B in the new position, the mutations that cause this to happen may not be in A or B. They may be in different genes altogether, and there may be multiple small changes rather than one large one. In general, because they have arisen by natural selection, developmental systems are much more complex than would seem necessary had they been designed by a conscious agent, and they do not necessarily change from one behavior to another using the most simple route.

This consideration also relates to the issue of **developmental constraints**. Biologists not concerned with developmental mechanisms often implicitly assume that there are no limits to possible mutational variation and with enough mutants, enough

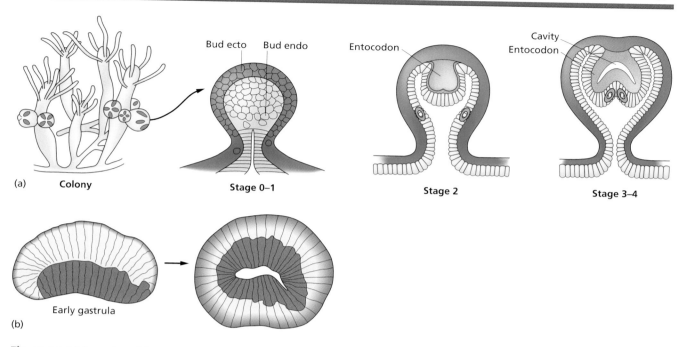

Fig. 22.10 (a) Formation of the entocodon in *Podocoryne* (a hydrozoan). Reproduced from Ball *et al. Nature Rev Genet* 2004; 5: 567–577, with permission from Nature Publishing Group. (b) Expression of *snail* in the endoderm of *Acropora* (a coral).

time, and enough selection it is possible to evolve anything from anything else. If you want to evolve a snail into an elephant: no problem, it is just a matter of time and selection. Developmental biologists often tend to fall into the other extreme of believing that the only mutational variation that is possible is that produced by gain or loss of function of each known component of a particular developmental system. This will confine the range of attainable morphologies to a relatively small set obtained by one-step changes away from the wild-type pattern. Mutagenesis screens on vertebrates such as the zebrafish or mouse generate only a modest range of phenotypes. The reason is that mutations in pleiotropic genes with multiple functions are often early lethals, and also that many morphological changes are not seen because they would need several mutations to be present simultaneously. In addition, it is possible to imagine all sorts of body plans that would probably confer greater fitness but have not evolved, such as angels which have a useful pair of wings in addition to arms and legs. On the other hand we know that it would indeed be possible to evolve a snail into an elephant given sufficient time and resources. It could be done by selecting initially for the loss of all morphology to convert the organism to a primitive worm-like creature, and then applying a second selective regime designed to culminate in large size, four legs, tusks, and a trunk. The potential importance of developmental constraints to evolutionary theory is immense. If constraints are very limited then the course of evolution is determined almost entirely by adaptation to a changing environment. But if developmental constraints are substantial then the course of evolution is determined mostly by what is possible given the current situation. At the present time, the true extent and role of developmental constraints remains unknown.

In general, the mutations of the known components of well-understood systems, such as those controlling segmental identity or limb patterning, will be **macromutations**: single mutations which bring about a significant morphological change in one step. Neo-Darwinists have always been uncomfortable with macromutations, believing that they will virtually always be deleterious and selected against. Developmental biologists are often more sympathetic to what has been called the "**hopeful monster**," the creature arising from a macromutation, because this type of mutational change is congruent with developmental explanations. Also some morphological characters seem intrinsically qualitative, and it is hard to imagine small, imperceptible, intermediate forms. For example if an organism adds a segment then the extra segment is either there or not there, and it seems more plausible that a mutant with an extra segment would arise in one step and then take its chances with natural selection rather than follow a trajectory involving a very small new segment, which gradually increased in size over many generations. But orthodox neo-Darwinists prefer to believe that evolution occurs by small, imperceptible steps and that many of the mutations that bring these about occur in "modifier genes," in other words genes outside the developmental machinery but which have quantitative effects on this machinery. Again, the truth is that we do not know the relative importance of macromutations versus those of small effect.

One principle that has become established, due to extensive genomic sequencing, is the idea that most mutations driving developmental changes occur in *cis*-regulatory regions of genes. They are typically creation or loss of transcription factor binding sites, change in affinity of such sites, or repositioning of regulatory sequences through the action of transposable elements. The reason for this is that most developmental genes are pleiotropic. If the gene itself loses function this will affect multiple systems and the result is likely to be lethality, or at least an organism of reduced fitness. But if just part of the regulatory region is altered, this can bring about a small change in the position or timing of expression, which may affect just one aspect of the morphology without deleterious effects elsewhere.

An interesting example is the changes of pattern of trichomes (small unicellular hairs, including the ventral denticles of the larva) between different species of *Drosophila*. Although hundreds of genes affect trichome pattern, the interspecies variation is all due to regulatory mutations at the *shavenbaby* locus (also known as *ovo*) (Fig. 22.11). This is because this gene integrates a large number of inputs from upstream developmental control genes, all of which have other functions, and it initiates the program for making a trichome. Significant gain or loss of function of upstream genes will have serious consequences for other systems, while mutations of downstream components can only affect the assembly of individual trichomes, but not their pattern.

New directions for research

EvoDevo is still a wide open field. We have very little idea what actually happened in evolution to create each of the key morphological novelties in animals. The clearest results are likely to be obtained by comparison of closely related species where the mutations responsible for a particular morphological change are not obscured by too many other changes.

Methods are now available to make the necessary investigations on nonmodel organisms:

1 complete genomic sequencing to make global comparisons;
2 *in situ* hybridization to analyze expression patterns;
3 RNAi to knock out gene activity;
4 electroporation to introduce genes for overexpression.

Even better, the best science journals have an insatiable appetite for stories about evolution!

Segmented body plans and Hox genes

The Arthropods are a very large phylum containing four classes: insects, crustaceans, myriapods (centipedes and millipedes), and chelicerates (spiders and scorpions). Each class is distinguished by a fundamentally different arrangement of segments in the body plan, but we know from both morphological and molecular phylogeny that the arthropods are a clade and have evolved from a common ancestor. Since the discovery of the Hox cluster and its role in controlling segment identity it has been natural to think that the changes in body pattern arose from changes in Hox gene expression. In particular, in *Drosophila* it is known that the genes *Ubx* and *abdA* effectively determine the boundary between thorax and abdomen, as they repress formation of legs in the abdomen by repressing an enhancer of the *distalless* gene, which is required for the outgrowth of all appendages. The expression of *Ubx* and *abdA* has been studied in a variety of arthropods (Fig. 22.12). This shows that there is a general association of Hox domains with body regions, but there is no specific association with the thorax or with the suppression of legs. In crustaceans these two Hox genes are expressed throughout the thoracic segments, all of which bear legs, but they are not expressed in head or genital regions. In myriapods *Ubx* and *abdA* are expressed in all but the first

thoracic segment. In chelicerates the anterior boundary occurs within the ophisthosoma, the posterior body region which does not bear any appendages. In crustaceans and myriapods *distalless* is co-expressed in the early legs along with *Ubx*, so is evidently not be repressed by *Ubx*. The general conclusion from this is that there are some associations of Hox boundaries with boundaries of segment type, but that specific functions such as leg suppression are not conserved. It follows that what the Hox genes must do is to provide a universal coordinate scheme, or **positional information**, for the organism, and that most of the evolutionary innovations occur in the regulatory connections downstream of them.

In one specific instance it is possible to track a close relationship between *Ubx* expression and a particular type of appendage. This relates to appendages called maxillipeds found in Crustacea. Different crustacean species may have one, two, or three pairs of maxillipeds, and this does correlate with the anterior boundary of *Ubx* expression, which lies just posterior to the maxilliped-bearing segments. It is therefore a reasonable hypothesis that the mutations that were responsible for the different maxilliped number were ones that changed the expression of *Ubx*. For this to be true a necessary condition is that changes of *Ubx* expression do, in fact, have the predicted consequences. The amphipod *Parhyale hawaiensis* can be reared in

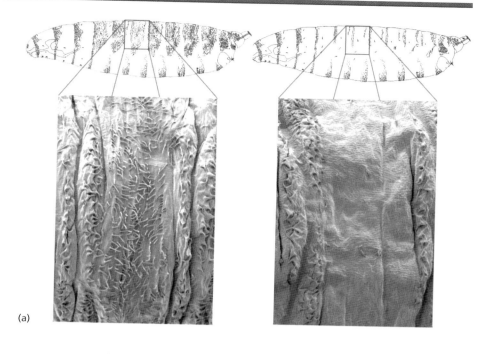

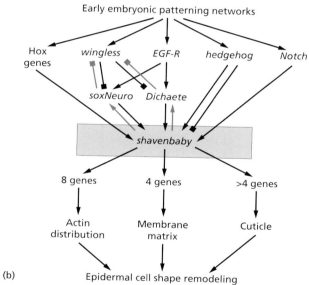

(a)

(b)

Fig. 22.11 Change in the pattern of trichomes between *Drosophila* species are due to mutations in the regulatory region of the *shavenbaby (ovo)* gene. (a) *D. melanogaster* and *D. sechellia* larvae. (b) The cis-regulatory region of the *shavenbaby* gene integrates information from many upstream genes to generate a prepattern of Shavenbaby protein, which controls the formation of trichomes. Reproduced from Stern and Orgogozo. *Science* 2009; 323: 746–751, with permission from American Association for the Advancement of Science.

the lab, and protocols have been developed for *in situ* hybridization, gene overexpression, and gene knock-down. Normally, the first thoracic segment (T1) is fused to the head, and bears maxillipeds. The next two thoracic segments (T2,3) bear grasping appendages, and the next five segments (T4–8) locomotory appendages. In *Parhyale* these three types of appendage correlate respectively with no expression, low expression, and high expression of *Ubx* during the establishment of the limb buds (Fig. 22.13). If Ubx protein synthesis is reduced by injection of siRNA into fertilized eggs, then the T2 and 3 appendages develop

as maxillipeds (Fig. 22.14b). Conversely, if *Ubx* is overexpressed, this causes a variety of posteriorizations in the head and thorax, including conversion of antennae, maxillae, and maxillipeds into thoracic legs (Fig. 22.14c). Transgenesis in *Parhyale* is achieved by injecting fertilized eggs with the *Ubx* gene driven by a heat-shock promoter, accompanied by RNA for a transposase that catalyzes integration into the genome. The application of a temperature shock then drives ectopic expression of *Ubx*. This set of experiments confirms that *Ubx* does control appendage character in this organism, but the nature of the

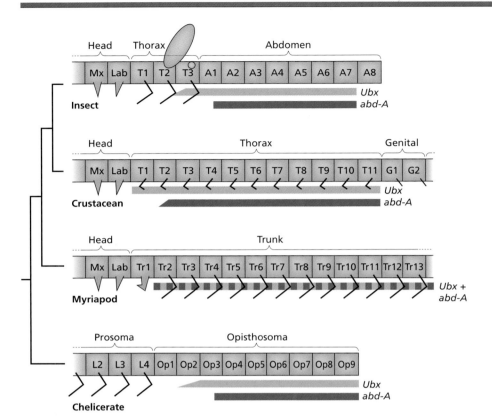

Fig. 22.12 Correlation of segmental pattern, appendage types and Hox gene expression in arthropods. Reproduced from Carroll, *From DNA to Diversity*, 2001: Blackwell, with permission from John Wiley & Sons Ltd.

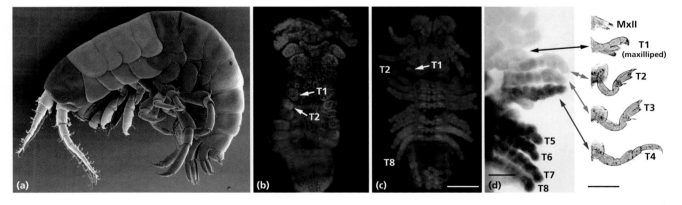

Fig. 22.13 Appendages in the amphipod *Parhyale*. (a) Scanning electron micrograph of a hatchling showing antennae in white, maxillae in blue, maxillipeds in green, grasping legs in yellow, and locomotory legs in magenta. (b–d) *In situ* hybridization of *Ubx* expression during appendage development. Scale bars 100 μm. *Ubx* is not expressed in maxilliped buds; is expressed at a low level in grasping leg buds; and at a high level in locomotory leg buds. Reproduced from Pavlopoulos *et al. Proc Natl Acad Sci USA* 2009;106:13897 (part a); and Liubicich *et al. Proc Natl Acad Sci USA* 2009;106:13892, with permission from National Academy of Sciences (parts b–d).

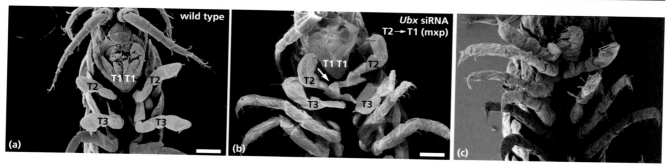

Fig. 22.14 Effect of Ubx on appendage character in *Parhyale*. (a) Wild-type hatchling showing maxillipeds in green. (b) Knock-down of *Ubx* translation with siRNA has transformed the T2 appendages toward a maxilliped morphology. (c) Gain of function of *Ubx* causes posteriorization of T1–4 appendages towards locomotory legs (blue). This specimen is a mosaic with ectopic expression of *Ubx* just on the affected side. Scale bars 50 μm. Reproduced from Liubicich *et al. Proc Natl Acad Sci USA* 2009; 106: 13892–13896, and Pavlopoulos *et al. Proc Natl Acad Sci USA* 2009; 106: 13897, with permission from National Academy of Sciences.

mutations that changed the expression of *Ubx*, and became fixed in the evolution of different crustacean groups, is not yet known.

Wings

It is known that some early fossil insects had larvae (nymphs) with wings on all segments, and that the adults were wingless. In modern insects wings reach their full development in the adult. This type of shift of relative timing of events in the life history, with resulting change in adult morphology, is known as a **heterochrony**, and is a common theme in evolution. In *Drosophila* the development of wings on all the nonwing-bearing segments is suppressed by Hox genes. In the first thoracic segment it is suppressed by *Sex combs reduced*, and in the abdomen by *Ubx*, *abdA*, and *AbdB*. Since *Drosophila* is a dipteran, it also lacks wings on the third thoracic segment where the dorsal imaginal disc forms a haltere and not a wing. This is controlled by *Ubx*, which prevents the activation of *vestigial* (see Chapter 17). By analogy with the crustacean example above, it might be supposed that non-Diptera have the anterior expression boundary of Ubx shifted into the abdomen, to allow wing formation on T3. But this is not the case. The expression of Hox genes relative to segments is fairly similar in all insects. So it seems that the wing suppression in Diptera must have arisen from the establishment of new regulatory connections downstream of *Ubx*, and not through changes in expression of the Hox genes themselves.

Atavisms

A mutation that yields a morphology characteristic of an evolutionary ancestor is known as an **atavism**. Since much of the diversification of arthropods seems to have depended on suppression of appendages by regulatory connections downstream of Hox activity, it is not surprising that loss-of-function mutations of Hox genes often have atavistic phenotypes. For example the four-winged *Drosophila*, resulting from complete loss of *Ubx* from the haltere disc, is an atavism.

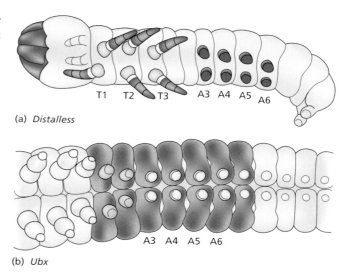

(a) *Distalless*

(b) *Ubx*

Fig. 22.15 Abdominal legs of the butterfly caterpillar. (a) *Distalless* expression. (b) *Ubx* expression at the embryonic stage.

Sometimes in the course of evolution atavisms may become established as the wild type and therefore represent partial reversals of evolution. For example the caterpillars of butterflies have legs on four of their abdominal segments (A3–A6; Fig. 22.15). It is thought that the primordial arthropod probably had legs on most of its segments, like modern myriapods. When the expression patterns of *Ubx*, *abdA*, and *Distalless* are examined in butterflies it can be seen that the Hox genes are actually turned off in the legs buds, and this presumably allows the expression of *Distalless* and the formation of the atavistic legs.

This example shows that similar evolutionary changes can be brought about in different ways. Changes of Hox expression can change the character of whole segments. Alternatively, as indicated by the butterfly, a local change of expression may modify structures within a segment. It is also possible for modifications to arise from mutations in the regulatory regions of genes downstream of the Hox genes. Just because one particular means of achieving a change appears elegant and simple to the developmental biologist, this does not mean that this particular means was adopted in the course of evolution.

Vertebrate limbs

The vertebrate limb has a distinguished history in evolutionary biology. It was early recognized as being a homologous structure across the tetrapods. Because of the presence of a complex skeleton it is often preserved in fossils, and it provides some striking examples of allometric change in evolution. **Allometry** means differential growth between parts (see Chapter 19). For example the middle digit of the horse leg has elongated during evolution until the modern horse has effectively just one digit with the others reduced to vestiges. By contrast, all the digits of the bat forelimb have become greatly elongated to bear the skinflaps that constitute the wings.

It is believed from the fossil record that the first tetrapods arose in the Devonian period and that their ancestors resembled a creature called *Panderichthys* (Fig. 22.16a). This was a lobe-finned fish belonging to the same order as the modern coelacanth. It is thought to have lived in shallow water and perhaps even be able to creep out of the water like some modern catfish. Until recently, the limbs of *Panderichthys* were thought to lack an **autopodium**. This fact was associated with the existence in tetrapods, but not in teleost fish, of an enhancer for the *Hoxd13* gene, which caused late expression in the region destined to become the autopodium. The idea arose that the origin of this genetic element was the key to the evolution of the tetrapod limb. However, it is now known from computed tomography scanning of *Panderichthys* fossils that the organism did, in fact, have distal skeletal elements resembling a primitive autopodium (Fig. 22.16a). Moreover, autopodial expression of *Hoxd13* has been detected in a basal bony fish, the paddlefish *Polyodon*, although it is not found in teleosts such as the zebrafish. This suggests that the basic developmental mechanisms for making a limb were present long before the actual limbs of land vertebrates appeared. The novelty of the tetrapod autopodium must instead derive from changes in regulation of downstream genes concerned with the pattern of skeletal differentiation.

Study of the development of the paired fins of zebrafish have provided further evidence for common developmental mechanisms underlying very different anatomy. In zebrafish embryos

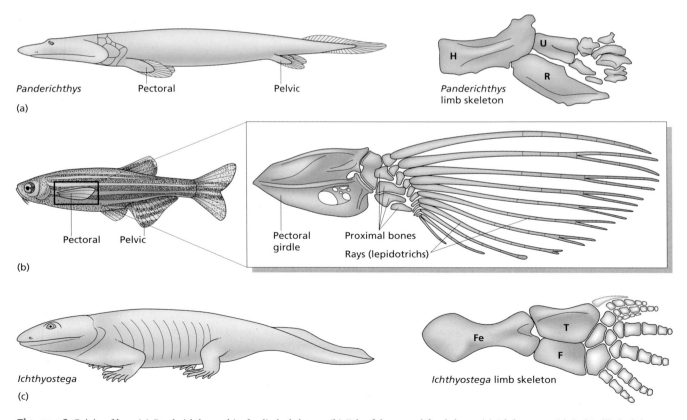

Fig. 22.16 Origin of legs. (a) *Panderichthys* and its forelimb skeletons. (b) Zebrafish pectoral fin skeleton. (c) *Ichthyostega* with its hindlimb skeleton.

injection of **morpholinos** against *Wnt2b* or *Tbx5* will both inhibit fin bud formation, as in higher vertebrates. During early fin bud outgrowth, there is an apical ectodermal ridge expressing *Fgf* genes and a posterior region expressing *Shh*, very similar to the tetrapod limb bud. But during this phase are laid down four parallel cartilage elements, which later ossify and serve as attachments for the fin rays. Both the pattern of these proximal bones and the fin rays themselves, which are formed by dermal ossification, are quite unlike a modern tetrapod limb, or even the *Panderichthys* limb (Fig. 22.16b). So again the critical evolutionary changes must have occurred in downstream genes.

One of the most obvious features of modern tetrapod limbs is that the number of digits very rarely exceeds five. The very name **pentadactyl** limb indicates that the primitive limb has five digits and that all types with reduced digit numbers are derived. It has been speculated that there is a developmental constraint preventing more than five digits because this is the number of Hox d genes available to create distinctions between the digit territories. Unfortunately, this theory does not stand up to the fossil evidence because there are some early Devonian tetrapods such as *Ichthyostega*, appearing soon after *Panderichthys*, having an autopodium which very obviously possessed more than five digits (Fig. 22.16c). How *Ichthyostega* achieved this is not known.

Limb presence and positioning

Some tetrapods have greatly reduced limbs or have lost them altogether. These include snakes, whales, and the forelimbs of flightless birds. Often a rudimentary girdle or portion of the proximal limb skeleton survives as a vestige, indicating that the ancestors had more substantial limbs than their modern descendants. The positions of the limbs on the lateral plate must be specified by the anteroposterior patterning system for the whole body. This is likely to involve the Hox genes although may include other genes as well, for example those of the NK group. Evidence of a role for Hox genes is provided by the effects of transgenic expression of *Hoxb8* at a more anterior level than usual in the mouse, which results in the formation of an extra zone of polarizing activity (ZPA) and duplication of the forelimb. In the mouse and chick the Hox genes *c6* and *c8* are expressed in the lateral plate between the two limb buds and it is believed that they repress limb formation in this region. The python is a type of snake that has no trace remaining of the forelimbs but does have a rudimentary pelvic girdle and a proximal vestige of the hindlimbs. It shows expression of *Hoxc6* and *c8* extending right up to the head, but posteriorly the expression border stops just short of the rudimentary hindlimbs (Fig. 22.17). Rather as in the crustacean maxilliped example given above, this suggests a hypothesis that the extension of Hox expression was a macromutation that suppressed limbs in evolutionary history. However, this would need to be confirmed both by experimental studies on the python to show that the change of Hox expression actually does suppress legs, and also

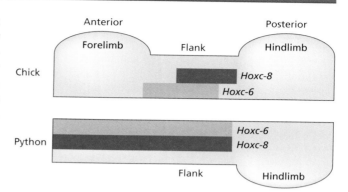

Fig. 22.17 Hox gene expression in the flank of chick and python embryos.

the genomes of many related organisms to find if there is any evidence for a one-time macromutation of the sort envisaged.

Vertebrate and insect wings

The vertebrate and the insect wing are, as mentioned above, considered analogous and not homologous, since the common ancestor of the two groups must have lived in the sea and cannot have possessed wings. However, there are a remarkable number of similarities in the expression patterns and the functions of transcription factors and signaling molecules in the two systems. Both have a dorsoventral pattern controlled by a homologous Lim transcription factor gene expressed on the dorsal side (*apterous* and *Lim1*). Both have an involvement of a *Fringe* gene in setting the dorsoventral boundary. Both have Hedgehog signaling controlling the anteroposterior pattern, and there is even an interaction of Hedgehog with BMP/Dpp in both cases. The distal outgrowth depends in both cases on the homeodomain factor Distalless. With so much developmental mechanism in common surely the insect and vertebrate wing have to be homologous! The solution to this mystery is not yet clear. But it is generally thought that the common features represent modules of developmental genetic circuitry. These may have evolved at an early stage, corresponding to the primordial bilaterian, and could then be co-opted for various purposes: in this case making appendages that needed to be asymmetrical in three dimensions. The lesson is that it is always essential to specify the level at which homology exists. The insect and vertebrate wings are not homologous as wings, but it is possible that they are homologous as appendages. Some appendage made with these systems may have existed in the common ancestor but probably had very different functions. It is certain that the two types of wing contain active genes and genetic pathways that are homologous. This example shows the importance of maintaining clear thinking about difficult concepts. This is valuable throughout developmental biology, but nowhere more than in the complex borderland between development and evolution.

Key points to remember

- Evolution involves both adaptive changes brought about by natural selection and neutral changes established by genetic drift. Neutral changes are most apparent in DNA and protein sequences.
- Classification of living organisms usually attempts to represent the evolutionary history of the groups concerned. This is called a phylogenetic or cladistic taxonomy.
- Homologous features are similar because they derive from a common ancestor. Analogous features are similar because they are driven together by natural selection.
- A phylogenetic tree for the animal phyla themselves can be created using information from descriptive embryology, molecular taxonomy, and the fossil record. It is thought that the common ancestor of all animals lived between 1000 and 600 million years ago.
- The time calibration for phylogenetic trees comes ultimately from radioactive dating of sedimentary rocks. This provides absolute dates for divergence times between lineages, which are observed in the fossil record.

- Animal taxa often have a phylotypic stage at which all groups show maximum similarity.
- All triploblastic animals (Bilateria) share expression domains, and probably function, of a basic set of genes responsible for setting up the general body plan. This configuration of gene activity comprises the Hox, ParaHox, and NK clusters, and certain other genes including *Otx*, *Emx*, and *Pax6*. This configuration, or zootype, was probably present in the common ancestor of all triploblastic animals.
- The importance in evolution of developmental constraints, and of macro- as opposed to micromutations, is still not known.
- Changes in the segmental organization of arthropods have arisen partly from changes in Hox expression but mostly from changes in downstream genes controlled by the Hox pattern.
- The basic developmental mechanisms of tetrapod limb development exist also in the paired fins of fish.
- Homology can exist at the level of genetic pathways and circuits even where there is no morphological homology.

Further reading

General

Gould, S.J. (1977) *Ontogeny and Phylogeny*. Cambridge, MA: Harvard University Press.

Raff, R.A. (1996) *The Shape of Life: Genes, Development and the Evolution of Animal Form*. Chicago: University of Chicago Press.

Gerhart, J. & Kirschner, M. (1997) *Cells, Embryos and Evolution*. Malden, MA: Blackwell Science.

Carroll, S.B., Grenier, J.K. & Weatherbee, S.D. (2001) *From DNA to Diversity*. Malden, MA: Blackwell Science.

Davidson, E.H. (2001) *Genomic Regulatory Systems in Development and Evolution*. New York: Academic Press.

Hall, B.K. & Olson W.M., eds. (2003) *Keywords and Concepts in Evolutionary Developmental Biology*. Cambridge, MA: Harvard University Press.

Arendt, D. (2008) The evolution of cell types in animals: emerging principles from molecular studies. *Nature Reviews Genetics* **9**, 868–882.

Carroll, S.B. (2008) Evo-Devo and an expanding evolutionary synthesis: a genetic theory of morphological evolution. *Cell* **134**, 25–36.

DeRobertis, E.M. (2008) Evo-devo: variations on ancestral themes. *Cell* **132**, 185–195.

Stern, D.L. & Orgogozo, V. (2009) Is genetic evolution predictable? *Science* **323**, 746–751.

Taxonomy

Holland, P.W.H., Garcia Fernàndez, J., Williams, N.A. & Sidow, A. (1994) Gene duplications and the origins of vertebrate development. *Development* Supplement, 125–133.

Bolker, J.A. & Raff, R.A. (1996) Developmental genetics and traditional homology. *Bioessays* **18**, 489–494.

Aguinaldo, A.M.A. & Lake, J.A. (1998) Evolution of the multicellular animals. *American Zoologist* **38**, 878–887.

Hedges, S.B. (2002) The origin and evolution of model organisms. *Nature Reviews Genetics* **3**, 838–849.

Telford, M.J. (2006) Animal phylogeny. *Current Biology* **16**, R981–R985.

DeSalle, R. & Schierwater, B. (2008) An even newer animal phylogeny. *BioEssays* **30**, 1043–1047.

Paleontology

Conway Morris, S. (1993) The fossil record and the early evolution of the metazoa. *Nature* **361**, 219–225.

Valentine, J.W., Jablonski, D. & Erwin, D.H. (1999) Fossils, molecules and embryos: new perspectives on the Cambrian explosion. *Development* **126**, 851–859.

Conway Morris, S. (2000) Evolution: bringing molecules into the fold. *Cell* **100**, 1–11.

Chen, J.Y., Bottjer, D.J., Oliveri, P., Dornbos, S.Q., Gao, F., Ruffins, S., Chi, H., Li, C.W. & Davidson, E.H. (2004) Small bilaterian fossils from 40 to 55 million years before the Cambrian. *Science* **305**, 218–222.

Raff, R.A. (2007) Written in stone: fossils, genes and evo-devo. *Nature Reviews Genetics* **8**, 911–920.

Body plans

Slack, J.M.W., Holland, P.W.H. & Graham, C.F. (1993) The zootype and the phylotypic stage. *Nature* **361**, 490–492.

Arendt, D. & Nübler-Jung, K. (1997) Dorsal or ventral: similarities in fate maps and gastrulation patterns in annelids, arthropods and chordates. *Mechanisms of Development* **61**, 7–21.

Ferrier, D.E.K. & Holland P.W.H. (2001) Ancient origin of the Hox gene cluster. *Nature Reviews Genetics* **2**, 33–38.

Erwin, D.H. & Davidson, E.H. (2002) The last common bilaterian ancestor. *Development* **129**, 3021–3032.

Mieko Mizutani, C. & Bier, E. (2008) EvoD/Vo: the origins of BMP signalling in the neuroectoderm. *Nature Reviews Genetics* **9**, 663–677.

Niehrs, C. (2010) On growth and form: a Cartesian coordinate system of Wnt and BMP signaling specifies bilaterian body axes. *Development* **137**, 845–857.

Case studies

Ahlberg, P.E. & Milner, A.R. (1994) The origin and early diversification of tetrapods. *Nature* **368**, 507–514.

Sordino, P. & Duboule, D. (1996) A molecular approach to the evolution of vertebrate paired appendages. *Trends in Ecology and Evolution* **11**, 114–119.

Shubin, N., Tabin, C. & Carroll, S. (1997) Fossils, genes and the evolution of animal limbs. *Nature* **388**, 639–648.

Weatherbee, S.D. & Carroll, S.B. (1999) Selector genes and limb identity in arthropods and vertebrates. *Cell* **97**, 283–286.

Akam, M. (2000) Arthropods: developmental diversity within a superphylum. *Proceedings of the National Academy of Sciences USA* **97**, 4438–4441.

Gehring, W.J. (2002) The genetic control of eye development and its implications for the evolution of the various eye types. *International Journal of Developmental Biology* **46**, 65–73.

Jordi, G.-F. & Èlia, B.-G. (2009) It's a long way from amphioxus: descendants of the earliest chordate. *BioEssays* **31**, 665–675.

Liubicich, D.M., Serano, J.M., Pavlopoulos, A., Kontarakis, Z., Protas, M.E., Kwan, E., Chatterjee, S., Tran, K.D., Averof, M. & Patel, N.H. (2009) Knockdown of Parhyale Ultrabithorax recapitulates evolutionary changes in crustacean appendage morphology. *Proceedings of the National Academy of Sciences* **106**, 13892–13896.

Technau, U. & Steele, R.E. (2011) Evolutionary crossroads in developmental biology: Cnidaria. *Development* **138**, 1447–1458.

Appendix

Key molecular components

This appendix contains a brief summary of the principal types of macromolecule and cell component involved in development. It is not intended as a substitute for a grounding in cell biology or molecular biology or biochemistry, but can be referred to if molecules or processes mentioned in the main text are unfamiliar.

Genes

A gene is a sequence of DNA that codes for a protein, or for a nontranslated RNA, and it is usually considered also to include the associated regulatory sequences as well as the coding region itself. The vast majority of eukaryotic genes are located in the nuclear chromosomes, but there are also a few genes carried in the DNA of mitochondria and chloroplasts. The genes encoding RNA include those for ribosomal or transfer RNAs, and also an indeterminate number of microRNAs that are probably involved in translational control. It is normal practice to italicize in lower case names of genes or their primary transcripts and to write the name of the protein product in Roman type, e.g. the *wingless* gene codes for the *wingless* messenger RNA and the wingless protein.

DNA is normally complexed into **chromatin** by the binding of basic proteins called histones. Protein-coding genes are transcribed into messenger RNA (mRNA) by **RNA polymerase II**. Transcription commences at a transcription start sequence and finishes at a transcription termination sequence. Genes are usually divided into several exons, each of which codes for a part of the mature mRNA. The primary RNA transcript is extensively processed before it moves from the nucleus to the cytoplasm. It acquires a "cap" of methyl guanosine at the 5′ end, and a polyA tail at the 3′ end, both of which stabilize the message

by protecting it from attack by exonucleases. The DNA sequences in between the exons are called introns and the portions of the initial transcript complementary to the introns are removed by splicing reactions catalyzed by "snRNPs" (small nuclear ribonucleoprotein particles). It is possible for the same gene to produce several different mRNAs as a result of alternative splicing, whereby different combinations of exons are spliced together from the primary transcript. In the cytoplasm the mature mRNA is translated into a polypeptide by the ribosomes. The mRNA still contains a 5′ leader sequence and a 3′ untranslated sequence flanking the protein-coding region, and these untranslated regions may contain specific sequences responsible for translational control or intracellular localization.

The total number of genes in vertebrate animals is about 25,000. In invertebrates it is rather fewer, about 19,000 in *Caenorhabditis elegans* and 13,000 in *Drosophila*. Of these, perhaps 1–2% are intimately involved in the processes of development, although many of the others are necessary for the formation of viable cells capable of developing.

Control of gene expression

There are many genes whose products are required in all tissues at all times, for example those concerned with basic cell structure, protein synthesis, or metabolism. These are referred to as "housekeeping" genes. But there are many others whose products are specific to particular cell types and indeed cell types differ from each other because they contain different repertoires of proteins. This means that the control of gene expression is central to developmental biology. Control may be exerted at several points. Most common is control of transcription, and we often speak of genes being "on" or "off" in particular

Essential Developmental Biology, Third Edition. Jonathan M.W. Slack.
© 2013 John Wiley & Sons, Ltd. Published 2013 by John Wiley & Sons, Ltd.

situations, meaning that they are or are not being transcribed. There are also many examples of translational regulation, where the mRNA exists in the cytoplasm but is not translated until some condition is satisfied. This can be particularly important in early embryos where mRNA present in the egg may not be translated until after fertilization. Control may also be exerted at the stage of nuclear RNA processing, or indirectly via the stability of individual mRNAs or proteins.

Control of transcription depends on regulatory sequences in the DNA, and on proteins called **transcription factors** that interact with these sequences. The **promoter** region of a gene is the region just upstream from the transcription start site to which the RNA polymerase binds. It binds assisted by a set of general transcription factors, which together make up a transcription complex. In addition to the general factors required for the assembly of the transcription complex, there are numerous specific transcription factors that bind to specific regulatory sequences that may be either adjacent to or at some distance from the promoter (Fig. A.1). Transcription factors usually contain a DNA-binding domain and a regulatory domain, which will either activate or repress transcription. Looping of the DNA may bring these regulatory domains into contact with the transcription complex and either promotes or inhibits its activity.

Genes involved in development often have very complex regulatory regions consisting of several sites. Each site may be capable of binding both positive and negative transcription factors. Whether the gene is or is not transcribed will depend on the combination of factors that is present in the cell. The same transcription factors will bind to regulatory sites controlling many different genes, thus ensuring that target genes are controlled in a coordinated manner. Regulatory sites at a distance from the gene are often called **enhancers**, although they may sometimes repress as well as enhance transcription. The term **promoter** is frequently used not just for the region binding the transcription complex but for a larger region adjacent to the 5′ end of the gene containing several regulatory sites. A promoter in this broad sense may act independently of its surroundings. If a specific promoter is joined to another gene and then introduced into the genome of an animal, it will often drive expression of the new gene with the appropriate tissue specificity, regardless of the position in the genome in which it integrates.

Transcription factors

Transcription factors are the proteins that regulate transcription. There are many families, classified by the type of DNA-binding domain they contain, such as the **homeodomain** or zinc-finger domain, and those most important in development are listed below. Most are nuclear proteins, although some exist in the cytoplasm until they are activated and then enter the nucleus. Activation often occurs in response to intercellular signaling (see below). One type of transcription factor, the nuclear receptor family, is directly activated by lipid-soluble signaling molecules such as retinoic acid or glucocorticoids.

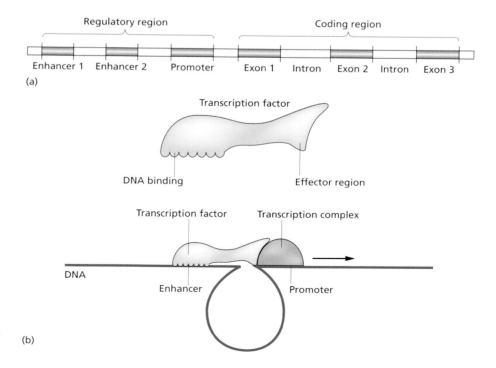

Fig. A.1 (a) Typical structure of a eukaryotic gene. (b) A transcription factor may bind to an enhancer and stimulate the transcription complex "at a distance."

Each type of DNA-binding domain in a protein has a corresponding type of target sequence in the DNA, usually 20 nucleotides or less. These can be identified by sequencing of the regulatory region of the target gene, although evidence that a particular transcription factor actually does regulate a given target gene *in vivo* needs to be established by experiment. Activation domains of transcription factors often contain many acidic amino acids forming an "acid blob," which accelerates the formation of the general transcription complex. Some transcription factors recruit histone acetylases which open up the chromatin by neutralizing amino groups on the histones, and allow access of other proteins to the DNA. In general, the molecular organization of transcription factors is modular, so new factors can be created in the laboratory by recombination of individual domains of existing factors. For example, a transcriptional activator may be converted to a repressor by swapping its acid blob for a repression domain. In the nuclear receptor family it is possible to combine the hormone-binding site of one factor with the DNA-binding and effector sites of another. For example if a retinoic acid-binding site in the receptor is replaced by a thyroid-binding site, then the genes normally responsive to retinoic acid would become responsive to thyroid hormone instead.

Although it is normal to classify transcription factors as activators or repressors of transcription, their action is also sensitive to context, and the presence of other factors may on occasion cause an activator to function as a repressor, or vice versa.

Transcription factor families

This lists some of the structural families of transcription factor most important in development. They are classified by DNA-binding domain.

Homeodomain factors

The **homeodomain** is an approximately 60 amino acid sequence containing many basic residues, and forms a helix-turn-helix structure that binds specific sites in DNA. The homeodomain sequence itself is coded by a corresponding **homeobox** in the gene. The homeobox was given its name because it was initially discovered in **homeotic genes** (see Chapter 4). However, there are many transcription factors that contain a homeodomain as their DNA-binding domain and although they are often involved in development, possession of a homeodomain does not guarantee a role in development, nor are mutants of homeobox genes necessarily homeotic. A very large number of homeodomain proteins have important developmental functions, e.g. Engrailed in *Drosophila* segmentation, Goosecoid in the vertebrate organizer, Cdx proteins in anteroposterior patterning. An important subset are the **HOX proteins** which have a special role in the control of anteroposterior pattern in animals. Homeobox genes are found in animals, plants, and fungi, but the **Hox** subset are found only in animals.

LIM-homeodomain proteins

The LIM domain is a cysteine-rich zinc-binding region responsible for protein–protein interactions, but is not itself a DNA-binding domain. LIM-homeoproteins possess two LIM domains together with the DNA-binding homeodomain. Examples are Lim-1 in the organizer, Islet-1 in motorneurons, Lhx factors in the limb bud, and Apterous in the *Drosophila* wing.

Pax proteins

These are characterized by a DNA-binding region called a paired domain with 6 α-helical segments. The name is derived from the paired protein in *Drosophila*. Many of the pax proteins also contain a homeodomain. Examples are Pax6 in the eye and Pax3 in the developing somite.

Zinc-finger proteins

This is a large and diverse group of proteins in which the DNA-binding region contains projections ("fingers") with Cys and/or His residues folding around a zinc atom. Some examples are the GATA factors important in the development of the blood and the gut, Krüppel in the early *Drosophila* embryo, Krox20 in the rhombomeres of the hindbrain, WT-1 in the kidney, and GAL4, the yeast transcription factor used for overexpression experiments.

Nuclear receptor superfamily

These are intracellular receptors that are also transcription factors. They are stimulated by lipophilic ligands including steroids, thyroid hormones, and retinoic acid. They have a hormone-binding domain, a DNA-binding domain, and a transcription-activation domain. Some types are normally complexed with a heat shock protein called hsp90 and thereby retained in the cell cytoplasm. On binding of the hormone, the hsp90 is displaced and the factors can form active dimers which move to the nucleus and bind to the target genes. Examples of nuclear receptors include the retinoic acid receptors (RARs, also now called NR1Bs) and the ecdysone receptor.

Basic helix–loop–helix (bHLH) proteins

These transcription factors are active as heterodimers. They contain a basic DNA-binding region and a hydrophobic helix-loop-helix region responsible for protein dimerization. One member of the dimer is found in all tissues of the organism and the other member is tissue specific. There are also proteins containing the HLH but not the basic part of the sequence. These form inactive dimers with other bHLH proteins and so inhibit their activity. Examples of bHLH proteins include E12, E47 which are ubiquitous in vertebrates, the myogenic factor MyoD, and the *Drosophila* pair-rule protein hairy. An example of an inhibitor with no basic region is Id, which is an inhibitor of myogenesis.

Winged helix proteins

These have a 100 amino acid winged helix domain which forms another type of DNA-binding region and are known as "Fox" proteins. Examples are Forkhead in *Drosophila* embryonic termini and FoxA2 (formerly HNF3 β) in the vertebrate main axis and gut.

T-box factors

These have a DNA-binding domain similar to the prototype gene product known as "T" in the mouse and as brachyury in other animals. They include the endodermal determinant VegT and the limb identity factors Tbx4 and Tbx5.

High mobility group (HMG)-box factors

These factors differ from most others because they do not have a specific activation or repression domain. Instead they work by bending the DNA to bring other regulatory sites into contact with the transcription complex. Examples are SRY, the testis-determining factor, Sox9, a "master switch" for cartilage differentiation, and the TCF and LEF factors whose activity is regulated by the Wnt pathway.

Other controls of gene activity

Some aspects of gene control are of a more stable and longer-term character than that exerted by combinations of positive and negative transcription factors. To some extent this depends on the remodeling of the **chromatin** structure, which is still poorly understood. Much of the DNA is complexed with histones into nucleosomes, and is then arranged as a 30-nm filament and higher structures. In much of the genome the nucleosomes are to some extent mobile, allowing access of transcription factors to the DNA. This is called **euchromatin**. In other regions it is highly condensed and inactive. This is called **heterochromatin**. In the extreme case of the nucleated red blood cells of nonmammalian vertebrates the entire genome is

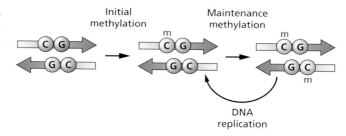

Fig. A.2 DNA methylation. Methylation is preserved through DNA replication by the action of the maintenance methylase.

heterochromatic and inactive. Chromatin structure is regulated to some degree by **chromobox** proteins, for example Polycomb in *Drosophila*, which affect the expression of many genes but are not themselves transcription factors.

An important element of the chromatin remodelling is the control through acetylation of lysines on the N-termini of histones. This opens up the chromatin structure enabling a transcription complex to assemble on the DNA. The degree of histone acetylation is controlled, at least partly, by DNA methylation, because histone deacetylases are recruited to methylated regions. Methylation occurs on cytosine residues in CG sequences of DNA. Because CG on one strand will pair with GC on the other, antiparallel, strand, potential methylation sites always lie opposite one another on the two strands. There are several DNA methyl transferase enzymes, including *de novo* methylases that methylate previously unmethylated CGs, and maintenance methylases that methylate the other CG of sites bearing a methyl group on only one strand. Once a site is methylated, it will be preserved through subsequent rounds of DNA replication, because the hemimethylated site resulting from replication will be a substrate for the maintenance methylase (Fig. A.2). There is an overall deficit of the dinucleotide CG in the vertebrate genome compared to the other 15 possible dinucleotides. Many of the CGs are clustered into about 40,000 regions near the 5′ ends of genes. These so-called CpG islands are predominantly unmethylated and are thought to indicate the positions of the "housekeeping" genes.

The state of histone acetylation may also, independently of methylation, be transmitted through DNA replication, perhaps by permanent association of the transacetylases with acetylated regions of chromatin. So both DNA methylation and histone acetylation provide means for maintaining the state of activity of genes in development even after the original signals for activation or repression have disappeared.

Signaling systems

Development depends to a large extent on the effects of one group of cells on another. These effects are exerted by **embryonic induction**, which involves the release of an **inducing factor** from a signaling center and its action on a responding cell group. There are several important molecular families of inducing factors, and these molecules may also be known as growth factors, cytokines, or hormones in other contexts. Inducing factors are mainly proteins, although a few are small lipid soluble molecules such as retinoic acid. Most are secreted from the signaling cells, although some are integral membrane proteins that can only interact with immediately adjacent cells. They bind to specific **receptors** and activate a **signal transduction pathway** within the cell. This may lead to activation or repression of specific genes or to changes in cell behavior mediated through the cytoskeleton or through changes to cellular metabolism. The repertoire of responses that a cell can show depends on which receptors it possesses, how these are coupled to signal transduction pathways, and how these pathways are coupled to gene regulation. An inducing factor that can evoke more than one response at different concentrations is called a **morphogen**. This is because a concentration gradient emitted from the signaling center will induce different responses at different distances from the source and thus create "morphology" where there was previously none (see Chapter 4).

Secretion of proteins

All secreted proteins, including the inducing factors, normally contain a signal peptide, which is a hydrophobic sequence of amino acids at the N-terminal end. During protein synthesis this is recognized by a signal recognition particle which binds the ribosome to a translocation channel in the endoplasmic reticulum. As synthesis proceeds the growing polypeptide is posted through this channel into the endoplasmic reticulum lumen. Finally the signal peptide is cleaved off and the protein released. Once in the endoplasmic reticulum, the protein becomes further processed. Any disulfide bonds that are required for the secondary structure are formed. These may be intramolecular or intermolecular and may lead to the formation of either homo- or heterodimers. Most cell-surface and extracellular proteins also become glycosylated by the addition of carbohydrate chains to serine, threonine, or asparagine residues. This usually does not affect the biological activity of the molecule but may affect its stability or interaction with other components. In some cases specificity is affected, for example in the case of Notch (see below). The proteins are carried in vesicles to the Golgi apparatus where further processing of the carbohydrates occurs, and then in exocytotic vesicles to the cell surface where they are released.

Signal transduction

The small molecule lipid soluble inducing factors, such as retinoic acid, can enter cells freely by diffusion. The retinoic acid receptor is itself a transcription factor, and ligand binding enables the receptor complex to activate its target genes (Fig. A.3a).

Protein-inducing factors cannot diffuse across the plasma membrane and so work by binding to specific cell-surface receptors. There are three main classes of these: enzyme-linked receptors, which are particularly important in development, G-protein-linked receptors, and ion channel receptors.

The main enzyme-linked receptors of importance in development are tyrosine kinases or Ser/Thr kinases (Fig. A.3b). All have a ligand-binding domain on the exterior of the cell, a single transmembrane domain, and the enzyme active site on the cytoplasmic domain. For receptor tyrosine kinases, the ligand-binding brings about dimerization of the receptor and this results in an autophosphorylation whereby each receptor molecule phosphorylates and activates the other. The phosphorylated receptors can then activate a variety of targets. Many of these are transcription factors that are activated by phosphorylation and move to the nucleus where they upregulate their target genes. In other cases there is a cascade of kinases that activate each other down the chain, culminating in the activation of a transcription factor. Roughly speaking, each class of factors has its own associated receptors and a specific signal transduction pathway; however, different receptors may be linked to the same signal transduction pathway, or one receptor may feed into more than one pathway. The effect of one pathway upon the others is often called "cross talk." For example, activated ERK (extracellular signal regulated kinase) can inhibit bone morphogenetic protein (BMP) signaling by phosphorylation of smad1 in its linker region. The significance of cross talk can be hard to assess from biochemical analysis alone, but is much easier using genetics. This is why the model organisms used for studying development are also useful for functional analysis of signal transduction and thereby useful for screening libraries of potential new drugs affecting signal transduction pathways. The inducing factor families and the signal transduction pathways most important in development are listed below.

There are several classes of G-protein-linked receptor (Fig. A.3c). The best known are seven-pass membrane proteins, meaning that they are composed of a single polypeptide chain crossing the membrane seven times. They are associated with trimeric G proteins composed of α-, β-, and γ-subunits. When the ligand binds, the activated receptor causes exchange of guanosine diphosphate (GDP) bound to the α-subunit for guanosine triphosphate (GTP), the activated α-subunit is released and can interact with other membrane components. The most common target is adenylyl cyclase, which converts adenosine triphosphate (ATP) to cyclic adenosine monophosphate (cAMP). Cyclic AMP activates protein kinase A (PKA) which phosphorylates various further target molecules affecting both intracellular metabolism and gene expression.

Another large group of G-protein-linked receptors use a different trimeric G protein to activate the inositol phospholipid pathway (Fig. A.3c). Here the G protein activates phospholipase C β which breaks down phosphatidylinositol bisphosphate

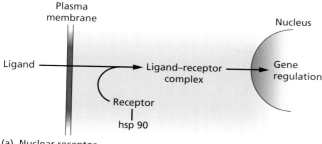

(a) Nuclear receptor

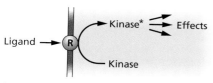

(b) Enzyme–linked receptor

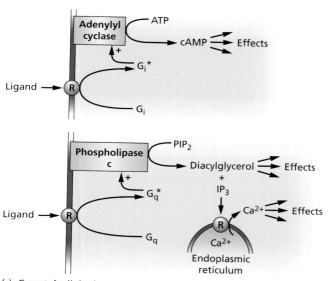

(c) G protein–linked receptors

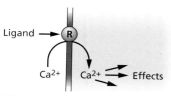

(d) Ion channel receptor

Fig. A.3 General modes of signal transduction. (a) Nuclear receptor mechanism used by steroids. (b) Enzyme-linked receptor. (c) G-protein-linked receptor. (d) Ligand gated ion channel.

(PIP$_2$) to diacylglycerol (DAG) and inositol trisphosphate (IP$_3$). The DAG activates an important membrane-bound kinase, protein kinase C. Like protein kinase A this has a large variety of possible targets in different contexts and can cause both metabolic responses and changes in gene expression. The IP$_3$

binds to an IP$_3$ receptor (IP$_3$R) in the endoplasmic reticulum and opens Ca channels which admit Ca ions into the cytoplasm. Normally cytoplasmic Ca is kept at a very low concentration of around 10^{-7} M. An increase caused either by opening of an ion channel in the plasma membrane, or as a result of IP$_3$ action, can again have a wide range of effects on diverse target molecules.

Ion channel receptors (Fig. A.3d) are not so prominent in development as the other types, but they play a key role in fertilization and their importance elsewhere may have been underestimated. They open on stimulation to allow passage of Na, K, Cl, or Ca.

Inducing factor families

Some important classes of protein-inducing factor are listed here. Most of these factors exist in both vertebrates and invertebrates but they often have different names if discovered by different routes. The normal vertebrate name is used here. Usually an invertebrate such as *C. elegans* or *Drosophila* will contain one gene for a particular factor, while the vertebrate species will contain a corresponding family of genes coding for proteins with similar, although not necessarily identical, biological activity. The usually accepted signal transduction mechanisms are shown in simplified form in Fig. A4, but the true picture is much more complex with many interactions between pathways.

Transforming growth factor β superfamily

Transforming growth factor (TGF) β was originally discovered as a mitogen (i.e. a factor promoting cell growth in tissue culture) secreted by "transformed" (cancer-like) cells. It has turned out to be the prototype for a large and diverse superfamily of signaling molecules, all of which share a number of basic structural characteristics. The mature factors are disulfide-bonded dimers of approximately 25 kDa. They are synthesized as longer pro-forms which need to be proteolytically cleaved to the mature form in order for biological activity to be shown.

The TGFβs themselves are in fact often inhibitory to cell division and promote the secretion of extracellular matrix materials. They are involved mainly in the organogenesis stages of development. The activin-like factors include the nodal-related family, which are all involved in induction and patterning of the mesoderm in vertebrate embryos. The bone morphogenetic proteins (BMPs) were discovered as factors promoting ectopic formation of cartilage and bone in rodents. They are involved in skeletal development, and also in the specification of the early body plan.

There are a number of receptors for the TGFβ superfamily. Their specificity for different factors is complex and overlapping, but in general different subsets of receptors bind to the

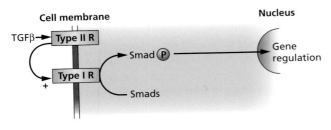

TGFβ pathway

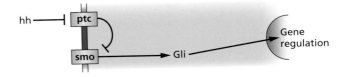

FGF pathway

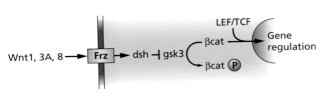

Insulin/IGF pathway

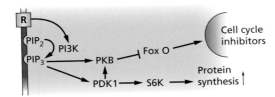

Hedgehog pathway

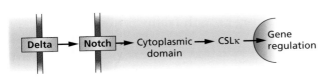

Canonical Wnt pathway

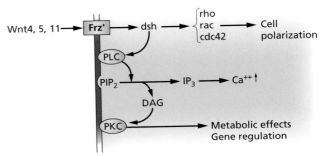

Other Wnt pathways

Notch pathway

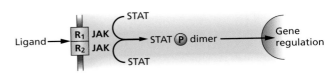

Cytokine system

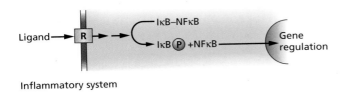

Inflammatory system

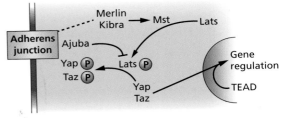

Hippo pathway

Fig. A.4 Various specific pathways of signal transduction.

TGFβs themselves, the activin-like factors, and the BMPs. In all cases the ligand binds first to a type II receptor and enables it to form a complex with a type I receptor. The type I receptor is a Ser-Thr kinase and becomes activated in the ternary complex. Activation causes phosphorylation of smad proteins in the cytoplasm. Smads 1, 5, and 8 are targets for BMP receptors; smads 2 and 3 for activin receptors. Smad 4 is required by both pathways, and smad 6 is inhibitory to both by displacing the binding of smad 4. Phosphorylation causes the smads to migrate to the nucleus where they function as transcription factors, regulating target genes (Fig. A.4a).

Fibroblast growth factor family

As their name suggests, fibroblast growth factors (FGFs) were first identified as mitogens for fibroblasts in tissue culture. They form a large family of factors with many important biological activities. They are monomeric polypeptides that bind tightly to heparan sulfate, a glycosaminoglycan found on all cell surfaces. FGFs have a very wide range of developmental functions including early anteroposterior patterning, regionalization of the brain, and promotion of limb outgrowth.

The FGF receptors are of the tyrosine kinase class. In vertebrates there are four receptors encoded by different genes and these each have splice forms that may show ligand specificity. In particular the R2b splice form is selective for FGF 7 and 10, while the R2c splice form is specific for FGF 4 and 8.

The complex of FGF and heparan sulfate binds to the receptors and brings about formation of receptor dimers. This leads to autophosphorylation and the activation of the extra-cellular signal regulated kinase (ERK) signal transduction pathway. This pathway involves a cascade of events. The first step is the activation of the GTP binding protein Ras, by exchange of a bound GTP for GDP. The activated Ras then activates the Raf protein by causing it to associate with the cell membrane. Raf is a kinase that phosphorylates another kinase called MEK (mitogen-activated, ERK-activating, kinase), which in turn phosphorylates ERK. Activated ERK enters the nucleus and activates various transcription factors by phosphorylation (Fig. A.4b). ERK used to be synonymous with MAP kinase, but this term now refers to a group of kinases which includes JNK and p38 as well as ERK itself.

Insulin family

Factors such as insulin and the insulin-like growth factors are important in growth control. They bind to tyrosine kinase-type receptors which are tetramers of two α and two β chains joined by disulfide bridges. These phosphorylate adapter proteins including IRS1 (insulin receptor substrate 1), which leads to activation of phosphatidylinositol 3-kinase (PI3 kinase). This acts in the membrane to phosphorylate various phosphatidyli-nositol substrates to generate a group of 3-phosphoinositides. These act as intact phospholipids and should not be confused with the free inositol 1,4,5-trisphosphate involved in the G-protein-linked pathway. The 3-phosphoinositides activate protein kinase B (also called Akt), and PDK1, which phosphorylate further substrates that have the effect of boosting protein synthesis and growth (Fig. A.4c).

The phosphatase PTEN dephosphorylates PKB and antagonizes insulin signaling.

Other tyrosine kinase-linked factors

Epidermal growth factor (EGF) was originally identified as a protein causing premature tooth eruption in newborn mice. It is structurally similar to TGFα, a factor isolated at the same time as TGFβ and hence with a similar name, although it is biochemically quite distinct. EGF/TGFα factors are very important in the maternal stage patterning of *Drosophila*. Like FGF receptors, EGF receptors are Tyr kinases and they also activate the ERK pathway.

A similar group of factors are the neuregulins, all formed by differential splicing from a single gene. These bind a different group of receptor tyrosine kinases (e.g. ErbB2) and are often secreted by neurons.

Platelet-derived growth factor (PDGF) is the major growth factor in fetal serum. There are A and B forms of the factor and α and β forms of the receptor. Knockout or inhibition of this system causes a number of developmental process to fail.

Hepatocyte growth factor (HGF) is a ligand distinct from the FGF and EGF classes. It binds to another tyrosine kinase type receptor called c-met. It is important in liver regeneration and for migration of myoblasts into the limb buds. It is also called scatter factor because its first discovered activity was to cause epithelial to mesenchymal transition of epithelial cells.

Neurotrophins are principally, although not exclusively, of importance in the nervous system. They are essential for survival of neurons and are often secreted by the target cells to which the axons project. The neurotrophins comprise nerve growth factor (NGF), brain-derived neurotrophic factor (BDNF), and neurotrophins 3 and 4 (NT3 and 4). The receptors are called Trk (pronounced "track") proteins, which are tyrosine kinase type receptors. Glial-derived neurotrophic factor (GDNF) has a different receptor, ret, also a tyrosine kinase.

Ephrins are ligands for the Eph class of tyrosine kinase receptors. The ephrin A subgroup remains attached to the producing cell by a glycerophosphoinositol (GPI) anchor and binds predominantly to the Eph A receptors. Members of the ephrin B subgroup are themselves transmembrane proteins and bind mainly to the Eph B receptors. So the ephrin system must operate between cells making contact with each other and signaling may occur in both directions where B-type ephrins are involved. The ephrins are known to be involved in gastrulation, the establishment of segmentation in the hindbrain, and establishment of topographic maps in the nervous system.

Hedgehog family

The hedgehogs were first identified because mutations of the gene in *Drosophila* disrupted the segmentation pattern and made the larvae look like hedgehogs. Sonic hedgehog is very important for the dorsoventral patterning of the neural tube and for anteroposterior patterning of the limbs. Indian hedgehog is important in skeletal development. The full-length hedgehog polypeptide is an autoprotease, cleaving itself into an active N-terminal and an inactive C-terminal part. The N-terminal fragment is normally modified by covalent addition of a fatty acyl chain and of cholesterol, which are needed for full activity.

The hedgehog receptor is called patched, again named after the phenotype of the gene mutation in *Drosophila*. This is of the G-protein-linked class. It is constitutively active and is repressed by ligand binding. When active it represses the activity of another cell membrane protein, smoothened, which in turn represses the proteolytic cleavage of Gli-type transcription factors. Full-length Gli factors are transcriptional activators that can move to the nucleus and turn on target genes, but the constitutive removal of the C-terminal region makes them into repressors. In the absence of hedgehog, patched is active, smoothened inactive, and Gli inactive. In the presence of hedgehog, patched is inhibited, smoothened is active, and Gli is active (Fig. A.4d). Activation of protein kinase A also represses Gli and hence antagonizes hedgehog signaling.

Wnt family

The founder member of the Wnt family was discovered through two routes, as an oncogene in mice and as the *wingless* mutation in *Drosophila*. Wnt factors are single-chain polypeptides containing a covalently linked fatty acyl group which is essential for activity and renders them insoluble in water.

The Wnt receptors are called frizzleds after another *Drosophila* mutation. There are several classes of receptor for different ligand types and they do not necessarily cross-react. Wnt 1, 3A, or 8 will activate frizzleds that cause the repression of a kinase, glycogen synthase kinase 3 (gsk3) via a multifunctional protein called dishevelled. When active, gsk3 phosphorylates β-catenin, an important molecule involved both in cell adhesion and in gene regulation. When gsk3 is repressed, β-catenin remains unphosphorylated and in this state can combine with a transcription factor, Tcf-1, and convey it into the nucleus (Fig. A.4e). This pathway is important in numerous developmental contexts, including early dorsoventral patterning in *Xenopus*, segmentation in *Drosophila*, and kidney development.

Other Wnts, including Wnts 4, 5, and 11, bind to a different subset of frizzled that activate two other signal transduction pathways. In the **planar cell polarity** pathway a domain of the dishevelled protein interacts with small GTPases and the cytoskeleton (see below) to bring about a polarization of the cell. In the Wnt-Ca pathway phospholipase C becomes activated by a frizzled. This then acts to generate diacylglycerol and inositol 1,4,5 trisphosphate, with consequent elevation of cytoplasmic calcium, as described above under G-protein-coupled receptors (Fig. A.4f).

Notch system

For the Delta–Notch system both the ligands (Delta, Jagged) and the receptor (Notch) are integral membrane proteins. Their interaction can therefore only take place if the cells making them are in contact, as for the ephrin-Eph system above. Binding of ligand to Notch causes cleavage of the cytoplasmic portion of Notch by an intramembranous protease, γ-secretase, and this causes release into the cytoplasm of a transcription factor, CSL-κ (**CBF1**, **SuCH**, and **Lag1** group, = suppressor of hairless in *Drosophila*). This migrates to the nucleus and activates target genes (Fig. A.4g). The γ-secretase is the same protease that generates the peptide whose accumulation in the brain leads to Alzheimer's disease.

Notch can carry O-linked tetrasaccharides and the presence of this carbohydrate chain can affect its specificity, increasing sensitivity to Delta and reducing sensitivity to Jagged (= Serrate in *Drosophila*). Control is often exercised through the activity of the glycosyl transferase Fringe, which adds a GlcNAc to the O-linked fucose.

The Delta–Notch system is important in numerous developmental situations, including neurogenesis, somitogenesis, and imaginal disc development.

Cytokine system

A somewhat heterogeneous group of signaling molecules including some classical cytokines (interferons), the hematopoietic growth factors, growth hormone, and leukemia inhibitory factor (LIF), work through this type of system. The receptors, which may be homodimeric or heterodimeric, bind intracellular tyrosine kinases called JAKs (Janus kinases). When the ligand brings receptor molecules together the JAKs phosphorylate and activate each other, and the receptor, and the active complex can then phosphorylate STAT-type transcription factors. Phosphorylation causes them to dissociate from the receptor and dimerize and they can then enter the nucleus and activate target genes (Fig. A.4h).

Inflammatory system

The proinflammatory cytokines comprising the interleukin 1 (IL1) and tumor necrosis factor families have receptors that are coupled to the activation of a transcription factor NFκB, which is normally complexed in an inactive form in the cytoplasm by another protein called IκB. When ligand binds receptor it activates a cytoplasmic kinase that phosphorylates IκB and causes it to dissociate from the complex, allowing NFκB to move to the nucleus and turn on target genes (Fig. A.4i). IκB can also be phosphorylated by protein kinase C.

IL1 and TNF are not prominent in developmental biology, but the same system acts in the dorsoventral patterning of *Drosophila* where the receptor called Toll is the homolog of the vertebrate IL1 receptor.

Axonal guidance systems

Diffusible factors primarily responsible for axonal guidance include the netrins and the semaphorins. The receptors for semaphorins are called neuropilins, and the receptors for netrins called DCC ("deleted in colon carcinoma").

Slit proteins are ligands for roundabout (robo)-like receptors. These mostly display axonal repulsion activity. They need heparan sulfate for activity and the intracellular effects work through the small GTPases (see below).

Cytoskeleton

The cytoskeleton is important in development for four distinct reasons. Firstly, eggs often have a significant internal regionalization, for example in the localization of particular mRNAs. Secondly, the orientation of cell division may be important. Thirdly, animal cells move around a lot, either as individuals or as part of moving cell sheets. Finally, the shape of cells is an essential part of their ability to carry out their functions. All of these activities are functions of the cytoskeleton.

The three main components of the cytoskeleton are:

1 **microfilaments**, made of actin;
2 **microtubules**, made of tubulin;
3 **intermediate filaments**, made of:
   (a) cytokeratins in epithelial cells
   (b) vimentin in mesenchymal cells
   (c) neurofilament proteins in neurons
   (d) glial fibrilliary acidic protein (GFAP) in glial cells.

Microtubules and microfilaments are universal constituents of eukaryotic cells, while intermediate filaments are found only in animals.

Microtubules

Microtubules (Fig. A.5) are hollow tubes of 25 nm diameter composed of tubulin. Tubulin is a generic name for a family of globular proteins which exist in solution as heterodimers of α- and β-type subunits and is one of the more abundant cytoplasmic proteins. The microtubules are polarized structures with a minus end anchored to the centrosome and a free plus end at which tubulin monomers are added or removed. Microtubules are not contractile but exert their effects through length changes based on polymerization and depolymerization. They are very dynamic, either growing by addition of tubulin monomers, or retracting by loss of monomers, and individual tubules can grow and shrink over a few minutes. The monomers contain GTP bound to the β-subunit and in a growing plus end this stabilizes the tubule. But if the rate of growth slows down, hydrolysis of GTP to GDP will catch up with the addition of monomers. The conversion of bound GTP to GDP renders the plus end of the tubule unstable and it will then start to depolymerize. The drugs colchicine and colcemid bind to monomeric tubulin and prevent polymerization. Among other effects this causes the disassembly of the mitotic spindle. These drugs

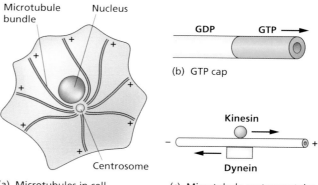

(a) Microtubules in cell

(b) GTP cap

(c) Microtubule motor proteins

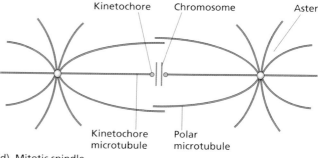

(d) Mitotic spindle

Fig. A.5 Microtubules. (a) Arrangement of microtubules in cell. (b) GTP cap at the plus end of microtubule. (c) Motor proteins associated with microtubules. (d) Arrangement of tubules in a mitotic spindle.

cause cells to become arrested in mitosis and are often used in studies of cell kinetics.

The shape and polarity of cells can be controlled by locating capping proteins in particular parts of the cell cortex which bind the free plus ends of the microtubules and stabilize them. The positioning of structures within the cell also depends largely on microtubules. There exist special **motor proteins** that can move along the tubules, powered by hydrolysis of ATP, and thereby transport other molecules to particular locations within the cell.

The kinesins move towards the plus ends of the tubules while the dyneins move towards the minus ends. In a developmental context they are important for the localization of substances in the egg, for example the dishevelled protein in *Xenopus*, or the bicoid mRNA in *Drosophila*.

Microtubules are prominent during cell division. The minus ends of the tubules originate in the centrosome, which is a microtubule-organizing center able to initiate the assembly of new tubules. In mitotic prophase the centrosome divides and each of the radiating sets of microtubules becomes known as an aster. The two asters move to the opposite sides of the nucleus to become the two poles of the mitotic spindle. The spindle contains two types of microtubules. The polar microtubules meet each other near the center and become linked by plus-directed motor proteins. These tend to drive the poles apart. Each chromosome has a special site called a kinetochore which binds another group of microtubules called kinetochore microtubules. At anaphase the kinetochores of homologous chromosomes separate. The polar microtubules continue to elongate while the kinetochore microtubules shorten by loss of tubulin from both ends and draw the chromosome sets into the opposite poles of the spindle.

Microfilaments

Microfilaments (Fig. A.6) are polymers of actin, which is the most abundant intracellular protein in most animal cells. In vertebrates there are several different gene products of which α actin is found in muscle and β/γ actins in the cytoskeleton of nonmuscle cells. For all actin types the monomeric soluble form is called G-actin. Actin filaments have an inert minus end, and a growing plus end to which new monomers are added. G-actin contains ATP and this becomes hydrolysed to ADP shortly after addition to the filament. As with tubules, a rapidly growing filament will bear an ATP cap which stabilizes the plus end. Microfilaments are often found to undergo "treadmilling" such that monomers are continuously added to the plus end and removed from the minus end while leaving the filament at the same overall length. Microfilament polymerization is prevented by a group of drugs called cytochalasins, and existing filaments are stabilized by another group called phalloidins. Like microtubules, microfilaments have associated motor proteins that will actively migrate along the fiber. The most abundant of these is myosin II, which moves toward the plus end of microfilaments, the process being driven

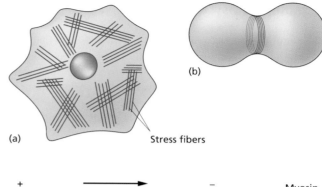
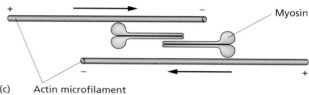

Fig. A.6 Microfilaments. (a) Arrangement of microfilaments in a fibroblast in tissue culture. (b) Contractile ring in a dividing cell. (c) Contraction of filament bundle driven by myosin.

by the hydrolysis of ATP. To bring about contraction of a filament bundle, the myosin is assembled as short bipolar filaments with motile centers at both ends. If neighboring actin filaments are arranged with opposite orientation then the motor activity of the myosin will draw the filaments past each other leading to a contraction of the filament bundle.

Microfilaments can be arranged in various different ways depending on the nature of the accessory proteins with which they are associated. Contractile assemblies contain microfilaments in antiparallel orientation associated with myosin. These are found in the contractile ring which is responsible for cell division, and in the stress fibers by which fibroblasts exert traction on their substratum. Parallel bundles are found in **filopodia** and other projections from the cell. Gels composed of short randomly orientated filaments are found in the cortical region of the cell.

Small GTPases

There are three well-known GTPases that activate cell movement in response to extracellular signals: Rho, Rac, and cdc42. They are activated by numerous tyrosine kinase-, G-coupled-, and cytokine-type receptors. Activation involves exchange of GDP for GTP and many downstream proteins can interact with the activated forms. Rho normally activates the assembly of stress fibers. Rac activates the formation of **lamellipodia** and ruffles. Cdc42 activates formation of filopodia. In addition all three promote the formation of focal adhesions, which are integrin-containing junctions to the extracellular matrix.

These proteins can also affect gene activity through the JNK and p38 MAP kinase pathways.

Cell adhesion molecules

Organisms are not just bags of cells, rather each tissue has a definite cellular composition and microarchitecture. This is determined partly by the cell-surface molecules by which cells interact with each other, and partly by the components of the **extracellular matrix** (**ECM**). Virtually all proteins on the cell surface or in the ECM are glycoproteins, containing oligosaccharide groups added in the endoplasmic reticulum or Golgi apparatus after translation and before secretion from the cell. These carbohydrate groups often have rather little effect on the biological activity of the protein but they may affect its physical properties and stability.

Cells are attached to each other by adhesion molecules (Fig. A.7). Among these are the cadherins, which stick cells together

in the presence of Ca, the cell adhesion molecules (CAMs), that do not require Ca, and the integrins that attach cells to the extracellular matrix. When cells come together they often form **gap junctions** at the region of contact. These consist of small pores joining the cytosol of the two cells. The pores, or connexons, are assembled from proteins called connexins. They can pass molecules up to about 1000 molecular weight by passive diffusion.

Cadherins

This is a family of single-pass transmembrane glycoproteins which can adhere tightly to similar molecules on other cells in the presence of calcium. Cadherins are the main factors attaching embryonic cells together, which is why embryonic tissues can often be caused to disaggregate simply by removal of calcium. The cytoplasmic tail of cadherins is anchored to actin bundles in the cytoskeleton by a complex including proteins called catenins. One of these, β-catenin, is also a component of the Wnt signaling pathway (see above), providing a potential link between cell signaling and cell association. Cadherins were first named for the tissues in which they were originally found, so E-cadherin occurs mainly in epithelia and N-cadherin occurs mainly in neural tissue.

Immunoglobulin superfamily

These are single-pass transmembrane glycoproteins with a number of disulfide-bonded loops on the extracellular region, similar to the loops found in antibody molecules. They also bind to similar molecules on other cells, but unlike the cadherins they do not need calcium to do so. The neural cell adhesion molecule (NCAM) is composed of a large family of different proteins formed by alternative splicing. It is most prevalent in the nervous system but also occurs elsewhere. It may carry a large amount of polysialic acid on the extracellular domain, and this can inhibit cell attachment because of the repulsion between the concentrations of negative charge on the two cells. Related molecules include L1 and ICAM (intercellular cell adhesion molecule).

Integrins

The integrins are cell-surface glycoproteins that interact mainly with components of the extracellular matrix. They are heterodimers of α- and β-subunits, and require either magnesium or calcium for binding. There are numerous different α and β chain types and so there is a very large number of potential heterodimers. Integrins are attached by their cytoplasmic domains to microfilament bundles, so, like cadherins, they provide a link between the outside world and the cytoskeleton.

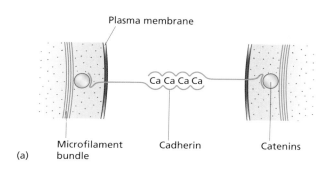

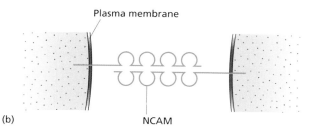

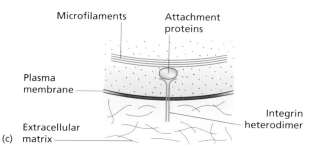

Fig. A.7 Cell adhesion molecules (CAMs). (a) Calcium-dependent adhesion via cadherins. (b) Calcium-independent adhesion via CAMs. (c) Adhesion to extracellular matrix via integrins.

They are also thought on occasion to be responsible for the activation of signal transduction pathways and new gene transcription following exposure to particular extracellular matrix components.

Extracellular matrix components

Glycosaminoglycans

Glycosaminoglycans (GAGs) are unbranched polysaccharides composed of repeating disaccharides of an amino sugar and a uronic acid, usually substituted with some sulfate groups. GAGs are constituents of proteoglycans, which have a protein core to which the GAG chains are added in the Golgi apparatus before secretion. One molecule of a proteoglycan may carry more than one type of GAG chain. GAGs have a high negative charge and a small amount can immobilize a large amount of water into a gel. Important GAGs, each of which have different component disaccharides, are heparan sulfate, chondroitin sulfate, and keratan sulfate. Heparan sulfate, closely related to the anticoagulant heparin, is particularly important for cell signaling as it is required to present various growth factors, such as the FGFs, to their receptors. Hyaluronic acid differs from other GAGs because it occurs free, and not as a constituent of a proteoglycan. It consists of repeating disaccharides of glucuronic acid and N-acetyl glucosamine, and is not sulfated. It is synthesized by enzymes at the cell surface and is abundant in early embryos.

Collagens

Collagens are the most abundant proteins by weight in most animals. The polypeptides, called α chains, are rich in proline and glycine. Before secretion, three α chains become twisted around each other to form a stiff triple helical structure. In the extracellular matrix, the triple helices become aggregated together to form the collagen fibrils visible in the electron micro- scope. There are many types of collagen, which may be composed of similar or of different α chains in the triple helix. Type I collagen is the most abundant and is a major constituent of most extracellular material. Type II collagen is found in cartilage and in the notochord of vertebrate embryos. Type IV collagen is a major constituent of the basal lamina underlying epithelial tissues. Collagen helices may become covalently crosslinked through their lysine residues, and this contributes to the changing mechanical properties of tissues with age.

Elastin

This is another extracellular protein with extensive intermolecular cross-linking. It confers the elasticity on tissues in which it is abundant, and also has some cell signaling functions.

Fibronectin

Fibronectin is composed of a large disulfide-bonded dimer. The polypeptides contain regions responsible for binding to collagen, to heparan sulfate, and to integrins on the cell surface. These latter, cell-binding, domains are characterized by the presence of the amino acid sequence Arg-Gly-Asp (= RGD). There are many different forms of fibronectin produced by alternative splicing.

Laminin

Laminin is a large extracellular glycoprotein, found particularly in basal laminae. It is composed of three disulfide-bonded polypeptides joined in a cross shape. It carries domains for binding to type IV collagen, heparan sulfate, and another matrix glycoprotein, entactin.

Further reading

General
Alberts, B., Johnson, A., Lewis, J., Raff, M., Roberts, K. & Walter, P. (2008) *Molecular Biology of the Cell*, 5th edn. New York: Garland Publishing.
Brown, T.A. (2010) *Gene Cloning and DNA Analysis: an introduction*, 6th edn. Oxford: Wiley-Blackwell.
Lodish, H. et al. (2008) *Molecular Cell Biology*, 6th edn. New York: W.H. Freeman.
Primrose, S.B. and Twyman, R.M. (2006) *Principles of Gene Manipulation*, 7th edn. Oxford: Wiley-Blackwell.

Gene regulation
Cross, S.H. & Bird, A.P. (1995) CpG islands and genes. *Current Opinion in Genetics and Development* **5**, 309–314.
Gould, A (1997) Functions of mammalian Polycomb group and trithorax group related genes. *Current Opinion in Genetics and Development* **7**, 488–494.
Wolffe, A. (1998) *Chromatin: structure and function*, 3rd edn. San Diego: Academic Press.
Latchman, D.S. (2003) *Eukaryotic Transcription Factors*. New York: Academic Press.

Signaling
Hancock, J.T. (1997) *Cell Signaling*. Harrow: Longman.
Hunter, T. (2000) Signaling – 2000 and beyond. *Cell* **100**, 113–127.
Downward, J. (2001) The ins and outs of signaling. *Nature* **411**, 759–762.
Heath, J.K. (2001) *Principles of Cell Proliferation*. Oxford: Blackwell Science.

Specific signaling pathways
Cadigan, K.M. & Nusse, R. (1997) Wnt signaling: a common theme in animal development. *Genes and Development* **11**, 3286–3305.
Holder, N. & Klein, R. (1999) Eph receptors and ephrins: effectors of morphogenesis. *Development* **126**, 2033–2044.

Szebenyi, G. & Fallon, J.F. (1999) Fibroblast growth factors as multi-functional signaling factors. *International Review of Cytology* **185**, 45–106.

Massagué, J. & Chen, Y.G. (2000) Controlling TGFβ signaling. *Genes and Development* **14**, 627–644.

Munn, J.S. & Kopan, R. (2000) Notch signaling: from the outside in. *Developmental Biology* **228**, 151–165.

Ingham, P.W. & McMahon, A.P. (2001) Hedgehog signaling in animal development: paradigms and principles. *Genes and Development* **15**, 3059–3087.

Von Bubnoff, A. & Cho, K.W.Y. (2001) Intracellular BMP signaling regulation in vertebrates: pathway or network? *Developmental Biology* **239**, 1–14.

Tata, J.R. (2002) Signaling through nuclear receptors. *Nature Reviews Molecular Cell Biology* **3**, 702–710.

Halder, G. & Johnson, R.L. (2011) Hippo signaling: growth control and beyond. *Development* **138**, 9–22.

Cytoskeleton, adhesion molecules, and extracellular matrix

Adams, J.C. & Watt, F.M. (1993) Regulation of development and differentiation by the extracellular-matrix. *Development* **117**, 1183–1198.

Kreis, T. & Vale, R. (1999) *Guidebook to the Cytoskeletal and Motor Proteins*, 2nd edn. Oxford: Oxford University Press.

Kreis, T. & Vale, R. (1999) *Guidebook to the Extracellular Matrix and Adhesion Proteins*, 2nd edn. Oxford: Oxford University Press.

Beckerle, M.C. (2002) *Cell Adhesion*. Oxford: Oxford University Press.

Thiery, J.P. (2003) Cell adhesion in development: a complex signaling network. *Current Opinion in Genetics and Development* **13**, 365–371.

Wheelock, M.J. & Johnson, K.R. (2003) Cadherins as modulators of cellular phenotype. *Annual Reviews of Cell and Developmental Biology* **19**, 207–235.

Glossary

To keep the glossary to a reasonable length it does not include the names of most animal taxa or individual genes or gene products. However the accounts of these in the text can easily be located from the index

acini (singular acinus): clusters of cells in a glandular epithelium.

acrosome: large secretory vesicle found in sperm.

adaptive evolution: evolutionary changes produced by natural selection.

adipose tissue: tissue storing and synthesizing fats.

adult stem cell: = tissue specific stem cell.

aggregation chimera: mouse made by aggregating two embryos of different genotypes.

AGMregion: aorta-gonad-mesonephros region of vertebrate embryo mesoderm.

agonist: a drug that will stimulate a particular cellular receptor or response, often mimicking the action of an endogenous substance.

allantois: extraembryonic structure of amniotes arising from posterior mesoderm.

allele: genes with alternative DNA sequences that occupy the same locus in the genome.

allelic series: a set of alleles showing a progressively stronger phenotype.

allograft: graft from one mature individual to another: likely to provoke an immune reaction unless the individuals are of the same inbred strain.

allometry: differential growth between two parts of an organism, or between homologous parts of different organisms.

amniocentesis: sampling of cells from within the amniotic cavity of a human embryo. This enables the early detection of chromosome abnormalities.

amnion: extraembryonic membrane characteristic of amniotes, consisting of ectoderm on the side facing the embryo and mesoderm on the side facing the chorion.

amniote: animal whose embryo has an amnion, i.e. a mammal, bird, or reptile.

analogy: structures that resemble each other for reasons other than homology.

anchor cell: cell acting as a signaling center for the *C. elegans* vulva.

androgenetic: an embryo whose genetic material derives only from the sperm.

angiogenesis: formation of new blood vessels by growth from pre-existing vessels.

animal cap: explanted region from the animal pole of an amphibian embryo, used to assay inducing factors.

animal hemisphere: the upper hemisphere of an egg or oocyte; after fertilization usually bears the polar bodies.

animal model: normally a transgenic or knockout mouse that is designed to have a defect similar to a human disease.

animal pole: the pole of an egg that normally lies uppermost, contains relatively yolk-free cytoplasm, and usually bears the polar bodies.

animal–vegetal axis: line running from animal to vegetal pole of an egg or oocyte.

antagonist: a substance that inhibits or blocks the action of another substance.

anterior intestinal portal: opening of the foregut into the subgerminal (midgut) cavity of an amniote embryo.

anterior midgut: the part of an insect embryo gut invaginationg from the anterior end.

anterior: the head end of an animal.

anteroposterior: the direction towards, or line joining, the anterior and posterior extremities of an animal.

antibody: immunoglobulin protein produced by B lymphocytes of the immune system, which can bind specifically and with high affinity to a specific substance of interest. Antibodies are made by immunizing an animal with the substance of interest, which is called the antigen.

antigen: any substance used to raise an anti-body by immunization of an animal.

antimorphic: = dominant negative.

antisense: sequence of DNA or RNA complementary to a target sequence and thus capable of hybridizing with it.

anurans: tailless amphibians: the frogs and toads.

apical ectodermal ridge (AER): epidermal ridge on limb bud, needed for distal outgrowth.

apical: of epithelia, the side away from the basement membrane, usually facing a lumen.

apoptosis: = programmed cell death.

appositional induction: induction between two cell sheets in contact.

Essential Developmental Biology, Third Edition. Jonathan M.W. Slack.
© 2013 John Wiley & Sons, Ltd. Published 2013 by John Wiley & Sons, Ltd.

archenteron: cavity formed as a result of gastrulation movements; becomes or contributes to the later gut lumen.

area opaca: outer region of the avian blastoderm which expands to form the yolk sac.

area pellucida: central region of an avian blastoderm from which the embryo itself develops.

asexual reproduction: reproduction without the fusion of two gametes to form a zygote. This can takes several forms including budding, fragmentation, or parthenogenesis from eggs.

aster: the microtubule array which serves to separate the chromosomes during cell division.

astrocyte: a type of glial cell in the central nervous system.

asymmetrical division: a situation where the two daughter cells resulting from a cell division are different from each other.

atavism: a morphological character, usually caused by a mutation, that resembles an ancestral organism.

autologous graft: graft from one part of the mature individual to another; avoids any immune rejection.

autonomous: of a process not requiring any input from surrounding cells.

autopod (= autopodium): the hand or foot of a vertebrate limb.

autoradiography: method for detecting radio-activity on a gel or a histological section by exposure to a photographic emulsion.

autosomes: chromosomes that are not sex chromosomes.

AVE region (= anterior visceral endoderm): an extraembryonic signaling center in the mouse conceptus.

axis: (1) a line or direction, as in anteroposterior axis; (2) the midline structures of a vertebrate embryo: notochord, somites, and neural tube.

B **β cells:** insulin-secreting endocrine cells of the islets of Langerhans, in the pancreas.

β-galactosidase: product of the *lacZ* gene. β-galactosidase enzymatic activity is easily detected histochemically and therefore the *lacZ* gene is often used as a reporter of gene expression.

β-geo: a fusion protein of β-galactosidase and neomycin resistance protein.

balancer chromosome: a chromosome that will not recombine with its homolog, and usually carries a recessive lethal mutation and a dominant marker. Especially used for simplifying mutagenesis screens and for maintaining stocks of recessive lethal mutants in heterozygous form.

band shift assay: gel electrophoretic technique for demonstrating the binding of a protein to a nucleic acid.

basal: (1) of epithelia, the side next to the basement membrane; (2) of taxa, those which are most similar to the common ancestor.

basement membrane: extracellular matrix layer underlying an epithelium.

bipolar form: a mirror symmetrical duplication of the whole body, especially encountered in the regeneration of planaria or hydroids.

birth defects (= congential defects): defects of a mammalian embryo apparent at the time of birth. These may arise from genetic causes, or chromosome abnormalities in the embryo, or teratological stimuli.

bistable switch: molecular mechanism which has two stable steady states that can be interconverted by some external signal.

bivalent: (1) the meiotic chromosomes containing four chromatids, formed by pairing of paternal and maternal chromosomes; (2) of a molecule, having two binding sites.

blastema: an undifferentiated bud of cells, dividing except in planaria, that gives rise to a structure.

blastocoel: fluid-filled cavity in the center of a blastula.

blastocyst: mammalian embryo after cavitation; differs from blastula because it contains differentiating extraembryonic structures.

blastoderm: similar to a blastula but arranged as a sheet rather than a ball of cells.

blastomeres: the large cells arising from the fertilized egg by cleavage divisions.

blastopore: depression or slit through which cells move to the interior during gastrulation.

blastula: early developmental stage; a ball of similar cells arising from repeated cleavage of a fertilized egg.

blot: a blot is made by transferring the products of a gel electrophoretic separation onto a membrane that is suitable for biochemical analysis by nucleic acid hybridization or antibody binding (see Northern, Southern, Western blot).

body plan (= Bauplan): the essential features of the whole body of a species or higher taxonomic group.

border cells: specialized ovarian follicle cells that control the polarity of the Drosophila oocyte.

brachial plexus: arrangement of nerves formed by the fusion and rebranching of the spinal nerves supplying the vertebrate forelimb.

branchial arches: (= pharyngeal arches) segmental structures between the pharyngeal pouches each comprising a cartilaginous arch, blood vessel, and cranial nerve.

branching morphogenesis: formation of a branched structure by cell movement and/or growth of an epithelium.

bromodeoxyuridine (BrdU): DNA nucleotide analog used for labeling cells in S phase.

C **cancer:** uncontrolled growth of the body's own cells; often displaying local invasive behavior and metastasis.

capacitation: of mammalian sperm, a process of conditioning in the female reproductive tract which makes them capable of fertilization.

carcinoma: a cancer derived from an epithelium.

cardia bifida: congenital abnormality: having two hearts.

carpals: bones of the wrist or ankle.

cartilage model: see model.

caudal: relating to the tail, often equivalent to posterior.

cavitation: formation of an internal cavity in a cell mass.

CCD (= charge coupled device): the optical sensor in a digital camera.

cell biology: approach to biology that focuses on the structure, function, and behavior of cells.

cell lineage: the "family tree" of a group of cells, showing which cell is the descendant of which other cell.

cellular blastoderm: stage of insect development at which the syncytial blastoderm becomes divided into cells.

centromere: the region of a chromosome that attaches to the spindle at cell division, attaches sister chromatids following DNA replication, and attaches homologous chromosome during meiosis.

cerebellum: part of the vertebrate brain controlling movement, formed from rhombomere 1.

chalone: a circulating, tissue-specific, inhibitor of growth.

checkpoint: time during the cell division cycle at which control is exerted on whether the cell shall continue in the cycle or not.

chemokine: class of extracellular signaling molecule particularly important for activation of white blood cells.

chim(a)era: (1) = genetic mosaic, especially a mammalian embryo in which cells of one genotype have been injected into an embryo of a different genotype; (2) also used of molecules assembled from more than one source by molecular cloning: see fusion protein and domain swap.

chondroblast: mitotically active cartilage cell, not yet a mature chondrocyte.

chondrocyte: mature cartilage cell.

chordates (= Chordata): phylum comprising the vertebrates together with the protochordates.

chorioallantoic membrane (CAM): extraembryonic membrane of avian embryos formed by fusion of the allantois with the chorion.

chorion: (1) extraembryonic membrane of amniotes, similar to the amnion but forming an outer layer; (2) an extracellular layer surrounding an insect or fish egg.

chromatin: DNA combined with chromosomal proteins in a manner which helps regulate its expression and behavior.

chromobox genes: genes encoding homologs of the *Polycomb* gene that regulate Hox gene expression via chromatin structure.

clade: a taxon which consists of all the descendants of a common ancestor.

cleavage: a type of cell division that occurs without growth, such that daughters are smaller than the mother; typically found in the early stages of development.

cloaca: common urogenital and anal aperture.

clonal analysis: information obtained about developmental mechanisms by studying the position and differentiation of the progeny of a single labeled cell.

clone: (1) population of cells derived from a single progenitor cell, either *in vivo* or *in vitro*; (2) DNA molecule prepared by molecular cloning.

cloning: (1) assembling a DNA sequence with the use of restriction enzymes and ligases followed by insertion into a cloning vector and growth to a useful quantity (= molecular cloning); (2) growth of a colony of cells from a single cell (= cell cloning); (3) formation of a whole organism from one cell (= whole animal cloning, or reproductive cloning of humans).

clonogenic: of cells, the ability to divide and form a clone in an appropriate assay system.

cnidarians (= Cnidaria): animal phylum including jellyfish and sea anemones.

coelom: body cavity lined with mesoderm.

combinatorial chemistry: methods for making large libraries of related compounds simultaneously instead of one at a time.

combinatorial: situation where different states are defined not by single elements but by a combination of elements. E.g. different states of cell commitment generally depend on the combination of transcription factors that is present, rather than a single factor.

commisuralneurons: neuronswhichconnect one side of the spinal cord to the other.

commitment: of a cell or tissue region indicates that it is programmed to follow a particular developmental pathway or fate.

compaction: process whereby cells of a mammalian morula become more adhesive and pack together more tightly.

compartment: a region of an embryo within which a clone of cells may move around but whose boundaries it does not cross.

competence: ability to respond to an inducing factor.

complementation test: genetic test to find if two recessive mutants lie in the same gene by introducing one mutation on the maternal and the other on the paternal chromosome. If they are in different genes then the wild-type alleles should complement and show no phenotype. But if they are in the same gene they will fail to complement and a phenotype will result.

compound microscope: type of visible light microscope used for looking at sections or other very thin objects.

conceptus: a mammalian embryo together with all the extraembryonic membranes formed from the zygote.

condensation: (1) a dense patch of cells within a mesenchyme; (2) formation of such a patch.

conditional knockout: a knockout mouse in which the ablation of the gene occurs under a particular set of circumstances controlled by the experimenter.

confocal (scanning) microscope: a microscope that scans the specimen with a laser to build up a digital image of a particular optical section.

connective tissue: strictly refers to tissues containing a substantial collagen-rich

extracellular matrix secreted by fibroblasts. Sometimes used more widely to refer to all tissues that are not epithelia.

constitutive: describes a gene or gene product that is active continuously.

convergentextension: morphogenetic movement in which a cell sheet elongates and narrows because of active movements of the constituent cells to alter the overall packing arrangement.

coronal section: = frontal section.

corpus luteum: structure formed in the ovary after ovulation of mammalian oocyte.

cortex: outer region of a cell, especially an egg or oocyte, comprising a few microns thickness beneath the plasma membrane.

cortical granules: secretory vesicles containing vitelline membrane components, released from the egg at fertilization.

cortical rotation: rotation of the zygote cortex relative to the internal cytoplasm.

cranial nerves: nerves coming from the brain.

cranial: relating to the head, often equivalent to anterior.

Cre-lox: system used to bring about DNA recombination or excision events at particular sites under controlled circumstances.

crossing over: recombination of paternal and maternal chromatids at meiosis; can also be artificially induced at four-strand stage of normal mitotic cycle.

cryostat (= cryotome): microtome operated below freezing point to cut frozen sections of a specimen.

cumulus: ovarian follicle cells released with a mammalian oocyte on maturation.

cystoblast: precursor cell to the oocyte+15 nurse cells in *Drosophila*.

cytoplasmic determinant: see determinant.

cytoskeleton: system of filaments and tubules that makes up the structural support for the cytoplasm of a cell.

D **dauer larva:** resistant dormant state of *C. elegans*.

deciduum: the swelling of a uterine crypt produced by its reaction to an implanted embryo.

deep cells: cells in the interior of an early embryo, especially in fish embryos.

definitive endoderm: the embryonic endoderm of an amniote embryo, as opposed to extraembryonic endoderm.

dehydration: removal of water.

denticles: small cellular appendages on the ventral side of a *Drosophila* larva.

derived: of taxa, those which are most different from the common ancestor.

dermal papilla: mound of dermal cells, especially in hair follicle.

dermomyotome: part of the somite forming the dermatome and myotome.

dermatome: part of the somite forming the dorsal dermis.

dermis: the connective tissue layer of the skin.

determinant (= cytoplasmic determinant): substance localized to part of an egg or blastomere that causes the cells that inherit it to acquire a particular developmental commitment.

determination: type of developmental commitment of cells or tissue explant that is irreversible following grafting to any different location in the embryo.

Deuterostomia: group of animal phyla characterized by radial cleavage, regulative development, and formation of the anus from the blastopore.

developmental constraint: aspect of the developmental mechanism that prevents the viability of some types of mutant.

diaphysis: the shaft of a long bone.

diencephalon: the posterior part of the vertebrate forebrain.

differentiation: the acquisition during development of the properties of mature functional cells. The term usually refers to the last step of development rather than to earlier events of commitment.

diI, diO: lipid-soluble fluorescent vital dyes used for fate mapping.

diploblasts: animal taxa with only two germ layers: the cnidarians and ctenophores.

diploid: having two sets of chromosomes, one from the father and one from the mother.

dissecting microscope: type of microscope for looking at, and manipulating, relatively large solid objects.

distal tip cell: cell acting as a signaling center for the *C. elegans* germ line.

distal: further from the body.

DNA sequencing: determination of the sequence of bases (A,T,C,G) in DNA.

domain swap: a gene encoding a protein in which two functional domains of other proteins are combined, for example a DNA binding and hormone binding region from different proteins.

dominant negative: type of mutation that antagonizes the effect of the wild-type allele.

dominant: type of mutation that produces a phenotype in the presence of the wild-type allele.

dorsal closure: morphogenetic movement of insect embryos leading to enclosure of the yolk mass by the ventrally located germ band.

dorsal lip: dorsal part of the ring-shaped blastopore of an amphibian embryo. The tissue just above the dorsal lip is Spemann's organizer.

dorsal root ganglia (DRG): the sensory ganglia on the dorsal branch of a spinal nerve.

dorsal: the upper surface (or back) of an animal.

dorsoventral: the direction towards, or line joining, the dorsal and ventral extremities of an animal.

dsRNA: double-stranded RNA, used in RNA interference.

ductus arteriosus: junction of aorta and pulmonary artery of mammalian fetus.

dysplasia: altered pattern of growth and differentiation.

 Ecdysozoa: a superphylum defined on the basis of primary sequence data; includes the arthropods and nematodes among other groups.

echinus rudiment: the bud in the sea urchin larva that gives rise to the adult during metamorphosis.

ectoderm: the outer of the three embryonic germ layers.

ectopic: not in its usual position; as in ectopic structures or ectopic gene expression.

egg chamber: structure in the female *Drosophila* consisting of the oocyte and its nurse cells surrounded by ovarian follicle cells.

egg cylinder: stage of rodent embryo in which the embryo consists of two layers arranged in a cup shape.

egg: strictly the female gamete after completion of the second meiotic division (= ovum). Also a fertilized egg (= zygote). Also often loosely used for secondary oocytes and early embryos.

electron microscope: microscope that forms an image using a beam of electrons and therefore has much higher resolving power, and possible magnification, than a light microscope.

electroporation: introduction of substances, usually DNA, into cells by means of electric field pulses.

embryoid bodies: structures resembling embryos which are formed by ES cells or teratocarcinoma cells when they are removed from a growth-promoting medium.

embryonic induction: the process whereby the development of one group of cells, called the competent region, is altered by an inducing factor from another group, called a signaling center or organizer.

embryonic shield: thickened dorsal part of the germ ring of a fish embryo.

embryonic stem cells (= ES cells): cells grown from early mammalian embryos that are developmentally pluripotent.

emission: the light of a particular characteristic wavelength that is emitted from a fluorochrome.

endocardial cushions: structures contributing to the ventricular septum and the atrioventricular valves of the heart.

endocardium: the inner, endothelial, layer of the heart.

endocrine: of a ductless gland, secreting substances into the blood.

endoderm: the inner of the three germ layers.

endothelial cells: the cells lining blood vessels and forming capillaries.

enhancer trap: a transgenic line containing a reporter gene whose activity is regulated by an endogenous enhancer.

enhancer: regulatory region of DNA that controls the expression of a gene; often independently of its precise position in relation to the gene.

enterocoely: formation of the coelom by budding of the mesodermal rudiment from the gut.

enterocytes: absorptive cells of small intestinal epithelium.

enteroendocrine cells: endocrine cells of the gut epithelium.

enveloping layer (EVL): outer layer of an early fish embryo, composed of flattened cells.

ependyma: the ventricular lining cells of the vertebrate central nervous system.

epiblast: upper layer of a mammalian or avian blastoderm, or fish germ ring.

epiboly: active spreading and increase in area of a cell sheet.

epidermis: the epithelial part of the skin.

epigenetic: may be used for anything to do with development, but nowadays more usually refers to mechanisms of gene control based on DNA methylation or chromatin structure.

epiphysis (plural epiphyses): (1) the expanded terminal regions of long bones; (2) = pineal gland.

epistasis: in general an effect exerted on the phenotype of one gene by the activity of another. In particular used to describe the suppression of one mutant phenotype by a mutation in a different gene.

epithelium (plural epithelia): a tissue type in which cells are arranged as a single or multilayered sheet lying on a basement membrane.

epitope: part of a molecule recognized by an antibody.

equivalence group: set of cells with the same competence.

erythrocyte: red blood cell.

ES cells: see embryonic stem cells.

euchromatin: decondensed, active chromatin.

excitation: the input of energy required to evoke fluorescence from a fluorochrome.

exocrine: of a gland secreting substances into a duct.

exogastrula: abnormal gastrula in which the endoderm + mesoderm separates from the ectoderm instead of invaginating within it.

experimental embryology: study of development by microsurgical methods.

extended germ band: the phylotypic stage of insect development.

extracellular matrix (ECM): material filling the space between cells, and also such structures as basement membranes, vitelline membranes, etc..

extraembryonic: relating to structures derived from the zygote which are not part of the embryo itself but serve to support or nourish it.

FACS: fluorescence-activated cell sorting. Enables separation of cell populations based on the attachment of different fluorescent labels.

fasicle: bundle of nerve axons, often surrounded by connective tissue.

fate map: diagram of an embryo or organ rudiment showing where each part will move and what it will become in the course of normal development.

fate: indicates what will happen to a region of the embryo. The fate can often be altered by experimental manipulation.

feeder layer: tissue culture cells treated to prevent division, which provide an environment enabling the growth of other cells that cannot be grown in standard media.

femur: bone of the vertebrate upper hindlimb (also a segment of insect leg).

fertilized egg: the egg from the time of sperm entry to first cleavage.

fibroblast: stellate cell secreting collagen, the main cell type of connective tissue.

fibula: posterior bone of vertebrate lower hind limb.

filopodium (plural filopodia): long thin finger-like extensions from cells.

fixation: treatment of a specimen to preserve it and enable it to withstand further treatments such as staining or sectioning.

fixative: chemical substance used for fixation, e.g. formaldehyde or glutaraldehyde.

floor plate: midventral part of the vertebrate embryo spinal cord.

floxed: a DNA sequence including loxP sites, which is a substrate for Cre-mediated recombination.

FLP system: method for inducing recombination at desired genomic sites, particularly used in Drosophila.

fluorescence: the emission of light of one wavelength following absorption of light of another, shorter, wavelength.

fluorochrome: fluorescent substance that can be attached to proteins, antibodies, etc..

follicle cells: somatic epithelial cells of the ovary.

foramen ovale: gap in the interatrial septum of the fetal mammalian heart.

forebrain: region of the vertebrate embryonic brain that will become the telencephalon and diencephalon.

foregut: region of the gut of a vertebrate embryo that lies anterior to the anterior intestinal portal. Later the region running from the pharynx to the common bile duct.

forward genetics: analysis of a biological phenomenon starting from a mutant phenotype.

foster mother: in mice, a female recipient to which a preimplantation embryo is transferred, to enable its further development.

frontal: relates to plane or section separating the dorsal and ventral parts of the body.

functional genomics: understanding gene function, especially in the context of total genome sequences where the function of most of the genes is not known.

functional proteomics: understanding protein function, especially in the context of comparison of complex populations of proteins.

fusion protein: protein composed of two or more functional domains, which is produced by molecular cloning and expression.

gain of function: of a mutation, confers additional or altered activity on the gene product.

gametes: haploid reproductive cells; sperm or eggs.

gametogenesis: the development of gametes from germ cells.

ganciclovir: drug used to select for desirable recombination events in tissue culture.

ganglion (plural ganglia): a structure in the peripheral nervous system containing neurons.

gap genes: genes whose expression defines a body region in the Drosophila embryo.

gap junctions: cell contacts that allow the passage of low molecular weight substances between the cells.

gastrula: stage of development in which gastrulation takes place.

gastrulation: phase of morphogenetic movements in early development that brings about the formation of three germ layers.

gene duplication: the appearance of a copy of an existing gene in the genome.

gene therapy: therapy involving the introduction of a gene into a patient.

gene trap: type of insertional mutation in which a reporter gene is introduced into the locus of an endogenous gene. The function of the endogenous gene is usually destroyed, but its regulatory elements drive the reporter in the normal expression pattern.

genetic marker: see marker gene.

genetic mosaic: an organism composed of two or more types of cell that are genetically different.

genital ridge: region of mesoderm of a vertebrate embryo that forms the gonads.

genome: the whole of the nuclear DNA of an organism.

genotype: relating to a single organism or to a genetic line, the genotype means the particular alleles present at specific genetic loci.

germ cells: cells belonging to the germ line.

germ layers: the ectoderm, mesoderm, and endoderm.

germ line: the cell lineage that will form the gametes, comprising the primordial germ cells, spermatogonia, and oogonia, together with other clonal progeny such as nurse cells. The germ line is often formed by the action of cytoplasmic determinants.

germ plasm: visible cytoplasmic determinant causing formation of the primordial germ cells.

germ ring: the thickened margin of the blastoderm of a fish embryo.

germinal vesicle: the nucleus of a primary oocyte.

GFP: see green fluorescent protein.

gizzard: muscular region of the avian stomach.

glial cells: the non-neuronal cells of the central nervous system and peripheral ganglia.

gnathal segments: the segments of an insect embryo lying in the posterior part of the head and bearing the mouthparts.

goblet cells: secretory cells containing a large mucinous vesicle: found in intestinal epithelium.

gonad: the somatic structure containing the germ cells.

gradient: a continuous change in some property with position. Often used to refer to concentration gradient of a morphogen.

graft: a piece of tissue transplanted from one place to another, or the act of carrying out a graft.

granulocytes: cells of the blood with characteristic granules, comprising eosinophils, neutrophils, and basophils.

gray (= grey) crescent: region of intermediate pigmentation on the dorsal side of an amphibian egg following the cortical rotation.

green fluorescent protein (GFP): a fluorescent protein often used as a reporter.

growth cone: structure at the leading tip of a growing nerve axon.

growth factors: generic name for biologically active extracellular proteins active at short range, includes many embryonic inducing factors.

growth plate: the zone of mitotic cartilage between the epiphysis and diaphysis of a developing vertebrate long bone.

gynogenetic diploid: an organism that is diploid by virtue of a duplication of the genome of the egg, and contains no paternal genome.

³H-thymidine (tritiated thymidine): radioactively labeled thymidine, used for labeling cells in S phase.

halteres: small balancing organs found on the metathorax of flies in place of the second pair of wings found in other insects.

haploid: having a single set of chromosomes, as after meiosis.

haploinsufficient: type of genetic dominance arising because the loss of one gene copy reduces the level of product enough to produce a phenotype.

Hayflick limit: the number of passages in tissue culture that a primary cell line can successfully be grown.

hemangioblast: pluripotent cell that forms both endothelial cells and hematopoietic stem cells.

hematopoietic (= hemopoietic): blood-forming.

Hemimetabola: the group of insects that undergo a gradual metamorphosis with nymphal stages separated by molts.

hemolymph: the "blood" of an arthropod.

Hensen's node: a condensation of cells at the anterior end of the primitive streak of an avian embryo.

hermaphrodite: individual organism producing both male and female gametes.

heterochromatic: condensed, inactive chromatin.

heterochrony: an evolutionary shift of the relative timing of events in development.

heterotopic: type of graft that is performed to a different position in the host to that from which it came in the donor.

hindbrain: region of the brain of a vertebrate embryo that will become the cerebellum and medulla oblongata.

hindgut: region of the gut of a vertebrate embryo that lies posterior to the posterior intestinal portal.

histological sections: thin slices of biological specimens that can be stained with a variety of techniques and enable visualization of cell arrangement and structure.

holoblastic: type of cleavage where the whole zygote becomes subdivided into blastomeres.

Holometabola: group of insects undergoing a complete metamorphosis.

homeobox: DNA sequence encoding a homeodomain, which is a 60-amino-acid DNA-binding domain defining an important class of transcription factors.

homeodomain: protein sequence encoded by a homeobox.

homeotic gene (= selector gene): a gene whose expression distinguishes two body parts. If mutated then one body part will be converted into the other.

homolog: (1) structures or genes which resemble each other because they are homologous, i.e. descended from a common ancestor; (2) the maternally and paternally derived chromosomes that pair with each other at meiosis.

homologous recombination: recombination of a transgene into a locus in the genome which has exactly the same sequence.

homology: related by descent from a common ancestor.

homophilic: binding of a molecule to others of the same type.

hopeful monster: individual organism arising as the result of a macromutation.

horseradish peroxidase (HRP): an enzyme used for cell labeling and neuronal tracing.

Hox genes: subset of homeobox genes that controls anteroposterior specification in animals.

humerus: bone of the upper vertebrate forelimb.

hybridization: (1) of nucleic acids, forming a double helix from two complementary sequences; (2) of animals, crossing one species with another to produce hybrid offspring, which are normally inviable or sterile.

hybridoma: cell line producing a particular monoclonal antibody.

hydroids: class of animal belonging to the phylum Cnidaria, usually sessile and colonial with a number of polyps connected by a stolon.

hyperdorsalized: having an excessive representation of dorsal-type structures.

hypermorph: mutant allele showing gain of function.

hyperplasia: excessive growth.

hypertrophy: excessive growth.

hypoblast: the lower layer of an avian or other amniote blastoderm, or of the germ ring of a fish embryo.

hypodermis: syncytial outer layer of C. elegans.

hypomorph: mutant allele showing partial loss of function.

imaginal discs: structures in the larva of Drosophila and other holometabolous insects that form the adult cuticle at the time of metamorphosis.

immunoprecipitation: isolation of a particular substance by incubating the sample with a specific antibody and then recovering the antibody–substance complex as a solid phase.

immunostaining: procedure to make visible the location of an antigen in a section or **wholemount** by using an **antibody** and a detection system.

imprinting: the expression of a gene from only one of the parental genomes.

in situ **hybridization:** detection of a specific RNA, or occasionally DNA, in a biological specimen by hybridizing to a specific **probe** and then visualizing the **probe** with a suitable detection method.

in vitro **fertilization (IVF):** creation of a fertilized egg outside the mother, typically by mixing eggs and sperm in a dish.

inducible: relating to a gene or gene product whose activity can be controlled by some experimental method.

inducing factor: a signal substance responsible for an embryonic **induction**.

induction: (1) = embryonic **induction**; (2) can refer to the regulated activation of transcription of a gene.

inner cell mass: the cells within a mammalian **blastocyst** that form the entire embryo and most of the **extraembryonic membranes**.

insertional mutagenesis: creation of mutations by means of a DNA element, such as a **transposon** or **retrovirus**, that will integrate at random into the genome and disrupt endogenous genes in the process.

instar: period in the life cycle of an organism in between two molts.

instructive induction: a process of induction that increases the complexity of the embryo because the responding cells can develop along two or more possible pathways depending on the concentration of the signal.

intercalary regeneration (= intercalation): regeneration at a junction between two body parts of the structures that would normally lie in between them.

intermediate filaments: components of the cytoskeleton; intracellular filaments of a size between **microfilaments** and **microtubules** and composed of various types of protein.

intermediate mesoderm: region of **mesoderm** in a vertebrate embryo lying between the **somites** and the **somatopleure**.

invagination: infolding of a cell sheet to form an internal protrusion or pocket.

involution: internalization of a cell sheet by movement led by a free edge.

isomerism: (1) loss of the normal left–right asymmetry of an animal; (2) in chemistry isomerism refers to two substances having the same molecular formula but different structures.

isthmus: constriction between vertebrate embryonic **midbrain** and **hindbrain**.

Kellerexplant: dorsalexplants from the *Xenopus* gastrula used to study morphogenetic movements.

keratinocyte: principal cell type of the **epidermis**.

knock-in: introduction of a transgene to a specific **locus** in the genome by **homologous recombination**.

knockout: null mutation made by targeted **mutagenesis**.

Kollar's sickle: thickened posterior region of **area pellucida** of early avian **blastoderm**.

lacZ: gene encoding *E.coli* β-galactosidase, often used as a **reporter gene**.

lamellipodium (plural lamellipodia): large flat extensions at the leading edge of a motile cell.

lampbrush chromosomes: large **bivalent** chromosomes in the **germinal vesicle** of an amphibian **oocyte**.

lateral inhibition: regional specification mechanism that involves isolated cells or cell clusters differentiating in a particular direction, and emitting an **inducing factor** that inhibits the surrounding cells from doing the same.

lateral plate: region of **mesoderm** of a vertebrate embryo lying lateral to the **somites**.

left–right asymmetry: differences between left and right sides of the body in terms of gene expression, morphogenetic movements, or differentiation.

Leydig cells: endocrine cells of the testis.

lineage label: method of labeling a cell or group of cells that will enable identification of all their descendants.

lineage: see **cell lineage**.

locus (plural loci): position in the **genome** occupied by a particular gene.

long terminal repeat (LTR): strong promoter of **retroviruses** often used to drive **transgenes**.

longitudinal section: a section made parallel to the long axis of the organism, often but not necessarily in the **sagittal** plane.

Lophotrochozoa: a superphylum defined on the basis of primary sequence data; includes the annelids, molluscs, and platyhelminths.

loss of function: usually describes a **mutation** which causes the protein product of the gene to be inactive or of lower activity than the **wild type**.

luciferase: enzyme catalyzing a phosphorescent reaction and often used as a **reporter**.

lumbosacral plexus: arrangement of nerves formed by the fusion and rebranching of the **spinal nerves** supplying the vertebrate hindlimb.

lumen: the cavity within an organ, often lined with an **epithelium**.

lymphocytes: cells responsible for specific immunity; T and B cells.

macroevolution: evolution of taxa above the species level.

macromeres: large **blastomeres**.

macromutation: mutation that produces a significant morphological effect.

macrophage: cell concerned with immunity and inflammation; originates from **hematopoietic stem cell** but found in tissues.

mantle layer: cell-rich layer adjacent to the **ventricular zone** of the embryonic vertebrate central nervous system.

marginal zone: (1) in an amphibian embryo the ring around the **blastula** that invaginates during **gastrulation**; (2) in a chick embryo the junction of the **area pellucida** and **area opaca**; (3) in the embryonic vertebrate central nervous system the cell-poor region near the **pial surface**.

marker gene: a **transgene**, or **allele** of endogenous gene, allowing visual identification of the cells possessing it.

mass spectrometry: technique for determining the molecular weight of a molecule, or of decomposition fragments of the molecule, to very high precision.

maternal effect: situation where the phenotype of the embryo corresponds to the **genotype** of the mother rather than its own **genotype**. This arises from defects in the assembly of the **oocyte**.

matrix: (1) the region of proliferating epidermal cells in the hair bulb; (2) = **extracellular matrix**.

medial (= median): relating to the midline of the organism.

mediolateral: the direction from, or line joining, the midline to the lateral extremities of an animal.

medusa: the jellyfish stage of the life cycle of a **cnidarian**.

megakaryocytes: bone marrow cells that form blood platelets.

meiosis: the final division leading to the formation of **gametes**, in which the chromosome complement is halved.

melanocytes: cells making the melanin pigments.

membrane bone: bone that develops directly from the dermis, with no cartilage model.

meroblastic: type of cleavage where only part of the zygote cleaves and the remainder, usually a yolk mass, does not.

mesenchyme: tissue type in which cells lie scattered within an extracellular matrix.

mesoderm induction: formation of the mesoderm in response to inducing factors.

mesoderm: the middle of the three germ layers.

mesometrium: membrane attaching the uterus of a mouse to the body wall.

mesonephros: the middle segments of the developing kidney.

mesothorax: the second thoracic segment of an insect.

metacarpals: proximal bones of the hand.

metamorphosis: major remodeling of the body at a late stage of development.

metanephros: the posterior region of the developing kidney, becomes the definitive kidney in mammals.

metaplasia: change of one tissue type into another.

metastasis (plural metastases): implantation and growth of cancer cells at remote sites in the body.

metatarsals: proximal bones of the vertebrate (hind)foot.

metathorax: the most posterior of the three thoracic segments of an insect.

microarray: a small slide containing a grid of dots of specific cDNAs or oligonucleotides. Used for establishing gene expression profiles by hybridization with cDNA from the cell types of interest.

microfilaments: components of the cytoskeleton; thin intracellular filaments composed of actin.

micromeres: small blastomeres.

microtome: machine for making thin sections of a biological specimen.

microtubule organizing center: a place in the cell, such as a centrosome or basal body, from which microtubules grow.

microtubules: components of the cytoskeleton; intracellular rod-like structures composed of tubulin, with a larger diameter than microfilaments or intermediate filaments.

microvilli: small projections from a cell, usually found on the apical surface of absorptive epithelia.

midblastula transition (MBT): set of coordinated changes in the late blastula including the onset of zygotic genome transcription.

midbrain: region of the embryonic vertebrate brain that will become the optic tecta, or equivalent, and some other structures.

midgut: region of an amniote embryo endoderm that lies between the anterior and posterior intestinal portals; eventually closes to the umbilicus.

mirror-symmetry: two sets of parts related such that one could be the reflection of the other in a mirror.

mitotic index: the proportion of cells in a specimen undergoing mitosis.

model: (1) model organism:species used for experimental work because of favorable technical considerations. In developmental biology the results obtained on a model species are considered to have a significance well beyond the species itself; (2) an animal model of a human disease; (3) a cartilage model is a skeletal element formed in cartilage that later becomes replaced by bone.

molecular clock: mechanism for the fixation in the population of neutral mutations: it leads to the prediction that the number of neutral mutations fixed in two lines of descent will be proportional to the time of divergence from a common ancestor.

molecular cloning: insertion of a nucleic acid sequence into a cloning vector, usually a bacterial plasmid, such that it can be amplified to a useful quantity.

monoclonal: consisting of or produced from a single clone of cells.

monocytes: macrophage-like cells in the blood.

morphogen: type of inducing factor to which competent cells can make at least two different responses at different threshold concentrations. This means that the responding cells will form a series of differently committed territories in response to a concentration gradient of the factor.

morphogenesis: the aspects of development involving movement of cells or of cell sheets, with the associated formation of structures.

morpholino: a chemical grouping that has given its name to a type of oligonucleotide analog which can hybridize with nucleic acids while being resistant to nuclease enzymes.

morphology: the structure of the organism (= anatomy).

morula: multicellular pre-blastula embryonic stage, usually of a mammalian embryo.

mosaic: (1) a type of embryo in which each part continues to develop in accordance with the fate map after separation from the rest of the embryo; (2) = genetic mosaic.

motor proteins: proteins that translocate along microtubules or microfilaments.

MSC: mesenchymal stem cell or marrow stromal cell.

mucosa: a moist internal epithelium together with its immediately underlying connective tissue.

Mullerian duct (= paramesonephric duct): duct that becomes the reproductive tract in females and regresses in males.

mutagen: substance inducing genetic mutations.

mutagenesis screens: the process for isolating a large number of mutants affecting some property of the organism. This may be the overall anatomy of the embryo, or some specific organ or cell type.

mutant: organism carrying a mutation.

mutation: change in the genomic DNA.

myeloid cells: nonlymphocytic cells of the blood comprising erythrocytes, monocytes, granulocytes, and megakaryocytes.

myoblast: mononucleate cell that is committed to differentiate into muscle.

myocardium: the muscular wall of the heart.

myoepithelial cell: epithelial cell capable of contraction.

myofiber: the multinucleate fibers of skeletal muscle.

myogenic: leading to formation of muscle.

myotome: the part of the somite forming striated muscle.

myotube: myofiber-like structures formed by myoblasts in tissue culture.

nauplius: a type of larva characteristic of crustacea.

neoblasts: the only dividing cells in planarian worms; required for tissue renewal and regeneration.

neocortex: additional layers of neurons in the cerebral cortex of mammals, as compared to lower vertebrates.

neomycin: drug used to select for desired recombination events in tissue culture.

neoplasm: any "new growth"; need not necessarily be malignant.

nephric duct (= Wolffian duct): duct forming the collecting system of the kidney, and, in males, the vas deferens.

nephrogenic: leading to formation of the kidney.

neural crest: migratory cells from the dorsal neural tube of a vertebrate embryo that form a variety of cell types.

neural plate: flat sheet of neuroepithelium that will roll up to form the neural tube of a vertebrate embryo.

neural stem cell: stem cell in the central nervous system that can produce neurons.

neural tube: the primordium for the vertebrate central nervous system.

neuroblast: a cell that divides to produce neurons.

neuroenteric canal: transient connection between the neural tube lumen and the hindgut of a vertebrate embryo.

neuroepithelium: the tissue comprising the neural tube.

neurogenesis: formation of neurons.

neurospheres: structures derived from the embryonic vertebrate central nervous system that contain neural stem cells and can be grown in suspension culture.

neurotrophin or neurotrophic factor: (1) class of growth factor that is particularly relevant to neuronal growth and survival; (2) substances released from nerves that are required for regeneration of other structures.

neurula: stage of vertebrate development during which the neural tube is formed.

neurulation: the morphogenetic movements of vertebrate embryos that create the neural plate and neural tube.

neutral evolution: evolution arising from the gradual accumulation of mutations which are neither beneficial nor deleterious.

neutral mutations: mutations which are neither beneficial nor deleterious.

neutralizing antibody: antibody that does not just bind to its target but also inhibits its biological activity.

Nieuwkoop center: region of the early amphibian embryo that induces Spemann's organizer.

node: condensation of cells at the anterior end of the primitive streak of a mouse embryo, similar to Hensen's node in the chick.

nonautonomous: process that is affected by events in surrounding cells.

nonpermissive: temperature at which a ts mutation causes a mutant phenotype.

normal development: events that occur in the course of development of the embryo undisturbed by any experimental manipulation.

Northern blot: method of detecting a specific mRNA by hybridization with a labeled probe following fractionation of the sample by gel electrophoresis, and transfer of the sample from the gel to a membrane.

notochord: cartilage-like rod which comprises the dorsal-most part of the mesoderm of a vertebrate or invertebrate chordate embryo.

nucleic acid hybridization: association of DNA or RNA strands through base complementarity (A binds to T/U and C binds to G). The specificity of hybridization enables the location or quantity of a particular sequence to be measured using a complementary labeled probe.

nucleus (plural nuclei): (1) of a cell; (2) a condensation of neurons in the central nervous system.

nude mouse: mouse with no thymus that does not reject tissue grafts.

null: relating to a mutation that shows a complete loss of function.

nurse cells: sister cells of the insect oocyte; 15 nurse cells and one oocyte are formed from female oogonia by four successive mitotic divisions.

oligodendrocyte: a type of glial cell in the central nervous system that forms myelin sheaths around axons.

oncogene: gene encoding a product that can confer cancer-like behavior if overexpressed in tissue culture cells.

oocyte: the female germ cell after completion of its mitotic divisions is a primary oocyte, and after completion of the first meiotic division is a secondary oocyte.

oogenesis: the development of oocytes.

oogonia: the mitotic germ cells that become oocytes.

optic tecta (singular tectum): regions of the midbrain where the axons from the optic nerve terminate.

optic vesicles: outgrowths of the forebrain of a vertebrate embryo that form the eyes.

organ culture: the growth of embryonic organ rudiments in vitro.

organizer (= signaling center): group of cells emitting an inducing factor.

organogenesis: the process, or stage, of formation of individual organs in vertebrate development.

ortholog: a gene is an ortholog of a gene in another species if it is the direct homolog, as opposed to a homolog to a duplicated locus that arose before the divergence of the two species.

orthotopic graft: type of graft that is performed to the same position in the host to that from which it came in the donor.

osteoblast: cell of the bone, strictly an immature mitotic cell.

osteoclast: multinucleate cell type of hematopoietic origin that resorbs bone and is essential for bone remodeling.

osteocyte: mature cell of the bone.

otic placodes: epidermal structures forming the ear vesicles.

oviduct: part of female reproductive tract: a tube conveying mature oocytes from the ovary to the uterus.

ovulation: the release of an oocyte, usually accompanied by the completion of first meiotic division.

P-element: transposable element of *Drosophila*.

P-granules: (= polar granules).

pair-rule genes: genes controlling the segmentation pattern of *Drosophila* that are expressed as one stripe for each prospective pair of segments.

Paneth cells: neutrophil-like cells at the base of small intestinal crypts that secrete a variety of antimicrobial substances.

parabiosis: two animals joined together in such a way that they share a common blood circulation.

paralog group: set of similar genes in the same genome, arising from a gene duplication. Especially used for the homologous sets of Hox genes within the vertebrate Hox clusters.

parasagittal: a plane parallel to the sagittal plane but displaced towards the right or left side.

parasegments: the initially formed segmental repeating structures in insect embryos.

paraxial mesoderm: plate of mesoderm that forms the somites of a vertebrate embryo.

parietal endoderm: extraembryonic endoderm lining the trophectoderm of a mouse embryo.

parthenogenesis: asexual development from oocytes, without a genetic contribution from the sperm.

pentadactyl: having five digits.

perdurance: persistance of a gene activity after the gene expression has stopped. It is due to continued presence of the protein product.

perichondrium: connective tissue sheath around a cartilage structure.

periosteum: connective tissue sheath around a bone.

permissive: (1) of inductions, a signal required for the continuation of a particular developmental pathway, but which does not control alternative developmental fates; (2) of temperatures, the temperature at which a ts mutant phenotype is not displayed.

phalanges: bones of the digits in vertebrate fore- or hindlimb.

pharyngeal arches: see branchial arches.

pharyngeal pouches: segmentally repeated pits or openings to the exterior in the pharyngeal region of a vertebrate embryo.

phenocopy: the same phenotype as that produced by a mutation, but produced by some other means.

phenotype: the collected characteristics of an organism, particularly in relation to other members of the same species. It is mostly determined by the individual's genotype.

phosphorescence: emission of light in the course of a chemical reaction.

phylogenetic tree: classification scheme which depicts the taxa in terms of their actual evolutionary relationships.

phylotypic stage: developmental stage at which members of an animal group most closely resemble each other.

phylum (plural phyla): the highest subdivision in animal taxonomy.

pial surface: the outer surface of the vertebrate central nervous system.

pigment epithelium: see retinal pigment epithelium.

placode: a thickening of the epidermis which is the rudiment for a structure, usually a sense organ.

planar cell polarity: polarity displayed by neighboring cells within a cell sheet.

Planaria: group of free-living flatworms belonging to the phylum Platyhelminthes. They have a blind-ended gut and no coelom.

Platyhelminthes: phylum including free-living and parasitic flatworms and tapeworms.

ploidy: indicates the number of parental chromosome sets per nucleus. Animals are normally diploid (two sets, one from each parent), but under special circumstances could be haploid (one set), triploid (three sets), tetraploid (four sets), etc..

pluripotent: able to form all or most of the differentiated cell types in the body.

polar body: chromosome set arising from a meiotic division of the oocyte and expelled from the oocyte as a small vesicle.

polar granules (= P-granules): associated with germ plasm of C. elegans.

polarity: (1) of cells, a difference between one end and the other, e.g. apico-basal polarity; (2) of an embryo or part of an embryo, a difference in commitment between the regions along a particular axis, e.g. animal–vegetal polarity.

pole cells: germ line cells of Drosophila, formed before cellularization of the rest of the embryo.

pole plasm: associated with germ plasm of Drosophila.

polyclonal: consisting of or formed from more than one clone of cells.

polydactyly: having too many digits.

polymerase chain reaction (PCR): amplification of a DNA sample by repeated cycles of strand separation, primer hybridization, and replication.

polyp: (1) the sessile phase of a cnidarian organism; (2) a projecting differentiated neoplasm, usually benign.

polyploidy: duplication or higher multiplication of the entire chromosome set.

polyspermy: fertilization of an egg by more than one sperm.

polytene, polyteny: repeated replication of DNA without separation of chromosomes; leads to giant chromosomes.

positional cloning: cloning of a gene starting from a mutation, which is mapped to very high resolution to find the gene within which it lies.

positional information: a hypothetical property of animal tissues that bears a one-to-one relationship to normal position in the body, and is called into play during regeneration.

posterior (= caudal) intestinal portal: the opening of the hindgut into the subgerminal cavity (midgut) of an amniote embryo.

posterior midgut: the part of an insect embryo gut invaginating from the posterior end.

posterior: the tail end of an animal.

potency: of a cell or tissue region indicates the full range of structures or cell types that it is capable of becoming if exposed to a range of different environments.

prenatal screening: a test to detect a developmental abnormality, which may be of genetic or spontaneous origin, before birth.

primary oocyte: see oocyte.

primitive ectoderm: = epiblast of mammalian embryo.

primitive endoderm: first formed layer of endoderm in mammalian embryo, contributes to extraembryonic membranes.

primitive streak: the region of convergence and invagination of cells in a gastrulating amniote embryo.

primordial germ cells (PGCs): germ line cells at a stage before they become oogonia or spermatogonia.

probe: a labeled and antisense nucleic acid that can be used to detect the complementary nucleic acid by hybridization, either in a biochemical method such as a blot, or in situ in a specimen.

procephalon: the part of an insect embryo forming the anterior head.

proctod(a)eum: most posterior part of the gut, nominally lined by ectoderm.

programmed cell death (= apoptosis): death of cells by a process involving the activation of caspase enzymes, which avoids the release of toxic products into the surroundings. Often occurs in embryonic development.

progress zone: the undetermined mesenchymal cells at the distal tip of a vertebrate limb bud.

promoter: the region of a gene 5′ to the coding sequence where the RNA polymerase binds. It may also contain regulatory sequences binding transcription factors that control gene expression.

pronephros: anterior segments of the developing kidney.

proneural cluster: group of cells which are competent to form neurons but of which only one or a few cells do so because of a lateral inhibition process.

pronucleus: transient nucleus-like structure of a fertilized egg containing the sperm or egg chromosomes.

proteomics: analysis of proteins, especially on a large scale using the techniques of 2D electrophoresis to resolve a complex mixture, followed by mass spectrometry to identify individual components.

prothorax: the most anterior of the three thoracic segments of an insect.

protochordate: member of the classes of the phylum Chordata that are not vertebrates, for example amphioxus is a cephalochordate.

Protostomia: group of animal **phyla** characterized by **spiral cleavage, mosaic** development, and **schizocoely.**

proventriculus: glandular region of the avian stomach.

proximal: nearer to the body.

pseudoalleles: genes that are so similar in sequence that they appear to be **alleles,** but actually belong at different genetic **loci.** Normally the result of **gene duplication.**

pseudocoelom: body cavity differing from a true **coelom** because it is not fully lined with **mesoderm.**

pseudopregnant: in mice, the hormonal state resulting from mating with a sterile male.

pseudotetraploid: a genome that has undergone complete duplication to the **tetraploid** state followed by some evolutionary divergence between the duplicated copies.

pupa: stage of **holometabolous** insect development at which **metamorphosis** occurs.

R

radial cleavage: symmetrical type of **cleavage.**

radial glia: cells spanning the full width of the developing brain, from the ventricular to the **pial surface,** now known to be neural **stem cells.**

radius: anterior bone of the vertebrate lower forelimb.

RCAS virus: replication-competent, avian specific **retrovirus,** used for overexpression experiments in chick embryos.

real time PCR: a quantitative type of PCR in which the formation of the product is continuously monitored during the amplification reaction.

receptor: type of molecule recognizing and activated by a specific ligand, often an **inducing factor.**

recessive: a mutation that produces no abnormal **phenotype** if it is present alongside the **wild type allele, but does so if present at both homologous loci.**

redundancy: partial or complete overlap of function of two or more genes.

regeneration: regrowth of a missing body part at an adult or late developmental stage.

regional specification: starting from a population of similar cells, the formation of a set of territories of cells each committed to become a different structure or cell type.

regulative: in addition to its general meaning of one thing controlling another, a **regulative embryo** is one where isolated parts do not necessarily develop in accordance with the **fate map,** usually forming more structures than predicted by the **fate map.**

renewal tissue: tissue undergoing continuous cell turnover.

reporter gene: usually a **transgene** that produces a product that is very easy to detect and hence "reports" on the nature and activity of its regulatory environment.

reproductive cloning: probably possible, but currently illegal, formation of a complete human from a single cell.

retinal pigment epithelium: pigmented layer of the eye derived from the optic cup.

retrovirus: a type of virus with an RNA genome that is reverse transcribed to DNA in the course of infection.

reverse genetics: investigations on a known gene, especially by creation of **loss-of-function** mutants.

reverse transcription polymerase chain reaction (RT-PCR): detection of a specific mRNA in a sample by reverse transcribing to DNA then amplifying the specific sequence of interest by the **polymerase chain reaction (PCR).**

rhombomeres: segmental structures in the **hindbrain** of vertebrate embryos.

RNA interference (= RNAi): method of inactivating a specific mRNA by introducing a short complementary double-stranded RNA sequence.

RNA polymerase II: the RNA polymerase responsible for transcription of proteincoding genes.

RNAse protection: method of detecting a specific mRNA by **hybridization** of the sample to a labeled **probe,** followed by RNAse digestion of unhybridized RNA, and separation of the protected product by electrophoresis.

roof plate: mid-dorsal part of the vertebrate embryo spinal cord.

Rosa26: **gene trap** mouse line that expresses β-**galactosidase** in all tissues.

S

S phase: the part of the cell cycle during which DNA is replicated.

sagittal: the **medial** plane of an organism, separating left and right halves of the body.

sarcoma: cancer derived from **connective tissue** (in broad sense).

satellite cells: mononuclear cells associated with vertebrate muscle that can become **myoblasts** and form new **myofibers** on appropriate stimulation.

scaffold: three-dimensional extracellular matrix used in **tissue engineering;** often made of synthetic polymers rather than natural extracellular components.

schizocoely: formation of the **coelom** by cavitation of a pre-existing mass of **mesoderm.**

Schwann cells: cells that form the myelin sheaths of peripheral nerve axons.

sclerotome: the part of the **somite** forming the vertebrae.

secondary oocyte: see **oocyte.**

sections: see **histological sections.**

segment polarity genes: genes in *Drosophila* that are expressed in one stripe per prospective segment.

segmental plate:= paraxial mesoderm.

segmentation: subdivision of the body, or of an appendage, into repeating structural units, usually involving all **germ layers.** In older literature can also be a synonym for **cleavage.**

selector gene:= homeotic gene.

septum intermedium: part of the interventricular septum of the heart.

septum transversum: region of **splanchnic mesoderm** into which the liver bud grows.

Sertoli cells: supporting cells of the testis.

sex determination: mechanism for ensuring that some individuals become male and some female.

sex linked: of genes or **mutations** that lie on a sex chromosome.

signal transduction pathways: metabolic pathway connecting a cell surface **receptor** to its intracellular effector (gene, cytoskeleton, secretory system). It often involves a sequence of protein phosphorylation events and movement of activated **transcription** factors into the nucleus where they can regulate gene activity.

signaling center (= organizer): group of cells emitting an **inducing factor.**

sink: region of cells destroying or otherwise removing an **inducing factor.**

situs inversus: inversion of the normal left–right asymmetry.

situs solitus: the normal asymmetrical arrangement of some organs on the left and right side of the body.

somatic cells: all cells of an organism that are not the **germ line;** may be collectively referred to as the **soma.**

somatic mesoderm: the layer of lateral **mesoderm** lying exterior to the **coelom.**

somatic: refers to all tissues that are not part of the **germ line.**

somatopleure: the lateral part of an amniote embryo consisting of somatic mesoderm overlain by epidermis.

somites: segmented mesodermal structures in vertebrate embryos that give rise to vertebrae and striated muscles.

source: region of cells producing an inducing factor.

Southern blot: method of detecting a specific DNA sequence by hybridization with a labeled probe following fractionation of the sample by gel electrophoresis, and transfer of the sample from the gel to a membrane (invented by Ed Southern).

specification: type of developmental commitment of cells or a tissue explant that is manifested on culture in isolation, but is not irreversible.

Spemann's organizer: the dorsal blastopore lip region of an amphibian embryo that emits BMP inhibitors and thereby dorsalizes the surrounding tissue.

spermatogonia: proliferating cells that give rise to sperm.

spinal nerves: segmental nerves coming from the spinal cord of a vertebrate.

spiral cleavage: asymmetrical type of cleavage in which quartets of micromeres are cut off in alternating directions.

splanchnic mesoderm: layer of lateral mesoderm lying internal to the coelom.

splanchnopleure: the splanchnic mesoderm together with the apposed endoderm.

squamous: describes a thin flat type of cell, usually found in a squamous epithelium.

stage series: description of normal development of a species divided into a number of standardized stages which can be identified by externally visible features.

stem cell niche: microenvironment that supports the survival and function of stem cells.

stem cell: a cell that is undifferentiated; survives a long time; divides to produce more copies of itself; and also divides to produce progeny destined to differentiate.

superficial cleavage: cleavage of nuclei without formation of cells.

symmetry breaking: process whereby two or more discrete cellular states arise starting from a uniform situation.

syncytium: a mass of cytoplasm containing many nuclei.

tailbud: (1) region at the posterior of a vertebrate embryo that generates part or all of the tail; (2) the phylotypic stage of vertebrate embryos.

tarsals: bones of the ankle.

taxon (plural taxa): a group of organisms.

taxonomy: the science of classification of organisms.

telencephalon: the anterior part of the vertebrate forebrain.

telomerase: enzyme-RNA complex that restores the telomeres of chromosomes after DNA replication.

telomeres: the ends of a chromosome, consisting of a short repeating sequence.

telson: unsegmented posterior extreme of an insect embryo.

temperature sensitive (ts): describes mutations that have an effect at one, usually high, temperature but not at another, usually lower, one.

teratocarcinoma: pluripotent tumor derived from germ cells.

teratogenic: leading to teratological effect.

teratology: the study of developmental abnormalities arising from exposure of the embryo to noxious stimuli such as toxic substances, radiation, or infection.

terminal system: system that activates certain genes at both termini of the Drosophila embryo.

tet-off: a system of inducible gene expression which is activated by withdrawal of tetracycline.

tet-on: a system of inducible gene expression activated by addition of tetracycline.

tetraploid: organism whose genome has become duplicated so that it contains two maternal copies and two paternal copies of every chromosome.

tetrapods: four-legged vertebrates: amphibians, reptiles, birds, and mammals.

thecal cells: endocrine cells of the ovary.

therapeutic cloning: growth in tissue culture of differentiated cells for transplantation, starting from ES cells made from a cloned human blastocyst.

therapeutic targets: gene products that may be suitable targets for the development of drugs to inhibit or enhance their activity.

threshold response: sharp and discontinuous change in cell state occurring at a particular concentration of an inducing factor.

tibia: anterior bone of the vertebrate lower hind limb (also a segment of insect leg).

tight junctions: band-like apical cell junctions within an epithelium that prevent substances crossing the epithelium through spaces between cells.

tissue culture: growth of cells in vitro, usually as a cell monolayer.

tissue engineering: the growth of tissues and organs in vitro, sometimes on a three-dimensional scaffold, either for transplantation or for temporary replacement of the function of a damaged organ.

tissue specific stem cell (= adult stem cell): stem cell found in a renewal tissue; normally committed to forming just cell types found in that tissue.

topographic: indicates a one-to-one mapping between parts of one structure and another, especially relevant to the formation of nerve connections.

totipotent: able to form a whole organism by itself. This term used to be used for embryonic stem cells but is now reserved for the fertilized egg alone.

tracheae: the respiratory system of insects.

transcription factor: protein controlling the transcription of specific genes.

transdetermination: change of the state of specification of a Drosophila imaginal disc to that of a different disc type.

transdifferentiation: strictly the change of one differentiated cell type to another, without passing through undifferentiated intermediate states. But often used as synonym for metaplasia.

transfilter: experimental setup whereby a signaling and responding tissue are separated by a membrane with defined permeability properties.

transgene: cloned gene introduced into an organism.

transgenesis: introduction of a new gene into an organism, normally into the germ line.

transit amplifying cells: dividing cells in a renewal tissue that are not stem cells. Normally they can only divide a limited number of times.

transplantation: the movement of tissue from one part of an organism to another, or from one individual to another. It can refer either to microscopic grafts carried out on embryos, or to organ transplants in adults.

transposable element or transposon: a piece of DNA that may occasionally move from one part of the genome to another.

transverse section: section made orthogonal to the anteroposterior axis of the organism.

triploblasts: animal taxa with three germ layers, ectoderm, mesoderm, and endoderm; most animals are triploblasts.

triploid: having three chromosome sets per nucleus.

trochophore: type of larva found among both annelids and molluscs.

trophectoderm: the outer layer of a mammalian blastocyst, forms part of the placenta.

trophoblast: extraembryonic structure formed from the **trophectoderm**.

truncoconal swellings: ingrowths of the outflow tract of the heart that separate it into pulmonary and systemic branches.

ts:= temperature sensitive.

tumor suppressor gene: a gene that encodes a growth inhibitor, such that its removal will provoke uncontrolled growth.

turning: morphogenetic movement whereby a rodent head-fold stage embryo moves and/or twists such that the dorsal midline structures come to lie in a C-shaped curve along the outside.

ulna: posterior bone of the vertebrate lower forelimb.

umbilical tube: the tube connecting an **amniote** embryo to the yolk mass or placenta.

unfertilized egg: strictly an egg arising from the second meiotic division. In fact most vertebrates are fertilized at the secondary oocyte stage, so the term is also often used of secondary oocytes.

ureteric bud: outgrowth of the nephric duct that forms the duct system of the **metanephros**.

urodeles: the tailed amphibians: newts and salamanders.

vasculogenesis: formation of new blood vessels by *de novo* differentiation from **mesoderm**.

vegetal hemisphere: the lower hemisphere of an egg or **oocyte**.

veliger: a type of larva characteristic of molluscan classes except cephalopods.

ventral furrow: invagination of prospective **mesoderm** into an insect embryo.

ventral: the lower surface of an animal.

ventricular septum: wall between the right and left ventricles of the heart.

ventricular zone: the cells next to the ventricles of the vertebrate central nervous system.

visceral endoderm: extraembryonic endoderm initially forming the lower layer of the egg cylinder stage embryo of rodents.

vital dye (= **vital stain**): stain that can be applied to a living specimen.

vitelline membrane: extracellular membrane around a **zygote**.

vitellointestinal duct: persistent projection of the **midgut** into the **umbilical tube** of an **amniote** embryo.

viviparous: where the embryo and fetus develops within the mother, nourished by a placenta.

Western blot: detection of a specific protein by use of an antibody, following fractionation of the sample by gel electrophoresis, and transfer of the sample from the gel to a membrane.

wholemount: a specimen that is viewed as a three-dimensional object.

wild type: the normally occurring **allele** at a genetic **locus**.

Wolffian duct: see **nephric duct**.

X-chromosome: sex-determining chromosome, in mammals females are XX and males are XY.

X-inactivation: shutting down of the activity of one of the two X chromosomes in female mammals.

yolk: granules of food reserve deposited in the **oocyte** and used for embryonic nutrition. Yolk granules contain a few principal proteins together with lipids.

yolk sac: in the avian embryo the **extraembryonic membrane** that adheres to the yolk mass, consisting of extraembryonic **mesoderm** and endoderm; in the rodent embryo the corresponding structure is the middle of three membranes eventually surrounding the fetus, derived from the endoderm and mesoderm connecting the egg cylinder with the ectoplacental cone.

yolk syncytial layer (YSL): syncytial region at the junction of the yolk cell and the cellular part of a fish embryo.

zona pellucida: transparent extracellular layer surrounding a mammalian early embryo.

zone of polarizing activity (ZPA): signaling center in the **posterior** region of the vertebrate limb bud that controls the **anteroposterior** pattern of **regional** specification.

zootype: configuration of gene expression domains at the **phylotypic stage** that defines a common pattern for all triploblastic animals.

zygote: fertilized egg, strictly after the stage of fusion of male and female pronuclei.

zygotic: relating to the **genome** of the embryo as opposed to that of the mother.

Index

Note: Terms in **bold** are listed in the Glossary. Page numbers in *italics* refer to figures (when outside text page ranges).

Essential Developmental Biology, Third Edition. Jonathan M.W. Slack.
© 2013 John Wiley & Sons, Ltd. Published 2013 by John Wiley & Sons, Ltd.